NORTH-HOLLAND
MATHEMATICS STUDIES 83

COMPUTATIONAL TECHNIQUES FOR DIFFERENTIAL EQUATIONS

Edited by

JOHN NOYE

Associate Dean
Faculty of Mathematical Sciences
The University of Adelaide
South Australia

1984

NORTH-HOLLAND – AMSTERDAM • NEW YORK • OXFORD

ISBN: 0 444 86783 x

Publishers:

ELSEVIER SCIENCE PUBLISHERS B.V.
P.O. Box Box 1991
1000 BZ Amsterdam
The Netherlands

Sole distributors for the U.S.A. and Canada:

ELSEVIER SCIENCE PUBLISHING COMPANY, INC.
52 Vanderbilt Avenue
New York, N.Y. 10017
U.S.A.

Library of Congress Cataloging in Publication Data
Main entry under title:

Computational techniques for differential equations.

(North-Holland mathematics studies ; 83)
Bibliography: p.
1. Differential equations--Numerical solutions--Congresses. 2. Differential equations, Partial--Numerical solutions--Congresses. I. Noye, John, 1930- . II. Series.
QA370.C626 1984 515.3'5 83-16370
ISBN 0-444-86783-X

PRINTED IN THE NETHERLANDS

PREFACE

Six invited papers on computational methods of solving partial differential equations were presented at the 1981 Conference on Numerical Solutions of Partial Differential Equations held at the University of Melbourne, Australia. They were also printed as part of the Conference Proceedings titled *Numerical Solutions of Partial Differential Equations*, edited by J. Noye and published by the North-Holland Publishing Company. The articles were so well received that it was decided to expand them and print them in a separate book so that the material they contained would be more readily available to postgraduate students and research workers in Universities and Institutes of Technology and to scientists and engineers in other establishments.

Because of the importance of ordinary differential equations and their use in the solution of partial differential equations, it was decided to include an additional article on this topic. Consequently, the first contribution, written by Robert May and John Noye, reviews the methods of solving initial value problems in ordinary differential equations. The next four articles are concerned with alternative techniques which may be used to solve problems involving partial differential equations: finite difference methods are described by John Noye of the University of Adelaide, Galerkin techniques by Clive Fletcher of the University of Sydney, finite element methods by Josef Tomas of the Royal Melbourne Institute of Technology, and boundary integral equation techniques by Leigh Wardle of the CSIRO Division of Applied Geomechanics. The first three of these are updated and extended revisions of the corresponding papers presented at the Melbourne conference; because the last mentioned author was unable to find time to revise his article, it has been reprinted in its original form from the 1981 Proceedings. The last two articles in this book describe the two basic methods of solving large sets of sparse linear algebraic equations: direct methods are presented by Ken Mann of the Chisholm Institute of Technology and iterative techniques by Len Colgan of the South Australian Institute of Technology. These methods are often incooperated in techniques for solving ordinary and partial differential equations.

My personal thanks go to the above-mentioned contributors for their cooperation in this venture, and to Drs. Arjen Sevenster (Mathematics Editor) and John Butterfield

(Technical Editor) of Elsevier Science Publishers B.V. (North-Holland), for their assistance with the printing of this book.

John Noye
The University of Adelaide
April, 1983

CONTENTS

Computational Techniques for Differential Equations
J. Noye (Editor)

THE NUMERICAL SOLUTION OF ORDINARY DIFFERENTIAL EQUATIONS: INITIAL VALUE PROBLEMS

ROBERT MAY
Royal Melbourne Institute of Technology, Victoria, Australia

JOHN NOYE
University of Adelaide, Adelaide, South Australia

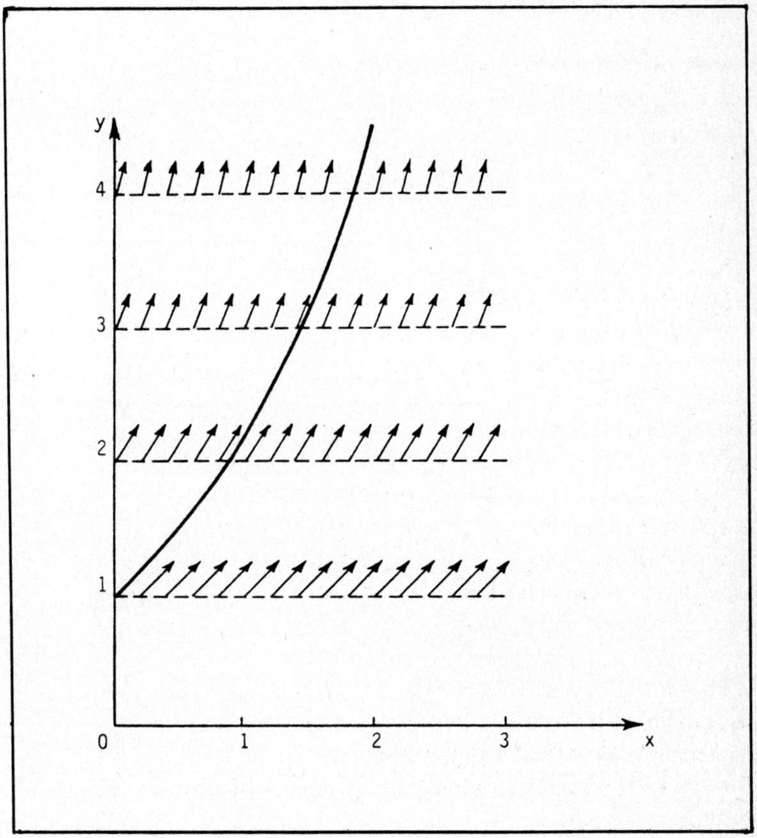

CONTENTS

1. INTRODUCTION

Mathematical models of problems in science and engineering often involve one or more ordinary differential equations. For instance, problems in mechanics such as the motion of projectiles or orbiting bodies, in population dynamics and in chemical kinetics may be modelled by ordinary differential equations.

Many clever methods of finding analytical solutions of ordinary differential equations are presented in elementary courses, but the majority of differential equations are not amenable to these methods, and unfortunately most of the differential equations which model practical problems fall into this category. For this reason many texts on mechanics, population dynamics and chemical kinetics develop elegant systems of ordinary differential equations, but give solutions to only very simple idealised problems.

Over the years various numerical methods have been devised, in general not by mathematicians but by people working in other fields for whom the method of solution was only incidental to the problem they were trying to solve. For instance, the technique now referred to as Adams method first appeared in an article on capillary action published by Bashforth and Adams (1883).

However, in the last thirty or so years mathematicians have put the subject on a much sounder theoretical basis, particularly in areas such as stability and the propagation of errors. Much work has been done on the implementation of methods and on their comparative testing. This has produced some agreement on what is the "best" method for a given non-stiff ordinary differential equation. Recently the problem of stiffness has received a lot of attention.

The aim of this article is to provide a practically oriented guide to help people with little or no previous knowledge to solve ordinary differential equations by numerical means. For this reason most of the theoretical results are merely described but not proved. For their proof, the interested reader is referred to the classical work of Henrici (1962) or the book by Stetter (1973). Section 11, on the choice of method and available software, is directed particularly to the beginner who must select a method to solve a particular problem and who wishes to find a suitable program to implement the method.

As the title of this article implies, only initial value problems have been considered. Boundary value problems for ordinary differential equations are also very important. A knowledge of the solution of initial value problems is useful when it comes to boundary value problems, which require more sophisticated numerical techniques for their solution. Keller (1968 and 1976) describes methods for solving the two-point boundary value problem, and Keller (1975) gives a general survey of boundary value problems in general. Unlike initial value problems, for which some well tried automatic computer codes are now readily available, the development of computer programs for boundary value problems is only in its infancy.

The main techniques considered in this article are those based on Taylor series (Section 3), the Runge-Kutta methods (Section 5), linear multi-step methods (Section 7) and extrapolation methods (Section 9).
All calculations made to demonstrate the relative accuracy of these methods were carried out on a CDC-CYBER 170-720, unless otherwise indicated. Besides going to the original articles for information, the

descriptions of these methods given in the books by Gear (1971), Henrici (1962), Ralston (1965), Shampine and Allen (1973), Shampine and Gordon (1975), and Stetter (1973) were used in the preparation of this article. Particular mention must be made of the excellent book by Lambert (1973).

Whenever an initial value problem in ordinary differential equations arises, a quick check to determine whether analytical techniques will give a general solution may be worthwhile. In this regard, books like Murphy (1960) are useful: they contain methods for solution of ordinary differential equations with a list of equations with known solutions. However, if a rapid search of this kind is not successful, the methods described here must be used.

Exact solutions for particular initial value problems are also useful for another reason. They can be used to check the accuracy of a numerical technique, and they are good indicators of possible coding errors.

2. PRELIMINARIES

2.1 Definitions

An equation of the form

$$F(x,y,y',y'',\ldots,y^{(m)}) = 0 \tag{2.1.1}$$

is called an *ordinary differential equation of order* m. A function $y(x)$ defined and m times differentiable on some interval I which satisfies (2.1.1) for all $x \in I$ is called a *solution* of the differential equation. Differential equations generally have many solutions, and extra conditions, known as *boundary conditions*, must be imposed to single out a particular solution. These boundary conditions usually take the form of the solution and/or its derivatives being specified for particular values of x, and it can be shown that a differential equation of order m requires m boundary conditions. If all the boundary conditions apply at one value of x they are called *initial conditions*, and the differential equation together with the initial conditions is termed an *initial value problem* - if more than one value of x is involved in the boundary conditions it is called a *boundary value problem*.

A differential equation is *linear* if y and its derivatives occur linearly, that is if the equation is of the form

$$a_m(x)y^{(m)} + a_{m-1}(x)y^{(m-1)} + \ldots + a_0(x)y = g(x). \tag{2.1.2}$$

In this paper we consider only *explicit differential equations*, that is differential equations which can be put in the form

$$y^{(m)} = f(x,y,y'',y''',\ldots,y^{(m-1)}). \tag{2.1.3}$$

Clearly all linear differential equations are in this category, as are the majority of non-linear equations. For the numerical solution of implicit differential equations see Fox and Mayers (1981).

2.2 Reduction of Higher-Order Differential Equations to First-Order Systems

Consider the mth-order initial value problem

$$\begin{aligned} &y^{(m)} = f(x,y,y',\ldots,y^{(m-1)}), \\ &y^{(i-1)}(a) = \eta_i, \quad i=1,2,\ldots,m. \end{aligned} \tag{2.2.1}$$

By introducing the variables y_i, $i=1,2,\ldots,m$, where

$$\begin{aligned} y_1 &\equiv y, \\ y_2 &\equiv y', \\ y_3 &\equiv y'', \\ &\vdots \\ y_m &\equiv y^{(m-1)}, \end{aligned} \tag{2.2.2}$$

equation (2.2.1) may be written as an initial value problem for a first-order system, namely

$$\begin{aligned} y_1' &= y_2, & y_1(a) &= \eta_1, \\ y_2' &= y_3, & y_2(a) &= \eta_2, \\ y_3' &= y_4, & y_3(a) &= \eta_3, \\ &\vdots & &\vdots \\ y_m' &= f(x,y_1,y_2,\ldots,y_m), & y_m(a) &= \eta_m. \end{aligned} \tag{2.2.3}$$

Using the matrix notation

$$\begin{aligned} \underset{\sim}{y} &= [y_1,y_2,\ldots,y_m]^T \\ \underset{\sim}{\eta} &= [\eta_1,\eta_2,\ldots,\eta_m]^T \\ \underset{\sim}{f} &= [y_2,y_3,\ldots,y_m,\ f(x,y_1,y_2,\ldots,y_m)]^T, \end{aligned} \tag{2.2.4}$$

the initial value problem becomes

$$\underset{\sim}{y}' = \underset{\sim}{f}(x,\underset{\sim}{y}), \quad \underset{\sim}{y}(a) = \underset{\sim}{\eta}\ . \tag{2.2.5}$$

In the same way an initial value problem involving a system of higher order equations can be put in the form (2.2.5).

Note that if the vector signs are omitted (2.2.5) defines a first-order initial value problem. This is particularly important because it means that most results which hold for a first-order initial value problem can be generalised to a system of m first-order equations and hence apply to an mth-order initial value problem. Similarly, any method of solution of a first-order initial value problem can be extended to a system of equations and thus may be used to solve an mth-order initial value problem. Throughout the rest of this article only first-order equations will be considered and from time-to-time it will be indicated how the result applies to the more general case.

Boundary value problems can also be reduced to a system of first-order equations, but the boundary conditions do not apply at the same value of x. Methods of solution of initial value problems can also be modified to solve boundary value problems.

2.3 Existence and Uniqueness of Solutions

The solutions of $y' = f(x,y)$ are generally a family of curves, and the initial condition $y(a) = \eta$ usually singles out one of these to give a unique solution. For example $y' = y$ has the solutions $y = ce^x$, and $y(0) = 1$ implies that $c = 1$, giving the unique solution $y = e^x$ (see Figure 2.1). However not all such problems have a unique solution. Consider

$$y' = \sqrt{y}\ , \quad y(0) = 0\ . \tag{2.3.1}$$

Clearly $y \equiv 0$ is a solution, but so is

$$y(x) = \begin{cases} 0 & 0 \le x \le c \\ \frac{1}{4}(x-c)^2 & x > c\ , \end{cases} \tag{2.3.2}$$

for any constant c. Thus this problem has infinitely many solutions (see Figure 2.2), and would obviously prove difficult to solve numerically.

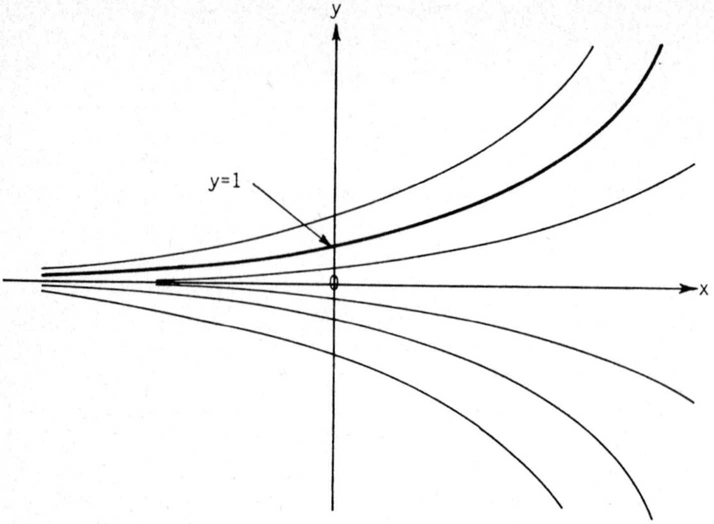

FIGURE 2.1: Solutions of $y'=y$ and the unique solution satisfying $y(0)=1$.

The initial value problem

$$y' = f(x,y), \quad y(a) = \eta, \tag{2.3.3}$$

is guaranteed a unique solution on some interval $[a,b]$ if $f(x,y)$ satisfies certain conditions, as the following theorem proved in Henrici (1962) shows.

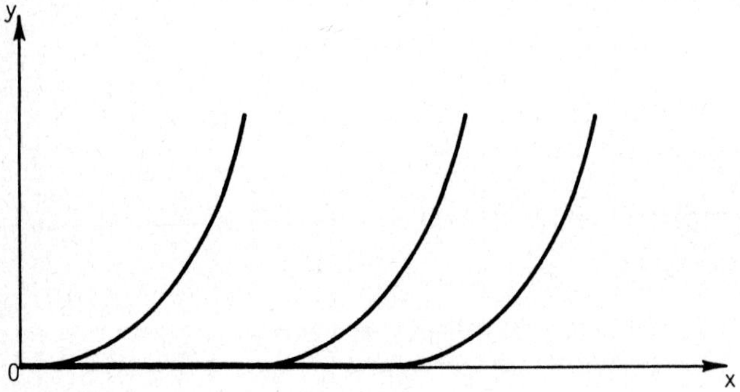

FIGURE 2.2: Some of the solutions of $y'=\sqrt{y}$, $y(0)=0$.

Theorem: If $f(x,y)$ *is defined and continuous for all points in the region*

$$D = \{(x,y) : a \le x \le b, \quad -\infty < y < \infty\} ,$$

and there exists a constant L *such that for all* (x,y) *and* (x,y^*) *in* D

$$|f(x,y) - f(x,y^*)| \le L\,|y-y^*| , \tag{2.3.4}$$

then the initial value problem (2.3.3) has a unique solution on $[a,b]$ *for any given number* η.

The constant L is called a *Lipschitz constant*, and the condition (2.3.4) is called the *Lipschitz condition*. This theorem applies to a system of equations - vector signs must be put under f, y and η and the absolute values in (2.3.4) replaced by vector norms.

In the case that $f(x,y)$ has a continuous partial derivative with respect to y, the mean value theorem gives

$$f(x,y) - f(x,y^*) = \partial f/\partial y\big|_{(x,\bar{y})} \quad (y-y^*) \tag{2.3.5}$$

where $\bar{y}$ lies between y and y^*. If $\partial f/\partial y$ is bounded by K on D, that is there exists some constant K such that

$$|\partial f/\partial y| < K \quad \text{for all} \quad (x,y) \in D , \tag{2.3.6}$$

we can take $L = K$. However, if $\partial f/\partial y$ is unbounded on D, then f does not satisfy a Lipschitz condition. For a system of m equations both f and y are m-vectors, and $\partial f/\partial y$ is the *Jacobian* of f with respect to y, that is a $m\times m$ matrix whose i,j element is $\partial f_i/\partial y_j$, so that a matrix norm subordinate to the vector norm used in (2.3.4) must replace the absolute value in (2.3.6).

Many problems do not satisfy the above theorem even though they have a unique solution. In this case it is often possible to modify the problem in such a way that the theorem is satisfied, but the solution is unchanged. Consider for example the initial value problem

$$y' = y^2, \quad y(0) = 1 . \tag{2.3.7}$$

On any interval $[0,c]$, where $c < 1$, this has the unique solution

$$y(x) = \frac{1}{1-x} , \tag{2.3.8}$$

but $f(x,y) = y^2$ giving $\partial f/\partial y = 2y$, and since this is unbounded as $y \to \pm\infty$, f does not satisfy a Lipschitz condition. However $\partial f/\partial y$ is bounded on any finite region, so if we define

$$f^*(x,y) = \begin{cases} y^2 & |y| \le M , \\ M^2 & |y| > M , \end{cases} \tag{2.3.9}$$

then $f^*(x,y)$ is continuous and satisfies a Lipschitz condition. Thus the initial value problem

$$y' = f^*(x,y), \quad y(0) = 1, \tag{2.3.10}$$

is guaranteed a unique solution, and the solution of (2.3.10) is identically equal to that of (2.3.7) for $|y| \le M$.

Another example is

$$y' = \sqrt{y}, \qquad y(0) = 1. \tag{2.3.11}$$

Here $\partial f/\partial y = 1/2\sqrt{y}$ is unbounded as $y \to 0$ and f is not defined for $y < 0$. We can define

$$f^*(x,y) = \begin{cases} \sqrt{y} & y \ge 1 , \\ 1 & y < 1 , \end{cases} \tag{2.3.12}$$

which is clearly continuous and satisfies a Lipschitz condition, and since the solution of (2.3.11) is monotone increasing $(y' > 0)$ then $y \ge 1$ for $x > 0$ so the modified problem has the same unique solution. Note that this example is nearly the same as the earlier example (2.3.1) of an initial value problem with more than one solution - only the initial value has been changed. The above modification cannot be carried out for (2.3.1) since the initial value $y(0) = 0$ is that at which f fails to satisfy a Lipschitz condition.

In many practical problems the function $f(x,y)$ is not continuous, but is only piecewise continuous. An example is the equations describing the motion of a multi-stage rocket - when a burnt out stage is detached the mass changes discontinuously. If the problem is split up into several problems, each corresponding to an interval on which f is continuous, then they may individually satisfy the theorem, and the end point of the solution on one interval is used as the initial value for the problem on the next interval.

A more complete and rigorous treatment of the uniqueness of solutions of ordinary differential equations is given by Coddington and Levinson (1955).

2.4 Autonomous Systems of Differential Equations

Some papers consider only the *autonomous system* of differential equations

$$\underset{\sim}{y}' = \underset{\sim}{f}(\underset{\sim}{y}), \qquad \underset{\sim}{y}(a) = \underset{\sim}{\eta} , \tag{2.4.1}$$

that is a system where the derivatives are independent of x. Any system of differential equations can be put into the form (2.4.1) with the addition of one equation. Consider the general system of m first-order differential equations

$$\begin{aligned} y_1' &= f_1(x,y_1,y_2,\ldots,y_m), & y_1(a) &= \eta_1, \\ y_2' &= f_2(x,y_1,y_2,\ldots,y_m), & y_2(a) &= \eta_2, \\ &\vdots & &\vdots \\ y_m' &= f_m(x,y_1,y_2,\ldots,y_m), & y_m(a) &= \eta_m. \end{aligned} \tag{2.4.2}$$

Introducing the variable $y_{m+1} = x$, we have $dy_{m+1}/dx = 1$ and $y_{m+1}(a) = a$ so that

$$
\begin{aligned}
y_1' &= f_1(y_{m+1}, y_1, y_2, \ldots, y_m), & y_1(a) &= \eta_1, \\
y_2' &= f_2(y_{m+1}, y_1, y_2, \ldots, y_m), & y_2(a) &= \eta_2, \\
&\vdots & &\vdots \\
y_m' &= f_m(y_{m+1}, y_1, y_2, \ldots, y_m), & y_m(a) &= \eta_m, \\
y_{m+1}' &= 1, & y_{m+1}(a) &= a,
\end{aligned} \tag{2.4.3}
$$

which is an autonomous system. Hence any method of solution of an autonomous system of differential equations can be used for a non-autonomous system.

2.5 Graphical Solution

A method of finding an approximate solution, but only to a single first-order equation, is the *graphical method*. If (x,y) is a point on the graph of $y(x)$, the solution of

$$y' = f(x,y), \quad y(a) = \eta, \tag{2.5.1}$$

then the value $f(x,y)$ is the slope of the tangent to the solution curve at the point (x,y). A *direction field* may be drawn by evaluating $f(x,y)$ at various points in the x-y plane and drawing a small arrow of slope $f(x,y)$ from (x,y). The approximate solution is then found by sketching a curve from the point (a,η) such that the arrows are tangential to it. Figure 2.3 shows the approximate solution to (2.3.11) obtained in this way.

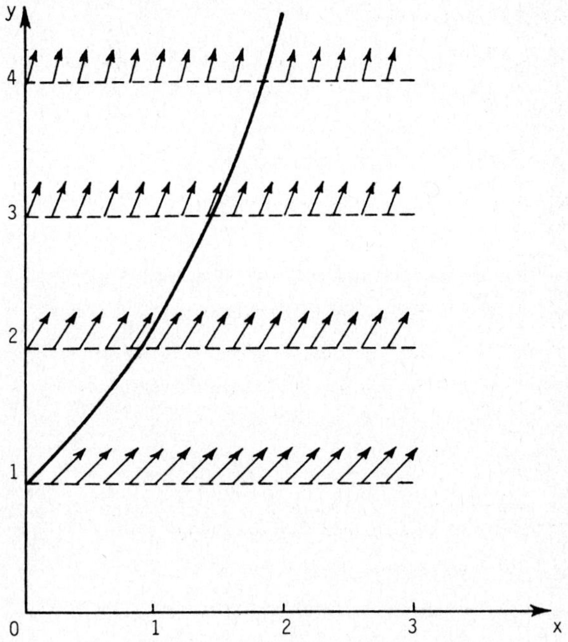

FIGURE 2.3: The graphical solution of $y'=\sqrt{y}$, $y(0)=1$.

3. TAYLOR SERIES METHODS

3.1 The Solution as a Taylor Series

The solution of the initial value problem

$$y' = f(x,y), \qquad y(a) = \eta \ , \tag{3.1.1}$$

may be expressed as the Taylor series

$$y(x) = y(a) + (x-a)y'(a) + \frac{(x-a)^2}{2!}\,y''(a) + \frac{(x-a)^3}{3!}\,y'''(a) + \ldots \tag{3.1.2}$$

provided it is infinitely differentiable at $x = a$. The second and higher derivatives in (3.1.2) may be obtained (if they exist) by repeatedly differentiating the differential equation using the chain rule. Thus

$$y(a) = \eta \ ,$$

$$y'(a) = f(x,y)\Big|_{\substack{x=a\\ y=\eta}} \ ,$$

$$y''(a) = \frac{d}{dx}\,f(x,y)\Big|_{\substack{x=a\\ y=\eta}}$$

$$= \left[\frac{\partial f}{\partial x} + \frac{\partial f}{\partial y}\frac{dy}{dx}\right]_{\substack{x=a\\ y=\eta}}$$

$$= \left[f_x + f_y\,f\right]_{\substack{x=a\\ y=\eta}} \ ,$$

$$y'''(a) = \ldots\ldots\ldots\ldots \quad \text{etc.} \tag{3.1.3}$$

In practice it is usually computationally more efficient to calculate the derivatives recursively as shown in the following example.

An example which will be used throughout this paper is the initial value problem

$$y' = 1 - 2xy, \qquad y(0) = 0 \ . \tag{3.1.4}$$

Using the linearity of the differential equation, it is easy to derive the solution

$$y = e^{-x^2}\int_0^x e^{t^2}\,dt. \tag{3.1.5}$$

This expression is known as *Dawson's integral*, and is graphed and tabulated in Abramowitz and Stegun (1965).

Now the initial condition is

$$y(0) = 0,$$

and substituting $x = 0$ into the differential equation gives

$y'(0) = 1.$

By successive differentiation of the differential equation followed by the substitution of $x = 0$, we have

$$y'' = -2y - 2xy' \quad \Rightarrow \quad y''(0) = 0,$$

$$y''' = -4y' - 2xy'' \quad \Rightarrow \quad y'''(0) = -4,$$

$$y^{(4)} = -6y'' - 2xy''' \quad \Rightarrow \quad y^{(4)}(0) = 0,$$

$$y^{(5)} = -8y''' - 2xy^{(4)} \Rightarrow \quad y^{(5)}(0) = 32. \qquad (3.1.6)$$

Substituting these values into (3.1.2) gives

$$y(x) = 0 + x(1) + \frac{x^2}{2!}(0) + \frac{x^3}{3!}(-4) + \frac{x^4}{4!}(0) + \frac{x^5}{5!}(32) + \ldots$$

$$= x - \frac{2}{3}x^3 + \frac{4}{15}x^5 + \ldots . \qquad (3.1.7)$$

An approximation to the solution at any value of x can be found by evaluating the truncated series. For example, if $x = 0.1$, we obtain

$$y(0.1) = 0.1 - \frac{2}{3}(0.001) + \frac{4}{15}(0.00001) + \ldots$$

$$\simeq 0.099336,$$

which agrees well with the tabulated value of 0.0993359924 (10D). However, many more terms would be needed to obtain the same accuracy for a much larger value of x.

This is the principal disadvantage of this method - to obtain accurate values of the solution well away from $x = a$ requires that a large number of terms of the Taylor series be used, which in turn requires that the solution must be many times differentiable at $x = a$, and that it is convergent for the particular value of x. The labour involved in calculating the derivatives may become prohibitive for more complicated problems unless computer software for symbolic manipulation including differentiation is available.

However, these problems can be overcome to some extent by using the method in a stepwise manner. No matter how few terms of the Taylor series are used (at least two always exist), an estimate of $y(a+h)$ can be obtained to any given accuracy simply by using a small enough value of h. This estimate can then be used as a new initial condition and a truncated Taylor series about $a+h$ calculated and used to advance the solution further. By repeating this procedure, the solution can be approximated for any value of x.

3.2 Euler's Method

For the present we assume that a constant steplength h is used so that an approximation to the solution of (3.1.1) is obtained at $x_n = a + nh$, $n=0,1,2,\ldots$. Usually the solution of a differential equation is required on some specific interval $[a,b]$, so we divide this interval into N equal subintervals of length h, namely

$$h = \frac{b - a}{N} . \qquad (3.2.1)$$

The approximation to $y(x_n)$ is denoted by y_n, and since $x_0 = a$ we take $y_0 = \eta$. Thus the initial value problem (3.1.1) may be written as

$$y' = f(x,y), \qquad y(x_0) = y_0 . \tag{3.2.2}$$

The solution of (3.2.2), evaluated at x_1, is

$$\begin{aligned} y(x_1) &= y(x_0 + h) \\ &= y(x_0) + hy'(x_0) + \frac{h^2}{2} y''(\xi) \\ &= y_0 + h\, f(x_0,y_0) + \frac{h^2}{2} y''(\xi) , \end{aligned} \tag{3.2.3}$$

where $x_0 < \xi < x_1$. If h is small the last term of (3.2.3) will be extremely small, so a good approximation to $y(x_1)$ is

$$y_1 = y_0 + h\, f(x_0,y_0). \tag{3.2.4}$$

Note that by using the first two terms of the Taylor series we have approximated the solution locally by its tangent at the point (x_0,y_0). The initial condition $y(x_1) = y_1$ is now used and the tangent at (x_1,y_1) to the solution of the differential equation passing through (x_1,y_1) used to approximate y_2. Repeating this procedure the approximation to $y(x_{n+1})$ is found from the tangent at (x_n,y_n) to the solution of the differential equation passing through (x_n,y_n), giving

$$y_{n+1} = y_n + h\, f(x_n,y_n). \tag{3.2.5}$$

The algorithm defined by (3.2.5) is known as *Euler's method* or the *first-order Taylor series method*, and gives rise to a polygonal approximation to the solution as shown in Figure 3.1.

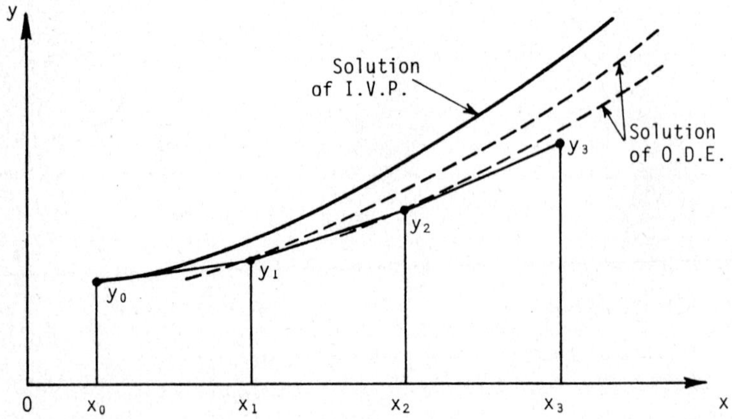

FIGURE 3.1: Polygonal approximation given by Euler's method.

Applying Euler's method to the initial value problem

$$y' = 1 - 2xy, \quad y(0) = 0 , \tag{3.2.6}$$

gives

$$y_{n+1} = y_n + h(1 - 2x_n y_n) , \tag{3.2.7}$$

with

$$x_0 = y_0 = 0$$

given by the initial condition. Table 3.1 shows a comparison of computed values for several values of h compared with the values of the solution given by Abramowitz and Stegun (1965). It can be seen that the solution improves as h is decreased - this is to be expected since the error of the approximation, the term dropped from (3.2.3), is proportional to h^2, although more steps are required to reach a particular value of x. A complete discussion of the errors is given in Sections 4.1 - 4.5.

TABLE 3.1: Numerical solutions of $y'=1-2xy$, $y(0)=0$ using Euler's method.

	h=0.1		h=0.01		h=0.001	
x_n	y_n	error	y_n	error	y_n	error
0.0	0.00000000	0.	0.00000000	0.	0.00000000	0.
0.1	0.10000000	-6.6E-04	0.09943166	-9.6E-05	0.09934584	-9.9E-06
0.2	0.19800000	-3.2E-03	0.19512735	-3.8E-04	0.19478915	-3.8E-05
0.3	0.29008000	-7.4E-03	0.28343576	-8.0E-04	0.28271262	-8.1E-05
0.4	0.37267520	-1.3E-02	0.36126548	-1.3E-03	0.36007610	-1.3E-04
0.5	0.44286118	-1.8E-02	0.42629936	-1.9E-03	0.42462282	-1.9E-04
0.6	0.49857507	-2.4E-02	0.47712328	-2.4E-03	0.47499895	-2.4E-04
0.7	0.53874606	-2.8E-02	0.51326004	-2.8E-03	0.51077896	-2.7E-04
0.8	0.56332161	-3.1E-02	0.53511193	-3.0E-03	0.53240165	-3.0E-04
0.9	0.57319015	-3.2E-02	0.54382743	-3.1E-03	0.54103328	-3.1E-04
1.0	0.57001592	-3.2E-02	0.54111554	-3.0E-03	0.53838165	-3.0E-04

Euler's method applied to the system of equations

$$\underset{\sim}{y}' = \underset{\sim}{f}(x,\underset{\sim}{y}), \quad \underset{\sim}{y}(a) = \underset{\sim}{\eta} \tag{3.2.8}$$

is simply

$$\underset{\sim}{y}_{n+1} = \underset{\sim}{y}_n + h\,\underset{\sim}{f}(x_n,\underset{\sim}{y}_n). \tag{3.2.9}$$

For example, consider the 2nd-order initial value problem

$$y'' + 2xy' + 2y = 0,$$

$$y(0) = 0, \quad y'(0) = 1. \tag{3.2.10}$$

The solution is once again Dawson's integral as this differential equation is the derivative of the previous differential equation (3.1.4). Setting $y_1 = y$ and $y_2 = y'$, (3.2.10) may be written as the system

$$y_1' = y_2, \qquad y_1(0) = 0,$$

$$y_2' = -2xy_2 - 2y_1, \qquad y_2(0) = 1, \tag{3.2.11}$$

so Euler's method is

$$\begin{bmatrix} y_{1,n+1} \\ y_{2,n+1} \end{bmatrix} = \begin{bmatrix} y_{2,n} \\ y_{2,n} \end{bmatrix} + h \begin{bmatrix} y_{2,n} \\ -2x_n y_{2,n} - 2y_{1,n} \end{bmatrix}, \tag{3.2.12}$$

with $x_0 = 0$, $y_{1,0} = 0$ and $y_{2,0} = 1$. The numerical solution of the second-order equation has exactly the same accuracy as that of the first-order equation, so to avoid duplicating the results of Table 3.1, numerical results for $h = 0.0001$ are shown in Table 3.2. Note that the tabulated values of y_2 are approximations to the first derivative of y and in many practical problems these can be most useful.

TABLE 3.2: The numerical solution of $y''+2xy'+2y = 0$, $y(0) = 0$, $y'(0) = 1$ using Euler's method with $h = 0.0001$.

x_n	$y_{1,n}$	error	$y_{2,n}$
0.0	0.00000000	0.	1.00000000
0.1	0.09933698	-9.9E-07	0.98013260
0.2	0.19475485	-3.8E-06	0.92209806
0.3	0.28263977	-8.1E-06	0.83041614
0.4	0.35995675	-1.3E-05	0.71203460
0.5	0.42445503	-1.9E-05	0.57554497
0.6	0.47478678	-2.4E-05	0.43025587
0.7	0.51053154	-2.7E-05	0.28525584
0.8	0.53213169	-3.0E-05	0.14858930
0.9	0.54075520	-3.1E-05	0.02664064
1.0	0.53810971	-3.0E-05	-0.07621941

3.3 Higher-Order Taylor Series Methods

In Euler's method, the approximation to $y(x_{n+1})$ is obtained by using the initial condition $y(x_n) = y_n$ and expanding the solution $y(x)$ as a Taylor series about x_n and retaining two terms. Taylor series methods of higher-order are obtained by using more terms. For example

$$\begin{aligned} y(x_{n+1}) &= y(x_n+h) \\ &= y(x_n) + h\,y'(x_n) + \frac{h^2}{2} y''(x_n) + \frac{h^3}{3!} y'''(\xi) \\ &= y_n + h\,y_n' + \frac{h^2}{2} y_n'' + \frac{h^3}{3!} y_n'''(\xi), \end{aligned} \tag{3.3.1}$$

where $x_n < \xi < x_{n+1}$ and the obvious notation $y_n' = y'(x_n), y_n'' = y''(x_n)$, ... etc. has been used. If h is small, the last term of Equation (3.3.1) will be negligible, so a good approximation to $y(x_{n+1})$ is

$$y_{n+1} = y_n + h\,y_n' + \frac{h^2}{2}\,y_n'' . \tag{3.3.2}$$

This is the *second-order Taylor series method*, and geometrically corresponds to approximating the solution curves locally by parabolas in place of the tangents of Euler's method. In general a pth*-order Taylor series method* is obtained by retaining the first $p+1$ terms of the Taylor series. These methods can be used for a system of differential equations in the same way as Euler's method.

The derivatives are most efficiently calculated recursively as before. For example, if the third-order Taylor series method is applied to

$$y' = 1 - 2xy, \qquad y(0) = 0, \tag{3.3.3}$$

we have

$$y_{n+1} = y_n + h\,y_n' + \frac{h^2}{2}\,y_n'' + \frac{h^3}{6}\,y_n''', \tag{3.3.4}$$

where from (3.1.6)

$$y_n' = 1 - 2x_n y_n ,$$

$$y_n'' = -2y_n - 2x_n y_n',$$

$$y_n''' = -4y_n' - 2x_n y_n''. \tag{3.3.5}$$

Table 3.3 shows the numerical solution of (3.3.3) obtained by Taylor series methods of orders two to four with $h = 0.1$. These can be compared with the numerical solution using Euler's method with $h = 0.1$ given in Table 3.1. Clearly the accuracy of the method increases with the order - this is to be expected since the term omitted is proportional to h^{p+1} for a pth-order method, and as $h < 1$ this decreases as p increases. For this example the fourth-order Taylor series method with $h = 0.1$ gives a more accurate result than Euler's method with $h = 0.00001$. Although each step of the fourth-order Taylor series method is more complicated, it is obviously computationally much more efficient than Euler's method.

TABLE 3.3: Numerical solutions of $y' = 1-2xy$, $y(0) = 0$ using Taylor series methods with $h = 0.1$.

	second-order		third-order		fourth-order	
x_n	y_n	error	y_n	error	y_n	error
0.0	0.00000000	0.	0.00000000	0.	0.00000000	0.
0.1	0.10000000	-6.6E-04	0.09933333	2.7E-06	0.09933333	2.7E-06
0.2	0.19602000	-1.3E-03	0.19473293	1.8E-05	0.19474600	5.0E-06
0.3	0.28437582	-1.7E-03	0.28258789	4.4E-05	0.28262492	6.7E-06
0.4	0.36198139	-2.0E-03	0.35986811	7.5E-05	0.35993593	7.6E-06
0.5	0.42656140	-2.1E-03	0.42432874	1.1E-04	0.42442902	7.4E-06
0.6	0.47677245	-2.0E-03	0.47462783	1.4E-04	0.47475693	6.3E-06
0.7	0.51222480	-1.7E-03	0.51034973	1.5E-04	0.51049955	4.5E-06
0.8	0.53341088	-1.3E-03	0.53193979	1.6E-04	0.53209931	2.4E-06
0.9	0.54155869	-8.3E-04	0.54056682	1.6E-04	0.54072404	2.8E-07
1.0	0.53843579	-3.6E-04	0.53793723	1.4E-04	0.53808104	-1.5E-06

4. ERRORS, CONVERGENCE, CONSISTENCY AND STABILITY

4.1 Local and Global Truncation Errors

The Taylor series methods considered in Section 3 belong to the class of *one-step methods* since only the values x_n, y_n are required to calculate y_{n+1}. A general explicit one step scheme may be written as

$$y_{n+1} = y_n + h\,\phi(x_n, y_n, h), \tag{4.1.1}$$

where $\phi(x_n, y_n, h)$ is called the *increment function*. For example, Euler's method has $\phi(x_n, y_n, h) = f(x_n, y_n)$, while the third-order Taylor series method has $\phi(x_n, y_n, h) = y_n' + h/2\, y_n'' + h^2/6\, y_n'''$, where y_n', y_n'' and y_n''' may be written as functions of x_n, y_n using (3.1.3).

If $y(x)$ is the unique solution of

$$y' = f(x,y), \qquad y(a) = \eta, \tag{4.1.2}$$

then we define the *local truncation error* T_{n+1} as

$$T_{n+1} = y(x_{n+1}) - y(x_n) - h\,\phi(x_n, y(x_n), h), \tag{4.1.3}$$

and the *global truncation error* e_{n+1} as

$$e_{n+1} = y(x_{n+1}) - y_{n+1}\ . \tag{4.1.4}$$

Many texts, including Henrici (1962) and Gear (1971), define these terms with the opposite sign, but this does not affect any of the results derived or stated here.

The truncation error T_{n+1} is local in the following sense: assuming that no previous errors have been made so that $y_n = y(x_n)$, we have from (4.1.3)

$$\begin{aligned} T_{n+1} &= y(x_{n+1}) - (y_n + h\,\phi(x_n, y_n, h)) \\ &= y(x_{n+1}) - y_{n+1}\ . \end{aligned} \tag{4.1.5}$$

Thus the local truncation error T_{n+1} is the error made in the step from x_n to x_{n+1} using the approximation (4.1.1) under the assumption of no previous errors. An alternative way of viewing the local truncation error is that it is the amount by which the solution of the initial value problem fails to satisfy the approximation (4.1.1). The global truncation error is the accumulation of the truncation errors made at each step. The local truncation error is easily found by expanding $y(x_{n+1})$ as a Taylor series about x_n. For example, the local truncation error for Euler's method is

$$\begin{aligned} T_{n+1} &= y(x_{n+1}) - y(x_n) - h\,f(x_n, y_n) \\ &= \{y(x_n) + h\,y'(x_n) + \frac{h^2}{2}\,y''(x_n) + \frac{h^3}{3!}\,y'''(x_n) + \ldots\} \\ &\qquad - y(x_n) - h\,f(x_n, y_n) \\ &= \frac{h^2}{2}\,y''(x_n) + \frac{h^2}{3!}\,y'''(x_n) + \ldots \\ &= O(h^2). \end{aligned} \tag{4.1.6}$$

The notation $T_{n+1} = O(h^2)$ means that there exists a constant K_{n+1} such that $|T_{n+1}| < K_{n+1} h^2$ for all h sufficiently small, and the local truncation error said to be of *order 2*. The first term of the truncation error, namely $\frac{1}{2}h^2y''(x_n)$, is called the principal local truncation error. Had a first degree Taylor polynomial with remainder been used in (4.1.6), we would have obtained

$$T_{n+1} = \tfrac{1}{2}h^2y''(\xi), \tag{4.1.7}$$

where $x_n < \xi < x_{n+1}$. This is precisely the term omitted in the derivation of Euler's method in Section 3.2. The local truncation errors for the Taylor series methods are simply the truncated terms of the Taylor series expansion. Thus for a pth-order Taylor series method

$$\begin{aligned} T_{n+1} &= \frac{h^{p+1}}{(p+1)!}\, y^{(p+1)}(x_n) + \ldots \\ &= O(h^{p+1}), \end{aligned} \tag{4.1.8}$$

so the local truncation error is of order $p+1$.

4.2 Convergence

A one-step method of solving the initial value problem (4.1.2) which satisfies the condition of the uniqueness theorem of Section 2.3 is said to be *convergent* if the numerical solution y_n approaches the analytic solution $y(x_n)$ at any fixed $x_n \in [a,b]$ as the step length h tends to zero and y_0 tends to η. The condition that x_n is fixed is necessary because if n was fixed $x_n = a + nh$ would approach a for all n as $h \to 0$. Since the global truncation error e_n is the difference between the numerical and analytic solution, a method is convergent if $e_n \to 0$ for all fixed $nh = x_n - a$ as $h \to 0$.

We can find the conditions for which the general one step scheme (4.1.1) is convergent as follows. Re-arranging (4.1.3) shows that the solution of the initial value problem satisfies

$$y(x_{n+1}) = y(x_n) + h\,\phi(x_n, y(x_n), h) + T_{n+1}, \tag{4.2.1}$$

and subtracting (4.1.1) gives

$$e_{n+1} = e_n + h[\phi(x_n, y(x_n), h) - \phi(x_n, y_n, h)] + T_{n+1}. \tag{4.2.2}$$

If $f(x,y)$ satisfies the conditions of the uniqueness theorem, it can be shown that $\phi(x,y,h)$ is continuous and satisfies a Lipschitz condition (Gear, 1971) so that for $x_n \in [a,b]$

$$|\phi(x_n, y(x_n), h) - \phi(x_n, y_n, h)| \le L|y(x_n) - y_n| = L|e_n|. \tag{4.2.3}$$

Assuming that the local truncation error is of order $p+1$, we have for $0 < h \le h_0$

$$|T_{n+1}| \le K\,h^{p+1}, \qquad n=0,1,2,\ldots,N-1, \tag{4.2.4}$$

where K is the maximum of $\{K_1, K_2, \ldots, K_N\}$. Therefore, from (4.2.2)

$$|e_{n+1}| \le |e_n| + L\,h|e_n| + K\,h^{p+1}, \tag{4.2.5}$$

or, putting $\alpha = 1 + Lh$ and $\beta = Kh^{p+1}$,

$$|e_{n+1}| \le \alpha|e_n| + \beta . \qquad (4.2.6)$$

Applying (4.2.6) recursively gives

$$|e_1| \le \alpha|e_0| + \beta,$$

$$|e_2| \le \alpha|e_1| + \beta \le \alpha^2|e_0| + \beta(1+\alpha),$$

$$|e_3| \le \alpha|e_2| + \beta \le \alpha^3|e_0| + \beta(1+\alpha+\alpha^2),$$

and so on, so clearly

$$|e_n| \le \alpha^n|e_0| + \beta(1+\alpha+\alpha^2+\ldots+\alpha^{n-1}). \qquad (4.2.7)$$

Now

$$1 + \alpha + \alpha^2 + \ldots + \alpha^{n-1} = \frac{\alpha^n - 1}{\alpha - 1}$$

and

$$e^{Lh} = 1 + Lh + \frac{(Lh)^2}{2!} + \ldots \ge 1 + Lh ,$$

which implies that

$$\alpha^n = (1+Lh)^n \le e^{Lnh} = \exp[L(x_n - a)],$$

so that (4.2.7) gives

$$|e_n| \le \exp[L(x_n - a)]|e_0| + Kh^p\left[\frac{\exp[L(x_n - a)] - 1}{L}\right]. \qquad (4.2.8)$$

Thus the bound on the global truncation error at x_n is made up of two parts, one being the propagated initial error e_0 and the other being the accumulation of the local truncation errors. Letting $e_0 \to 0$ and $h \to 0$ while x_n is fixed in (4.2.8), the right-hand side tends to zero if $p \ge 1$, so that $e_n \to 0$ if $p \ge 1$. Hence the one-step scheme (4.1.1) is convergent if $p + 1 \ge 2$, that is, if its local truncation error is at least order two. If $e_0 = 0$, then for $0 < h \le h_0$

$$|e_n| \le K\left[\frac{\exp[L(x_n - a)] - 1}{L}\right]h^p , \qquad (4.2.9)$$

or

$$e_n = O(h^p), \qquad (4.2.10)$$

and we say that the numerical scheme *has order* p. Note that the order of the scheme is one less than the order of the truncation error. This explains the naming of the Taylor series methods of Section 3.3. When the terms of $O(h^3)$ were omitted, that is when the local truncation error was of order 3, we called the resulting scheme the second-order Taylor series in anticipation of the above result.

The bound (4.2.8) sometimes grossly over estimates the actual error. This is for three reasons, the first being that all errors are assumed to be additive, whereas in practice they may cancel. The second reason can be illustrated by considering the numerical solution of $y' = \lambda y$, $y(0) = A$, using Euler's method. If an error e_0 is made in the initial condition, so it becomes $y(0) = A - e_0$, the solution is $y(x) = (A - e_0)e^{\lambda x}$, and the resulting error at x_n is $e_0 e^{\lambda x_n}$. Hence the error decays exponentially as it propagates if $\lambda < 0$, and grows exponentially if $\lambda > 0$. However, since $\phi(x,y,h) = f(x,y)$ for Euler's method, L is the Lipschitz constant of $f(x,y) = \lambda y$ giving $L = |\lambda|$, so that (4.2.8) puts a bound of $|e_0| e^{|\lambda| x_n}$ on the error at x_n. Thus the sign of λ is disregarded in (4.2.8) and the worst case is assumed, namely that the error grows exponentially as it propagates. Similarly the sign of λ or, more generally, the sign of $\partial f/\partial y$ is ignored in the term arising from the accumulation of the truncation errors. The third reason is that the constants K and L are chosen so that the inequalities (4.2.3) and (4.2.4) hold over the whole interval $[a,b]$, even though the local truncation error may be large in magnitude on only a very small part of the interval, while (4.2.8) assumes that this large error is made at every step.

We now apply (4.2.8) to the numerical solution of $y' = 1 - 2xy$, $y(0) = 0$ on $[0,1]$ by Euler's method. Since $\phi(x,y,h) = f(x,y)$ for Euler's method, L is a bound of $\partial f/\partial y = -2x$ so we take $L = 2$. The truncation error of Euler's method is $\frac{1}{2}h^2 y''(\xi)$, and by evaluating the second derivative from the numerical solution we found $|y''(x)| < 1.5$ on $[0,1]$, so we use $K = 0.75$. Assuming $e_0 = 0$, Equation (4.2.8) gives

$$|e_n| \le 0.75\ h\left[\frac{\exp(2x_n)-1}{2}\right], \qquad (4.2.11)$$

so at $x_n = 1$

$$|e_n| \le 2.4h\ . \qquad (4.2.12)$$

Table 3.1 shows that the errors are actually about $0.3h$ in magnitude, so in this example the bound has over estimated the error by a factor of about 8.

Gear (1971) has shown that (4.2.8), with the moduli replaced by vector norms, applies to the numerical solution of a system of ordinary differential equations by a one-step method. He has also proved that this result holds when variable length steps are taken if h is interpreted as the maximum step length.

4.3 Consistency

In the previous section it was shown that a one-step method of order $p \ge 1$ is convergent when applied to an initial value problem having a unique solution. Having its order $p \ge 1$ is equivalent to having the order of its local truncation error $p + 1 \ge 2$. If a numerical method has an order $p \ge 1$ then it is said to be *consistent*, so that a consistent method is convergent when used to solve an initial value problem which has a unique solution.

We will now see why the term "consistent" is used. The local truncation error is defined to be

$$T_{n+1} = y(x_{n+1}) - y(x_n) - h\ \phi(x_n, y(x_n), h), \qquad (4.3.1)$$

so expanding $y(x_{n+1})$ as a Taylor series about x_n and $\phi(x_n, y_n, h)$ as a Taylor series in h about $h = 0$ gives

$$\begin{aligned} T_{n+1} &= [y(x_n) + h\ y'(x_n) + \ldots] - y(x_n) - h[\phi(x_n, y(x_n), 0) + \ldots] \\ &= h[y'(x_n) - \phi(x_n, y(x_n), 0)] + \ldots \qquad (4.3.2) \end{aligned}$$

so the order of T_{n+1} is ≥ 2 if and only if

$$y'(x_n) = \phi(x_n, y(x_n), 0) = f(x_n, y(x_n)). \quad (4.3.3)$$

Thus a one-step scheme is consistent if and only if $\phi(x,y,0) = f(x,y)$. Suppose the general one-step scheme defined by (4.1.1) is consistent and hence convergent. Rearranging (4.1.1) gives

$$\frac{y_{n+1} - y_n}{h} = \phi(x_n, y_n, h), \quad (4.3.4)$$

and taking the limit as $h \to 0$ gives, since $y_n \to y(x_n)$,

$$y'(x_n) = \phi(x_n, y_n, 0) = f(x_n, y_n). \quad (4.3.5)$$

Thus in the limit $h \to 0$ the numerical scheme tends to the differential equation, so we say the numerical scheme is *consistent with the differential equation*. However, we generally use only the word "consistent" as above.

4.4 Round-Off Errors

In the analysis of errors in Section 4.2 we assumed that the values y_n satisfied the difference equation (4.1.1) exactly. However, when the y_n are calculated using a fixed number of digits, errors called round-off errors are introduced. If $\tilde{y}_n$ is the calculated solution of (4.1.1), δ_n is the error in evaluating $\phi(x_n, \tilde{y}_n, h)$, and the error in multiplying this result by h and adding it to $\tilde{y}_n$ is ρ_n, then

$$\tilde{y}_{n+1} = \tilde{y}_n + h[\phi(x_n, \tilde{y}_n, h) + \delta_n] + \rho_n . \quad (4.4.1)$$

Defining the *global error* by

$$\tilde{e}_n = y(x_n) - \tilde{y}_n , \quad (4.4.2)$$

and repeating the working of Section 4.2 with the additional bounds $|\delta_n| \leq \delta$ and $|\rho_n| \leq \rho$ leads to

$$|\tilde{e}_n| \leq |\tilde{e}_0| \exp[L(x_n - a)] + [K h^p + \delta + \frac{\rho}{h}] \left[\frac{\exp[L(x_n - a)] - 1}{L}\right]. \quad (4.4.3)$$

Note that the first term in the right-hand side of (4.4.3) is attributable to round-off - it is the error in representing the initial value η to a fixed number of digits. The error bound (4.4.3) shows that for fixed x_n the maximum global truncation error decreases as h decreases, but the maximum round-off error increases as h decreases. The latter is reasonable since decreasing h increases the number of steps to reach x_n, so that the accumulated round-off error is likely to increase. The overall effect is that the maximum error will decrease with h until at some value h_{min} it reaches a minimum, after which the maximum error increases as h is decreased (see Figure 4.1).

Although (4.2.3) is a rather crude bound,as it assumes that the round-off errors are always additive both to themselves and to the local truncation errors, it is found in practice that the errors do behave as predicted by this result. In Figure 4.2 we have plotted the errors at $x_n = 1$ in the numerical solution of (3.2.6) using Euler's method with various values of h. Results are given for two machines, a CDC CYBER 170-720 computer and a Hitachi PEACH MB-6890 minicomputer.

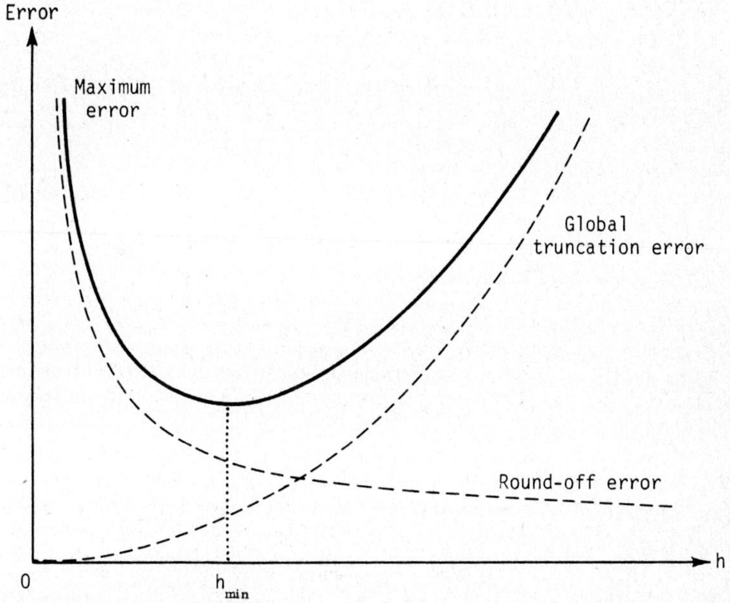

FIGURE 4.1: The maximum error as a function of h.

For $h \geq 10^{-3}$ the errors almost lie on a straight line of slope one; the global truncation error dominates for large h and is proportional to h since Euler's method is first-order. For $h \leq 10^{-5}$ the errors given by the PEACH become larger due to the build up of round-off error. The error actually changes sign between $h = 10^{-4}$ and $h = 10^{-5}$, showing that the global truncation error and round-off error have different signs which explains the smaller than expected error at $h = 10^{-4}$. The CYBER has about 14 significant digits compared to about 7 for the PEACH, so the errors given by the CYBER remain close to the line for much smaller h due to the smaller round-off errors, and only start to increase at $h = 10^{-8}$.

Whether or not the round-off errors become a problem depends on two things, the required accuracy of the solution and the number of significant digits used in the computation. Presumably the accuracy requirement is fixed, so all one can do is to reduce the round-off error by using more accurate arithmetic. Double-precision arithmetic can be used if it is available, but this will decrease the speed of the computation considerably. A much less expensive alternative is to use *partial double precision*. It can be seen from (4.4.3) that the trouble-some term is ρ/h which arises from the error in the multiplication of $\phi(x_n, y_n, h)$ by h and adding the result to y_n. If these operations only are performed in double precision, then ρ and hence the total round-off error will be reduced considerably, while the evaluation of $\phi(x_n, y_n, h)$, which is the bulk of the computation, is still performed in single precision.

Unless they are specifically mentioned, round-off errors are ignored in the rest of this paper. Nevertheless, one must be mindful of their presence in any calculated result.

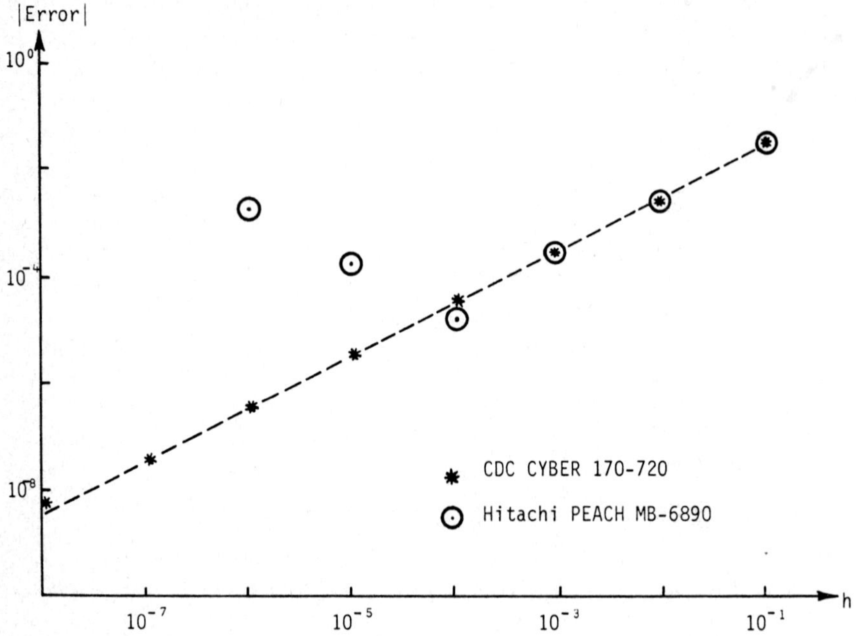

FIGURE 4.2: Errors at $x=1$ in the numerical solution of $y' = 1-2xy$, $y(0) = 0$ using Euler's method.

4.5 Stability

In general terms, a numerical process is said to be *unstable* if an error, such as a round-off error, introduced at some stage of the calculation becomes unbounded as the calculation proceeds, and is *stable* otherwise. Recall that when we are calculating y_{n+1} we are trying to follow the solution $y^*(x)$ of the differential equation which passes through the previously calculated point y_n (see Figure 4.3).

Now

$$\begin{aligned} e_{n+1} &= y(x_{n+1}) - y_{n+1} \\ &= [y(x_{n+1}) - y^*(x_{n+1})] + [y^*(x_{n+1}) - y_{n+1}] . \qquad (4.5.1) \end{aligned}$$

The second term in (4.5.1) is the error made in trying to follow $y^*(x)$ from x_n to x_{n+1} and is called the *local error or one-step error*.

The local error is not quite the same as the local truncation error, the former being the amount by which $y^*(x)$ does not satisfy the approximation (4.1.1), while the latter is the amount by which the analytic solution $y(x)$ does not satisfy (4.1.1). However, in general they will have the same order. Unless the numerical solution is very inaccurate, they will be approximately equal since $y(x)$ and $y^*(x)$ will not be very different.

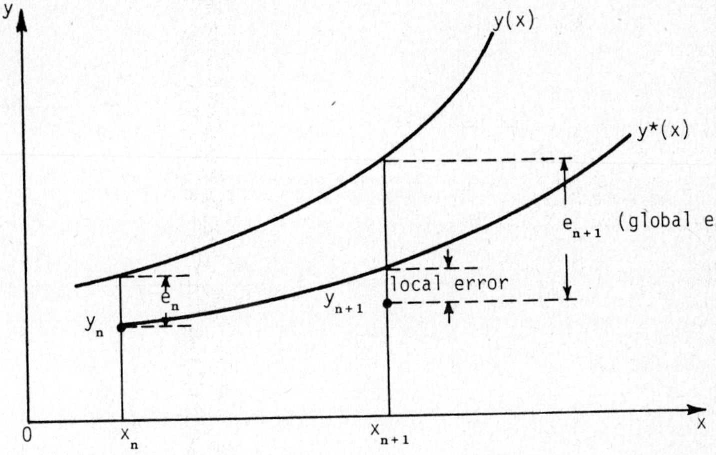

FIGURE 4.3: Error propagation in the $(n+1)^{th}$ step.

The first term of (4.5.1) is the propagated value of the global error e_n, and whether or not it is propagated in a stable manner depends on the solutions $y(x)$ and $y^*(x)$. If $\partial f/\partial y > 0$, the two solutions continue to diverge, so the error continues to grow and the method is unstable. If $\partial f/\partial y > 0$ the differential equation is termed *unstable*, and any method used to solve it is *inherently unstable*. However, if $\partial f/\partial y < 0$ the solution curves converge (see Figure 4.4) and the differential equation is *stable*, and a numerical method used to solve it may or may not be stable.

Obviously stability, as defined above, is not only a property of the numerical method but also of the differential equation it is used to solve. In order that we can talk about the stability of the numerical method, we always examine the stability when it is used to solve the test problem $y' = \lambda y$, where λ is a complex constant. There are two reasons why this differential equation is chosen. Firstly, if a numerical scheme is unstable when used to solve such a simple problem it is unlikely to be stable for more complicated problems. Secondly, any differential equation can be locally linearized and put into this form. Now

$$\begin{aligned} y' &= f(x,y) \\ &= f(a,b) + (x-a)f_x(a,b) + (y-b)f_y(a,b) + \dots , \end{aligned} \tag{4.5.2}$$

and in a neighbourhood of (a,b) the higher order terms can be omitted. A simple change of variables then brings the equation to the form $y = \lambda y$. Linearizing a system of m equations in this way leads to the linear system $\underset{\sim}{Y}' = \underline{A}\underset{\sim}{Y}$, where $\underline{A}$ is the $m \times m$ Jacobian matrix

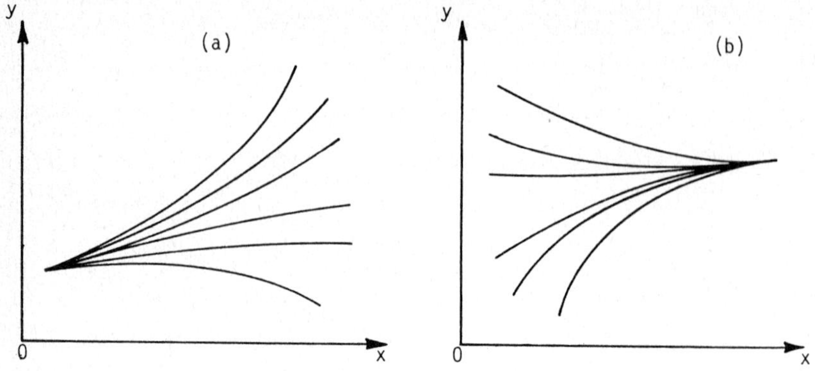

FIGURE 4.4: Solutions of (a) an unstable differential equation ($\partial f/\partial y>0$), and (b) a stable differential equation ($\partial f/\partial y<0$).

$[\partial f_i/\partial y_j]$. If $\lambda_1,\lambda_2,\ldots,\lambda_m$ are distinct eigenvalues of the matrix $\underline{\underline{A}}$ there exists an orthogonal matrix $\underline{\underline{H}}$ so that $\underline{\underline{H}}^T\underline{\underline{A}}\underline{\underline{H}} = \underline{\underline{\Lambda}} = \text{diag}(\lambda_1,\lambda_2,\ldots,\lambda_m)$. Making the transformation $\underset{\sim}{Y} = \underline{\underline{H}}\underset{\sim}{y}$ gives $\underline{\underline{H}}\underset{\sim}{y}' = \underline{\underline{A}}\underline{\underline{H}}\underset{\sim}{y}$, and premultiplying by $\underline{\underline{H}}^T$ leads to $\underset{\sim}{y}' = \underline{\underline{\Lambda}}\underset{\sim}{y}$. Thus we have m equations of the form $y_i' = \lambda_i y_i$, where the eigenvalues λ_i will be complex in general, which is the reason for permitting λ to be complex in our test problem. Thus the constant λ in the differential equation $y' = \lambda y$ corresponds either to $\partial f/\partial y$ for a single equation or the eigenvalues of the Jacobian matrix for a system of equations.

When a one-step method is used to solve $y' = \lambda y$ it gives

$$y_{n+1} = r_1(\lambda h)y_n \,, \tag{4.5.3}$$

where the quantity $r_1(\lambda h)$ depends on the particular method. Since the solution of $y' = \lambda y$ which passes through y_n is $y^*(x) = y_n e^{\lambda x}$, a method of order p (which has a local truncation error of order $p+1$) will give

$$y_{n+1} = y_n e^{\lambda h} + O(h^{p+1}) \,. \tag{4.5.4}$$

Equating (4.5.3) and (4.5.4) gives

$$r_1(\lambda h) = e^{\lambda h} + O(h^{p+1}) \,. \tag{4.5.5}$$

Referring to (4.5.3) we see that introducing an error ε into y_{n+1} results in an error $r_1(\lambda h)\varepsilon$ in y_{n+1}, and in every subsequent step the error is again multiplied by the factor $r_1(\lambda h)$. Hence the error will remain bounded if $|r_1(\lambda h)| \le 1$. However, to conform with the definition of stability for a linear multistep method, we exclude the equality. The error will therefore decay, and we therefore say the scheme is *absolutely stable* if $|r_1(\lambda h)| < 1$. The region in the complex plane in which λh must lie so that the scheme is absolutely stable is called the *region of absolute stability* and its intersection with the real axis is called the *interval of absolute stability*.

Applying Euler's method to $y' = \lambda y$ gives

$$y_{n+1} = y_n + h(\lambda y_n) = (1 + \lambda h)y_n , \tag{4.5.6}$$

so $r_1 = 1 + \lambda h$. Equation (4.5.5) is satisfied since r_1 equals the first two terms of the Taylor series for $e^{\lambda h}$. Hence Euler's method is absolutely stable if $|1 + \lambda h| < 1$, which implies that the region of absolute stability is the interior of the unit circle centred at $(-1,0)$, and the interval of convergence is $(-2,0)$. It may be easily verified that for the pth-order Taylor series method r_1 is equal to the first $p+1$ terms of the Taylor series for $e^{\lambda h}$. The intervals of absolute stability for Taylor series method of orders 1 to 6 are given in Table 4.1.

TABLE 4.1: Intervals of absolute stability of Taylor series methods

Order	Interval of absolute stability
1	(-2,0)
2	(-2,0)
3	(-2.51,0)
4	(-2.78,0)
5	(-3.21,0)
6	(-3.55,0)

It can be seen from Table 4.1 that none of the intervals of absolute stability contain any part of the positive real axis. This is not surprising, for if λ is real and positive then the differential equation is unstable and any numerical method is inherently unstable. Nevertheless it is found that these methods do give good solutions for $\lambda > 0$. This is because the solution of $y' = \lambda y$ is increasing, as is the numerical solution, and although errors will grow, they only grow at the same rate as the numerical solution. However, if the method is not absolutely stable when $\lambda < 0$ the error will increase and soon "swamp" the decreasing solution of the differential equation.

Note that absolute stability is not the only criterion for success of a method. For example, Euler's method with $\lambda h = -3/2$ will give a numerical solution which decays in an oscillatory way, whereas the analytic solution decays monotonically. We will see later that absolute stability is important for stiff differential equations, but for non-stiff equations it is accuracy which is of primary importance, and this forces the use of a small enough step length that the method is absolutely stable when $\text{Re}(\lambda h) < 0$.

5. RUNGE-KUTTA METHODS

5.1 Second-Order Runge-Kutta Methods

The simplest one-step method is Euler's method, but its low order means that it is not particularly accurate and so not of much practical use. The higher-order Taylor series methods overcome the accuracy problem, but at a cost of the calculation of higher derivatives which tend to become quite complicated for all but the simplest differential equation. The Runge-Kutta methods attain higher-order accuracy at a lower cost by taking the increment function $\phi(x_n, y_n, h)$ to be a weighted average of first derivatives at points in the interval $[x_n, x_{n+1}]$.

Suppose we take

$$\phi(x_n, y_n, h) = c_1 f(x_n, y_n) + c_2 f(x_n + a_2 h, y_n + b_{21} h\, f(x_n, y_n)) \qquad (5.1.1)$$

where c_1, c_2, a_2 and b_{21} are constants to be determined. This expression can be calculated by evaluating $f(x,y)$ twice. Expanding the second term about (x_n, y_n) gives

$$\phi(x_n, y_n, h) = c_1 f + c_2 f + c_2 a_2\, h f_x + c_2 b_{21} h\, f\, f_y + O(h^2), \qquad (5.1.2)$$

where f and its partial derivatives are all evaluated at (x_n, y_n). Thus the local truncation error is

$$\begin{aligned} T_{n+1} &= y(x_{n+1}) - y(x_n) - h\,\phi(x_n, y_n, h) \\ &= [y + hf + \frac{h^2}{2}(f_x + ff_y) + O(h^3)] - y \\ &\qquad - h[(c_1+c_2)\,f + c_2 a_2 h\, f_x + c_2 b_{21} h\, ff_y + O(h^2)] \\ &= hf(1-c_1-c_2) + h^2 f_x(\tfrac{1}{2}-c_2 a_2) + h^2 ff_y(\tfrac{1}{2}-c_2 b_{21}) + O(h^3). \qquad (5.1.3) \end{aligned}$$

Here the arguments (x_n) and (x_n, y_n) have again been omitted, $y(x_{n+1})$ has been expanded about x_n and the derivatives of $y(x_n)$ have been expressed in terms of f and its partial derivatives (see Equations (3.1.3)). The local truncation error is third-order if

$$\begin{aligned} c_1 + c_2 &= 1, \\ c_2 a_2 &= 1/2, \\ c_2 b_{21} &= 1/2. \end{aligned} \qquad (5.1.4)$$

Note that the first of these conditions ensures that the local truncation error is at least second-order and hence that the scheme is consistent, a result which could have been deduced directly from (5.1.1) using the condition $\phi(x,y,0) = f(x,y)$. As there are four unknowns in the three equations (5.1.4), there are infinitely many solutions given by

$$\begin{aligned} c_1 &= 1 - \gamma, \\ c_2 &= \gamma, \\ a_2 &= b_{21} = \frac{1}{2\gamma}, \end{aligned} \qquad (5.1.5)$$

for any value $\gamma \neq 0$.

Two particular values of γ yield well known methods. If $\gamma = 1$ the resulting method is

$$y_{n+1} = y_n + h\, f(x_n + \tfrac{1}{2}h,\ y_n + \tfrac{1}{2}h\, f(x_n, y_n)), \tag{5.1.6}$$

or, as it is more commonly written,

$$k_1 = f(x_n, y_n),$$

$$k_2 = f(x_n + \tfrac{1}{2}h,\ y_n + \tfrac{1}{2}h\, k_1),$$

$$y_{n+1} = y_n + h\, k_2. \tag{5.1.7}$$

This method is referred to as the *modified Euler or improved polygon method.*

To illustrate how this method approximates y_{n+1} (see Figure 5.1), we write (5.1.6) as

$$\bar{y}_{n+\frac{1}{2}} = y_n + \tfrac{1}{2}h\, f(x_n, y_n), \tag{5.1.8a}$$

$$y_{n+1} = y_n + h\, f(x_n + \tfrac{1}{2}h,\ \bar{y}_{n+\frac{1}{2}}). \tag{5.1.8b}$$

Euler's method is used with a step length of $\tfrac{1}{2}h$ to approximate $y^*(x_n + \tfrac{1}{2}h)$ by $\bar{y}_{n+\frac{1}{2}}$ using Equation (5.1.8a), and y_{n+1} is found by moving along a line parallel to the tangent to the solution passing through $(x_n + \tfrac{1}{2}h,\ \bar{y}_{n+\frac{1}{2}})$ using Equation (5.1.8b).

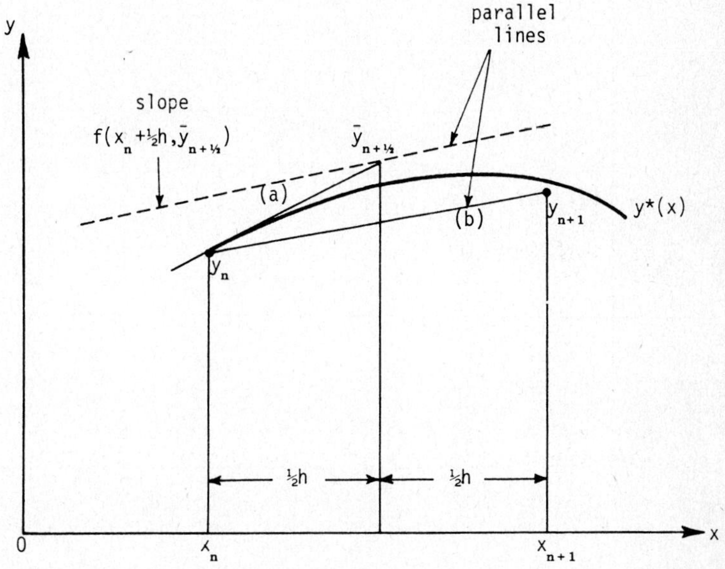

FIGURE 5.1: Geometrical interpretation of the modified Euler method.

Another choice is $\gamma = \tfrac{1}{2}$ which gives the *improved Euler method*

$$y_{n+1} = y_n + \tfrac{1}{2}h[f(x_n, y_n) + f(x_n + h,\ y_n + h\, f(x_n, y_n))], \tag{5.1.9}$$

which may be written

$$k_1 = f(x_n, y_n),$$

$$k_2 = f(x_n + h,\ y_n + h\,k_1),$$

$$y_{n+1} = y_n + \tfrac{1}{2}h\,(k_1 + k_2). \qquad (5.1.10)$$

This may be written as

$$\bar{y}_{n+1} = y_n + h\,f(x_n, y_n), \qquad (5.1.11a)$$

$$y_{n+1} = y_n + \tfrac{1}{2}h[f(x_n, y_n) + f(x_n + h, \bar{y}_{n+1})] \qquad (5.1.11b)$$

where $\bar{y}_{n+1}$ is an approximate value of $y^*(x_n)$ calculated by Euler's method, and y_{n+1} lies on a line whose slope is the average of the slopes of the tangents to solutions through (x_n, y_n) and $(x_{n+1}, \bar{y}_{n+1})$. Thus Equation (5.1.11a) may be thought of as a predictor for y_{n+1}, and (5.1.11b) as a corrector as it gives a corrected value for y_{n+1} (see Figure 5.2). Predictor-corrector methods are an important class of methods which will be examined later.

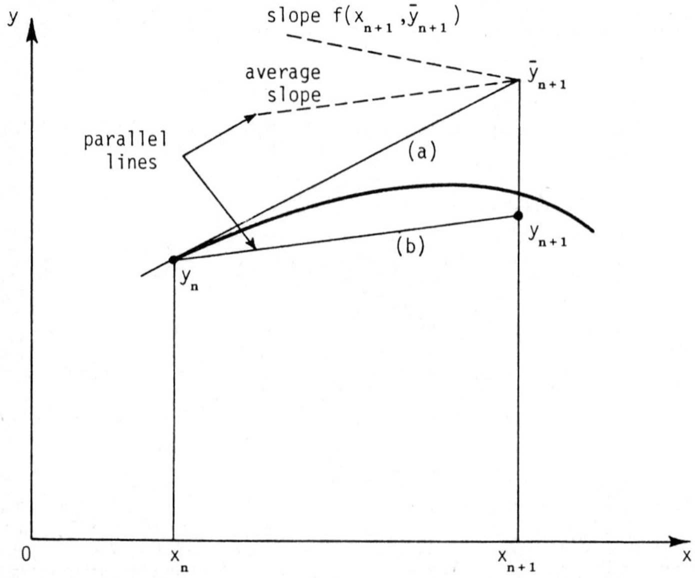

FIGURE 5.2: Geometrical interpretation of the improved Euler method.

The above choices of γ were made to give simple coefficients suitable for hand calculations. If terms had been retained to one order higher in h in the expansions leading to the expression for the local truncation error (5.1.3) and then c_1, c_2, a_2 and b_{21} eliminated using (5.1.5), it would have been found that

$$T_{n+1} = \frac{h^3}{6}\left[(1 - \frac{3}{4\gamma})(f_{xx} + f^2 f_{yy} + 2ff_{xy}) + f_x f_y + ff_y^2\right] + O(h^4). \qquad (5.1.12)$$

Clearly no choice of γ can make the local truncation error fourth-order, nor can γ be chosen to minimise the local truncation error for a general function $f(x,y)$. Ralston (1965) proposed that γ could be specified so as to minimise a bound on T_{n+1}. He used bounds suggested by Lotkin (1951) where constants P and Q are found so that f and its derivatives satisfy

$$|f(x,y)| < Q,$$

$$\left|\frac{\partial^{i+j} f}{\partial x^i \partial y^j}\right| < \frac{P^{i+j}}{Q^{j-1}}, \qquad (5.1.13)$$

for $x \in [a,b]$, $y \in (-\infty,\infty)$ and $i+j<p$, where p is the order of the method (in this case 2). This form of bound was chosen so that the terms in the local truncation error are bounded by the same expression of P and Q. Thus in Equation (5.1.12) f_{xx}, $f^2 f_{yy}$,... and so on, are all bounded by P^2Q. Consequently, for any second-order Runge-Kutta method

$$|T_{n+1}| < \frac{h^3}{6}\left[4\left|1 - \frac{3}{4\gamma}\right|P^2Q + 2P^2Q\right], \qquad (5.1.14)$$

so the local truncation error bound is a minimum if $\gamma = 3/4$, when

$$|T_{n+1}| < \frac{1}{3} h^3P^2Q. \qquad (5.1.15)$$

Thus the second-order Runge-Kutta scheme which has the minimum error bound is

$$y_{n+1} = y_n + \frac{h}{4}[f(x_n,y_n)+3f(x_n + \tfrac{2}{3}h,\ y_n + \tfrac{2}{3}h\ f(x_n,y_n))], \qquad (5.1.16)$$

or, alternatively

$$k_1 = f(x_n,y_n),$$

$$k_2 = f(x_n + \tfrac{2}{3}h,\ y_n + \tfrac{2}{3}h\ k_1),$$

$$y_{n+1} = y_n + \frac{h}{4}(k_1 + 3k_2). \qquad (5.1.17)$$

Table 5.1 shows the numerical solution of the standard example

$$y' = 1 - 2xy, \qquad y(0) = 0,$$

by the above methods using $h = 0.1$. For this problem the modified Euler method has given the best results. Note that this is not a contradiction to the above result that (5.1.17) has the minimum error bound, for it is the bound that is minimized, not the actual local truncation error.

However, the minimum error bound scheme is probably the best scheme to use for the following reasons. Equation (5.1.12) is of the form

$$T_{n+1} = \frac{h^3}{6}\left[(1 - \frac{3}{4\gamma})A+B\right] + O(h^4) \qquad (5.1.19)$$

so that, if the higher order terms are ignored, either

$$|T_{n+1}(\gamma < \tfrac{3}{4})| < |T_{n+1}(\gamma = \tfrac{3}{4})| < |T_{n+1}(\gamma > \tfrac{3}{4})|, \qquad (5.1.20)$$

TABLE 5.1: The numerical solution of $y' = 1-2xy$, $y(0) = 0$ using second-order Runge-Kutta methods with $h = 0.1$.

	Modified Euler		Improved Euler		Minimum Error Bound	
x_n	y_n	error	y_n	error	y_n	error
0.0	0.00000000	0.	0.00000000	0.	0.00000000	0.
0.1	0.09950000	-1.6E-04	0.09900000	3.4E-04	0.09933333	2.7E-06
0.2	0.19504485	-2.9E-04	0.19406960	6.8E-04	0.19471978	3.1E-05
0.3	0.28298765	-3.6E-04	0.28159900	1.0E-03	0.28252482	1.1E-04
0.4	0.36027279	-3.3E-04	0.35856291	1.4E-03	0.35970297	2.4E-04
0.5	0.42464522	-2.1E-04	0.42272650	1.7E-03	0.42400593	4.3E-04
0.6	0.47476980	-6.6E-06	0.47276294	2.0E-03	0.47410131	6.6E-04
0.7	0.51025293	2.5E-04	0.50827497	2.2E-03	0.50959429	9.1E-04
0.8	0.53157264	5.3E-04	0.52972640	2.4E-03	0.53095812	1.1E-03
0.9	0.53993468	7.9E-04	0.53830098	2.4E-03	0.53939119	1.3E-03
1.0	0.53707998	1.0E-03	0.53571321	2.4E-03	0.53662561	1.5E-03

or

$$|T_{n+1}(\gamma < \tfrac{3}{4})| > |T_{n+1}(\gamma = \tfrac{3}{4})| > |T_{n+1}(\gamma > \tfrac{3}{4})| , \qquad (5.1.21)$$

unless $A = 0$ when all schemes have the same error. We do not know which of these situations occurs, and it probably will alter as we work our way across the interval, so we might as well use the scheme with $\gamma = 3/4$ as this cannot produce the least accurate scheme in any circumstance. It also has the added advantage of producing a third-order scheme when $f(x,y)$ is independent of y, since then $B = 0$ (see Equation (5.1.12)).

5.2 Higher-Order Runge-Kutta Methods

The general *R-stage Runge-Kutta method* is defined by

$$k_r = f(x_n + ha_r , y_n + h \sum_{s=1}^{R} b_{rs} k_s), \qquad r=1,2,3,\ldots,R,$$

$$y_{n+1} = y_n + h \sum_{r=1}^{R} c_r k_r . \qquad (5.2.1)$$

It is *explicit* if $b_{rs} = 0$ for $s \geq r$ and *implicit* otherwise, although the explicit Runge-Kutta methods are usually referred to simply as "Runge-Kutta methods", a practice we adopt. In the previous section, we considered a 2-stage method and found that a second-order scheme resulted if three equations in four unknowns were satisfied.

As the number of stages increases, so does the complexity of the algebra. Ralston (1965) shows that to obtain a fourth-order 4-stage method, 13 equations in 11 unknowns must be satisfied, so once again there are infinitely many schemes. The most common scheme is [see Runge (1895), and Kutta (1901)]

$$k_1 = f(x_n, y_n),$$

$$k_2 = f(x_n + \tfrac{1}{2}h,\ y_n + \tfrac{1}{2}h\ k_1),$$

$$k_3 = f(x_n + \tfrac{1}{2}h,\ y_n + \tfrac{1}{2}h\ k_2),$$

$$k_4 = f(x_n + h,\ y_n + h\ k_3),$$

$$y_{n+1} = y_n + \frac{h}{6}(k_1 + 2k_2 + 2k_3 + k_4), \qquad (5.2.2)$$

which is usually referred to as the *classical Runge-Kutta method*. Note that if $f(x,y)$ is independent of y then $k_2 = k_3$ and (5.2.1) is equivalent to integrating the differential equation using Simpson's rule for quadrature.

Another fourth-order Runge-Kutta method is due to Gill (1951). The scheme, which we refer to as the *Runge-Kutta-Gill method*, is

$$k_1 = f(x_n, y_n),$$

$$k_2 = f(x_n + \tfrac{1}{2}h,\ y_n + \tfrac{1}{2}h\ k_1),$$

$$k_3 = f(x_n + \tfrac{1}{2}h,\ y_n + (-\tfrac{1}{2} + 1/\sqrt{2})h\ k_1 + (1 - 1/\sqrt{2})h\ k_2),$$

$$k_4 = f(x_n + h,\ y_n - hk_2/\sqrt{2} + (1 + 1/\sqrt{2})h\ k_3),$$

$$y_{n+1} = y_n + \frac{h}{6}[k_1 + 2(1 - 1/\sqrt{2})k_2 + 2(1 + 1/\sqrt{2})k_3 + k_4]. \qquad (5.2.3)$$

This scheme also reduces to Simpson's rule if $f(x,y)$ is independent of y.

The computational form of Gill's scheme is

$$k_1 = f(x_n, y_n),$$

$$w_1 = y_n + \tfrac{1}{2}(hk_1 - 2q_4),$$

$$q_1 = q_4 + \frac{3}{2}(hk_1 - 2q_4) - \tfrac{1}{2}hk_1\ ,$$

$$k_2 = f(x_n + \tfrac{1}{2}h, w_1),$$

$$w_2 = w_1 + (1 - 1/\sqrt{2})(hk_2 - q_1),$$

$$q_2 = q_1 + 3(1 - 1/\sqrt{2})(hk_2 - q_1) - (1 - 1/\sqrt{2})hk_2,$$

$$k_3 = f(x_n + \tfrac{1}{2}h, w_2),$$

$$w_3 = w_2 + (1 + 1/\sqrt{2})(hk_3 - q_2),$$

$$q_3 = q_2 + 3(1 + 1/\sqrt{2})(hk_3 - q_2) - (1 + 1/\sqrt{2})hk_3,$$

$$k_4 = f(x_n + h, w_3),$$

$$y_{n+1} = w_3 + \frac{1}{6}(hk_4 - 2q_3),$$

$$q_4 = q_3 + \tfrac{1}{2}(hk_4 - 2q_3) - \tfrac{1}{2}hk_4, \qquad (5.2.4)$$

where $q_4 = 0$ is used for the first step. This scheme allows the k's and q's to be overwritten, and the w_1, w_2, w_3, y_{n+1} to overwrite y_n. Thus for a system of equations only three arrays need storing, compared with four for the classical Runge-Kutta method. The role of the q's is to reduce round-off error. If exact arithmetic was used q_4 would be zero, but in practice it will be non-zero and a correction is made for this error in the next step. The lower storage requirement and reduced round-off errors made this scheme extremely popular in the early days of digital computers, and these features make it a good scheme for a microcomputer, a type of machine increasing in popularity. Blum (1962) has shown that the classical Runge-Kutta method can be organised in such a way that the same advantages are realised. Shampine (1979) discusses techniques for minimising the storage requirements of general Runge-Kutta methods.

The local truncation error of the classical Runge-Kutta method is bounded by

$$|T_{n+1}| < \frac{73}{720} h^5 P^4 Q, \tag{5.2.5}$$

where P and Q are constants chosen so that the bounds (5.1.13) hold (Lotkin, 1951). Ralston (1965) has shown that the minimum error bound

$$|T_{n+1}| < 0.0546\, h^5 P^4 Q \tag{5.2.6}$$

is achieved by the scheme

$$\begin{aligned}
k_1 &= f(x_n, y_n),\\
k_2 &= f(x_n + 0.4h,\ y_n + 0.4h\,k_1),\\
k_3 &= f(x_n + 0.45573725h,\ y_n + 0.29697761h\,k_1 + 0.15875964h\,k_2)\\
k_4 &= f(x_n + h,\ y_n + 0.21810040h\,k_1 - 3.05096516h\,k_2\\
&\qquad + 3.83286476h\,k_3),\\
y_{n+1} &= y_n + h(0.17476028k_1 - 0.55148066k_2\\
&\qquad + 1.20553560k_3 + 0.17118478k_4).
\end{aligned} \tag{5.2.7}$$

Because of the large coefficients, some of which are necessarily negative, this scheme is more susceptible to round-off error growth than the other schemes.

The numerical solution of

$$y' = 1 - 2xy, \qquad y(0) = 0, \tag{5.2.8}$$

using the above fourth-order Runge-Kutta methods with $h = 0.1$ are shown in Table 5.2. The classical method and Gill's method have produced identical results because the round-off errors are insignificant, while the minimum error bound scheme of Ralston has produced slightly more accurate values.

So far we have considered 2-stage and 4-stage Runge-Kutta methods and found that schemes of order two and four respectively can be obtained. However, it is not always possible to derive an Rth-order scheme using R-stages. Butcher (1965) has shown that if $p^*(R)$ is the maximum

TABLE 5.2: The numerical solution of $y' = 1 - 2xy$, $y(0) = 0$ using fourth-order Runge-Kutta methods with $h = 0.1$

	Classical Runge-Kutta		Runge-Kutta-Gill		Minimum error bound	
x_n	y_n	error	y_n	error	y_n	error
0.0	0.00000000	0.	0.00000000	0.	0.00000000	0.
0.1	0.09933582	1.7E-07	0.09933583	1.7E-07	0.09933599	-1.5E-09
0.2	0.19475069	3.4E-07	0.19475069	3.4E-07	0.19475103	-1.3E-09
0.3	0.28263113	5.4E-07	0.28263113	5.4E-07	0.28263165	1.5E-08
0.4	0.35994271	7.7E-07	0.35994271	7.7E-07	0.35994341	6.9E-08
0.5	0.42443534	1.0E-06	0.42443534	1.0E-06	0.42443619	1.9E-07
0.6	0.47476181	1.4E-06	0.47476181	1.4E-06	0.47476281	4.0E-07
0.7	0.51050226	1.8E-06	0.51050226	1.8E-06	0.51050335	7.1E-07
0.8	0.53209947	2.2E-06	0.53209947	2.2E-06	0.53210060	1.1E-06
0.9	0.54072164	2.7E-06	0.54072164	2.7E-06	0.54072274	1.6E-06
1.0	0.53807645	3.1E-06	0.53807645	3.1E-06	0.53807746	2.0E-06

attainable order for a R-stage Runge-Kutta method, then

$$p^*(R) \begin{cases} = R & R \le 4 \\ = R-1 & R = 5,6,7 \\ \le R-2 & R \ge 8\ . \end{cases} \tag{5.2.9}$$

Hence two extra stages (and function evaluations) are required to increase the order of a scheme from four to five, a result which probably explains the popularity of fourth-order methods. However, the important criterion is to minimize the total number of function evaluations required to find the solution to a given accuracy over an interval, and schemes of order greater than four may be better in this regard; in fact, for sufficiently high accuracy, high order schemes must be more efficient. High order Runge-Kutta formulas are given in Butcher (1964), Luther and Konen (1965) and Luther (1966, 1968) - see also the references in Sections 5.3 and 6.2.

It is quite straightforward to use Runge-Kutta methods to solve a system of m ordinary differential equations, as the y_n's and k_r's are then just m-vectors. However, if the order of the method for a single equation is p then the order for a system may be less than p if $p > 4$, but the order will be unchanged if $p \le 4$. Due to the freedom of choice in the coefficients of Runge-Kutta methods, any extra conditions that arise when solving a system of equations can usually be satisfied, so that most high order schemes do not suffer any loss of order when used to solve such systems. However, note that the error bounds quoted above do not apply to systems of equations. Similar bounds can be derived - see for example Henrici (1962).

5.3 Stability of Runge-Kutta Methods

It can be shown that when an R-stage (explicit) Runge-Kutta method is used to solve $y' = \lambda y$ the scheme reduces to

$$y_{n+1} = r_1(\lambda h)y_n, \qquad (5.3.1)$$

where r_1 is a polynomial of degree R. In Section 4.5 it was deduced that

$$r_1(\lambda h) = e^{\lambda h} + O(h^{p+1})$$

for a method of order p. Consequently if an R-stage Runge-Kutta has order R (possible for $R \le 4$) then

$$r_1(\lambda h) = 1 + \lambda h + \ldots + \frac{(\lambda h)^R}{R!}, \qquad (5.3.2)$$

a result which is identical to that for a Taylor series method of order R. It follows that for any given $R \le 4$, all stage Runge-Kutta methods of order R have the same region of absolute stability given by $|r_1(\lambda h)| < 1$. Plots of these regions may be found in Lambert (1973) page 227, and the corresponding intervals of absolute stability are given in Table 4.1.

If an R-stage method has order $p < R$ (which is always the case for $p > 4$), then

$$r_1(\lambda h) = 1 + \lambda h + \ldots + \frac{(\lambda h)^P}{P!} + \gamma_{p+1}(\lambda h)^{p+1} + \ldots + \gamma_R(\lambda h)^R,$$

where $\gamma_{p+1}, \gamma_{p+2}, \ldots, \gamma_R$ depend on the parameters in terms of which the coefficients of the Runge-Kutta method may be written once the order requirements have been satisfied. One criterion for choosing the parameters has been seen, that of minimizing the local truncation error bound. Another is to choose the parameters so the region of absolute stability is maximized. Lambert (1973) considers the example of a three-stage order two method for which

$$r_1 = 1 + \lambda h + \frac{(\lambda h)^2}{2} + \gamma_3(\lambda h)^3.$$

Choosing the parameters so that γ_3 takes the values 0, 1/6 and 1/12 gives intervals of absolute stability of (-2,0), (-2.51,0) and (-4.52,0) respectively. In this example the larger interval is obtained at a cost of lower order (and hence lower accuracy), but if $R > 4$ then R is larger than the maximum attainable order, so the region of stability may be maximized while the maximum possible order is maintained. Lawson (1966, 1967) has derived Runge-Kutta methods of order five and six with extended regions of absolute stability.

5.4 Implicit Runge-Kutta Methods

The general R-stage Runge-Kutta method was defined in Section 5.2 (Equation (5.2.1)), and is *implicit* if there is at least one $b_{rs} \neq 0$ for $s \ge r$ so that at least one k_r is defined implicitly. Butcher (1963) has shown that for any $R \ge 2$ there exists an R-stage implicit Runge-Kutta method of order $2R$. For example, the following 2-stage scheme of Hammer and Hollingsworth (1955) has order four:

$$k_1 = f(x_n + (\tfrac{1}{2} + \tfrac{\sqrt{3}}{6})h,\ y_n + \tfrac{1}{4}h\,k_1 + (\tfrac{1}{4} + \tfrac{\sqrt{3}}{6})h\,k_2)$$

$$k_2 = f(x_n + (\tfrac{1}{2} - \tfrac{\sqrt{3}}{6})h,\ y_n + (\tfrac{1}{4} - \tfrac{\sqrt{3}}{6})h\,k_1 + \tfrac{1}{4}h\,k_2)$$

$$y_{n+1} = y_n + \tfrac{h}{2}(k_1 + k_2). \qquad (5.4.1)$$

In this case both k_1 and k_2 are defined implicitly by a system of two equations which will be non-linear unless $f(x,y)$ is linear in y. An R-stage implicit Runge-Kutta method results in the system of R generally non-linear equations. They may be solved by the iteration

$$k_r^{[t+1]} = f(x_n + ha_r, y_n + h\sum_{s=1}^{r-1} b_{rs} k_s^{[t+1]} + h\sum_{s=r}^{R} b_{rs} k_s^{[t]}),$$

$$r=1,2,\ldots,R, \qquad (5.4.2)$$

where $k_r^{[t]}$, $r=1,2,\ldots,R$ is the approximation at the tth stage. Starting with $k_r^{[0]}$, $r=1,2,\ldots,R$, the above scheme is used for $t=0,1,2,\ldots$ until some convergence criterion is satisfied. Butcher (1964a) has proved that the iteration will converge for any choice of $k_r^{[0]}$, $r=1,2,\ldots,R$, provided

$$h < \left\{L\left[\max_r \sum_{s=r}^{R} |b_{rs}| + \max_r \sum_{s=1}^{r-1} |b_{rs}|\right]\right\}^{-1}, \qquad (5.4.3)$$

where L is the Lipschitz constant of $f(x,y)$. Thus an R-stage implicit Runge-Kutta method will in general require many more function evaluations than the R appearing in (5.2.1).

The main advantage of implicit schemes is their improved stability characteristics. Lambert (1973) has shown that for the fourth-order scheme (5.4.1),

$$r_1(\lambda h) = \frac{1 + \frac{1}{2}\lambda h + \frac{1}{12}(\lambda h)^2}{1 - \frac{1}{2}\lambda h + \frac{1}{12}(\lambda h)^2}. \qquad (5.4.4)$$

This is the fourth-order (2,2) Padé approximation to $\exp(\lambda h)$. It follows from (5.4.4) that the interval of absolute stability is $(-\infty,0)$ compared with the stability interval of $(-2.78,0)$ of the 4-stage fourth-order explicit Runge-Kutta methods.

The difficulty of finding the values of k_r is reduced in methods which are termed *semi-explicit* by Butcher (1964a). Semi-explicit Runge-Kutta methods have $b_{rs} = 0$ for $s > r$, and since $b_{rs} = 0$ for $s \geq r$ implies the method is explicit, at least one b_{rr} must be non-zero. The k_r, $r=1,2,\ldots,R$, may be found in that order using the iteration

$$k_r^{[t+1]} = f(x_n + ha_r, y_n + h\sum_{s=1}^{r-1} b_{rs} k_s + h b_{rr} k_r^{[t]}), \qquad (5.4.5)$$

if $b_{rr} \neq 0$. This iteration will converge for any starting values $k_r^{[0]}$ if

$$h < \frac{1}{L \max_r |b_{rr}|}. \qquad (5.4.6)$$

An example of a 3-stage fourth-order semi-explicit method given by Butcher (1964a) is

$$k_1 = f(x_n, y_n),$$

$$k_2 = f(x_n + \tfrac{1}{2}h,\ y_n + \tfrac{1}{4}h\,k_1 + \tfrac{1}{4}h\,k_2),$$

$$k_3 = f(x_n + h,\ y_n + h\,k_2),$$

$$y_{n+1} = y_n + \frac{h}{6}(k_1 + 4k_2 + k_3). \qquad (5.4.7)$$

This method also has a larger stability interval than the 4-stage fourth-order explicit Runge-Kutta methods. It is easily shown that

$$r_1(\lambda h) = \frac{1 + \frac{3}{4}\lambda h + \frac{1}{4}(\lambda h)^2 + \frac{1}{24}(\lambda h)^3}{1 - \frac{1}{4}\lambda h}, \qquad (5.4.8)$$

which is the (3,1) Padé approximation to $\exp(\lambda h)$, and by computing the real roots of $|r_1(\lambda h)| = 1$, one finds the interval of absolute stability is $(-5.41,0)$, compared with $(-2.78,0)$ for the explicit scheme.

Table 5.3 shows numerical solutions of

$$y' = 1 - 2xy, \quad y(0) = 0, \qquad (5.4.9)$$

obtained by using the implicit scheme (5.4.1) and the semi-explicit scheme (5.4.7). They have similar accuracies, both being slightly more accurate than the fourth-order explicit schemes (see Table 5.2).

TABLE 5.3: The numerical solution of $y' = 1 - 2xy$, $y(0) = 0$ using an implicit and a semi-explicit Runge-Kutta method of order 4 with $h=0.1$.

	Implicit		Semi-explicit	
x_n	y_n	error	y_n	error
0.0	0.00000000	0.	0.00000000	0.
0.1	0.09933610	-1.1E-07	0.09933583	1.7E-07
0.2	0.19475123	-2.0E-07	0.19475073	3.1E-07
0.3	0.28263192	-2.6E-07	0.28263126	4.0E-07
0.4	0.35994375	-2.7E-07	0.35994305	4.3E-07
0.5	0.42443662	-2.3E-07	0.42443600	3.8E-07
0.6	0.47476335	-1.4E-07	0.47476296	2.4E-07
0.7	0.51050406	-4.5E-09	0.51050403	3.0E-08
0.8	0.53210154	1.7E-07	0.53210195	-2.4E-07
0.9	0.54072396	3.5E-07	0.54072486	-5.5E-07
1.0	0.53807897	5.4E-07	0.53808036	-8.5E-07

6. IMPLEMENTATION OF ONE-STEP METHODS

6.1 Variable Step Length

To this point we have assumed that the step length is constant. However, it is desirable to use a variable step length, choosing it so that certain accuracy requirements are met. Many solutions vary in character over the interval on which they are sought, being smooth in places so that a large step length is adequate, while in other places they change very rapidly so that a small step length is necessary. Using the small step length over the whole interval is not only inefficient, but also may lead to larger global errors due to the build-up of round-off errors.

Of course, the error that we wish to control is the global error and this can be achieved by controlling the local error, that is by controlling the error made in each step. If h_n is the length of the step from x_n to x_{n+1}, then

$$x_{n+1} = x_n + h_n , \tag{6.1.1}$$

and

$$y_{n+1} = y_n + h_n \phi(x_n, y_n, h_n). \tag{6.1.2}$$

The local error in this step is (see Figure 4.3)

$$\begin{aligned} d_{n+1} &= y^*(x_{n+1}) - y_{n+1} \\ &= y^*(x_{n+1}) - y_n - h_n \phi(x_n, y_n, h_n) \\ &= y^*(x_{n+1}) - y^*(x_n) - h_n \phi(x_n, y^*(x_n), h_n) \\ &= C(x_n) h_n^{p+1} + O(h_n^{p+2}) \end{aligned} \tag{6.1.3}$$

for a scheme of order p. If the local error per unit length d_{n+1}/h_n is less in magnitude than ε and the conditions of the uniqueness theorem hold, then the global truncation error satisfies the inequality

$$|e_n| < \varepsilon \left| \frac{\exp[L(x_n - a)] - 1}{L} \right| \tag{6.1.4}$$

for any $x_n \in [a,b]$. A proof of this result may be found in Birkhoff and Rota (1978). Thus the strategy is to choose the step length h_n so that for some prescribed value of ε,

$$\left| \frac{d_{n+1}}{h_n} \right| < \varepsilon . \tag{6.1.5}$$

We neglect all but the leading term in (6.1.3), but even so cannot in general calculate the required step length directly from it since the function $C(x)$ is usually unknown. The way around this is to take a step of length H, estimate the local error at $x_n + H$ (this is discussed in the next section), and use the estimate to calculate the step length $h_n = \theta H$ that should have been used. If d_{n+1}^* is the estimated local error associated with the step of length H, then

$$d_{n+1}^* = C(x_n) H^{p+1} \tag{6.1.6}$$

and the local error produced by a step of θH is

$$d_{n+1} = C(x_n)(\theta H)^{p+1} . \tag{6.1.7}$$

The inequality (6.1.5) will be satisfied if

$$|C(x_n)(\theta H)^p| < \varepsilon , \tag{6.1.8}$$

and using (6.1.6) to eliminate $C(x_n)$, we see that (6.1.5) is satisfied if

$$\theta < \left|\frac{\varepsilon H}{d^*_{n+1}}\right|^{1/p} . \tag{6.1.9}$$

We want to take the largest step length possible, but because a rejected step is very costly we take

$$\theta = 0.8\left|\frac{\varepsilon H}{d^*_{n+1}}\right|^{1/p} . \tag{6.1.10}$$

The factor 0.8 compensates in part for the approximations made in deducing the result (6.1.9).

If the value of θ given by (6.1.10) is greater than one then H is smaller than necessary, but rather than recalculate the step we accept the value y_{n+1} with $h_n = H$. If $\theta < 1$ then the step must be repeated with H replaced by θH, a new value of θ calculated and that step accepted if θ is greater than one, and so on. Once a step is accepted we start the next step using θH for H. Here we are assuming that $C(x_{n+1})$ is not very different from $C(x_n)$; the factor 0.8 in (6.1.10) allows for some variation.

There are some other constraints which should be put on the step length h_n. If the step length becomes too small, the different values of x used in the algorithm may in fact be made equal due to the finite precision of the computer. If ℓh_n is the minimum distance between two points used in the algorithm, and u is the unit-roundoff, the smallest quantity for which the computed value of $1+u$ is greater than one, then we require $\ell h_n \geq x_n u$, so $h_{min} = x_n u/\ell$. If β is the base used by the computer and s is the number of digits in the mantissa, then

$$u = \begin{cases} \beta^{1-s} & \text{(chopped or truncated arithmetic)} \\ \tfrac{1}{2}\beta^{1-s} & \text{(rounded arithmetic).} \end{cases} \tag{6.1.11}$$

If the subroutine tries to take a step of length $h_n < h_{min}$ then the problem cannot be solved to the requested accuracy and the subroutine should exit with a message to that effect.

Too large a step should not be used, since then some feature of the solution could be skipped over. The maximum step length h_{max} is obviously dependent on the problem so should be supplied by the user.

It is also undesirable for the program to change step length too abruptly. Suppose the solution is very smooth up to some point where it has a rapid change. The program will approach this point with a large h, and when it steps past, it will find a very large error, causing it to take a very small step. This small step will be taken, and since the solution is smooth, will result in a very small error, so the program will attempt to take a large step again and so the cycle will be repeated. The end result is that the point will be approached in a series of unnecessarily small steps. If we restrict an increase in step length to a factor of say 2 the step length can still increase rapidly over a few steps. A factor by which the step length may decrease is more difficult to assign - if it is too large then several consecutive step rejections may occur. Shampine and Allen (1973) suggest the value 1/10 and allow an increase in step length by a factor of 5.

When the end of the interval [a,b] on which the solution is sought is approached, the step length should be changed so that the numerical solution at b is calculated. If at some stage $x_n + H > b$, the value of H must be reduced to $b - x_n$. Often the numerical solution is required at equal intervals on [a,b]. The usual procedure is to divide [a,b] into the appropriate subintervals and repeatedly call the integrating routine. This has the effect of often causing unnecessarily small steps to be taken at the end of each interval, so it may be more efficient to interpolate the required values if the solution is required at a large number of points. Gladwell (1979) uses Hermite interpolation, because not only are the solution values known, but also their derivatives. He uses three points in the interpolation, and an interesting point is that he concludes that the step length should not be increased by more than a factor of 2 to control the errors in the interpolation.

A flow chart of the step length selection procedure is shown in Figure 6.1. For simplicity the calculation of the value "y_{new}" is shown in box (A), but, to make the program as efficient as possible, only those calculations required to calculate d* should be performed there, the rest being carried out in box (B) after the step has been accepted. In practice some of the calculations of a rejected step may be used for the step chosen. An example of this is the evaluation of k_1 in a Runge-Kutta method since it is independent of the step length.

The strategy that we have described controls the global error or absolute error of the approximation. Some authors, for example Shampine and Gordon (1975), advocate a mixed relative-absolute error criterion and let

$$\varepsilon = \varepsilon_a + \varepsilon_r |y_n| . \tag{6.1.12}$$

Setting $\varepsilon_r = 0$ gives the criterion we have described, while setting $\varepsilon_a = 0$ gives a relative error control which may be appropriate for solutions which become large in magnitude. However, this relative error criterion could lead to excessively small steps in the region of a zero of the solution, which could be avoided if a mixed criterion was used instead.

6.2 Estimation of the Local Error

The selection of the step length is dependent on a good estimate being made of the local error. First we consider using the Taylor series method of order p. The local error at x_{n+1} is given by

$$d_{n+1} = H^{p+1} \frac{y^{*(p+1)}(x_n)}{(p+1)!} + O(H^{p+2}), \tag{6.2.1}$$

where H is the step length used. In this case $y^{*(p+1)}(x_n)/(p+1)!$

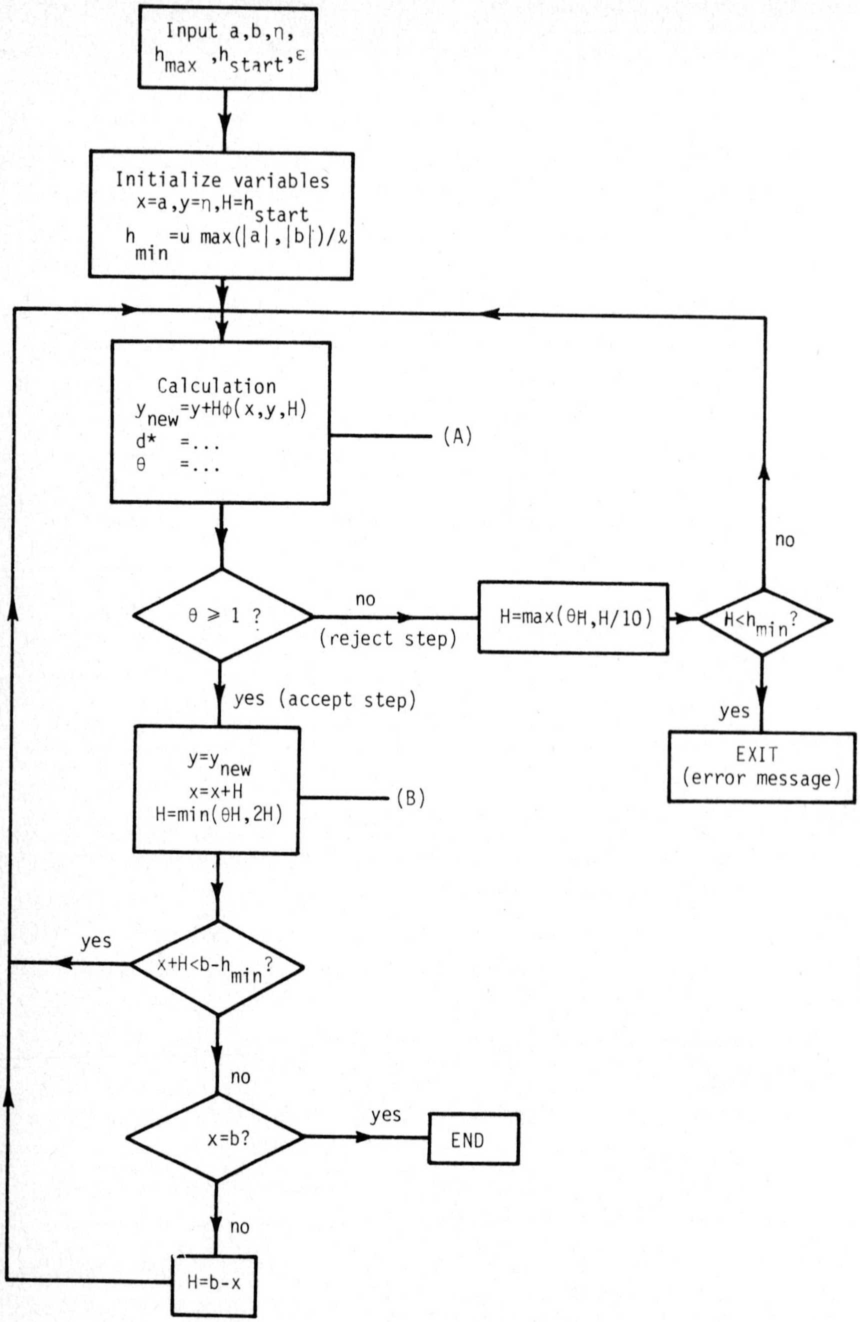

FIGURE 6.1: *A flow chart for a one-step method with variable step length.*

can be calculated directly and, considering the leading term, we see that (6.1.5) is satisfied if

$$h_n < \left| \frac{\varepsilon(p+1)!}{y^{*(p+1)}(x_n)} \right|^{1/p} \qquad (6.2.2)$$

To allow for the other terms in (6.2.1) it would be advisable to use a value of h_n somewhat less than the maximum value allowed by (6.2.2). Note that the leading term of d_{n+1} is the difference between the values obtained by Taylor series methods of order $p+1$ and p, and consequently very little extra work is required to obtain the result of the order $p+1$ scheme. This more accurate result should be used.

The basic ideas considered above apply to a general one-step method. If the step from x_n to x_{n+1} is performed by methods of order p and $p+1$, then an approximation to the local error in the pth-order result is the difference between the computed values of y_{n+1}. This is easily shown as follows. Using the overbar ($\bar{\ }$) and circumflex ($\hat{\ }$) to denote the order p and $p+1$ results respectively, then

$$y^*(x_{n+1}) - \hat{y}_{n+1} = O(h_n^{p+2}) , \qquad (6.2.3)$$

and

$$y^*(x_{n+1}) - \bar{y}_{n+1} = C(x_n)h_n^{p+1} + O(h_n^{p+2}). \qquad (6.2.4)$$

Subtracting these two equations gives

$$\hat{y}_{n+1} - \bar{y}_{n+1} = C(x_n)h_n^{p+1} + O(h_n^{p+2}), \qquad (6.2.5)$$

so an estimate of the local error is

$$d^*_{n+1} = \hat{y}_{n+1} - \bar{y}_{n+1} . \qquad (6.2.6)$$

Note that

$$\hat{y}_{n+1} = \bar{y}_{n+1} + d^*_{n+1} , \qquad (6.2.7)$$

and no matter how the local error estimate is found, $\hat{y}_{n+1}$ will be generally a better approximation to $y^*(x_{n+1})$ than $\bar{y}_{n+1}$ and so should be used as y_{n+1}. This process is called *local extrapolation*. In this case y_{n+1} is just the result obtained using the method of order $p+1$. Thus the high-order result is used in the calculation, but as the step length selection is based on the lower order scheme the computed solution should be more accurate than predicted by (6.1.4).

When a pth order scheme is obtained as a by-product of a scheme of order $p+1$, it is known as an *embedded method*. This idea was first proposed by Merson (1957), who derived a 5-stage Runge-Kutta method of order four with a local error estimate. The scheme is

$$k_1 = f(x_n, y_n),$$

$$k_2 = f(x_n + \tfrac{1}{3}h,\ y_n + \tfrac{1}{3}h\ k_1),$$

$$k_3 = f(x_n + \tfrac{1}{3}h,\ y_n + \tfrac{1}{6}h\ k_1 + \tfrac{1}{6}h\ k_2),$$

$$k_4 = f(x\ + \tfrac{1}{2}h,\ y\ + \tfrac{1}{8}h\ k_1 + \tfrac{3}{8}h\ k_3),$$

$$k_5 = f(x_n + h,\ y_n + \tfrac{1}{2}h\,k_1 - \tfrac{3}{2}h\,k_3 + 2h\,k_4),$$

$$y_{n+1} = y_n + \frac{h}{6}(k_1 + 4k_4 + k_5), \qquad (6.2.8)$$

and its local error estimate is

$$d^*_{n+1} = \frac{h}{30}(-2k_1 + 9k_3 - 8k_4 + k_5). \qquad (6.2.9)$$

Scraton (1964) has shown that this error estimate is valid only if $f(x,y)$ is linear in both x and y and that it often grossly over-estimates the error in a non-linear equation. It also occasionally under-estimates the error (England, 1969), but nevertheless it is commonly used; for example it is still used in the NAG library (Gladwell, 1979).

Embedded Runge-Kutta methods which are applicable to a general system of ordinary differential equations are given by England (1969), Shintani (1965, 1966, 1966a) and Fehlberg (1964, 1969, 1970). Perhaps the most popular of these is the Fehlberg's (1970) 6-stage method of order 5(4), given by

$$k_1 = f(x_n, y_n),$$

$$k_2 = f(x_n + \tfrac{1}{4}h,\ y_n + \tfrac{1}{4}h\,k_1),$$

$$k_3 = f(x_n + \tfrac{3}{8}h,\ y_n + \tfrac{3}{32}h\,k_1 + \tfrac{9}{32}h\,k_2),$$

$$k_4 = f(x_n + \tfrac{12}{13}h,\ y_n + \tfrac{1932}{2197}h\,k_1 - \tfrac{7200}{2197}h\,k_2 + \tfrac{7296}{2197}h\,k_3),$$

$$k_5 = f(x_n + h,\ y_n + \tfrac{439}{216}h\,k_1 - 8h\,k_2 + \tfrac{3680}{513}h\,k_3 - \tfrac{845}{4104}h\,k_4),$$

$$k_6 = f(x_n + \tfrac{h}{2},\ y_n - \tfrac{8}{27}h\,k_1 + 2h\,k_2 - \tfrac{3544}{2565}h\,k_3 + \tfrac{1859}{4104}h\,k_4 - \tfrac{11}{40}h\,k_5),$$

$$y_{n+1} = y_n + h(\tfrac{16}{135}k_1 + \tfrac{6656}{12825}k_3 + \tfrac{28561}{56430}k_4 - \tfrac{9}{50}k_5 + \tfrac{2}{55}k_6). \qquad (6.2.10)$$

The local error estimate is

$$d^*_{n+1} = h(\tfrac{1}{360}k_1 - \tfrac{128}{4275}k_3 - \tfrac{2197}{75240}k_4 + \tfrac{1}{50}k_5 + \tfrac{2}{55}k_6). \qquad (6.2.11)$$

A disadvantage of Fehlberg's schemes is that they give error estimates of zero when y' depends only on x. Verner (1978, 1979) has developed embedded Runge-Kutta methods which overcome this disadvantage. One of his schemes, an 8-stage method of order 6(5) is used in the IMSL Library routine DVERK (Verner, 1978).

Another way of estimating the local error is termed *extrapolation* and is a particular example of Richardson extrapolation (Richardson, 1927). The step from x_n to x_{n+1} is performed twice, firstly in one step of length H and secondly by two steps of length $H/2$. If the method has order p and we denote the results by $\bar{y}_{n+1}$ and $\hat{y}_{n+1}$ respectively, then

$$y^*(x_{n+1}) - \bar{y}_{n+1} = C(x_n)H^{p+1} + O(H^{p+2}), \qquad (6.2.12)$$

and it can be shown that

$$y^*(x_{n+1}) - \hat{y}_{n+1} = 2C(x_n)(H/2)^{p+1} + O(H^{p+2}). \qquad (6.2.13)$$

Subtracting (6.2.13) from (6.2.12) gives

$$\hat{y}_{n+1} - \bar{y}_{n+1} = 2C(x_n)(H/2)^{p+1}[2^p-1] + O(H^{p+1}), \qquad (6.2.14)$$

so that an estimate of the local error of $\hat{y}_{n+1}$ is

$$d^*_{n+1} = \frac{\hat{y}_{n+1} - \bar{y}_{n+1}}{2^p - 1} \; . \qquad (2.6.15)$$

Local extrapolation gives

$$y_{n+1} = \hat{y}_{n+1} + d^*_{n+1} , \qquad (6.2.16)$$

and because the term of order $p+1$ has been eliminated, (6.2.16) is essentially a $(p+1)^{th}$ order result. If a fourth-order Runge-Kutta scheme is used with extrapolation, then 11 function evaluations are required for each step as the scheme is used three times but the value k_1 can be used for the full-step and the first half-step.

When local extrapolation is used, the local error of the extrapolated value is one order higher than the local error estimated by d^*_{n+1}. If the step length selection is based on (6.1.5) the extra accuracy is a bonus, but an alternative is to use the criterion

$$|d^*_{n+1}| < \varepsilon , \qquad (6.2.17)$$

in which case the actual local error per unit length is bounded by some unknown multiple of ε. This is the approach used by Shampine and Gordon (1975). Other authors, for example Gear (1971, 1971a), use the criterion (6.2.17) even though local extrapolation is not used.

6.3 Estimation of the Global Error

The bounds on the global error given by (4.2.8) and (6.1.4) are of no practical use as they usually grossly over-estimate the error. A common method of estimating the global error of a numerical solution is to solve the problem a second time with a smaller error tolerance and then compare the two results, assuming that the second solution is much more accurate. Shampine (1980) reports that this procedure is hazardous with a variable step method, and gives an example where a reduction of the error tolerance actually increases the global error.

However, a good estimate of the global error can be obtained using Richardson extrapolation if a constant step length is used. If $\hat{y}$ and $\bar{y}$ denote the numerical solutions at some given value of x obtained by a method of order p using step lengths $\hat{h}$ and $\bar{h}$ respectively, then it is easily shown that

$$\hat{e} \simeq \frac{(\hat{h})^p(\bar{y}-\hat{y})}{(\hat{h})^p-(\bar{h})^p} , \qquad (6.3.1)$$

and

$$\bar{e} \simeq \frac{(\bar{h})^p(\bar{y}-\hat{y})}{(\hat{h})^p-(\bar{h})^p} . \qquad (6.3.2)$$

Of course these error estimates can be used to obtain a more accurate result, either

$$\hat{y} + \hat{e} \quad \text{or} \quad \bar{y} + \bar{e}.$$

In Section 3.2 the solution of

$$y' = 1 - 2xy, \qquad y(0) = 0,$$

was found using Euler's method, and at $x = 1$ the results $\hat{y} = 0.57001592$ and $\bar{y} = 0.54111554$ were obtained with $\hat{h} = 0.1$ and $\bar{h} = 0.01$ (see Table 3.1). Using the above approximations (with $p=1$ since Euler's method is first-order) the values $\hat{e} = -0.032$ and $\bar{e} = -0.0032$ are obtained, which are in good agreement with the actual errors $\hat{e} = -0.032$ and $\bar{e} = -0.0030$.

In arriving at the above estimates of the global error and in the discussion of Sections 6.1 and 6.2, it has been assumed that a method of order p produces numerical results which have global errors of order p and local errors of order $p+1$. This is only true if the solution of the initial value problem is sufficiently well behaved. It is certainly true if the solution has continuous derivatives to order $p+1$, but such a stringent condition is not necessary. A simple discontinuity in a derivative at some points does not affect the order of the errors provided the points are not in the interior of any interval $[x_n, x_{n+1}]$, and this can be guaranteed by splitting the original problem into a number of smaller problems. However, other discontinuities can decrease the effective order of the scheme. Gear (1971) has given an example where the third derivative of the solution becomes infinite, and has found that a second-order method gives global errors which are $O(h^2)$ but the global errors of a fourth-order method are not $O(h^4)$. Even so, the fourth-order method has produced more accurate results for all h.

6.4 Choice of Method and Order

We have considered three distinct methods, namely Taylor series methods, Runge-Kutta methods and implicit Runge-Kutta methods. The implicit Runge-Kutta methods require the iterative solution of non-linear equations and so are nowhere near as efficient as the other two methods - they are only useful for the solution of stiff equations which will be considered later.

When compared with Runge-Kutta methods, Taylor series methods have the disadvantage of requiring the calculation of higher derivatives which generally become increasingly complicated as the order is increased. The user must supply algebraic expressions for these derivatives. The advantage of Taylor series methods over Runge-Kutta methods is that the step-length can be calculated a priori, so no computation is wasted from rejected steps. If the differentiation required by Taylor series methods is performed automatically, then most of the disadvantage is overcome and it allows the use of schemes of very high order (typically orders of around 30). Barton et al (1971) and Corliss and Chang (1982) have implemented such methods and conclude that they compare well with other methods.

Nevertheless Runge-Kutta methods are generally considered to be the most efficient one-step methods. Embedded methods are usually preferred to schemes which use extrapolation to obtain a local error estimate. The order of the method must be a compromise unless a variable order scheme is used, because low order methods are generally more efficient when the accuracy requirement is low, while high order methods are superior for high accuracy. Variable order Runge-Kutta schemes are not very commonly used, probably because of their higher overheads and larger storage requirements than fixed order schemes. However, Shampine, Gordon and Wisniewski (1980) conclude that they are worthy of consideration. A detailed discussion on the choice of a fixed order Runge-Kutta scheme is given by Shampine and Watts (1977). They choose one of Fehlberg's schemes of order 5(4), and give details of the writing of the code RKF45 in a later paper, Shampine and Watts (1979).

7. LINEAR MULTISTEP METHODS

7.1 The General Linear k-step Method

In the preceding Sections we have considered one-step methods where the numerical solution for some value x_{n+1} of the independent variable is calculated using information from only the previous value, x_n. In a k-step method information is used from the previous k equispaced values, so that y_{n+k} is calculated using values of y computed at $x_{n+j} = x_n + jh$, $j=0,1,2,\ldots,k-1$. Note that a special starting procedure is needed for $k \geq 2$ as to begin with only one value is known from the initial condition. Any suitable one-step method can be used to calculate the necessary values. Using the notation $f_n = f(x_n, y_n)$, the general *linear k-step method* may be written

$$\sum_{j=0}^{k} \alpha_j y_{n+j} = h \sum_{j=0}^{k} \beta_j f_{n+j} , \tag{7.1.1}$$

where α_j and β_j are constants with $\alpha_k=1$ and α_0 and β_0 not both zero. Equation (7.1.1) can be rearranged to give

$$y_{n+k} = h\beta_k f(x_{n+k}, y_{n+k}) + \sum_{j=0}^{k-1} (h\beta_j f_{n+j} - \alpha_j y_{n+j}). \tag{7.1.2}$$

If $\beta_k=0$ then the right hand side of (7.1.2) is known so the method is *explicit*, but if $\beta_k \neq 0$ the scheme is *implicit*. An implicit method requires the solution of the generally non-linear Equation (7.1.2) at every step, and this may be accomplished by the iteration

$$y_{n+k}^{[t+1]} = h\beta_k f(x_{n+k}, y_{n+k}^{[t]}) + B, \tag{7.1.3}$$

where B is the known sum in (7.1.2). The iteration (7.1.3) is carried out for $t=0,1,2,\ldots$ until some convergence criterion is satisfied. Convergence is guaranteed for any starting value $y_{n+k}^{[0]}$ if

$$h < 1/L|\beta_k| , \tag{7.1.4}$$

where L is the Lipschitz constant of $f(x,y)$.

Most of the terms defined in connection with one-step methods, such as the global truncation error, the local error, etc., apply to any method, and in particular to linear k-step methods. The local truncation error is the error in calculating y_{n+k} assuming that y_{n+j}, $j=0,1,\ldots,k-1$, are all exact, or alternatively, is the amount by which the exact solution fails to satisfy (7.1.1). Thus the *local truncation error* T_{n+k} is defined by

$$T_{n+k} = \sum_{j=0}^{k} [\alpha_j y(x_{n+j}) - h\beta_j y'(x_{n+j})] . \tag{7.1.5}$$

Expanding $y(x_{n+j})$ and $y'(x_{n+j})$ in Taylor series about x_n gives

$$T_{n+k} = \sum_{\ell=0}^{\infty} C_\ell h^\ell y^{(\ell)}(x_n), \tag{7.1.6}$$

and if $C_0 = C_1 = C_2 = \ldots = C_p = 0$ and $C_{p+1} \neq 0$ then we say the linear k-step method (7.1.1) is of *order* p, and if $p \geq 1$ then the k-step method (7.1.1) is *consistent with the differential equation* $y' = f(x,y)$. The following formulae for $C_0, C_1, \ldots$ are easily derived:

$$C_0 = \sum_{j=0}^{k} \alpha_j ,$$

$$C_1 = \sum_{j=1}^{k} j\alpha_j - \sum_{j=0}^{k} \beta_j ,$$

$$C_\ell = \frac{1}{\ell!} \sum_{j=1}^{k} (j^\ell \alpha_j - \ell j^{\ell-1} \beta_j), \qquad \ell=2,3,\ldots \tag{7.1.6}$$

Thus a linear k-step method is consistent if and only if

$$\sum_{j=0}^{k} \alpha_j = 0 \text{ and } \sum_{j=1}^{k} j\alpha_j = \sum_{j=0}^{k} \beta_j . \tag{7.1.7}$$

7.2 Derivation of Linear Multistep Methods

There are several different ways of deriving linear multistep methods. One way is to obtain equations for the coefficients α_j and β_j by using Taylor series expansions - we can use the general results quoted above. For example, suppose we want to derive the explicit two-step method of highest order. Then we have $\alpha_2=1$, $\beta_2=0$ and four unknowns $\alpha_0, \alpha_1, \beta_0$ and β_1, so it should be possible to choose these so that $C_0 = C_1 = C_2 = C_3 = 0$. From (7.1.6) with $k=2$ we obtain the following equations:

$$\begin{aligned}
\alpha_0 + \alpha_1 + 1 &= 0,\\
\alpha_1 + 2 - \beta_0 - \beta_1 &= 0,\\
\alpha_1 + 4 \qquad - 2\beta_1 &= 0,\\
\alpha_1 + 8 \qquad - 3\beta_1 &= 0.
\end{aligned} \tag{7.2.1}$$

Solving these equations gives $\alpha_0 = -5$, $\alpha_1 = 4$, $\beta_0 = 2$ and $\beta_1 = 4$, so the resulting multistep method is

$$y_{n+2} = -4y_{n+1} + 5y_n + h(4f_{n+1}+2f_n). \tag{7.2.2}$$

Since

$$C_4 = \frac{1}{4!}(\alpha_1+16\alpha_2-4\beta_1-32\beta_2) = \frac{1}{6} \tag{7.2.3}$$

the method is third-order with the local truncation error

$$T_{n+2} = \frac{1}{6} h^4 y^{(4)}(x_n) + \ldots . \tag{7.2.4}$$

Another way of deriving a linear multistep method is to use numerical integration. Integrating the differential equation from x_n to x_{n+2} yields

$$y(x_{n+2})-y(x_n) = \int_{x_n}^{x_{n+2}} f(x,y(x))dx. \tag{7.2.5}$$

The integral may be approximated by Simpson's rule to give

$$y(x_{n+2})-y(x_n) = \frac{h}{3}[f(x_n,y(x_n))+4f(x_{n+1},y(x_{n+1}))+f(x_{n+2},y(x_{n+2}))]$$

$$- \frac{h^5}{90}\frac{d^4}{dx^4}[f(x,y(x))]_{x=\xi_n}, \qquad (7.2.6)$$

where ξ_n is some constant in the range (x_n, x_{n+2}).

Dropping the error term gives the implicit two-step method

$$y_{n+2} = y_n + \frac{h}{3}(f_n + 4f_{n+1} + f_{n+2}). \qquad (7.2.7)$$

The error term is of course the local truncation error, and since $y' = f(x,y)$ it follows that

$$T_{n+2} = -\frac{h^5}{90}y^{(5)}(\xi_n), \qquad (7.2.8)$$

which demonstrates that the method is fourth-order. Other methods can be found by using other Newton-Cotes quadrature formulas to approximate the integral - the closed formulas give implicit methods while the open formulas give explicit methods.

The Newton-Cotes formulas are derived by integrating interpolating polynomials, and this method can be used to derive multistep methods which are not identical to those obtained using Newton-Cotes formulas. For example, integrating the differential equation from x_{n+1} to x_{n+2} gives

$$y(x_{n+2})-y(x_{n+1}) = \int_{x_{n+1}}^{x_{n+2}} f(x,y(x))dx. \qquad (7.2.9)$$

If the integral is approximated by the integral of the quadratic which interpolates f_n, f_{n+1} and f_{n+2}, we obtain the scheme

$$y_{n+2} = y_{n+1} + \frac{h}{12}(5f_{n+2} + 8f_{n+1} - f_n). \qquad (7.2.10)$$

This is another two-step implicit scheme, and from (7.1.6) the values $C_0 = C_1 = C_2 = C_3 = 0$ and $C_4 = -1/24$ are obtained, so the method is third-order with a local truncation error of

$$T_{n+2} = -\frac{1}{24}h^4y^{(4)}(x_n) + \ldots . \qquad (7.2.11)$$

Other ways of deriving multistep methods include the use of Hermite interpolation or splines instead of polynomial interpolation (see Lambert, 1973).

7.3 Convergence and Stability of Linear Multistep Methods

A linear k-step method is said to be *convergent* if the numerical solution y_n approaches the analytic solution $y(x_n)$ for any fixed $x_n \in [a,b]$ and the starting values $y_0, y_1, \ldots, y_{k-1}$ approach η as the step length h tends to zero. This definition only differs from the definition of convergence of a one-step method by having an extra condition for the starting values. If $y_0 = \eta$ and the other starting values are found from a convergent one-step method this condition will be satisfied.

Table 7.1 shows the solution of the initial value problem

$$y' = 1 - 2xy, \quad y(0) = 0, \qquad (7.3.1)$$

Table 7.1: The numerical solution of $y' = 1-2xy$, $y(0) = 0$ using the multistep method (7.2.2).

	h = 0.1		h = 0.01	
x_n	y_n	error	y_n	error
0.0	0.	0.	0.	0.
0.1	9.933E-02	2.7E-06	9.929E-02	4.9E-05
0.2	1.947E-01	3.1E-05	-4.819E+02	4.8E+02
0.3	2.827E-01	-2.6E-05	-4.850E+09	4.8E+09
0.4	3.596E-01	3.9E-04	-4.938E+16	4.9E+16
0.5	4.261E-01	-1.7E-03	-5.088E+23	5.1E+23
0.6	4.654E-01	9.4E-03	-5.306E+30	5.3E+30
0.7	5.605E-01	-5.0E-02	-5.599E+37	5.6E+37
0.8	2.592E-01	2.7E-01	-5.980E+44	6.0E+44
0.9	2.043E+00	-1.5E+00	-6.464E+51	6.5E+51
1.0	-7.830E+00	8.4E+00	-7.071E+58	7.1E+58

by the third-order explicit scheme (7.2.2). The extra starting value needed by this scheme was calculated using the third-order Taylor series method. With a step length $h=0.1$ the errors grow quite rapidly, but decreasing h to 0.01 had a disastrous effect. To find the cause of this, we examine the stability of the scheme. If the scheme is used to solve the test problem $y' = \lambda y$ we have $f_n = \lambda y_n$, so (7.2.2) becomes

$$y_{n+2} + 4(1-\lambda h)y_{n+1} - (5+2\lambda h)y_n = 0. \tag{7.3.1}$$

This is a linear difference equation with constant coefficients so it is easy to solve. Trying for a solution of the form $y_n = Br^n$, gives the characteristic equation

$$r^2 + 4(1-\lambda h)r - (5+2\lambda h) = 0 . \tag{7.3.2}$$

Solving this for r gives

$$r = -2(1-\lambda h) \pm \sqrt{4(1-\lambda h)^2+(5+2\lambda h)} \tag{7.3.3}$$

from which we can deduce

$$r_1 = 1 + \lambda h + O(h^2),$$

$$r_2 = -5 + 3\lambda h + O(h^2). \tag{7.3.4}$$

Since (7.3.2) is linear, its solution is

$$y_n = B_1(r_1)^n + B_2(r_2)^n . \tag{7.3.5}$$

The first term represents the solution of $y' = \lambda y$ as $(r_1)^n \to e^{\lambda nh}$ as $h \to 0$ with nh fixed, but the second term is spurious and arises from the approximation of a first-order differential equation by a second-order difference equation. Even if the values y_0 and y_1 were such that $B_2 = 0$, round-off errors would soon have the same effect as making B_2 non-zero, and hence any errors will propagate in an unstable manner through this term. In the above example $\partial f/\partial y = -2$ at $x=1$, so taking $\lambda = -2$ gives $r_2 = -5.6$ if $h = 0.1$, and this is the factor by which the errors grew (see Table 7.1). When $h = 0.01$, we obtain $r_2 = -5.06$, but as ten steps are required to move from 0.9 to 1.0 the error will grow by $(-5.06)^{10} = 1.1 \times 10^7$, a result which is again in agreement with the actual errors. Even in the limit $h \to 0$ the difference equation has the roots $r_1 = 1$ and $r_2 = -5$, so this particular multistep method is obviously not convergent.

Using the general linear k-step scheme defined by (7.1.1) to solve the differential equation $y' = \lambda y$ gives the kth-order difference equation

$$\sum_{j=0}^{k} (\alpha_j - h\lambda\beta_j) y_{n+j} = 0, \tag{7.3.6}$$

which has the characteristic equation

$$\pi(r,h) = \sum_{j=0}^{k} (\alpha_j - h\lambda\beta_j) r^j = 0. \tag{7.3.7}$$

The polynomial $\pi(r,h)$ is often referred to as the *characteristic polynomial* or *stability polynomial* of the method. If r_ℓ, $\ell=1,2,\ldots,k$, are the roots of the characteristic polynomial, then the solution of the difference equation (7.3.6) is, for distinct roots,

$$y_n = \sum_{\ell=1}^{k} B_\ell (r_\ell)^n . \tag{7.3.8}$$

If a root r_ℓ has multiplicity q, then it gives rise to the q terms

$$[B_{\ell 1} + nB_{\ell 2} + n(n-1)B_{\ell 3} + \ldots + n(n-1) \ldots (n-q+2)B_{\ell q}](r_\ell)^n \tag{7.3.9}$$

in the sum (7.3.8). Because of the linearity of the difference equation (7.3.6), any errors in $y_0, y_1, y_2, \ldots$ satisfy the difference equation and hence the error in y_n may be expressed in the form (7.3.8). Consequently the errors can not become unbounded as n increases if $|r_\ell| < 1$, $\ell=1,2,\ldots,k$. Note that we can not allow $|r_\ell| = 1$, because if it is a multiple root (7.3.9) shows that the error would become unbounded. Thus we say a linear k-step method is *absolutely stable* if the roots r_ℓ of its characteristic polynomial satisfy $|r_\ell| < 1$, $\ell=1,2,\ldots,k$. The *region of absolute stability* is the region in the complex λh plane in which the method is absolutely stable, and the *interval of absolute stability* its intersection with the real axis.

It can be shown (Lambert, 1973) that one of the roots, say r_1, is an approximation to $\exp(\lambda h)$ and in particular

$$r_1 = e^{\lambda h} + O(h^{p+1}), \tag{7.3.10}$$

where p is the order of the method. Hence the term $B_1(r_1)^n$ approximates the solution of the differential equation, while the other terms of (7.3.8), often called *parasitic solutions*, arise because the first-order differential equation is approximated by a difference equation of order k. It would seem desirable to have $|r_\ell| < |r_1|$, $\ell=2,3,\ldots,k$, so that the parasitic solutions decay relative to the approximating solution $B_1(r_1)^n$. If $|r_\ell| < |r_1|$, $\ell=2,3,\ldots,k$, the errors are dominated by the term involving $(r_1)^n$, and even if $|r_1| > 1$, which must happen for small positive λh, the errors will grow at about the same rate as the numerical solution and the relative error will remain small. For this reason we say that a linear k-step method is *relatively stable* if the roots r_ℓ of its characteristic polynomial satisfy $|r_\ell| < |r_1|$, $\ell=2,3,\ldots,k$. The region in the complex λh plane for which the method is relatively stable is called the *region of relative stability*, and the intersection of this with the real axis is called the *interval of relative stability*.

We now return to the question of convergence. A fundamental result, the proof of which can be found in Henrici (1962), is that a linear multi-step scheme is convergent if and only if it is zero-stable and consistent. A linear k-step method is defined to be *zero-stable* if the roots r_ℓ of the polynomial $\pi(r,0)$ satisfy $|r_\ell| \le 1$, $\ell=1,2,\ldots,k$, with the equality only holding for simple roots. Thus a linear multi-step method is zero-stable if for $h=0$ all the roots of the characteristic polynomial lie in or on the unit circle, with those on the unit circle being simple. Henrici has also derived a bound on the global truncation error e_n - he has shown that $e_n = O(h^p)$ for a convergent p^{th}-order linear multistep method provided that the starting values have errors of $O(h^q)$ where $q \ge p$. This justifies our earlier definition of order.

While convergence is essential, it is clearly not enough on its own. If $Re(\lambda) > 0$ the solution of the differential equation is increasing so it is not possible for a method to be absolutely stable (for small h anyway), but it is necessary that the method is relatively stable. Similarly, if $Re(\lambda) < 0$ the solution of the differential equation decreases so absolute stability is imperative. In this case relative stability is also desirable, since without it the errors will not decay as quickly as the solution, causing the relative error to increase.

As an example consider Simpson's rule (7.2.7)

$$y_{n+2} = y_n + \frac{h}{3}(f_{n+2}+4f_{n+1}+f_n). \tag{7.3.11}$$

Proceeding in the same way as the previous example, it is easy to show that the roots of the characteristic polynomial are

$$r_1 = 1 + \lambda h + O(h^2),$$

$$r_2 = -1 + \tfrac{1}{3}\lambda h + O(h^2). \tag{7.3.12}$$

When $h=0$ we have $r_1 = 1$ and $r_2 = -1$ so the method is zero-stable, and as it is consistent it is convergent. Suppose that a single equation is solved, so that we can assume that λ is real and that it corresponds to $\partial f/\partial y$. It can be seen from (7.3.12) that if λh is small and positive then $|r_1| > 1$, and if λh is small and negative $|r_2| > 1$, so that the errors grow no matter what is the sign of $\partial f/\partial y$. Decreasing the step length h reduces the error, but even so the error will continue to grow and a very small value of h may be needed, particularly if the solution is required on a large interval. Consequently Simpson's rule can not be recommended.

Convergent methods which have an empty region of absolute stability are sometimes called *weakly stable*, and those with a non-empty region of absolute stability are called *strongly stable*. Simpson's rule belongs to the former category, and the instability described above is called *weak instability*. There are many slightly different definitions of absolute, relative, weak, ..., etc. stability which can be important when comparing results from different sources. Note that regions and intervals of stability can not be calculated from expressions for the roots of the characteristic polynomial like (7.3.12). The methods by which they may be found are described in Lambert (1973).

7.4 The Adams Methods

Linear k-step methods with characteristic polynomials such that $\pi(r,0) = r^k - r^{k-1}$ are called *Adams methods*. In particular, those which are explicit are called *Adams-Bashforth methods*, while those which are implicit are called *Adams-Moulton methods*. The roots of $\pi(r,0)$ are $r_1 = 1$, $r_\ell = 0$, $\ell=2,3,\ldots,k$, and since the roots of a polynomial depend continuously on its coefficients, the parasitic roots of $\pi(r,h)$ will be small in magnitude for small $|\lambda h|$. Consequently these methods have good stability properties.

The Adams methods are derived from the integrated differential equation

$$y(x_{n+k})-y(x_{n+k-1}) = \int_{x_{n+k-1}}^{x_{n+k}} f(x,y(x))dx. \tag{7.4.1}$$

If the integral is approximated by integrating the polynomial interpolating the points $f_n, f_{n+1}, \ldots, f_{n+k-1}$, the following Adams-Bashforth methods are obtained for $k=1,2,3,4$:

$$\begin{aligned}
y_{n+1} &= y_n + hf_n, \\
y_{n+2} &= y_{n+1} + \frac{h}{2}(3f_{n+1}-f_n), \\
y_{n+3} &= y_{n+2} + \frac{h}{12}(23f_{n+2}-16f_{n+1}+5f_n), \\
y_{n+4} &= y_{n+3} + \frac{h}{24}(55f_{n+3}-59f_{n+2}+37f_{n+1}-9f_n).
\end{aligned} \tag{7.4.2}$$

These methods have orders of 1,2,3 and 4 respectively, and of course the first method is none other than Euler's method. To derive the Adams-Moulton methods, a polynomial which interpolates the points $f_n, f_{n+1}, \ldots, f_{n+k}$ is used, resulting in the following formulas for $k=1,2,3,4$:

$$\begin{aligned}
y_{n+1} &= y_n + \frac{h}{2}(f_{n+1}+f_n), \\
y_{n+2} &= y_{n+1} + \frac{h}{12}(5f_{n+2}+8f_{n+1}-f_n), \\
y_{n+3} &= y_{n+2} + \frac{h}{24}(9f_{n+3}+19f_{n+2}-5f_{n+1}+f_n), \\
y_{n+4} &= y_{n+3} + \frac{h}{720}(251f_{n+4}+646f_{n+3}-264f_{n+2}+106f_{n+1}-19f_n).
\end{aligned} \tag{7.4.3}$$

The orders of these formulas are 2,3,4 and 5 respectively. The first formula is known as the *trapezoidal method* as it is clearly just the trapezoidal quadrature formula. For completeness we add one further method,

$$y_{n+1} = y_n + hf_{n+1}. \tag{7.4.4}$$

This is an Adams-Moulton method but does not have the highest attainable order possible for a one-step method. It is called the *backward Euler method* and is first-order.

The two one-step Adams-Moulton methods have extremely large regions of absolute stability. For the backward Euler method

$$r_1 = \frac{1}{1-\lambda h}, \tag{7.4.5}$$

so the method is absolutely stable for λh outside of the unit circle centred at $(1,0)$. The trapezoidal method has

$$r_1 = \frac{1+\frac{1}{2}\lambda h}{1-\frac{1}{2}\lambda h}, \tag{7.4.6}$$

so that $|r_1| < 1$ for $\mathrm{Re}(\lambda h) < 0$, giving a region of absolute stability of the entire left half-plane. As the order increases the regions of absolute stability of the Adams methods decrease in size, a behaviour which is the opposite to that of the Runge-Kutta methods. Plots of the absolute stability regions of the Adams methods can be found in Gear (1971), page 131, and intervals of absolute stability are given in Table 7.2.

TABLE 7.2: A comparison of the Adams-Bashforth and Adams-Moulton methods of step-numbers 1-4.

	Adams-Bashforth				Adams-Moulton			
Step-number (k)	1	2	3	4	1	2	3	4
Order (p)	1	2	3	4	2	3	4	5
C_{p+1}	$\frac{1}{2}$	$\frac{5}{12}$	$\frac{3}{8}$	$\frac{251}{720}$	$-\frac{1}{12}$	$-\frac{1}{24}$	$-\frac{19}{720}$	$-\frac{3}{160}$
Interval of absolute convergence	$(-2,0)$	$(-1,0)$	$(-\frac{6}{11},0)$	$(-\frac{3}{10},0)$	$(-\infty,0)$	$(-6,0)$	$(-3,0)$	$(-\frac{90}{49},0)$

It can be seen from Table 7.2 that the (implicit) Adams-Moulton methods have several advantages over the (explicit) Adams-Bashforth methods of the same order. Comparing the values of the *error constant* C_{p+1} which is the coefficient of the principal truncation error $C_{p+1}h^{p+1}y^{(p+1)}(x_n)$, we see that the implicit method is much more accurate. For example, the error constant of the implicit fourth-order method is smaller by a factor of about 1/13 than the error constant of the fourth-order explicit method. This is not surprising because the explicit Adams methods were derived by integrating a polynomial which extrapolated $f(x,y(x))$ on the interval over which it was integrated, whereas when the implicit methods were derived the polynomial interpolated $f(x,y(x))$ on the interval of integration. This probably also explains the better stability properties of the implicit methods. For the fourth-order methods the absolute stability interval of the Adams-Moulton method is ten times larger than that of the Adams-Bashforth method. A minor advantage of the implicit methods is that they have a smaller step-number for a given order. The above comparison is typical of all explicit and implicit methods.

Implicit schemes do have the major disadvantage that the value y_{n+k} is defined implicitly, so that in general it must be found by an iterative scheme such as the one defined in (7.1.3). An initial estimate of $y_{n+k}^{[0]}$ is needed to start the iteration, and we would like to make this estimate as accurate as possible so that the number of iterations required is minimised. An explicit method is ideally suited to provide the value $y_{n+k}^{[0]}$; it is called a *predictor* while the implicit scheme which is used to improve the estimate of y_{n+k} is called a *corrector* - together they make a *predictor-corrector pair*.

TABLE 7.3: The numerical solution of $y' = 1-2xy$, $y(0) = 0$ using Adams-Bashforth and Adams-Moulton methods of order 4 with $h=0.1$.

	Adams-Bashforth		Adams-Moulton	
x_n	y_n	error	y_n	error
0.0	0.00000000	0.	0.00000000	0.
0.1	0.09933333	2.7E-06	0.09933333	2.7E-06
0.2	0.19474600	5.0E-06	0.19474600	5.0E-06
0.3	0.28262492	6.7E-06	0.28262492	6.7E-06
0.4	0.35985124	9.2E-05	0.35994208	1.4E-06
0.5	0.42430135	1.4E-04	0.42443757	-1.2E-06
0.6	0.47461475	1.5E-04	0.47476434	-1.1E-06
0.7	0.51038052	1.2E-04	0.51050297	1.1E-06
0.8	0.53202720	7.5E-05	0.53209702	4.7E-06
0.9	0.54071409	1.0E-05	0.54071561	8.7E-06
1.0	0.53813410	-5.5E-05	0.53806723	1.2E-05

The numerical solution of

$$y' = 1 - 2xy, \quad y(0) = 0 , \tag{7.4.7}$$

by the fourth-order Adams-Bashforth and Adams-Moulton methods for $h = 0.1$ are shown in Table 7.3. The necessary starting values were found using the fourth-order Taylor series method, and to make the comparison fair three values were calculated in each case even though the Adams-Moulton method only needs two values. As expected the Adams-Moulton method has given better results, but it is not as accurate as the fourth-order Runge-Kutta methods (see Table 5.2).

8. IMPLEMENTATION OF LINEAR MULTISTEP METHODS

8.1 Predictor-Corrector Modes

For the reasons discussed in the previous Section, linear multistep methods are almost always used as predictor-corrector methods. There are two basic approaches in the implementation of a predictor-corrector method. The first is to perform the iteration (7.1.3) of the corrector until some convergence criterion such as

$$|y_{n+k}^{[t+1]} - y_{n+k}^{[t]}| < \delta , \tag{8.1.1}$$

where δ is a preassigned tolerance, is satisfied. This mode of operation is called *correcting to convergence*. Each iteration requires the evaluation of $f(x,y)$, so the number of function evaluations will vary from step to step. As the error in the iterate $y_{n+k}^{[t]}$ is reduced by a factor of approximately $h|\beta_k|L$ in each iteration, reducing h will not only improve the accuracy of the initial estimate $y_{n+k}^{[0]}$ given by the predictor but also increase the rate of convergence of the iteration. Thus the convergence criterion (8.1.1) can always be satisfied in less than a fixed number of iterations if h is made small enough.

The second approach is to specify in advance the number of iterations, ν, we will use at each step, and take $y_{n+k} = y_{n+k}^{[\nu]}$. This process is best described in the standard notation introduced by Hull and Creemer (1963), where P denotes the application of the predictor, C a single iteration of the corrector and E an evaluation of f. First we compute $y_{n+k}^{[0]}$ from the predictor, next we calculate $f(x_{n+k},y_{n+k}^{[0]})$ as this is needed on the right hand side of (7.1.3), and then we obtain $y_{n+k}^{[1]}$ from the corrector. To this point we have carried out the operations PEC. Another evaluation $f(x_{n+k},y_{n+k}^{[1]})$, followed by an iteration of the corrector, gives $y_{n+k}^{[2]}$ and the calculation thus far is denoted by PECEC or $P(EC)^2$. After ν iterations we have $y_{n+k} = y_{n+k}^{[\nu]}$ and the calculation is denoted by $P(EC)^{\nu}$. We now must decide on a value for f_{n+k} as this is needed in subsequent steps. There are two choices which can be made: we can use the previously calculated value $f(x_{n+k},y_{n+k}^{[\nu-1]})$ or we can compute $f(x_{n+k},y_{n+k}^{[\nu]})$. If the former value is used we have $P(EC)^{\nu}$, but if the last evaluation is carried out the scheme is denoted by $P(EC)^{\nu}E$.

To illustrate the above modes we use the fourth-order Adams predictor-corrector pair defined in (7.4.2) and (7.4.3). Because the step-number of the Adams-Moulton corrector is less than that of the Adams-Bashforth predictor, its formula must be "shifted". The $P(EC)^{\nu}E$ mode scheme is given by

$$\text{P}: y_{n+4}^{[0]} = y_{n+3} + \frac{h}{24}(55f_{n+3} - 59f_{n+2} + 37f_{n+1} - 9f_n),$$

$$\left.\begin{aligned} \text{E}: f_{n+4}^{[t]} &= f(x_{n+4},y_{n+4}^{[t]}),\\ \text{C}: y_{n+4}^{[t+1]} &= y_{n+3} + \frac{h}{24}(9f_{n+4}^{[t]} + 19f_{n+3} - 5f_{n+2} + f_{n+1}), \end{aligned}\right\} \quad t = 0,1,.,\nu-1,$$

$$\text{E}: f_{n+4}^{[\nu]} = f(x_{n+4},y_{n+4}^{[\nu]}),$$

$$y_{n+4} = y_{n+4}^{[\nu]} ,$$

$$f_{n+4} = f_{n+4}^{[\nu]} . \qquad (8.1.2)$$

In practice the values y_{n+4} and f_{n+4} are overwritten by $y_{n+4}^{[t]}$ and $f_{n+4}^{[t]}$ respectively, so that the last two "replacements" do not have to be carried out. In the $P(EC)^\nu$ mode, the above scheme is modified by removing the last evaluation and replacing the ν by ν-1 in the last line of (8.1.2).

We have used the fourth-order Adams predictor-corrector pair in both $P(EC)^\nu$ and $P(EC)^\nu E$ modes to solve

$$y' = 1 - 2xy, \quad y(0) = 0 , \qquad (8.1.3)$$

for $h = 0.1$ and ν=1,2,3. The results are given in Tables 8.1 and 8.2. The iteration of the corrector has almost converged for ν=3 : this can be seen by comparing these results with those in Table 7.3 where the Adams-Moulton scheme was iterated to convergence.

TABLE 8.1: The numerical solution of $y' = 1-2xy$, $y(0) = 0$, using the fourth-order Adams predictor-corrector pair in $P(EC)^\nu$ mode with $h=0.1$.

	PEC		$P(EC)^2$		$P(EC)^3$	
x_n	y_n	error	y_n	error	y_n	error
0.0	0.00000000	0.	0.00000000	0.	0.00000000	0.
0.1	0.09933333	2.7E-06	0.09933333	2.7E-06	0.09933333	2.7E-06
0.2	0.19474600	5.0E-06	0.19474600	5.0E-06	0.19474600	5.0E-06
0.3	0.28262492	6.7E-06	0.28262492	6.7E-06	0.28262492	6.7E-06
0.4	0.35994481	-1.3E-06	0.35994200	1.5E-06	0.35994208	1.4E-06
0.5	0.42444765	-1.1E-05	0.42443724	-8.6E-07	0.42443758	-1.2E-06
0.6	0.47477740	-1.4E-05	0.47476386	-6.5E-07	0.47476436	-1.2E-06
0.7	0.51051627	-1.2E-05	0.51050249	1.6E-06	0.51050298	1.1E-06
0.8	0.53210429	-2.6E-06	0.53209682	4.9E-06	0.53209702	4.7E-06
0.9	0.54071423	1.0E-05	0.54071600	8.3E-06	0.54071558	8.7E-06
1.0	0.53805325	2.6E-05	0.53806848	1.1E-05	0.53806713	1.2E-05

If a predictor-corrector method is iterated to convergence, the predictor has no effect on the computed approximation and therefore the stability of the method depends only on the stability of the corrector. However, if a fixed number of iterations are used the stability depends on both the predictor and corrector and also on the actual mode used. For example, Brown, Riley and Bennett (1965) have computed intervals of absolute stability for the fourth-order Adams predictor-corrector pair and obtained the following results:

$$\begin{aligned} \text{Correcting to convergence} &: (-3.00,0), \\ PEC &: (-0.16,0), \\ PECE &: (-1.25,0), \\ P(EC)^2 &: (-0.90,0). \end{aligned} \qquad (8.1.4)$$

TABLE 8.2: The numerical solution of $y' = 1-2xy$, $y(0) = 0$, using the fourth-order Adams predictor-corrector pair in $P(EC)^{\nu}E$ mode with $h = 0.1$.

	PECE		$P(EC)^2E$		$P(EC)^3E$	
x_n	y_n	error	y_n	error	y_n	error
0.0	0.00000000	0.	0.00000000	0.	0.00000000	0.
0.1	0.09933333	2.7E-06	0.09933333	2.7E-06	0.09933333	2.7E-06
0.2	0.19474600	5.0E-06	0.19474600	5.0E-06	0.19474600	5.0E-06
0.3	0.28262492	6.7E-06	0.28262492	6.7E-06	0.28262492	6.7E-06
0.4	0.35994481	-1.3E-06	0.35994200	1.5E-06	0.35994208	1.4E-06
0.5	0.42444237	-6.0E-06	0.42443741	-1.0E-06	0.42443758	-1.2E-06
0.6	0.47476982	-6.6E-06	0.47476415	-9.4E-07	0.47476435	-1.1E-06
0.7	0.51050732	-3.3E-06	0.51050282	1.2E-06	0.51050297	1.1E-06
0.8	0.53209842	3.3E-06	0.53209703	4.7E-06	0.53209702	4.7E-06
0.9	0.54071262	1.2E-05	0.54071590	8.4E-06	0.54071559	8.7E-06
1.0	0.53805917	2.0E-05	0.53806788	1.2E-05	0.53806718	1.2E-05

The modes which use a fixed number of corrections have stability intervals which are much smaller than does the mode of correcting to convergence, but this is not always the case for predictor-corrector methods. However, it is generally found that the PEC mode has a very small interval of convergence compared to the PECE or $P(EC)^2$ modes.

Crane and Klopfenstein (1965) have derived a fourth-order predictor which gives the large interval of absolute stability of $(-2.48,0)$ when used in PECE mode with the fourth-order Adams-Moulton corrector. The predictor is

$$y_{n+4} = 1.547652y_{n+3} - 1.867503y_{n+2} + 2.017204y_{n+1} - 0.697353y_n$$
$$+h(2.002247f_{n+3} - 2.031690f_{n+2} + 1.818609f_{n+1} - 0.714320f_n). \quad (8.1.5)$$

A disadvantage of this predictor is that it needs much more storage than the fourth-order Adams-Bashforth predictor, requiring the additional storage of y_{n+2}, y_{n+1} and y_n, where these are m-vectors for a system of m equations. The amount of storage required by a predictor-corrector method can become large for high-order schemes, but Krogh (1966) has developed predictors of order 4 to 8 which only use the values y_{n+k-1} and y_{n+k-2} (one more than the Adams-Bashforth predictors) and which increase the size of the region of absolute stability when used in PECE mode with Adams-Moulton correctors. Approximate intervals of absolute stability, obtained from diagrams in Krogh's paper for the Adams pair corrected to convergence and in PECE mode, and Krogh's predictor with the Adams-Moulton corrector in PECE mode, are given in Table 8.3. The stability decreases with increasing order in all cases, but the stability of the PECE modes do not decrease as rapidly as that of the mode of correcting to convergence, and for order 8 the scheme using Krogh's predictor in PECE mode is the most stable. An example of Krogh's predictors is the fourth-order predictor

$$y_{n+4} = \tfrac{1}{2}y_{n+3} + \tfrac{1}{2}y_{n+2} + \frac{h}{48}(119f_{n+3} - 99f_{n+2} + 69f_{n+1} - 17f_n). \quad (8.1.6)$$

TABLE 8.3: Approximate intervals of absolute stability using Adams-Moulton correctors.

Order	Correcting to convergence	Adams-Bashforth predictor (PECE)	Krogh's predictor (PECE)
4	(-3,0)	(-1.3,0)	(-1.8,0)
5	(-1.8,0)	(-1.0,0)	(-1.4,0)
6	(-1.2,0)	(-0.7,0)	(-1.0,0)
7	(-0.8,0)	(-0.5,0)	(-0.8,0)
8	(-0.5,0)	(-0.4,0)	(-0.6,0)

If the evaluation of $f(x,y)$ is very time consuming a PEC mode scheme may be more economical. A fourth-order predictor which gives the reasonable interval of absolute stability of $(-0.78,0)$ when used with the Adams-Moulton corrector in PEC mode, is given by Klopfenstein and Millman (1968). It is

$$\begin{aligned} y_{n+4} &= -0.29y_{n+3}-15.39y_{n+2}+12.13y_{n+1}+4.55y_n \\ &\quad +h(2.27f_{n+3}+6.65f_{n+2}+13.91f_{n+1}+0.69f_n). \end{aligned} \tag{8.1.7}$$

8.2 Estimation of the Local Error

Suppose that the predictor and corrector are both of order p and have error constants C_{p+1} and $\hat{C}_{p+1}$ respectively. We assume that $y_n, y_{n+1}, \ldots, y_{n+k-1}$ all lie on a solution $y^*(x)$ of the differential equation. The local error of the predicted value $y_{n+k}^{[0]}$ is then

$$y^*(x_{n+k}) - y_{n+k}^{[0]} = C_{p+1}h^{p+1}y^{*(p+1)}(x_n) + O(h^{p+2}), \tag{8.2.1}$$

and it can be shown that (Lambert, 1973) the local error of the corrected value $y_{n+k}^{[t]}$, $t=1,2,\ldots,$ is

$$y^*(x_{n+k}) - y_{n+k}^{[t]} = \hat{C}_{p+1}h^{p+1}y^{*(p+1)}(x_n) + O(h^{p+2}). \tag{8.2.2}$$

On subtracting (8.2.2) from (8.2.1) we obtain

$$y_{n+k}^{[t]} - y_{n+k}^{[0]} = (C_{p+1}-\hat{C}_{p+1})h^{p+1}y^{*(p+1)}(x_n) + O(h^{p+2}), \tag{8.2.3}$$

and using this expression we eliminate $y^{*(p+1)}(x_n)$ from (8.2.2) to give

$$y^*(x_{n+k}) - y_{n+k}^{[t]} = \frac{\hat{C}_{p+1}}{C_{p+1}-\hat{C}_{p+1}}(y_{n+k}^{[t]}-y_{n+k}^{[0]}) + O(h^{p+2}). \tag{8.2.4}$$

Thus an estimate of the local error in the last iterate $y_{n+k}^{[\nu]}$ is

$$d^*_{n+k} = \frac{\hat{C}_{p+1}}{C_{p+1}-\hat{C}_{p+1}}(y_{n+k}^{[\nu]}-y_{n+k}^{[0]}). \tag{8.2.6}$$

This technique of estimating the local error was originated by

W.E. Milne and is known as *Milne's device*. Note that in practice $y_n, y_{n+1}, \ldots, y_{n+k-1}$ will not lie exactly on a solution of the differential equation, but it should be remembered that (8.2.6) is only an estimate and other approximations such as ignoring the higher-order terms have been made.

From Table 7.2 we see that the error constants of the fourth-order Adams predictor-corrector are $C_{p+1} = 251/720$ and $\hat{C}_{p+1} = -19/720$, giving the local error estimate

$$d^*_{n+4} = -\frac{19}{270}(y^{[\nu]}_{n+4} - y^{[0]}_{n+4}). \qquad (8.2.7)$$

When the other predictors given in Section 8.1 are used with the fourth-order Adams-Moulton corrector the following local error estimates apply:

$$\text{Predictor (8.1.5)} : d_{n+4} = -(y^{[\nu]}_{n+4} - y^{[0]}_{n+4})/16.21966,$$

$$\text{Predictor (8.1.6)} : d_{n+4} = -(y^{[\nu]}_{n+4} - y^{[0]}_{n+4})/13.7105,$$

$$\text{Predictor (8.1.7)} : d_{n+4} = -(y^{[\nu]}_{n+4} - y^{[0]}_{n+4})/18.0274. \qquad (8.2.8)$$

If d^*_{n+k} is a good estimate of the local error of $y^{[\nu]}_{n+k}$, then rather than taking $y_{n+k} = y^{[\nu]}_{n+k}$ we can obtain a more accurate approximation by letting

$$y_{n+k} = y^{[\nu]}_{n+k} + \frac{\hat{C}_{p+1}}{C_{p+1} - \hat{C}_{p+1}}(y^{[\nu]}_{n+k} - y^{[0]}_{n+k}). \qquad (8.2.9)$$

This is just the process we called local extrapolation, but in the present context (8.2.9) is known as a *modifier* and denoted by M. In a similar way the predicted value can be modified. An expression for the local error in $y^{[0]}_{n+k}$ may be found by eliminating $y^{*(p+1)}(x_n)$ from (8.2.1) using (8.2.3), giving

$$y^*(x_{n+k}) - y^{[0]}_{n+k} = \frac{C_{p+1}}{C_{p+1} - \hat{C}_{p+1}}(y^{[\nu]}_{n+k} - y^{[0]}_{n+k}) + O(h^{p+2}), \qquad (8.2.10)$$

but when the predicted value $y^{[0]}_{n+k}$ is calculated $y^{[\nu]}_{n+k}$ is not known. However, since

$$C_{p+1}h^{p+1}y^{*(p+1)}(x_n) = C_{p+1}h^{p+1}y^{*(p+1)}(x_{n-1}) + O(h^{p+2})$$

$$= \frac{C_{p+1}}{C_{p+1} - \hat{C}_{p+1}}(y^{[\nu]}_{n+k-1} - y^{[0]}_{n+k-1}) + O(h^{p+2}), \qquad (8.2.11)$$

we replace $y^{[0]}_{n+k}$ by the modified value $\bar{y}^{[0]}_{n+k}$, where

$$\bar{y}^{[0]}_{n+k} = y^{[0]}_{n+k} + \frac{C_{p+1}}{C_{p+1} - \hat{C}_{p+1}}(y^{[\nu]}_{n+k-1} - y^{[0]}_{n+k-1}), \qquad (8.2.12)$$

and this value is used to start the iteration of the corrector.

Modifiers can be incorporated into $P(EC)^{\nu}$ or $P(EC)^{\nu}E$ modes to give $PM(EC)^{\nu}$, $PM(EC)^{\nu}M$,.etc. It should be noted that the use of modifiers will change the stability characteristics of the method. An example of a method which uses modifiers is *Hamming's method* (Hamming, 1959), a PMECME mode method given by

$$\begin{aligned}
P &: y_{n+4}^{[0]} = y_n + \tfrac{4}{3}h(2f_{n+3} - f_{n+2} + 2f_{n+1}),\\
M &: \bar{y}_{n+4}^{[0]} = y_{n+4}^{[0]} + \tfrac{112}{121}(y_{n+3}^{[1]} - y_{n+3}^{[0]}),\\
E &: f_{n+4}^{[0]} = f(x_{n+4}, \bar{y}_{n+4}^{[0]}),\\
C &: y_{n+4}^{[1]} = \tfrac{9}{8}y_{n+3} - \tfrac{1}{8}y_{n+1} + \tfrac{3}{8}h(f_{n+4}^{[0]} + 2f_{n+3} - f_{n+2}),\\
M &: y_{n+4} = y_{n+4}^{[1]} - \tfrac{9}{121}(y_{n+4}^{[1]} - y_{n+4}^{[0]}),\\
E &: f_{n+4} = f(x_{n+4}, y_{n+4}). \qquad (8.2.13)
\end{aligned}$$

8.3 Variable Step Length

For one-step methods the step length is determined solely on the basis of controlling the global error by ensuring the local error per unit length is less than some prescribed value ε. The step length of a predictor-corrector method is chosen in the same way, using Milne's device to estimate the local error, but two other conditions must be satisfied. Firstly, h must satisfy (7.1.4) so that the iteration of the corrector is convergent, and secondly, h must be small enough that the method is stable. Both these conditions require a knowledge of the value of $\partial f/\partial y$ for a single equation or the eigenvalues of the Jacobian matrix for a system of equations, but for non-stiff differential equations it is generally the accuracy consideration which determines h. Thus most programs select h in a similar way to that described for one-step methods.

However it is not easy to change step length with a multistep scheme because a k-step method needs in general *back values* of y and f at k equispaced points and it is unlikely that these are available. If an increase in length is restricted to doubling h the back values have previously been computed (but twice as many must be stored), and if the step length is halved then half of the back values have been calculated and are available. Thus many programs restrict changes in step length to halving or doubling h. Note that the Adams predictor-corrector methods only use one previous value of y which is always available, but of course they require back values of f.

One method of calculating the back values is to use a one-step method, the method used to calculate the starting values. Another way is to interpolate the stored values of y to obtain the required back values and then evaluate $f(x,y)$ to find the back values of f. For methods using Adams predictor-corrector pairs, the back values of f can be found directly by interpolation. When interpolation is used the error of the interpolation formula should be of the same order as the local truncation error of the method of solving the initial value problem. No matter which of these methods are used, changes in step length should not be made too often because of the quite considerable computation required to do so.

An entirely different approach is due to Nordsieck (1962). He proposed saving y_{n+k-1} and the first $k-1$ derivatives of the polynomial which interpolates $y_n, y_{n+1}, \ldots, y_{n+k-1}$ evaluated at x_{n+k-1}, resulting in a one-step method which is equivalent to a multistep predictor-corrector method. This is essentially the method described by Gear (1971,1971a) who uses an Adams-Bashforth predictor and an Adams-Moulton corrector of the same order corrected to convergence (the original Nordsieck method used an Adams predictor and corrector of different order). If the corrector does not converge in three iterations the step length is reduced by a factor of 4 and, instead of using Milne's device to estimate the local error, an estimate based on difference of the highest stored derivative is employed. Gear (1971, 1971b) has written the program DIFSUB which incorporates this method with orders of one to seven. It is self-starting, using the first-order method initially, and the order is chosen so that the step length is maximized. This minimizes the computational effort, since the amount of computation per step is virtually independent of the order used.

Another way around the problem of changing step lengths is to remove the restriction that the back values must be equispaced. The Adams methods were derived by integration of a polynomial which interpolated the values $f_n, f_{n+1}, \ldots, f_{n+k-1}$ (and f_{n+k} for an implicit scheme), and this can be carried out when the points are not equispaced. It means that the coefficients of the predictor and corrector are not constant but will change at each step, but they can be efficiently calculated using divided differences. This is the basic method used by Shampine and Gordon (1975), who have developed a variable order scheme with orders one to twelve. An interesting feature of their method is that a predictor of order p is used with a corrector of order $p+1$ in PECE mode, which has improved the stability of the method. Plots of the stability regions are given in the above mentioned reference. They obtain an estimate of the local error from the difference between the predicted and corrected values.

8.4 Comparison of Predictor-Corrector Methods with Runge-Kutta Methods

It is difficult to compare two different classes of method such as the predictor-corrector methods and the Runge-Kutta methods. To simplify matters, we first consider the methods when a constant step length is used. It is not possible to compare the accuracies of the methods for an arbitrary problem because of the different form of the local truncation error, but it would seem that for a given order the Runge-Kutta methods may be slightly more accurate. However, while a Runge-Kutta method requires more function evaluations as the order is increased, a predictor-corrector method in $P(EC)^{\nu}$ or $P(EC)^{\nu}E$ mode needs ν or $\nu+1$ function evaluations no matter what the order, and in most implementations the number of evaluations is limited to a maximum of three even if the corrector is iterated to convergence. Thus if a Runge-Kutta method of order p is compared with a predictor-corrector method of order p on the basis of equal work, that is if the step lengths are adjusted so the same number of function evaluations are required on a given interval, we would expect that for $p \geq 4$ the predictor-corrector method would be more accurate. Alternatively, if the step lengths are kept equal a predictor-corrector method can be made more accurate than a Runge-Kutta method without increased computational cost, simply by increasing its order.

Another property which should be compared is the stability of the methods. Once again the behaviour is quite different; the absolute stability region of the Runge-Kutta methods grow as the order is increased while the regions of absolute stability of the predictor-corrector methods become smaller as the order is increased. However, if methods of equal order are compared on the basis of equal computation, the smaller step length allowed by the predictor-corrector method more than compensates for the smaller region of absolute stability.

Runge-Kutta methods do have some advantages. Their storage requirements and their computational overheads are generally lower. Predictor-corrector methods require the storage of back values which must be updated after each step, and the resulting shifting operations can form a significant part of the computation in the step. Runge-Kutta methods are self-starting, whereas predictor-corrector methods require the use of some one-step method to calculate sufficient values for them to proceed.

Varying the step length is no problem for the one-step Runge-Kutta methods - in fact the only difficulty is to determine what the new step length should be. It is much more difficult to change the step length with a predictor-corrector method, and it requires much extra computation. Predictor-corrector methods do have the extremely cheap local error estimate of Milne's device, although an embedded Runge-Kutta method provides an estimate of the local error at reasonable cost. There is no doubt that a Runge-Kutta method is easier to program than a predictor-corrector method for either a fixed step length scheme or a variable step scheme.

The optimum predictor-corrector method is the variable step variable order method. Codes of this type have been compared with variable step Runge-Kutta codes, such as those based on the Fehlberg formulas, by Hull et al (1972) and they conclude that when the derivative function is expensive to calculate (that is, requires about 25 or more arithmetic operations) the variable order Adams method is superior. Runge-Kutta methods are preferred for low accuracy requirements when the derivative evaluation is cheap, and a third class of methods, the extrapolation methods (see Sections 9.1, 9.2), is recommended when the function evaluation is cheap but a high accuracy is required. A more recent comparison has been made by Shampine et al (1976), and although they claim to test codes rather than methods they reach the same conclusions.

9. EXTRAPOLATION METHODS AND OTHER METHODS

9.1 Polynomial extrapolation

Suppose that an approximation to $y^*(x_n+H)$, where $y^*(x)$ is the solution of $y' = f(x,y)$, $y(x_n) = y_n$, is obtained by taking N_i steps of length h_i where $N_i h_i = H$. We denote the approximation by $y(x_n+H; h_i)$, and assume that for some integer γ

$$y(x_n+H; h_i) = y^*(x_n+H) + A_1 h_i^{\gamma} + A_2 h_i^{2\gamma} + A_3 h_i^{3\gamma} + \ldots \quad (9.1.1)$$

Note that the terms $A_1 h_i^{\gamma} + \ldots$ represent the global truncation error. If the calculations are performed with step lengths h_0 and h_1, $h_0 > h_1$ then

$$y(x_n+H; h_0) + \frac{y(x_n+H; h_0) - y(x_n+H; h_1)}{(h_1/h_0)^{\gamma} - 1} = y^*(x_n+H) - A_2 h_0^{\gamma} h_1^{\gamma} + \ldots, \quad (9.1.2)$$

which is clearly a better estimate of $y^*(x_n+H)$. This is essentially Richardson extrapolation (see Section 6.3), and if $y(x_n+H; h_i)$ is calculated for three values $h_0 > h_1 > h_2$ the process can be repeated to eliminate the A_2 term and so on. This repeated Richardson extrapolation is most efficiently carried out using an algorithm due to Neville (1934) by which the following tableau is constructed for $h_0 > h_1 > h_2 > h_3 > \ldots$:

$$\begin{array}{lllll}
y(x_n+H; h_0) = P_0^{(0)} & & & & \\
y(x_n+H; h_1) = P_1^{(0)} & P_0^{(1)} & & & \\
y(x_n+H; h_2) = P_2^{(0)} & P_1^{(1)} & P_0^{(2)} & & \\
y(x_n+H; h_3) = P_3^{(0)} & P_2^{(1)} & P_1^{(2)} & P_0^{(3)} & \\
\vdots & \vdots & \vdots & \vdots & \ddots
\end{array} \quad (9.1.3)$$

The tableau may be determined row by row from the formula

$$P_i^{(j)} = P_{i+1}^{(j-1)} + \frac{P_{i+1}^{(j-1)} - P_i^{(j-1)}}{(h_i/h_{i+j})^{\gamma} - 1}, \quad j=1,2,\ldots . \quad (9.1.4)$$

The value $P_i^{(j)}$ is the value at $h=0$ of the polynomial of h^{γ} which interpolates $y(x_n+H; h_i), y(x_n+H; h_{i+1}),\ldots,y(x_n+H; h_{i+j})$.

It can be shown (Gragg, 1965) that

$$P_i^{(j)} = y^*(x_n+H) + O(h_i^{\gamma} h_{i+1}^{\gamma} \ldots h_{i+j}^{\gamma}) \quad (9.1.5)$$

so that each column converges to $y^*(x_n+H)$ faster than the columns to the left, each row converges to $y^*(x_n+H)$ faster than the rows above, and the upper diagonal converges faster than any row or column.

An expansion of the form (9.1.1) is necessary for an extrapolation method, and the larger the value of γ, the faster the method will converge. Gragg (1965) has investigated methods which have expansions with $\gamma=2$ and has found that the explicit two-step *mid-point rule*

$$y_{n+2} = y_n + 2hf_{n+1} \tag{9.1.6}$$

gives such an expansion, provided the starting value is found from Euler's method and the N_i are either all odd or all even, although there are some advantages in using even values. It is easy to show the mid-point rule is only weakly stable (that is, it has an empty region of absolute stability) but the weak instability can be controlled by applying a smoothing procedure at the end of the basic step without destroying the form of the expansion. The resulting method is known as *Gragg's method* or the *modified mid-point method* and is defined by

$$\begin{aligned}
&h_i = H/N_i, \qquad N_i \text{ even},\\
&Y_0 = y_n, \qquad X_0 = x_n,\\
&Y_1 = Y_0 + h_i f(X_0, Y_0),\\
&\left.\begin{aligned} X_{j+1} &= X_j + h_i,\\ Y_{j+2} &= Y_j + 2h_i f(X_{j+1}, Y_{j+1}), \end{aligned}\right\} \quad j=0,1,2,\ldots,N_i-1,\\
&y(x_n+H;\, h_i) = \tfrac{1}{4}Y_{N_i-1} + \tfrac{1}{2}Y_{N_i} + \tfrac{1}{4}Y_{N_i+1}\,.
\end{aligned} \tag{9.1.7}$$

This procedure is carried out for an increasing sequence $\{N_i\}$ of even integers, enabling polynomial extrapolation as defined by (9.1.3)-(9.1.5) to be performed. The sequence $\{2,4,8,16,32,64,\ldots\}$ is sometimes used, but although very accurate it is expensive since the amount of computation is doubled for each stage. The sequence $\{2,4,6,8,10,12,..\}$ is much cheaper to calculate, but gives poor results since the denominator $(h_i/h_{i+1}-1)$ becomes very small. The most popular sequence is $\{2,4,6,8,12,16,24,32,\ldots\}$, which is a compromise between efficiency and accuracy.

Gragg (1965) has observed that using (9.1.7) with a fixed number of extrapolations is equivalent to a Runge-Kutta method so that stability is guaranteed. Stetter (1969) has shown (Lambert, 1973) that the stability of Gragg's method with polynomial extrapolation compares favourably with that of other methods.

The numerical solution of

$$y' = 1-2xy, \qquad y(0) = 0, \tag{9.1.8}$$

using Gragg's method with $H = 0.1$ followed by polynomial extrapolation, is shown in Table 9.1. The results in the first column are from Gragg's method which is a second-order method, but it should be noted that the step length was actually $H/2 = 0.05$. Clearly each extrapolation has improved the accuracy, and when one extrapolation is used, giving a fourth-order method, the results are better than those given by the fourth-order Runge-Kutta methods (Table 5.2).

Gragg's method with polynomial extrapolation can be used with a variable step length, but a discussion of this is left to the next Section where rational extrapolation is considered since the two methods are implemented in exactly the same way. Rational extrapolation is found to generally give better results than polynomial extrapolation.

TABLE 9.1: The numerical solution of $y'=1-2xy$, $y(0)=0$ using Gragg's method with polynomial extrapolation with $H=0.1$.

	no extrapolations		one extrapolation		two extrapolations	
x_n	y_n	error	y_n	error	y_n	error
0.0	0.00000000	0.	0.00000000	0.	0.00000000	0.
0.1	0.09925250	8.3E-05	0.09933603	-4.1E-08	0.09933599	2.0E-11
0.2	0.19458429	1.7E-04	0.19475110	-7.1E-08	0.19475103	3.8E-11
0.3	0.28238289	2.5E-04	0.28263174	-7.2E-08	0.28263166	5.5E-11
0.4	0.35961606	3.3E-04	0.35994351	-3.2E-08	0.35994348	7.6E-11
0.5	0.42403720	4.0E-04	0.42443632	6.1E-08	0.42443638	1.1E-10
0.6	0.47430356	4.6E-04	0.47476299	2.1E-07	0.47476320	1.7E-10
0.7	0.50999998	5.0E-04	0.51050364	4.1E-07	0.51050406	2.8E-10
0.8	0.53157328	5.3E-04	0.53210105	6.6E-07	0.53210171	4.6E-10
0.9	0.54019424	5.3E-04	0.54072340	9.2E-07	0.54072432	7.1E-10
1.0	0.53757097	5.1E-04	0.53807835	1.2E-06	0.53807951	1.0E-09

9.2 Rational extrapolation

Stoer (1961) and Bulirsch and Stoer (1964) have devised a tableau similar to (9.1.3) which corresponds to extrapolation to $h=0$ using rational functions $P(h^\gamma)/Q(h^\gamma)$, where $P(h^\gamma)$ and $Q(h^\gamma)$ are polynomials. This technique was used with Gragg's method by Bulirsch and Stoer (1966), giving a rational extrapolation method of solving initial value problems which is now known as the *Gragg-Bulirsch-Stoer method*. Assuming the values $y(x_n+H;\, h_i)$ are found by Gragg's method, the tableau is

$$\begin{array}{llllll}
y(x_n+H;\, h_0) = R_0^{(0)} & & & & & \\
y(x_n+H;\, h_1) = R_1^{(0)} & R_0^{(1)} & & & & \\
y(x_n+H;\, h_2) = R_2^{(0)} & R_1^{(1)} & R_0^{(2)} & & & \\
y(x_n+H;\, h_3) = R_3^{(0)} & R_2^{(1)} & R_1^{(2)} & R_0^{(3)} & & \\
\quad\vdots & \vdots & \vdots & \vdots & \vdots & \vdots \quad \ddots,
\end{array} \tag{9.2.1}$$

where the $R_i^{(j)}$ are found from the recurrence relation

$$R_i^{(-1)} = 0,$$

$$R_i^{(j)} = R_{i+1}^{(j-1)} + \frac{R_{i+1}^{(j-1)} - R_i^{(j-1)}}{(h_i/h_{i+j})^2[1-(R_{i+1}^{(j-1)}-R_i^{(j-1)})/(R_{i+1}^{(j-1)}-R_{i+1}^{(j-2)})]-1}. \tag{9.2.2}$$

In this case $R_i^{(j)}$ is the value obtained by interpolating the points $y(x_n+H;\, h_i)$, $y(x_n+H;\, h_{i+1})$ ... $y(x_n+H;\, h_{i+j})$ by the rational function of h^2 defined by

$$\frac{P(h^2)}{Q(h^2)} = \begin{cases} \dfrac{a_0+a_2h^2+\ldots+a_j h^j}{b_0+b_2h^2+\ldots+b_j h^j}, & j \text{ even}, \\[2ex] \dfrac{a_0+a_2h^2+\ldots+a_{j-1} h^{j-1}}{b_0+b_2h^2+\ldots+b_{j-1} h^{j-1}+b_{j+1} h^{j+1}}, & j \text{ odd}, \end{cases} \qquad (9.2.3)$$

and evaluating it at $h=0$. Gragg (1965) has shown that

$$R_i^{(j)} = y^*(x_n+H) + O(h_i^2 h_{i+1}^2, \ldots, h_{i+j}^2), \qquad (9.2.4)$$

a result identical to that for polynomial extrapolation. Nevertheless, rational extrapolation is found to generally give better results than polynomial extrapolation.

The formula (9.2.2) involves calculating differences of numbers which are nearly equal so is prone to the build-up of round-off errors. Consequently it is replaced by the equivalent algorithm defined below:

$$R_i^{(0)} = C_i^{(0)} = D_i^{(0)} = y(x_n+H;\ h_i),$$

$$W_i^{(0)} = R_i^{(0)} - D_{i-1}^{(0)},$$

$$\left.\begin{aligned} D_{i-j}^{(j)} &= \frac{C_{i-j+1}^{(j-1)}\, W_{i-j+1}^{(j-1)}}{(h_{i-j}/h_i)^2 D_{i-j}^{(j-1)} - C_{i-j+1}^{(j-1)}}, \\ R_{i-j}^{(j)} &= R_{i-j+1}^{(j-1)} + D_{i-j}^{(j)}, \\ C_{i-j}^{(j)} &= \frac{(h_{i-j}/h_i)^2 D_{i-j}^{(j-1)}\, W_{i-j+1}^{(j-1)}}{(h_{i-j}/h_i)^2 D_{i-j}^{(j-1)} - C_{i-j+1}^{(j-1)}}, \\ W_{i-j}^{(j)} &= C_{i-j}^{(j)} - D_{i-j-1}^{(j)}. \end{aligned}\right\} \quad j=1,2,\ldots,i. \qquad (9.2.5)$$

Thus the tableau is formed row by row $(i=0,1,2,\ldots)$. The only values that need to be stored at each stage are the values $D_{i-j}^{(j)}$, $j=0,1,2,\ldots,i$. Usually only the latest estimate $R_{i-j}^{(j)}$ is kept.

The algorithm (9.2.5) or (9.2.3) is one of the few that can not be modified to apply to the system $\underset{\sim}{y} = \underset{\sim}{f}(x,\underset{\sim}{y})$ simply by putting vector signs under $R_i^{(j)}$ and so on, although they are m-vectors. The equations apply to components of the vectors. This is no disadvantage, for even if an equation is written in vector form, the actual calculation is performed component by component.

The results obtained by using Gragg's method with rational extrapolation to solve

$$y' = 1-2xy, \qquad y(0) = 0, \qquad (9.2.6)$$

with $H = 0.1$ are given in Table 9.2. For this problem there is little difference between the approximations obtained using rational extrapolation and polynomial extrapolation (see Table 9.1).

TABLE 9.2: The numerical solution of $y'=1-2xy$, $y(0)=0$ using the Gragg-Bulirsch-Stoer method with $H=0.1$.

	no extrapolation		one extrapolation		two extrapolations	
x_n	y_n	error	y_n	error	y_n	error
0.0	0.00000000	0.	0.00000000	0.	0.00000000	0.
0.1	0.09925250	8.3E-05	0.09933605	-5.9E-08	0.09933599	1.1E-11
0.2	0.19458429	1.7E-04	0.19475113	-9.7E-08	0.19475103	2.4E-11
0.3	0.28238289	2.5E-04	0.28263177	-1.0E-07	0.28263166	4.1E-11
0.4	0.35961606	3.3E-04	0.35994355	-6.9E-08	0.35994348	5.8E-11
0.5	0.42403720	4.0E-04	0.42443636	2.2E-08	0.42443638	5.7E-11
0.6	0.47430356	4.6E-04	0.47476303	1.7E-07	0.47476320	1.7E-11
0.7	0.50999998	5.0E-04	0.51050368	3.7E-07	0.51050406	-8.9E-11
0.8	0.53157328	5.3E-04	0.53210109	6.2E-07	0.53210171	-2.9E-10
0.9	0.54019424	5.3E-04	0.54072344	8.8E-07	0.54072432	-6.1E-10
1.0	0.53757097	5.1E-04	0.53807838	1.1E-06	0.53807951	-1.1E-09

Extrapolation can be thought of as a variable order method, since the result $R_i^{(j)}$ essentially comes from a method of order $2j+2$, and the basic step length H may also be changed. A further variable is the number of stages or rows in the tableau, and the practical implementation must involve a choice between increasing the order or the number of stages or decreasing the step length to achieve the desired accuracy.

Since the accumulation of round-off errors increases both with the number of extrapolations and the number of stages, they were limited to six and ten respectively in the original implementation by Bulirsch and Stoer (1966). The local error was estimated by

$$d_{n+1}^* = \begin{cases} R_0^{(i)} - R_0^{(i-1)}, & i \le 6 \\ R_{i-6}^{(6)} - R_{i-7}^{(6)}, & 7 \le i \le 9, \end{cases} \tag{9.2.7}$$

and a relative error test was used so that the step was terminated when, for $i \ge 3$,

$$|d_{n+1}^*| \le \varepsilon y_{max}, \tag{9.2.8}$$

where

$$y_{max} = \max_{x\in[x_n, x_n+H]} |y(x)| . \tag{9.2.9}$$

The local error estimate (9.2.7) is based on a comparison of approximations of different orders (see (6.2.6)), and the higher-order result is used as the approximation to $y(x)$ at $x_{n+1} = x_n+H$.

If the error test (9.2.8) is satisfied for $i=3,4,5,6$, then

$$y_{n+1} = P_0^{(i)} \tag{9.2.10}$$

and θH used as the basic step length in the next step, where the recommended rule of thumb is

$$\theta = 1.5. \tag{9.2.11}$$

When (9.2.8) is satisfied for $i=7,8,9$ the aim is to reduce the step length by a factor $\theta < 1$ so that $R_0^{(6)}$ will be sufficiently accurate in the next step. Assuming the implied function in (9.2.4) varies slowly from step to step, this will be achieved if

$$(\theta h_0)^2(\theta h_1)^2 \ldots (\theta h_6)^2 = h_{i-6}^2 h_{i-5}^2 \ldots h_i^2 . \tag{9.2.12}$$

Rather than use this equation directly to find θ, the rule

$$\theta = 0.9(0.6)^{i-7} \tag{9.2.13}$$

is used. This is derived from (9.1.12) using the approximation $h_{i+1}/h_i \simeq 0.6$. If the convergence is so slow that (9.2.7) is not satisfied for $i \le 9$ the tableau is abandoned and the step repeated with a basic step length $H/2$. In the interests of efficiency values at $x_n+H/2$ are stored when each $y(x_n+H;\, h_i)$ is calculated, so one rejected step is not too costly.

The paper of Bulirsch and Stoer (1966) contains an Algol program which was converted to Fortran by N. Clark of Argonne National Laboratories. Error test options and printing options were added to this routine for use at Bell Telephone Laboratories by P. Crane, and a version of this code is presented by Fox (1971). Gear (1971) contains a Fortran program derived from the Argonne National Laboratories routine, but unlike the other codes the number of extrapolations and number of stages are not limited to six and ten respectively. Both of these programs differ from the original implementation of Bulirsch and Stoer in the way the step length is controlled. Rather than increase the step length by the fixed factor $\theta = 1.5$ when the error test is satisfied for $3 \le i \le i_{max}$, where i_{max} is the maximum number of extrapolations allowed ($i_{max} = 6$ for Fox's program), they use

$$\theta = 1 + (i_{max} - i)/i_{max} , \tag{9.2.14}$$

and in place of (9.2.13) they have

$$\theta = (\sqrt{2})^{i - i_{max} - 1} , \qquad i > i_{max} . \tag{9.2.15}$$

Gear's program also gives the user the option of using polynomial extrapolation or rational extrapolation.

The heuristic step control described above was found to perform well in the tests of Hull et al (1972) and Shampine et al (1976), and in both cases the Gragg-Bulirsch-Stoer method was found to be the best method when the function evaluations are inexpensive and the accuracy requirements are high. Stoer (1974) has proposed a rather complicated strategy for choosing the basic step length H which leads to a significant improvement in performance.

The step control strategy described above strongly biases the order to the highest permitted order. Murphy and Evans (1981) have proposed a scheme by which the order is chosen to minimize the computational effort of the step. They use only the values $R_0^{(j)}$ as approximations, with a local error estimate of

$$d^{*(j)}_{n+1} = R_0^{(j+1)} - R_0^{(j)} . \tag{9.2.16}$$

They show that the local error per unit length of $R_0^{(j)}$ in the next step will be not greater than ε if the current step length H is changed by the factor θ_j where

$$\theta_j = \left| \frac{H\varepsilon}{d^{*(j)}_{n+1}} \right|^{\frac{1}{2j}} \tag{9.2.17}$$

(compare with Equation (6.1.9)). For the sequence $\{N_j\} = \{2,4,6,8,12,16,...\}$ the number of function evaluations required is $\{A_j\} = \{3,7,13,21,33,49,...\}$, since $f(x_n, y_n)$ needs only to be evaluated once. Therefore, the cost of using j extrapolations with a step length $\theta_j H$ is proportional to A_j/θ_j, so j is chosen to minimize this quantity. Normally this information is only available for j less than the current number of extrapolations, but after k steps using k extrapolations an extra extrapolation is carried out to determine whether an increase in order is warranted. If the error test is not satisfied within the predicted number of extrapolations a test for the convergence of the tableau is carried out before it is extended. In this way the abandonment of large tableau is avoided.

9.3 Other methods

In this Section some of the less popular methods are briefly presented. More details may be found in Lambert (1973).

Block methods

The idea of block methods is to simultaneously produce a "block" of approximations $y_{n+1}, y_{n+2}, ..., y_{n+N}$. Block methods can generally be written either in terms of linear multistep methods or as an equivalent Runge-Kutta method. Block methods that are equivalent to explicit Runge-Kutta methods of step length Nh are given by Rosser (1967). An example with $N=2$ is

$$\begin{aligned}
k_1 &= f(x_n, y_n), \\
k_2 &= f(x_n+h,\ y_n+hk_1), \\
k_3 &= f(x_n+h,\ y_n+\tfrac{1}{2}hk_1+\tfrac{1}{2}hk_2), \\
k_4 &= f(x_n+2h,\ y_n+2hk_3), \\
y_{n+1} &= y_n + \frac{h}{12}(5k_1+8k_3-k_4), \\
k_5 &= f(x_n+h,\ y_{n+1}), \\
k_6 &= f(x_n+2h,\ y_n + \frac{h}{3}(k_1+k_4+4k_5)), \\
y_{n+2} &= y_n + \frac{h}{3}(k_1+4k_5+k_6).
\end{aligned} \tag{9.3.1}$$

The local truncation error of y_{n+2} is order five, so the scheme can be thought of as a six-stage explicit Runge-Kutta method of order four with a step length $2h$. The approximation y_{n+1} has a global error of order three, but as only y_{n+2} is used in the next step this does not affect the overall accuracy of the scheme. This lower order accuracy at the "interior" points is a feature of block methods. It can be shown that k_6 is a third-order approximation to $f(x_{n+2}, y^*(x_{n+2}))$ so it can be used for k_1 in the next step without lowering the order. Thus this method requires five function evaluations per step, compared with eight function evaluations for a conventional fourth-order Runge-Kutta method to take two steps of length h. Explicit block methods equivalent to implicit Runge-Kutta methods may also be derived, and as expected they have good stability properties.

Hybrid methods

A hybrid method is a linear multistep method that also uses information from an "off-step" point like a Runge-Kutta method. Thus a k-step hybrid method can be written

$$\sum_{j=0}^{k} \alpha_j y_{n+j} = h \sum_{j=0}^{k} \beta_j f_{n+j} + h\beta_\nu f_{n+\nu} \tag{9.3.2}$$

where $\alpha_k = 1$, α_0 and β_0 are not both zero, $\beta_\nu \neq 0$ and $\nu \neq 0,1,2,\ldots,k$. A predictor is also needed to provide a value for $y_{n+\nu}$ which is needed to evaluate $f_{n+\nu} = f(x_{n+\nu}, y_{n+\nu})$. The predictor may be of the form

$$y_{n+\nu} + \sum_{j=0}^{k-1} \bar{\alpha}_j y_{n+j} = h \sum_{j=0}^{k-1} \bar{\beta}_j f_{n+j} \ . \tag{9.3.3}$$

The constants in these formulas can be found by replacing y_{n+j} by $y(x_{n+j})$ and f_{n+j} by $y'(x_{n+j})$ and expanding them as Taylor series about x_n, as was done for linear multistep methods. Many of the properties of linear multistep methods apply to hybrid methods. This is hardly surprising, for if $\nu = k-\frac{1}{2}$ (a common choice) the hybrid method can be considered as a multistep method with a step length $h/2$. The advantage of hybrid methods is their greater accuracy for a given step length than linear multistep methods, but whether or not this compensates for the extra work is an interesting question.

Multiderivative multistep methods

Another modification of linear multistep methods is to use higher derivatives like the Taylor series methods. The resulting method is known as a multiderivative multistep method or an *Obrechkoff method*, and a general k-step method takes the form

$$\sum_{j=0}^{k} \alpha_j y_{n+j} = \sum_{i=1}^{\ell} h^i \left(\sum_{j=0}^{k} \beta_{ij} y_{n+j}^{(i)} \right), \tag{9.3.4}$$

where $\alpha_k = 1$ and one of α_0, β_{i0}, $i=1,2,\ldots,\ell$ is non-zero. Note that this method is actually an extension of the Taylor series method since they are given by $k=1$ and $\beta_{1i} = 0$, $i=1,2,\ldots,\ell$, in (9.3.4). As for linear multistep methods, implicit schemes are found to be more accurate and have better stability properties than explicit schemes of the same order. In addition, explicit and implicit methods can be combined to give predictor-corrector methods.

10. STIFF INITIAL VALUE PROBLEMS

10.1 Stiffness

Consider the system

$$y_1' = -1001y_1 + 999y_2 + 2; \qquad y_1(0) = 3,$$

$$y_2' = 999y_1 - 1001y_2 + 2; \qquad y_2(0) = 1, \tag{10.1.1}$$

which has the solution

$$y_1(x) = e^{-2000x} + e^{-2x} + 1,$$

$$y_2(x) = -e^{-2000x} + e^{-2x} + 1. \tag{10.1.2}$$

The terms e^{-2000x} and e^{-2x} may be classified as fast and slow transients respectively, and the fast transient will almost die out by $x = 0.01$ and the slow transient by about $x = 10$ leaving the steady-state solution $y_1(x) = 1$, $y_2(x) = 1$. If we were to use the fourth order Runge-Kutta method to solve (10.1.1) it could be expected that we would have to use a small step for $0 \le x \le 0.01$ so that the rapidly decaying fast transient would be accurately represented. In fact we require $\lambda h \in (-2.78, 0)$ for absolute stability, and since $\lambda = -2000$ this implies that $h < 0.00139$. Once $x = 0.01$ has been reached, and the fast transient has died away, we might expect to be able to take a larger step length. However, this is not the case since the stability requirement must still be satisfied. Thus the small step length must be used for any x, and to reach the steady state solution at $x = 10$ will take at least 7,200 steps. Had it been possible to change the system of differential equations for $x > 0.01$ so that the solution did not contain the fast transient, a step length up to $h = 1.39$ would have been possible, although a smaller value would be required for accurate results. This leads to one definition of stiffness, that "stiffness occurs when stability rather than accuracy dictates the choice of step length".

Consider the general linear constant coefficient system

$$\underset{\sim}{y}' = \underline{\underline{A}}\underset{\sim}{y} + \underset{\sim}{\psi}(x), \tag{10.1.3}$$

where $\underline{\underline{A}}$ is an $m \times m$ matrix whose eigenvalues λ_j, $j=1,2,\ldots,m$, are assumed distinct. If the eigenvectors of $\underline{\underline{A}}$ are $\underset{\sim}{u}_j$, $j=1,2,\ldots,m$, then the general solution of (10.1.3) is

$$\underset{\sim}{y}(x) = \sum_{j=1}^{m} c_j \underset{\sim}{u}_j e^{\lambda_j x} + \underset{\sim}{\Psi}(x), \tag{10.1.4}$$

where the c_j are constants and $\Psi(x)$ is the particular integral of (10.1.3).

A more formal definition of stiffness (Lambert, 1980) is that (10.1.3) is said to be *stiff* if:

(i) $\mathrm{Re}(\lambda_j) < 0$ for $j=1,2,\ldots,m$,

(ii) $\max_j |\mathrm{Re}(\lambda_j)| / \min_j |\mathrm{Re}(\lambda_j)| = S \gg 1.$ (10.1.5)

S is called the *stiffness ratio*. The general system $\underset{\sim}{y} = \underset{\sim}{f}(x,\underset{\sim}{y})$ is said to be stiff on an interval I if the eigenvalues $\lambda(x)$ of the Jacobian matrix satisfy (i) and (ii) for all $x \in I$. Note that in the above example $S = 1000$; in many practical problems stiffness ratios of 10^{10} or higher are not uncommon (Curtis, 1978). Stiff initial value problems arise in areas such as chemical engineering, chemical kinetics, control theory, networks, and so on.

The definition (10.1.5) is perhaps too restrictive in demanding that all the eigenvalues should have negative real parts. Stiffness really arises because of the different scales of terms which make up the solution, so that a system of differential equations is stiff if the Jacobian matrix has at least one eigenvalue whose real part is negative and large in magnitude compared to the overall scale of the solution. Since the independent variable will often denote time, a system is stiff if its solution contains transients which decay in a time which is short compared to the time-scale of the solution. This is essentially the definition given by Gear (1971) and Curtis (1978), and as the time scale of the solution may be determined by a forcing term, such as a diurnal temperature change in a heating problem, it does not require that $m > 1$.

An example of a single stiff equation is

$$y' = \lambda(y-F(x)) + F'(x), \tag{10.1.6}$$

where $\lambda \ll 0$ and $F(x)$ varies slowly. The solutions of this differential equation are $y(x) = Ae^{\lambda x} + F(x)$, so that all the solution curves rapidly coalesce. This behaviour is typical of stiff systems and is sometimes termed *overstability*. Figure 10.1 shows the effect of using Euler's method with a large step length to solve (10.1.6). If

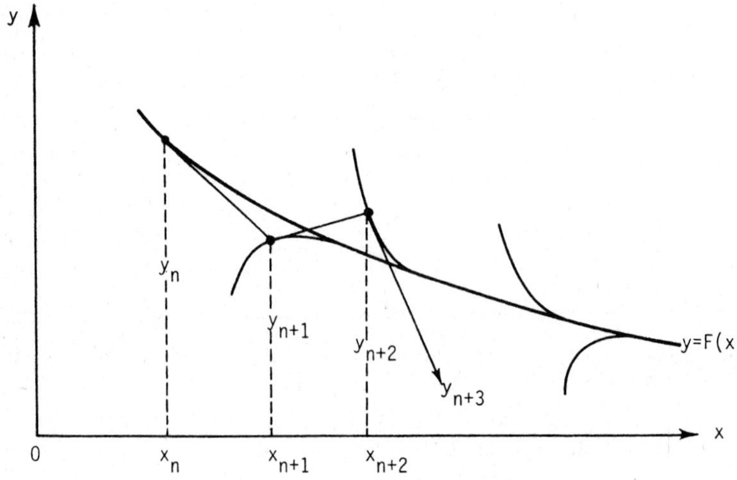

FIGURE 10.1: The numerical solution of $y' = \lambda(y-F(x)) + F'(x)$ using Euler's method.

$|1+\lambda h| > 1$ the error is amplified at each step even though the transient has died away completely. However, if the backward Euler method (7.4.4) is used the numerical solution converges to $y(x_n) = F(x_n)$ even with a large step length (see Figure 10.2). In this case the error is multiplied by $(1-\lambda h)^{-1}$ at each step (see Equations (7.4.5)) but for $\lambda < 0$ this factor is smaller than 1. A small step length is needed to accurately represent the transient $Ae^{\lambda x}$, but once this has died away the large step length can be used.

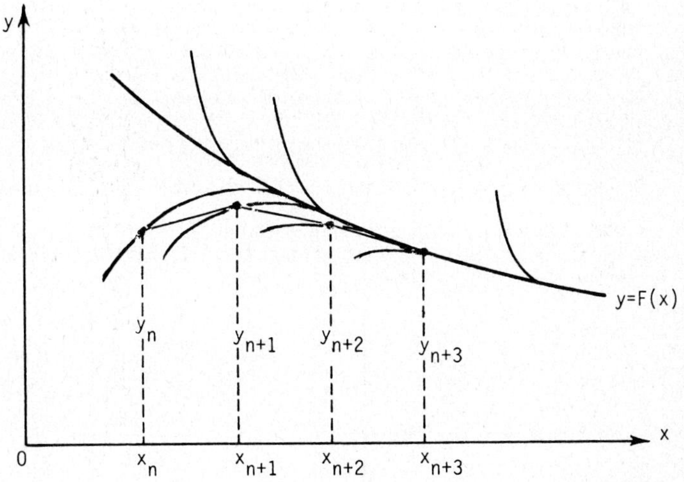

FIGURE 10.2: The numerical solution of $y' = \lambda(y-F(x)) + F'(x)$ *using the backward Euler method.*

It is generally difficult to determine if a system is stiff or not. Lambert (1980) has given some examples of linear systems where even a knowledge of the eigenvalues of the Jacobian matrix is not sufficient to determine the behaviour of the solution. Stiffness should be suspected if a fixed step length method gives very poor results - or even diverges - for "reasonable" step lengths, or if a program with variable step length can not meet the accuracy requirement or takes excessively small steps. Some work has been done by Shampine (1977) and Shampine and Hiebert (1977) on the automatic detection of stiffness in a non-stiff code. One of the NAG routines includes such a test (Gladwell, 1979).

10.2 Stability considerations

When the system (10.1.3) is solved numerically the step length h must be chosen according to two criteria: firstly, the non-decreasing part of the solution must be accurately represented and, secondly, $h\lambda_j$ must be within the region of absolute stability for all j such that $Re(\lambda_j) < 0$. If the latter condition determines h the system is stiff. Clearly the problem is overcome if the region of absolute stability contains the entire left-half complex plane $\{\lambda h | \text{Re}(\lambda h) < 0\}$. Methods with this property are called *A-stable* (Dahlquist, 1963), but Dahlquist proved that a linear multistep method of order greater than two cannot be A-stable and that the most accurate A-stable linear multistep method is the trapezoidal method.

In view of this restriction, relaxed stability conditions have been postulated. Widlund (1967) defines a method as *A(α)-stable* if its region of absolute stability contains the region $\{\lambda h; -\alpha < \pi - \text{Arg}(\lambda h) < \alpha\}$ for $\alpha \in (0,\pi/2)$. He also defines *A(0)-stability* as A(α)-stability for some sufficiently small $\alpha \in (0,\pi/2)$. Gear (1969) defines a method as *stiffly-stable* if it is absolutely stable in the region $\{\lambda h: \text{Re}(\lambda h) \le -a\}$ and is accurate in the region $\{\lambda h; -a < \text{Re}(\lambda h) < b, -c < \text{Im}(\lambda h) \le c\}$ where a,b,c are positive constants. This last condition of accuracy requires absolute stability to the left of the imaginary axis and relative stability to the right of the imaginary axis. A weaker condition is given by Cryer (1973) who calls a method *A_0-stable* if its region of absolute stability contains the whole negative real axis. Since

$$\text{A-stability} \Rightarrow \text{stiff-stability} \Rightarrow \text{A}(\alpha)\text{-stability} \Rightarrow A_0\text{-stability}$$

(see Figure 10.3) it is easier to find methods with a stability property nearer the end of this list, but they will prove unsatisfactory for a larger class of problems.

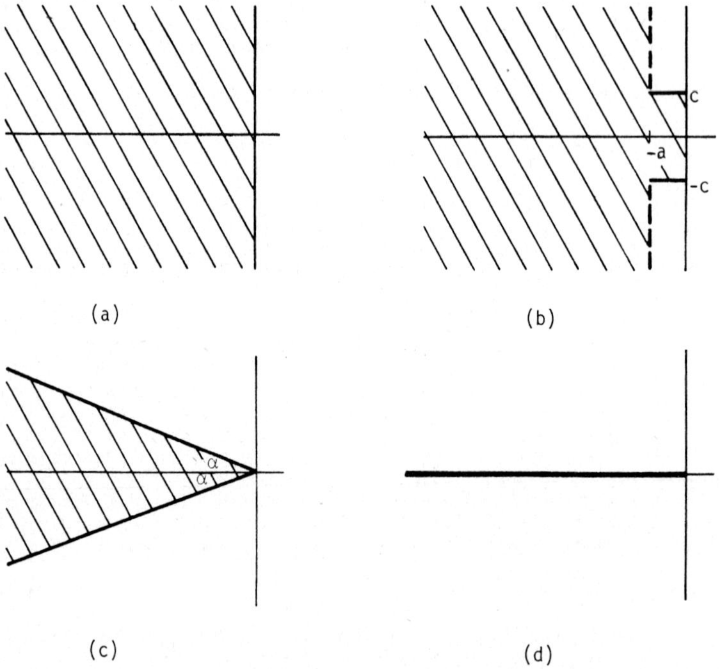

FIGURE 10.3: If the region of absolute stability includes the shaded region then the method is (a) A-stable, (b) stiffly-stable, (c) A(α)-stable, (d) A_0-stable.

While A-stability is a very severe requirement, an even stronger stability condition is defined for one-step methods. For a one step method $y_{n+1}/y_n = r_1(\lambda h)$, and for the trapezoidal method, it is seen that

$$r_1(\lambda h) = \frac{1+\lambda h}{1-\lambda h} \to -1 \quad \text{as} \quad Re(\lambda h) \to -\infty . \tag{10.2.1}$$

Thus any decaying components of the solution will do so slowly, in an oscillatory manner, for a large step length. In contrast, the backward Euler method has

$$r_1(\lambda h) = \frac{1}{1-\lambda h} \to 0 \quad \text{as} \quad Re(\lambda h) \to -\infty , \tag{10.2.2}$$

so decaying components will be rapidly damped out in a monotonic way. An A-stable one-step method such that $r_1(\lambda h) \to 0$ as $Re(\lambda h) \to -\infty$ is said to be *L-stable, strongly A-stable,* or *stiffly A-stable*. While this property may be desirable, the region of absolute stability of an L-stable method must contain part of the right half-plane and hence may damp out a component of the solution which should be growing. Lambert (1980) gives an example where this happens. Thus, a desirable property of a method is that its region of absolute stability does not encroach on the right half-plane. The trapezoidal method is such a method, as its absolute stability region is precisely the left-plane.

The oscillation produced by the trapezoidal method may be overcome simply by using a small step length until the fast transients have almost died away, then increasing the step length which will only cause an oscillation due to the small remnant of the transients. This procedure has the advantage of accurately following the transient, but when this is not required an alternative approach is to apply the smoothing formula used in Gragg's method (see (9.1.7)).

10.3 Solving the implicit equations

For most classes of method, certainly for linear multistep methods and Runge-Kutta methods, the requirement of even A_0-stability can only be met by an implicit method. For a linear multistep method this means that at each step we must solve a set of simultaneous, generally non-linear, equations of the form

$$\underset{\sim}{y}_{n+k} = h\beta_k \underset{\sim}{f}(x_{n+k}, \underset{\sim}{y}_{n+k}) + \underset{\sim}{B} , \tag{10.3.1}$$

where $\underset{\sim}{B}$ is the known vector

$$\underset{\sim}{B} = \sum_{j=0}^{k-1} (h\beta_j \underset{\sim}{f}_{n+j} - \alpha_j \underset{\sim}{y}_{n+j}). \tag{10.3.2}$$

The iteration suggested earlier (Equation (7.1.3)) to solve this equation is not feasible for a stiff system of equations because it requires

$$h < 1/L|\beta_k| \tag{10.3.3}$$

for convergence. Since $L = \| \partial f_i / \partial y_j \| \geq \max|\lambda_j|$, L will be very large and application of (10.3.3) will require choice of an extremely small h. Similarly, the conditions (5.4.3) and (5.4.6), which guarantee convergence of the iterations for the implicit and semi-implicit Runge-Kutta methods, both require that h is less than a multiple of L^{-1}. Therefore choice of a small h is required when the system is stiff.

This limitation on h is overcome by using Newton's method to solve the non-linear equations. To solve the vector equation $\underset{\sim}{F}(\underset{\sim}{y}) = \underset{\sim}{0}$, Newton's method is

$$\underset{\sim}{y}^{[t+1]} = \underset{\sim}{y}^{[t]} - \underset{\sim}{F}(\underset{\sim}{y}^{[t]})\underset{=}{J}^{-1}(\underset{\sim}{y}^{[t]}), \qquad t=0,1,\dots, \tag{10.3.4}$$

where $\underset{=}{J}$ is the Jacobian matrix $\partial\underset{\sim}{F}/\partial\underset{\sim}{y} = [\partial F_i/\partial y_j]$. Applying this to Equation (10.3.1) gives

$$\underset{\sim}{y}_{n+k}^{[t+1]} = \underset{\sim}{y}_{n+k}^{[t]} - [\underset{\sim}{y}_{n+k}^{[t]} - h\beta_k \underset{\sim}{f}(x_{n+k}, \underset{\sim}{y}_{n+k}^{[t]}) - \underset{\sim}{B}] \times$$

$$[\underset{=}{I} - h\beta_k \partial\underset{\sim}{f}(x_{n+k}, \underset{\sim}{y}_{n+k}^{[t]})/\partial\underset{\sim}{y}]^{-1}, \qquad t=0,1,2,\dots, \tag{10.3.5}$$

where $\underset{=}{I}$ is the $m\times m$ unit matrix. This iteration will converge provided the initial estimate $\underset{\sim}{y}_{n+k}^{[0]}$ is sufficiently accurate, and a suitable predictor can be used to provide this value. In practice the matrix $\underset{=}{I} - h\beta_k \partial\underset{\sim}{f}/\partial\underset{\sim}{y}$ is not inverted, but (10.3.5) is multiplied by this matrix and the resulting set of equations solved while making use of any sparseness of the Jacobian. Although (10.3.5) calls for the evaluation of the Jacobian at every iteration, it is usual to hold it constant for the iteration unless the iteration does not converge in, say, three steps. If the Jacobian does not vary rapidly it is also possible for it to keep the same value over several integration steps. Note that the corrector must be iterated to convergence for stiff systems, otherwise the absolute stability region will not be that of the corrector alone.

The non-linear equations arising from the use of an implicit or semi-implicit Runge-Kutta method must also be solved by Newton's method.

10.4 Methods for stiff initial value problems

By far the most common method of solution of stiff systems is the use of backward differentiation formulas. They are linear multistep methods of the form

$$\sum_{j=0}^{k} \alpha_j y_{n+j} = h\beta_k f_{n+k} \tag{10.4.1}$$

where $\alpha_k = 1$, $\alpha_0 \neq 0$ and $\beta_k \neq 0$. Their order, p, is equal to the step number, k, and for orders of one to six they are stiffly stable. Regions of absolute stability are given in Gear (1971), pages 215 and 216, and the coefficients are given in Table 10.1. This approximate value of the parameter a used in the definition of stiff-stability is also given in the table (also see Figure 10.3(b)). The first-order method is the backward Euler method, and it, together with the second and third-order methods, is absolutely stable in the right half-plane not particularly far from the origin. The higher-order methods do not suffer this problem, but instead are not absolutely stable in a region of the left half-plane near the imaginary axis. Thus these methods may give poor results for a system with an eigenvalue near the imaginary axis, either damping a solution which should be increasing in value or allowing a solution to grow when it should decay.

TABLE 10.1: Coefficients of the backward differentiation formulas.

Order	β_k	α_0	α_1	α_2	α_3	α_4	α_5	α_6	-a
1	1	-1	1						0
2	2/3	1/3	-4/3	1					0
3	6/11	-2/11	9/11	-18/11	1				0.1
4	12/25	3/25	-16/25	36/25	-48/25	1			0.7
5	60/137	-12/137	75/137	-200/137	300/137	-300/137	1		2.4
6	60/147	10/147	-72/147	225/147	-400/147	450/147	-360/147	1	6.1

Gear uses these formulas in his program DIFSUB (Gear, 1971, 1971b). This program has a stiff or non-stiff option, the non-stiff method using Adams predictor-corrector pairs of order one to seven. The backward differentiation formulas are implemented in the same variable-order variable-step manner as the Adams formulas, using the method of Nordsieck (1962) to store prior information. A pth order predictor of the form

$$y^{[0]}_{n+k} = h\beta^*_k f_{n+k-1} - \sum_{j=0}^{k-1} \alpha^*_j y_{n+j} \qquad (10.4.2)$$

is used to provide the initial estimate for the Newton iteration of the pth order corrector (10.4.1). The Jacobian matrix can either be supplied as a subroutine by the user, or else the program calculates it using differences. The program DIFSUB has been modified by A.C. Hindmarsh of Lawrence Livermore Laboratory and the resulting program is the basis of various routines in both the NAG and IMSL libraries.

The implicit Runge-Kutta methods are another class of methods which have suitable stability characteristics for use on stiff systems. Ehle (1969) has proved that the R-stage implicit Runge-Kutta method of order $2R$ is A-stable. L-stable methods are also possible. The difficulty with this procedure is that for a system of m equations there are mR simultaneous non-linear equations to solve at each step, and Newton's method requires sufficiently accurate estimates of $\boldsymbol{k}_r$, $r=1,2,\ldots,R$, to ensure convergence. Even an R-stage semi-explicit method requires the solution of R sets of m non-linear equations at each step. Consequently, these methods are not very practical.

The Jacobian matrix $\partial\boldsymbol{f}/\partial\boldsymbol{y}$ is required for the Newton iteration of the non-linear equations and, since $\boldsymbol{y}'' = (\partial\boldsymbol{f}/\partial\boldsymbol{y})\boldsymbol{f}$ for an autonomous system $\boldsymbol{y}' = \boldsymbol{f}(\boldsymbol{y})$, Obrechkoff methods seem attractive. Enright (1974) uses stiffly-stable Obrechkoff methods with step number $k \le 7$ and order $p = k+2$ in a variable-step variable-order scheme. Newton's method now involves $\partial\boldsymbol{y}''/\partial\boldsymbol{y}$ but, even though the second order derivatives are ignored, the coefficient matrix of the equations involves the square of the Jacobian so some sparseness is lost. Therefore the iteration is more costly than that of the backward differentiation formulas.

Rosenbrock (1963) proposed that the Jacobian might be used in the coefficients of Runge-Kutta methods, giving the general R-stage method

$$\left.\begin{aligned} \underset{\sim}{k}_r &= h\underset{\sim}{f}(\underset{\sim}{y}_n + \sum_{s=1}^{r-1} b_{rs}\underset{\sim}{k}_s) + h\gamma_r \frac{\partial \underset{\sim}{f}(\underset{\sim}{\eta}_r)}{\partial \underset{\sim}{y}} \underset{\sim}{k}_r , \\ \underset{\sim}{\eta}_r &= \underset{\sim}{y}_n + \sum_{s=1}^{r-1} \beta_{rs}\underset{\sim}{k}_s , \end{aligned}\right\} \quad r=1,2,\ldots,R ,$$

$$\underset{\sim}{y}_{n+1} = \underset{\sim}{y}_n + \sum_{r=1}^{R} c_r \underset{\sim}{k}_r , \qquad (10.4.3)$$

of solving the autonomous system $\underset{\sim}{y}' = \underset{\sim}{f}(\underset{\sim}{y})$. This may be thought of as an extension of the explicit Runge-Kutta method or a linearization of the semi-explicit Runge-Kutta method. It is implicit but only requires the solution of R sets of m linear equations at each step. In general the R-stage method requires the evaluation of the Jacobian matrix R times per step, but Bui (1979) has developed A-stable and L-stable four-stage methods of order four with $\beta_{rs} = 0$ and $\gamma_r = \gamma$ for all r and s so that only one Jacobian evaluation per step is needed. The four sets of equations also have the same coefficient matrix, so only one matrix decomposition per step is required.

The regions of absolute stability of Bui's A-stable and L-stable methods are almost identical. The simplified algorithm is

$$\left[\underset{=}{I} - \gamma h \frac{\partial \underset{\sim}{f}(\underset{\sim}{y}_n)}{\partial \underset{\sim}{y}}\right] \underset{\sim}{k}_r = h\underset{\sim}{f}(\underset{\sim}{y}_n + \sum_{s=1}^{r-1} b_{rs}\underset{\sim}{k}_s), \quad r=1,2,3,4,$$

$$\underset{\sim}{y}_{n+1} = \underset{\sim}{y}_n + \sum_{r=1}^{4} c_r \underset{\sim}{k}_r , \qquad (10.4.4)$$

and the constants for the L-stable method are

$\gamma = 0.5728160625,$

$b_{21} = -0.5,$

$b_{31} = -0.1012236115,$ $b_{32} = 0.9762236115,$

$b_{41} = -0.3922096763,$ $b_{42} = 0.7151140251,$ $b_{43} = 0.1430371625,$

$c_1 = 0.9451564786,$ $c_2 = 0.341323172,$

$c_3 = 0.5655139575,$ $c_4 = -0.8519936081.$ (10.4.5)

A variation of (10.4.3) called the *Rosenbrock-Wanner method* (see Kaps and Rentrop, 1979) follows:

$$\left[\underset{=}{I} - \gamma h \frac{\partial \underset{\sim}{f}(\underset{\sim}{y}_n)}{\partial \underset{\sim}{y}}\right] \underset{\sim}{k}_r = h\underset{\sim}{f}(\underset{\sim}{y}_n + \sum_{s=1}^{r-1} \alpha_{rs}\underset{\sim}{k}_s) + h \frac{\partial \underset{\sim}{f}(\underset{\sim}{y}_n)}{\partial \underset{\sim}{y}} \sum_{s=1}^{r-1} \gamma_{rs}\underset{\sim}{k}_s , \quad r=1,2,\ldots,R,$$

$$\underset{\sim}{y}_{n+1} = \underset{\sim}{y}_n + \sum_{r=1}^{R} c_r \underset{\sim}{k}_r . \qquad (10.4.6)$$

This method has the same advantage as Bui's in needing only one Jacobian evaluation and one matrix factorization per step. Kaps and Rentrop (1979) have derived an A-stable and an A(89.3°)-stable four-stage method of this type which is of order four and has an order three method embedded in it. The third-order and fourth-order results are denoted by $\bar{y}_n$ and $\hat{y}_n$ respectively, and because the fourth-order result is used as the approximation to $y(x_n)$ we have

$$\bar{y}_{n+1} = \hat{y}_n + \sum_{r=1}^{3} \bar{c}_r k_r ,$$

$$\hat{y}_{n+1} = \hat{y}_n + \sum_{r=1}^{4} \hat{c}_r k_r . \qquad (10.4.7)$$

To avoid the matrix multiplications in (10.4.6) the k_r's are found by solving

$$\left[I - \gamma h \frac{\partial f(y_n)}{\partial y} \right] (k_r + \sum_{s=1}^{r-1} \bar{\gamma}_{rs} k_s)$$

$$= hf(y_n + \sum_{s=1}^{r-1} \alpha_{rs} k_s) + \sum_{s=1}^{r-1} \bar{\gamma}_{rs} k_s , \quad r=1,2,3,4, \qquad (10.4.8)$$

where

$$\bar{\gamma}_{rs} = \gamma_{rs}/\gamma . \qquad (10.4.9)$$

The coefficients for the A-stable scheme are given below:

$\gamma = 0.395,$

$\gamma_{21} = -0.767672395484,$

$\gamma_{31} = -0.851675323742,$ $\gamma_{32} = 0.522967289188,$

$\gamma_{41} = 0.288463109545,$ $\gamma_{42} = 0.0880214273381,$ $\gamma_{43} = -0.337389840627,$

$\alpha_{21} = 0.438,$

$\alpha_{31} = 0.796920457938,$ $\alpha_{32} = 0.0730795420615,$

$\alpha_{41} = \alpha_{31},$ $\alpha_{42} = \alpha_{32},$ $\alpha_{43} = 0,$

$\bar{c}_1 = 0.346325833758,$ $\bar{c}_2 = 0.285693175712,$

$\bar{c}_3 = 0.367980990530,$

$\hat{c}_1 = 0.199293275701,$ $\hat{c}_2 = 0.482645235674,$

$\hat{c}_3 = 0.0680614886256,$ $\hat{c}_4 = 0.25.$ (10.4.10)

An estimate of the local error of the third-order result is

$$d^*_{n+1} = \hat{y}_{n+1} - \bar{y}_{n+1} , \qquad (10.4.11)$$

from which the scaled error E is calculated:

$$E = \max_{i=1,2,\ldots,m} |d^*_{i,n+1}/S_i| , \qquad (10.4.12)$$

where $d^*_{i,n+1}$ is the ith component of d^*_{n+1} and $S = [S_1, S_2, \ldots, S_m]^T$ is a suitable scaling vector. The step control procedure advised by

Kaps and Rentrop is essentially the same as that described in Section 6.1. The step length is chosen so that the scaled error E per step is not greater than some prescribed value ε, giving

$$\theta = 0.9\left(\frac{\varepsilon}{E}\right)^{1/4} . \qquad (10.4.13)$$

This equation is the analogue of (6.1.10), and θ is limited by 0.5 and 1.5. As the step size is chosen from the error in the third-order result, the error in the fourth-order result is controlled on an error per unit step basis. Kaps and Rentrop recommend this method for ε between 10^{-2} and 10^{-4} and over a set of test problems it compares favourably with a slight modification of Gear's backward differentiation formula.

If $\underset{\sim}{S} = [1,1,\ldots,1]^T$ then E is the maximum absolute error, but taking $\underset{\sim}{S} = \hat{\underset{\sim}{y}}_{n+1}$ makes E the maximum relative error. Curtis (1978) advocates that relative error control should be used for stiff systems, and Curtis (1980) gives a scheme so that $\underset{\sim}{S} \approx \hat{\underset{\sim}{y}}_{n+1}$ while guarding against the occurrence of zero components.

10.5 Solution of partial differential equations

In this section we show how the methods of solution of an ordinary differential equation can be used in the solution of some partial differential equations. As an illustration we consider the one-dimensional diffusion equation and use a notation which is consistent, as far as possible, with that of Noye (1983). Suppose the problem to be solved is

$$\frac{\partial\bar{\tau}}{\partial t} = \alpha\,\frac{\partial^2\bar{\tau}}{\partial x^2} ,$$

$$\bar{\tau}(x,0) = f(x), \qquad 0 \le x \le 1,$$

$$\bar{\tau}(0,t) = g_0(t), \qquad t \ge 0,$$

$$\bar{\tau}(1,t) = g_1(t), \qquad t \ge 0. \qquad (10.5.1)$$

If the spatial derivative is approximated by the second-order central difference approximation, we have

$$\frac{d\tau_j}{dt} = \alpha\left[\frac{\tau_{j+1} - 2\tau_j + \tau_{j-1}}{\Delta x^2}\right] , \qquad j=1,2,\ldots,M-1, \qquad (10.5.2)$$

where $\tau_j(t)$ is an approximation to $\bar{\tau}(j\Delta x,t)$ and $\Delta x = 1/M$. Thus, we have a system of first-order differential equations with $\tau_0(t) = g_0(t)$ and $\tau_M(t) = g_1(t)$ and the initial conditions $\tau_j(0) = f(j\Delta x)$. The reduction of (10.5.1) to (10.5.2) is known as the *method of lines*. Using the notation $\underset{\sim}{\tau} = [\tau_1,\tau_2,\ldots,\tau_{M-1}]^T$, (10.5.2) may be written as

$$\frac{d\underset{\sim}{\tau}}{dt} = \underset{=}{A}\,\underset{\sim}{\tau} + \underset{\sim}{\phi}(t) , \qquad (10.5.3)$$

where $\underset{=}{A}$ is the $(M-1) \times (M-1)$ tridiagonal matrix

$$\underline{\underline{A}} = \frac{\alpha}{\Delta x^2}\begin{bmatrix} -2 & 1 & 0 & . & . & . & . & 0 \\ 1 & -2 & 1 & . & . & . & . & 0 \\ . & & & & & & & . \\ . & & & & & & & . \\ . & & & & & & & . \\ 0 & . & . & . & . & 1 & -2 & 1 \\ 0 & . & . & . & . & 0 & 1 & -2 \end{bmatrix}, \tag{10.5.4}$$

and $\phi(t) = [g_0(t),0,0,\ldots,0,g_1(t)]^T$. The eigenvalues of $\underline{\underline{A}}$ are $\lambda_k = \alpha(-2+2\cos(k\pi/M))/\Delta x^2$, $k=1,2,\ldots,M-1$, so that $-4\alpha/\Delta x^2 < \lambda_k < 0$. The stiffness ratio is $S = |\lambda_{M-1}/\lambda_1|$ and varies from about 360 for $M = 30$ to about 4050 for $M = 100$, so the system is moderately stiff.

If Euler's method is used to solve (10.4.2), we obtain

$$\tau_j^{n+1} = \tau_j^n + \frac{\alpha\Delta t}{(\Delta x)^2}[\tau_{j+1}^n - 2\tau_j^n + \tau_{j-1}^n], \tag{10.5.5}$$

where τ_j^n is an approximation to $\tau_j(n\Delta t)$ and is therefore an approximation to $\bar{\tau}(j\Delta x, n\Delta t)$. Since the interval of absolute stability is $(-2,0)$ for Euler's method, the method will be stable if

$$-2 \le -\frac{4\alpha}{(\Delta x)^2}\Delta t,$$

or

$$\frac{\alpha\Delta t}{(\Delta x)^2} \le \tfrac{1}{2}. \tag{10.5.6}$$

This is precisely the stability condition obtained by Noye (1983) and the forward-time centred space approximation (10.5.5) is said to be *conditionally stable*.

If the trapezoidal method is used to solve (10.5.2) the resulting scheme is

$$\tau_j^{n+1} = \tau_j^n + \frac{\alpha\Delta t}{2\Delta x^2}[(\tau_{j+1}^{n+1} - 2\tau_j^{n+1} + \tau_{j-1}^{n+1}) + (\tau_{j+1}^n - 2\tau_j^n + \tau_{j-1}^n)]. \tag{10.5.7}$$

This is the well known Crank-Nicolson method, and since the trapezoidal rule is A-stable there is no restriction on Δt and the method is said to be *unconditionally stable*. Thus, stability of these finite difference methods of solution of the partial differential equation (10.5.1) is related to absolute stability of the system (10.5.2). As the eigenvalues of $\underline{\underline{A}}$ are real any A_0-stable method of solution of (10.5.2) will give an unconditionally stable method of solution of (10.5.1). The method of lines can also be used for hyperbolic partial differential equations with a similar correspondence between the stability of the partial differential equation and the approximating system of ordinary differential equations.

11. CHOICE OF METHOD AND AVAILABLE SOFTWARE

11.1 Choice of method

In this section we suggest some methods which are relatively easy to program and are suitable for the solution of "simple" problems. If the problem is such that these methods take a large amount of computer time to run and library routines such as those discussed in Section 11.2 are not available, then one can do little better than use or adapt one of the programs referred to in Section 11.3. The various program libraries give advice on which method should be used, and as for the other programs we can do no more than repeat the advice given earlier that predictor-corrector methods are superior when the function evaluation is expensive, but when the evaluation is cheap Runge-Kutta methods are best if accuracy requirements are low and extrapolation methods best for high accuracy requirements. Factors such as output frequency influence these "rankings", extrapolation in particular being less advantageous if the solution is required at closely spaced values of the dependent variable.

By far the easiest method to program is a Runge-Kutta method with a fixed step length. A method of order four is a good compromise between accuracy and amount of work per step, and the Runge-Kutta-Gill method has much to recommend it. Another possibility, although more complicated, is the Gragg-Bulirsch-Stoer method in fixed step length form. It has the advantage that the order of the method is simply changed by altering the number of extrapolations that are performed. If the function evaluation is particularly expensive a predictor-corrector method is more efficient, but a one-step method is needed to provide the starting values. The predictor (8.1.7) used in PEC mode with the fourth-order Adams-Moulton corrector (7.4.3) gives a particularly economical scheme.

For stiff differential equations the trapezoidal method has the advantage of being precisely A-stable, but as it is only of order two it requires small step lengths. Another disadvantage is that a system of non-linear equations must be solved at each step unless $f(x,y)$ is linear in y. The fourth-order L-stable method developed by Bui (1979) (see Section 10.4) is relatively straight-forward and only requires the solution of linear equations. The A-stable scheme of Kaps and Rentrop (1979) described in Section 10.4 is only slightly more complicated and may be programmed in variable step form in much the same way as shown in the flow-chart of Figure 6.1.

Note that the fixed step length schemes have the advantage of allowing easy calculation of the global error by means of Richardson extrapolation (see Section 6.3).

11.2 IMSL and NAG libraries

Many computer installations have libraries of subroutines available to the user, perhaps the most common being the International Mathematics and Statistics Library (IMSL) and the Numerical Algorithms Group (NAG) library. Both of these libraries contain sections on the numerical solution of ordinary differential equations.

The IMSL library, Edition 8 (1980) contains three subroutines, DVERK, DGEAR and DREBS for the numerical solution of initial value problems.

The subroutine DVERK uses an embedded Runge-Kutta method of order 6(5) which was developed by Verner (1978) and is based on the code of Hull, Enright and Jackson (1976). DREBS, an extrapolation routine using the Gragg-Bulirsch-Stoer method, is adapted from the program of Fox (1971). The third subroutine DGEAR is based on Hindmarsh's version (Hindmarsh, 1974) of Gear's subroutine DIFSUB (Gear, 1971, 1971a) and it is a variable order method which uses Adams formulas of order one to twelve for non-stiff differential equations and backward differentiation formulas of order one to five for stiff equations. When the stiff option is chosen the user can specify whether the Newton iteration of the corrector employs a user supplied Jacobian, a numerically calculated Jacobian or a diagonal approximation to the Jacobian.

The NAG library, Mark 7 (1979) contains about seventeen subroutines for the solution of initial value problems. Three methods are used, Runge-Kutta, variable-order Adams and variable-order backward differentiation formula methods; the latter two methods are also based on the subroutine GEAR of Hindmarsh (1974) while the Runge-Kutta methods are based on the Merson (1957) algorithm. For each method subroutines are provided to integrate over a domain, integrate over a domain with intermediate output, integrate until a component of the solution becomes zero, or integrate until a function of the solution becomes zero. Other subroutines include a Runge-Kutta-Merson subroutine with a global error estimate and stiffness check, and subroutines for the interpolation of the solution. More details of the initial value subroutines of the NAG library may be found in Gladwell (1979).

11.3 Other software

A number of listings of FORTRAN programs for the numerical solution of initial value problems appear in the literature. We only consider programs which may be used to solve a general system of equations, and this has resulted in all programs having automatic step length selection. Some minor changes may have to be made to the code, particularly with regard to the type declaration of variables.

Runge-Kutta methods

Gear (1971) contains the listing of a program which uses the classical (fourth-order) Runge-Kutta method. Richardson extrapolation is used to estimate the local error, and step length chosen in order that the local error at each step is controlled. A program based on a Runge-Kutta-Fehlberg method or order 5(4) is listed in Shampine and Allen (1973). This program uses a local error per unit step criterion to determine the step length. A more sophisticated version of this program is described in Shampine and Watts (1976); a listing of this program may be found in Shampine and Watts (1976a). A feature of this program is that it provides an estimate of the global error, but it should be noted that the step selection is based on limiting the local error per step.

Predictor-corrector methods

A program incorporating a variable-order predictor-corrector method may be found in Gear (1971 and 1971b). For non-stiff differential equations Adams methods of order one to seven are used, and for stiff equations backward differentiation formulas of order one to six are employed, with local error per step control. A correction to the program in Gear (1971b) has been made - the coefficient A(3) in the seventh order Adams formula should be changed from -1.235 to -1.225.

Another variable-order Adams method program is given by Shampine and Gordon (1975). This uses formulas of order one to twelve with the step length determined by the magnitude of the local error per unit step. As a complete book is devoted to the development and implementation of the method, it is the best documented program available. Because much higher order Adams formulas are used, this program is found to perform better than Gear's for many problems.

Extrapolation methods

Extrapolation programs are found in Gear (1971) and Fox (1971). Both programs essentially come from the same source so they are very similar, except that Gear's version allows the user to specify the maximum number of extrapolations and stages and it also has the option of polynomial extrapolation as well as rational extrapolation. Both programs select the step length or the basis of error per step.

Methods for stiff differential equations

The programs of Gear (1971 and 1971b) described previously have an option for solving stiff differential equations. Another program for stiff equations is given by Tendler, Bickart and Picel (1978, 1978a). This program uses stiffly-stable cyclic composite methods of order one to seven with local error per step control, and is reported to perform better than Gear's method when eigenvalues of the Jacobian matrix lie close to the imaginary axis.

APPENDIX - Notation

a	- value of the independent variable at which the initial condition is given.
$[a,b]$	- interval on which the solution is sought.
C_{p+1}	- error constant of a linear multistep method of order p (local truncation error is $C_{p+1} h_n^{p+1} y^{(p+1)}(x_n)$).
$C(x_n)$	- function such that the local truncation error at x_{n+1} is $C(x_n)h_n^{p+1}$ for a method of order p.
d_{n+1}	- local error in the nth step.
d^*_{n+1}	- estimate of d_{n+1}.
e_n	- global truncation error at x_n.
$f(x,y)$	- right-hand side of ordinary differential equation $y' = f(x,y)$.
h	- fixed step length.
h_n	- step length from x_n to x_{n+1} (variable).
L	- Lipschitz constant of $f(x,y)$ with respect to y.
m	- order of differential equation or number of equations in a first order system.
p	- order of numerical method.
$P(EC)^{\nu}$	- mode of predictor-corrector method where the corrector is iterated ν times (see Section 8.1) - also $P(EC)^{\nu}E$.
$P_i^{(j)}$	- extrapolated polynomial approximation to $y(x_n+H)$.
r_ℓ	- roots of characteristic polynomial of method where r_1 is an approximation to $\exp(\lambda h)$.
$R_i^{(j)}$	- extrapolated rational polynomial approximation to $y(x_n+H)$.
S	- stiffness ratio.
T_n	- local truncation error at x_n.
x	- independent variable.
x_n	- discrete value of independent variable with $x_{n+1} = x_n + h_n$ and $x_0 = a$.
y	- dependent variable (may be an m-vector).
$y(x)$	- analytic solution of initial value problem.
$y^*(x)$	- solution of differential equation safisfying the initial condition $y(x_n) = y_n$.
y_n	- approximation to $y(x_n)$.
$\tilde{y}_n$	- calculated value of y_n (includes round-off errors).
$\bar{y}_n, \hat{y}_n$	- values of y_n obtained in some particular way (see text).
$y_{n+k}^{[t]}$	- t^{th} iterate of y_{n+k}.
$y(x_n+H;h_i)$	- estimate of $y(x_n+H)$ using Gragg's method with N_i steps of length h_i from x_n.
δ, ρ	- round-off errors.
ε	- specified bound on the local error per unit step (or per step).
η	- value of the dependent variable given by the initial condition $y(a)=\eta$.

Θ	-	factor by which the step length should be multiplied to satisfy the local error criterion.
λ	-	value of $\partial f/\partial y$ for a single differential equation or eigenvalues of Jacobian matrix $\partial \tilde{f}/\partial \tilde{y}$ for a system of equations.
ν	-	number of corrector iterations in a predictor-corrector method.
$\Pi(r,h)$	-	characteristic polynomial of linear multistep method.
$\phi(x,y,h)$	-	increment function for an explicit one-step method.

REFERENCES

Abramowitz, M. and Stegun, I.A. (1965), *Handbook of Mathematical Functions*, Dover Publications.

Barton, D., Willers, I.M. and Zahar, I.V.M. (1971), *Taylor Series Methods for Ordinary Differential Equations - An Evaluation*, Mathematical Software, ed. J.R. Rice, Academic Press, pp. 369-390.

Bashforth, F and Adams, J.C. (1883), *An attempt to test the theories of capillary action ... with an explanation of the method of integration employed*, Cambridge University Press .

Birkhoff, G. and Rota, G. (1978), *Ordinary Differential Equations*, Third ed., John Wiley and Sons.

Blum, E.K. (1962), *A modification of the Runge-Kutta fourth-order method*, Math. Comp., 16, pp. 176-187.

Brown, R. R., Riley, J.D. and Bennett, P.Q. (1965), *Stability Properties of Adams-Moulton Type Methods*, Math. Comp. 19, pp. 90-96.

Bui, T.D. (1979), *Some A-Stable and L-Stable Methods for the Numerical Integration of Stiff Ordinary Differential Equations*, J. Assoc. Comp. Mach., 26, pp. 483-493.

Bulirsch, R. and Stoer, J. (1964), *Fehlerabschätzungen und Extrapolation mit rationalen Funktionen bei Verfahren vom Richardson-Typus*, Numer. Math. 6, pp.413-427.

Bulirsch, R. and Stoer, J. (1966), *Numerical Treatment of Ordinary Differential Equations by Extrapolation Methods*, Numer. Math., 8, pp. 1-13.

Butcher, J.C. (1963), *Coefficients for the Study of Runge-Kutta Integration Processes*, J. Austral. Math. Soc., 3, pp. 185-201.

Butcher, J.C. (1964), *On Runge-Kutta Processes of High Order*, J. Austral. Math. Soc., 4, pp. 179-194.

Butcher, J.C. (1964a), *Implicit Runge-Kutta Processes*, Math. Comp., 18, pp. 50-64.

Butcher, J.C. (1965), *On the Attainable Order of Runge-Kutta Methods*, Math. Comp., 19, pp. 408-417.

Coddington, E.A. and Levinson, N. (1955), *Theory of Ordinary Differential Equations*, McGraw-Hill.

Corliss, G. and Chang, Y.F. (1982), *Solving Ordinary Differential Equations Using Taylor Series*, A.C.M. Trans. Math. Software, 8, pp. 114-144.

Crane, R.L. and Klopfenstein, R.W. (1965), *A Predictor-Corrector Algorithm With an Increased Range of Stability*, J. Assoc. Comp. Mach., 12, pp. 227-241.

Cryer, C.W. (1973), *A New Class of Highly Stable Methods: A_0-Stable Methods*, B.I.T., 13, pp. 153-159.

Curtis, A.R. (1978), *Solution of Large, Stiff Initial Value Problems - The State of the Art*, Numerical Software - Needs and Availability, ed. D. Jacobs, Academic Press, pp. 257-278.

Curtis, A.R. (1980), *The FACSIMILE Numerical Integrator for Stiff Initial Value Problems*, Computational Techniques for Ordinary Differential Equations, ed. I. Gladwell and D.K. Sayers, Academic Press, pp. 47 - 82.

Dahlquist, G. (1963), *A Special Stability Problem for Linear Multistep Methods*, B.I.T., 3, pp. 27-43.

Ehle, B.L. (1969), *On Padé Approximations to the Exponential Function and A-Stable Methods for the Numerical Solution of Initial Value Problems*, University of Waterloo, Department of Applied Analysis and Computer Science, Research Report No. CSRR 2010.

England, R. (1969), *Error Estimates for Runge-Kutta Type Solutions to Systems of Ordinary Differential Equations*, Comput. J., 12, pp. 166-170.

Enright, W.H. (1974), *Optimal Second Derivative Methods for Stiff Systems*, Stiff Differential Systems, ed. R.A. Willoughby, Plenum Press, pp. 95-109.

Fehlberg, E. (1964), *New High-Order Runge-Kutta Formulas with Step Size Control for Systems of First- and Second-Order Differential Equations*, Z. Agnew. Math. Mech., 44, pp. 17-29.

Fehlberg, E. (1969), *Klassische Runge-Kutta Formeln fünfter und siebenter Ordnung mit Schrittweiten-Kontrolle*, Computing, 4, pp. 93-106.

Fehlberg, E. (1970), *Klassische Runge-Kutta Formeln vierter und niedrigerer Ordnung mit Schrittweiten-Kontrolle und ihre Anwendung auf Wärmeleitungs - probleme*, Computing, 6, pp. 61-71.

Fox L. and Mayers, D.F. (1981), *On the Numerical Solution of Implicit Ordinary Differential Equations*, I.M.A. J. Numer. Anal., 1, pp. 377-401.

Fox, P.A. (1971), *DESUB: Integration of a First-Order System of Ordinary Differential Equations*, Mathematical Software, ed. J. Rice, Academic Press, pp. 477-507.

Gear, C.W. (1969), *The Automatic Integration of Stiff Ordinary Differential Equations*, Information Processing 68, ed. A.J.H. Morrel, North-Holland, pp. 187-193.

Gear, C.W. (1971), *Numerical Initial Value Problems in Ordinary Differential Equations*, Prentice-Hall.

Gear, C.W. (1971a), *The Automatic Integration of Ordinary Differential Equations*, Comm. A.C.M., 14, pp. 176-179.

Gear, C.W. (1971b), *Algorithm 407, DIFSUB for Solution of Ordinary Differential Equations*, Comm. A.C.M., 14, pp. 185-190.

Gill, S. (1951), *A process for the step-by-step integration of differential equations in an automatic digital computing machine*, Proc. Cambridge Philos. Soc., 47, pp. 95-108.

Gladwell, I. (1979), *Initial Value Routines in the NAG Library*, A.C.M. Trans. Math. Software, 5, pp. 386-400.

Gragg, W.B. (1965), *On Extrapolation Algorithms for Ordinary Initial Value Problems*, SIAM J. Numer. Anal., 2, pp. 384-403.

Hammer, P.C. and Hollingsworth, J.W. (1955), *Trapezoidal Methods of Approximating Solutions of Differential Equations*, M.T.A.C., 9, pp. 92-96.

Hamming, R.W. (1959), *Stable Predictor-Corrector Methods for Ordinary Differential Equations*, J. Assoc. Comp. Mach., 6, pp. 37-47.

Henrici, P. (1962), *Discrete Variable Methods for Ordinary Differential Equations*, John Wiley and Sons.

Hindmarsh, A.C. (1974), *GEAR: Ordinary Differential Equation Solver*, Report UCID-30001, Revision 3, Lawrence Livermore Laboratory.

Hull, T.E. and Creemer, A.L. (1963), *Efficiency of Predictor-Corrector Procedures*, J. Assoc. Comp. Mach., 10, pp. 291-301.

Hull, T.E., Enright, W.H., Fellen, B.M., and Sedgwick, A.E. (1972), *Comparing Numerical Methods for Ordinary Differential Equations*, SIAM J. Numer. Anal., 9, pp. 603-637.

Hull, T.E., Enright, W.H. and Jackson, K.R. (1976), *Users Guide for DVERK-A Subroutine for Solving Non-Stiff O.D.E.'s*, TR No. 100, Department of Computer Science, University of Toronto.

IMSL (1980), Reference Manual, Edition 8, IMSL Inc., 7500 Bellaire Boulevard, Houston, Texas, U.S.A.

Kaps, P. and Rentrop, P. (1979), *Generalized Runge-Kutta Methods of Order Four with Step Control for Stiff Ordinary Differential Equations*, Numer. Math., 33, pp. 55-68.

Keller, H.B. (1968), *Numerical Methods for Two-point Boundary Value Problems*, Blaisdell Publishing Company.

Keller, H.B. (1975), *Numerical Solution of Boundary Value Problems for Ordinary Differential Equations: Survey and some recent results on difference methods*, Numerical Solutions of Boundary Value Problems for Ordinary Differential Equations, edited A. Aziz, Academic Press, pp. 27-88.

Keller, H.B. (1976), *Numerical Solution of Two-point Boundary Value Problems*, Regional Conf. Series in Appl. Math. No. 24, SIAM.

Klopfenstein, R.W. and Millman, R.S. (1968), *Numerical Stability of a One Evaluation Predictor-Corrector Algorithm for Numerical Solution of Ordinary Differential Equations*, Math. Comp., 22, pp. 557-564.

Krogh, F.T. (1966), *Predictor-Corrector Methods of High Order With Improved Stability Characteristics*, J. Assoc. Comp. Mach., 13, pp. 374-385.

Kutta, W. (1901), *Beitrag zur näherungsweisen Integration totaler Differentialgleichungen*, Z. Math. Phys., 46, pp. 435-453.

Lambert, J.D. (1973), *Computational Methods in Ordinary Differential Equations*, John Wiley and Sons.

Lambert, J.D. (1980), *Stiffness*, Computational Techniques for Ordinary Differential Equations, ed. I. Gladwell and D.K. Sayers, Academic Press, pp. 19- 46.

Lawson, J.D. (1966), *An Order Five Runge-Kutta Process with Extended Region of Stability*, SIAM J. Numer. Anal., 3, pp. 593-597.

Lawson, J.D. (1967), *An Order Six Runge-Kutta Process with Extended Region of Stability,* SIAM J. Numer. Anal., 4, pp. 620-625.

Lotkin, M. (1951), *On the Accuracy of Runge-Kutta's Method,* M.T.A.C., 5, pp. 128-132.

Luther, H.A. (1966), *Further Explicit Fifth-Order Runge-Kutta Formulas,* SIAM Rev., 8, pp. 374-380.

Luther, H.A. (1968), *An Explicit Sixth-Order Runge-Kutta Formula,* Math. Comp., 22, pp. 434-436.

Luther, H.A. and Konen, H.P. (1965), *Some Fifth-Order Classical Runge-Kutta Formulas,* SIAM Rev., 7, pp. 551-558.

Merson, R.H. (1957), *An Operational Method for the Study of Integration Processes,* Proc. Symp. Data Processing, Weapons Research Establishment, Salisbury, South Australia .

Murphy, C.P. and Evans, D.J. (1981), *A Flexible Variable Order Extrapolation Technique for Solving Non-Stiff Ordinary Differential Equations,* Int. J. Comp. Math., 10, pp. 63-75.

Murphy, G.M. (1960), *Ordinary Differential Equations and Their Solutions,* Van Nostrand-Reinhold.

NAG (1979), Manual, Mark 7, NAG Central Office, 7 Banbury Road, Oxford, England.

Neville, E.H. (1934), *Iterative Interpolation,* J. Ind. Math. Soc., 20, pp. 87-120.

Nordsieck, A. (1962), *On Numerical Integration of Ordinary Differential Equations,* Math. Comp., 16, pp. 22-49.

Noye, J. (1983), *Finite Difference Techniques for Partial Differential Equations*, Computational Techniques for Differential Equations, edited J. Noye, North-Holland Publishing Coy., pp.95-354.

Ralston, A. (1965), *A First Course in Numerical Analysis,* McGraw-Hill.

Richardson, L.F. (1927), *The Deferred Approach to the Limit, I-Single Lattice,* Trans. Roy. Soc. London, 226, pp. 299-349.

Rosenbrock, H.H. (1963), *Some General Implicit Processes for the Numerical Solution of Differential Equations,* Comput. J., 5, pp. 329-330.

Rosser, J.B. (1967), *A Runge-Kutta For All Seasons,* SIAM Rev., 9, pp. 417-452.

Runge, C. (1895), *Über die numerische Auflösung von Differentialgleichungen,* Math. Ann., 46, pp. 167-178.

Scraton, R.E. (1964), *Estimation of the Truncation Error in Runge-Kutta and Allied Processes,* Comput. J., 7, pp. 246-248.

Shampine, L.F. (1977), *Stiffness and Non-Stiff Differential Equation Solvers, II: Detecting Stiffness with Runge-Kutta Methods,* A.C.M. Trans. Math. Software, 3, pp. 44-53.

Shampine, L.F. (1979), *Storage Reduction for Runge-Kutta Codes,* A.C.M. Trans. Math. Software, 5, pp. 245-250.

Shampine, L.F. (1980), *What Everyone Solving Differential Equations Should Know,* Computational Techniques for Ordinary Differential Equations, ed. I. Gladwell and D.K. Sayers, Academic Press, pp. 1-17.

Shampine, L.F. and Allen, R.C. (1973), *Numerical Computing: An Introduction,* W.B. Saunders Company.

Shampine, L.F. and Gordon, M.K. (1975), *Computer Solution of Ordinary Differential Equations: The Initial Value Problem,* W.H. Freeman and Company.

Shampine, L.F., Gordon, M.K. and Wisniewski, J.A. (1980), *Variable Order Runge-Kutta Codes,* Computational Techniques for Ordinary Differential Equations, ed. I. Gladwell and D.K. Sayers, Academic Press, pp. 83-101.

Shampine, L.F. and Hiebert, K.L. (1977), *Detecting Stiffness with the Fehlberg (4,5) Formulas,* Comput. Math. Appl., 3, pp. 41-46.

Shampine, L.F. and Watts, H.A. (1976), *Global Error Estimation for Ordinary Differential Equations,* A.C.M. Trans. Math. Software, 2, pp. 172-186.

Shampine, L.F. and Watts, H.A. (1976a), *Algorithm 504, GERK: Global Error Estimation for Ordinary Differential Equations,* A.C.M. Trans. Math. Software, 2, pp. 200-203.

Shampine, L.F. and Watts, H.A. (1977), *The Art of Writing a Runge-Kutta Code, Part I,* Mathematical Software III, ed. J.R. Rice, Academic Press, pp. 257-275.

Shampine, L.F. and Watts, H.A. (1979), *The Art of Writing a Runge-Kutta Code, Part II,* Appl. Math. Comp., 5, pp. 93-121.

Shampine, L.F., Watts, H.A. and Davenport, S.M. (1976), *Solving Nonstiff Ordinary Differential Equations - The State of the Art,* SIAM Rev., 18, pp. 376-411.

Shintani, H. (1965), *Approximate Computation of Errors in Numerical Integration of Ordinary Differential Equations by One-Step Methods,* J. Sci. Hiroshima Univ. Ser. A-1 Math.,29, pp. 97-120.

Shintani, H. (1966), *On a One-Step Method of Order 4,* J. Sci. Hiroshima Univ. Ser. A-1 Math., 30, pp. 91-107.

Shintani, H. (1966a), *Two-Step Processes by One-Step Methods of Order 3 and of Order 4,* J. Sci. Hiroshima Univ. Ser. A-1 Math., 30, pp. 183-195.

Stetter, H.J. (1969), *Stability Properties of the Extrapolation Methods,* Conference on the Numerical Solution of Differential Equations, Dundee, editor J. Ll. Morris, Springer-Verlag.

Stetter, H. (1973), *Analysis of Discretization Methods for Ordinary Differential Equations,* Springer-Verlag.

Stoer, J. (1961), *Über zwei Algorithm zur Interpolation mit rationalen Funktionen,* Numer. Math., 3, pp. 285-304.

Stoer, J. (1974), *Extrapolation Methods for The Solution of Initial Value Problems and Their Practical Realisation,* Proceedings of the Conference on Numerical Solutions of Ordinary Differential Equations, ed. D. Bettis, Springer-Verlag.

Tendler, J.M., Bickart, T.A. and Picel, Z. (1978), *A Stiffly Stable Integration Process Using Cyclic Composite Methods,* A.C.M. Trans. Math. Software, 4, pp. 339-368.

Tendler, J.M., Bickart, T.A. and Picel, Z. (1978a), *Algorithm 534, STINT: STiff (differential equations)INTegrator,* A.C.M. Trans. Math. Software, 4, pp. 399-403.

Verner, J.H. (1978), *Explicit Runge-Kutta Methods with Estimates of the Local Truncation Error,* SIAM J. of Numer. Anal., 15, pp. 772-790.

Verner, J.H. (1979), *Families of Imbedded Runge-Kutta Methods,* SIAM J. Numer. Anal., 16, pp. 857-875.

Widlund, O.B. (1967), *A Note on Unconditionally Stable Linear Multistep Methods,* B.I.T., 7, pp. 65-70.

Computational Techniques for Differential Equations
J. Noye (Editor)

FINITE DIFFERENCE TECHNIQUES
FOR
PARTIAL DIFFERENTIAL EQUATIONS

JOHN NOYE
The University of Adelaide, South Australia

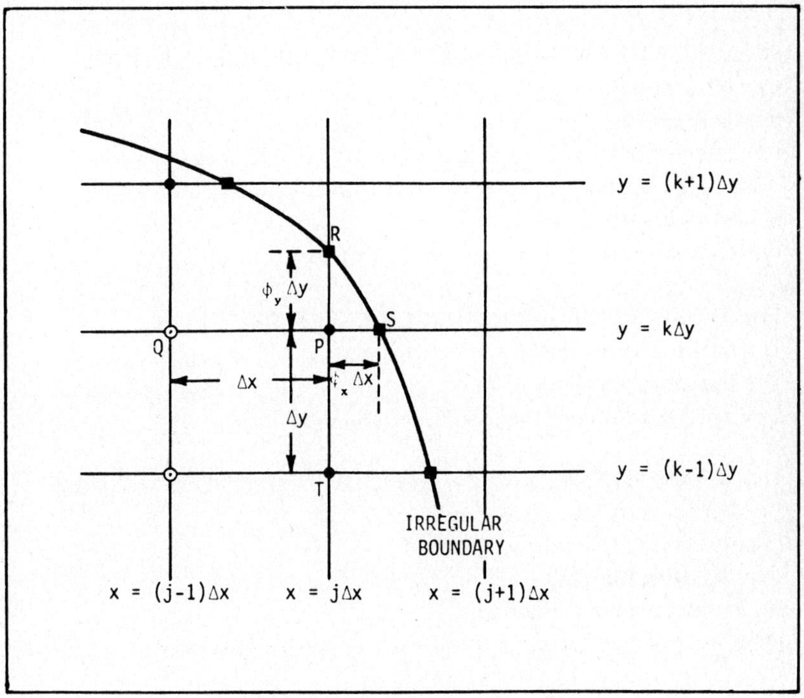

CONTENTS

1. INTRODUCTION

1.1 General Considerations

This article describes some of the finite difference methods available for solving partial differential equations, with particular application to those equations which model heat and mass transfer in fluids. It also discusses the more common difficulties encountered by the mathematician, physicist or engineer when he first enters this field of numerical simulation.

Section 1.2 introduces the transport equation which is the example used to illustrate most of the various finite difference techniques in the following sections. For some initial and boundary conditions, analytic solutions can be found for this particular equation. These solutions are very useful for comparison, in order to get an idea of the accuracy of any numerical method.

However, since a numerical technique is employed only to solve problems to which an analytic solution is either very difficult or impossible to obtain, even with greatly simplified initial and boundary conditions, it is useful to know of methods which prove stability and convergence of such a technique. Unfortunately, such methods have not been developed for the more complicated (generally non-linear) problems and *ad hoc* approaches must be employed in which numerical methods become an experimental tool. This work deals mainly with linear partial differential equations; non-linear cases will be dealt with in detail in a future article.

In Section 2, the various finite difference approximations to derivatives at a point in the solution domain are described, and it is shown how to apply the forward time and centred space approximations to the derivatives in the one-dimensional diffusion equation, to obtain approximate solutions of this equation at certain points in the region being considered.

In Section 3 the concepts of convergence, consistency and stability are considered in some detail, while in Section 4 some other finite difference methods, both explicit and implicit, of solving the one-dimensional diffusion equation are developed. Ways of incorporating derivative boundary conditions and of dealing with discontinuities between initial and boundary values are also discussed. Section 5 deals with methods of solving the transport equation in one space dimension; the first part considers the convection equation and wave propagation characteristics, the second part deals with the complete transport equation. Section 6 deals with the transport equation in two and three space dimensions, and describes ways of dealing with boundary conditions on irregular boundaries. Methods of solving the wave equation are considered in Section 7. Section 8 considers methods of varying the grid spacing and transforming the coordinate system used. Improving the accuracy of solutions is discussed in Section 9, especially Richardson's method of extrapolating the approximations found for different sized grid spacings to obtain the limiting value for a zero grid spacing. It should also be noted that, for the readers' reference, Appendix 4 lists in aphabetical order all the symbols used in the following, with their meaning.

It must be emphasised that the theoretical foundation of finite difference methods has not kept pace with their application. It is generally impossible to obtain a theoretical estimate of the accuracy of a numerical solution of a problem in which the non-linear terms in the equations are not small. Users of finite difference techniques must therefore apply every method available, such as successive grid refinement and comparison of the numerical solution with analytical solutions to simplified versions of the problem in order to verify the quality of their results.

Readers who wish to use the methods described in this article should also study the original papers referenced and they should read more recent literature which discuss these methods. Publications such as Mathematics of Computation, Journal of Computational Physics, Computers and Fluids, Computer Methods in Applied Mechanics and Engineering, Numerical Methods in Engineering and the Journal of Fluid Mechanics, regularly include articles of relevance to finite difference techniques.

1.2 The Transport Equation

The behaviour of a fluid undergoing mass, vorticity or forced convective heat transfer, is described by a set of partial differential equations which are mathematical formulations of one or more of the conservation laws of physics. These laws include those of conservation of mass, momentum and energy.

For instance, if the fluid is incompressible and has constant thermal conductivity κ, the *heat equation*, a mathematical formulation of the law of conservation of thermal energy, is

$$\rho C_p \frac{DT}{Dt} = \kappa \nabla^2 T, \tag{1.2.1}$$

where ρ is the density, C_p is the specific heat of the fluid at constant pressure, D/Dt is the operation of differentiation following the motion of the fluid, and T is its temperature. In Cartesian coordinates this can be written

$$\frac{\partial T}{\partial t} + u\frac{\partial T}{\partial x} + v\frac{\partial T}{\partial y} + w\frac{\partial T}{\partial z} = a\left(\frac{\partial^2 T}{\partial x^2} + \frac{\partial^2 T}{\partial y^2} + \frac{\partial^2 T}{\partial z^2}\right) \tag{1.2.1a}$$

where $a = \kappa/\rho C_p$ and u,v,w are the velocity components of the fluid at the point (x,y,z) at time t.

For fluids at rest, and for solids, u=v=w=0 and Equation (1.2.1) reduces to the *diffusion (conduction) equation*

$$\frac{\partial T}{\partial t} = a\left(\frac{\partial^2 T}{\partial x^2} + \frac{\partial^2 T}{\partial y^2} + \frac{\partial^2 T}{\partial z^2}\right). \tag{1.2.1b}$$

Equation (1.2.1a) is a form of the transport equation: it involves transfer of heat due to the motion of the fluid (convection) and due to diffusion (conduction). The three terms $u\partial T/\partial x$, $v\partial T/\partial y$ and $w\partial T/\partial z$ are called convective terms, and the three terms $a\partial^2 T/\partial x^2$, $a\partial^2 T/\partial y^2$ and $a\partial^2 T/\partial z^2$ are usually called diffusive terms.

Similar cylindrical and spherical coordinate forms of Equation (1.2.1a) exist, but for the purpose of describing methods for their solution only the rectangular Cartesian forms will be considered here.

In general, the transport equation (1.2.1a) may also describe the transfer of dissolved material or vorticity, as well as heat, to mention just three examples. In heat transfer problems T is the temperature, in material transfer problems it is the concentration and for vorticity transfer T represents the vorticity.

Throughout this article, Equation (1.2.1a) is used in its various forms, such as:

the one-dimensional diffusion equation

$$\frac{\partial T}{\partial t} = a\frac{\partial^2 T}{\partial x^2}, \tag{1.2.2a}$$

the one-dimensional convection equation

$$\frac{\partial T}{\partial t} + u\frac{\partial T}{\partial x} = 0, \tag{1.2.2b}$$

the one-dimensional steady transport equation

$$u\frac{\partial T}{\partial x} = a\frac{\partial^2 T}{\partial x^2}, \tag{1.2.2c}$$

and so on. These forms will be used to illustrate the finite difference techniques described in later sections. For some initial and boundary value conditions, analytic solutions (usually in terms of infinite series) can be found for these equations. The accuracy of the finite difference methods developed in this article can be determined by comparison with these exact solutions.

In the examples which follow, the coordinates will be non-dimensionalised so that the boundaries in each direction will be located at 0 and 1, the time will be non-dimensionalised so that a typical time period in the solution T will become unity, and the speed of the fluid flow will be non-dimensionalised accordingly.

1.3 Non-Dimensionalisation of the Transport Equation

The transport equation (1.2.1a) can have the space and time variables non-dimensionalised in the following way.

Let L, B, H be typical lengths, such as the distances between boundaries in the x, y and z directions; for instance, the space region *R* over which (1.2.1a) is to be solved may be defined by $0 \le x \le L$, $0 \le y \le B$, $0 \le z \le H$. Replace the x,y,z variables by the dimensionless space variables

$$\left.\begin{aligned} x_* &= x/L, \\ y_* &= y/B, \\ z_* &= z/H, \end{aligned}\right\} \tag{1.3.1a}$$

so that the region R is now defined by $0 \le x_* \le 1$, $0 \le y_* \le 1$, $0 \le z_* \le 1$.

Let P be a characteristic interval of time associated with the given problem, such as the period of an oscillation in the dependent variable T. Replace the variable t by

$$t_* = t/P, \tag{1.3.1b}$$

so that in terms of the dimensionless time variable t_* the period of the oscillation is now unity.

Substitution of the new independent variables x_*, y_*, z_* and t_* into the transport equation (1.2.1a) gives

$$\frac{\partial T}{\partial t_*} + \frac{uP}{L}\frac{\partial T}{\partial x_*} + \frac{vP}{B}\frac{\partial T}{\partial y_*} + \frac{wP}{H}\frac{\partial T}{\partial z_*} = \frac{aP}{L^2}\frac{\partial^2 T}{\partial x_*^2} + \frac{aP}{B^2}\frac{\partial^2 T}{\partial y_*^2} + \frac{aP}{H^2}\frac{\partial^2 T}{\partial z_*^2}. \tag{1.3.2}$$

If T_0 is a typical value of the dependent variable T, this may also be non-dimensionalised by means of the substitution

$$T_* = T/T_0. \tag{1.3.3}$$

Substitution of (1.3.3) into Equation (1.3.2) yields

$$\frac{\partial T_*}{\partial t_*} + u_*\frac{\partial T_*}{\partial x_*} + v_*\frac{\partial T_*}{\partial y_*} + w_*\frac{\partial T_*}{\partial z_*} = \alpha_x\frac{\partial^2 T_*}{\partial x_*^2} + \alpha_y\frac{\partial^2 T_*}{\partial y_*^2} + \alpha_z\frac{\partial^2 T_*}{\partial z_*^2}, \qquad (1.3.4a)$$

where

$$\left.\begin{array}{l} u_* = uP/L,\ v_* = vP/B,\ w_* = wP/H, \\ \alpha_x = \dfrac{aP}{L^2},\ \alpha_y = \dfrac{aP}{B^2},\ \alpha_z = \dfrac{aP}{H^2}. \end{array}\right\} \qquad (1.3.4b)$$

Equation (1.3.4) is the non-dimensional form of the transport equation and appears in the same form as Equation (1.2.1a). It is to be solved on the space region *R* now defined by

$$0 \le x_* \le 1,\ 0 \le y_* \le 1,\ 0 \le z_* \le 1.$$

For convenience, in the following it will be assumed that all equations have been non-dimensionalised. The asterisks in Equation (1.3.4) will therefore be omitted. In general, we will consider the transport equation

$$\frac{\partial \bar{\tau}}{\partial t} + u\frac{\partial \bar{\tau}}{\partial x} + v\frac{\partial \bar{\tau}}{\partial y} + w\frac{\partial \bar{\tau}}{\partial z} = \alpha_x\frac{\partial^2 \bar{\tau}}{\partial x^2} + \alpha_y\frac{\partial^2 \bar{\tau}}{\partial y^2} + \alpha_z\frac{\partial^2 \bar{\tau}}{\partial z^2}, \qquad (1.3.5)$$

in which $\bar{\tau}$ is some suitably non-dimensionalised property such as temperature, concentration, or vorticity, t is a non-dimensional time variable with a characteristic value of 1, and x,y,z are non-dimensional space variables which range from 0 to 1.

2. FINITE DIFFERENCE APPROXIMATIONS

2.1 The One-Dimensional Diffusion Equation

Many of the concepts associated with the numerical solution of partial differential equations by finite difference methods can be illustrated by considering the one-dimensional diffusion equation

$$\frac{\partial \bar{\tau}}{\partial t} = \alpha \frac{\partial^2 \bar{\tau}}{\partial x^2}, \tag{2.1.1}$$

assumed to have been suitably non-dimensionalised so that $0 \le \bar{\tau} \le 1$ and $0 \le x \le 1$. For instance, the dependent variable $\bar{\tau}(x,t)$ may represent the non-dimensional temperature at time t at position x along a thin insulated rod of unit length with ends located at x=0 and x=1, as in Figure 2.1. If the rod is composed of homogeneous material then α, the non-dimensional coefficient of heat diffusion, is constant.

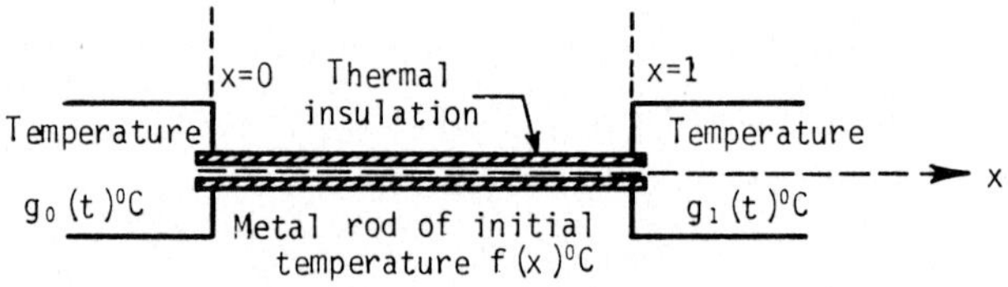

Figure 2.1 : The insulated homogeneous metal rod.

Certain initial and boundary conditions are required before Equation (2.1.1) can be solved, for example

$$\left.\begin{aligned} \bar{\tau}(x,0) &= f(x)\ ,\ 0 \le x \le 1 \\ \bar{\tau}(0,t) &= g_0(t),\ t \ge 0 \\ \bar{\tau}(1,t) &= g_1(t),\ t \ge 0 \end{aligned}\right\} . \tag{2.1.2}$$

Interpreted in terms of the problem defined by Figure 2.1, the first of these conditions is a statement that the temperature everywhere along the rod is known to be f(x) degrees initially, the second is a statement that the temperature at the left end of the rod (that is at x=0) is prescribed for all time by $g_0(t)$ degrees, and the third is a statement that at the right end of the rod (that is at x=1) the temperature is given by $g_1(t)$ degrees.

Given the description of f(x), $g_0(t)$ and $g_1(t)$, the solution $\bar{\tau}(x,t)$ is to be determined over the semi-infinite rectangle in x-t space defined by $0 < x < 1$, $t > 0$ (see Figure 2.2). This region, $\mathcal{D}$, on which the solution of the given differential equation (2.1.1) is to be found subject to the initial and boundary conditions (2.1.2), is called the solution domain.

For example, a thin insulated rod initially at a temperature T of 0°C suddenly has the temperature at each end raised to 100°C and held at that value for all time $t > 0$. Dividing the temperature in degrees C by 100°C gives a suitable non-dimensional temperature $\bar{\tau}(x,t)$, for which the initial and boundary conditions (2.1.2) become

$$\left.\begin{aligned} \bar{\tau}(x,0) &= 0,\ 0 \le x \le 1 \\ \bar{\tau}(0,t) &= 1,\ t \ge 0 \\ \bar{\tau}(1,t) &= 1,\ t \ge 0 \end{aligned}\right\} . \tag{2.1.3}$$

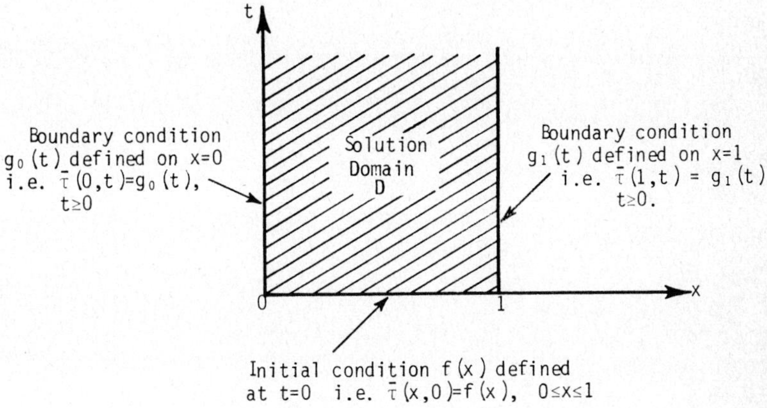

Figure 2.2 : The solution domain in x-t space for the one-dimensional diffusion equation.

A problem arises when the initial temperatures at x=0 and x=1, namely $\bar{\tau}(0,0)$ and $\bar{\tau}(1,0)$, are to be prescribed. In practice the temperature at x=0 and x=1 cannot be altered instantaneously, but (2.1.3) requires

$$\bar{\tau}(0,0-) = 0 \text{ while } \bar{\tau}(0,0+) = 1,$$

$$\bar{\tau}(1,0-) = 0 \text{ while } \bar{\tau}(1,0+) = 1.$$

The simplest way of defining $\bar{\tau}(0,0)$ and $\bar{\tau}(1,0)$ so the resulting problem is tractable mathematically is to give $\bar{\tau}(0,0)$ and $\bar{\tau}(1,0)$ the value of 1.

Equation (2.1.1) with initial and boundary conditions as now defined can be solved analytically by substituting $T = 1 - \bar{\tau}$ in (2.1.1) and (2.1.3), then applying the method of separation of variables (see Kreyszig, 1979, p.513) giving the solution

$$\bar{\tau}(x,t) = 1 - \sum_{m=1}^{\infty} \frac{4}{(2m-1)\pi} \sin\{(2m-1)\pi x\}\exp\{-\alpha(2m-1)^2\pi^2 t\}, \qquad (2.1.4)$$

which permits the value of the non-dimensional temperature $\bar{\tau}$ to be calculated at any point in the solution domain $0 < x < 1$, $t > 0$. For a given x and t, this requires the summation of the series (2.1.4) on a computer. This is done by adding more and more terms of the infinite series until successive partial sums remain constant with the required accuracy.

It should be noted that the method of separation of variables referred to above is restricted to the solution of linear partial differential equations with constant coefficients, and boundary values which are equivalent to reflection or periodicity conditions. It is difficult, and in some cases impossible, to find exact solutions to non-linear partial differential equations with variable coefficients and complicated boundary conditions.

An alternative method of solving the given partial differential equation is to use finite difference methods which are not restricted by such criteria as linearity, constant coefficients, and so forth.

2.2 The Finite Difference Grid

The solution domain in x-t space is covered by a rectangular grid (sometimes called a mesh or net) with grid spacings of Δx, Δt in the x,t directions respectively, the values of Δx and Δt being assumed uniform for the present. The grid consists of the set of lines parallel to the t-axis given by

$$x = x_j \,,\ j=0(1)J, \tag{2.2.1}$$

where $x_j = j\Delta x$ and $\Delta x = 1/J$, and the set of lines parallel to the x-axis given by

$$t = t_n \,,\ n=0,1,2,\ldots \tag{2.2.2}$$

where $t_n = n\Delta t$, as shown in Figure 2.3.

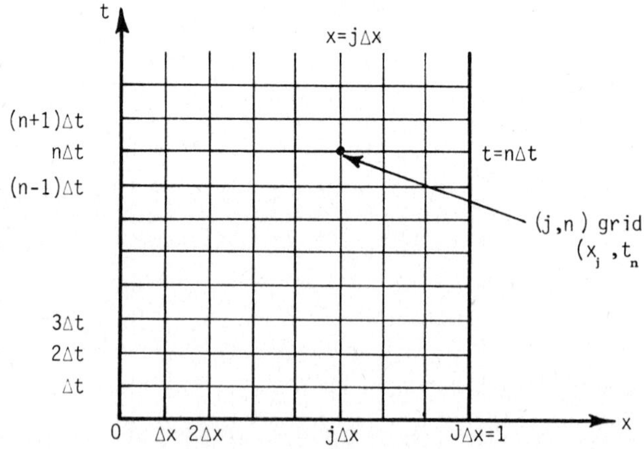

Figure 2.3 : The finite difference grid in the solution region for the one-dimensional diffusion equation.

Finite difference methods will now be developed which determine approximately the values of $\bar{\tau}$ at interior points within the solution domain where the lines defined by (2.2.1) and (2.2.2) intersect. These points, with coordinates $(x_j\,,t_n)$, are called *grid-points* and are often denoted (j,n). The terms mesh-point or node are sometimes used instead of grid-point. Grid-points within the solution domain are termed interior; those on the boundary of the solution domain, that is those for which j=0 or J or n=0, are called boundary grid-points. In the diffusion problem under consideration, the values of $\bar{\tau}$ on the boundaries of the solution region are defined by the initial and boundary conditions (2.1.3) (see Figure 2.2).

The following notation will be used for values of $\bar{\tau}$ and its derivatives at the (j,n) grid-point:

$$\left.\begin{aligned} \bar{\tau}_j^n &= \bar{\tau}(x_j\,,t_n), \\ \left.\frac{\partial\bar{\tau}}{\partial t}\right|_j^n &= \frac{\partial\bar{\tau}(x_j\,,t_n)}{\partial t} \end{aligned}\right\}, \tag{2.2.3}$$

and so on. Thus we denote the second spatial derivative at $x_{j-1}=(j-1)\Delta x$, $t_{n+1}=(n+1)\Delta t$, that is at the $(j-1,n+1)$ grid-point, by

$$\left.\frac{\partial^2\bar{\tau}}{\partial x^2}\right|_{j-1}^{n+1}.$$

Note that the superscript n in the left side of Equations (2.2.3) indicates the time level. In order to avoid confusion, if a superscript is to indicate a power, the number to be raised to that power will be enclosed in brackets. Thus, $(G)^n$ is the nth power of the number G. The brackets will be omitted when there is no possibility of misinterpretation.

2.3 Finite Difference Approximations to Derivatives

As a first step in developing a method of calculating the values of $\bar{\tau}$ at each interior grid-point, the space and time derivatives of $\bar{\tau}$ at the (j,n) grid-point must be expressed in terms of values of $\bar{\tau}$ at nearby grid-points. Taylor series expansions of $\bar{\tau}$ about the (j,n) grid-point will be used in this process, namely

$$\bar{\tau}(x_j+\Delta x,t_n) = \sum_{m=0}^{\infty}\frac{(\Delta x)^m}{m!}\frac{\partial^m\bar{\tau}(x_j,t_n)}{\partial x^m},$$

or, in the notation defined by (2.2.3),

$$\bar{\tau}_{j+1}^{n} = \sum_{m=0}^{\infty}\frac{(\Delta x)^m}{m!}\left.\frac{\partial^m\bar{\tau}}{\partial x^m}\right|_j^n$$

$$= \bar{\tau}_j^n + \Delta x\left.\frac{\partial\bar{\tau}}{\partial x}\right|_j^n + \frac{(\Delta x)^2}{2!}\left.\frac{\partial^2\bar{\tau}}{\partial x^2}\right|_j^n + \frac{(\Delta x)^3}{3!}\left.\frac{\partial^3\bar{\tau}}{\partial x^3}\right|_j^n + \dots \quad . \qquad (2.3.1)$$

Similarly

$$\bar{\tau}(x_j,t_n+\Delta t) = \sum_{m=0}^{\infty}\frac{(\Delta t)^m}{m!}\frac{\partial^m\bar{\tau}(x_j,t_n)}{\partial t^m},$$

which may be written

$$\bar{\tau}_j^{n+1} = \sum_{m=0}^{\infty}\frac{(\Delta t)^m}{m!}\left.\frac{\partial^m\bar{\tau}}{\partial t^m}\right|_j^n$$

$$= \bar{\tau}_j^n + \Delta t\left.\frac{\partial\bar{\tau}}{\partial t}\right|_j^n + \frac{(\Delta t)^2}{2!}\left.\frac{\partial^2\bar{\tau}}{\partial t^2}\right|_j^n + \frac{(\Delta t)^3}{3!}\left.\frac{\partial^3\bar{\tau}}{\partial t^3}\right|_j^n + \dots \quad . \qquad (2.3.2)$$

These Taylor series expansions may be truncated after any number of terms, the resulting error being dominated by the next term in the expansion if $\Delta x \ll 1$ in (2.3.1), or if $\Delta t \ll 1$ in (2.3.2). Thus we may write

$$\bar{\tau}_{j+1}^{n} = \bar{\tau}_j^n + \Delta x\left.\frac{\partial\bar{\tau}}{\partial x}\right|_j^n + \frac{(\Delta x)^2}{2!}\left.\frac{\partial^2\bar{\tau}}{\partial x^2}\right|_j^n + O\{(\Delta x)^3\}, \qquad (2.3.3)$$

and

$$\bar{\tau}_j^{n+1} = \bar{\tau}_j^n + \Delta t\left.\frac{\partial\bar{\tau}}{\partial t}\right|_j^n + \frac{(\Delta t)^2}{2!}\left.\frac{\partial^2\bar{\tau}}{\partial t^2}\right|_j^n + O\{(\Delta t)^3\}. \qquad (2.3.4)$$

The term $O\{(\Delta t)^3\}$, for example, is interpreted as meaning there exists a positive constant κ, depending on $\bar{\tau}$, such that the difference between $\bar{\tau}$ at the $(j,n+1)$ grid-point and the first three terms on the right side of (2.3.4), all evaluated at the (j,n) grid-point, is numerically less than $\kappa(\Delta t)^3$ for all sufficiently small Δt. When there is no possibility of confusion, the qualifying statement "as $\Delta t \to 0$" is omitted. It is clear that the error involved in this approximation reduces in magnitude as the grid-spacing Δt decreases.

The value of the continuous derivative $\partial\bar{\tau}/\partial t$ at the (j,n) grid-point may be approximated in terms of values of $\bar{\tau}$ at nearby grid-points in a number of ways. For instance, suppose it is desired to express the derivative $\partial\bar{\tau}/\partial t|_j^n$ in terms of the values of $\bar{\tau}_j^n$ and $\bar{\tau}_j^{n+1}$, the latter being the value of $\bar{\tau}$ at the corresponding spatial position but one time level ahead. We therefore write

$$\left.\frac{\partial\bar{\tau}}{\partial t}\right|_j^n = a_0\bar{\tau}_j^n + a_1\bar{\tau}_j^{n+1} + O\{(\Delta t)^m\}. \tag{2.3.5}$$

The term $O\{(\Delta t)^m\}$ will indicate the accuracy of the resulting approximation for $\partial\bar{\tau}/\partial t|_j^n$.

Since, by (2.3.4)

$$a_1\bar{\tau}_j^{n+1} = a_1\bar{\tau}_j^n + a_1\Delta t\left.\frac{\partial\bar{\tau}}{\partial t}\right|_j^n + a_1\frac{(\Delta t)^2}{2}\left.\frac{\partial^2\bar{\tau}}{\partial t^2}\right|_j^n + \ldots,$$

addition of $a_0\bar{\tau}_j^n$ gives

$$a_1\bar{\tau}_j^{n+1} + a_0\bar{\tau}_j^n = (a_1+a_0)\bar{\tau}_j^n + a_1\Delta t\left.\frac{\partial\bar{\tau}}{\partial t}\right|_j^n + a_1\frac{(\Delta t)^2}{2}\left.\frac{\partial^2\bar{\tau}}{\partial t^2}\right|_j^n + \ldots \tag{2.3.6}$$

Clearly, we require $a_1\Delta t = 1$ and $a_1+a_0 = 0$ on the right side of this equation. Thus $a_1 = 1/\Delta t$ and $a_0 = -a_1 = -1/\Delta t$. Substitution in Equation (2.3.6) yields

$$\left.\frac{\partial\bar{\tau}}{\partial t}\right|_j^n = \frac{\bar{\tau}_j^{n+1} - \bar{\tau}_j^n}{\Delta t} + O\{\Delta t\}. \tag{2.3.7a}$$

Thus we may write

$$\left.\frac{\partial\bar{\tau}}{\partial t}\right|_j^n \simeq \frac{\bar{\tau}_j^{n+1} - \bar{\tau}_j^n}{\Delta t}, \tag{2.3.7b}$$

which has an error of $O\{\Delta t\}$. Because the right hand side of this expression uses the value $\bar{\tau}_j^{n+1}$ which is forward of $\bar{\tau}_j^n$ in time, it is called the forward difference approximation to the time derivative evaluated at the (j,n) grid-point.

An alternative way of approximating the value of $\partial\bar{\tau}/\partial t$ at the (j,n) grid-point is to use the relation

$$\left.\frac{\partial\bar{\tau}}{\partial t}\right|_j^n = \frac{\bar{\tau}_j^n - \bar{\tau}_j^{n-1}}{\Delta t} + O\{\Delta t\}, \tag{2.3.8a}$$

which yields the approximation

$$\left.\frac{\partial\bar{\tau}}{\partial t}\right|_j^n \simeq \frac{\bar{\tau}_j^n - \bar{\tau}_j^{n-1}}{\Delta t}. \tag{2.3.8b}$$

Because the right side of (2.3.8b) uses the value $\bar{\tau}_j^{n-1}$ at a previous time level to $\bar{\tau}_j^n$ it is called the backward difference approximation to the time derivative. Like the forward difference approximation it has an error of $O\{\Delta t\}$.

Another way of approximating the value of $\partial\bar{\tau}/\partial t$ at the (j,n) grid-point is to use the relation

$$\left.\frac{\partial\bar{\tau}}{\partial t}\right|_j^n = \frac{\bar{\tau}_j^{n+1} - \bar{\tau}_j^{n-1}}{2\Delta t} + O\{(\Delta t)^2\}, \tag{2.3.9a}$$

which leads to the approximation

$$\left.\frac{\partial\bar{\tau}}{\partial t}\right|_j^n \simeq \frac{\bar{\tau}_j^{n+1} - \bar{\tau}_j^{n-1}}{2\Delta t} . \tag{2.3.9b}$$

The error in using the right side of (2.3.9b) for the value of $\partial\bar{\tau}/\partial t|_j^n$ is $O\{(\Delta t)^2\}$, which is smaller than $O\{\Delta t\}$ for $\Delta t << 1$, so this is a more accurate approximation than either the forward or backward difference approximations to $\partial\bar{\tau}/\partial t$. Because the time level $n\Delta t$ is centred between the time levels $(n+1)\Delta t$ and $(n-1)\Delta t$ at which the values $\bar{\tau}_j^{n+1}$ and $\bar{\tau}_j^{n-1}$ occur, the right hand side of (2.3.9b) is called the central difference approximation to the first-order time derivative $\partial\bar{\tau}/\partial t$ at the (j,n) grid-point.

In a similar way, finite difference approximations to the second-order time derivative $\partial^2\bar{\tau}/\partial t^2|_j^n$ can be derived. For example, suppose it is desired to express the derivative at the (j,n) grid-point in terms of the values of $\bar{\tau}$ at that grid-point and also at grid-points one time level ahead and two time levels behind.

We therefore write

$$\left.\frac{\partial^2\bar{\tau}}{\partial t^2}\right|_j^n = a_{-2}\bar{\tau}_j^{n-2} + a_{-1}\bar{\tau}_j^{n-1} + a_0\bar{\tau}_j^n + a_1\bar{\tau}_j^{n+1} + O\{(\Delta t)^m\}, \tag{2.3.10}$$

where the coefficients a_{-2}, a_{-1}, a_0 and a_1 are to be determined, as well as the order of the error term.

Since

$$a_{-2}\bar{\tau}_j^{n-2} = a_{-2}\{\bar{\tau}_j^n - 2\Delta t\left.\frac{\partial\bar{\tau}}{\partial t}\right|_j^n + \frac{4(\Delta t)^2}{2!}\left.\frac{\partial^2\bar{\tau}}{\partial t^2}\right|_j^n - \frac{8(\Delta t)^3}{3!}\left.\frac{\partial^3\bar{\tau}}{\partial t^3}\right|_j^n + \frac{16(\Delta t)^4}{4!}\left.\frac{\partial^4\bar{\tau}}{\partial t^4}\right|_j^n + ..\},$$

$$a_{-1}\bar{\tau}_j^{n-1} = a_{-1}\{\bar{\tau}_j^n - \Delta t\left.\frac{\partial\bar{\tau}}{\partial t}\right|_j^n + \frac{(\Delta t)^2}{2!}\left.\frac{\partial^2\bar{\tau}}{\partial t^2}\right|_j^n - \frac{(\Delta t)^3}{3!}\left.\frac{\partial^3\bar{\tau}}{\partial t^3}\right|_j^n + \frac{(\Delta t)^4}{4!}\left.\frac{\partial^4\bar{\tau}}{\partial t^4}\right|_j^n + ..\},$$

$$a_0\bar{\tau}_j^n = a_0\bar{\tau}_j^n,$$

$$a_1\bar{\tau}_j^{n+1} = a_1\{\bar{\tau}_j^n + \Delta t\left.\frac{\partial\bar{\tau}}{\partial t}\right|_j^n + \frac{(\Delta t)^2}{2!}\left.\frac{\partial^2\bar{\tau}}{\partial t^2}\right|_j^n + \frac{(\Delta t)^3}{3!}\left.\frac{\partial^3\bar{\tau}}{\partial t^3}\right|_j^n + \frac{(\Delta t)^4}{4!}\left.\frac{\partial^4\bar{\tau}}{\partial t^4}\right|_j^n + ..\},$$

then their sum is

$$a_{-2}\bar{\tau}_j^{n-2} + a_{-1}\bar{\tau}_j^{n-1} + a_0\bar{\tau}_j^n + a_1\bar{\tau}_j^{n+1}$$

$$= (a_{-2}+a_{-1}+a_0+a_1)\bar{\tau}_j^n + \Delta t(-2a_{-2}-a_{-1}+a_1)\left.\frac{\partial\bar{\tau}}{\partial t}\right|_j^n$$

$$+ \frac{(\Delta t)^2}{2}(4a_{-2}+a_{-1}+a_1)\left.\frac{\partial^2\bar{\tau}}{\partial t^2}\right|_j^n + \frac{(\Delta t)^3}{6}(-8a_{-2}-a_{-1}+a_1)\left.\frac{\partial^3\bar{\tau}}{\partial t^3}\right|_j^n$$

$$+ \frac{(\Delta t)^4}{24}(16a_{-2}+a_{-1}+a_1)\left.\frac{\partial^4\bar{\tau}}{\partial t^4}\right|_j^n + \ldots \qquad (2.3.11)$$

From (2.3.10) it is clear that the right side of (2.3.11) is to become

$$\left.\frac{\partial^2\bar{\tau}}{\partial t^2}\right|_j^n + O\{(\Delta t)^m\}, \qquad (2.3.12)$$

so we require

$$\left.\begin{aligned} a_{-2} + a_{-1} + a_0 + a_1 &= 0, \\ -2a_{-2} - a_{-1} + a_1 &= 0, \\ 4a_{-2} + a_{-1} + a_1 &= 2(\Delta t)^{-2}. \end{aligned}\right\} \qquad (2.3.13a)$$

Also, as there is still one degree of freedom left, the term in $(\Delta t)^3$ can be eliminated, giving

$$-8a_{-2} - a_{-1} + a_1 = 0. \qquad (2.3.13b)$$

Solving the set of linear algebraic equations (2.3.13) gives

$$a_{-2} = 0,\ a_{-1} = (\Delta t)^{-2},\ a_0 = -2(\Delta t)^{-2},\ a_1 = (\Delta t)^{-2}, \qquad (2.3.14)$$

so that (2.3.11) becomes

$$\left.\frac{\partial^2\bar{\tau}}{\partial t^2}\right|_j^n = \frac{\bar{\tau}_j^{n-1} - 2\bar{\tau}_j^n + \bar{\tau}_j^{n+1}}{(\Delta t)^2} + O\{(\Delta t)^2\}. \qquad (2.3.15)$$

Note that a more accurate approximation to $\partial^2\bar{\tau}/\partial t^2$ is obtained by using values of $\bar{\tau}$ at only three consecutive time-levels in the manner of (2.3.15), than by using values of $\bar{\tau}$ at four consecutive time-levels.

In a similar manner finite difference approximations to the spatial derivatives of $\bar{\tau}$ at the (j,n) grid-point may be obtained. For the first-order spatial derivative the appropriate finite difference forms are:

$$\left.\frac{\partial\bar{\tau}}{\partial x}\right|_j^n = \frac{\bar{\tau}_{j+1}^n - \bar{\tau}_j^n}{\Delta x} + O\{\Delta x\}, \qquad (2.3.16)$$

which leads to a *forward difference* approximation;

$$\left.\frac{\partial\bar{\tau}}{\partial x}\right|_j^n = \frac{\bar{\tau}_j^n - \bar{\tau}_{j-1}^n}{\Delta x} + O\{\Delta x\}, \qquad (2.3.17)$$

which leads to a *backward difference* approximation;

$$\left.\frac{\partial\bar{\tau}}{\partial x}\right|_j^n = \frac{\bar{\tau}_{j+1}^n - \bar{\tau}_{j-1}^n}{2\Delta x} + O\{(\Delta x)^2\}, \tag{2.3.18}$$

which leads to a *central difference* approximation.

The central difference approximation to the second-order spatial derivative is based on the relation

$$\left.\frac{\partial^2\bar{\tau}}{\partial x^2}\right|_j^n = \frac{\bar{\tau}_{j+1}^n - 2\bar{\tau}_j^n + \bar{\tau}_{j-1}^n}{(\Delta x)^2} + O\{(\Delta x)^2\}. \tag{2.3.19}$$

ASSESSING FINITE DIFFERENCE APPROXIMATIONS TO DERIVATIVES OF WAVE-LIKE MOTIONS

Since wave-like motions are of interest in many problems, and since any motion may be considered as a sum of wave-like motions (that is, its functional form can be expressed as a Fourier series) it is of interest to determine the accuracy of finite difference approximations to a simple harmonic wave.

Consider the progressive wave

$$\bar{\tau}(x,t) = \cos(\kappa x - \sigma t), \tag{2.3.20}$$

κ and σ being constants. This is a sinusoidal wave which moves in the positive x-direction with wave speed σ/κ (see Figure 2.4) and at a fixed point x_j it is periodic in time with period $P = 2\pi/\sigma$ (see Figure 2.5). It is also periodic in space with wave-length $\lambda = 2\pi/\kappa$.

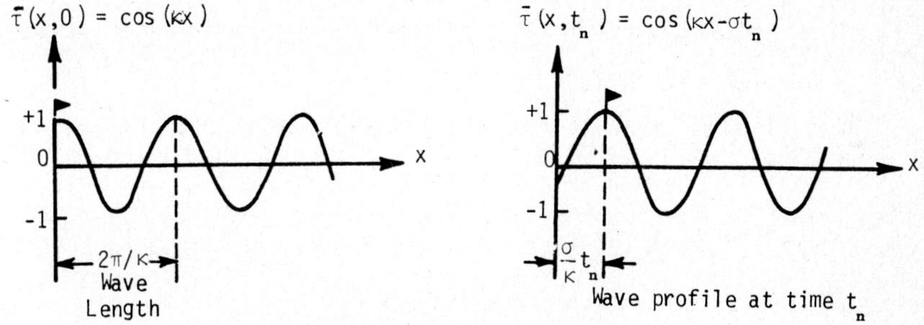

Figure 2.4 : The sinusoidal wave $\tau(x,t) = \cos(\kappa x - \sigma t)$ has wavelength $2\pi/\kappa$ and moves with speed σ/κ.

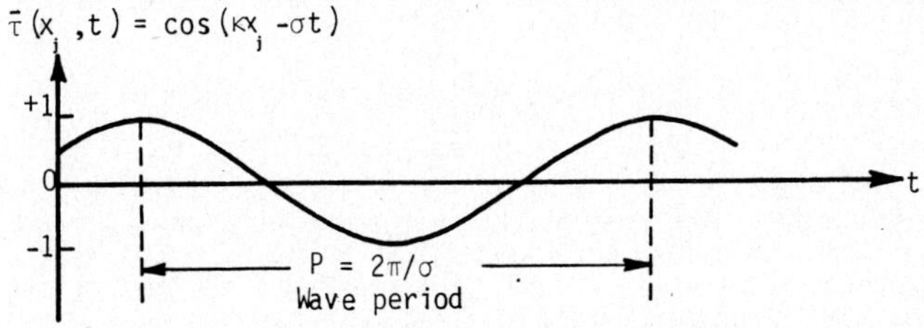

Figure 2.5 : The wave $\cos(\kappa x_j - \sigma t)$ is periodic in time with period $P = 2\pi/\sigma$.

At the point x_j at the n-thtime level, the exact value of the time derivative of $\bar{\tau}$ is

$$\left.\frac{\partial\bar{\tau}}{\partial t}\right|_j^n = \sigma \sin(\kappa x_j - \sigma t_n). \qquad (2.3.21)$$

However, the forward-time finite difference approximation to this derivative is

$$\frac{\bar{\tau}_j^{n+1} - \bar{\tau}_j^n}{\Delta t} = \frac{\cos(\kappa x_j - \sigma t_n - \sigma\Delta t) - \cos(\kappa x_j - \sigma t_n)}{\Delta t}$$

$$= \sigma\left\{\frac{\sin(\sigma\Delta t/2)}{(\sigma\Delta t/2)}\right\} \sin(\kappa x_j - \sigma t_n - \frac{\sigma\Delta t}{2}). \qquad (2.3.22)$$

Clearly an error has been introduced in the amplitude and phase of the derivative if (2.3.22) is used as an approximation for (2.3.21). That is, the forward-time approximation for the time derivative of (2.3.20) is equal to the true value of the derivative given by Equation (2.3.21) multiplied by the factor $\{\sin(\sigma\Delta t/2)/(\sigma\Delta t/2)\}$ and altered in phase by $-\sigma\Delta t/2$, which is equivalent to a time lead of $\Delta t/2$. Clearly, as $\Delta t \to 0$ the factor multiplying the amplitude tends to 1 and the phase error tends to zero. Therefore the accuracy of the forward difference approximation in time increases with decreasing values of Δt. If the ratio of the time step to the period is such that $P \geq 20\Delta t$, so there are at least 20 time steps in one period, then $\sigma\Delta t/2 \leq \pi/20$ and the phase lag is less than $\pi/20$ radians or 9^0. Also the amplitude of the derivative is multiplied by a factor between 0.9959 and 1, and the forward time finite difference approximation is a good approximation to the time derivative. On the other hand, when $\Delta t \geq P/2$, so there are two or less time steps in one period and $\sigma\Delta t/2 \geq \pi/2$, there is a phase error greater than one quarter of a cycle and the amplitude is multiplied by a factor less than 0.64. These errors are unacceptably large.

The effect of using the central difference approximation for the second space derivative $\partial^2\bar{\tau}/\partial x^2|_j^n$ when $\bar{\tau}$ is given by the progressive wave form (2.3.20), is seen by comparing the exact value

$$\left.\frac{\partial^2\bar{\tau}}{\partial x^2}\right|_j^n = -\kappa^2\cos(\kappa x_j - \sigma t_n) \qquad (2.3.23)$$

and the central difference approximation

$$\frac{\bar{\tau}_{j+1}^n - 2\bar{\tau}_j^n + \bar{\tau}_{j-1}^n}{(\Delta x)^2} = \frac{\cos(\kappa x_j + \kappa\Delta x - \sigma t_n) - 2\cos(\kappa x_j - \sigma t_n) + \cos(\kappa x_j - \kappa\Delta x - \sigma t_n)}{(\Delta x)^2}$$

$$= -\kappa^2\left\{\frac{\sin(\kappa\Delta x/2)}{(\kappa\Delta x/2)}\right\}^2 \cos(\kappa x_j - \sigma t_n). \qquad (2.3.24)$$

This comparison shows no phase error is introduced, only a change to the amplitude of the derivative. Since the wavelength of (2.3.20) is $\lambda = 2\pi/\kappa$, then for small space steps so there are at least 20 in one wavelength, that is $\lambda \geq 20\Delta x$ and $\kappa\Delta x/2 \leq \pi/20$, the amplitude of the derivative is multiplied by a factor between 0.992 and 1. However, when there are fewer than 2 grid spacings in one wavelength, so that $\Delta x \geq \lambda/2$ and $\kappa\Delta x/2 \geq \pi/2$, the amplitude of the derivative is less than 0.42 of its correct value.

2.4 An Approximate Solution to the Diffusion Equation

Approximate values τ_j^n to the true values $\bar{\tau}_j^n$ at the interior grid-points, that is at the (j,n) grid-points for $j=1(1)J-1$ and $n=1,2,3,..$, may be found for the diffusion equation (2.1.1) by substituting finite difference approximations to the derivatives involved. At the (j,n) grid-point, the given equation is

$$\left.\frac{\partial\bar{\tau}}{\partial t}\right|_j^n = \alpha \left.\frac{\partial^2\bar{\tau}}{\partial x^2}\right|_j^n , \tag{2.4.1}$$

and any of the various approximations for $\partial\bar{\tau}/\partial t$ and $\partial^2\bar{\tau}/\partial x^2$ described in Section 2.3 can be used. One of the simplest finite difference equations which approximates the partial differential equation (2.4.1) is obtained by using the forward difference approximation to the time derivative with the central difference approximation to the space derivative.

Substitution from Equations (2.3.7a) and (2.3.19) into Equation (2.4.1) gives

$$\frac{\bar{\tau}_j^{n+1} - \bar{\tau}_j^n}{\Delta t} + O\{\Delta t\} = \alpha\{\frac{\bar{\tau}_{j+1}^n - 2\bar{\tau}_j^n + \bar{\tau}_{j-1}^n}{(\Delta x)^2} + O\{(\Delta x)^2\}\}. \tag{2.4.2}$$

Solving the finite difference equation

$$\frac{\tau_j^{n+1} - \tau_j^n}{\Delta t} = \alpha\{\frac{\tau_{j+1}^n - 2\tau_j^n + \tau_{j-1}^n}{(\Delta x)^2}\}, \tag{2.4.3}$$

with the given initial and boundary conditions therefore gives approximate values τ to the true solution $\bar{\tau}$ of the partial differential equation (2.4.1). Equation (2.4.3) is called the Forward-Time Centred-Space (FTCS) finite difference approximation to Equation (2.4.1). Clearly, the use of the finite difference approximation (2.4.3) involves a leading error of $O\{\Delta t,(\Delta x)^2\}$. The last statement is to be interpreted as meaning that there exist two positive constants κ_1 and κ_2, both depending on $\bar{\tau}$, such that the absolute value of the error in assuming (2.4.3) is the same as (2.4.1), is less than or equal to $\kappa_1\Delta t+\kappa_2(\Delta x)^2$, for all Δt and Δx which are sufficiently small.

Equation (2.4.3) may be rewritten in the form

$$\tau_j^{n+1} = s\tau_{j-1}^n + (1-2s)\tau_j^n + s\tau_{j+1}^n \tag{2.4.4}$$

where

$$s = \alpha\ \Delta t/(\Delta x)^2, \tag{2.4.5}$$

which is a formula giving the approximate value τ_j^{n+1} at the $(n+1)$th time level in terms of the approximate values τ_{j-1}^n, τ_j^n and τ_{j+1}^n at the nth time level.

Application of Equation (2.4.4) with $n=0$ gives the approximate values τ_j^1 at grid-points along the first time row $t=\Delta t$, in terms of the known boundary and initial values $\tau_j^0 = \bar{\tau}_j^0$, $j=0(1)J$. Application of Equation (2.4.4) with $n=1$ then gives the approximate values τ_j^2 to the true solution $\bar{\tau}_j^2$ at grid-points along the second time level in terms of the computed values τ_j^1 at the grid-points along the first time level. An equation such as (2.4.4), which expresses one unknown value τ_j^{n+1} directly in terms of known values of τ_j^n, τ_j^{n-1}, and so on, is an explicit finite difference equation. Also, because it involves values along only two time levels, it is called a two-level formula.

Returning to the diffusion Equation (2.1.1) with $\alpha = 10^{-2}$ and boundary conditions given by (2.1.3), the explicit FTCS finite difference formula (2.4.4) can be used to obtain approximate solutions to the given problem at all interior grid-points. Experience shows that at least ten intervals are necessary to give reasonably accurate solutions to this problem, so $\Delta x=0.1$ will be chosen to illustrate the method.

Noting that Equation (2.4.4) takes the simple form

$$\tau_j^{n+1} = \tfrac{1}{2}(\tau_{j-1}^n + \tau_{j+1}^n), \qquad (2.4.6)$$

when $s=\tfrac{1}{2}$, choose

$$\Delta t = \frac{(\Delta x)^2}{2\alpha} = 0.5.$$

In accordance with the initial conditions (2.1.3), the values of τ at grid-points along the time level n=0 are $\tau_0^0 = 1$, $\tau_1^0 = \tau_2^0 = \tau_3^0 = \tau_4^0 = \tau_5^0 = \tau_6^0 = \tau_7^0 = \tau_8^0 = \tau_9^0 = 0$, $\tau_{10}^0 = 1$. Equation (2.4.4) is then applied with n=0 and j=1,2,..,9 to calculate the approximate solutions τ_j^1 at each of the internal grid-points along the time level n=1; that is,

$$\tau_j^1 = \tfrac{1}{2}(\tau_{j-1}^0 + \tau_{j+1}^0),\ j=1(1)9,$$

is used. Values of τ_0^1 and τ_J^1, that is at grid-points on the boundaries j=0 and J, are specified by the boundary conditions of this problem and need not be calculated. The order in which the approximations τ_j^1 are calculated at the internal grid-points is immaterial when using this method, since Equation (2.4.4) gives τ_j^1 explicitly in terms of known values from previous time levels. The approximate values τ calculated at the first time level t=0.5 are therefore

$$\tau_0^1 = 1.0,\ \tau_1^1 = 0.5,\ \tau_2^1 = \tau_3^1 = \tau_4^1 = \tau_5^1 = \tau_6^1 = \tau_7^1 = \tau_8^1 = 0,$$

$$\tau_9^1 = 0.5,\ \tau_{10}^1 = 1.0.$$

Table 2.1 shows the results of these and calculations at further time-levels. As the time increases, heat diffuses inwards and the internal temperatures are seen to be increasing steadily and it would be found, if the calculations were continued, that the temperatures gradually approach the equilibrium value of 1 at all internal grid-points. This is physically realistic. Note that it takes five time steps for the boundary temperature of 1 to have any effect on the temperature at the centre of the rod using this finite difference scheme with J=10.

Clearly, a large number of time steps will be needed before the values calculated at internal grid-points in the solution domain are close to 1. In order to reduce the number of arithmetic operations required, a larger time step is preferable. Choosing $\Delta t = 1$, for which $s = 1$, Equation (2.4.4) becomes

$$\tau_j^{n+1} = \tau_{j-1}^n - \tau_j^n + \tau_{j+1}^n. \qquad (2.4.7)$$

Besides requiring only half the number of time levels to reach a particular time, this formula uses less computer time than application of (2.4.6). This is because multiplications and divisions take longer to carry out on a computer than do additions and subtractions. For instance, a multiplication or division on a CYBER 173 takes approximately

TABLE 2.1

Approximate solution of the diffusion problem using (2.4.4) with s=½

t

n	$n\Delta t$	Boundary Values										Boundary Values
6	3.0	1										1
5	2.5	1	0.6875	0.375	0.21875	0.0625	0.0625	0.0625	0.21875	0.375	0.6875	1
4	2.0	1	0.625	0.375	0.125	0.0625	0.0	0.0625	0.125	0.375	0.625	1
3	1.5	1	0.625	0.25	0.125	0.0	0.0	0.0	0.125	0.25	0.625	1
2	1.0	1	0.5	0.25	0.0	0.0	0.0	0.0	0.0	0.25	0.5	1
1	0.5	1	0.5	0.0	0.0	0.0	0.0	0.0	0.0	0.0	0.5	1
→ 0	0.0	1	0	0	0	0	0	0	0	0	0	1

x

Initial Values

$j\Delta x$	0.0	0.1	0.2	0.3	0.4	0.5	0.6	0.7	0.8	0.9	1.0
j	0	1	2	3	4	5	6	7	8	9	10

TABLE 2.2

Approximate solution of the diffusion problem using (2.4.4) with s=1

t

n	$n\Delta t$	Boundary Values										Boundary Values
6	6	1										1
5	5	1	7	-8	7	-3	2	-3	7	-8	7	1
4	4	1	-2	4	-2	1	0	1	-2	4	-2	1
3	3	1	2	-1	1	0	0	0	1	-1	2	1
2	2	1	0	1	0	0	0	0	0	1	0	1
1	1	1	1	0	0	0	0	0	0	0	1	1
→ 0	0	1	0	0	0	0	0	0	0	0	0	1

x

Initial Values

$j\Delta x$	0.0	0.1	0.2	0.3	0.4	0.5	0.6	0.7	0.8	0.9	1.0
j	0	1	2	3	4	5	6	7	8	9	10

3 micro-seconds (micro equals one-millionth) whereas an addition or subtraction takes approximately ½ micro-second. Therefore the arithmetic operations in (2.4.7) take 1 micro-second to perform, while those in (2.4.6) take about 4 micro-seconds, four times as long.

Table 2.2 lists the results obtained by applying (2.4.7). Obviously the values shown are physically impossible; the effect of changing from s=½ to s=1 has been disastrous. Clearly the restrictions which apply to a particular finite difference method must be known to the user, otherwise he may accept unrealistic values as being accurate solutions to the given partial differential equation. In order to establish these restrictions three important concepts will be studied in Section 3, those of convergence, consistency and stability.

2.5 The Finite Difference Method - a Summary

The process of solving an initial value problem involving partial differential equations, using finite difference techniques, may be summarised as follows.

Firstly, a grid is defined which covers the solution domain. At the grid-points, the approximate solution to the problem will be found by algebraic processes.

As a second step, all dependent variables possible are initialised. These initial values may be exact, corresponding to a real initial situation for a transient problem, or they may be an approximation (perhaps a rough guess) for a steady (equilibrium) solution if the transient part is not of interest.

Once these two steps have been completed, the computational cycle begins. A finite difference equation, which is an analogue of the continuous partial differential equation is constructed. This is developed by replacing the continuous derivatives in the equation applied to an interior grid-point by approximations in terms of values of the unknown dependent variable at nearby grid-points. The finite difference equation is used to calculate approximate values of the dependent variables at grid-points in the solution domain.

For instance, with the heat diffusion equation, the new temperature τ_j^{n+1} can be calculated approximately from the previously calculated temperatures τ_{j-1}^n, τ_j^n and τ_{j+1}^n using the FTCS formula (2.4.4). In this process, with time held constant at $t = n\Delta t$, j is varied from 1 (1)J-1 until the temperature at all points $j\Delta x$ in the grid are computed at the next time level $t = (n+1)\Delta t$. Once these values have been found, the time is again increased by Δt and the process repeated to get the temperatures at the new time level, $t = (n+2)\Delta t$, using the computed values at the time level $t = (n+1)\Delta t$.

This is continued until a predetermined time is reached, or until the dependent variables at each space position $j\Delta x$ remain unchanged when computed at successive time levels if an equilibrium solution is required.

However, the finite difference analogue of the partial differential equation must satisfy certain conditions, or its solution will not approximate the required solution of the given partial differential equation. These conditions are examined in the next Section.

3. CONVERGENCE, CONSISTENCY AND STABILITY

3.1 Convergence

A solution to a finite difference equation which approximates a given partial differential equation is said to be *convergent* if, at each grid-point in the solution domain, the finite difference solution approaches the solution of the partial differential equation as the grid-spacings (Δx and Δt for the FTCS method of solving the one-dimensional diffusion equation) tend to zero. Thus, if we let $\bar{\tau}_j^n$ represent the exact solution of the one-dimensional diffusion equation at (x_*, t_*) and τ_j^n the exact solution of the approximating finite difference equation at the same point, then the finite difference solution is said to be convergent if for $j\Delta x = x_*$, $n\Delta t = t_*$, then $\tau_j^n \to \bar{\tau}_j^n$ as $\Delta x \to 0$, $\Delta t \to 0$.

The difference between the true solution of the partial differential equation and the exact solution of the approximating finite difference equation is called the *discretization error*, which will be denoted e_j^n; that is

$$e_j^n = \bar{\tau}_j^n - \tau_j^n. \tag{3.1.1}$$

The exact solution of the finite difference equation is obtained when no numerical errors of any sort, such as round-off errors, are introduced into the computations.

The magnitude of the error e_j^n at the (j,n) grid-point depends on the size of the grid-spacings Δx and Δt, and on the magnitude of the higher-order derivatives omitted in the finite difference approximations to the derivatives in the given differential equation. For instance, in the case of the FTCS method being used to solve Equation (2.1.1), rearrangement of Equation (2.4.2) gives

$$\bar{\tau}_j^{n+1} = s\,\bar{\tau}_{j-1}^n + (1-2s)\bar{\tau}_j^n + s\,\bar{\tau}_{j+1}^n + O\{(\Delta t)^2, \Delta t(\Delta x)^2\}. \tag{3.1.2}$$

By comparing Equation (2.4.4) with Equation (3.1.2) it is clear that the error term of $O\{(\Delta t)^2, \Delta t(\Delta x)^2\}$ is the contribution to the discretisation error e_j^{n+1} in the process of stepping from time level n to time level (n+1) using the FTCS equation. However, it must be remembered that this contribution to e_j^{n+1} also depends on the magnitude of the higher-order derivatives associated with the terms which contain $(\Delta t)^2$, $\Delta t(\Delta x)^2$.

Comparison of Exact Solutions to the Diffusion Equation and FTCS Equation

Guidance concerning the convergence of finite difference solutions calculated using a particular method, such as the FTCS approximation (2.4.4) to the diffusion equation, can be obtained by considering special cases in which corresponding exact solutions of the approximating finite difference equation and the original partial differential equation can be found. For instance, with initial condition $f(x) = a\sin(m\pi x)$, where m is an integer, and boundary conditions $g_1(t) = g_2(t) \equiv 0$ in (2.1.2), the diffusion equation (2.1.1) has the exact solution

$$\bar{\tau}(x,t) = a\sin(m\pi x).\exp(-\alpha m^2\pi^2 t), \tag{3.1.3}$$

while the approximating finite difference equation (2.4.4) has the exact solution

$$\tau_j^n = a\{1 - 4s\sin^2(m\pi\Delta x/2)\}^n \sin(m\pi j\Delta x). \tag{3.1.4}$$

These solutions may be verified by substitution of (3.1.3), (3.1.4) into the appropriate equations (2.1.1), (2.4.4) respectively, and the given boundary and initial conditions. Similar results for other linear partial differential equations may be obtained using the method of separation of variables (see Kreyszig, 1979). For other finite difference equations, solutions may be obtained by using methods such as those described in Levy and Lessman (1959).

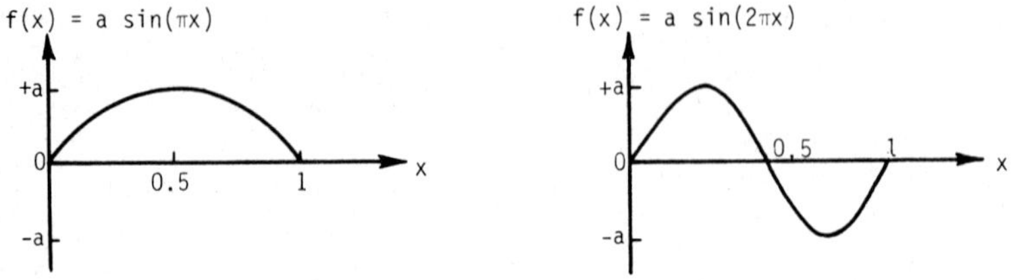

Figure 3.1 : The initial conditions $f(x) = a\sin(m\pi x)$, $m=1,2$, $0 \le x \le 1$.

Whatever the value of m, Equation (3.1.3) shows that $\bar{\tau}(x,t)$ decreases in size as t increases. However, if $s > \frac{1}{2}$, then for some values of m

$$1 - 4s\sin^2(m\pi\Delta x/2) < -1, \tag{3.1.5}$$

so that for these m the magnitude of τ_j^n increases exponentially and its sign alternates as n (and therefore t) increases. Clearly, in these cases the value τ_j^n obtained at the (j,n) grid-point bears little resemblance to the solution of the diffusion equation at this point.

It should be noted that the exact solution (2.1.4) of the original heat diffusion problem (2.1.1) with initial and boundary conditions (2.1.3), contains terms of the form (3.1.3) in which m is an odd integer and $a = 4/m\pi$. Therefore, for $s > \frac{1}{2}$ the relation (3.1.5) holds for some of these terms and a divergent oscillating solution to the approximating finite difference equation (2.4.3) is to be expected. The manner in which the exact solution of equation (2.4.3) with s = 1 diverges from the true solution of the given partial differential equation was shown in Table 2.2.

For $s < \frac{1}{2}$, the discretization error in this example is the difference between (3.1.3) evaluated at $(j\Delta x, n\Delta t)$, namely

$$\bar{\tau}_j^n = a\sin(m\pi j\Delta x).\exp(-\alpha m^2\pi^2 n\Delta t), \tag{3.1.6a}$$

and (3.1.4), which can be expanded as follows. Since $4s\sin^2(m\pi\Delta x/2)<1$ for $s < \frac{1}{2}$,

$$\begin{aligned}\tau_j^n &= a\sin(m\pi j\Delta x).\exp\{n\,ln[1 - 4s\sin^2(m\pi\Delta x/2)]\},\\ &= a\sin(m\pi j\Delta x).\exp n\{-4s\sin^2(m\pi\Delta x/2) - 8s^2\sin^4(m\pi\Delta x/2)\\ &\qquad - \frac{64}{3}s^3\sin^6(m\pi\Delta x/2) - \ldots\},\end{aligned}$$

where all indices represent powers. Furthermore, as

$$\begin{aligned}\sin^2(m\pi\Delta x/2) &= \tfrac{1}{2}\{1 - \cos(m\pi\Delta x)\}\\ &= m^2\pi^2(\Delta x)^2/4 - m^4\pi^4(\Delta x)^4/48 + O\{(\Delta x)^6\},\end{aligned}$$

it follows that, for fixed s and m,

$$\tau_j^n = a\sin(m\pi j\Delta x).\exp[-\alpha m^2\pi^2 n\Delta t + \tfrac{1}{2}\alpha m^4\pi^4 n\Delta t(s - \tfrac{1}{6})(\Delta x)^2 + O\{(\Delta x)^4\}], \tag{3.1.6b}$$

Substituting (3.1.6a,b) into Equation (3.1.1) yields the discretization error

$$e_j^n = a\sin(m\pi j\Delta x)\exp(-\alpha m^2\pi^2 n\Delta t)[1-\exp\{\tfrac{1}{2}\alpha m^4\pi^4(\Delta x)^2 n\Delta t(s-1/6)+O\{(\Delta x)^4\}\}] \tag{3.1.7}$$

Therefore the approximation (3.1.4) will be very accurate until the number of time steps, n, becomes large enough that

$$\exp\{\tfrac{1}{2}\alpha m^4\pi^4(\Delta x)^2 n\Delta t(s - 1/6) + O\{(\Delta x)^4\}\} \tag{3.1.8}$$

differs significantly from 1. Clearly, the choice $s = 1/6$ should lead to particularly accurate results.

A Proof of Convergence of the FTCS Method for Solving the Diffusion Equation

Exact solutions for suitable special initial and boundary conditions applied to the given partial differential equation and its finite difference analogue are often difficult to find. In such cases, an alternative procedure is to examine the behaviour of the discretization error e_j^n at the (j,n) grid-point and show that $|e_j^n|$ is bounded by some number κ which tends to zero as $\Delta x \to 0$, $\Delta t \to 0$.

Again, considering the diffusion equation (2.1.1) with exact solution $\bar{\tau}_j^n$ at the (j,n) grid-point, and the explicit FTCS finite difference approximation (2.4.3) which has exact solution τ_j^n at the same point, Equation (3.1.1) gives

$$\tau_j^n = \bar{\tau}_j^n - e_j^n. \tag{3.1.9}$$

Substitution of (3.1.9) into Equation (2.4.4) yields

$$e_j^{n+1} = s\,e_{j-1}^n + (1-2s)e_j^n + s\,e_{j+1}^n + \bar{\tau}_j^{n+1} - \bar{\tau}_j^n - s\{\bar{\tau}_{j+1}^n + \bar{\tau}_{j-1}^n - 2\bar{\tau}_j^n\} \tag{3.1.10}$$

Application of the truncated Taylor series expansion about the (j,n) grid-point, with a remainder term expressed in Lagrange's form (see Thomas, 1968,p.639), gives

$$\bar{\tau}_{j+1}^n = \bar{\tau}_j^n + \Delta x \left.\frac{\partial\bar{\tau}}{\partial x}\right|_j^n + \frac{(\Delta x)^2}{2!}\left.\frac{\partial^2\bar{\tau}}{\partial x^2}\right|_{j+\theta_1}^n,$$

$$\bar{\tau}_{j-1}^n = \bar{\tau}_j^n - \Delta x \left.\frac{\partial\bar{\tau}}{\partial x}\right|_j^n + \frac{(\Delta x)^2}{2!}\left.\frac{\partial^2\bar{\tau}}{\partial x^2}\right|_{j-\theta_2}^n,$$

$$\bar{\tau}_j^{n+1} = \bar{\tau}_j^n + \Delta t \left.\frac{\partial\bar{\tau}}{\partial t}\right|_j^{n+\theta_3},$$

where $0 < \theta_1 < 1$, $0 < \theta_2 < 1$, $0 < \theta_3 < 1$.

Substitution of these expansions into Equation (3.1.10) gives

$$e_j^{n+1} = s\,e_{j-1}^n + (1-2s)e_j^n + s\,e_{j+1}^n$$
$$+ \Delta t \left\{ \frac{\partial \bar{\tau}}{\partial t}\Big|_j^{n+\theta_3} - \frac{s(\Delta x)^2}{2}\left\{\frac{\partial^2 \bar{\tau}}{\partial x^2}\Big|_{j+\theta_1}^n + \frac{\partial^2 \bar{\tau}}{\partial x^2}\Big|_{j-\theta_2}^n\right\}\right. . \qquad (3.1.11)$$

If $s \le \frac{1}{2}$, so that $1-2s \ge 0$, and noting that $s = \alpha\Delta t/(\Delta x)^2 > 0$ since $\Delta t > 0$ and $\alpha > 0$ in any practical problem, Equation (3.1.11) may be written

$$|e_j^{n+1}| \le s|e_{j-1}^n| + (1-2s)|e_j^n| + s|e_{j+1}^n|$$
$$+ \Delta t \left| \frac{\partial \bar{\tau}}{\partial t}\Big|_j^{n+\theta_3} - \frac{\alpha}{2}\left\{\frac{\partial^2 \bar{\tau}}{\partial x^2}\Big|_{j+\theta_1}^n + \frac{\partial^2 \bar{\tau}}{\partial x^2}\Big|_{j-\theta_2}^n\right\}\right| \qquad (3.1.12)$$

Let

$$M = \underset{\substack{j=1(1)J-1\\ n=0(1)N-1}}{\text{Maximum}} \left| \frac{\partial \bar{\tau}}{\partial t}\Big|_j^{n+\theta_3} - \frac{\alpha}{2}\left\{\frac{\partial^2 \bar{\tau}}{\partial x^2}\Big|_{j+\theta_1}^n + \frac{\partial^2 \bar{\tau}}{\partial x^2}\Big|_{j-\theta_2}^n\right\}\right|,$$

where $t_N = N\Delta t$ is the last time step to be considered, and let

$$e_{max}^n = \underset{j=1(1)J-1}{\text{Maximum}} |e_j^n|.$$

That is, e_{max}^n is the maximum absolute value of the discretization error at interior grid-points at the nth time level. Then Equation (3.1.12) becomes

$$|e_j^{n+1}| \le s\,e_{max}^n + (1-2s)e_{max}^n + s\,e_{max}^n + \Delta t\, M,$$

or

$$|e_j^{n+1}| \le e_{max}^n + \Delta t\, M.$$

Therefore, as this relation applies for all j,

$$e_{max}^{n+1} \le e_{max}^n + \Delta t\, M. \qquad (3.1.13)$$

Repeated application of (3.1.13) with $n = 0(1)N-1$ yields

$$e_{max}^N \le e_{max}^0 + N\,\Delta t\, M. \qquad (3.1.14)$$

But there is no initial discretization error because the initial values are known exactly. Therefore $e_{max}^0 = 0$, giving

$$e_{max}^N \le M\, t_N. \qquad (3.1.15)$$

Now as $\Delta x \to 0$, $\Delta t = s(\Delta x)^2/\alpha \to 0$ for fixed s, and

$$M \to \underset{\substack{j=1(1)J-1\\ n=0(1)N-1}}{\text{Maximum}} \left| \frac{\partial \bar{\tau}}{\partial t}\Big|_j^n - \alpha \frac{\partial^2 \bar{\tau}}{\partial x^2}\Big|_j^n \right|$$
$$\to 0,$$

because $\bar{\tau}_j^n$ is the exact solution to the diffusion equation (2.1.1). It there fore follows that, as $\Delta x \to 0$ for fixed values of s in the range $0 < s \le \frac{1}{2}$,

$$|e_j^N| \to 0, \tag{3.1.16}$$

for fixed N. Therefore the solution to the finite difference approximation (2.4.4) converges to the exact solution of the given diffusion equation (2.1.1) as the grid becomes finer, provided $s \le \frac{1}{2}$.

This method, unfortunately, is very difficult to apply when the given partial differential equation is even slightly more complicated than the diffusion equation (2.1.1). In such cases the only indication of convergence may be the accuracy of the finite difference solutions of simplified problems, similar to the one in question, for which exact solutions are known, or by the application of Lax's equivalence theorem if consistency and stability can be determined (see Sections 3.2 to 3.4).

3.2 Consistency

A finite difference equation is said to be *consistent* with a partial differential equation if in the limit as the grid-spacings tend to zero, the difference equation becomes the same as the partial differential equation at each point in the solution domain.

Consider the FTCS finite difference approximation to the one-dimensional diffusion equation. The substitution of $\bar{\tau}_j^n$, the exact solution of the diffusion equation at the (j,n) grid-point, for the approximation τ_j^n in the FTCS formula gives

$$\bar{\tau}_j^{n+1} = s\,\bar{\tau}_{j-1}^n + (1-2s)\bar{\tau}_j^n + s\,\bar{\tau}_{j+1}^n . \tag{3.2.1}$$

We wish to determine how closely Equation (3.2.1) corresponds to the diffusion equation (2.4.1) at the (j,n) grid-point, that is, how close it is to

$$\left.\frac{\partial\bar{\tau}}{\partial t}\right|_j^n = \alpha \left.\frac{\partial^2\bar{\tau}}{\partial x^2}\right|_j^n .$$

Substitution of the Taylor series expansion about the (j,n) grid-point for each term of Equation (3.2.1) gives

$$\begin{aligned}
&\bar{\tau}_j^n + \Delta t \left.\frac{\partial\bar{\tau}}{\partial t}\right|_j^n + \frac{(\Delta t)^2}{2!}\left.\frac{\partial^2\bar{\tau}}{\partial t^2}\right|_j^n + \frac{(\Delta t)^3}{3!}\left.\frac{\partial^3\bar{\tau}}{\partial t^3}\right|_j^n + \ldots \\
&\quad = s\{\bar{\tau}_j^n + \Delta x\left.\frac{\partial\bar{\tau}}{\partial x}\right|_j^n + \frac{(\Delta x)^2}{2!}\left.\frac{\partial^2\bar{\tau}}{\partial x^2}\right|_j^n + \frac{(\Delta x)^3}{3!}\left.\frac{\partial^3\bar{\tau}}{\partial x^3}\right|_j^n + \frac{(\Delta x)^4}{4!}\left.\frac{\partial^4\bar{\tau}}{\partial x^4}\right|_j^n \\
&\qquad + \frac{(\Delta x)^5}{5!}\left.\frac{\partial^5\bar{\tau}}{\partial x^5}\right|_j^n + \frac{(\Delta x)^6}{6!}\left.\frac{\partial^6\bar{\tau}}{\partial x^6}\right|_j^n + \ldots\} \\
&\quad + s\{\bar{\tau}_j^n - \Delta x\left.\frac{\partial\bar{\tau}}{\partial x}\right|_j^n + \frac{(\Delta x)^2}{2!}\left.\frac{\partial^2\bar{\tau}}{\partial x^2}\right|_j^n - \frac{(\Delta x)^3}{3!}\left.\frac{\partial^3\bar{\tau}}{\partial x^3}\right|_j^n + \frac{(\Delta x)^4}{4!}\left.\frac{\partial^4\bar{\tau}}{\partial x^4}\right|_j^n \\
&\qquad - \frac{(\Delta x)^5}{5!}\left.\frac{\partial^5\bar{\tau}}{\partial x^5}\right|_j^n + \frac{(\Delta x)^6}{6!}\left.\frac{\partial^6\bar{\tau}}{\partial x^6}\right|_j^n - \ldots\} \\
&\quad + (1-2s)\,\bar{\tau}_j^n .
\end{aligned} \tag{3.2.2}$$

Upon simplification, using $s = \alpha\Delta t/(\Delta x)^2$, Equation (3.2.2) becomes

$$\left.\frac{\partial\bar{\tau}}{\partial t}\right|_j^n = \alpha \left.\frac{\partial^2\bar{\tau}}{\partial x^2}\right|_j^n + E_j^n , \tag{3.2.3}$$

where

$$E_j^n = -\frac{\Delta t}{2}\left.\frac{\partial^2\bar{\tau}}{\partial t^2}\right|_j^n + \frac{\alpha(\Delta x)^2}{12}\left.\frac{\partial^4\bar{\tau}}{\partial x^4}\right|_j^n + O\{(\Delta t)^2,(\Delta x)^4\}. \tag{3.2.4}$$

The exact solution of the FTCS finite difference approximation to the diffusion equation (2.1.1) is therefore the solution of the partial differential equation (3.2.3), which differs from the diffusion equation by the inclusion of the extra terms contained in E_j^n.

This error E_j^n is called the *truncation error:* it is the discrepancy between the approximating finite difference equation with the exact values $\bar{\tau}_j^n$ substituted for τ_j^n, and the partial differential equation which the difference equation replaces at the (j,n) grid-point. Clearly, as the grid spacing gets smaller and smaller with the FTCS method, the truncation error gets smaller and smaller at a fixed point (x,t) in the solution domain. In the limit as $\Delta x \to 0$, $\Delta t \to 0$, the finite difference equation (3.2.1) is equivalent to the partial differential equation (2.4.1). This property is called *consistency* (or compatability).

For the FTCS finite difference approximation to the diffusion equation, the truncation error (3.2.4) is seen to be generally $O\{\Delta t,(\Delta x)^2\}$. However, since $\bar{\tau}$ satisfies the diffusion equation (2.1.1), it also satisfies the equation

$$\frac{\partial}{\partial t}\left(\frac{\partial\bar{\tau}}{\partial t}\right) = \frac{\partial}{\partial t}\left(\alpha\,\frac{\partial^2\bar{\tau}}{\partial x^2}\right) = \alpha\,\frac{\partial^2}{\partial x^2}\left(\frac{\partial\bar{\tau}}{\partial t}\right) = \alpha\,\frac{\partial^2}{\partial x^2}\left(\alpha\,\frac{\partial^2\bar{\tau}}{\partial x^2}\right)$$

or

$$\frac{\partial^2\bar{\tau}}{\partial t^2} = \alpha^2\,\frac{\partial^4\bar{\tau}}{\partial x^4}. \tag{3.2.5}$$

Therefore the truncation error (3.2.4) may be written

$$E_j^n = -\frac{\alpha(\Delta x)^2}{2}\left(s - \frac{1}{6}\right)\left.\frac{\partial^4\bar{\tau}}{\partial x^4}\right|_j^n + O\{(\Delta t)^2,(\Delta x)^4\}. \tag{3.2.6}$$

If $s = 1/6$, the first term in the above expression vanishes and the truncation error is now $O\{(\Delta t)^2,(\Delta x)^4\}$, or, what is the same thing for fixed s, $O\{(\Delta x)^4\}$. In this case, the truncation error goes to zero faster than for any other value of s when the grid spacing is made smaller and smaller. Therefore, the solution of the FTCS finite difference equation approaches the solution of the diffusion equation more rapidly when $s = 1/6$ than for other values of $s \le \frac{1}{2}$, as the grid-spacing is reduced.

Clearly, consistency is necessary if the finite difference solution is to converge to the solution of the partial differential equation being approximated. However, this is not a sufficient condition, for even though the finite difference equation might become equivalent to a certain partial differential equation as the grid spacing tends to zero, it does not follow that the *solution* of the finite difference equation approaches the *solution* of the partial differential equation. For instance, if $\alpha = 10^{-2}$, choosing $\Delta x = 0.1$ and $\Delta t = 1$ so $s = 1$, the FTCS explicit finite difference equation simplifies to (2.4.7) which can be solved as in Table 2.2. Continuing to reduce Δx and Δt, so that s remains unchanged at the value 1, the exact calculation of the finite difference solution at specific points in the solution domain rapidly diverge, as shown in Table 3.1, in contrast with solutions using $s=\frac{1}{2}$, $s = 1/6$.

TABLE 3.1

Finite difference approximations to $\bar{\tau}(0.4,8) = 0.4503963$

J	Δx	$\Delta t/s$	s = 1	s = 1/2	s = 1/6
5	0.2	4	$\tau_2^2 = 1.0000$	$\tau_2^4 = 0.5000000$	$\tau_2^{12} = 0.4684500$
10	0.1	1	$\tau_4^8 = 122.00$	$\tau_4^{16} = 0.4754791$	$\tau_4^{48} = 0.4549047$
15	$0.06\dot{6}$	$0.44\dot{4}$	$\tau_6^{18} = 197\times10^6$	$\tau_6^{36} = 0.4561530$	$\tau_6^{108} = 0.4523997$
20	0.05	0.25	∞	$\tau_8^{64} = 0.4566436$	$\tau_8^{192} = 0.4515231$
25	0.04	0.16	∞	$\tau_{10}^{100} = 0.4524738$	$\tau_{10}^{300} = 0.4511174$
40	0.025	0.0625	∞	$\tau_{16}^{256} = 0.4519564$	$\tau_{16}^{768} = 0.4506780$
50	0.02	0.04	∞	$\tau_{20}^{400} = 0.4513947$	$\tau_{20}^{1200} = 0.4505766$
100	0.01	0.01	∞	$\tau_{40}^{1600} = 0.4506459$	$\tau_{40}^{4800} = 0.4504414$

Note that the functional dependence of the truncation error E_j^n on Δt and Δx not only indicates how closely the solution of the differential equation satisfies the approximating finite difference equation (that is, it shows *consistency* if it exists), it also indicates how rapidly the solution of the finite difference equation approaches the solution of the given differential equation as $\Delta x \to 0$, $\Delta t \to 0$ if the solution is convergent. For instance, with $s = 1/6$ the truncation error for the FTCS scheme is $O\{(\Delta t)^2,(\Delta x)^4\}$, whereas with $s = 1/2$ it is $O\{\Delta t,(\Delta x)^2\}$. Reversing the procedures involved in deriving (3.2.3) from Equation (3.2.1), we find that

$$\bar{\tau}_j^{n+1} = s\,\bar{\tau}_{j-1}^n + (1-2s)\,\bar{\tau}_j^n + s\,\bar{\tau}_{j+1}^n + \Delta t\,E_j^n, \tag{3.2.7}$$

which is equivalent to (3.1.2). Clearly, the contribution to the discretization error in going from one time level to the next is $\Delta t E_j^n$, so that

$$e_j^n = O\{n\Delta t\,E_j^n\}. \tag{3.2.8}$$

The contribution to the discretisation error at each time step is therefore $O\{(\Delta t)^2,\Delta t(\Delta x)^2\}$ in general, but is $O\{(\Delta t)^3,\Delta t(\Delta x)^4\}$ when $s = 1/6$. Therefore, there is a much greater rate of convergence of the FTCS finite difference equation to the solution of the diffusion equation, if $s = 1/6$, than for other values of $0 < s \le \frac{1}{2}$. This greater rate of convergence is seen in Table 3.1. However, note that with $s = 1$, even though $E_j^n = O\{\Delta t,(\Delta x)^2\}$ and the finite difference equation *is* consistent with the diffusion equation, the solution of the difference equation does *not* converge to the solution of the diffusion equation.

That the discretisation error behaves according to (3.2.8) is seen in the following. The approximate value of $\bar{\tau}(0.4,8)$ obtained by using the FTCS method to solve the diffusion equation (2.1.1) with initial condition

$$\bar{\tau}(x,0) = \sin(\pi x), \quad 0 \le x \le 1, \tag{3.2.9a}$$

and boundary conditions

$$\bar{\tau}(0,t) = \bar{\tau}(1,t) = 0, \quad t \ge 0, \tag{3.2.9b}$$

has been computed to 14 significant figures, for various values of Δx and $s \le \frac{1}{2}$. The exact solution to this problem is

$$\bar{\tau}(x,t) = \sin(\pi x)\exp(-\pi^2 \alpha t). \tag{3.2.10}$$

Table 3.2 lists the discretisation errors given by (3.1.1), namely

$$e_j^n = \bar{\tau}_j^n - \tau_j^n, \tag{3.2.11}$$

for values of $s = \frac{1}{2}$, 1/3, 1/6, 1/10.

TABLE 3.2

Discretisation errors e in the calculation of $\bar{\tau}(0.4,8) = 0.4318184$ using the FTCS method to solve the diffusion equation (2.1.1) with initial and boundary conditions (3.2.9).

Δx	$s = \frac{1}{2}$	$s = 1/3$	$s = 1/6$	$s = 1/10$
1/20	1.41×10^{-3}	7.03×10^{-4}	3.85×10^{-7}	-2.81×10^{-4}
1/30	6.25×10^{-4}	3.12×10^{-4}	7.60×10^{-8}	-1.25×10^{-4}
1/40	3.51×10^{-4}	1.75×10^{-4}	2.40×10^{-8}	-7.01×10^{-5}
1/50	2.25×10^{-4}	1.12×10^{-4}	9.84×10^{-9}	-4.49×10^{-5}
1/60	1.56×10^{-4}	7.79×10^{-5}	4.75×10^{-9}	-3.12×10^{-5}
1/70	1.15×10^{-4}	5.72×10^{-5}	2.56×10^{-9}	-2.29×10^{-5}
1/80	8.77×10^{-5}	4.38×10^{-5}	1.50×10^{-9}	-1.75×10^{-5}
1/90	6.93×10^{-5}	3.46×10^{-5}	9.38×10^{-10}	-1.38×10^{-5}
1/100	5.61×10^{-5}	2.80×10^{-5}	6.15×10^{-10}	-1.12×10^{-5}

The values of $\log_{10}|e|$ have been plotted against $\log_{10}(\Delta x)$, for each value of s, in Figure 3.2. The equation of the lines of best fit for each s are:

$$\left.\begin{aligned} s &= 1/2, \; e = 0.57(\Delta x)^{2.00}, \\ s &= 1/3, \; e = 0.282(\Delta x)^{2.00}, \\ s &= 1/6, \; e = 0.062(\Delta x)^{4.00}, \\ s &= 1/10, e = -0.112(\Delta x)^{2.00}. \end{aligned}\right\} \tag{3.2.12}$$

The results for s = 1/6 reflect the higher accuracy of the FTCS method for that value, and the higher rate of convergence to the exact solution as $\Delta x \to 0$.

From equations (3.2.8) and (3.2.6) we see that

$$e_j^n = \begin{cases} O\{n(\Delta t)^2,\ n\Delta t(\Delta x)^2\}, & s \neq 1/6, \\ O\{n(\Delta t)^3,\ n\Delta t(\Delta x)^4\}, & s = 1/6. \end{cases} \qquad (3.2.13)$$

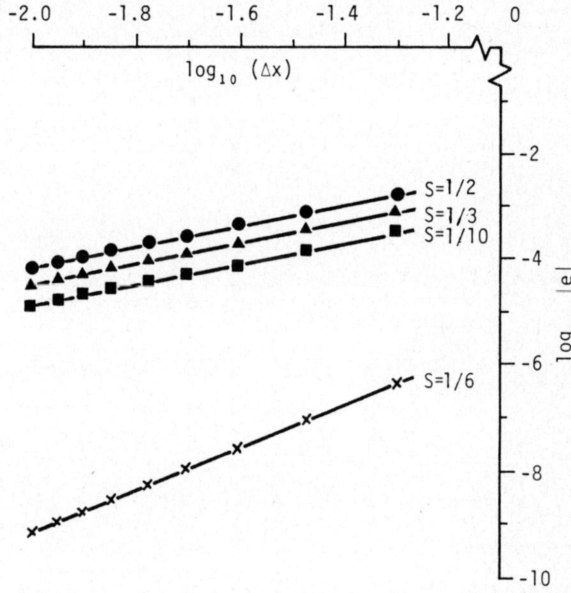

Figure 3.2 : The discretisation error $|e|$ plotted against grid-spacing Δx for the FTCS method of solving the diffusion equation

However, in order to reach the fixed time t = 8, the value of n, the number of time-steps taken, will be inversely proportional to Δt, that is $n \alpha 1/\Delta t$. Therefore, in Table 3.2

$$e = \begin{cases} O\{\Delta t,(\Delta x)^2\}, & s \neq 1/6, \\ O\{(\Delta t)^2,(\Delta x)^4\}, & s = 1/6, \end{cases} \qquad (3.2.14)$$

which is the same as the order of the truncation error E. Furthermore, if s is *fixed*, varying Δx changes Δt in a manner proportional to $(\Delta x)^2$, so we may write

$$e = \begin{cases} O\{(\Delta x)^2\}, & s \neq 1/6, \\ O\{(\Delta x)^4\}, & s = 1/6. \end{cases} \qquad (3.2.15)$$

This is consistent with the results of (3.2.12).

However, when assessing the economy of any particular numerical scheme, it is necessary to take into account the computational effort involved with each application of the formula, as well as the total number of applications. For example, even though the truncation and hence the discretization error is minimised by choosing $s = 1/6$, as evident in Table 3.1, the reduced number of applications of the FTCS formula to achieve a given accuracy may be more than offset by an increased computational effort at each application.

Consider the results obtained in Table 3.1 using the CYBER 173 computer, for which multiplications and divisions take very nearly 3 micro-seconds while additions and subtractions take approximately ½ micro-second. With $s = 1/6$, the result τ_6^{108} is slightly more accurate than τ_{10}^{100} obtained with $s = \frac{1}{2}$. In the former case 14 applications of the FTCS formula (3.2.1) must be made at each of 108 time levels, a total of 1,512 applications. As each application of (3.2.1) requires 7 micro-seconds if the calculations are carried out in the quickest way, which is to write it as

$$\tau_j^{n+1} = \frac{1}{6}(\tau_{j-1}^n + 4\tau_j^n + \tau_{j+1}^n), \tag{3.2.16}$$

then the computer time used in obtaining this result is 10,584 micro-seconds. In the latter case, with $s = \frac{1}{2}$, 24 applications of (2.4.6), each taking 3½ micro-seconds must be made at each of 100 time-levels. Thus many more applications of the formula are required, namely 2,400 but the required computer time is actually less than the previous case, namely 8,400 micro-seconds.

The economy in choosing the FTCS scheme with $s = 1/6$ is really apparent in the evaluation of τ_{20}^{1200}, which required 58,800 applications of the formula in a time of 411,600 micro-seconds. This is less than the time taken to obtain the less accurate value τ_{40}^{1600} with $s = \frac{1}{2}$, which required 158,400 applications of the FTCS formula in a time of 554,400 micro-seconds

3.3 Stability

A third important feature of a finite difference method of solving a partial differential equation is the stability of the associated finite difference equation which must be solved. The stability of such an equation is concerned with the growth, or decay, of errors produced in the finite difference solution by errors introduced in previously calculated values. In this context, the errors referred to are not those caused by incorrect logic but those which occur because the computer produces truncated or rounded-off results.

If it were possible to carry out all numerical operations to an infinite number of decimal places, the exact solution to the finite difference equation would be found. In practice, however, each calculation made by the computer is carried out to a finite number of significant figures which introduces a "round-off" or truncation error at every step of a computation. Hence the finite difference solution found to Equation (2.4.4) is not τ_j^{n+1}, but $^*\tau_j^{n+1}$ (say). $^*\tau_j^n$ is called the *numerical solution* of the finite difference equation in contrast with its exact solution τ_j^n.

A set of finite difference equations is said to be stable if the cumulative effect of all the rounding off errors is negligible. More specifically, consider the errors

$$\xi_j^n = \tau_j^n - {}^*\tau_j^n \tag{3.3.1}$$

introduced at the (j,n) grid-points for $j = 1(1)J-1$ and for time-levels up to $N\Delta t$, all of which have absolute value $|\xi_j^n|$ less than some positive number δ (say). Then the finite difference equations are said to be *stable* if the maximum value of $|\xi_j^{n+1}|$ is bounded or tends to zero as $\delta \to 0$. That is, the difference between the numerical solution $*\tau_j^{n+1}$ and the exact solution τ_j^{n+1} does not increase exponentially as the numbers of rows of calculations at successive time levels in the solution domain is increased. If the errors do not increase exponentially as n increases, but persist as linear combinations of the initial errors, they are usually numerically tolerable provided their sum remains much smaller than the exact finite difference solution τ_j^n.

It is usually not possible to determine the exact value of the numerical error ($\tau_j^n - *\tau_j^n$) at the (j,n) grid-point for an arbitrary distribution of errors at other grid-points. However, it can be estimated using certain standard methods, some of which will be discussed in this section. Note that the numerical solutions are invariably more accurate than these estimates indicate, because stability analyses always assume the worst possible combination of individual errors. For instance, it may be assumed that all errors have a distribution of signs so their total effect is additive, which is not always the case.

Note that stability does not involve the difference between the exact solutions of the finite difference equation and the given partial differential equation. This difference is the discretization error, which is the province of convergence. Stability concerns only the calculation of the solution of the finite difference equation. The phenomenon shown in Table 2.2 is not due to instability, because the calculations were all performed exactly - no errors were introduced. Table 2.2 is an example of a non-convergent solution.

Each of the methods of stability analysis to be described give further insight into how a solution of a finite difference equation behaves when round-off errors effect the calculations. These techniques are:

(1) Discrete perturbation stability analysis;
(2) The matrix method;
(3) von Neumann's stability analysis.

These methods will be illustrated by applications to the FTCS finite difference formula (2.4.4) which approximates the one-dimensional diffusion equation.

Note that in the calculation of τ_j^{n+1} using the FTCS formula we actually compute the numerical approximation

$$*\tau_j^{n+1} = \tau_j^{n+1} - \xi_j^{n+1} \tag{3.3.2a}$$

using numerical approximations from the previous time level $t = n\Delta t$, namely

$$*\tau_{j-1}^{n} = \tau_{j-1}^{n} - \xi_{j-1}^{n}, \tag{3.3.2b}$$

$$*\tau_{j}^{n} = \tau_{j}^{n} - \xi_{j}^{n}, \tag{3.3.2c}$$

$$*\tau_{j+1}^{n} = \tau_{j+1}^{n} - \xi_{j+1}^{n}. \tag{3.3.2d}$$

It can be shown that, for linear finite difference equations, the corresponding error terms satisfy the same *homogeneous* finite difference equation as the values of τ. For instance, using Equation

(2.4.4) means that we are actually calculating $*\tau_j^{n+1}$ using $*\tau_{j-1}^n$, $*\tau_j^n$ and $*\tau_{j+1}^n$, so that

$$*\tau_j^{n+1} = s(*\tau_{j-1}^n) + (1-2s)(*\tau_j^n) + s(*\tau_{j+1}^n). \tag{3.3.3}$$

Substitution of Equations (3.3.2) into (3.3.3), followed by application of Equation (2.4.4) which applies since the exact numerical solutions τ satisfy the FTCS finite difference equation, yields the homogeneous difference equation

$$\xi_j^{n+1} = s\,\xi_{j-1}^n + (1-2s)\xi_j^n + s\,\xi_{j+1}^n. \tag{3.3.4}$$

The initial errors, ξ_j^0, j=0(1)J, and the boundary errors, ξ_0^n and ξ_J^n, n=1,2,3,... for this equation, assuming given boundary and initial values of the form (2.1.2), will all be zero and, unless some error is introduced in calculating the finite difference solution τ_j^n at some interior grid-point, the resulting errors in the solution will remain zero.

(1) Discrete Perturbation Stability Analysis

In this method a discrete perturbation, or error, ε is introduced into the finite difference scheme in the determination of τ_j^n at an arbitrary (j,n) grid-point and its effect on the computation of subsequent values of the solution of the finite difference equation is examined. Stability is indicated if the perturbation eventually dies out.

For a given value of s, application of (3.3.4) with zero errors up to some time level (n-1) and zero errors at grid-points everywhere along the nth time-level except one, where an error ε is introduced, will show whether the introduced perturbation increases without bound and the scheme is unstable. For example, Table 3.3 shows that the introduction of an error $\xi_4^1 = \varepsilon$ in the value of τ_4^1 in the FTCS finite difference scheme considered in Section 2.4, with $s = \frac{1}{2}$, produces errors which decrease in magnitude at successive time-levels.

TABLE 3.3
Error propagation due to introduction of error $\xi_4^1 = \varepsilon$ with $s = \frac{1}{2}$

n	nΔt	t										Errors in Boundary Values
7	3.5	0										
6	3.0	0	$\frac{1}{8}\varepsilon$	0	$\frac{7}{16}\varepsilon$	0	$\frac{7}{16}\varepsilon$	0	$\frac{5}{32}\varepsilon$	0	$\frac{1}{32}\varepsilon$	0
5	2.5	0	0	$\frac{1}{4}\varepsilon$	0	$\frac{3}{8}\varepsilon$	0	$\frac{1}{4}\varepsilon$	0	$\frac{1}{16}\varepsilon$	0	0
4	2.0	0	$\frac{1}{8}\varepsilon$	0	$\frac{3}{8}\varepsilon$	0	$\frac{3}{8}\varepsilon$	0	$\frac{1}{8}\varepsilon$	0	0	0
3	1.5	0	0	$\frac{1}{4}\varepsilon$	0	$\frac{1}{2}\varepsilon$	0	$\frac{1}{4}\varepsilon$	0	0	0	0
2	1.0	0	0	0	$\frac{1}{2}\varepsilon$	0	$\frac{1}{2}\varepsilon$	0	0	0	0	0
1	0.5	0	0	0	0	ε	0	0	0	0	0	0
0	0.0	0	0	0	0	0	0	0	0	0	0	0 → x
	jΔx	0.0	0.1	0.2	0.3	0.4	0.5	0.6	0.7	0.8	0.9	1.0
	j	0	1	2	3	4	5	6	7	8	9	10

Errors in Initial Values

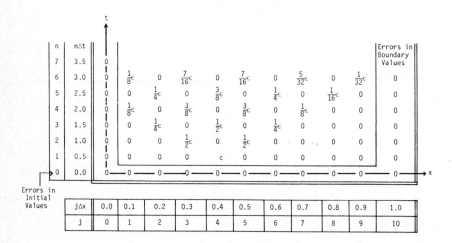

However, with s = 1, the introduction of the error $\xi_4^1 = \varepsilon$ into the FTCS finite difference equation produces a perturbation which rapidly increases in numerical value as in Table 3.4. Clearly, the finite difference equation (2.4.4) is unstable with s = 1.

TABLE 3.4

Error propagation due to introduction of error $\xi_4^1 = \varepsilon$ with s = 1

n	$n\Delta t$											Errors in Boundary Values
7	7	0										0
6	6	0	14ε	-30ε	45ε	-51ε	45ε	-30ε	15ε	-5ε	ε	0
5	5	0	-4ε	10ε	-16ε	19ε	-16ε	10ε	-4ε	ε	0	0
4	4	0	ε	-3ε	6ε	-7ε	6ε	-3ε	ε	0	0	0
3	3	0	0	ε	-2ε	3ε	-2ε	ε	0	0	0	0
2	2	0	0	0	ε	$-\varepsilon$	ε	0	0	0	0	0
1	1	0	0	0	0	ε	0	0	0	0	0	0
0 (Errors in Initial Values)	0	0	0	0	0	0	0	0	0	0	0	0
	$j\Delta x$	0.0	0.1	0.2	0.3	0.4	0.5	0.6	0.7	0.8	0.9	1.0
	j	0	1	2	3	4	5	6	7	8	9	10

In general, if we wish to determine the values of s for which the finite difference equation (2.4.4) is stable, we must sweep through the grid as in Tables 3.3 and 3.4 using the formula for error propagation, namely Equation (3.3.4), with arbitrary s. The only permissible values of s are those for which the resulting errors produced at higher time levels have moduli less than or equal to $|\varepsilon|$. With the only error at the nth time level being $\xi_j^n = \varepsilon$, j=1(1)J-1, then the resulting errors at the (n+1)th time level are $\xi_{j-1}^{n+1} = s\varepsilon$, $\xi_j^{n+1} = (1-2s)\varepsilon$ and $\xi_{j+1}^{n+1} = s\varepsilon$. These errors are not greater in magnitude than ε if

$$|s| \leq 1 \text{ and } |1-2s| \leq 1,$$

both of which are satisfied if

$$0 \leq s \leq 1. \tag{3.3.5}$$

Since s = 0 is impractical, as it implies $\Delta t = 0$, then the requirement for no increase in error from the nth to the (n+1)th time-level is

$$0 < s \leq 1. \tag{3.3.6}$$

The errors at the (n+2)th time level produced by the perturbation $\xi_j^n = \varepsilon$ are

$$\xi_{j-2}^{n+2} = s^2\varepsilon, \quad \xi_{j-1}^{n+2} = (2s-4s^2)\varepsilon, \quad \varepsilon_j^{n+2} = (1-4s+6s^2)\varepsilon,$$

$$\varepsilon_{j+1}^{n+2} = (2s-4s^2)\varepsilon, \quad \xi_{j+2}^{n+2} = s^2\varepsilon,$$

where the superscripts on s indicate powers.

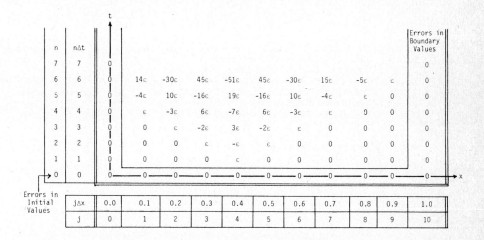

These errors are equal to or less than ε in magnitude if

$$|s^2| \le 1, \ |2s-4s^2| \le 1 \text{ and } |1-4s+6s^2| \le 1;$$

that is, if

$$0 < s \le 2/3. \tag{3.3.7}$$

This process may be continued to higher time levels, but the algebra involved soon becomes prohibitively complicated. Consideration of the errors at time level (n+1) lead to the requirement $0 < s \le 1$ for errors not to increase in magnitude; consideration of the errors at time level (n+2) leads to the requirement $0 < s \le 2/3$ for errors not to increase in magnitude. The asymptotic limit of this restriction is required as higher time levels are considered.

The initial set of round-off errors at the first time level will propagate through the grid as described by Equation (3.3.4). To these errors will be added new errors produced by round-off in further calculations. It can be expected after many time steps that errors at any time level will be approximately the same in magnitude, either having the same sign or an alternating sign distribution along that time level.

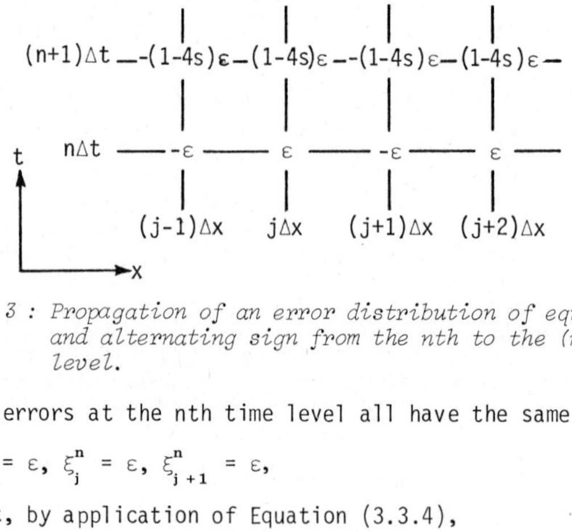

Figure 3.3 : Propagation of an error distribution of equal magnitude and alternating sign from the nth to the (n+1)th time-level.

If the errors at the nth time level all have the same sign, then

$$\xi^n_{j-1} = \varepsilon, \ \xi^n_j = \varepsilon, \ \xi^n_{j+1} = \varepsilon,$$

so that, by application of Equation (3.3.4),

$$\xi^{n+1}_j = \varepsilon,$$

which yields no restriction on s for stability.

If the distribution of errors at the nth time level has alternate positive and negative signs, such as

$$\xi^n_{j-1} = -\varepsilon, \ \xi^n_j = \varepsilon, \ \xi^n_{j+1} = -\varepsilon,$$

application of (3.3.3) yields $\varepsilon^{n+1}_j = (1-4s)\varepsilon$. Stability therefore requires that

$$|1-4s| \le 1$$

which yields the condition

$$0 < s \le \tfrac{1}{2}. \qquad (3.3.8)$$

Examination of the error propagation at succeeding time levels does not change this restriction on s. The FTCS finite difference scheme is therefore both convergent and stable if (3.3.8) is satisfied.

For a given α and a fixed space interval Δx, condition (3.3.8) places a limitation on the time step Δt, namely

$$\Delta t \le (\Delta x)^2/2\alpha. \qquad (3.3.9)$$

The inequality (3.3.9) has important implications about the amount of computer time required to solve the FTCS finite difference approximation to the diffusion equation. Suppose the FTCS equation has been solved up to a certain time T with a mesh size Δx_1, using the maximum possible time step of $\Delta t_1 = (\Delta x_1)^2/2\alpha$. If the computation is repeated with a halved grid-spacing, say $\Delta x_2 = \Delta x_1/2$, in order to improve accuracy for example, then

$$\Delta t_2 = (\Delta x_2)^2/2\alpha = (\Delta x_1)^2/8\alpha = \Delta t_1/4.$$

To find solutions up to the same time T, the second case requires four times as many time levels in the solution domain. Furthermore, at each time level the calculations take twice as long to perform, since there are twice as many interior grid-points at one time level. The required computer time is therefore increased eight-fold. Obviously, finite difference methods of solving the diffusion equation which do not have a stability restriction of the type (3.3.8) are highly desirable.

(2) The matrix method

A more rigorous treatment of the manner in which errors propagate through the solution domain is to express the governing equation (3.3.4) in matrix form and examine the eigenvalues of the associated matrix.

Substituting j = 1(1)J-1 into (3.3.4) and noting that errors at the boundary are zero, so that $\xi_0^n = 0$ and $\xi_J^n = 0$ for all n, gives

$$\left.\begin{aligned}
\xi_1^{n+1} &= (1-2s)\xi_1^n + s\,\xi_2^n \\
\xi_2^{n+1} &= s\,\xi_1^n + (1-2s)\xi_2^n + s\,\xi_3^n \\
\xi_3^{n+1} &= \qquad s\,\xi_2^n + (1-2s)\xi_3^n + s\,\xi_4^n \\
&\ \ \vdots \\
\xi_{J-1}^{n+1} &= \qquad\qquad s\,\xi_{J-2}^n + (1-2s)\xi_{J-1}^n
\end{aligned}\right\} \qquad (3.3.10)$$

Equation (3.3.10) may be written in matrix form as

$$\underset{\sim}{\xi}^{n+1} = \underline{\underline{A}}\,\underset{\sim}{\xi}^n, \quad n = 1,2,\ldots, \qquad (3.3.11)$$

where $\underline{\underline{A}}$ is a (J-1) square matrix and $\underset{\sim}{\xi}^n$ is a vector of length (J-1), defined by

$$\underline{\underline{A}} = \begin{bmatrix} (1-2s) & s & & & & 0 \\ s & (1-2s) & s & & & \\ & s & (1-2s) & s & & \\ & & \cdot & \cdot & \cdot & \\ & & & \cdot & \cdot & \cdot \\ 0 & & & & s & (1-2s) \end{bmatrix}, \quad \underset{\sim}{\xi}^n = \begin{bmatrix} \xi_1^n \\ \xi_2^n \\ \xi_3^n \\ \cdot \\ \cdot \\ \cdot \\ \cdot \\ \cdot \\ \xi_{J-1}^n \end{bmatrix}$$

Suppose the (J-1) eigenvalues of A are λ_m, m=1(1)J-1 and that they are all different. Then the corresponding (J-1) eigenvectors $\underset{\sim m}{\nu}$ satisfy the relation

$$\underline{\underline{A}}\, \underset{\sim}{\nu}_m = \lambda_m \underset{\sim}{\nu}_m, \tag{3.3.12}$$

by definition of an eigenvector (see Wade, 1951). Also because the λ_m are distinct, the eigenvectors $\underset{\sim}{\nu}_m$ form a linearly independent set, so that the initial error vector $\underset{\sim}{\xi}^1$ can be expressed uniquely in terms of the eigenvectors. That is, we can write

$$\underset{\sim}{\xi}^1 = \sum_{m=1}^{J-1} C_m \underset{\sim}{\nu}_m, \tag{3.3.13}$$

where the C_m are constants which are the solutions of the (J-1) linear algebraic equations obtained by writing (3.3.13) in full, namely

$$\begin{bmatrix} \xi_1^1 \\ \xi_2^1 \\ \cdot \\ \cdot \\ \xi_{J-1}^1 \end{bmatrix} = C_1 \begin{bmatrix} \nu_{1,1} \\ \nu_{1,2} \\ \cdot \\ \cdot \\ \nu_{1,J-1} \end{bmatrix} + C_2 \begin{bmatrix} \nu_{2,1} \\ \nu_{2,2} \\ \cdot \\ \cdot \\ \nu_{2,J-1} \end{bmatrix} + \dots + C_{J-1} \begin{bmatrix} \nu_{J-1,1} \\ \nu_{J-1,2} \\ \cdot \\ \cdot \\ \nu_{J-1,J-1} \end{bmatrix}.$$

Applying Equation (3.3.11) with n=1 gives

$$\underset{\sim}{\xi}^2 = \underline{\underline{A}}\,\underset{\sim}{\xi}^1 = \sum_{m=1}^{J-1} C_m(\underline{\underline{A}}\,\underset{\sim}{\nu}_m) = \sum_{m=1}^{J-1} \lambda_m C_m \underset{\sim}{\nu}_m;$$

then with n=2 it is found that

$$\underset{\sim}{\xi}^3 = \underline{\underline{A}}\,\underset{\sim}{\xi}^2 = \sum_{m=1}^{J-1} \lambda_m C_m(\underline{\underline{A}}\,\underset{\sim}{\nu}_m) = \sum_{m=1}^{J-1} (\lambda_m)^2 C_m \underset{\sim}{\nu}_m;$$

and so on, until

$$\underset{\sim}{\xi}^{n+1} = \sum_{m=1}^{J-1} (\lambda_m)^n C_m \underset{\sim}{\nu}_m. \tag{3.3.14}$$

It follows that the errors are bounded as n increases if the eigenvalues λ_m all have absolute values less than or equal to one; that is, if

$$|\lambda_m| \le 1, \; m=1(1)J-1. \tag{3.3.15}$$

Now, the eigenvalues of the tri-diagonal M-square matrix

$$\underline{\underline{B}} = \begin{bmatrix} b & c & & & & 0 \\ a & b & c & & & \\ & a & b & c & & \\ & & \cdot & \cdot & \cdot & \\ & & & a & b & c \\ 0 & & & & a & b \end{bmatrix}$$

are given by the formula

$$\lambda_m = b + 2\sqrt{ac}\cos\{\frac{m\pi}{M+1}\}, \quad m=1(1)M. \tag{3.3.16}$$

This may be verified by substitution of the given eigenvalues λ_m into the characteristic equation $|\underline{\underline{B}} - \lambda\underline{\underline{I}}| = 0$. Therefore, the eigenvalues of matrix $\underline{\underline{A}}$ are

$$\lambda_m = (1-2s) + 2s\cos(m\pi/J), \quad m=1(1)J-1,$$

which simplifies to

$$\lambda_m = 1 - 4s\sin^2(m\pi/2J). \tag{3.3.17}$$

Application of the stability condition (3.3.15) allows only values of s for which

$$|1 - 4s\sin^2(m\pi/2J)| \le 1,$$

that is

$$-1 \le 1 - 4s\sin^2(m\pi/2J) \le 1. \tag{3.3.18}$$

The right side of this inequality is satisfied for all m and s, while the left side requires

$$s\sin^2(m\pi/2J) \le \tfrac{1}{2},$$

which is true for all m if $s \le \frac{1}{2}$. The FTCS finite difference equation is therefore stable for $s \le \frac{1}{2}$.

It must be pointed out that if the eigenvalues λ_m are not distinct, the corresponding eigenvectors $\underset{\sim}{v}_m$ may not form a linearly independent set, in which case Equation (3.3.13) will not hold for an arbitrary initial error vector $\underset{\sim}{\xi}^0$. If the eigenvectors of $\underline{\underline{A}}$ are not distinct, the matrix method can be used only if a corresponding set of linearly independent eigenvectors $\underset{\sim}{v}_m$ can be found to satisfy (3.3.13).

Alteration of the nature of the boundary conditions at x = 0 and/or x = 1 only slightly modifies this method (see Section 4.4).

(3) von Neumann Stability Analysis

This is the most commonly used method of stability analysis. However, it only provides information about the influence of derivative boundary conditions on the numerical stability of the finite difference scheme if it is applied separately to the algorithms used at the boundaries. This information is automatically obtained if the matrix method is used. The von Neumann (or Fourier series) method was developed in 1944 and it was rapidly revised and improved by von Neumann and his co-workers. The earliest complete description was given by O'Brien et al (1950).

In this method, the errors distributed along grid-points at one time level are expanded as a finite Fourier series. Then, the stability or instability of the finite difference equation is determined by considering whether separate Fourier components of the error distribution decay or amplify on progressing to the next time level. Thus the initial error vector $\underset{\sim}{\xi}^0$ is expressed as a finite complex Fourier series, so that at $x = j\Delta x$ the error is

$$\xi_j^0 = \sum_{m=1}^{J-1} a_m e^{i(m\pi j\Delta x)},$$

where $e^{i(m\pi j\Delta x)} = \exp\{i(m\pi j\Delta x)\}$ and $i = \sqrt{-1}$. That is,

$$\xi_j^0 = \sum_{m=1}^{J-1} a_m e^{i\beta_m j}, \quad j=0(1)J, \tag{3.3.19}$$

where $\beta_m = m\pi\Delta x$. Because the finite difference equation being studied here is linear, then so is Equation (3.3.4) governing the error propagation, and it is sufficient to study the propagation of the error due to just the single term $e^{i\beta j}$ of the Fourier series representation (3.3.19). In the following, the coefficient a_m is a constant, so it has been omitted, and the subscript m has been dropped from β_m.

A solution of the error equation (3.3.4) is therefore sought in the variables separable form

$$\xi_j^n = (G)^n e^{i\beta j}, \tag{3.3.20}$$

where the time dependence of this Fourier component of the error at $x = j\Delta x$ is contained in the coefficient $(G)^n$, which is the nth power of the complex number G.

Substitution of the mth Fourier component of the errors at the nth time level into the error equation (3.3.4) gives the mth Fourier component of the errors at the (n+1)th time level from the equation

$$(G)^{n+1} e^{i\beta j} = s(G)^n e^{i\beta(j-1)} + (1-2s)(G)^n e^{i\beta j} + s(G)^n e^{i\beta(j+1)}.$$

This gives, on dividing throughout by $(G)^n e^{i\beta j}$,

$$G = se^{-i\beta} + (1-2s) + s\, e^{i\beta}. \tag{3.3.21}$$

Clearly, G is the amplification factor for the mth Fourier mode of the error distribution, as it propagates one step forward in time, since

$$\xi_j^{n+1}/\xi_j^n = G. \tag{3.3.22}$$

Note that G is a function of s and β. That is, $G = G(s,\beta)$ depends on the size of the grid-spacings and the particular Fourier mode being considered since $s = \alpha\Delta t/(\Delta x)^2$ and $\beta = m\pi\Delta x$. The errors will remain bounded if the absolute value of G, called the gain, is never greater than unity for all Fourier modes, that is, for all β. The stability requirement is therefore that

$$|G| \le 1 \text{ for all } \beta. \tag{3.3.23}$$

Note that, in general the amplification factor G is a complex number and the amplitude of the mth Fourier component of the error is multiplied by $|G|$ and its phase is increased by arg G as the method steps from one time level to the next.

On use of the relations $e^{i\beta} = \cos\beta + i\sin\beta$ and $\cos\beta = 1-2\sin^2(\beta/2)$, Equation (3.3.21) gives

$$G = 1 - 4s\sin^2(\beta/2). \tag{3.3.24}$$

Clearly, this amplification factor has a different value for each β, that is, for each Fourier component.

Our stability requirement for the FTCS finite difference equation is therefore, by (3.3.23),

$$|1 - 4s\sin^2(\beta/2)| \leq 1,$$

or

$$-1 \leq 1 - 4s\sin^2(\beta/2) \leq 1, \text{ for all } \beta. \tag{3.3.25}$$

The right-hand inequality is satisfied for all s and β, whereas the left-hand inequality requires

$$s\sin^2(\beta/2) \leq \tfrac{1}{2},$$

which is true for all β if

$$s \leq \tfrac{1}{2}. \tag{3.3.26}$$

This result is identical to that found by both discrete perturbation analysis and the matrix method.

The von Neumann analysis is the most commonly used method of determining stability criteria as it is generally the easiest to apply, the most straightforward and the most dependable. Unfortunately, like the matrix method, it can only be used to establish necessary and sufficient conditions for stability of linear initial value problems with constant coefficients. Practical problems typically involve variable coefficients, non-linearities and complicated types of boundary conditions, in which case these methods can only be applied locally. If the complete non-linear difference equation is linearized in a small part of the solution domain where its coefficients may be considered constant, then the conditions for the applicability of these methods are satisfied locally even though they are not satisfied over the whole solution domain. And just as the von Neumann method can provide useful stability information when applied locally at interior grid-points, similar local application of the method at the boundaries can provide useful boundary stability information (see Trapp and Ramshaw, 1976).

Comparison of Stability Methods

There are many other methods of stability analysis in addition to the three described. Four others which deserve mention those due to Hirt (1968), Friedrichs (see Hahn, 1958), Keller and Lax (see Richtmeyer and Morton, 1967) and Eddy (1949). In the first of these, the terms of the finite difference equation are expanded in a Taylor series in order to develop a corresponding continuum partial differential equation. Stability of the difference equation is then determined from known stability properties of the resulting differential equation. The next two of these apply in practice only to the simplest finite difference equations. All four methods consider directly the amplification properties of the finite difference equations on introduced errors, rather than on their discrete Fourier components as in von Neumann's method.

Each of the three methods described in this article has its own particular merits.

The discrete perturbation method of stability analysis is a rather hit or miss affair compared with the more methodical matrix or von Neumann methods, and for this reason it is not widely used. However, in certain cases (see Section 6), it readily produces stability criteria obtained only by much effort using other methods. It focusses attention on the individual calculations performed, rather than on related abstractions, so that the magnitude of the errors can be estimated. It also provides insight about the best way of incorporating boundary conditions.

Like the von Neumann method, the matrix method is clearly formulated. It has the advantage that it can include the effect of boundary conditions which must be treated separately in the von Neumann method. However, the matrix method is more difficult to apply and requires considerable knowledge of matrix algebra (for example, as in Wade, 1951). The von Neumann analysis provides information not only on attenuation (or damping) produced by the finite difference approximation, but also on any phase effects produced. It therefore gives an insight into resulting dispersion errors (see Section 5).

In summary, all three methods provide valuable information about the stability of a finite difference equation. At the present time the von Neumann method is favoured. However, none of them can be considered completely adequate, so when actually computing solutions to a finite difference equation they should be supplemented by numerical experimentation. Kusic and Lavi (1968) and Kusic(1969) present a non-iterative method of evaluating stability during the process of computing the solutions to a finite difference equation.

It must be noted that the application of the discrete perturbation technique can be extended to non-linear problems in order to get an estimate of stability, but the basis of both the matrix method and the von Neumann stability analysis is that they apply only to linear systems with constant coefficients. They can however, give a rough estimate if applied to locally linearised forms of non-linear finite difference equations. John (1952) has shown that a mildly strengthened form of the von Neumann condition for boundaries is sufficient for stability of linear parabolic differential equations with variable coefficients. Lax (see Lax and Richtmeyer, 1956) has shown that a similar result holds for variable-coefficient hyperbolic differential equations.

3.4 Lax's Equivalence Theorem

There exist important connections between the stability of linear finite difference schemes and the convergence of their solutions to that of the linear partial differential equations to which they are consistent. An example of this connection is given by Lax's equivalence theorem.

This theorem, proved by Lax and described in Lax and Richtmeyer (1956), states: "Given a properly posed linear initial value problem and a finite difference approximation to it that satisfies the consistency condition, stability is the necessary and sufficient condition for convergence." This result is of great practical importance, for while it is relatively easy to show stability of a finite difference equation and its consistency with a partial differential equation, is usually very difficult to show convergence of its solution to that of the partial differential equation.

The conditions under which the theorem applies include:

(a) the initial-value problem must be well posed. This requires that the solution of the partial differential equation depends continuously on the given initial conditions.

(b) the problem must be linear. Because of this the errors propagate according to the homogeneous form of the given difference equation. It should be noted that attempts to find an equivalent theorem for non-linear problems have been unsuccessful.

Lax's equivalence theorem by-passes the need to prove convergence of the solution of the FTCS finite difference equation (2.4.4) to the solution of the partial differential equation (2.4.1). Because the diffusion equation and its associated initial and boundary conditions are linear and well-posed, and the FTCS finite difference equation is stable and consistent with the diffusion equation, it follows that the FTCS method is convergent.

4. SOLVING THE ONE-DIMENSIONAL DIFFUSION EQUATION

4.1 Explicit Methods

The FTCS Method

By substituting the forward difference approximation for the time derivative and the central difference approximation for the space derivative in Equation (2.4.1), namely

$$\left.\frac{\partial\bar{\tau}}{\partial t}\right|_j^n = \alpha \left.\frac{\partial^2\bar{\tau}}{\partial x^2}\right|_j^n ,$$

the classical FTCS explicit finite difference equation (2.4.4) for solving the one-dimensional diffusion equation was obtained,

$$\tau_j^{n+1} = s\,\tau_{j-1}^n + (1-2s)\tau_j^n + s\,\tau_{j+1}^n$$

with s defined by Equation (2.4.5),

$$s = \alpha\Delta t/(\Delta x)^2 .$$

(see Section 2.4.)

Then, in Section 3.1 it was shown that the solution τ_j^n of the finite difference equation (2.4.4) *converged* to the solution $\bar{\tau}_j^n$ of the diffusion equation (2.4.1) in the limit $\Delta x \to 0$, $\Delta t \to 0$, provided $s \le \frac{1}{2}$. In Section 3.2 the finite difference equation (2.4.4), with $\bar{\tau}_j^n$ substituted for τ_j^n, was shown to be *consistent* with the one-dimensional diffusion equation; that is

$$\bar{\tau}_j^{n+1} = s\,\bar{\tau}_{j-1}^n + (1-2s)\bar{\tau}_j^n + s\,\bar{\tau}_{j+1}^n ,$$

was shown to be equivalent to (2.4.1) in the limit as $\Delta x \to 0$, $\Delta t \to 0$. Finally, in Section 3.3, it was shown that the finite difference equation (2.4.4) was *stable* provided $s \le \frac{1}{2}$; that is, round-off errors in the calculation of τ_j^n at any time level produced bounded errors in values of τ computed at later times.

Equation (2.4.4) simplifies to $\tau_j^{n+1} = \frac{1}{2}(\tau_{j-1}^n + \tau_{j+1}^n)$ when $s = \frac{1}{2}$. That is, the value of τ at $x = j\Delta x$ at the (n+1)th time level is the arithmetic mean of the values of τ at $(j-1)\Delta x$ and $(j+1)\Delta x$ at the nth time level. The process of taking the arithmetic mean can be carried out either numerically or graphically, and in the latter form has been familiar for many years in works on heat transfer under the name of Schmidt's method (see Carslaw and Jaeger, 1962).

Since the one-dimensional diffusion equation is linear and is well-posed, Lax's equivalence theorem indicates that consistency and stability of the FTCS finite difference approximation is necessary and sufficient for the finite difference solution to converge to the solution of the diffusion equation. This theorem is also used to establish convergence of other finite difference methods of solving the diffusion equation developed in this Section, since stability and consistency are much easier to prove than convergence.

Once convergence has been proved, the solution to the given partial differential equation can be obtained to any desired degree of accuracy, provided the spacings Δx and Δt of the finite difference grid are made sufficiently small.

The FTCS method (2.4.4) is classified as explicit, because the value of τ_j^{n+1} at the (n+1)th time level may be calculated directly from known values of τ_j^n at the previous time level. It is a two-level method because values of τ at only two levels of time are involved in the approximating finite difference equation. Two other explicit methods will now be examined, those developed by Richardson (1910) and DuFort and Frankel (1953).

Richardson's Method

This is a modification of the finite difference approximation (2.4.3) to Equation (2.4.1), in which the forward difference approximation for $\partial\bar{\tau}/\partial t$ is replaced by the central difference form. Equation (2.4.1) is written

$$\frac{\bar{\tau}_j^{n+1} - \bar{\tau}_j^{n-1}}{2\Delta t} + O\{(\Delta t)^2\} = \alpha\{\frac{\bar{\tau}_{j+1}^n - 2\bar{\tau}_j^n + \bar{\tau}_{j-1}^n}{(\Delta x)^2} + O\{(\Delta x)^2\}\}, \qquad (4.1.1)$$

which yields, on rearrangement,

$$\bar{\tau}_j^{n+1} = \bar{\tau}_j^{n-1} + 2s(\bar{\tau}_{j-1}^n - 2\bar{\tau}_j^n + \bar{\tau}_{j+1}^n) + O\{(\Delta t)^3, \Delta t\ (\Delta x)^2\}. \qquad (4.1.2)$$

This yields the explicit three-level finite difference equation

$$\tau_j^{n+1} = \tau_j^{n-1} + 2s(\tau_{j-1}^n - 2\tau_j^n + \tau_{j+1}^n), \qquad (4.1.3)$$

when the terms of $O\{(\Delta t)^3, \Delta t(\Delta x)^2\}$ are omitted from Equation (4.1.2).

Because τ_j^{n+1} is calculated using values at the *two* previous time levels, an alternative procedure, such as the FTCS method with n = 0, is required to start the computations and calculate the approximations τ_j^1, j=1(1)J-1, at time level n = 1 from the known initial values τ_j^0.

Comparison of Equations (4.1.2) and (4.1.3) indicates that there is a contribution of $O\{(\Delta t)^3, \Delta t(\Delta x)^2\}$ to the discretisation error in stepping from time levels (n-1) and n to the time level (n+1) with Equation (4.1.3). Also, Equation (4.1.1) indicates that the truncation error is generally $O\{(\Delta t)^2, (\Delta x)^2\}$. In order to determine the exact nature of the truncation error it is necessary to go through a formal consistency analysis of the method. For instance, without carrying out the formal consistency analysis of the FTCS method, the optimal case with minimal truncation and discretisation error for s = 1/6 would not have been found.

To determine the *consistency* of the finite difference equation (4.1.2), the values τ_j^n, which satisfy this equation are replaced by values $\bar{\tau}_j^n$ which are solutions of the diffusion equation (2.4.1). Replacing these terms by their truncated Taylor series expanded about the (j,n) grid-point yields

$$\bar{\tau}_j^n + \Delta t \left.\frac{\partial\bar{\tau}}{\partial t}\right|_j^n + \frac{(\Delta t)^2}{2!}\left.\frac{\partial^2\bar{\tau}}{\partial t^2}\right|_j^n + \frac{(\Delta t)^3}{3!}\left.\frac{\partial^3\bar{\tau}}{\partial t^3}\right|_j^n + O\{(\Delta t)^4\}$$

$$= \bar{\tau}_j^n - \Delta t \left.\frac{\partial\bar{\tau}}{\partial t}\right|_j^n + \frac{(\Delta t)^2}{2!}\left.\frac{\partial^2\bar{\tau}}{\partial t^2}\right|_j^n - \frac{(\Delta t)^3}{3!}\left.\frac{\partial^3\bar{\tau}}{\partial t^3}\right|_j^n + O\{(\Delta t)^4\}$$

$$+ 2s\begin{Bmatrix} \bar{\tau}_j^n - \Delta x \left.\frac{\partial\tau}{\partial x}\right|_j^n + \frac{(\Delta x)^2}{2!}\left.\frac{\partial^2\tau}{\partial x^2}\right|_j^n - \frac{(\Delta x)}{3!}\left.\frac{\partial^3\tau}{\partial x^3}\right|_j^n + \frac{(\Delta x)^4}{4!}\left.\frac{\partial^4\bar{\tau}}{\partial x^4}\right|_j^n + O\{(\Delta x)^5\} \\ -2\ \bar{\tau}_j^n \\ + \bar{\tau} + \Delta x \left.\frac{\partial\bar{\tau}}{\partial x}\right|_j^n + \frac{(\Delta x)^2}{2!}\left.\frac{\partial^2\bar{\tau}}{\partial x^2}\right|_j^n + \frac{(\Delta x)}{3!}\left.\frac{\partial^3\bar{\tau}}{\partial x^3}\right|_j^n + \frac{(\Delta x)^4}{4!}\left.\frac{\partial^4\bar{\tau}}{\partial x^4}\right|_j^n + O\{(\Delta x)^5\} \end{Bmatrix}.$$

This equation simplifies to

$$\left.\frac{\partial\bar{\tau}}{\partial t}\right|_j^n = \alpha \left.\frac{\partial^2\bar{\tau}}{\partial x^2}\right|_j^n + E_j^n \tag{4.1.4a}$$

where

$$E_j^n = \frac{(\Delta t)^2}{6}\left.\frac{\partial^3\bar{\tau}}{\partial t^3}\right|_j^n + \frac{\alpha(\Delta x)^2}{12}\left.\frac{\partial^4\bar{\tau}}{\partial x^4}\right|_j^n + O\{(\Delta t)^3,(\Delta x)^3\}. \tag{4.1.4b}$$

Richardson's method is therefore consistent, because the truncation error $E_j^n \to 0$ as $\Delta x \to 0$, $\Delta t \to 0$, and it is of a higher order of accuracy than the FTCS method because it has truncation error $O\{(\Delta t)^2,(\Delta x)^2\}$ compared with $O\{\Delta t,(\Delta x)^2\}$.

In order to establish whether there is an optimum value of s for which E_j^n is a minimum, we can follow the same procedure that was used for the FTCS method in Section 3.2. In this case, the term $\partial^3\bar{\tau}/\partial t^3|_j^n$ may be replaced by $\alpha^3\, \partial^6\bar{\tau}/\partial x^6|_j^n$, which follows from (3.2.5) because $\bar{\tau}$ satisfies the diffusion equation (2.4.1). However, no matter what value of s is chosen, the term in $(\Delta x)^2$ cannot be eliminated.

To investigate the *stability* of Richardson's method the mth Fourier component of the error distribution at the nth time-level, namely $\xi_j^n = (G)^n e^{i\beta j}$ at the (j,n) grid-point, is substituted into the error equation corresponding to (4.1.3), namely

$$\xi_j^{n+1} = \xi_j^{n-1} + 2s\,(\xi_{j-1}^n - 2\xi_j^n + \xi_{j+1}^n), \tag{4.1.5}$$

giving

$$(G)^{n+1}e^{i\beta j} = (G)^{n-1}e^{i\beta j} + 2s\{(G)^n e^{i\beta(j-1)} - 2(G)^n e^{i\beta j} + (G)^n e^{i\beta(j+1)}\}. \tag{4.1.6}$$

Here G is the amplification factor by which the mth Fourier component of the error distribution at any time level is multiplied to get the corresponding Fourier component at the next time level. Dividing (4.1.6) by $(G)^n e^{i\beta j}$ yields

$$G - G^{-1} = 4s(\cos\beta - 1), \tag{4.1.7}$$

which gives the equation to be solved for G, namely

$$G^2 + 8s\gamma G - 1 = 0, \tag{4.1.8}$$

where $\gamma = \sin^2(\beta/2)$.

Since the discriminant of the quadratic equation (4.1.8) is positive for all s and β, then it has two real roots, G_1 and G_2, which satisfy the pair of equations

$$G_1G_2 = -1 \tag{4.1.9a}$$

and

$$G_1 + G_2 = -8\gamma s. \tag{4.1.9b}$$

It follows from (4.1.9a) that either

$$|G_1| > 1 \quad\text{or}\quad |G_2| > 1,$$

in which case the errors at a given time level will contain some Fourier components which grow without bound as n increases; otherwise

$$G_1 = -G_2 = 1 \text{ (say)},$$

in which case (4.1.9b) indicates that $s = 0$, or $\Delta t = 0$, when the method has no practical value. Richardson's method is therefore unstable for all $s > 0$ and errors in the values computed using Equation (4.1.3) become as large as the true values after only a few time steps. This equation therefore cannot be used to approximately solve the diffusion equation.

But, it must not be assumed that all methods which may involve steps in time or space for which the errors magnify, are of no practical use. Such steps may be part of a time-splitting method for an initial value problem, or a space-stepping procedure in a boundary value problem in which the errors are kept small relative to the true values (see Section 6).

DuFort-Frankel Method

Another explicit method for solving the diffusion equation is that developed by DuFort and Frankel (1953). This is a modification of Richardson's method in which the central grid-point value $\bar{\tau}^n_j$ in the finite difference approximation for the diffusion term $\alpha\partial^2\bar{\tau}/\partial x^2$ in Equation (4.1.1) is replaced by its average at the (n-1) and (n+1) time levels. Since

$$\bar{\tau}^n_j = \tfrac{1}{2}(\bar{\tau}^{n+1}_j + \bar{\tau}^{n-1}_j) + O\{(\Delta t)^2\}, \tag{4.1.10}$$

it follows that Equation (4.1.1) can be rewritten

$$\frac{\bar{\tau}^{n+1}_j - \bar{\tau}^{n-1}_j}{2\Delta t} + O\{(\Delta t)^2\} = \alpha\{\frac{\bar{\tau}^n_{j+1} - (\bar{\tau}^{n+1}_j + \bar{\tau}^{n-1}_j) + \bar{\tau}^n_{j-1}}{(\Delta x)^2} + O\{(\Delta x)^2, (\frac{\Delta t}{\Delta x})^2\}\}. \tag{4.1.11}$$

Although a value of $\bar{\tau}$ at the (n+1)th time level appears on the right-hand side of this equation, it is at the jth space position, so Equation (4.1.11) can be rearranged to give

$$\bar{\tau}^n_j = \frac{2s}{1+2s}(\bar{\tau}^n_{j-1} + \bar{\tau}^n_{j+1}) + \frac{1-2s}{1+2s}\bar{\tau}^{n-1}_j + O\{(\Delta t)^3, \Delta t(\Delta x)^2, \frac{(\Delta t)^3}{(\Delta x)^2}\}. \tag{4.1.12}$$

Dropping the terms of $O\{(\Delta t)^3, \Delta t(\Delta x)^2, (\Delta t)^3/(\Delta x)^2\}$ gives the explicit finite difference equation

$$\tau^{n+1}_j = \frac{2s}{1+2s}(\tau^n_{j-1} + \tau^n_{j+1}) + \frac{1-2s}{1+2s}\tau^{n-1}_j. \tag{4.1.13}$$

Like Richardson's formula, the DuFort-Frankel formula involves values of the dependent variable at grid-points along three time levels and it requires special starting procedures. These two methods are three level methods, in contrast with the FTCS explicit method which is a two level method.

A stability analysis of the error equation corresponding to (4.1.13), namely

$$\xi^{n+1}_j = \frac{2s}{1+2s}(\xi^n_{j-1} + \xi^n_{j+1}) + \frac{1-2s}{1+2s}\xi^{n-1}_j, \tag{4.1.14}$$

by means of the von Neumann method, gives the amplification factor from one time level to the next for the mth Fourier component of the error distribution as

$$G = \frac{2s\cos\beta \pm (1-4s^2\sin^2\beta)^{1/2}}{1+2s} . \qquad (4.1.15)$$

If s and β are such that $1 - 4s^2\sin^2\beta \geq 0$, then both terms in the numerator of (4.1.15) are real. Also, $1+2s > 0$ so that

$$|G| \leq \frac{|2s\cos\beta| + (1-4s^2\sin^2\beta)^{1/2}}{1+2s},$$

and as

$$|2s\cos\beta| \leq 2s,\ 0 \leq 1-4s^2\sin^2\beta \leq 1, \text{ for all } \beta,$$

it follows that

$$|G| \leq \frac{2s+1}{1+2s} = 1. \qquad (4.1.16)$$

Otherwise, $1-4s^2\sin^2\beta < 0$ so that

$$G = \frac{2s\cos\beta \pm i(4s^2\sin^2\beta-1)^{1/2}}{1+2s},$$

and

$$|G|^2 = \frac{4s^2\cos^2\beta+(4s^2\sin^2\beta-1)}{(1+2s)^2}$$

$$= \frac{4s^2-1}{4s^2+4s+1}$$

$$\leq 1. \qquad (4.1.17)$$

Therefore it follows that this finite difference method is stable for all positive s.

Alternatively, the stability of the DuFort-Frankel formula may be investigated by the matrix method. Since (4.1.11) is a three-level formula, the matrix method for a two-level formula described in Section 3.3 requires modification.

If $\underset{\sim}{\xi}^n$ represents the vector of error values at interior grid-points along the nth time level, with zero errors at the boundaries, then

$$\underset{\sim}{\xi}^{n+1} = \left(\frac{2s}{1+2s}\right)\underline{\underline{A}}\underset{\sim}{\xi}^n + \left(\frac{1-2s}{1+2s}\right)\underset{\sim}{\xi}^{n-1}, \quad n=1,2,3,\ldots, \qquad (4.1.18)$$

where

$$\underline{\underline{A}} = \begin{bmatrix} 0 & 1 & & & 0 \\ 1 & 0 & 1 & & \\ & 1 & 0 & 1 & \\ & & & \ddots & \\ 0 & & & 1 & 0 \end{bmatrix}, \quad \underset{\sim}{\xi}^n = \begin{bmatrix} \xi_1^n \\ \xi_2^n \\ \xi_3^n \\ \vdots \\ \xi_{J-1}^n \end{bmatrix}.$$

Equation (4.1.18) can be written in partitioned form as

$$\begin{bmatrix} \underset{\sim}{\xi}^{n+1} \\ \\ \underset{\sim}{\xi}^{n} \end{bmatrix} = \begin{bmatrix} \frac{2s}{1+2s}\underline{\underline{A}} & \frac{1-2s}{1+2s}\underline{\underline{I}} \\ \\ \underline{\underline{I}} & \underline{\underline{0}} \end{bmatrix} \begin{bmatrix} \underset{\sim}{\xi}^{n} \\ \\ \underset{\sim}{\xi}^{n-1} \end{bmatrix}, \quad n=1,2,3,.., \tag{4.1.19}$$

where $\underline{\underline{I}}$ is the unit matrix and $\underline{\underline{0}}$ the zero matrix, both of order (J-1). Writing

$$\underset{\sim}{\zeta}^{n} = \begin{bmatrix} \underset{\sim}{\xi}^{n} \\ \\ \underset{\sim}{\xi}^{n-1} \end{bmatrix}, \quad \underline{\underline{B}} = \begin{bmatrix} \frac{2s}{1+2s}\underline{\underline{A}} & \frac{1-2s}{1+2s}\underline{\underline{I}} \\ \\ \underline{\underline{I}} & \underline{\underline{0}} \end{bmatrix}, \tag{4.1.20}$$

then Equation (4.1.19) can be written

$$\underset{\sim}{\zeta}^{n+1} = \underline{\underline{B}}\,\underset{\sim}{\zeta}^{n}, \quad n=1,2,3,.. \,. \tag{4.1.21}$$

If the eigenvalues λ_m of $\underline{\underline{B}}$ are distinct then the corresponding eigenvectors $\underset{\sim}{\nu}_m$ form a linearly independent set and $\underset{\sim}{\zeta}^1$, any error distribution at the time levels 0 and 1, can always be written as a linear combination of ν_m; that is

$$\underset{\sim}{\zeta}^{1} = \sum_{m=1}^{2J-2} C_m \underset{\sim}{\nu}_m. \tag{4.1.22}$$

On application of Equation (4.1.21) with n=1,2,3,.., it follows that

$$\underset{\sim}{\zeta}^{2} = \underline{\underline{B}}\,\underset{\sim}{\zeta}^{1} = \sum_{m=1}^{2J-2} C_m(\underline{\underline{B}}\underset{\sim}{\nu}_m) = \sum_{m=1}^{2J-2} C_m\lambda_m\underset{\sim}{\nu}_m,$$

$$\underset{\sim}{\zeta}^{3} = \underline{\underline{B}}\,\underset{\sim}{\zeta}^{2} = \sum_{m=1}^{2J-2} C_m(\lambda_m)^2\underset{\sim}{\nu}_m,$$

and so on, until finally

$$\underset{\sim}{\zeta}^{n} = \sum_{m=1}^{2J-2} C_m(\lambda_m)^{n-1}\underset{\sim}{\nu}_m. \tag{4.1.23}$$

Clearly, Equation (4.1.23) indicates that, for stability, the (2J-2) eigenvalues λ of $\underline{\underline{B}}$ must all satisfy the inequality $|\lambda| \le 1$.

The eigenvalues of $\underline{\underline{B}}$ are, by definition, the roots of the characteristic equation $|\underline{\underline{B}}-\lambda\underline{\underline{I}}^*| = 0$, where $\underline{\underline{I}}^*$ is the unit matrix of order (2J-2). Using (4.1.20) the characteristic equation can be rewritten

$$\begin{vmatrix} \frac{2s}{1+2s}\underline{\underline{A}}-\lambda\underline{\underline{I}} & \frac{1-2s}{1+2s}\underline{\underline{I}} \\ \\ \underline{\underline{I}} & -\lambda\underline{\underline{I}} \end{vmatrix} = 0,$$

so the eigenvalues of $\underline{\underline{B}}$ are the roots of

$$|\underline{\underline{A}} - \frac{(2s+1)\lambda^2+(2s-1)}{2s\lambda}\underline{\underline{I}}| = 0.$$

This equation is satisfied if the coefficient of $\underline{\underline{I}}$ is an eigenvalue of $\underline{\underline{A}}$, which may be found using (3.3.16). The values of λ therefore satisfy

$$\frac{(2s+1)\lambda^2+(2s-1)}{2s\lambda} = 2\cos(m\pi/J), \quad m=1(1)J-1. \tag{4.1.24}$$

The required eigenvalues of $\underline{\underline{B}}$ are the (2J-2) roots of

$$(2s+1)\lambda^2 - 4s\cos(m\pi/J)\lambda + (2s-1) = 0, \quad m=1(1)J-1.$$

Solving this equation gives

$$\lambda = \frac{2s\cos(m\pi/J) \pm \{1-4s^2\sin^2(m\pi/J)\}^{1/2}}{1+2s}, \quad m=1(1)J-1. \tag{4.1.25}$$

This equation is analogous to Equation (4.1.15), and using the same reasoning as then, it can be shown that $|\lambda| \le 1$ for all $s > 0$ and all m. This shows that the DuFort-Frankel method is unconditionally stable.

Therefore Δt may be chosen as large as we like, provided the discretisation error does not become too large. The discretisation error depends on the truncation error, which can be established by a consistency analysis.

On checking the consistency of the finite difference equation (4.1.13) by substitution of $\bar{\tau}$, the solution of the diffusion equation, for the finite difference solution τ, followed by expansion of each term in truncated Taylor series about the (j,n) grid-point, the following relation is obtained:

$$\left.\frac{\partial\bar{\tau}}{\partial t}\right|_j^n = \alpha\left.\frac{\partial^2\bar{\tau}}{\partial x^2}\right|_j^n - \alpha\left(\frac{\Delta t}{\Delta x}\right)^2\left.\frac{\partial^2\bar{\tau}}{\partial t^2}\right|_j^n - \frac{(\Delta t)^2}{6}\left.\frac{\partial^3\bar{\tau}}{\partial t^3}\right|_j^n + \frac{(\Delta x)^2}{12}\left.\frac{\partial^4\bar{\tau}}{\partial x^4}\right|_j^n$$

$$+ O\{(\Delta t)^3,(\Delta x)^3\}. \tag{4.1.26}$$

The DuFort-Frankel method is therefore consistent with the one-dimensional diffusion equation, with truncation error going to zero, only if $\Delta t/\Delta x \to 0$ as both Δx and $\Delta t \to 0$. Therefore, even though this method is unconditionally stable, $\Delta t \ll \Delta x$ is required for consistency. If Δt and $\Delta x \to 0$ in such a way that $\Delta t/\Delta x \to K$, a constant, then the finite difference equation (4.1.13) is consistent with the hyperbolic equation

$$\frac{\partial\bar{\tau}}{\partial t} = \alpha\left(\frac{\partial^2\bar{\tau}}{\partial x^2} - K^2\frac{\partial^2\bar{\tau}}{\partial t^2}\right), \tag{4.1.27}$$

and not with the parabolic one-dimensional diffusion equation (2.1.1).

In spite of the restriction $\Delta t \ll \Delta x$ required for consistency, the stablising effect of the substitution of the average value of τ_j^{n+1} and τ_j^{n-1} for τ_j^n in the finite difference approximation of the diffusion terms in various equations of fluid dynamics has been used successfully by many workers including Payne (1958), Fromm and Harlow (1963), Amsden and Harlow (1964), Hung and Macagno (1966), and Torrance (1968).

4.2 Implicit Methods

Each of the methods described previously is explicit. At the new time level, the finite difference equation contains only the one unknown value τ_j^{n+1}, which is calculated explicitly from values of τ known at previous time levels. These methods are easy to program and require few computations to determine the values of τ at each new time level. Unfortunately, each of the two usable methods already described is restricted to very small time steps Δt, the classical FTCS method because of stability requirements and the DuFort-Frankel method because of consistency requirements. For instance, if a numerical approximation at t = 10 is required and Δx must take the value 10^{-2}, then use of the FTCS method with $\alpha = 5$ implies that Δt must be chosen to satisfy

$$s = \frac{5\Delta t}{(10^{-2})^2} \le \tfrac{1}{2}$$

so that

$$\Delta t \le 10^{-5} .$$

To generate a numerical solution at t = 10 would require the computation of τ on at least one million rows of grid-points in the solution domain.

The Classical Implicit Method

By substituting the backward difference form (2.3.8a) for the time derivative and the central difference form (2.3.19) for the space derivative in the diffusion equation (2.1.1) evaluated at the (j,n+1) grid-point, namely

$$\left.\frac{\partial\bar{\tau}}{\partial t}\right|_j^{n+1} = \alpha \left.\frac{\partial^2\bar{\tau}}{\partial x^2}\right|_j^{n+1} , \tag{4.2.1a}$$

gives

$$\frac{\bar{\tau}_j^{n+1} - \bar{\tau}_j^n}{\Delta t} + O\{\Delta t\} = \alpha\{\frac{\bar{\tau}_{j+1}^{n+1} - 2\bar{\tau}_j^{n+1} + \bar{\tau}_{j-1}^{n+1}}{(\Delta x)^2} + O\{(\Delta x)^2\}\}. \tag{4.2.1b}$$

The analogous finite difference approximation to (4.2.1a) is therefore

$$\frac{\tau_j^{n+1} - \tau_j^n}{\Delta t} = \alpha\{\frac{\tau_{j+1}^{n+1} - 2\tau_j^{n+1} + \tau_{j-1}^{n+1}}{(\Delta x)^2}\} , \tag{4.2.2a}$$

with an apparent truncation error of $O\{\Delta t, (\Delta x)^2\}$. Equation (4.2.2a) can be rearranged to yield the finite difference equation

$$-s\,\tau_{j-1}^{n+1} + (1+2s)\tau_j^{n+1} - s\tau_{j+1}^{n+1} = \tau_j^n . \tag{4.2.2b}$$

This equation is not explicit, since there are three unknown values of τ at the (n+1)th time level. With n = 0, substitution of j=1(1)J-1 in Equation (4.2.2b) gives (J-1) simultaneous linear algebraic equations for the (J-1) unknown values of τ at internal grid-points along the first time level, in terms of known initial and boundary values. These equations can be solved to give the values of τ_j^1, j=1(1)J-1. Similarly, putting n = 1 gives (J-1) equations for the (J-1) unknown values of τ at the second time level in terms of boundary values and the computed values of τ at the first time level, and so on. Methods such as this, in which the unknown values of τ at any time level are found by solving a set of algebraic equations are described as *implicit*. Equation (4.2.2b) defines what is known as the classical implicit method for solving the one-dimensional diffusion equation (see Laasonen, 1949, and O'Brien et al, (1950).

Assuming values are known at the nth time level we may substitute j=1(1)J-1 in Equation (4.2.2b) to obtain the following set of simultaneous linear algebraic equations with unknowns τ_j^{n+1}, j=1(1)J-1:

$$\left.\begin{array}{rl}
(1+2s)\tau_1^{n+1} - s\,\tau_2^{n+1} & = d_1^n \\
-s\,\tau_1^{n+1} + (1+2s)\tau_2^{n+2} - s\,\tau_3^{n+1} & = d_2^n \\
-s\,\tau_2^{n+1} + (1+2s)\tau_3^{n+1} - s\,\tau_4^{n+1} & = d_3^n \\
\ddots \qquad & \vdots \\
-s\,\tau_{J-2}^{n+1} + (1+2s)\tau_{J-1}^{n+1} & = d_{J-1}^n
\end{array}\right\} \quad (4.2.3)$$

where

$$d_1^n = \tau_1^n + s\,\bar{\tau}_0^{n+1},$$

$$d_j^n = \tau_j^n,\ j=2(1)J-2,$$

$$d_{J-1}^n = \tau_{J-1}^n + s\,\bar{\tau}_J^{n+1}.$$

This tri-diagonal system may be solved by the elimination process described by Thomas (1949) and outlined in Appendix 1. The Thomas algorithm makes full use of the fact that most of the coefficients of the unknown values of τ at time level (n+1) are zero's. It is much faster than using a Gauss elimination "package" which automatically goes through a process of making the coefficient of many terms zero, when they are already so.

The process of solving the set of Equations (4.2.3) must be repeated at every time step. However, compensation for the extra arithmetic required at each time step compared with that required for an explicit method occurs because the finite difference equation (4.2.2) is stable for all values of s, so that larger values of Δt may be used. In fact, application of the von Neumann stability analysis to the corresponding error equation

$$-s\xi_{j-1}^{n+1} + (1+2s)\xi_j^{n+1} - s\xi_{j+1}^{n+1} = \xi_j^n,$$

yields the amplification factor

$$G = \{1 + 2s(1 - \cos\beta)\}^{-1},$$

for the propagation of the mth Fourier mode of the error distribution from one time level to the next. Since $(1 - \cos\beta) \geq 0$, then

$$1+2s(1-\cos\beta) \geq 1 \text{ for all } s > 0 \text{ and all } \beta.$$

Therefore $|G| \leq 1$ for all $s > 0$ and all β, so the classical implicit method is unconditionally stable.

The consistency of this implicit finite difference method clearly follows from the manner in which the finite difference equation (4.2.2b) was constructed from Equation (4.2.1b), the truncation error normally being of $O\{\Delta t,(\Delta x)^2\}$. However, the consistency should be formally established to determine whether any values of s occur for which the truncation error, and therefore the discretisation error, is of optimal order. This is done by substituting the solutions $\bar{\tau}$ of the given partial differential equation for the values of τ in the finite difference approximation (4.2.2), to obtain the relation

$$-s\,\bar{\tau}^{n+1}_{j-1} + (1+2s)\bar{\tau}^{n+1}_{j} - s\,\bar{\tau}^{n+1}_{j+1} = \bar{\tau}^{n}_{j}. \tag{4.2.4}$$

The difference between this equation and the one-dimensional diffusion equation is determined by expanding each term in a Taylor series about the (j,n) grid-point. The Taylor series expansion of $\bar{\tau}^{n+1}_{j+1}$ is

$$\bar{\tau}^{n+1}_{j+1} = \sum_{m=0}^{\infty} \frac{1}{m!}\left[\Delta t\frac{\partial}{\partial t} + \Delta x\frac{\partial}{\partial x}\right]^m \bar{\tau}\Big|^n_j, \tag{4.2.5}$$

which becomes, on expansion,

$$\begin{aligned}\bar{\tau}^{n+1}_{j+1} &= \bar{\tau}^n_j + \left[\Delta t\frac{\partial}{\partial t} + \Delta x\frac{\partial}{\partial x}\right]\bar{\tau}\Big|^n_j \\ &+ \frac{1}{2!}\left[\Delta t\frac{\partial}{\partial t} + \Delta x\frac{\partial}{\partial x}\right]^2\bar{\tau}\Big|^n_j \\ &+ \frac{1}{3!}\left[\Delta t\frac{\partial}{\partial t} + \Delta x\frac{\partial}{\partial x}\right]^3\bar{\tau}\Big|^n_j + \dots \\ &= \bar{\tau}^n_j + \Delta t\frac{\partial\bar{\tau}}{\partial t}\Big|^n_j + \Delta x\frac{\partial\bar{\tau}}{\partial x}\Big|^n_j \\ &+ \frac{(\Delta t)^2}{2}\frac{\partial^2\bar{\tau}}{\partial t^2}\Big|^n_j + \Delta t\,\Delta x\frac{\partial^2\bar{\tau}}{\partial t\partial x}\Big|^n_j + \frac{(\Delta x)^2}{2}\frac{\partial^2\bar{\tau}}{\partial x^2}\Big|^n_j \\ &+ \frac{(\Delta t)^3}{6}\frac{\partial^3\bar{\tau}}{\partial t^3}\Big|^n_j + \frac{(\Delta t)^2\Delta x}{2}\frac{\partial^3\bar{\tau}}{\partial t^2\partial x}\Big|^n_j + \frac{\Delta t(\Delta x)^2}{2}\frac{\partial^3\bar{\tau}}{\partial t\partial x^2}\Big|^n_j \\ &+ \frac{(\Delta x)^3}{6}\frac{\partial^3\bar{\tau}}{\partial x^3}\Big|^n_j + \dots\end{aligned} \tag{4.2.6a}$$

The Taylor series expansion of $\bar{\tau}^{n+1}_{j-1}$ may be found by substituting $-\Delta x$ for Δx in the above equation, giving

$$\begin{aligned}\bar{\tau}^{n+1}_{j-1} &= \bar{\tau}^n_j + \Delta t\frac{\partial\bar{\tau}}{\partial t}\Big|^n_j - \Delta x\frac{\partial\bar{\tau}}{\partial x}\Big|^n_j \\ &+ \frac{(\Delta t)^2}{2}\frac{\partial^2\bar{\tau}}{\partial t^2}\Big|^n_j - \Delta t\,\Delta x\frac{\partial^2\bar{\tau}}{\partial t\partial x}\Big|^n_j + \frac{(\Delta x)^2}{2}\frac{\partial^2\bar{\tau}}{\partial x^2}\Big|^n_j \\ &+ \frac{(\Delta t)^3}{6}\frac{\partial^3\bar{\tau}}{\partial t^3}\Big|^n_j - \frac{(\Delta t)^2\Delta x}{2}\frac{\partial^3\bar{\tau}}{\partial t^2\partial x}\Big|^n_j + \frac{(\Delta t)^2\Delta x}{2}\frac{\partial^3\bar{\tau}}{\partial t\partial x^2}\Big|^n_j \\ &- \frac{(\Delta x)^3}{6}\frac{\partial^3\bar{\tau}}{\partial x^3}\Big|^n_j + \dots\ .\end{aligned} \tag{4.2.6b}$$

Finally, by (2.3.2),

$$\bar{\tau}^{n+1}_j = \bar{\tau}^n_j + \Delta t\frac{\partial\bar{\tau}}{\partial t}\Big|^n_j + \frac{(\Delta t)^2}{2}\frac{\partial^2\bar{\tau}}{\partial t^2}\Big|^n_j + \frac{(\Delta t)^3}{6}\frac{\partial^3\bar{\tau}}{\partial t^3}\Big|^n_j + \dots\ . \tag{4.2.6c}$$

Substitution of the expansions (4.2.6) into Equation (4.2.4) yields the equivalent relation

$$\left.\frac{\partial\bar{\tau}}{\partial t}\right|_j^n = \alpha \left.\frac{\partial^2\bar{\tau}}{\partial x^2}\right|_j^n + E_j^n \qquad (4.2.7a)$$

where E_j^n is given by

$$E_j^n = -\frac{\Delta t}{2}\left.\frac{\partial^2\bar{\tau}}{\partial t^2}\right|_j^n + \alpha\Delta t \left.\frac{\partial^3\bar{\tau}}{\partial t\partial x^2}\right|_j^n + \frac{\alpha(\Delta x)^2}{12}\left.\frac{\partial^4\bar{\tau}}{\partial x^4}\right|_j^n + O\{(\Delta t)^2,\Delta t(\Delta x)^2\}. \qquad (4.2.7b)$$

Since $\bar{\tau}$ is a solution of the one-dimensional diffusion equation (2.1.1) then

$$\frac{\partial^2\bar{\tau}}{\partial t^2} = \alpha\frac{\partial^3\bar{\tau}}{\partial t\partial x^2} = \alpha^2\frac{\partial^4\bar{\tau}}{\partial x^4}.$$

Hence, since $(\Delta x)^2 = s\Delta t/\alpha$, it follows that E_j^n may be expressed as

$$E_j^n = \frac{\Delta t}{2}\{1 + \frac{1}{6s}\}\left.\frac{\partial^2\bar{\tau}}{\partial t^2}\right|_j^n + O\{(\Delta t)^2\}, \text{ for fixed } s. \qquad (4.2.8)$$

Clearly, as $\Delta t \to 0$ for a fixed value of s, $E_j^n \to 0$ and formula (4.2.4) approaches Equation (4.1.1), so the finite difference equation (4.2.2b) is consistent with the one-dimensional diffusion equation. Also, there is no optimal value of s for which the order of the truncation error may be reduced; for all $s > 0$, the truncation error E_j^n remains $O\{\Delta t\}$. Therefore, for a given s, taking a large value of Δt increases the error E_j^n in a manner approximately proportional to Δt. We must therefore compromise between taking large time steps, to save on the number of arithmetic operations needed, and increased inaccuracy produced by the larger truncation error and associated discretisation error which occurs. "Stability without accuracy has little to commend it" (Fox, 1962, p.243).

An aspect of implicit finite difference methods seldom considered, is whether the resulting system of linear algebraic equations can be solved in practice. If the coefficient matrix of the system is *diagonally dominant* then any direct method of solving the corresponding linear algebraic system, such as that due to Thomas for tridiagonal systems or the Gauss elimination method for more general systems, is stable. That is, a round-off error in any computation does not magnify during the following calculations to make the results so inaccurate they cannot be taken as the true solutions of the original linear algebraic system of equations.

A system of linear algebraic equations is said to be diagonally dominant if along each row of the coefficient matrix the absolute value of the coefficient on the leading diagonal is greater than or equal to, the sum of the absolute values of the remaining coefficients along that row, and along at least one row there is strict inequality.

For the classical implicit method, in which the set of equations (4.2.3) must be solved, we have for the rows indicated

$$j=1,J-1 : |1+2s| > |-s|,$$

$$j=2(1)J-2: |1+2s| > |-s| + |-s|,$$

since $s > 0$. Therefore the system (4.2.3) is *solvable* for all s.

TABLE 4.1

Discretisation errors in the computation of $\bar{\tau}(0.4,8) = 0.4318184$, using the classical implicit method to solve the diffusion equation (2.1.1) with initial and boundary conditions (3.2.9).

Δx	s = 1/2	s = 1	s = 2	s = 4
1/20	-2.79×10^{-3}	-4.85×10^{-3}	-8.90×10^{-3}	-1.67×10^{-2}
1/30	-1.24×10^{-3}	-2.17×10^{-3}	-4.01×10^{-3}	-7.63×10^{-3}
1/40	-7.00×10^{-4}	-1.22×10^{-3}	-2.27×10^{-3}	-4.33×10^{-3}
1/50	-4.49×10^{-4}	-7.81×10^{-4}	-1.45×10^{-3}	-2.78×10^{-3}
1/60	-3.12×10^{-4}	-5.40×10^{-4}	-1.01×10^{-3}	-1.93×10^{-3}
1/70	-2.30×10^{-4}	-3.94×10^{-4}	-7.43×10^{-4}	-1.42×10^{-3}
1/80	-1.76×10^{-4}	-2.99×10^{-4}	-5.69×10^{-4}	-1.08×10^{-3}
1/90	-1.40×10^{-4}	-2.32×10^{-4}	-4.50×10^{-4}	-8.51×10^{-4}
1/100	-1.14×10^{-4}	-1.84×10^{-4}	-3.64×10^{-4}	-6.84×10^{-4}

The behaviour of the discretisation error of the classical implicit method is evident in Table 4.1 and Figure 4.1. Using the same reasoning as that given for the FTCS method in Section 3.2, it is seen that the order of the discretisation error e at t = 8, and the truncation error E are both the same, namely $O\{(\Delta x)^2\}$ for a fixed value of s. This is evident in Figure 4.1, since the slope of each line is very nearly 2. In fact, the straight lines of best fit for each s are:

$$s = 1/2, \quad e = -1.12(\Delta x)^{1.99},$$

$$s = 1\ , \quad e = -2.30(\Delta x)^{2.01},$$

$$s = 2\ , \quad e = -3.52(\Delta x)^{1.99},$$

$$s = 4\ , \quad e = -6.98(\Delta x)^{2.00}.$$

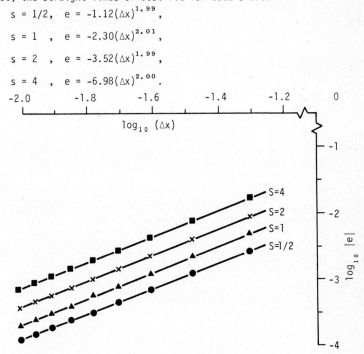

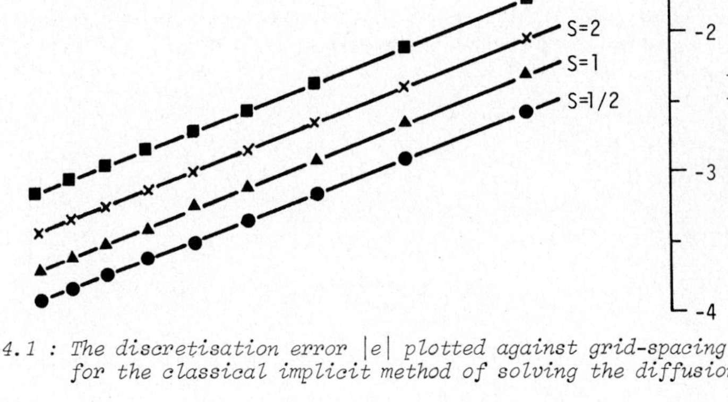

Figure 4.1 : The discretisation error |e| plotted against grid-spacing Δx for the classical implicit method of solving the diffusion equation.

The Crank-Nicolson method

Another implicit method used to solve the one-dimensional diffusion equation is due to Crank and Nicolson (1947). This method uses centred finite difference approximations for both time and space derivatives at the point $(j\Delta x, n\Delta t+\frac{1}{2}\Delta t)$ which is halfway between the (j,n) and $(j,n+1)$ grid-points. The central difference approximation for the time derivative $\partial\bar{\tau}/\partial t$ at this point in the solution domain is the same as the forward difference approximation for this derivative at the (j,n) grid-point but it is a more accurate approximation, the error being of $O\{(\Delta t)^2\}$. The spatial derivative $\partial^2\bar{\tau}/\partial x^2$ at $(j\Delta x, n\Delta t+\frac{1}{2}\Delta t)$ is approximated by the average of the central difference approximations of these spatial derivatives at the (j,n) and $(j,n+1)$ grid-points.

The diffusion equation applied at the point $(j\Delta x, n\Delta t+\frac{1}{2}\Delta t)$ is

$$\left.\frac{\partial\bar{\tau}}{\partial t}\right|_j^{n+\frac{1}{2}} = \alpha\left.\frac{\partial^2\bar{\tau}}{\partial x^2}\right|_j^{n+\frac{1}{2}}. \tag{4.2.9}$$

Replacing the right side of this equation by the average of the values at time-levels n and (n+1) gives

$$\left.\frac{\partial\bar{\tau}}{\partial t}\right|_j^{n+\frac{1}{2}} = \frac{\alpha}{2}\left\{\left.\frac{\partial^2\bar{\tau}}{\partial x^2}\right|_j^{n} + \left.\frac{\partial^2\bar{\tau}}{\partial x^2}\right|_j^{n+1} + O\{(\Delta t)^2\}\right\},$$

which becomes, on applying central difference forms for each derivative,

$$\frac{\bar{\tau}_j^{n+1}-\bar{\tau}_j^{n}}{\Delta t} + O\{(\Delta t)^2\} = \frac{\alpha}{2}\left\{\frac{\bar{\tau}_{j+1}^{n+1}-2\bar{\tau}_j^{n+1}+\bar{\tau}_{j-1}^{n+1}}{(\Delta x)^2} + \frac{\bar{\tau}_{j+1}^{n}-2\bar{\tau}_j^{n}+\bar{\tau}_{j-1}^{n}}{(\Delta x)^2} + O\{(\Delta x)^2\}\right.$$

$$\left. + O\{(\Delta t)^2\}\right\}. \tag{4.2.10}$$

Neglecting the terms of $O\{(\Delta t)^2,(\Delta x)^2\}$, which is therefore the usual order of the truncation error, this becomes, on writing the values of τ at the (n+1)th time level on the left side and the known values of τ at the nth time level on the right side,

$$-\tfrac{1}{2}s\tau_{j-1}^{n+1} + (1+s)\tau_j^{n+1} - \tfrac{1}{2}s\tau_{j+1}^{n+1} = \tfrac{1}{2}s\tau_{j-1}^{n} + (1-s)\tau_j^{n} + \tfrac{1}{2}s\tau_{j+1}^{n}. \tag{4.2.11}$$

Clearly, the error involved in using Equation (4.2.11) as an approximation for (4.2.10) is $O\{(\Delta t)^3,\Delta t(\Delta x)^2\}$. As the values of τ at the (n+1) time level are computed using this equation, the contribution to the discretisation error in stepping from one time level to the next is $O\{(\Delta t)^3,\Delta t(\Delta x)^2\}$.

Assuming values of τ_j^n, $j=0(1)J$, are known, the right side of this equation is known for $j=1(1)J-1$. Application of the boundary conditions which define the values of τ_0^{n+1}, τ_J^{n+1} then yields a set of simultaneous linear algebraic equations in the unknowns τ_j^{n+1}, $j=1(1)J-1$, similar to Equations (4.2.3). These are

$$\left.\begin{aligned}
(1+s)\tau_1^{n+1} - \tfrac{1}{2}s\tau_2^{n+1} &= d_1^n\\
-\tfrac{1}{2}s\tau_1^{n+1} + (1+s)\tau_2^{n+1} - \tfrac{1}{2}s\tau_3^{n+1} &= d_2^n\\
-\tfrac{1}{2}s\tau_2^{n+1} + (1+s)\tau_3^{n+1} - \tfrac{1}{2}s\tau_4^{n+1} &= d_3^n\\
&\vdots\\
-\tfrac{1}{2}s\tau_{j-1}^{n+1} + (1+s)\tau_j^{n+1} - \tfrac{1}{2}s\tau_{j+1}^{n+1} &= d_j^n\\
&\vdots\\
-\tfrac{1}{2}s\tau_{J-2}^{n+1} + (1+s)\tau_{J-1}^{n+1} &= d_{J-1}
\end{aligned}\right\} \tag{4.2.12}$$

where $d_1^n = \frac{1}{2}s\tau_0^n + (1-s)\tau_1^n + \frac{1}{2}s\tau_2^n + \frac{1}{2}s\bar{\tau}_0^{n+1}$,

$d_j^n = \frac{1}{2}s\tau_{j-1}^n + (1-s)\tau_j^n + \frac{1}{2}s\tau_{j+1}^n$, j=2(1)J-2,

$d_{J-1}^n = \frac{1}{2}s\tau_{J-2}^n + (1-s)\tau_{J-1}^n + \frac{1}{2}s\tau_J^n + \frac{1}{2}s\bar{\tau}_J^{n+1}$,

are all known. Because it is tridiagonal, this system can be solved efficiently using the Thomas algorithm. Commencing with n=0, then taking n=1,2,3,... in turn, values of τ_j^1, τ_j^2, τ_j^3, τ_j^4,.., j=1(1)J-1, are found.

When the stability of the finite difference equation (4.2.11) is investigated, it is found that errors ξ_j^n in the computation of τ_j^n propagate according to the relation

$$-\tfrac{1}{2}s\xi_{j-1}^{n+1} + (1+s)\xi_j^{n+1} - \tfrac{1}{2}s\xi_{j+1}^{n+1} = \tfrac{1}{2}s\xi_{j-1}^n + (1-s)\xi_j^n + \tfrac{1}{2}s\xi_{j+1}^n, \qquad (4.2.13)$$

for j=1(1)J-1. Application of the von Neumann method yields the amplification factor G for the mth Fourier component of the error distribution at any time level

$$G = \frac{1-2s\,\sin^2(\beta/2)}{1+2s\,\sin^2(\beta/2)}.$$

Because

$$|G| \leq 1$$

for all positive s and all β, it is clear that this method applied to the diffusion problem with given boundary values is *unconditionally stable*.

Alternatively, the stability criteria for the Crank-Nicolson method may be found by the matrix method. Since no errors occur at the boundaries j=0,J then $\xi_0^n = \xi_0^{n+1} = 0$ and $\xi_J^n = \xi_J^{n+1} = 0$. Therefore, when stepping from the nth time level to the (n+1)th time level, the error propagation may be represented by the matrix equation

$$\begin{bmatrix} (1+s) & -\frac{1}{2}s & & 0 \\ -\frac{1}{2}s & (1+s) & -\frac{1}{2}s & \\ & \cdot & \cdot & \cdot \\ 0 & & -\frac{1}{2}s & (1+s) \end{bmatrix} \begin{bmatrix} \xi_1^{n+1} \\ \xi_2^{n+1} \\ \vdots \\ \xi_{J-1}^{n+1} \end{bmatrix}$$

$$= \begin{bmatrix} (1-s) & \frac{1}{2}s & & 0 \\ \frac{1}{2}s & (1-s) & \frac{1}{2}s & \\ & \cdot & \cdot & \cdot \\ 0 & & \frac{1}{2}s & (1-s) \end{bmatrix} \begin{bmatrix} \xi_1^{n} \\ \xi_2^{n} \\ \vdots \\ \xi_{J-1}^{n} \end{bmatrix},$$

that is, by

$$(\underline{\underline{I}} + s\underline{\underline{B}})\underset{\sim}{\xi}^{n+1} = (\underline{\underline{I}} - s\underline{\underline{B}})\underset{\sim}{\xi}^{n},$$

where

$$\underset{\approx}{B} = \begin{bmatrix} 1 & -\frac{1}{2} & & 0 \\ -\frac{1}{2} & 1 & -\frac{1}{2} & \\ & \cdot & \cdot & \cdot \\ 0 & & -\frac{1}{2} & 1 \end{bmatrix}, \quad \underset{\sim}{\xi}^n = \begin{bmatrix} \xi_1^n \\ \xi_2^n \\ \vdots \\ \xi_{J-1}^n \end{bmatrix}.$$

Hence

$$\underset{\sim}{\xi}^{n+1} = (\underset{\approx}{I} + s\underset{\approx}{B})^{-1}(\underset{\approx}{I} - s\underset{\approx}{B})\underset{\sim}{\xi}^n. \tag{4.2.14}$$

Applying the same argument as that described in Section 3.3, it is seen that the finite difference scheme is stable for values of s for which the moduli of the eigenvalues of

$$\underset{\approx}{A} = (\underset{\approx}{I} + s\underset{\approx}{B})^{-1}(\underset{\approx}{I} - s\underset{\approx}{B})$$

are each less than one. Application of relation (3.3.16) then gives

$$(\lambda_B)_m = 2\sin^2(m\pi/2J),$$

$m=1(1)J-1$, as the eigenvalues of $\underset{\approx}{B}$. If f is an elementary function, and the eigenvalues of $\underset{\approx}{B}$ are λ_B then the eigenvalues of $f(\underset{\approx}{B})$ are $f(\lambda_B)$ (see Wade, 1951, p.109). It follows that the eigenvalues of $\underset{\approx}{A}$ are

$$(\lambda_A)_m = \{1 + 2s\sin^2(m\pi/2J)\}^{-1}\{1 - 2s\sin^2(m\pi/2J)\}, \tag{4.2.15}$$

$m=1(1)J-1$. Since the numerical values of $(\lambda_A)_m$ are clearly less than unity for all positive values of s, it follows that the Crank-Nicolson finite difference scheme has unrestricted stability.

The consistency of this finite difference scheme is investigated by substituting the exact solution $\bar{\tau}$ for the approximation τ in (4.2.11) and then expanding each term of the resulting equation in a Taylor series about the (j,n) grid-point, yielding

$$\left.\frac{\partial\bar{\tau}}{\partial t}\right|_j^n = \alpha\left.\frac{\partial^2\bar{\tau}}{\partial t^2}\right|_j^n + O\{(\Delta t)^2,(\Delta x)^2\}.$$

This gives Equation (4.1.1) in the limit as $\Delta x \to 0$, $\Delta t \to 0$, so the Crank-Nicolson equation is consistent with the original diffusion equation. It also has, as suggested previously, a truncation error which is $O\{(\Delta t)^2,(\Delta x)^2\}$.

As mentioned earlier, the tri-diagonal set of equations (4.2.12) can readily be solved by the Thomas algorithm so long as the coefficient matrix of the linear algebraic system is diagonally dominant. This is true for all $s > 0$, since for the rows indicated, it is true that

$$j=1,J-1 \quad : \quad |1+s| > |-\tfrac{1}{2}s|,$$

$$j=2(1)J-2 \quad : \quad |1+s| > |-\tfrac{1}{2}s| + |-\tfrac{1}{2}s|.$$

Thus both the classical and Crank-Nicolson implicit methods are *solvable* for all $s > 0$.

Clearly, the solution of the set of Crank-Nicolson equations (4.2.12) requires more calculations per time step than the classical implicit method to evaluate the d_j^n on the right side of the equations. Each d_j^n in Equation (4.2.12) includes two values of τ more than in the d_j^n of the classical implicit equations (4.2.3). What then is the advantage of the Crank-Nicolson scheme over the classical implicit method? Because at each time step the contribution to the discretisation errors are $O\{(\Delta t)^3, \Delta t(\Delta x)^2\}$ then the Crank-Nicholson method is more accurate than the classical implicit method which contributes at each step an error of $O\{(\Delta t)^2, \Delta t(\Delta x)^2$. For the same degree of accuracy, a larger time increment can be used for the Crank-Nicolson method and fewer time steps are necessary to compute values of τ up to a given time level.

It is of interest to note that the unconditionally stable Crank-Nicolson scheme may be considered as a combination of the conditionally stable FTCS explicit method and the unconditionally stable classical implicit method. Consider the FTCS method applied over the first half of the time interval from $n\Delta t$ to $(n+1)\Delta t$, producing values τ_j^* at the intermediate time $(n\Delta t + \tfrac{1}{2}\Delta t)$. The values of τ_j^* are therefore given in terms of τ_j^n by means of (2.4.4), namely

$$\tau_j^* = s^*\tau_{j-1}^n + (1-2s^*)\tau_j^n + s^*\tau_{j+1}^n \tag{4.2.16a}$$

where $s^* = \alpha(\Delta t/2)/(\Delta x)^2 = s/2$. Application of the classical implicit equation over the second half of the time interval Δt gives the values of τ_j^{n+1} in terms of τ_j^* according to the relation

$$-s^*\tau_{j-1}^{n+1} + (1+2s^*)\tau_j^{n+1} - s^*\tau_{j+1}^{n+1} = \tau_j^*. \tag{4.2.16b}$$

The result of adding these two equations is to give the finite difference equation,

$$-s^*\tau_{j-1}^{n+1} + (1+2s^*)\tau_j^{n+1} - s^*\tau_{j+1}^{n+1} = s^*\tau_{j-1}^n + (1-2s^*)\tau_j^n + s^*\tau_{j+1}^n, \tag{4.2.17}$$

which directly connects values of τ at time level $(n+1)$ with those at time level n. Substitution of $s/2$ for s^* then yields the Crank-Nicolson finite difference equation for stepping the one-dimensional diffusion equation from one time level to the next.

Therefore, every application of the Crank-Nicolson finite difference equation is, in effect, an application of the FTCS explicit equation over the first half of each time interval followed by the classical implicit equation over the second half of the interval. This time-splitting of a stable finite difference equation into two equations, one of which may be unstable by itself, is used extensively to solve parabolic partial differential equations in two and three space dimensions. This is the basis of the Alternating Direction Implicit methods (see Section 6.3).

The Crank-Nicolson method has been used to solve the problem presented in Section 2.1 and the results obtained are displayed in Table 4.2. Comparing these results with those in Table 3.1, obtained using the FTCS explicit method, it is clear that this implicit method gives more accurate answers. For the same grid-spacing Δx, a relatively large value of Δt and therefore of s, say $s = 4.0$, still yields a better answer than the FTCS method for $s = 1/2$, and for the optimal value $s = 1/6$.

TABLE 4.2

Crank-Nicolson approximations to $\bar{\tau}(0.4,8) = 0.4503963$

J	Δx	Δt/s	s = 1/2	s = 1	s = 2	s = 4
10	0.1	1	τ_4^{16} =.4515527	τ_4^{8} =.4517348	τ_4^{4} =.4518655	τ_4^{2} =.4568709
20	0.05	0.25	τ_8^{64} =.4506646	τ_8^{32} =.4506763	τ_8^{16} =.4507236	τ_8^{8} =.4508537
40	0.025	0.0625	τ_{16}^{256} =.4504620	τ_{16}^{128}=.4504628	τ_{16}^{64} =.4504658	τ_{16}^{32} =.4504777
50	0.02	0.04	τ_{20}^{400} =.4504383	τ_{20}^{200}=.4504386	τ_{20}^{100}=.4504129	τ_{20}^{50} =.4504447
100	0.01	0.01	τ_{40}^{1600}=.4504068	τ_{40}^{800}=.4504068	τ_{40}^{400}=.4504069	τ_{40}^{200}=.4504072

However, this increased accuracy is acquired only with an increase in the number of arithmetic operations required to obtain the result. Since the time taken to perform a multiplication or division on a computer is generally much longer than that required to perform an addition or subtraction - for example, 2.9×10^{-6} sec compared with 0.55×10^{-6} sec on the CYBER 173 - only the number of multiplications and divisions required will be considered when comparing the computational times required for each of the above examples.

If the formula used with the FTCS method is written

$$\tau_j^{n+1} = s(\tau_{j-1}^n + \tau_{j+1}^n) + (1-2s)\tau_j^n,$$

then only two multiplications are used for every application of the formula, unless $s = \frac{1}{2}$ when only one division is required. Therefore, computing all the values of τ at each time level requires $2(J-1)$, or approximately $2J$ multiplications. For example, the answer obtained in Table 4.2 when J=20, $s=\frac{1}{2}$, required $64 \times 1 \times 20 = 1{,}280$ such operations. When using the Crank-Nicolson method with the Thomas algorithm, to compute values of τ at each new time level required approximately 5J multiplications or divisions. In addition, at each time step the values d_j^n, $j=1(1)J-1$, which appear on the right side of Equation (4.2.12), must be obtained and this requires approximately 2J multiplications, unless $s = 1$ when the number of multiplications is only J. Hence, a total of about 7J multiplications or divisions are required to advance one time step if $s \neq 1$, or approximately 6J multiplications if $s = 1$. For example, when J=20 and $s = \frac{1}{2}$, 64 time steps had to be made to reach $t = 8$ and so about $64 \times 7 \times 20 = 8{,}960$ multiplications or divisions were required. Therefore, using the Crank-Nicolson method required about 7 times the computational effort to reach $t = 8$ compared with the FTCS explicit method. However, the answer obtained is much more accurate.

A comparison of the computational efficiency of the explicit FTCS method and the implicit Crank-Nicolson method, is given in Table 4.3. With a fixed grid-spacing in the x-direction, the value of Δt was varied, and the absolute error in the value of the approximation τ computed for $\bar{\tau}(0.4,8)$ was compared with the approximate number of arithmetic operations required to achieve the result. It took 160,000 such operations to obtain a result with an error of 25×10^{-5} with $s = \frac{1}{2}$ in the FTCS method; with s=1/6 six times as many operations (960,000) reduced the error to one-sixth (4.5×10^{-5}). But using s=1 with the Crank-Nicholson method took only 480,000 operations to give an error one-quarter of this, namely 1.05×10^{-5}.

With s = 8, the number of operations required with the implicit method decreased to 70,000 while the error increased only slightly. A direct comparison of the FTCS method with s = ½ and the Crank-Nicolson method with s = 8 indicates the superiority of the latter; less than half the number of computations produced a result with less than one-twentieth of the error produced by the FTCS scheme.

TABLE 4.3

Comparison of computational efficiency of the FTCS and Crank-Nicolson methods for calculating $\bar{\tau}(0.4,8)$ = 0.4503963 for the diffusion problem.

	FTCS Method		Crank-Nicolson Method	
J=100	s = 1/6	s = ½	s = 1	s = 8
Mult. and Div. per time-level	200	100	600	700
Number of time-levels	4800	1600	800	100
Total number of Mult. and Div.	960,000	160,000	480,000	70,000
ABSOLUTE ERROR	4.51×10^{-5}	24.96×10^{-5}	1.05×10^{-5}	1.21×10^{-5}

The method of weighted averages

Crandall (1955) used a weighted average instead of the Crank-Nicolson simple average of the central difference approximations to $\partial^2\bar{\tau}/\partial x^2$ at time levels n and (n+1) in Equation (4.2.10). The diffusion equation (2.1.1) is evaluated at the point $(j\Delta x, n\Delta t+\theta\Delta t)$, $0 \le \theta \le 1$, namely

$$\left.\frac{\partial\bar{\tau}}{\partial t}\right|_j^{n+\theta} = \alpha\left.\frac{\partial^2\bar{\tau}}{\partial x^2}\right|_j^{n+\theta} . \tag{4.2.18}$$

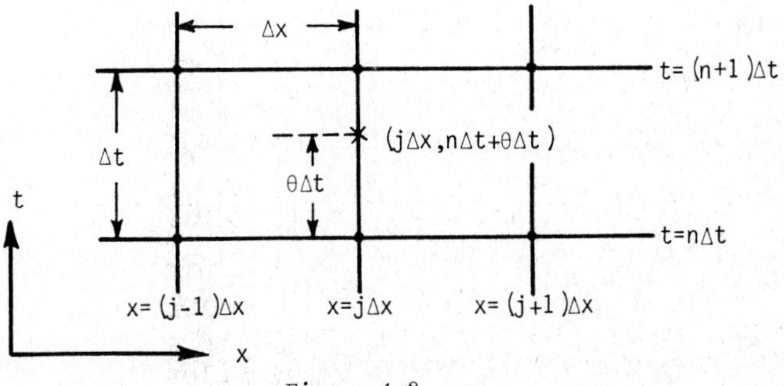

Figure 4. 2

We now express the derivatives in Equation (4.2.18) in terms of values of $\bar{\tau}$ at surrounding gridpoints (see Figure 4.2). The time derivative can be written

$$\left.\frac{\partial\bar{\tau}}{\partial t}\right|_j^{n+\theta} = \frac{\bar{\tau}_j^{n+1} - \bar{\tau}_j^{n}}{\Delta t} + O\{(1-2\theta)\Delta t,(\Delta t)^2\}, \tag{4.2.19}$$

where the additional terms on the right side are $O\{\Delta t\}$, except for $\theta=\frac{1}{2}$ when they are $O\{(\Delta t)^2\}$. The space derivative in Equation (4.2.18) can be written as a weighted average of the values at time levels n and (n+1) using the relation

$$\bar{\tau}_j^{n+\theta} = (1-\theta)\bar{\tau}_j^n + \theta\bar{\tau}_j^{n+1} + O\{(\Delta t)^2\}, \quad 0 \le \theta \le 1. \tag{4.2.20}$$

If θ = 0,1 the terms of $O\{(\Delta t)^2\}$ are omitted because (4.2.20) is exact. Therefore (4.2.18) becomes

$$\frac{\bar{\tau}_j^{n+1} - \bar{\tau}_j^n}{\Delta t} = \alpha\left\{(1-\theta)\frac{\partial^2\bar{\tau}}{\partial x^2}\bigg|_j^n + \theta\,\frac{\partial^2\bar{\tau}}{\partial x^2}\bigg|_j^{n+1}\right\} + O\{(1-2\theta)\Delta t,(\Delta t)^2\}. \tag{4.2.21}$$

Writing the space derivatives on the right side of (4.2.21) in central difference form yields

$$\frac{\bar{\tau}_j^{n+1} - \bar{\tau}_j^n}{\Delta t} = \alpha(1-\theta)\left\{\frac{\bar{\tau}_{j+1}^n - 2\bar{\tau}_j^n + \bar{\tau}_{j-1}^n}{(\Delta x)^2}\right\} + \alpha\theta\left\{\frac{\bar{\tau}_{j+1}^{n+1} - 2\bar{\tau}_j^{n+1} + \bar{\tau}_{j-1}^{n+1}}{(\Delta x)^2}\right\}$$

$$+ O\{(1-2\theta)\Delta t,(\Delta t)^2,(\Delta x)^2\}. \tag{4.2.22}$$

The finite difference approximation to the diffusion equation is therefore

$$\frac{\tau_j^{n+1} - \tau_j^n}{\Delta t} = \alpha(1-\theta)\left\{\frac{\tau_{j+1}^n - 2\tau_j^n + \tau_{j-1}^n}{(\Delta x)^2}\right\} + \alpha\theta\left\{\frac{\tau_{j+1}^{n+1} - 2\tau_j^{n+1} + \tau_{j-1}^{n+1}}{(\Delta x)^2}\right\}, \tag{4.2.23}$$

where $0 \le \theta \le 1$, the truncation error being $O\{(1-2\theta)\Delta t,(\Delta t)^2,(\Delta x)^2\}$. As in the Crank-Nicolson method, three values of τ at the (n+1)th time level are expressed in terms of the known values of τ at the nth time level, in the following way:

$$-\theta s\tau_{j-1}^{n+1} + (1+2\theta s)\tau_j^{n+1} - \theta s\tau_{j+1}^{n+1}$$

$$= s(1-\theta)\tau_{j-1}^n + [1-2s(1-\theta)]\tau_j^n + s(1-\theta)\tau_{j+1}^n, \tag{4.2.24}$$

for j=1(1)J-1.

The contributions to the discretisation error in the approximation to $\bar{\tau}_j^{n+1}$ is clearly of $O\{(1-2\theta)(\Delta t)^2,(\Delta t)^3,\Delta t(\Delta x)^2\}$ in the step from the time level n to the time level (n+1).

If the weighting factor is θ = 1, Equation (4.2.24) reduces to the classical implicit method considered at the start of this section. If $\theta = \frac{1}{2}$ the Crank-Nicolson method is obtained, and if θ = 0 the FTCS explicit scheme results.

If $\theta \neq 0$, this scheme results in the following tri-diagonal set of (J-1) simultaneous linear algebraic equations for the unknown values τ_j^{n+1}, j=1(1)J-1, in terms of the known values of τ_j^n, j=0(1)J, at the nth time level and the known boundary values τ_0^{n+1}, τ_J^{n+1} at the (n+1)th time level:

$$\left.\begin{aligned}
&(1+2\theta s)\tau_1^{n+1} - \theta s\tau_2^{n+1} &&= d_1^n\\
&-\theta s\tau_1^{n+1} + (1+2\theta s)\tau_2^{n+1} - \theta s\tau_3^{n+1} &&= d_2^n\\
&\quad -\theta s\tau_2^{n+1} + (1+2\theta s)\tau_3^{n+1} - \theta s\tau_4^{n+1} &&= d_3^n\\
&\qquad \ddots &&\vdots\\
&\qquad -\theta s\tau_{j-1}^{n+1} + (1+2\theta s)\tau_j^{n+1} - \theta s\tau_{j+1}^{n+1} &&= d_j^n\\
&\qquad\qquad \ddots &&\vdots\\
&\qquad\qquad -\theta s\tau_{J-2}^{n+1} + (1+2\theta s)\tau_{J-1}^{n+1} &&= d_{J-1}^n
\end{aligned}\right\} \qquad (4.2.25)$$

where

$$d_1^n = s(1-\theta)\tau_0^n + [1-2s(1-\theta)]\tau_1^n + s(1-\theta)\tau_2^n + \theta s\tau_0^{n+1},$$

$$d_j^n = s(1-\theta)\tau_{j-1}^n + [1-2s(1-\theta)]\tau_j^n + s(1-\theta)\tau_{j+1}^n, \quad j=2(1)J-2$$

$$d_{J-1}^n = s(1-\theta)\tau_{J-2}^n + [1-2s(1-\theta)]\tau_{J-1}^n + s(1-\theta)\tau_J^n + \theta s\tau_J^{n+1},$$

are all known. With $\theta = 1$ this reduces to the system of Equations (4.2.3), and with $\theta = \frac{1}{2}$ this becomes the system (4.2.12). The system of Equations (4.2.25) can be solved using the Thomas algorithm provided the coefficient matrix is diagonally dominant. As $\theta \geq 0$, $s > 0$, then

$$|1+2\theta s| > |-\theta s| + |-\theta s|$$

so the system is diagonally dominant and is *solvable*.

Application of the von Neumann stability analysis to Equation (4.2.24) yields the amplification factor for the mth Fourier component of the error distribution,

$$G = \frac{1-4s(1-\theta)\sin^2(\beta/2)}{1+4s\theta\sin^2(\beta/2)} .$$

With $s > 0$ and $\theta \geq 0$ the denominator of G is always positive, so that, as G is real, the stability condition $|G| \leq 1$ requires

$$-1 - 4s\theta\sin^2(\beta/2) \leq 1 - 4s(1-\theta)\sin^2(\beta/2) \leq 1 + 4s\theta\sin^2(\beta/2). \qquad (4.2.26)$$

The right-hand inequality of (4.2.26) requires

$$4s\sin^2(\beta/2) \geq 0,$$

which is always satisfied because $s > 0$. The left-hand inequality of (4.2.26) requires

$$2s(1-2\theta)\sin^2(\beta/2) \leq 1. \qquad (4.2.27)$$

If $\theta < \frac{1}{2}$, this is true if

$$s \leq \frac{1}{2(1-2\theta)\sin^2(\beta/2)} ,$$

which is satisfied for all β if

$$s \le \frac{1}{2(1-2\theta)}\,. \tag{4.2.28}$$

If $\theta \ge \frac{1}{2}$, since $s > 0$ the left side of condition (4.2.27) is always negative or zero and the inequality is satisfied. The weighted average scheme defined by Equation (4.2.24) is therefore *conditionally stable*, requiring

$$\begin{aligned} 0 < s \le \frac{1}{2(1-2\theta)} \quad & \text{for } 0 \le \theta < \tfrac{1}{2}, \\ s > 0 \quad & \text{for } \tfrac{1}{2} \le \theta \le 1. \end{aligned} \tag{4.2.29}$$

The consistency of Equation (4.2.24) is formally established by substituting $\bar{\tau}$, the solution of the diffusion equation, for the approximate value τ, and expanding each term in a Taylor series about the (j,n) grid-point. This yields, on application of the relation $\partial\bar{\tau}/\partial t = \alpha\, \partial^2\bar{\tau}/\partial x^2$ and using $s = \alpha\, \Delta t/(\Delta x)^2$,

$$\left.\frac{\partial\bar{\tau}}{\partial t}\right|_j^n = \alpha \left.\frac{\partial^2\bar{\tau}}{\partial x^2}\right|_j^n + E_j^n \tag{4.2.30a}$$

where

$$E_j^n = -\,\Delta t \left.\frac{\partial^2\bar{\tau}}{\partial t^2}\right|_j^n (\tfrac{1}{2}-\theta) + \frac{(\Delta x)^2}{12\alpha} \left.\frac{\partial^2\bar{\tau}}{\partial t^2}\right|_j^n + O\{(\Delta t)^2,(\Delta x)^4\}. \tag{4.2.30b}$$

Clearly, $E_j^n \to 0$ and (4.2.30a) yields diffusion equation in the limit as $\Delta t \to 0$, $\Delta x \to 0$. Thus the finite difference equation (4.2.24) is consistent with the one-dimensional diffusion equation and generally has truncation error of $O\{\Delta t,(\Delta x)^2\}$. The truncation error reduces to $O\{(\Delta t)^2,(\Delta x)^2\}$ when $\theta = \frac{1}{2}$, as then the term in Δt vanishes. This is the result established previously, that the truncation error for the Crank-Nicolson method is second order in both Δt and Δx.

The accuracy of (4.2.24) can be further improved by suitably choosing θ and s so both the terms in Δt and $(\Delta x)^2$ disappear from Equation (4.2.30b), since the truncation error may be rewritten

$$E_j^n = \Delta t \left.\frac{\partial^2\bar{\tau}}{\partial t^2}\right|_j^n \{-\tfrac{1}{2} + \theta + \frac{1}{12s}\} + O\{(\Delta t)^2,(\Delta x)^4\}. \tag{4.2.31}$$

The error E_j^n will be of $O\{(\Delta t)^2,(\Delta x)^4\}$ if θ and s are chosen to satisfy

$$\theta = \tfrac{1}{2} - \frac{1}{12s}. \tag{4.2.32}$$

Since $\Delta t = s(\Delta x)^2/\alpha$, this is an error of $O\{(\Delta x)^4\}$ if s is fixed.

Equation (4.2.32) is compatible with the stability relation

$$s < \frac{1}{2(1-2\theta)}, \tag{4.2.33}$$

which is necessary for $\theta < \frac{1}{2}$, since the inequality (4.2.33) may be rewritten as

$$\theta > \tfrac{1}{2} - \frac{1}{4s},$$

which is true for all θ given by the relation (4.2.32).

The implicit methods are only advantageous if a value of s much greater than one-half is chosen, otherwise the FTCS explicit method is more suitable. Equation (4.2.32) suggests that suitable values of s and θ for a stable method with minimal truncation error might be $s = 1$ with $\theta = 5/12$, $s = 2$ with $\theta = 11/24$, $s = 3$ with $\theta = 17/36$, and so on.

These results for the stability and accuracy of the weighted average method are summarised in Figure 4.3. Each point in the (s,θ) plane represents a different weighted-average finite difference scheme for solving the one-dimensional diffusion equation. All are implicit except those with $\theta = 0$. The range of stability is shown together with the family of formulae for which the truncation error is $O\{(\Delta t)^2,(\Delta x)^2\}$ and $O\{(\Delta t)^2,(\Delta x)^4\}$ instead of the usual $O\{\Delta t,(\Delta x)^2\}$. The actual stability limits depend on the size of the grid-spacing Δx and on the nature of the boundary conditions, but as Δx decreases there is a rapid approach to the limiting values shown. Note that, for formulae that have a truncation error $O\{(\Delta x)^4\}$, better overall accuracy is obtained if the boundary conditions are incorporated with a contribution to the discretisation error of the same order as the contribution made by the formulae which apply at grid-points in the interior of the solution domain.

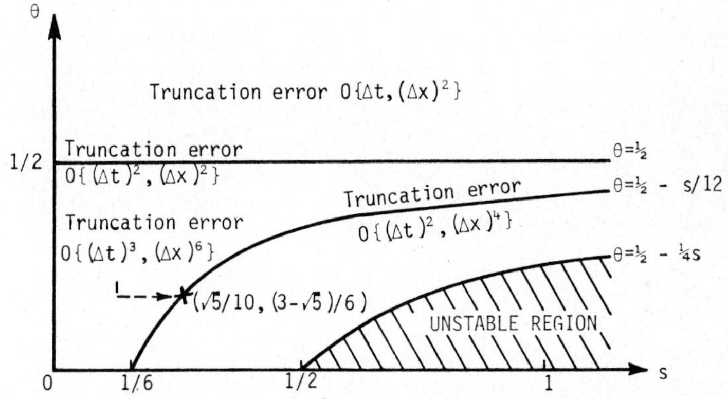

Figure 4.3 : Properties of the weighted average implicit finite-difference approximations for the one-dimensional diffusion equation.

The problem presented in Section 2.1 has been solved using the method of weighted averages with $J = 40$ and various s, and the results are presented in Table 4.4. As expected, the Crank-Nicolson method ($\theta = \frac{1}{2}$) gives more accurate results than either the explicit ($\theta = 0$) or classical implicit ($\theta = 1$) methods. However, the Crank-Nicolson results are also more accurate than those obtained using the Crandall method in which an optimal θ is chosen for a given value of s according to Equation (4.2.32). This apparent disagreement with the theoretical result obtained above has also been noted by other workers in this field. For example, Moore, Kaplan and Mitchell (1975) found that the Crank-Nicolson method gave more accurate answers to the boundary value problem they were modelling than did the Crandall method.

TABLE 4.4

Numerical approximations to $\bar{\tau}(0.4,8) = 0.4503963$ of the diffusion equation using the method of weighted averages with J=40 and various s

s	Explicit $\theta=0$	Fully Implicit $\theta=1$	Crank-Nicolson $\theta=\frac{1}{2}$	Crandall's Method $\theta=\frac{1}{2}-1/12s$
½	0.4519564	0.4498152	0.4504620	0.4506782
1	-	0.4491719	0.4504628	0.4506790
2	-	0.4478948	0.4504658	0.4506820
4	-	0.4453794	0.4504777	0.4506941
8	-	0.4405031	0.4505257	0.4507418
16	-	0.4313556	0.4506036	0.4508018
32	-	0.4152484	0.4506501	0.4508143

It is also interesting to note that the terms of order $(\Delta t)^2$ in Equation (4.2.31) can be made zero for a particular value of s and θ as well as the term of order Δt. Crandall (1955) showed that choosing $s = \sqrt{5}/10$ and $\theta = (3-\sqrt{5})/6$ gives a finite difference approximation which has a truncation error of $O\{(\Delta x)^6\}$. These values of s and θ have also been used to solve the problem presented in Section 2. For J=100 the result $\tau(0.4,8) = 0.4503557$ was obtained; for J=50, 0.4503700 and for J=40 the answer obtained was 0.4503505. A disadvantage of this method is that $\sqrt{5}$ is irrational and so it is impossible to find τ exactly at the time t = 8.0 secs. For example, the above results are those values calculated at times t = 7.9984, t = 7.9962 and t = 7.9939 respectively. Such higher order methods seem to be of little practical use since there are methods available such as Richardson's extrapolation (see Section 9) which give results at some grid points just as accurately as the above but with a greatly reduced number of calculations.

A *semi-implicit* method, one which is implicit and therefore has the advantage of being very stable but which does not have the disadvantage of requiring the solution of a set of linear algebraic equations, has been described by Saul'yev (1964). The method is based on the following discretisation of the one dimensional diffusion equation (2.1.1) at the point $(j\Delta x, n\Delta t+\frac{1}{2}\Delta t)$, using only values of the dependent variable at gridpoints (j-1,n+1), (j,n+1), (j,n) and (j+1,n) - see Figure 4.4.

The time derivative in the differential equation

$$\left.\frac{\partial\bar{\tau}}{\partial t}\right|_j^{n+\frac{1}{2}} = \alpha \left.\frac{\partial^2\bar{\tau}}{\partial x^2}\right|_j^{n+\frac{1}{2}}$$

is replaced by the central difference form

$$\left.\frac{\partial\bar{\tau}}{\partial t}\right|_j^{n+\frac{1}{2}} = \frac{\bar{\tau}_j^{n+1} - \bar{\tau}_j^{n}}{\Delta t} + O\{(\Delta t)^2\}. \quad (4.2.34)$$

The spatial derivative is replaced as follows:

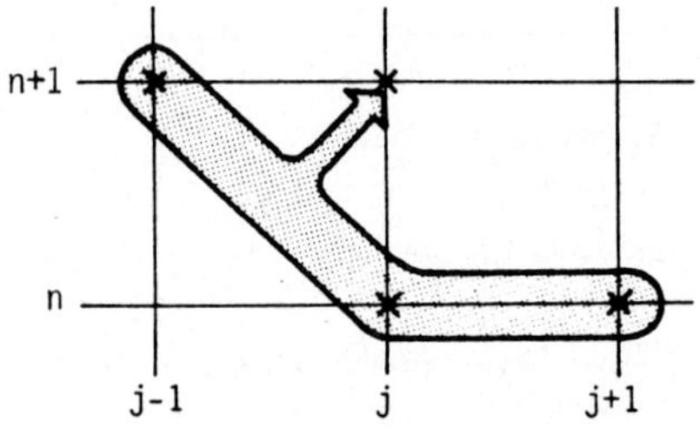

Figure 4.4

$$\left.\frac{\partial^2\bar{\tau}}{\partial x^2}\right|_j^{n+\frac{1}{2}} = \left.\frac{\partial}{\partial x}\left[\frac{\partial\bar{\tau}}{\partial x}\right]\right|_j^{n+\frac{1}{2}}$$

$$= \frac{1}{\Delta x}\left\{\left.\frac{\partial\bar{\tau}}{\partial x}\right|_{j+\frac{1}{2}}^{n+\frac{1}{2}} - \left.\frac{\partial\bar{\tau}}{\partial x}\right|_{j-\frac{1}{2}}^{n+\frac{1}{2}}\right\} + O\{(\Delta x)^2\}$$

$$= \frac{1}{\Delta x}\left\{\left.\frac{\partial\bar{\tau}}{\partial x}\right|_{j+\frac{1}{2}}^{n} - \left.\frac{\partial\bar{\tau}}{\partial x}\right|_{j-\frac{1}{2}}^{n+1} + O\{\Delta t\}\right\} + O\{(\Delta x)^2\}$$

$$= \frac{1}{\Delta x}\left\{\frac{\bar{\tau}_{j+1}^{n} - \bar{\tau}_{j}^{n}}{\Delta x} - \frac{\bar{\tau}_{j}^{n+1} - \bar{\tau}_{j-1}^{n+1}}{\Delta x} + O\{(\Delta x)^2\}\right\} + O\{\Delta t/\Delta x,(\Delta x)^2\} \tag{4.2.35}$$

Substitution in the diffusion equation, dropping terms of $O\{\Delta x,\Delta t/\Delta x\}$ and rearranging, gives the difference equation

$$\tau_j^{n+1} = \frac{1-s}{1+s}\tau_j^n + \frac{s}{1+s}(\tau_{j-1}^{n+1} + \tau_{j+1}^n). \tag{4.2.36}$$

Commencing with a given boundary value τ_0^{n+1}, the values of τ_j^{n+1} may be computed in the order $j=1(1)J-1$, since τ_{j-1}^{n+1} is known at each application of (4.2.36). In this manner the implicit finite difference equation (4.2.36) is computed as though it is explicit in nature.

A similar form

$$\tau_j^{n+1} = \frac{1-s}{1+s}\tau_j^n + \frac{s}{1+s}(\tau_{j+1}^{n+1} + \tau_{j-1}^n),\ j=J-1(1)1, \tag{4.2.37}$$

may be derived, using values of τ at the gridpoints shown in Figure 4.5. This form sweeps from right to left across the spatial grid commencing with the known boundary value τ_J^{n+1}.

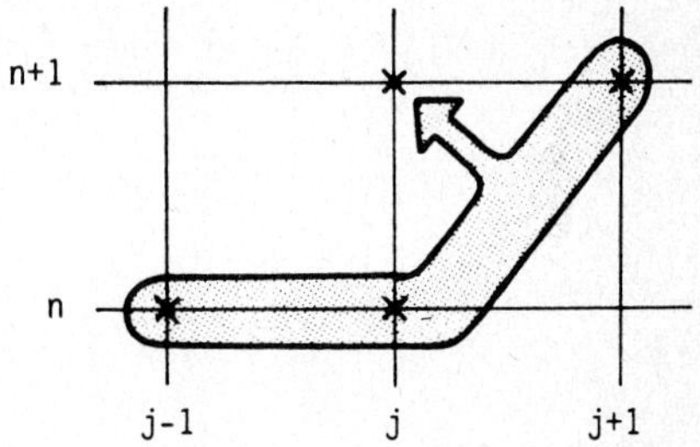

Figure 4.5

Applications of (4.2.36), which sweeps from left to right, for $n=0,2,4,\ldots$ with (4.2.37), which sweeps from right to left, used for $n=1,3,5,\ldots$, produces a method which is more accurate than the individual equations alone. This is because the truncation error of each of the separate steps is $O\{\Delta x,\Delta t/\Delta x\}$, whereas for the two step procedure it is $O\{(\Delta x)^4,(\Delta t)^2/\Delta x\}$.

A consistency analysis of this pair of equations may be carried out by considering them centred on the (j,n) gridpoint. Taking the first of these to be (4.2.37) with $n+1$ replaced by n, namely

$$\frac{1}{\Delta t}(\bar{\tau}_j^n - \bar{\tau}_j^{n-1}) = \frac{\alpha}{(\Delta x)^2}\{(\bar{\tau}_{j+1}^n - \bar{\tau}_j^n) - (\bar{\tau}_j^{n-1} - \bar{\tau}_j^{n-1})\}, \tag{4.2.38}$$

which is obtained by rearrangement and division by Δt. Substitution of the following forms given in Appendix 3,

$$\bar{\tau}_j^n - \bar{\tau}_j^{n-1} = \Delta t\left.\frac{\partial\bar{\tau}}{\partial t}\right|_j^n - \frac{(\Delta t)^2}{2}\left.\frac{\partial^2\bar{\tau}}{\partial t^2}\right|_j^n + \frac{(\Delta t)^3}{6}\left.\frac{\partial^3\bar{\tau}}{\partial t^3}\right|_j^n - \ldots$$

$$\bar{\tau}_{j+1}^n - \bar{\tau}_j^n = \Delta x\left.\frac{\partial\bar{\tau}}{\partial x}\right|_j^n + \frac{(\Delta x)^2}{2}\left.\frac{\partial^2\bar{\tau}}{\partial x^2}\right|_j^n + \frac{(\Delta x)^3}{6}\left.\frac{\partial^3\bar{\tau}}{\partial x^3}\right|_j^n + \frac{(\Delta x)^4}{24}\left.\frac{\partial^4\bar{\tau}}{\partial x^4}\right|_j^n + \ldots$$

$$\bar{\tau}_j^{n-1} - \bar{\tau}_{j-1}^{n-1} = \Delta x\frac{\partial\bar{\tau}}{\partial x}\bigg|_j^n - \frac{(\Delta x)^2}{2}\frac{\partial^2\bar{\tau}}{\partial x^2}\bigg|_j^n + \frac{(\Delta x)^3}{6}\frac{\partial^3\bar{\tau}}{\partial x^3}\bigg|_j^n - \frac{(\Delta x)^4}{24}\frac{\partial^4\bar{\tau}}{\partial x^4}\bigg|_j^n$$

$$- \Delta x\Delta t\frac{\partial^2\bar{\tau}}{\partial x\partial t}\bigg|_j^n + \frac{(\Delta x)^2\Delta t}{2}\frac{\partial^3\bar{\tau}}{\partial x^2\partial t} + \frac{\Delta x(\Delta t)^2}{2}\frac{\partial^3\bar{\tau}}{\partial x\partial t^2}\bigg|_j^n$$

$$- \frac{(\Delta x)^3\Delta t}{6}\frac{\partial^4\bar{\tau}}{\partial x^3\partial t}\bigg|_j^n - \frac{(\Delta x)^2(\Delta t)^2}{4}\frac{\partial^4\bar{\tau}}{\partial x^2\partial t^2}\bigg|_j^n - \frac{\Delta x(\Delta t)^3}{6}\frac{\partial^4\bar{\tau}}{\partial x\partial t^3}\bigg|_j^n + \ldots,$$

into Equation (4.2.38) gives

$$\frac{\partial\bar{\tau}}{\partial t}\bigg|_j^n = \alpha\frac{\partial^2\bar{\tau}}{\partial x^2}\bigg|_j^n + E^{(1)}\big|_j^n , \qquad (4.2.39)$$

where

$$E^{(1)} = \frac{\Delta t}{2}\frac{\partial^2\bar{\tau}}{\partial t^2} - \frac{(\Delta t)^2}{6}\frac{\partial^3\bar{\tau}}{\partial t^3} + \ldots$$

$$+ \frac{\alpha(\Delta x)^2}{12}\frac{\partial^4\bar{\tau}}{\partial x^4} + \alpha\frac{\Delta t}{\Delta x}\frac{\partial^2\bar{\tau}}{\partial x\partial t} - \frac{\alpha\Delta t}{2}\frac{\partial^3\bar{\tau}}{\partial x^2\partial t} - \frac{\alpha(\Delta t)^2}{2\Delta x}\frac{\partial^3\bar{\tau}}{\partial x\partial t^2} +$$

$$+ \frac{\alpha\Delta x\Delta t}{6}\frac{\partial^4\bar{\tau}}{\partial x^3\partial t} + \frac{\alpha(\Delta t)^2}{4}\frac{\partial^4\bar{\tau}}{\partial x^2\partial t^2} + \frac{\alpha(\Delta t)^3}{6\Delta x}\frac{\partial^4\bar{\tau}}{\partial x\partial t^3} - \ldots \qquad (4.2.40)$$

This step of the Saul'yev method is consistent with the transport equation only if $E^{(1)} \to 0$ as $\Delta t \to 0$, $\Delta x \to 0$, which requires the additional condition that $\Delta t/\Delta x \to 0$. It has a truncation error of $O\{\Delta t/\Delta x, (\Delta x)^2\}$.

The step which follows Equation (4.2.37) will use Equation (4.2.36). A similar rearrangement and substitution to that carried out for the first step, yields the corresponding partial differential equation

$$\frac{\partial\bar{\tau}}{\partial t}\bigg|_j^n = \alpha\frac{\partial^2\bar{\tau}}{\partial x^2}\bigg|_j^n + E^{(2)}\big|_j^n , \qquad (4.2.41)$$

where

$$E^{(2)} = - \frac{\Delta t}{2}\frac{\partial^2\bar{\tau}}{\partial t^2} - \frac{(\Delta t)^2}{6}\frac{\partial^3\bar{\tau}}{\partial t^3} + \ldots$$

$$+ \frac{\alpha(\Delta x)^4}{12}\frac{\partial^4\bar{\tau}}{\partial x^4} - \frac{\alpha\Delta t}{\Delta x}\frac{\partial^2\bar{\tau}}{\partial x\partial t} + \frac{\alpha\Delta t}{2}\frac{\partial^3\bar{\tau}}{\partial x^2\partial t} - \frac{\alpha(\Delta t)^2}{2\Delta x}\frac{\partial^4\bar{\tau}}{\partial x\partial t^2}$$

$$- \frac{\alpha\Delta x\Delta t}{6}\frac{\partial^4\bar{\tau}}{\partial x^3\partial t} + \frac{\alpha(\Delta t)^2}{4}\frac{\partial^4\bar{\tau}}{\partial x^2\partial t^2} - \frac{\alpha(\Delta t)^3}{6\Delta x}\frac{\partial^4\bar{\tau}}{\partial x\partial t^3} + \ldots \qquad (4.2.42)$$

Again, as with the other step of the method, there is a truncation error of $O\{(\Delta x)^2, \Delta t/\Delta x\}$, so consistency requires $\Delta t/\Delta x \to 0$ as well as $\Delta t \to 0$, $\Delta x \to 0$.

However, treated as a two-step method the truncation error will be the average of $E^{(1)}$ and $E^{(2)}$, namely

$$E(x,t) = - \frac{(\Delta t)^2}{6}\frac{\partial^3\bar{\tau}}{\partial t^3} + \frac{\alpha(\Delta x)^4}{12}\frac{\partial^4\bar{\tau}}{\partial x^4} - \frac{\alpha(\Delta t)^2}{2\Delta x}\frac{\partial^3\bar{\tau}}{\partial x\partial t^2} + \frac{\alpha(\Delta t)^2}{4}\frac{\partial^4\bar{\tau}}{\partial x^2\partial t^2} + \ldots, \qquad (4.2.43)$$

which is $O\{(\Delta x)^4, (\Delta t)^2/\Delta x\}$. Many of the terms in the truncation errors for the separate steps have cancelled. Consistency for the two-step process requires $(\Delta t)^2/\Delta x \to 0$ as $\Delta t \to 0$, $\Delta x \to 0$. This is much less restrictive than the consistency requirement for the separate equations, merely requiring Δt, Δx to be of the same order of magnitude as they tend to zero.

A von Neumann stability analysis of Equation (4.2.36) gives the amplification factor

$$G^{(1)} = \frac{1 - s + s\cos\beta + i\, s\sin\beta}{1 + s - s\cos\beta + i\, s\sin\beta}, \tag{4.2.44}$$

for that step. Then

$$|G^{(1)}|^2 = \frac{(1 - s + s\cos\beta)^2 + (s\sin\beta)^2}{(1 + s - s\cos\beta)^2 + (s\sin\beta)^2}$$

$$= \frac{1 - 4(s - s^2)\sin^2(\beta/2)}{1 + 4(s + s^2)\sin^2(\beta/2)} \tag{4.2.45}$$

so that

$$|G^{(1)}|^2 - 1 = \frac{-8s\sin^2(\beta/2)}{1 + 4(s + s^2)\sin^2(\beta/2)} \le 0$$

for all β so long as $s > 0$. Therefore $|G^{(1)}| \le 1$ for all $s > 0$, so (4.2.36) is unconditionally stable.

Similarly, for Equation (4.2.37) the amplification factor of the von Neumann method is

$$G^{(2)} = \frac{1 - s + s\cos\beta - i\, s\sin\beta}{1 + s - s\cos\beta - i\, s\sin\beta}, \tag{4.2.46}$$

which is the complex conjugate of $G^{(1)}$. Therefore

$$|G^{(2)}| = |G^{(1)}|,$$

and (4.2.37) is unconditionally stable.

Clearly, the two-step method defined by a left to right sweep across the spatial grid using (4.2.36) followed by a right to left sweep using (4.2.37) is unconditionally stable.

It is difficult to determine the order of the discretisation errors, e, in the double sweep of the Saul'yev method, from the truncation error (4.2.43). However, they are very nearly $O\{(\Delta x)^2\}$ in the computed approximations to the value of $\bar{\tau}(0.4,8) = 0.4318184$ using this numerical technique.

In Figure 4.6, graphs of $\log_{10}|e|$ against $\log_{10}(\Delta x)$ have slopes which are close to 2 for each s. The lines of best fit gave the following formulae for e in terms of Δx:

$$s = 1/2,\quad e = -0.95(\Delta x)^{1.97},$$

$$s = 1\ ,\quad e = -2.93(\Delta x)^{1.96},$$

$$s = 2\ ,\quad e = -10.3(\Delta x)^{1.94},$$

$$s = 4\ ,\quad e = -31.9(\Delta x)^{1.89}.$$

It is clearly more important to make s small than to make Δx small.

The various methods of solving the diffusion equation, with comments on their accuracy, stability and other limitations, is given in Table 4.5.

TABLE 4.5

Finite difference techniques for solving the diffusion equation $\frac{\partial \bar{\tau}}{\partial t} = \alpha \frac{\partial^2 \bar{\tau}}{\partial x^2}$.

Method	Finite difference equation	Truncation error E for fixed s	Amplification factor G	Stability restriction
FTCS Explicit	Eq. (2.4.4) $\tau_j^{n+1} = s\tau_{j-1}^n + (1-2s)\tau_j^n + s\tau_{j+1}^n$	Eq. (3.2.6) $O\{(\Delta x)^2\}$ for $s \neq 1/6$; $O\{(\Delta x)^4\}$ for $s = 1/6$	$1 - 4s\sin^2(\frac{\beta}{2})$	$s \leq \frac{1}{2}$
DuFort-Frankel Explicit	Eq. (4.1.13) $\tau_j^{n+1} = \frac{2s}{1+2s}\{\tau_{j-1}^n + \tau_{j+1}^n\} + \frac{1-2s}{1+2s}\tau_j^{n-1}$	Eq. (4.1.26) $O\{(\Delta x)^2\}$ for $s \neq 1/2\sqrt{3}$; $O\{(\Delta x)^4\}$ for $s = 1/2\sqrt{3}$	$\frac{2s\cos\beta \pm (1-4s^2\sin^2\beta)^{\frac{1}{2}}}{1+2s}$	None
Classical Implicit	Eq. (4.2.2b) $-s\tau_{j-1}^{n+1} + (1+2s)\tau_j^{n+1} - s\tau_{j+1}^{n+1} = \tau_j^n$	Eq. (4.2.8) $O\{(\Delta x)^2\}$	$\frac{1}{1+4s\sin^2(\beta/2)}$	None

Method	Finite difference equation	Truncation error E for fixed s	Amplification factor G	Stability restriction
Crank-Nicolson Implicit	Eq. (4.2.11) $-\frac{1}{2}s\tau_{j-1}^{n+1} + (1+s)\tau_j^{n+1} - \frac{1}{2}s\tau_{j+1}^{n+1} = \frac{1}{2}s\tau_{j-1}^{n} + (1-s)\tau_j^{n} + \frac{1}{2}s\tau_{j+1}^{n}$	$O\{(\Delta t)^2,(\Delta x)^2\}$ $\equiv O\{(\Delta x)^2\}$	$\frac{1-2s\sin^2(\beta/2)}{1+2s\sin^2(\beta/2)}$	None
Richtmyer Implicit ($\theta=\frac{1}{2}$)	Eq. (4.6.5) $-s\tau_{j-1}^{n+1} + (\frac{3}{2}+2s)\tau_j^{n+1} - s\tau_{j+1}^{n+1} = 2\tau_j^{n} - \frac{1}{2}\tau_j^{n-1}$	$O\{(\Delta t)^2,(\Delta x)^2\}$ $\equiv O\{(\Delta x)^2\}$	$\frac{\frac{1}{2} \pm \frac{2}{3}i(\frac{3}{16} + 2s\sin^2(\frac{\beta}{2}))^{\frac{1}{2}}}{1+\frac{8s}{3}\sin^2(\frac{\beta}{2})}$	None
Saul'yev	Eq. (4.2.36) $\tau_j^{n+1} = \frac{1-s}{1+s}\tau_j^{n} + \frac{s}{1+s}(\tau_{j-1}^{n+1}+\tau_{j+1}^{n})$ Eq. (4.2.37) $\tau_j^{n+1} = \frac{1-s}{1+s}\tau_j^{n} + \frac{s}{1+s}(\tau_{j+1}^{n+1}+\tau_{j-1}^{n})$	Eq. (4.2.43) $O\{\frac{(\Delta t)^2}{\Delta x},(\Delta x)^4\}$ $\equiv O\{(\Delta x)^3\}$	$\sqrt{\frac{\{1-2s\sin^2(\frac{\beta}{2})\}^2+s^2\sin^2\beta}{\{1+2s\sin^2(\frac{\beta}{2})\}^2+s^2\sin^2\beta}}$	None

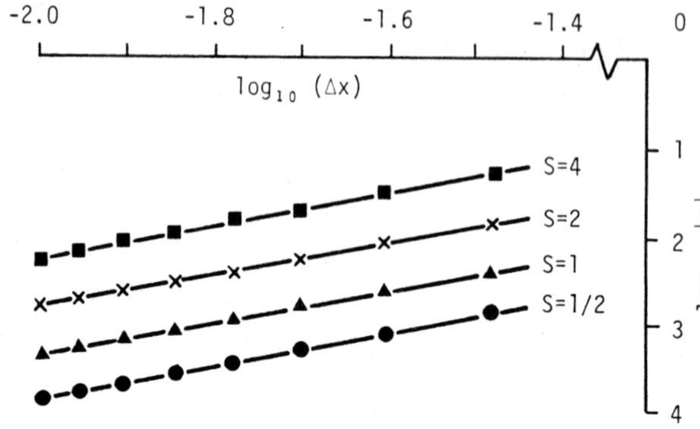

Figure 4.6 : The discretisation error e plotted against grid-spacing Δx for the Saul'yev method of solving the diffusion equation.

4.3 Iterative Techniques for Solving Implicit Finite Difference Equations

The stability restrictions which applied to the explicit methods described in Section 4.1 are largely overcome if implicit methods such as those described in Section 4.2 are used, thereby permitting the use of much larger time steps when using finite difference methods. For example, the Crank-Nicolson method for the one dimensional diffusion equation (2.4.1) is given by Equation (4.2.11), namely

$$-\tfrac{1}{2}s\tau_{j-1}^{n+1} + (1+s)\tau_j^{n+1} - \tfrac{1}{2}s\tau_{j+1}^{n+1} = \tfrac{1}{2}s\tau_{j-1}^{n} + (1-s)\tau_j^{n} + \tfrac{1}{2}s\tau_{j+1}^{n}, \; j=1(1)J-1. \tag{4.3.1}$$

This can be written as the set of linear algebraic equations (4.2.12) in which the values of τ on the left side are at the (n+1)th time level, and the right side contains all the terms which are known, namely those from the previous time level and from the boundary at the (n+1) time level. Therefore, knowing the values of τ at the nth time level, those at the (n+1)th level can be computed. The best way of doing this is to use the very efficient Thomas algorithm for inverting tri-diagonal systems of linear algebraic equations. However, if the Crank-Nicolson method is applied to the two-dimensional diffusion equation (see Section 6.2), a penta-diagonal system of linear algebraic equations results. Such a system is much more difficult to solve than the tri-diagonal system of the one-dimensional case, since two of the non-zero diagonals in the coefficient matrix are separated from the three along the leading diagonal by a wide band of zero coefficients. Any direct elimination method is therefore much less economical computationally than the Thomas method. For such equations, iterative techniques may be employed. This method is illustrated by application to the Crank-Nicolson formula (4.3.1).

An iterative method for solving an equation or a set of equations is one in which a first approximation to the unknowns is used to determine a second approximation, which in turn is used to calculate a third estimate, and so on. Any iterative method is said to be convergent when the difference between the exact solution and the successive approximations tends to zero as the number of iterations increase.

Iterative methods may be used to advantage when solutions are to be found for a set of linear algebraic equations which contain a large number of unknowns, but in which the individual equations involve only a few of these unknowns. The main advantage is speed. This is because iterative methods automatically take advantage of the zeros which occur naturally in the matrix of coefficients for the system of equations, whereas many of the calculations of general elimination methods are concerned with introducing zero coefficients which may already exist.

In order to solve the set of equations (4.3.1) by iteration, it is rewritten in the form

$$\tau_j^{n+1} = \frac{1}{2(1+s)}\{s\tau_{j-1}^{n+1} + s\tau_{j+1}^{n+1} + 2D_j^n\}, \tag{4.3.2}$$

where

$$D_j^n = \tfrac{1}{2}s\tau_{j-1}^n + (1-s)\tau_j^n + \tfrac{1}{2}s\tau_{j+1}^n,$$

for all interior grid points, that is, for j=1(1)J-1. For j=0,J the value of τ_j^{n+1} is either prescribed as the boundary conditions or must be found from some alternative scheme as described in Section 4.5 if derivative boundary conditions are prescribed. At any time level n, the value of D_j^n is known, while the remaining terms on the right hand side of Equation (4.3.2) can be taken, at a first estimate, to be their values at the time level n. Equation (4.3.2) is therefore used for all j for which τ is unknown at the (n+1)th time level to produce, successively, a set of values for τ of improving accuracy. The iterative form of this equation is

$$\tau_j^{(p+1)} = \frac{1}{2(1+s)}\{s\tau_{j-1}^{(p)} + s\tau_{j+1}^{(p)} + 2D_j^n\}, \tag{4.3.3}$$

for p=0,1,2,.., with $\tau_j^{(0)} = \tau_j^n$. For convenience, the superscript indicating the time level has been omitted. The bracketed superscript p denotes the iteration number. However, it is important to remember that the values of $\tau_j^{(p)}$, p=0,1,2,.., are successive estimates of the value of τ_j^{n+1}. A suitable solution is assumed to be found when successive estimates of τ_j^{n+1} differ by less than a specified small quantity for all j. It is then assumed that the process has converged sufficiently.

The manner in which the Jacobi iterations proceed to find estimates $\tau_j^{(p+1)}$ at the (n+1)th time level is illustrated in Figure 4.7. Commencing with the known boundary value g_0^{n+1}, and the estimate $\tau_2^{(0)}$, a better estimate $\tau_1^{(1)}$ for τ_1^{n+1} is found by applying (4.3.3) with p=0, j=1. Then a better estimate for τ_2^{n+1} is found by applying (4.3.3) with p=0, j=2, giving $\tau_2^{(1)}$ in terms of $\tau_1^{(0)}$ and $\tau_3^{(0)}$. Continuing with p=0, j=3(1)J-1, the improved estimates $\tau_j^{(1)}$ are found. Then, with p=1, j=1 in Equation (4.3.3), an improved estimate for τ_1^{n+1} is found from g_0^{n+1} and $\tau_2^{(1)}$, namely $\tau_1^{(2)}$. Similarly, the improved estimates $\tau_j^{(1)}$, j=2(1)J-1 are obtained from (4.3.3) with p=1. This process continues until, say

$$|\tau_j^{(p+1)} - \tau_j^{(p)}| < \varepsilon,$$

for all j, where ε is some small number, say 10^{-6}.

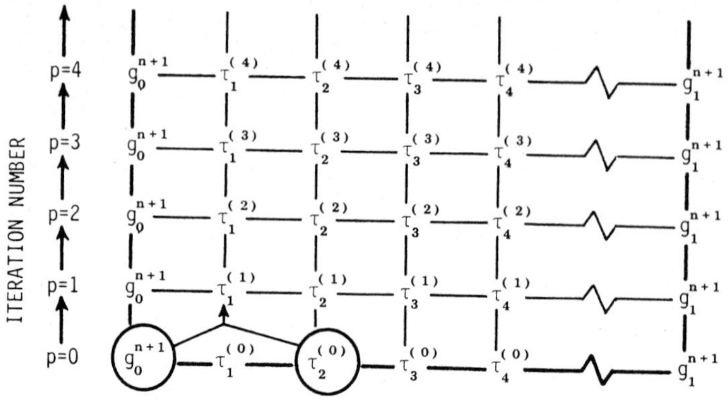

Figure 4.7 : The way in which Jacobi iterations proceed

This particular process is an example of the *Jacobi method.* The order, in terms of j, in which the (p+1)th iterates for τ_j^{n+1} are calculated will not affect the values obtained or their rate of convergence, since they are calculated only in terms of the pth iterates.

However, this method is rather inefficient because it does not use the most up-to-date approximations for the values of τ in the right side of Equation (4.3.3). A better method is to use the most recent iterates as soon as they are available, that is to replace $\tau^{(p)}$ by $\tau^{(p+1)}$ as soon as it is calculated. When a computer is used, this replacement may be made systematically. For instance, if the next iterates are computed in order of increasing j then $\tau_{j-1}^{(p+1)}$ will have been found before $\tau_j^{(p+1)}$ is evaluated. Therefore it is more efficient to replace Equation (4.3.3) by

$$\tau_j^{(p+1)} = \frac{1}{2(1+s)}\{s\tau_{j-1}^{(p+1)} + s\tau_{j+1}^{(p)} + 2D_j^n\}, \tag{4.3.4}$$

with j=1(1)J-1, for each p. This procedure is called the *Gauss-Seidel* or *unextrapolated Liebmann* method and in this example it converges in about half the number of iterations for the Jacobi method. Unlike the Jacobi method the order, in terms of j, of calculating the (p+1)th iterates $\tau_j^{(p+1)}$ is fixed.

It can be shown that the rate of convergence of the Gauss-Seidel scheme is about twice that of the Jacobi method in general.

This is illustrated by the following. The diffusion equation (2.4.1) with $\alpha = 1$, and initial conditions $\bar{\tau}(x,0) = \sin(\pi x)$, $0 \le x \le 1$, and boundary conditions $\bar{\tau}(0,t) = \bar{\tau}(1,t) = 0$, $t \ge 0$, was solved using $\Delta x=0.1$, $s = 2$, for one time step using the Crank-Nicolson equation solved by both the Jacobi and Gauss-Seidel iterative methods respectively.

The Jacobi iteration for the values of τ at the (n+1)th time-level is defined by Equation (4.3.3), which with s = 2 in this case reduces to

$$\tau_j^{(p+1)} = \frac{1}{3}\{\tau_{j-1}^{(p)} + \tau_{j+1}^{(p)} + D_j^n\} \tag{4.3.5}$$

where

$$D_j^n = \tau_{j-1}^n - \tau_j^n + \tau_{j+1}^n .$$

That is,

$$\tau_j^{(p+1)} = \frac{1}{3}\{\tau_{j-1}^{(p)} + \tau_{j+1}^{(p)} + \tau_{j-1}^n - \tau_j^n + \tau_{j+1}^n\},\ j=1(1)J-1, \qquad (4.3.6)$$

with starting values $\tau_j^{(0)} = \tau_j^n$.

At the first time level the starting iterates are $\tau_j^{(0)} = \tau_j^0 = \sin(\pi j/10)$, $j=1(1)9$. The solutions of Equations (4.3.6) with $n=0$ are given in Table 4.6 for the first 27 iterations.

The last line is the solution obtained to the corresponding finite difference scheme (4.2.12) using the Thomas method. Clearly, the final row of iterates is correct to five significant figures.

TABLE 4.6

Jacobi Iterative Method applied to Equation (4.3.1) with $\Delta x = 0.1$, $s = 2$. Results are shown only for $x = j/10$, $j=1(1)5$.

$j\Delta x$	0	0.1	0.2	0.3	0.4	0.5
$\tau_j^0 = \tau(j\Delta x,0)$	0.000	0.30902	0.58779	0.80902	0.95106	1.00000
p = 0	0.000	0.30902	0.58779	0.80902	0.95106	1.00000
p = 1	0.000	0.28885	0.54943	0.75622	0.88899	0.93474
p = 2	0.000	0.27607	0.52511	0.72275	0.84964	0.89337
⋮	⋮	⋮	⋮	⋮	⋮	⋮
p = 26	0.000	0.25391	0.48297	0.66476	0.78147	0.82168
p = 27	0.000	0.25391	0.48297	0.66475	0.78147	0.82168
τ_j^1 (Thomas)	0.000	0.25391	0.48297	0.66475	0.78147	0.82168

The Gauss-Seidel iteration for values of τ at the (n+1)th time level is defined by Equation (4.3.4), which in this case reduces to

$$\tau_j^{(p+1)} = \frac{1}{3}\{\tau_{j-1}^{(p+1)} + \tau_{j+1}^{(p)} + \tau_{j-1}^n - \tau_j^n + \tau_{j+1}^n\},\ j=1(1)J-1, \qquad (4.3.7)$$

with starting iterates $\tau_j^{(0)} = \tau_j^n$. The iteration equations for the values of τ at the first time level are the same as those for the Jacobi process, except $\tau_1^{(p+1)}$ is used instead of $\tau_1^{(p)}$ in the evaluation of $\tau_2^{(p+1)}$, $\tau_2^{(p+1)}$ is used for $\tau_2^{(p)}$ in the evaluation of $\tau_3^{(p+1)}$, $\tau_3^{(p+1)}$ is used instead of $\tau_3^{(p)}$ in the evaluation of $\tau_4^{(p+1)}$, $\tau_4^{(p+1)}$ is used in place of $\tau_4^{(p)}$ in the evaluation of $\tau_5^{(p+1)}$, and so on.

The solutions of these equations for p=0(1)15 are given in Table 4.7. In this case, only 15 iterations were required to achieve the same accuracy as 27 Jacobi iterations shown in Table 4.3.

There is an even better iterative method available, based on a modification of the Gauss-Seidel method. Equation (4.3.4) is rewritten as

$$\tau_j^{(p+1)} = \tau_j^{(p)} + R_j^{(p)} \tag{4.3.8a}$$

where

$$R_j^{(p)} = \frac{1}{2(1+s)} \{s\tau_{j-1}^{(p+1)} + s\tau_{j+1}^{(p)} - 2(1+s)\tau_j^{(p)} + 2D_j^n\}. \tag{4.3.8b}$$

The term $R_j^{(p)}$ may be considered as a correction to be added to $\tau_j^{(p)}$ to make it more exact, and as the process converges $R_j^{(p)}$ tends to zero. The improvement to the Gauss-Seidel method is achieved by "over-relaxing"; that is, in Equation (4.3.8a) the correction term is multiplied by a factor ω, called the relaxation factor, where usually $\omega > 1$, producing

$$\tau_j^{(p+1)} = \tau_j^{(p)} + \omega R_j^{(p)} . \tag{4.3.9}$$

TABLE 4.7

Gauss-Seidel Iterative Method applied to Equation (4.3.1) with $\Delta x = 0.1$, $s = 2$. Results are shown only for $x = j/10$, $j=1(1)5$.

$j\Delta x$	0	0.1	0.2	0.3	0.4	0.5
$\tau_j^0 = (j\Delta x, 0)$	0.000	0.30902	0.58779	0.80902	0.95106	1.00000
p = 0	0.000	0.30902	0.58779	0.80902	0.95106	1.00000
p = 1	0.000	0.28885	0.54271	0.74120	0.86639	0.90652
p = 2	0.000	0.27382	0.51509	0.70377	0.82275	0.86090
⋮	⋮	⋮	⋮	⋮	⋮	⋮
p = 14	0.000	0.25392	0.48297	0.66476	0.78147	0.82168
p = 15	0.000	0.25391	0.48297	0.66475	0.78147	0.82168
τ_j^1 (Thomas)	0.000	0.25391	0.48297	0.66475	0.78147	0.82168

The improved Gauss-Seidel method is generally called the *successive over-relaxation method* (after Young, 1954), or, alternatively, the *extrapolated Liebmann* method (Frankel, 1950). If the value of ω lies in the range $0 < \omega < 1$, then the process is called successive under-relaxation. This is not as widely used as successive over-relaxation.

The manner in which the Gauss-Seidel iteration is improved by the use of the relaxation factor ω is illustrated in Figure 4.8.

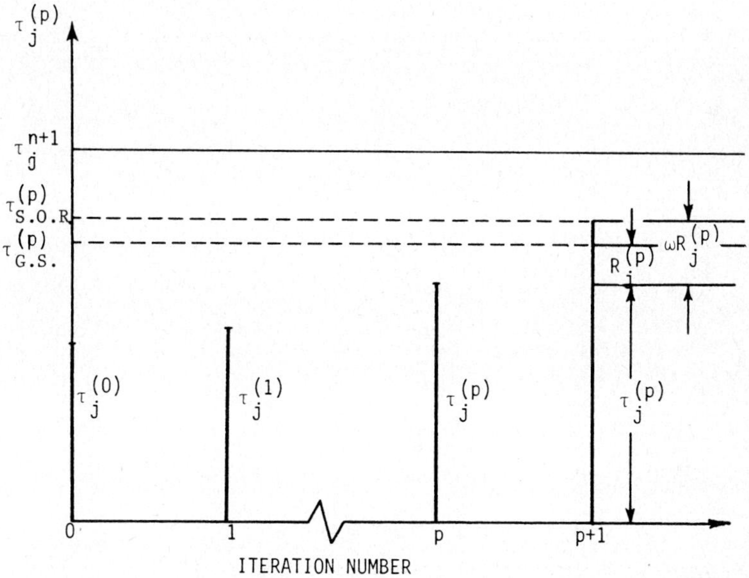

Figure 4.8 : GS indicates the new iterate if the Gauss-Seidel method is used to compute $\tau_i^{(p+1)}$ from $\tau_j^{(p)}$; SOR indicates the new iterate if Successive Over-Relaxation is used.

With the successive over-relaxation method much computing time can be saved by determining ω_0 the optimum value of ω for convergence of the iterative scheme. A practical method for finding ω_0 has been given by Isenberg and deVahl Davis (1975); they chose different values for ω, carried out five to ten test iterations, each time starting from the same initial set of values of τ, and selected the value of ω which gave the most rapid decrease in the difference between successive iterates.

It has been shown that using ω_0, the successive over-relaxation method is faster than the Gauss-Seidel method by a factor of the order of $2/\Delta x$. From this it is obvious that for small Δx it is important to be able to find ω_0.

Formulae for determining ω_0 may be found for certain finite difference methods. For example, solving the one-dimensional diffusion problem using successive over-relaxation on the Crank-Nicolson formula (4.2.11), which is analogous to the one-dimensional Poisson problem, it is found that

$$\omega_0 = 2/(1 + \sqrt{1-\mu^2}) ,$$

where μ is the spectral radius of the corresponding Jacobi method, which in this case is given by

$$\mu = \frac{s}{1+s} \cos(\pi/J).$$

Here (J-1) is the total number of internal grid-points along a time level, the boundary values being known.

For equations whose solutions require a *large number* of iterations, the rate of convergence of the Jacobi, Gauss-Seidel and successive over-relaxation method with optimal ω are found theoretically to be proportional to

$$|\ln \mu|,\ 2|\ln \mu| \quad \text{and} \quad |\ln(\omega_0 - 1)|,$$

respectively. In the illustrations of the Jacobi and Gauss-Seidel methods considered previously, with J=10 and s = 2, then μ = 0.63404 and ω_0 = 1.1278. Therefore $|\ln \mu|$ = 0.4556, $2|\ln \mu|$ = 0.9112, $|\ln(\omega_0-1)|$ = 2.0573, showing that, in this case, the successive over-relaxation method converges more than twice as fast as the Gauss-Seidel method.

This is illustrated by applying the successive over-relaxation method to solve the finite difference equation (4.3.1), for one time step with Δx = 0.1, as before, and comparing the results with those obtained for the Jacobi and Gauss-Seidel methods. The optimum relaxation factor ω_0 = 1.1278 is used in the equation

$$\tau_j^{(p+1)} = \tau_j^{(p)} + \omega_0 R_j^{(p)}, \tag{4.3.10a}$$

where

$$R_j^{(p)} = \frac{1}{3}\{\tau_{j-1}^{(p+1)} + \tau_{j+1}^{(p)} - 3\tau_j^{(p)} + D_j^n\} \tag{4.3.10b}$$

and

$$D_j^n = \tau_{j-1}^n - \tau_j^n + \tau_{j+1}^n . \tag{4.3.10c}$$

The solution for successive iterates p=0(1)9 for the first time step is shown in Table 4.8 and compared with the results obtained by the Thomas method. Results correct to five significant figures are obtained after nine iterations, showing that convergence is approximately twice as fast as for the Gauss-Seidel method (see Table 4.7).

TABLE 4.8

Successive Over-Relaxation Method applied to Equation (4.3.1) with Δx=0.1, s = 2, ω_0 = 1.1278. Results are shown for only x=j/10, j=1(1)5.

$j\Delta x$	0	0.1	0.2	0.3	0.4	0.5
$\tau_j^0 = \tau(j\Delta x, 0)$	0.000	0.30902	0.58779	0.80902	0.95106	1.00000
p = 0	0.000	0.30902	0.58779	0.80902	0.95106	1.00000
p = 1	0.000	0.28627	0.53597	0.72999	0.85135	0.88892
p = 2	0.000	0.26970	0.50666	0.69159	0.80790	0.84476
⋮	⋮	⋮	⋮	⋮	⋮	⋮
p = 8	0.000	0.25393	0.48298	0.66476	0.78147	0.82168
p = 9	0.000	0.25391	0.48297	0.66475	0.78147	0.82168
τ_j^1 (Thomas)	0.000	0.25391	0.48297	0.66475	0.78147	0.82168

If possible, the successive over-relaxation method should be used with a value of ω as near as possible to the optimum value ω_0, because the rate of convergence can be very sensitive to small changes in ω. This is illustrated by a problem considered by Carre (1961) in which $\omega_0=1.9$. This value of the relaxation factor ω gave a rate of convergence forty times faster than that obtained using the Gauss-Seidel method ($\omega=1$) and twice as fast as that given by reducing ω very slightly from 1.9 to 1.875.

Table 4.9 compares the number of iterations required for the three iterative methods described to reach seven figure accuracy in the previous example, and the computation time on the University of Adelaide Cyber 173 computer. Firstly, it is notable that the successive over-relaxation method is less sensitive to a small over-estimate of the optimal relaxation parameter ω_0, say $\omega = \omega_0 + 0.0722 = 1.2000$ when 1 extra iteration is required, than to an even smaller under-estimate, say $\omega = \omega_0 - 0.0278 = 1.100$ when 2 extra iterations are required. Secondly, it is clear that Thomas's direct method is superior to all of the iterative methods described for solving this spatially one-dimensional problem. However, in problems involving more than one space dimension direct methods become relatively more cumbersome and time consuming and it is often preferable to use iterative methods.

TABLE 4.9

Comparison of numbers of iterations and computer time for 7 figure accuracy, for various numerical methods of solving the implicit equation (4.3.1) to find τ_j^1.

Method	No. of iterations	Computer time (secs)
Jacobi iterative method	34	0.105
Gauss-Seidel ($\omega=1$)	19	0.085
S.O.R. ($\omega=1.1000$)	13	0.052
S.O.R. ($\omega=\omega_0=1.1278$)	11	0.045
S.O.R. ($\omega=1.2000$)	12	0.053
Thomas Direct Method	-	0.012

Note that in these iterative methods, each approximation is found by a single calculation, and errors introduced do not prevent convergence of the process since the use of any incorrect values is equivalent to beginning with a new set of starting values. Hence the methods, if convergent, will converge to the correct values whatever the starting values.

4.4 A Marching Method for Solving Implicit Finite Difference Equations

Consider the solution of the one dimensional diffusion equation (2.4.1) by means of the Crank-Nicolson equation (4.2.11), namely

$$-\tfrac{1}{2}s\tau_{j-1}^{n+1} + (1+s)\tau_j^{n+1} - \tfrac{1}{2}s\tau_{j+1}^{n+1} = D_j^n, \qquad j=1(1)J-1, \tag{4.4.1}$$

where

$$D_j^n = \tfrac{1}{2}s\tau_{j-1}^n + (1-s)\tau_j^n + \tfrac{1}{2}s\tau_{j+1}^n. \tag{4.4.2}$$

The set of linear algebraic equations (4.4.1) must be solved to obtain τ_j^{n+1}, j=1(1)J-1, using the known values of D_j^n, j=1(1)J-1, which are calculated from values at the previous time level, and the boundary values $\tau_0^{n+1} = g_0^{n+1}$, $\tau_J^{n+1} = g_J^{n+1}$ which are known at the new time level.

Rearrangement of (4.4.1) gives the equation

$$\tau_{j+1}^{n+1} = -\tau_{j-1}^{n+1} + 2(1+s^{-1})\tau_j^{n+1} - 2s^{-1}D_j^n, \qquad j=1(1)J-1. \qquad (4.4.3)$$

Clearly, if τ_1^{n+1} was known, then (4.4.3) could be used to march from $\tau_0^{n+1} = g_0^{n+1}$ across the interval $0 < x < 1$, giving the unknown values of τ_j^{n+1}, j=2(1)J-1. However, τ_1^{n+1} is not known, but it can be found from the known boundary value τ_J^{n+1} in the following way.

Since (4.4.3) is linear, then τ_J^{n+1} computed in this way depends on τ_1^{n+1} in a linear manner. Thus

$$\tau_J^{n+1} = a\tau_1^{n+1} + b, \qquad (4.4.4)$$

where a and b are constants.

So long as a and b can be found, the required value of τ_1^{n+1} which yields $\tau_J^{n+1} = g_1^{n+1}$ can be obtained by rearrangement of (4.4.4), giving

$$\tau_1^{n+1} = a^{-1}\{g_1^{n+1} - b\}. \qquad (4.4.5)$$

In order to find the values of a and b, proceed in the following way. Set $\tau_1^{n+1} = 0$ in (4.4.3), and march across thegrid at the (n+1)th time level until the corresponding value of τ_J^{n+1} is reached. Let this be $\tau_J^{(0)}$. Substitution in (4.4.4) gives

$$b = \tau_J^{(0)}. \qquad (4.4.6)$$

Then set $\tau_1^{n+1} = 1$, and march across the grid to obtain the final value $\tau_J^{(1)}$. Equation (4.4.4) then yields

$$\tau_J^{(1)} = a.1 + b,$$

so that

$$a = \tau_J^{(1)} - \tau_J^{(0)}. \qquad (4.4.7)$$

A final march across the grid using (4.4.3), j=1(1)J-2, commencing with

$$\begin{aligned} \tau_0^{n+1} &= g_0^{n+1}, \\ \tau_1^{n+1} &= \{\tau_J^{(1)} - \tau_J^{(0)}\}^{-1}\{g_1^{n+1} - \tau_J^{(0)}\}, \end{aligned} \qquad (4.4.8)$$

gives the required values of τ_j^{n+1}, j=2(1)J-1. A check of the accuracy of this procedure can be made by carrying out an additional step with j=J-1, which gives the value of τ_J^{n+1} which should equal g_1^{n+1}.

This technique of marching from x = 0 to x = 1, used to solve the implicit Crank-Nicolson equations, is similar to the time stepping procedure of initial value problems. Therefore, it should be investigated to determine whether errors introduced in the value of τ_1^{n+1} may increase as j increases, when it is possible for them to become so large they could render the value of τ_j^{n+1} obtained so inaccurate that it would be of no use. Suppose an error $\varepsilon > 0$ is introduced in the value of τ_1^{n+1}. Since the propagation of errors ξ_j^{n+1} by (4.4.1) is according to the homogenous difference equation

$$\xi_{j+1}^{n+1} = - \xi_{j-1}^{n+1} + 2(1+s^{-1})\xi_j^{n+1}, \tag{4.4.9}$$

then commencing with

$$\xi_0^{n+1} = 0,$$
$$\xi_1^{n+1} = \varepsilon,$$

we obtain

$$\begin{aligned} \xi_2^{n+1} &= 2(1+s^{-1})\varepsilon > 0, \\ \xi_3^{n+1} &= 4(1+s^{-1})^2\varepsilon - \xi_1^{n+1} > 0, \\ \xi_4^{n+1} &= 8(1+s^{-1})^3\varepsilon - \xi_2^{n+1} > 0, \end{aligned} \tag{4.4.10}$$

and so on. Clearly, these errors increase in value as we march across the grid. This is equivalent to the unstable situation in the perturbation stability analysis for the FTCS method. Continuing in this manner, it can be shown that

$$|\xi_j^{n+1}| < [2(1+s^{-1})]^{j-1}\varepsilon, \tag{4.4.11}$$

with a similar relation obtained for $\varepsilon < 0$.

For a computer which truncates the numbers it stores, ε is less than a unit in the last decimal place of τ_1^{n+1}, and the loss of significant figures in the value of τ_J^{n+1} then depends on the value of $[2(1+s^{-1})]^{j-1}$. The greatest possible loss of significant figures is

$$L < (j-1)\log_{10}[2(1+s^{-1})]. \tag{4.4.12}$$

For $s=4$, $L < 0.4(j-1)$.

The loss of significant figures with $j = 21$ is therefore less than eight. On a computer such as the CDC CYBER-173, which carries out calculations to 14 figures, the answers will be accurate to *at least* six figures so long as $J \le 23$. Use of double precision on this computer permits J to be as large as 58 before the accuracy of the answers is less than six figures.

In fact, because the very worst case was considered above, the loss is usually much less than indicated. The accuracy of this method can always be tested by comparing the computed value of τ_J^{n+1} and the given boundary value g_1^{n+1}.

It is worthy of note that, in this case, a basically unstable scheme has been useful in practice. This method has been used by Noye (1970) to solve a one-dimensional wave-equation governing wind forced motions in lakes. It is also the basis of a method of solving the resulting implicit finite difference equations obtained by discretising the two dimensional diffusion equation (see Section 6.4). It converts the problem of solving a very large sparse set of linear algebraic equations, to one of solving a very small dense system.

4.5 Boundary Conditions

In the diffusion example considered so far for numerical solution, the boundary conditions consisted of known values of the dependent variable $\bar{\tau}$ at all time t. This is only one type of boundary condition that may apply in a physical problem.

Consider the case of heat flow across the boundary B of a solid S, surrounded by some medium M at a different temperature $\bar{\tau}_M$, as in Figure 4.9.

Assuming that the rate at which heat is transferred from the surface to the surrounding medium is proportional to their difference in temperature, then

$$\alpha \left.\frac{\partial\bar{\tau}}{\partial\eta}\right|_B = H(\bar{\tau}_M - \bar{\tau}_B) \quad (4.5.1)$$

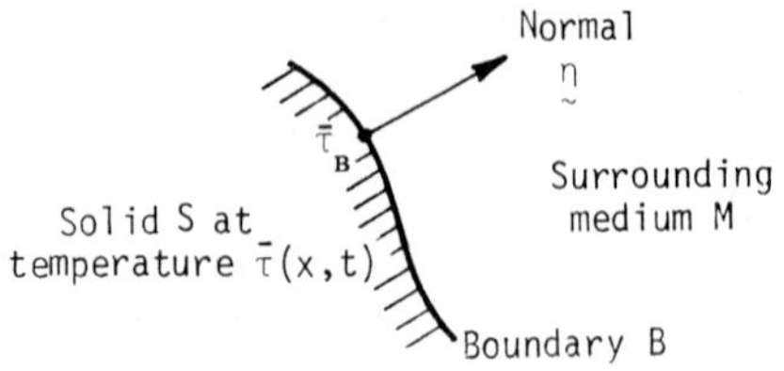

Figure 4.9

where $\bar{\tau}_B$ is the temperature of the surface of the solid S at time t, η is the measure of distance in the direction of the normal pointing outward from the surface of the body into the surrounding medium, and α and H are positive constants of thermal conductivity and surface heat transfer respectively. Equation (4.5.1) is called a mixed boundary condition due to the fact that both unknown quantities $\partial\bar{\tau}/\partial\eta|_B$ and $\bar{\tau}_B$ are involved in a condition which must be used with the partial differential equation which applies to the interior of the solid in order to compute values of $\bar{\tau}$ within the solid and on its surface.

Boundary conditions which can occur with a diffusion problem similar to the one described may be classified into three mathematical types.

Firstly, there is the Dirichlet problem. In this problem the dependent variable $\bar{\tau}$ is given explicitly on all boundaries of the space region involved, even though the boundaries may not be regular. For the flow of heat along a thin insulated metal rod, this is the case where the temperature is given at each end, as described in Section 2.1. For instance, at x = 0 the Dirichlet boundary condition for the diffusion problem may be written

$$\bar{\tau}(0,t) = c_0(t), \quad (4.5.2)$$

where $c_0(t)$ is a known function of t.

Secondly, there is the Neumann problem. In this case the normal derivative of $\bar{\tau}$ out from the boundary of the space region is given everywhere on the boundary. In general it may be written

$$\left.\frac{\partial\bar{\tau}}{\partial\eta}\right|_B = c_B(t) \quad (4.5.3)$$

where η is the measure in the direction of the outward normal at B. For heat flow along the thin rod, this is the case where $\partial\bar{\tau}/\partial x$ is given at each end of the rod. At x = 0 the condition would be written

$$\left.\frac{\partial\bar{\tau}}{\partial(-x)}\right|_{(0,t)} = c_0(t). \quad (4.5.4a)$$

The negative sign appears with the x coordinate, because at x = 0 the direction of the outward normal from the insulated rod is the negative x-direction. At x = 1, Equation (4.5.3) would be written

$$\left.\frac{\partial\bar{\tau}}{\partial x}\right|_{(1,t)} = c_J(t) \tag{4.5.4b}$$

(see Figure 4.10).

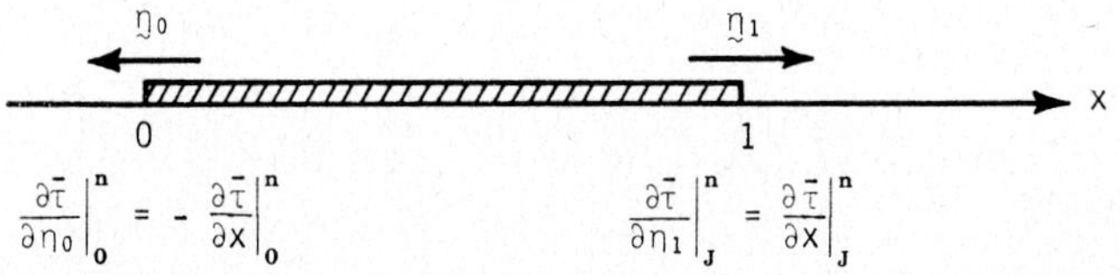

Figure 4.10

Thirdly, there is the mixed problem in which the value of a linear combination of $\bar{\tau}$ and $\partial\bar{\tau}/\partial\eta$ is known on the boundary B of the space region. This may be written

$$a_B(t)\bar{\tau}_B + b_B(t)\left.\frac{\partial\bar{\tau}}{\partial\eta}\right|_B = c_B(t), \tag{4.5.5}$$

where the functions a_B, b_B and c_B may depend on the position B on the boundary.

Equation (4.5.1), which applied to the temperature $\bar{\tau}_B$ of the surface of a solid in a medium at a different temperature $\bar{\tau}_M$, is an example of a mixed boundary condition. Rewriting Equation (4.5.1) gives

$$H\bar{\tau}_B + \alpha\left.\frac{\partial\tau}{\partial\eta}\right|_B = H\bar{\tau}_M$$

which is the same as (4.5.5) with

$$a_B(t) \equiv H,\ b_B(t) \equiv \alpha,\ c_B(t) = H\bar{\tau}_M.$$

Clearly, in practical problems both $a_B(t)$ and $b_B(t)$ have the same sign for all t, and in general are constants.

In particular, for the case of heat flow at the end x = 0 of the insulated rod of unit length, the boundary condition (4.5.5) becomes

$$a_0(t)\bar{\tau}(0,t) + b_0(t)\left.\frac{\partial\bar{\tau}}{\partial(-x)}\right|_{(0,t)} = c_0(t), \tag{4.5.6}$$

where a_0, b_0, c_0 are the coefficients applicable at x = 0. At the time level $t = n\Delta t$ this may be written

$$a_0^n\bar{\tau}_0^n - b_0^n\left.\frac{\partial\bar{\tau}}{\partial x}\right|_0^n = c_0^n \tag{4.5.7a}$$

where $a_0^n = a_0(n\Delta t)$, $b_0^n = b_0(n\Delta t)$, $c_0^n = c_0(n\Delta t)$.

At the end x = 1, the corresponding mixed boundary condition has the form

$$a_J^n \bar{\tau}_J^n + b_J^n \left.\frac{\partial \bar{\tau}}{\partial x}\right|_J^n = c_J^n, \tag{4.5.7b}$$

where a_J^n, b_J^n have the same sign.

Derivative Boundary Conditions

A typical derivative boundary condition at the end x = 0 for the diffusion problem (2.1.1), is

$$\left.\frac{\partial \bar{\tau}}{\partial x}\right|_{(0,t)} = c_0(t), \tag{4.5.8}$$

which is a particular example of the mixed boundary condition (4.5.6) with $a_0(t) \equiv 0$, $b_0(t) \equiv -1$.

Using the forward difference approximation in space at the (0,n+1) grid-point,Equation (4.5.8) becomes

$$\frac{\bar{\tau}_1^{n+1} - \bar{\tau}_0^{n+1}}{\Delta x} = c_0^{n+1} + O\{\Delta x\}, \tag{4.5.9}$$

where $c_0^{n+1} = c_0(n\Delta t+\Delta t)$. This may be rewritten

$$\bar{\tau}_0^{n+1} = \bar{\tau}_1^{n+1} - \Delta x c_0^{n+1} + O\{(\Delta x)^2\}. \tag{4.5.10}$$

The corresponding finite difference approximation to (4.5.10) is

$$\tau_0^{n+1} = \tau_1^{n+1} - \Delta x c_0^{n+1}. \tag{4.5.11}$$

Equation (4.5.11) can be used with the FTCS or any other finite difference approximation to (2.1.1) in order to obtain the boundary value τ_0^{n+1} once the interior value τ_1^{n+1} is known. Equation (4.5.9) shows that the truncation error of the finite difference approximation of the derivative boundary condition is $O\{\Delta x\}$, while a comparison of Equations (4.5.10) and (4.5.11) shows that the approximation for τ_0^{n+1} has a second order error in Δx which contributes to the discretisation error at each time step. In the case of the FTCS scheme, the truncation error is $O\{\Delta t,(\Delta x)^2\}$ and the contribution to the discretisation error at each time step is $O\{(\Delta t)^2,\Delta t(\Delta x)^2\}$, which is fourth order in Δx for a given value of $s \neq 1/6$. Therefore, no matter how accurate is the finite difference method used in the interior of the solution domain, the overall accuracy is governed by the low order of accuracy of the boundary condition approximation.

Matching Truncation Errors

To overcome this anomaly a method of higher order accuracy, such as one using the central difference approximation for the space derivative, could be used. Applying this method with Equation (4.5.8) at the (0,n) grid-point gives

$$\frac{\bar{\tau}_1^n - \bar{\tau}_{-1}^n}{2\Delta x} = c_0^n + O\{(\Delta x)^2\}, \tag{4.5.12}$$

where $\bar{\tau}_{-1}^n$ is a fictitious value of $\bar{\tau}$ prescribed at the exterior point $(-\Delta x, n\Delta t)$ to the solution domain. Rearranging Equation (4.5.12) gives

$$\bar{\tau}_{-1}^n = \bar{\tau}_1^n - 2\Delta x c_0^n + O\{(\Delta x)^3\}, \tag{4.5.13}$$

the corresponding finite difference approximation to the value $\bar{\tau}^n_{-1}$ being

$$\tau^n_{-1} = \tau^n_1 - 2\Delta x c^n_0. \tag{4.5.14}$$

The computation of τ^n_{-1} by (4.5.14) has truncation error $O\{(\Delta x)^2\}$ which is consistent with the truncation error of the FTCS finite difference equation (2.4.4) used at interior grid points, provided $s \neq 1/6$.

It is assumed that the solution domain is extended to include the point $(-\Delta x, n\Delta t)$ so that fictitious values such as τ^n_{-1} can be incorporated with the FTCS method in the following way. Substitution of Equation (4.5.14) into Equation (2.4.4) with j=0 gives

$$\tau^{n+1}_0 = (1-2s)\tau^n_0 + 2s\tau^n_1 - 2s\Delta x\, c^n_0. \tag{4.5.15}$$

The boundary value τ^{n+1}_0 is therefore given in terms of the known values τ^n_0, τ^n_1 found at the previous time level, and the known derivative boundary value c^n_0.

The matching of truncation errors of the finite difference equation which represents the partial differential equation at gridpoints in the interior of the space region and the finite difference equation which represents the conditions applying at gridpoints on the boundary, is the usual criteria taken in order to obtain optimal finite difference solutions to a particular order of accuracy (see Smith, 1969, or Carnahan, Luther and Wilkes, 1969).

Results obtained in this way are better than those obtained using a lower order, and therefore larger, truncation error when approximating the boundary equations.

As an example, consider the insulated rod depicted in Figure 2.1. The temperature, denoted by $\bar{\tau}(x,t)$, is initially 0°C throughout the rod. The right hand end of the rod is kept at 0°C for all time and at $t = 0$ a constant heat flux is suddenly imposed at the left hand end. The mathematical description of this problem is given by Equation (2.1.1) subject to the following boundary and initial conditions:

$$\begin{aligned} \partial\bar{\tau}/\partial x|_{(0,t)} &= -1, \quad t > 0, \\ \bar{\tau}(1,t) &= 0, \quad t \geq 0, \\ \bar{\tau}(x,0) &= 0, \quad 0 \leq x \leq 1. \end{aligned} \tag{4.5.16}$$

The negative sign on the right side of the first equation indicates that there is a steady flow of heat into the rod at $x = 0$.

The analytical solution to this non-homogeneous problem may be obtained using the method of separation of variables (see Myers, 1971, p.103) and is

$$\bar{\tau}(x,t) = 1 - x - \frac{8}{\pi^2} \sum_{m=1}^{\infty} (2m-1)^{-2}\cos\{(2m-1)\pi x/2\}\exp\{-(2m-1)^2\pi^2\alpha t/4\}. \tag{4.5.17}$$

This problem has also been solved using the FTCS explicit method, firstly with the forward difference approximation to the spatial derivative in the boundary condition at x=0 as given by Equation (4.5.11) and then with the central difference approximation which is given by Equation (4.5.15). The value of $\bar{\tau}(0.4,4)$ obtained using the analytic result is 0.090711; the approximate value obtained using the FTCS finite difference method with J=20, s=0.2, and incorporating Equation (4.5.11) is 0.099872 which has a relative error of 10.1%; the value obtained using the FTCS method with Equation (4.5.15) is 0.090468 which is in error by approximately 0.3%. Clearly, the use of the central difference approximation for the derivative boundary condition at x=0 gives a much more accurate result than using the forward difference approximation.

The relative accuracy achieved using the two approximations to the derivative boundary condition considered is more clearly shown in Figure 4.11 in which the distribution of percentage errors along the bar at time t=4.0 is shown for both schemes. In each case the percentage error introduced at x=0 increases from left to right along the bar. Also, for a fixed s it is found that the relative error decreases at a given position on the bar as time passes. This is due to the values of $\bar{\tau}(x,t)$ increasing as t increases, for a given x, while the errors remain about the same. For instance, using the central difference approximation to the boundary derivative at x=0, the errors at x=0.4 at the end of the fourth and sixteenth seconds are respectively 0.268% and 0.028%.

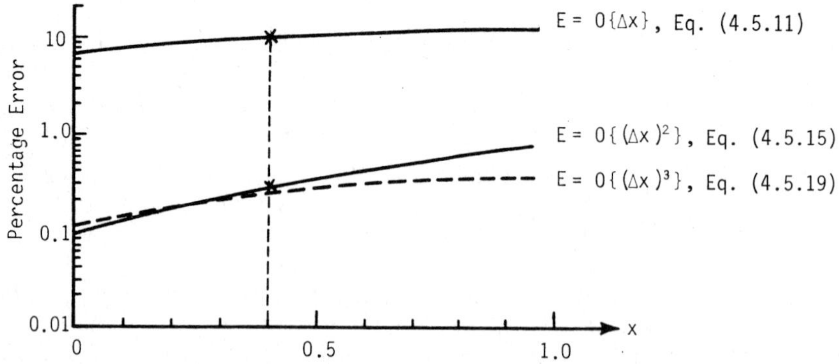

Figure 4.11: Absolute error on $0 \le x < 1$ using different finite difference approximations to a derivative boundary condition at $x = 0$.

Matching Discretisation Errors

There is another criterion which can be used for matching the accuracy of the finite difference approximations to the conditions given on the boundary and the partial differential equation which applies to the interior of the space region. Although truncation errors may be of the same order of accuracy, such as the finite difference equations (4.5.16) and (2.4.4), both of which have $E_j^n = 0\{(\Delta x)^2\}$, the discretisation errors may be of quite different orders. Calculation of τ_0^{n+1} by means of (4.5.16) contributes an error of $0\{(\Delta x)^3\}$ to the discretisation error at each time step, wheras the calculation of τ_j^{n+1}, j=1(1)J-1, by means of (2.4.4) only contributes an error of $0\{(\Delta x)^4\}$. Since τ_0^{n+1} is used in the calculation of τ_j^{n+2}, j=0 and 1, which are in turn used in the calculation of τ_j^{n+3}, j=0,1 and 2, etc., then the contributions to the discretisation error at each time step must overall be considered of $0\{(\Delta x)^3\}$.

Finite difference approximations to the boundary equations which produce boundary values with discretisation errors of the same order of magnitude as the discretisation errors of the finite difference equations used at interior gridpoints, should give optimal results. Boundary conditions usually involve partial differential equations with derivatives of lower order than those in the partial differential equations which apply to the interior of the region. Therefore the truncation errors associated with the boundary equation approximations may need to be of a higher order than the truncation error associated with finite difference equations used at interior gridpoints if the discretisation errors are to be of the same order.

For instance, consider the relation

$$\left.\frac{\partial\bar{\tau}}{\partial x}\right|_j^n = \frac{-2\bar{\tau}_{j-1}^n - 3\bar{\tau}_j^n + 6\bar{\tau}_{j+1}^n - \bar{\tau}_{j+2}^n}{6\Delta x} + O\{(\Delta x)^3\}, \qquad (4.5.18)$$

and the equation (4.5.8). From these we may obtain the finite difference equation for the fictitious value of τ at $(-\Delta x, n\Delta t)$,

$$\tau_{-1}^n = -3/2\tau_0^n + 3\tau_1^n - \tfrac{1}{2}\tau_2^n - 3\Delta x\, c_0^n,$$

with an $O\{(\Delta x)^4\}$ contribution to the discretisation error at each time step.

Substitution of this value of τ_{-1}^n into Equation (2.4.4) with j=0, gives the finite difference equation,

$$\tau_0^{n+1} = (1 - 7s/2)\tau_0^n + 4s\tau_1^n - \tfrac{1}{2}s\tau_2^n - 3s\Delta x\, c_0^n, \qquad (4.5.19)$$

with the same contribution to the discretisation error as τ_j^{n+1}, $j=1(1)J-1$, computed using Equation (2.4.4), if $s \neq 1/6$.

Results obtained using Equation (4.5.19) with (2.4.4) for $j=1(1)J-1$ to solve Equation (2.1.1) subject to conditions (4.5.16) are compared with those obtained using the less accurate Equations (4.5.15) and (4.5.11) in Figure 4.11.

A series of numerical experiments has been carried out, in order to determine which type of finite difference approximation is the best to use when incorporating derivative boundary conditions into a numerical method. The finite difference equations which the various boundary approximations supplemented, were the FTCS explicit method (2.4.4) for solving the one-dimensional diffusion equation, the classical implicit equation (4.2.2b), the Crank-Nicolson equation (4.2.11) and the Saul'yev method (4.2.36-37). Boundary approximations with truncation errors of $O\{\Delta x\}$, $O\{(\Delta x)^2\}$ and $O\{(\Delta x)^3\}$ were tested, and variations of the first two were considered in which fictitious values at exterior gridpoints were, or were not, involved.

The one-dimensional diffusion equation (2.1.1) was solved, with parameters $\alpha = 0.1$, $\Delta x = 0.02$, $s = 0.5$, initial condition

$$\bar{\tau}(x,0) = \sin(\pi x/2), \quad 0 \le x \le 1,$$

and boundary conditions

$$\bar{\tau}(0,t) = 0, \quad \partial\bar{\tau}/\partial x|_{(1,t)} = 0, \quad t \ge 0.$$

The exact solution to this problem is

$$\bar{\tau}(x,t) = \sin(\pi x/2).\exp(-\pi^2\alpha t/4).$$

The average of the modulus of the numerical error across the grid-line at t = 4 was used as a measure of the accuracy of the method. Results are shown in Table 4.10.

TABLE 4.10

Comparison of average error $\frac{1}{J}\sum_{j=1}^{J}|e_j^n|$, J=50, in $\bar{\tau}(x,4)$, for various ways of incorporating a derivative boundary condition at x = 1 into some methods of solving the diffusion equation (2.1.1) with s = 0.5.

Truncation error of boundary approx.	$O\{\Delta x\}$	$O\{\Delta x\}$	$O\{(\Delta x)^2\}$	$O\{(\Delta x)^2\}$	$O\{(\Delta x)^3\}$
Fictitious exterior values?	No	Yes	No	Yes	Yes
FTCS explicit method	1.00×10^{-3}	0.93×10^{-3}	1.00×10^{-5}	0.95×10^{-5}	Unstable
Classical implicit method	1.10×10^{-3}	0.98×10^{-3}	1.90×10^{-5}	1.93×10^{-5}	1.91×10^{-5}
Crank-Nicolson method	0.93×10^{-3}	0.83×10^{-3}	0.42×10^{-5}	0.47×10^{-5}	0.45×10^{-5}
Saul'yev method	1.03×10^{-3}	0.96×10^{-3}	1.75×10^{-5}	1.78×10^{-5}	1.77×10^{-5}

The results obtained with a finite difference approximation of truncation error $O\{\Delta x\}$ to the derivative on the boundary, were much worse than those obtained with truncation errors of $O\{(\Delta x)^2\}$ or $O\{(\Delta x)^3\}$. In general, the errors were 100 times larger. The use of fictitious values allotted to exterior gridpoints did not affect the errors very much, nor did the use of the more complicated approximations with truncation error $O\{(\Delta x)^3\}$ when compared with the simpler forms with an error of $O\{(\Delta x)^2\}$. The greater complexity of the forms with $E = O\{(\Delta x)^3\}$, and the fact that they usually involve more restrictive stability criteria than the finite difference equation used at interior gridpoints, leads to the conclusion that forms with $E = O\{(\Delta x)^2\}$ on the boundary are probably the best.

Mixed Boundary Conditions

A general example of a mixed boundary condition at x = 0 for the heat conduction problem is Equation (4.5.7a), namely

$$a_0^n \bar{\tau}_0^n - b_0^n \left.\frac{\partial \bar{\tau}}{\partial x}\right|_0^n = c_0^n .$$

Approximating this derivative boundary condition using the forward difference equation for the spatial derivative at the $(0,n\Delta t)$ gridpoint gives (see Equation (4.5.9))

$$a_0^n \bar{\tau}_0^n - b_0^n \left(\frac{\bar{\tau}_1^n - \bar{\tau}_0^n}{\Delta x}\right) = c_0^n + O\{\Delta x\}. \tag{4.5.20}$$

Rearranging Equation (4.5.20) yields the finite difference approximation

$$\tau_0^n = (b_0^n \tau_1^n + c_0^n \Delta x)/(b_0^n + a_0^n \Delta x), \tag{4.5.21}$$

with an $O\{(\Delta x)^2\}$ contribution to the discretisation error at each time step.

Equation (4.5.21) can now be incorporated into the finite difference scheme being used so it covers the whole solution domain including gridpoints on the boundary. However, even though the FTCS equation only contributes an error of $O\{(\Delta t)^2, \Delta t(\Delta x)^2\}$ to the discretisation error at each time step, overall the results will be governed by the lower accuracy of the difference equations used at the boundary.

Matching Truncation Errors

To increase the order of accuracy when incorporating the boundary condition (4.5.6) at x=0, the central difference approximation, which has a second order truncation error, may be used for the same derivative. Equation (4.5.7a) then yields the fictitious value of τ^n_{-1} for the point $(-\Delta x, n\Delta t)$,

$$\tau^n_{-1} = \tau^n_1 - (2a^n_0\,\Delta x/b^n_0)\tau^n_0 + 2c^n_0\,\Delta x/b^n_0, \tag{4.5.22}$$

provided $b^n_0 \neq 0$. This equation contributes an error of $O\{(\Delta x)^3\}$ at each time step. Using this expression for τ^n_{-1} the value of τ at the boundary x=0 can be calculated using any of the finite difference schemes previously described, with j=0. For example, substitution of Equation (4.5.22) into the FTCS equation (2.4.4) with j=0, yields

$$\tau^{n+1}_0 = \{1-2s(1+a^n_0\Delta x/b^n_0)\}\tau^n_0 + 2s\tau^n_1 + 2s\Delta x\; c^n_0/b^n_0. \tag{4.5.23}$$

The boundary value τ^{n+1}_0 is therefore represented in terms of values of τ known at the previous time level and the known coefficients a^n_0, b^n_0, c^n_0 of Equation (4.5.7a) with a contribution of $O\{(\Delta x)^3\}$ to the discretisation error at each time step.

When $a_0(t) \equiv 0$ and $b_0(t) \equiv -1$, the mixed boundary condition (4.5.7a) reduces to the derivative boundary condition (4.5.8) and the resulting equations involving τ^{n+1}_0 reduce to the corresponding equations in the derivative boundary case; for example, Equation (4.5.23) reduces to Equation (4.5.15)

At x=1, the relation corresponding to (4.5.23) is obtained from Equation (4.5.7b), namely

$$\tau^{n+1}_J = 2s\tau^n_{J-1} + \{1-2s(1+a^n_J\Delta x/b^n_J\}\tau^n_J + 2s\; x\Delta c^n_J/b^n_J, \tag{4.5.24}$$

$b^n_J \neq 0$. Again, the contribution to the discretisation error is $O\{(\Delta x)^3\}$.

Effect on Stability

Consider the diffusion equation being solved with the mixed boundary conditions (4.5.7) at x=0,1 where the a, b and c's are constants, that is

$$a^n_0 \equiv a_0,\; b^n_0 \equiv b_0,\; c^n_0 \equiv c_0,$$
$$a^n_J \equiv a_J,\; b^n_J \equiv b_J,\; c^n_J \equiv c_J.$$

If Equation (4.5.23) is used to find τ^{n+1}_0, the FTCS equation (2.4.4) is used to find τ^{n+1}_j, j=1(1)J-1, and Equation (4.5.24) is used to find τ^n_J, then the propagation of the errors ξ^n_j is given by the matrix equation (3.3.9), namely

$$\underset{\sim}{\xi}^{n+1} = \underline{\underline{A}}\,\underset{\sim}{\xi}^n,\; n=0,1,2,\ldots,$$

where $\underline{\underline{A}}$ is the (J+1) square matrix, and $\underset{\sim}{\xi}^n$ the (J+1) vector:

$$\underline{\underline{A}} = \begin{pmatrix} \{1-2s(1+a_0b_0^{-1}\Delta x)\} & 2s & & 0 \\ s & (1-2s) & s & \\ & \cdot & \cdot & \cdot \\ 0 & & 2s & \{1-2s(1+a_Jb_J^{-1}\Delta x)\} \end{pmatrix},\quad \underset{\sim}{\xi}^n = \begin{pmatrix} \xi^n_0 \\ \xi^n_1 \\ \vdots \\ \xi^n_J \end{pmatrix}.$$

Note that the errors satisfy the *homogeneous* forms of Equations (4.5.23) and (4.5.24).

As previously described in the matrix method of stability analysis (Section 3.3) so long as the eigenvalue of $\underline{\underline{A}}$ with the largest modulus has a numerical value less than or equal to unity, then the errors will not increase exponentially as n increases.

Application of Brauer's theorem assists with the determination of the conditions which result. This theorem states:

> "If R_m is the sum of the moduli of the terms along the mth row of a matrix $\underline{\underline{A}}$, excluding the diagonal element $a_{m,m}$, then every eigenvalue λ of $\underline{\underline{A}}$ lies inside or on the boundary of at least one of the circles $|\lambda - a_{m,m}| = R_m$."

For m=2(1)J, $a_{m,m} = 1-2s$, $R_m = 2s$, and application of Brauer's theorem yields

$$|\lambda - (1-2s)| \le 2s, \tag{4.5.25}$$

or

$$-2s \le \lambda - 1 + 2s \le 2s.$$

It follows that

$$1 - 4s \le \lambda \le 1.$$

Therefore $|\lambda| \le 1$ if

$$1 - 4s \ge -1,$$

which yields the same restriction as that found previously for non-derivative boundary conditions, namely

$$0 < s \le \tfrac{1}{2}. \tag{4.5.26}$$

For m=1, $a_{m,m} = 1-2s(1+a_0 b_0^{-1}\Delta x)$, $R_m = 2s$, and Brauer's theorem indicates that the eigenvalues λ of $\underline{\underline{A}}$, lie in the range

$$|\lambda - \{1 - 2s(1 + a_0 b_0^{-1}\Delta x)\}| \le 2s, \tag{4.5.27}$$

so that

$$1 - 2s(2 + a_0 b_0^{-1}\Delta x) \le \lambda \le 1 - 2s\, a_0 b_0^{-1}\Delta x.$$

Again, for stability $|\lambda| \le 1$ is required. Since $1 - 2s\, a_0 b_0^{-1}\Delta x < 1$ as s, Δx and $a_0 b_0^{-1}$ are positive, the stability requirement is therefore

$$-1 \le 1 - 2s(2 + a_0 b_0^{-1}\Delta x)$$

or

$$s \le 1/(2 + a_0 b_0^{-1}\Delta x). \tag{4.5.28a}$$

For m=J+1, the corresponding condition on s is

$$s \le 1/(2 + a_J b_J^{-1}\Delta x). \tag{4.5.28b}$$

These three conditions on s may be summarised by the statement

$$0 < s \leq \text{Min}\ \{1/(2+a_0 b_0^{-1} \Delta x),\ 1/(2+a_J b_J^{-1} \Delta x)\}. \qquad (4.5.29)$$

This condition on s for the FTCS method incorporates the effect of the inclusion of the mixed boundary conditions with constant coefficients when the central difference formulation for the spatial derivatives is used in the boundary conditions.

Alternatively, the von Neumann method may be used to determine the stability criterion in the following way.

Firstly, we recall that, in Section 3.3, application of this method to the error equation corresponding to the FTCS finite difference equation assumed to apply to all grid-points $j=0(1)J$ gave the result that, for

$$s \leq \tfrac{1}{2},$$

the equation was stable.

Application of this method to the error equation corresponding to Equation (4.5.23), namely

$$\xi_0^{n+1} = 1-2s(1+a_0 b_0^{-1} \Delta x)\}\xi_0^n + 2s\ \xi_1^n, \qquad (4.5.30)$$

assuming it applies to all the gridpoints $j=0(1)J-1$, gives an amplification factor of

$$G = \{1-2s(1+a_0 b_0^{-1} \Delta x)\} + 2se^{i\beta}$$

for the propagation of the m-th Fourier component of an error distribution from one time level to the next. That is

$$G = \{1-2(1+\gamma)s + 2s \cos \beta\} - i\{2s \sin \beta\}, \qquad (4.5.31)$$

where $\gamma = a_0 b_0^{-1} \Delta x$, which is positive as a_0, b_0 have the same sign. Therefore

$$|G|^2 - 1 = 4s\{[(2+2\gamma+\gamma^2)s - (1+\gamma)] + [1-2(1+\gamma)s]\cos \beta\}. \qquad (4.5.32)$$

Defining the linear function

$$F(\chi) = [1-2(1+\gamma)s]\chi + [(2+2\gamma+\gamma^2)s - (1+\gamma)], \qquad (4.5.33)$$

then the stability requirement $|G| \leq 1$ becomes the condition $F(\chi) \leq 0$ for all χ in the closed interval $<-1,1>$

If $1-2(1+\gamma)s \geq 0$, so that $s \leq 1/(2+2\gamma)$, then $F(\chi) \leq 0$ on $<-1,1>$ so long as $F(1) \leq 0$, that is if

$$\gamma^2 s - \gamma \leq 0$$

or

$$s \leq 1/\gamma.$$

Since $1/\gamma > 1/(2+2\gamma)$ for all $\gamma > 0$, it follows that $F(\chi) \leq 0$ on $<-1,1>$ if

$$s \leq 1/(2+2\gamma). \qquad (4.5.34)$$

If $1-2(1+\gamma)s \le 0$, so that $s \ge 1/(2+2\gamma)$, then $F(\chi) \le 0$ on $<-1,1>$ so long as $F(-1) \le 0$, that is if

$$(\gamma+2)^2 s - (\gamma+2) \le 0$$

or

$$s \le 1/(2+\gamma). \tag{4.5.35}$$

Since for $\gamma > 0$ it follows that $1/(2+\gamma) > 1/(2+2\gamma)$, combining (4.5.34) and (4.5.35) gives $F(\chi) \le 0$ on $<-1,1>$ provided

$$0 < s \le 1/(2+\gamma).$$

The finite difference equation (4.5.23) is therefore locally stable if

$$s \le 1/(2+a_0 b_0^{-1} \Delta x). \tag{4.5.36}$$

Similarly, application of the von Neumann method to the error equation

$$\xi_J^{n+1} = 2s\ \xi_{J-1}^n + \{1-2s(1+a_J b_J^{-1} \Delta x)\}\xi_J^n, \tag{4.5.37}$$

indicates that the finite difference equation (4.5.24) is locally stable if

$$s \le 1/(2+a_J b_J^{-1} \Delta x). \tag{4.5.38}$$

Combining relations (4.5.36) and (4.5.38) yields the same stability condition on s as the matrix method, that is inequality (4.5.29).

At this point we consider an example which illustrates the points discussed so far.

Consider the problem

$$\frac{\partial \bar{\tau}}{\partial t} = \frac{\partial^2 \bar{\tau}}{\partial x^2},\ 0 \le x \le 1, \tag{4.5.39}$$

satisfying the initial condition $\tau(x,0) = 1$ for $0 \le x \le 1$. Let the boundary condition at x=0 be

$$\bar{\tau}_0^n - \left.\frac{\partial \bar{\tau}}{\partial x}\right|_0^n = 0 \text{ for all } t; \tag{4.5.40}$$

that is, $a_0^n = \alpha \equiv 1$, $b_0^n \equiv 1$, and $c_0^n = 0$, in Equation (4.5.7a). At x=1 let

$$\bar{\tau}_J^n + \left.\frac{\partial \bar{\tau}}{\partial x}\right|_J^n = 0 \text{ for all } t; \tag{4.5.41}$$

that is, $a_J^n = \alpha \equiv 1$, $b_J^n \equiv 1$, and $c_J^n = 0$, in Equation (4.5.7b). Using the FTCS Equation (2.4.4) with $s \ne 1/6$, and corresponding boundary approximations with the same truncation error, we have (from Equation (4.5.23))

$$\tau_0^{n+1} = \{1 - 2s(1 + \Delta x)\}\tau_0^n + 2s\tau_1^n. \tag{4.5.42}$$

With J = 10, so that $\Delta x = 0.1$, we obtain

$$\tau_0^{n+1} = (1 - 2.2s)\tau_0^n + 2s\tau_1^n, \tag{4.5.43}$$

with the corresponding boundary approximation at x = 1, from Equation (4.5.24),

$$\tau_{10}^{n+1} = 2s\tau_9^n + \{1 - 2.2s\}\tau_{10}^n. \tag{4.5.44}$$

This system of finite difference equations is stable if the inequality (4.5.29) is satisfied by s, that is if

$$s \le 1/2.1 \simeq 0.476. \tag{4.5.45}$$

Table 4.11 lists the error $\bar{\tau} - \tau$ between the finite difference solution and the analytic solution at x = 0.2 for various values of t. This particular numerical solutions was calculated with $\Delta t = 1/800$, so that s=1/8.

The analytic solution to this problem is

$$\bar{\tau}(x,t) = \sum_{m=1}^{\infty} \frac{\cos\{\gamma_m(2x-1)\}\exp(-4\gamma_m^2 t)}{(\gamma_m^2 + 3/4)\cos\gamma_m}, \tag{4.5.46}$$

where γ_m, m=1,2,3,..., are the roots of $\gamma_m \tan\gamma_m = \frac{1}{2}, (m-1)\pi < \gamma_m < m\pi$. When (4.5.43) and (4.5.44) are used at the boundaries, the truncation error is the same as that in the FTCS explicit equation (2.4.4) used in the interior of the solution domain. However, in this case the boundary approximations contribute terms of $O\{(\Delta x)^3\}$ to the discretisation error, which is larger than that for the FTCS equation, namely $O\{\Delta t^2, \Delta t(\Delta x)^2\}$ or $O\{(\Delta x)^4\}$ for fixed s.

Finite difference equations to compute boundary values at x = 0 with a contribution of $O\{(\Delta x)^4\}$ to the discretisation error, are obtained by substituting (4.5.18), with j=0, into (4.5.40), giving

$$\bar{\tau}_{-1}^n = -\frac{3}{2}(1 + 2\Delta x)\bar{\tau}_0^n + 3\bar{\tau}_1^n - \tfrac{1}{2}\bar{\tau}_2^n + O\{(\Delta x)^4\}. \tag{4.5.47}$$

Substitution of this relation for $\bar{\tau}_{-1}^n$ into (2.4.4) with j=0, gives the finite difference equation

$$\tau_0^{n+1} = (1 - \frac{7s}{2} - 3s\Delta x)\tau_0^n + 4s\tau_1^n - \frac{s}{2}\tau_2^n, \tag{4.5.48}$$

with a contribution of $O\{(\Delta t)^2, \Delta t(\Delta x)^2, (\Delta x)^4\}$ to the discretisation error. Similarly

$$\tau_J^{n+1} = -\frac{s}{2}\tau_{J-2}^n + 4s\tau_{J-1}^n + (1 - \frac{7s}{2} - 3s\Delta x)\tau_J^n \tag{4.5.49}$$

may be used to compute values of τ on the boundary x = 1 with the same accuracy. Both Equations (4.5.48) and (4.5.49) contribute terms of $O\{(\Delta x)^4\}$ to the discretisation error, if s is fixed.

The range of values of s for which the two boundary equations are stable is given by a von Neumann analysis. Equation (4.5.48) has an amplification factor of

$$G = (1 - \frac{7s}{2} - 3s\Delta x) + 4se^{i\beta} - \frac{s}{2}e^{i.2\beta},$$

which becomes, with $\Delta x = 1/10$,

$$G = (1 - \frac{33s}{5} + 4s\cos\beta - s\cos^2\beta) + i\, s\sin\beta(4 - \cos\beta). \tag{4.5.50}$$

It follows that

$$|G|^2 - 1 = (\frac{38s^2}{5} - 2s)\cos^2\beta - (\frac{172s^2}{5} - 8s)\cos\beta + (\frac{2689s^2}{100} - \frac{33s}{5}).$$

The required condition for stability, $|G| \le 1$, is therefore the same as

$$F(\chi) \le 0, \ -1 \le \chi \le 1, \tag{4.5.51}$$

where

$$F(\chi) = (38s^2 - 10s)\chi^2 - (172s^2 - 40s)\chi + (\frac{2689s^2}{20} - 33s). \qquad (4.5.52)$$

The quadratic function $F(\chi)$ has a maximum value if

$$38s(s - 5/19) \le 0$$

or

$$0 < s \le \frac{5}{19} \simeq 0.26315 \ldots . \qquad (4.5.53)$$

Since we also require

$$F(-1) \le 0 \quad \text{or} \quad s(s - \frac{1660}{6889}) \le 0,$$

namely

$$0 < s \le \frac{1660}{6889} \simeq 0.24096\ldots, \qquad (4.5.54)$$

then we need only find values of s satisfying (4.5.51) in the range (4.5.54).

Firstly, $F(\chi) \le 0$ for all χ if the zeros of (4.5.52) are not real, or real and equal, that is if

$$\Delta = (172s^2 - 40s)^2 - 4(38s^2 - 10s)(\frac{2689s^2}{20} - 33s) \le 0, \qquad (4.5.55)$$

or

$$0.12705\ldots \simeq \frac{255 - 5\sqrt{249}}{1386} \le s \le \frac{255 + 5\sqrt{249}}{1386} \simeq 0.24090 \ldots . \qquad (4.5.56)$$

This lies in the range (4.5.54).

Secondly, $F(\chi) \le 0$ for all $\chi \le 1$ if $F(1) \le 0$ together with $F'(1) \ge 0$, even when (4.5.55) is not true. This requires

$$F(1) = \frac{9}{20}s(s - \frac{20}{3}) \le 0$$

or

$$0 < s \le 6\frac{2}{3}, \qquad (4.5.57)$$

and

$$F'(1) = [2(38s^2-10s)\chi-(172s^2-40s)]_{\chi=-1} = -96\,s(s - \frac{5}{24}) \ge 0$$

or

$$0 < s \le 5/24 \simeq 0.20833 \ldots . \qquad (4.5.58)$$

The inequality (4.5.58) satisfies (4.5.57) and is in the range (4.5.54).

It follows that the finite difference equation (4.5.48) is stable if both (4.5.56) and (4.5.58) are satisfied; that is, if

$$0 < s \le 0.24090\ldots . \qquad (4.5.59)$$

For these values of s, Equation (4.5.49) is also stable. Therefore, $s = 0.125$ is a suitable choice for a comparison of boundary approximations with truncation errors of $O\{(\Delta x)^2\}$ and $O\{(\Delta x)^3\}$, since it satisfies both (4.5.45) and (4.5.59). The results are shown in Table 4.11.

When constructing approximations to the boundary conditions, the results in Table 4.11 indicate that making the contribution to the discretisation error the same as that of the finite difference equation used at interior grid points does not always produce better results than merely matching the truncation errors. In the above table they are often worse. Since the more accurate boundary approximations are complicated to develop and are more restrictive on values of s for stability, they appear unwarranted.

TABLE 4.11
Finite difference approximations to solutions of the one-dimensional diffusion equation (2.1.1) with boundary conditions (4.5.40-41), using $\Delta x=0.1$, $s=0.125$.

Time t	Analytic solution $\bar{\tau}(0.2, t)$	Errors obtained using boundary approximations with truncation errors	
		$O\{(\Delta x)^2\}$	$O\{(\Delta x)^3\}$
0.005	0.998357784	-3.82×10^{-4}	1.30×10^{-4}
0.050	0.911986738	-9.26×10^{-4}	11.39×10^{-4}
0.100	0.834221375	-5.34×10^{-4}	10.21×10^{-4}
0.250	0.645420847	-0.49×10^{-4}	7.62×10^{-4}
0.500	0.421212852	3.40×10^{-4}	4.72×10^{-4}
1.000	0.179398897	4.61×10^{-4}	1.79×10^{-4}

4.6 Initial Conditions

Starting Procedures

Given the initial conditions at $t = 0$ there is no difficulty in starting the FTCS method, since going from the time level $n = 0$ to the level $n = 1$ is explicitly determined by Equation (2.4.4). However, when choosing an explicit method of solving the diffusion equation (2.1.1) it may be preferable to use the DuFort-Frankel method, for which much larger values of Δt may be used as it is stable for all $s > 0$. Because this is a three-level method, in which a value of τ at one time level is calculated using values from the previous two time levels, there is a problem with starting, because initial conditions are only given at the time level $t = 0$ and there are no values of $\bar{\tau}$ available for $t = -\Delta t$. To overcome this difficulty the FTCS method can be used to obtain values at the time level $n = 1$ from those at $n = 0$. Then, by using values at both these time levels, the DuFort-Frankel method can be used to continue the process. Since the FTCS method is used only once, it is possible to use a value of $s > \frac{1}{2}$, say $s = 2$; use of the DuFort-Frankel method at succeeding steps then controls the growth of round-off errors.

The classical implicit method and the Crank-Nicolson method are procedures by which values at any time level are computed in terms of known values at only one previous time level, so they require no special starting procedures. They may also be used for the first time step to begin the DuFort-Frankel method, as they are stable and solvable for all values of $s > 0$.

However, not all implicit methods are "two-level" methods such as these. Consider the following finite difference formulation (Richtmeyer, 1957) for the diffusion equation, in which a weighted average of finite difference approximations to the time derivative is used for the derivative $\partial\bar{\tau}/\partial t$.

Since for $0 < \theta \le 1$,

$$\bar{\tau}_j^n = (1+\theta)\bar{\tau}_j^n - \theta\bar{\tau}_j^{n-1} + O\{\Delta t\}, \tag{4.6.1}$$

then the diffusion equation at the (j,n+1) grid-point, namely

$$\left.\frac{\partial\bar{\tau}}{\partial t}\right|_j^{n+1} = \alpha \left.\frac{\partial^2\bar{\tau}}{\partial x^2}\right|_j^{n+1},$$

may be written

$$(1+\theta)\left.\frac{\partial\bar{\tau}}{\partial t}\right|_j^{n+1} - \theta \left.\frac{\partial\bar{\tau}}{\partial t}\right|_j^{n} + O\{\Delta t\} = \alpha \left.\frac{\partial^2\bar{\tau}}{\partial x^2}\right|_j^{n+1}. \tag{4.6.2}$$

Using the backward difference forms for each of the time derivatives on the left side of this equation and the centred difference form for the second order space derivative on the right side, one obtains

$$(1+\theta)\left(\frac{\bar{\tau}_j^{n+1} - \bar{\tau}_j^{n}}{\Delta t}\right) - \theta\left(\frac{\bar{\tau}_j^{n} - \bar{\tau}_j^{n-1}}{\Delta t}\right) + O\{\Delta t\}$$
$$= \alpha\left\{\frac{\bar{\tau}_{j+1}^{n+1} - 2\bar{\tau}_j^{n+1} + \bar{\tau}_{j-1}^{n+1}}{(\Delta x)^2}\right\} + O\{(\Delta x)^2\}. \tag{4.6.3}$$

Dropping the terms of $O\{\Delta t,(\Delta x)^2\}$ gives a finite difference approximation to (4.6.3) which has a truncation error of that order. Rearrangement then gives the implicit "three-level" equation

$$-s\tau_{j-1}^{n+1} + (1 + 2s + \theta)\tau_j^{n+1} - s\tau_{j+1}^{n+1} = (2\theta+1)\tau_j^{n} - \theta\tau_j^{n-1}, \tag{4.6.4}$$

which, in general, contributes an error of $O\{(\Delta t)^2,\Delta t(\Delta x)^2\}$ to the discretisation error at each time step. The equation is also stable for all $s > 0$.

Richtmeyer has shown that this equation is extremely useful when dealing with rapidly varing or discontinuous initial data because it damps the resulting spurious short wave-length components very rapidly.

Choosing $\theta = 0$ yields the classical implicit equation (4.2.2b); choosing $\theta = \frac{1}{2}$ gives the formula

$$-s\tau_{j-1}^{n+1} + (3/2 + 2s)\tau_j^{n+1} - s\tau_{j+1}^{n+1} = 2\tau_j^{n} - 1/2\tau_j^{n-1}, \tag{4.6.5}$$

which is a special case with greater accuracy since it has truncation error $O\{(\Delta t)^2,(\Delta x)^2\}$ and contributes an error of only $O\{(\Delta t)^3,\Delta t(\Delta x)^2\}$ to the approximation for $\bar{\tau}_j^n$ at each time step.

Once again, because it is a three-level finite difference equation, there is a problem when starting the method. However, Equation (4.6.4) can be used after applying the FTCS method at the first time step and there is no loss of accuracy, unless $\theta = \frac{1}{2}$, when a more accurate method such as that due to Crank-Nicolson could be used at the first time step to retain an overall truncation error of $O\{(\Delta t)^2,(\Delta x)^2\}$.

Initial-value, boundary-value singularities

As noted in Section 2.1, singularities in the value of $\bar{\tau}$ may be encountered at the intersection of lines where initial and boundary values are specified in the solution domain. If the singularity is ignored in the application of a finite difference method, which was the case for the example involving heat diffusion considered in Section 2.4, then values of $\partial^4\bar{\tau}/\partial x^4$ and $\partial^2\bar{\tau}/\partial t^2$ are relatively very large near $x = 0$ and $x = 1$, so that, for a particular pair of values of Δx and Δt the truncation error of the finite difference approximation is much larger near the singularity than near the centre of the rod. As a result, the method is less accurate near the ends of the rod, and much smaller values of Δx may be required there than in the centre of the rod. The use of variable size grids to permit this is described in Section 8.

Application of special procedures for a short initial time

A better approach, however, is to employ an asymptotic expansion near the singularity, which usually results in the application of a simple analytic expression for the solution which is valid near the singularity.

Consider the example of the insulated rod, at an initial temperature of zero with its ends suddenly heated to $\bar{\tau} = 1$ as described in Section 2.1. Due to symmetry, the region of interest is the interval $0 \le x \le \frac{1}{2}$. Initially only the region near $x = 0$ "feels" the sudden temperature disturbance so that, to start with, the problem is the same as that of a semi-infinite insulated rod initially at 0°C and suddenly heated to $\bar{\tau} = 1$ at $x = 0$. Thus, for a short time the temperature may be considered to satisfy the diffusion equation (2.1.1) with initial condition

$$\bar{\tau}_L(x,0) = 0, \quad 0 < x < \infty, \tag{4.6.6a}$$

and boundary conditions

$$\bar{\tau}_L(0,t) = 1, \quad \bar{\tau}_L(\infty,t) = 0, \quad t \ge 0. \tag{4.6.6b}$$

With these conditions the diffusion equation yields the solution

$$\bar{\tau}_L(x,t) = \text{erfc}\{x/2\sqrt{\alpha t}\}, \tag{4.6.7}$$

where erfc(x) is the complementary error function (see Crank, 1975). This solution is valid for the original diffusion problem until the centre of the rod, at $x = 1/2$, begins to experience a significant increase in temperature. At this point in time, approximately $t=(100\alpha)^{-1}$, the finite difference method can be started using values for τ obtained by application of Equation (4.6.7) for $0 < x \le \frac{1}{2}$, and of its mirror image in $x = \frac{1}{2}$, namely

$$\bar{\tau}_R(x,t) = \text{erfc}\{(1-x)/2\sqrt{\alpha t}\}, \tag{4.6.8}$$

for $\frac{1}{2} < x < 1$.

Alternatively, we may use a very fine uniform grid to start with, and when $\partial^4\bar{\tau}/\partial x^4$ and $\partial^2\bar{\tau}/\partial t^2$ reduce to reasonable values, revert to a coarse grid. It may be noted that with the coarse grid, larger time spacings may be used with explicit techniques such as the FTCS method.

5. THE ONE-DIMENSIONAL TRANSPORT EQUATION

5.1 The Convection Equation

The one dimensional transport equation

$$\frac{\partial\bar{\tau}}{\partial t} + u\frac{\partial\bar{\tau}}{\partial x} = \alpha\,\frac{\partial^2\bar{\tau}}{\partial x^2}\,, \qquad (5.1.1)$$

governs the change in the scalar property $\bar{\tau}$ caused by diffusion governed by the coefficient $\alpha > 0$ and convection in a fluid moving with a speed u parallel to the x-axis. The term $u\,\partial\bar{\tau}/\partial x$ is the convection term and $\alpha\,\partial^2\bar{\tau}/\partial x^2$ is the diffusion term. In the following it will be assumed that the fluid moves with a *constant* speed $u > 0$, that is, with a steady speed in the positive x direction. In order to solve (5.1.1) on some interval of the x axis, such as $0 \le x \le 1$, for $t \ge 0$, it is necessary to prescribe one set of initial conditions at $t = 0$ and two sets of boundary conditions, one at $x = 0$ and the other at $x = 1$. The solution domain would then be the same as Figure 2.2 for the one dimensional diffusion equation.

Previously, in Sections 2, 3 and 4, the pure diffusion problem in one dimension was considered, namely Equation (2.1.1)

$$\frac{\partial\bar{\tau}}{\partial t} = \alpha\,\frac{\partial^2\bar{\tau}}{\partial x^2}\,.$$

This equation is the same as (5.1.1) with $u = 0$; that is, the diffusion is taking place in a stationary fluid or a solid.

Pure convection is described by the partial differential equation

$$\frac{\partial\bar{\tau}}{\partial t} + u\frac{\partial\bar{\tau}}{\partial x} = 0\,. \qquad (5.1.2)$$

In order to solve (5.1.2) on the interval $0 \le x \le 1$, for $t \ge 0$, one set of initial conditions at $t = 0$ must be prescribed, and *either* one set of boundary conditions (say at $x = 0$) or the fact that the problem is cyclic in space with periodicity 1 must be given. The latter condition requires $\bar{\tau}(x,t) = \bar{\tau}(1+x,t), t \ge 0$, for all x.

For a fixed value of u, the general solution of Equation (%.1.2.) is

$$\bar{\tau}(x,t) = f(x-ut)\,, \quad -\infty < x < \infty, \qquad (5.1.3)$$

where

$$\bar{\tau}(x,0) = f(x) \qquad (5.1.4)$$

is the initial distribution of $\bar{\tau}$ for $-\infty < x < \infty$. That is, the initial distribution of $\bar{\tau}$ along the x-axis is translated without change as it travels along the axis at the constant speed u.

Evaluated at the (j,n) grid-point, the convection equation becomes

$$\left.\frac{\partial\bar{\tau}}{\partial t}\right|_j^n + u\left.\frac{\partial\bar{\tau}}{\partial x}\right|_j^n = 0. \qquad (5.1.5)$$

Using the forward difference form (2.3.7a) for the time derivative and the central difference form (2.3.18) for the space derivative, Equation (5.1.5) becomes

$$\frac{\bar{\tau}_j^{n+1} - \bar{\tau}_j^n}{\Delta t} + O\{\Delta t\} + u\,[\frac{\bar{\tau}_{j+1}^n - \bar{\tau}_{j-1}^n}{2\Delta x} + O\{(\Delta x)^2\}] = 0. \tag{5.1.6}$$

Rearranging this equation gives

$$\bar{\tau}_j^{n+1} = \bar{\tau}_j^n - \tfrac{1}{2}c(\bar{\tau}_{j+1}^n - \bar{\tau}_{j-1}^n) + O\{(\Delta t)^2, \Delta t(\Delta x)^2\}, \tag{5.1.7}$$

where the constant

$$c = u\Delta t/\Delta x \tag{5.1.8}$$

is called the Courant number.

The explicit finite difference equation

$$\tau_j^{n+1} = \tau_j^n - \tfrac{1}{2}c(\tau_{j+1}^n - \tau_{j-1}^n), \tag{5.1.9}$$

which approximates (5.1.5) with truncation error $O\{\Delta t, (\Delta x)^2\}$, may then be used to compute approximate solutions to the convection Equation (5.1.2).

If Equation (5.1.9) is analysed for stability using the von Neumann method, it is found that the amplification factor for the mth Fourier component of any error distribution propogating from one time level to the next, is

$$G = 1 - ic \sin \beta. \tag{5.1.10}$$

Hence the gain $|G|$ is found by solving

$$|G|^2 = 1 + (c \sin \beta)^2 \geq 1, \tag{5.1.11}$$

where the superscripts indicate powers. This result implies that Equation (5.1.9) is unstable for all values of Δx and Δt chosen. This method (which is analogous to the FTCS method for the pure diffusion problem) is therefore of no practical use.

An alternative method which uses *upwind differencing* instead of central differencing for the spatial derivative in (5.1.5) was first used by Courant, Isaacson and Rees (1952) in order to overcome the instability of Equation (5.1.9). Since then, the method has been used under various names with different rationales. For example, for some time meteorologists have known of the stabilising effect of "upwind" (Forsythe and Wasow, 1960) or "weather" (Frankel, 1956) differencing and have applied it to incompressible fluid flow problems. Mathematicians generally refer to the finite difference equations as having "positive coefficients" (Forsythe and Wasow, 1960).

In the upwind differencing method a one sided, rather than space centred, differencing is used for the space derivative in the convection term $u\,\partial\bar{\tau}/\partial x$, the direction of the difference being "upwind". That is, the backward-space finite difference form (2.3.17) is used for $\partial\bar{\tau}/\partial x$ at the (j,n) grid-point when $u > 0$. This has the following physical basis. When determining the value of τ_j^{n+1} from known values of τ at the nth time level, it must be remembered that the fluid flow carries information from the position x_{j-1} to x_j, so a backward difference form which involves values of τ at the (j-1,n) and the (j,n) grid-points is chosen for the spatial derivative. To use a forward difference approximation would be unrealistic, since information cannot be carried from the position x_{j+1}

to x_j in such a case. However, this choice would be appropriate if u was negative.

For $u > 0$, the forward time, backward space form of Equation (5.1.5) is

$$\frac{\bar{\tau}_j^{n+1} - \bar{\tau}_j^n}{\Delta t} + O\{\Delta t\} + u\left(\frac{\bar{\tau}_j^n - \bar{\tau}_{j-1}^n}{\Delta x} + O\{\Delta x\}\right) = 0. \qquad (5.1.12)$$

This equation can be rewritten

$$\bar{\tau}_j^{n+1} = (1-c)\bar{\tau}_j^n + c\bar{\tau}_{j-1}^n + O\{(\Delta t)^2, \Delta t \Delta x\}, \qquad (5.1.13)$$

which leads to the explicit finite difference equation for solving the convection equation,

$$\tau_j^{n+1} = (1-c)\tau_j^n + c\tau_{j-1}^n, \; u > 0, \qquad (5.1.14)$$

with an error of $O\{(\Delta t)^2, \Delta t\Delta x\}$ contributed to the discretisation error e_j^n at each of the n time steps.

Therefore when u is a positive constant, the errors ξ_j^n propagate according to the relation

$$\xi_j^{n+1} = (1-c)\xi_j^n + c\xi_{j-1}^n. \qquad (5.1.15)$$

Application of the von Neumann method of stability analysis to this error equation yields the amplification factor

$$G = \{1 - c(1-\cos\beta)\} - i\{c\sin\beta\}, \qquad (5.1.16)$$

so the gain $|G|$ for the propagation of the mth Fourier mode of any error distribution from one time level to the next is given by

$$|G| = \{1 - 4c(1-c)\sin^2(\beta/2)\}^{1/2}. \qquad (5.1.17)$$

When $c \le 1$, $|G|^2$ has a maximum value of 1 when $\sin^2(\beta/2) = 0$, since the second term on the right hand side of Equation (5.1.17) is always positive. The method is then stable. When $c > 1$, $|G|^2$ has a maximum value of $1 + 4c(c-1) = (2c-1)^2 > 1$ when $\sin^2(\beta/2) = 1$, and the method is unstable. The finite difference Equation (5.1.14) is therefore stable so long as the Courant number is not greater than 1, that is if

$$c \le 1. \qquad (5.1.18)$$

The relation (5.1.18) is usually referred to as the Courant-Friedrichs-Lewy (CFL) condition. This condition applies generally to explicit finite difference approximations to hyperbolic partial differential equations. It requires $u\Delta t \le \Delta x$, which means that the fluid should not travel more than one grid spacing in the x-direction in one time step.

If $c = 1$ when $u > 0$, Equation (5.1.14) becomes

$$\tau_j^{n+1} = \tau_{j-1}^n, \qquad (5.1.19)$$

which is the exact solution to the convection equation. It is the exact solution, since by (5.1.3)

$$\bar{\tau}(x_{j-1},t_n) = f(x_{j-1}-ut_n)$$
$$= f((j-1)\Delta x-un\Delta t) \quad (5.1.20)$$

and

$$\bar{\tau}(x_j,t_{n+1}) = f(x_j-ut_{n+1})$$
$$= f(j\Delta x-u(n+1)\Delta t)$$
$$= f((j-1)\Delta x+\Delta x-un\Delta t-u\Delta t)$$
$$= \bar{\tau}(x_{j-1},t_n), \quad (5.1.21)$$

as $c = 1$ implies that $\Delta x = u\Delta t$.

Further for $c < 1$, the method introduces an artificial damping, the values of τ_j^n being reduced in magnitude at successive time levels in the same way as the errors ξ_j^n are reduced, since Equations (5.1.14) and (5.1.15) are analogous.

The amplification factor G gives a quantitative estimate of the damping and phase shift produced in any infinitely long wave train propagating according to the partial differential equation (5.1.2). Consider the wave train given initially as

$$\bar{\tau}(x,0) = e^{i(\pi mx)} \quad \text{(real part intended)}, \quad (5.1.22)$$

where

$$e^{i(\pi mx)} = \exp(i\pi mx).$$

This wave train has wave number πm and wavelength

$$\lambda = 2/m. \quad (5.1.23)$$

Therefore, at grid-points along the initial time level, the values of $\bar{\tau}$ are given by

$$\bar{\tau}_j^0 = e^{i\beta j}, \quad (5.1.24)$$

where, as in Section 3.3,

$$\beta = \pi m\Delta x. \quad (5.1.25)$$

An exact solution to the finite difference equation (5.1.14) may be found in the variables separable form

$$\tau_j^n = (G)^n e^{i\beta j}, \quad (5.1.26)$$

which clearly satisfies the initial condition (5.1.24). Solving Equation (5.1.14) for G yields the same result as (5.1.16), namely

$$G = \{1 - c(1-\cos\beta)\} - ic\sin\beta.$$

This happens because the form assumed for τ_j^n in (5.1.26) is the same as the error form $\xi_j^n = (G)^n e^{i\beta j}$ used in the von Neumann method, and the error equation (5.1.15) is the same as the homogeneous finite difference equation (5.1.14).

We may write the complex number G in polar form as

$$G = |G|\, e^{-i\psi} \tag{5.1.27}$$

where $|G|$ is given by (5.1.17) and

$$\psi = -\mathrm{Arg}\{G\}. \tag{5.1.28}$$

For the upwind method of solving the convection equation G is given by (5.1.16), and

$$\begin{aligned} \mathrm{Arg}\ G &= -\arctan\{\frac{c \sin \beta}{1-2c\sin^2(\beta/2)}\}, && \text{if } 2c\sin^2(\beta/2) < 1. \\ &= -\pi/2, && \text{if } 2c\sin^2(\beta/2) = 1, \\ &= -\pi + \arctan\{\frac{c \sin \beta}{1-2c\sin^2(\beta/2)}\}, && \text{if } 2c\sin^2(\beta/2) > 1. \end{aligned} \tag{5.1.29}$$

Both $|G|$ and ψ are functions of β and c. Therefore, using the upwind differencing method, the exact finite difference solution to (5.1.2) with initial condition (5.1.22) is obtained by substituting (5.1.27) into (5.1.26). This leads to the value of G given by

$$\tau_j^n = |G|^n e^{i\pi m(j\Delta x - n\psi/\pi m)} \tag{5.1.30}$$

on application of (5.1.25), where the superscripts on the right side represent powers.

However, by (5.1.3), the exact solution to the partial differential equation (5.1.2) with initial condition (5.1.22) is

$$\bar{\tau}(x,t) = e^{i\pi m(x-ut)},$$

so that the exact solution at the (j,n) grid-point is

$$\bar{\tau}(x_j, t_n) = e^{i\pi m(j\Delta x - n\, u\Delta t)}. \tag{5.1.31}$$

Clearly, (5.1.30) differs from the exact solution (5.1.31) by a factor of $|G|^n$ in amplitude. The wave speed has also changed from u to u_N, since $\psi/\pi m$ in (5.1.30) may be written as $u_N \Delta t$, similar to (5.1.31), where

$$u_N = \psi/(\pi m\Delta t). \tag{5.1.32}$$

Substitution of $\Delta t = c\Delta x/u$, from (5.1.8), into this result then yields the ratio of the wave speed of the numerical solution relative to the true wave speed: that is,

$$\frac{u_N}{u} = \frac{\psi}{\pi m\, c\Delta x} = \frac{\psi}{\beta c}. \tag{5.1.33}$$

Clearly, $|G|$ indicates the damping of an infinite wave-train produced in one time step by any two level finite difference method used to solve (5.1.2), and $\psi = -\mathrm{Arg}\{G\}$ may be used to determine the error in the wave speed by substitution into Equation (5.1.33). In an ideal numerical method, the value of $|G|$ should be 1 for all β and c and the argument of G should be given by

$$\text{Arg}\{G\} = -\beta c, \tag{5.1.34}$$

since then $u_N/u = 1$.

For $c = 1$ with upwind differencing, Equation (5.1.17) gives $|G| = 1$, so the method propagates a wave train with the correct amplitude. Also, Equation (5.1.29) gives $\text{Arg}\{G\} = -\arctan\{\tan\beta\} = -\beta$, so that (5.1.34) is satisfied. The method therefore propagates a wave train without change of wave speed.

Consider a case when the wavelength λ of a travelling wave is large compared with the grid spacing Δx. Suppose $\lambda = 20\Delta x$, then as $m = 2/\lambda$ from (5.1.23) it follows that $\beta = \pi m\Delta x = 2\pi\Delta x/\lambda = \pi/10$.

If c is less than 1, say $c = 0.8$, then from (5.1.17)

$$|G| = \{1 - 0.64 \sin^2(\pi/20)\}^{\frac{1}{2}} = 0.9921.$$

Also, from (5.1.28) and (5.1.29)

$$\psi = \arctan\{0.8 \sin(\pi/10)/(0.2 + 0.8 \cos(\pi/10))\} = 0.0802\pi,$$

so that, from (5.1.33)

$$u_N/u = 0.0802\pi/(0.8 \times \pi/10) = 1.002.$$

In this case, the numerical method transmits the wave in slightly "damped" form, that is with its amplitude reduced to 0.9921 of its value at the previous time step. The method also propagates the wave train with a slightly increased speed, namely 1.002 of its true value.

By increasing the value of c to 0.9 while retaining 20 grid-spacings in one wavelength, it is found that $|G| = 0.9996$ and $u_N/u = 1.0013$. Therefore the wave amplitude is less damped than with $c = 0.8$ as the method proceeds from one time level to the next, and the speed of the wave train is closer to the true speed.

If, on the other hand, the wavelength is represented by only three grid-spacings, so that $\lambda = 3\Delta x$, then $\beta = 2\pi/3$. With $c=0.8$, it is found that $|G| = 0.7211$ and $u_N = 1.105u$.

Clearly there is considerable damping of the wave in this case since the amplitude decreases to 0.7211 of its value as the upwind difference method progresses through one time step. The wave speed is also increased, by the factor 1.105. These results indicate that with the upwind difference method the Courant number should be made as near to 1 as practicable and each wavelength should be represented by at least 20 grid-spacings.

Knowing G as a function of c and β, the change of amplitude and speed of an infinitely long travelling wave can be expressed as a function of the number of grid spacings in one wavelength,

$$N_\lambda = \lambda/\Delta x, \tag{5.1.35}$$

for each Courant number c, in the following way.

Firstly, denote the ratio of the speed of the numerically computed wave to the true wave by μ. That is,

$$\mu = u_N/u. \tag{5.1.36}$$

Then from (5.1.33) and (5.1.28)

$$\mu = -\mathrm{Arg}\{G(c,\beta)\}/\beta c. \tag{5.1.37}$$

Substitution of Δx from (5.1.35), and m from (5.1.23), into (5.1.25) gives

$$\beta = 2\pi/N_\lambda. \tag{5.1.38}$$

Therefore μ may be written as a function of N_λ and c, since (5.1.37) becomes

$$\mu = -\frac{N_\lambda}{2\pi c}\,\mathrm{Arg}\{G(c,N_\lambda)\}. \tag{5.1.39}$$

For the upwind method of solving the convection equation, Arg{G} is given by (5.1.29), so

$$\mu = \frac{N_\lambda}{2\pi c}\;\arctan\left\{\frac{c\,\sin\,(2\pi/N_\lambda)}{1-2c\,\sin^2\,(\pi/N_\lambda)}\right\},\quad 2c\,\sin^2\,(\pi/N_\lambda) < 1 \tag{5.1.40}$$

with similar expressions for $2c\,\sin^2\,(\pi/N_\lambda) \geq 1$.

Plots of the relative wave speed μ against the number of grid spacings per wavelength are shown in Figure 5.1. The relative wave speed is very close to 1 so long as $N_\lambda \geq 20$, for all values of c. For c near 1, such as c = 0.9, the relative wave speed is close to 1 but greater than 1 for all values of N_λ. For small values of c, such as c = 0.1 and 0.4, the relative wave speed is very small for $N_\lambda \leq 4$; in fact, for $N_\lambda = 2$, for these values of c the numerical method produces a wave which is not moving, as $\mu = 0$.

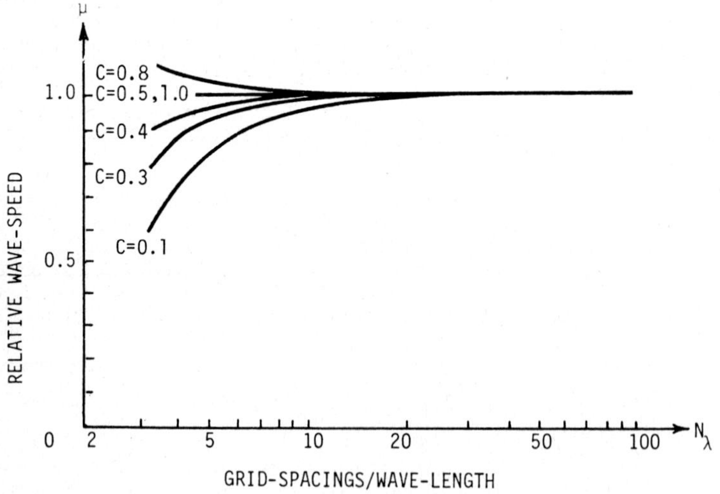

Figure 5.1 : Relative wave speed using the upwind method of solving the convection equation.

The true wave (5.1.31) has a period P which is given by the time taken for the wave to trave one wavelength, namely

$$P = \lambda/u. \tag{5.1.41}$$

The number of time steps in one period is therefore

$$N_P = \frac{\lambda}{u\Delta t}. \tag{5.1.42}$$

Substitution of $u\Delta t = c\Delta x$ from (5.1.8) and use of (5.1.35), then gives

$$N_P = N_\lambda/c. \tag{5.1.43}$$

Therefore, in the time taken for the wave to travel one wavelength, the ratio of the amplitude of the numerically computed wave to the true amplitude becomes

$$\gamma = |G|^{N_P} = |G|^{N_\lambda/c}. \tag{5.1.44}$$

For the upwind method of solving the convection equation $|G|$ is given by (5.1.17), so the relative amplitude is

$$\gamma = \{1 - 4c(1-c)\sin^2(\pi/N_\lambda)\}^{N_\lambda/c}. \tag{5.1.45}$$

Plots of γ against N_λ, the number of grid spacings per wavelength, are shown in Figure 5.2. Even for very large values of N_λ, say $N_\lambda = 100$, the amplitude of the true wave has been noticeably attenuated by the numerical method, unless $c = 1$ when $\gamma = 1$ for all N_λ.

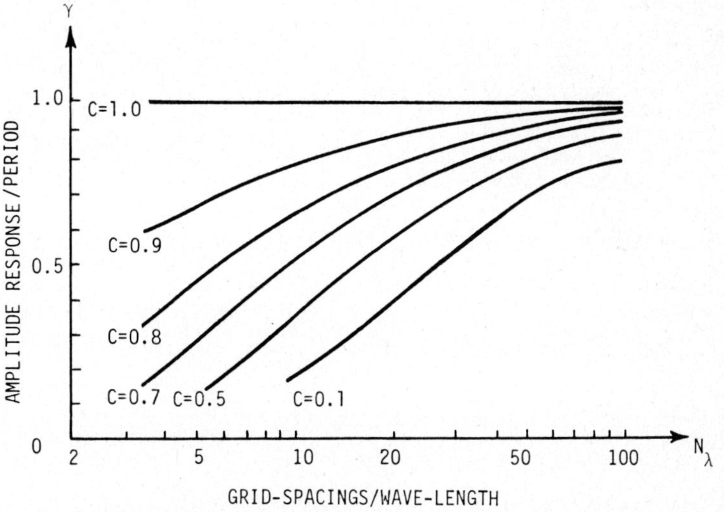

Figure 5.2 : Amplitude attenuation produced by the upwind method of solving the convection equation.

The consistency of the difference equation (5.1.14) with the convection equation (5.1.2) follows from the manner in which it was derived from (5.1.12): the truncation error is $O\{\Delta t, \Delta x\}$ and tends to zero as $\Delta t \to 0$, $\Delta x \to 0$. Let us now formally check the consistency by deriving the actual truncation error E_j^n, in order to determine if there are ways of choosing the Courant number c so that E_j^n and therefore the contribution to the discretisation error at each time step, $\Delta t\, E_j^n$, can be minimised.

Substitution of the exact solution $\bar{\tau}$ for τ in (5.1.14) and expanding each term as a Taylor series evaluated at the (j,n) grid-point, yields the equivalent partial differential equation

$$\left.\frac{\partial\bar{\tau}}{\partial t}\right|_j^n + u\left.\frac{\partial\bar{\tau}}{\partial x}\right|_j^n = \tfrac{1}{2}u\Delta x\left.\frac{\partial^2\bar{\tau}}{\partial x^2}\right|_j^n - \tfrac{1}{2}\Delta t\left.\frac{\partial^2\bar{\tau}}{\partial t^2}\right|_j^n + O\{(\Delta t)^2,(\Delta x)^2\}. \qquad (5.1.46)$$

Clearly, this is consistent with the pure convection equation (5.1.2) as $\Delta t \to 0$, $\Delta x \to 0$. However, in any practical situation $\frac{1}{2}u\Delta x$ is finite, so there is a diffusion type term, namely $\frac{1}{2}u\Delta x\ \partial^2\bar{\tau}/\partial x^2|_j^n$ in Equation (5.1.46). This is not the only such term as the following shows. Since $\bar{\tau}$ satisfies (5.1.2), then

$$\frac{\partial\bar{\tau}}{\partial t} = -u\frac{\partial\bar{\tau}}{\partial x},$$

so that

$$\frac{\partial^2\bar{\tau}}{\partial t^2} = u^2\frac{\partial^2\bar{\tau}}{\partial x^2}. \qquad (5.1.47)$$

Substitution of (5.1.47) into Equation (5.1.46) yields

$$\left.\frac{\partial\bar{\tau}}{\partial t}\right|_j^n + u\left.\frac{\partial\bar{\tau}}{\partial x}\right|_j^n = \alpha'\left.\frac{\partial^2\bar{\tau}}{\partial x^2}\right|_j^n + O\{(\Delta t)^2,(\Delta x)^2\}, \qquad (5.1.48)$$

where

$$\alpha' = \tfrac{1}{2}u\Delta x(1-c). \qquad (5.1.49)$$

Therefore the upwind differencing method has introduced an artificial (or numerical) diffusion of $O\{\Delta t,\Delta x\}$ because of the introduction of the non-physical coefficient α' of $\partial^2\bar{\tau}/\partial x^2$. For a given value of $c \neq 1$, large values of the speed u and of the space step Δx result in large values of α' and hence considerable artificial diffusion.

An example which illustrates this artificial diffusion and damping is shown in Figure 5.3. The function $\bar{\tau}(x,t)$ which is governed by the convection equation (5.1.2) is initially given by

$$\bar{\tau}(x,0) = \begin{cases} 20x & , 0 \le x \le 0.05 \\ 20(0.1-x) & , 0.05 \le x \le 0.1, \\ 0 & , 0.1 \le x \le 1.0, \end{cases}$$

and satisfies boundary conditions

$$\bar{\tau}(0,t) = \bar{\tau}(1,t) = 0 \quad , t \ge 0.$$

On the interval $0 \le x \le 1$, the exact solution to this problem up to $t = 0.9u^{-1}$ is given by

$$\bar{\tau}(x,t) = \begin{cases} 0 & , 0 \le x \le ut, \\ 20(x-ut) & , ut \le x \le ut+0.05, \\ 20(0.1-x+ut) & , ut+0.05 \le x \le ut+0.1, \\ 0 & , ut+0.1 \le x \le 1.0 \end{cases}$$

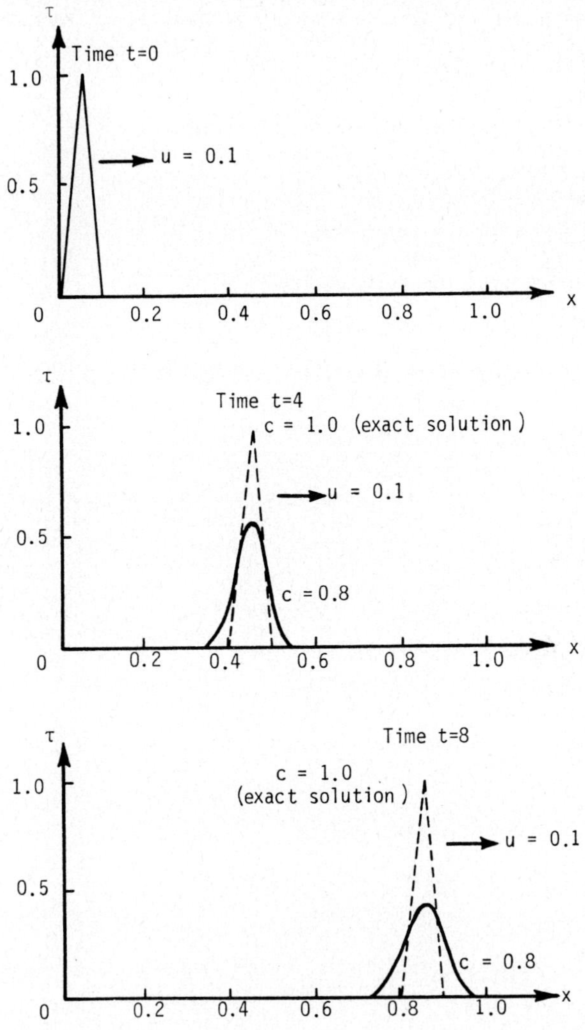

Figure 5.3 : Solutions to the convection equation using upstream differencing with the Courant numbers c = 1 and 0.8.

That is, the triangular pulse with a peak at $x = 0.05$ at time $t = 0$ in Figure 5.3 simply moves to the right with speed u. In the example considered, u was taken to be 0.1. Clearly, when $c = 1$, the correct solution was obtained using the upwind finite difference method. However, for other values of $c < 1$, this numerical method has the peak moving to the right with the correct speed, but it has undergone artificial diffusion. That is, the triangular pulse has spread out at its base and its height has decreased as it moved downstream; this effect increases as c gets smaller.

Another finite difference method of solving the convection equation is obtained if the central difference form (2.3.9a) is used for the time derivative in Equation (5.1.5) with the central difference form (2.3.18) for the space derivative, giving

$$\frac{\bar{\tau}_j^{n+1} - \bar{\tau}_j^{n-1}}{2\Delta t} + O\{(\Delta t)^2\} + u\left(\frac{\bar{\tau}_{j+1}^{n} - \bar{\tau}_{j-1}^{n}}{2\Delta x} + O\{(\Delta x)^2\}\right) = 0. \qquad (5.1.50)$$

Rearrangement of this equation then yields

$$\bar{\tau}_j^{n+1} = \bar{\tau}_j^{n-1} + c(\bar{\tau}_{j-1}^{n} - \bar{\tau}_{j+1}^{n}) + O\{(\Delta t^3, \Delta t(\Delta x)^2\}, \qquad (5.1.51)$$

which can be approximated by the three level explicit finite difference equation

$$\tau_j^{n+1} = \tau_j^{n-1} + c(\tau_{j-1}^{n} - \tau_{j+1}^{n}), \qquad (5.1.52)$$

with a contribution of $O\{(\Delta t)^3, \Delta t(\Delta x)^2\}$ to the discretisation error at each time step. This is much smaller than the contribution of $O\{(\Delta t)^2, \Delta t\Delta x\}$ to the discretisation error for each application of the upwind differencing method.

Because the formula (5.1.52) determines the value of τ_j^{n+1} from the corresponding value at τ_j^{n-1} by adding a term computed from values at the nth time level, this difference scheme is often called the *leapfrog method.*

The stability of (5.1.52) can be established using the von Neumann analysis. As the method proceeds from one time level to the next, the amplification factor G of the mth Fourier mode of the error distribution is found to satisfy

$$(G)^2 + (2ic \sin\beta)G - 1 = 0. \qquad (5.1.53)$$

Solving the quadratic equation (5.1.53) gives

$$G_1, G_2 = -ic \sin\beta \pm \sqrt{1-(c \sin\beta)^2}. \qquad (5.1.54)$$

If $c \le 1$, then the quantity under the square root sign is real, and

$$|G_1| = |G_2| = (c \sin\beta)^2 + \{1-(c \sin\beta)^2\} = 1 \qquad (5.1.55)$$

for all β. If $c > 1$, then for some β we have $c \sin\beta > 1$ when the quantity under the square root sign is negative and

$$G_1, G_2 = -i\{c \sin\beta \pm \sqrt{(c \sin\beta)^2-1}\}. \qquad (5.1.56)$$

At least one of these values G_1, G_2 has a modulus greater than 1, so the corresponding Fourier modes of the error distribution magnify exponentially in modulus as values of τ at successive time levels are computed. This results in computational instability. The leap-frog method is therefore stable only if the Courant number $c \leq 1$.

As was seen with the upwind difference method for solving the convection equation, the amplification factor G of the von Neumann method gives more information than just a qualitative determination of stability of the finite difference equation. Because τ propagates according to the same difference equation as the errors, then seeking a solution to (5.1.52) with $c \leq 1$ and initial condition $\tau_j^0 = e^{i\beta j}$, in the variables separable form

$$\tau_j^n = (G)^n e^{i\beta j}, \text{ (real part intended)},$$

it is found that G has the two possible values G_1, G_2 given by Equation (5.1.54). In its most general form, the exact solution to the leap-frog finite difference equation is therefore

$$\tau_j^n = a(G_1)^n e^{i\beta j} + b(G_2)^n e^{i\beta j}, \tag{5.1.57}$$

where a and b are arbitrary constants.

Firstly, substitution of $n = 0$ in (5.1.57) gives the initial condition, so that

$$(a+b)e^{i\beta j} = e^{i\beta j}$$

whence

$$a = 1 - b. \tag{5.1.58}$$

Also, for $c \leq 1$,

$$|G_1| = |G_2| = 1,$$

and

$$\arg\{G_1\} = -\psi, \; \arg\{G_2\} = \pi + \psi,$$

where

$$\psi = \arcsin\{c \sin\beta\}. \tag{5.1.59}$$

Therefore

$$\tau_j^n = (1-b)(e^{-i\psi})^n e^{i\beta j} + b(e^{i(\pi+\psi)})^n e^{i\beta j}, \tag{5.1.60}$$

and substitution of $\beta = \pi m\Delta x$, $e^{i\pi} = -1$, yields

$$\tau_j^n = (1-b)e^{i\pi m(j\Delta x - n\psi/\pi m)} + (-1)^n b e^{i\pi m(j\Delta x + n\psi/\pi m)}. \tag{5.1.61}$$

The real part of (5.1.61) is the exact solution of the leap-frog difference equation on the infinite interval $-\infty < x < \infty$ with initial condition $\tau_j^0 = e^{i\pi mj\Delta x}$, and consists of two waves rather than the one wave (5.1.31), namely

$$\bar{\tau}_j^n = e^{i\pi m(j\Delta x - nu\Delta t)},$$

which is the exact solution of the convection equation (5.1.2) with the given initial condition. The two waves occur in the solution of the leap-frog method because it involves three time levels, so the corresponding difference equation is second order.

Comparison of the exact finite difference solution (5.1.61) and the exact solution (5.1.31) of the convection equation shows that the amplitude of the first wave in the numerical solution is (1-b), that of the second wave is b, while the exact solution to the convection equation has an amplitude of 1. Secondly, the speed of the first numerical wave compared with the true wave speed u is given by the ratio $\psi/(\beta c)$ (see Equation (5.1.33)), whereas that of the second wave is $-\psi/(\beta c)$; the first wave moves in the positive x-direction, the second wave moves in the negative x-direction. Thirdly, the second numerical wave undergoes a phase change every time step, because of the factor $(-1)^n$.

When c = 1, the speed of the first wave is the same as that of the physical wave which is the true solution of (5.1.2) because

$$\psi/\beta = \beta^{-1}\arcsin(\sin\beta) = 1, \text{ for all } \beta.$$

For this reason, the first wave in the finite difference solution (5.1.61) is referred to as the *physical mode*. The second wave in the finite difference solution travels in the opposite direction to the true solution and changes phase at every time step. Because it has no counterpart in the exact solution and arises only from the finite difference method used, it is called the *computational mode*. The effect of the computational mode, which appears as a wave with period 2Δt, on the solution is shown in Figure 5.4.

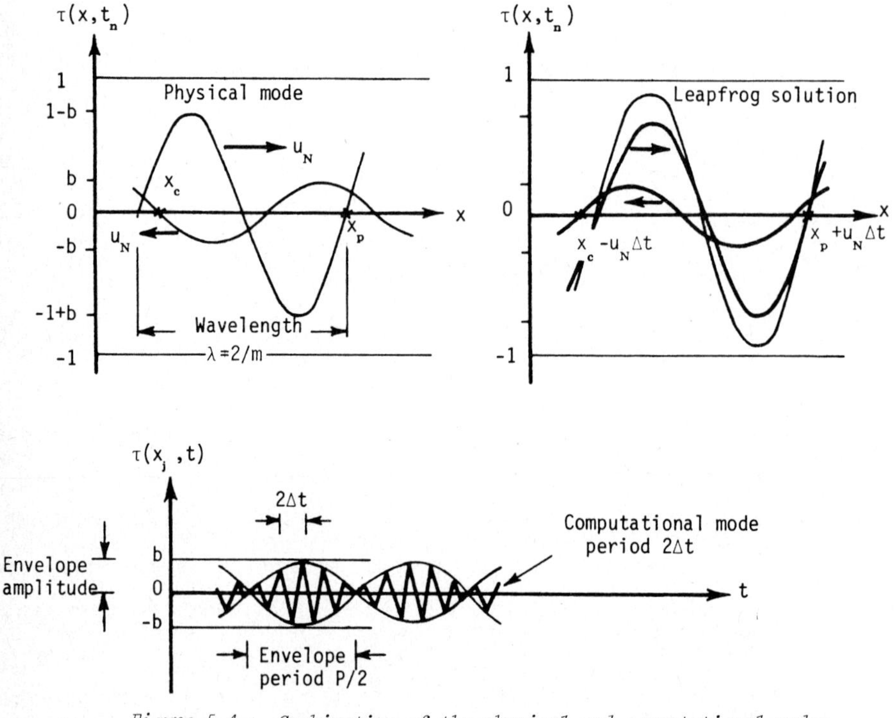

Figure 5.4 : Combination of the physical and computational modes in the leapfrog solution of the convection equation.

The same as any of the three level schemes used to solve the diffusion equation, a special technique must be used to start this method. With $n = 0$, τ_j^1 cannot be computed using (5.1.52) since the values of τ_j^{-1} are not known.

For instance, any two level method may be used for the first time step. Suppose it is the upwind differencing scheme (5.1.14) with $c \le 1$, which gives for the initial condition $\tau_j^0 = e^{i\beta j}$, the numerical solution

$$\tau_j^1 = Ge^{i\beta j}, \tag{5.1.62}$$

namely (5.1.26) with $n = 1$, where the amplification factor G is given by (5.1.16), namely

$$G = \{1 - c(1-\cos\beta)\} - ic \sin \beta.$$

After the first time step the leap-frog method (5.1.52) is used. Therefore the constant b must be such that (5.1.61) with $n = 1$ must have the same value as (5.1.62). Thus

$$(1-b)e^{i\pi mj\Delta x - i\psi} - be^{i\pi mj\Delta x + i\psi}$$

$$= \{(1-c + c\cos\beta) - ic \sin\beta\}e^{i\pi mj\Delta x}, \tag{5.1.63}$$

where ψ for the leap-frog scheme is given by (5.1.59). Division by $e^{i\pi mj\Delta x}$, and rearrangement gives

$$(1-2b)\cos\psi - i\sin\psi = (1-c+c\cos\beta) - ic \sin\beta. \tag{5.1.64}$$

Since from (5.1.59),

$$\sin\psi = c \sin\beta,$$

it follows that the imaginary parts of (5.1.64) are equal. Also,

$$\cos\psi = \{1-(c \sin\beta)^2\}^{\frac{1}{2}}.$$

Equating the real parts of (5.1.64) gives

$$(1-2b)\{1-(c \sin\beta)^2\}^{\frac{1}{2}} = 1 - c + c\cos\beta, \tag{5.1.65}$$

whence

$$b = \tfrac{1}{2}\{1 - \frac{1-2c\sin^2(\beta/2)}{\sqrt{1-(c\sin\beta)^2}}\}. \tag{5.1.66}$$

Clearly, as $\Delta x \to 0$ then $\beta \to 0$ and hence $b \to 0$. The amplitude of the physical mode then approaches the amplitude of the exact solution, while the amplitude of the computational mode tends to zero.

Now, for a given value of c, Equation (5.1.59) shows that the value of ψ will decrease with decreasing values of $\beta = \pi m\Delta x = 2\pi\Delta x/\lambda$. Therefore, when the wavelength λ is large compared to the grid-spacing Δx, β will be small and from (5.1.66) the value of b is small so the amplitude of the physical mode will be nearly that of the exact solution of the partial differential equation and the amplitude of the computational mode will be small. For example, if the wavelength equals 20 grid-spacings, then $\lambda = 20\Delta x$ and $\beta = \pi/10$. Choosing $c = 0.8$, then $b = 0.0042$, so the amplitude of the physical mode is about 0.9958 while that of the computational mode is only 0.0042. Also, the speed of the physical mode relative to the true wave speed is given by (5.1.33), namely,

$$u_N/u = \beta^{-1}c^{-1}\arcsin(c \sin \beta).$$

Here, the value of u_N/u is 0.9939. Therefore, in this example the amplitude and speed of the physical mode in the finite difference solution is very close to that of the true solution of this problem. On the other hand, if the grid-spacing is a large fraction of the wave length, say $\Delta x/\lambda=0.5$ when $\beta = \pi$ and $b = 0.8$, then the amplitude of the spurious computational mode is 0.8 while the physical mode has an amplitude of only 0.2. Clearly the numerical solution bears little resemblance to the true solution of the convection problem in this case.

The ratio of the amplitude of the physical mode to that of the true wave is

$$a = 1 - b < 1 \tag{5.1.67}$$

where b is given by (5.1.66). This amplitude ratio may be considered as a function of the Courant number c and the number of grid spacings per wavelength N_λ by substitution of Equation (5.1.38) in (5.1.67), giving

$$a = \tfrac{1}{2}\{1 + \frac{1-2c\sin^2(\pi/N_\lambda)}{\sqrt{1-c^2\sin^2(2\pi/N_\lambda)}}\}. \tag{5.1.68}$$

This ratio is plotted against N_λ for different values of c in Figure 5.5. However, it must be remembered that this ratio is determined only by the starting procedure, and remains constant after the first time step. This is in contrast with the result for the upwind method, when the amplitude attenuation applies to each wave period, causing the amplitude of the numerical solution to tend to zero with increasing numbers of time steps.

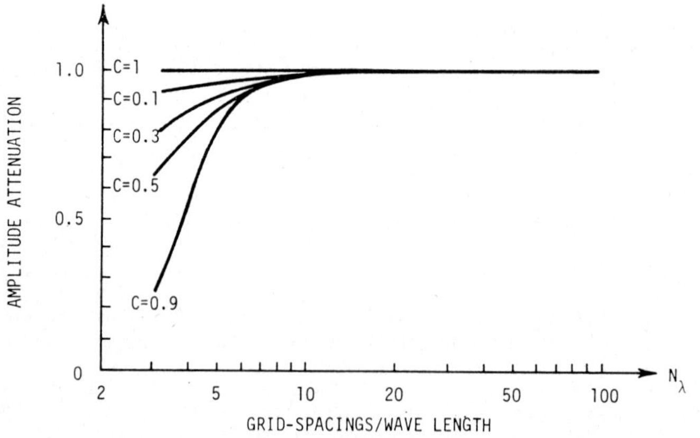

Figure 5.5 : The ratio of the amplitude of the physical mode in the leap-frog solution to the amplitude of the true wave, using the upwind method at the first time step.

The relative wave speed, μ, of the computational mode is given by (5.1.39) with $\mathrm{Arg}\{G\} = -\psi$ where ψ is given by (5.1.59). Therefore

$$\mu = \frac{N_\lambda}{2\pi c} \arcsin\{c \sin(2\pi/N_\lambda)\}. \tag{5.1.69}$$

Graphs of μ against N_λ for various values of c are shown in Figure 5.6 In general, larger values of c produce a relative speed μ nearer to 1 than smaller c, and for all c the value of $\mu \to 1$ as N_λ increases.

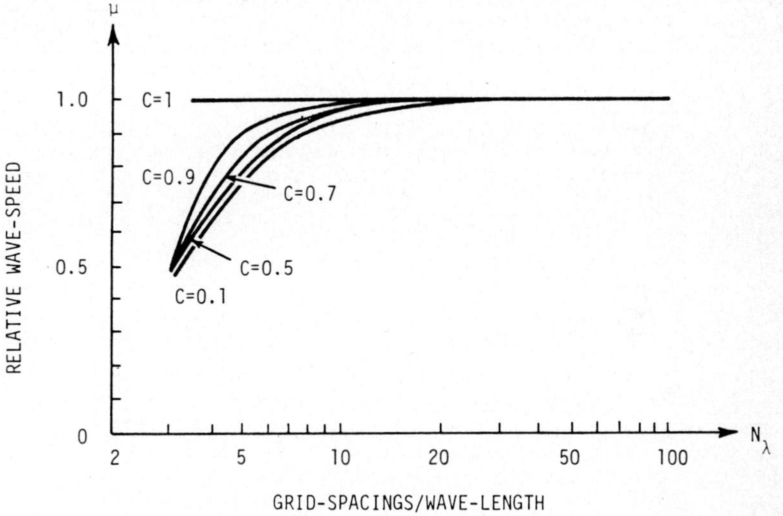

Figure 5.6 : The relative wave speed for the physical mode in the leap-frog solution to the convection equation.

Implicit finite difference methods may also be used to solve the one-dimensional convection equation (5.1.2). Consider the following scheme, in which the partial differential equation is evaluated at the point $(j\Delta x, n\Delta t + \frac{1}{2}\Delta t)$ in the solution domain, giving

$$\left.\frac{\partial\bar{\tau}}{\partial t}\right|_j^{n+\frac{1}{2}} + u\left.\frac{\partial\bar{\tau}}{\partial x}\right|_j^{n+\frac{1}{2}} = 0. \tag{5.1.70}$$

This equation is now approximated in terms of values of $\bar{\tau}$ at nearby grid points. On writing the space derivative in terms of its average value at the nth and (n+1)th time levels, (5.1.70) becomes

$$\left.\frac{\partial\bar{\tau}}{\partial t}\right|_j^{n+\frac{1}{2}} + \frac{u}{2}\left\{\left.\frac{\partial\bar{\tau}}{\partial x}\right|_j^{n} + \left.\frac{\partial\bar{\tau}}{\partial x}\right|_j^{n+1} + O\{(\Delta t)^2\}\right\} = 0. \tag{5.1.71}$$

Writing the derivatives in this equation in their central-difference form gives

$$\frac{\bar{\tau}_j^{n+1} - \bar{\tau}_j^{n}}{\Delta t} + O\{(\Delta t)^2\} + \frac{u}{2}\left\{\frac{\bar{\tau}_{j+1}^{n} - \bar{\tau}_{j-1}^{n}}{2\Delta x} + \frac{\bar{\tau}_{j+1}^{n+1} - \bar{\tau}_{j-1}^{n+1}}{2\Delta x} + O\{(\Delta t)^2, (\Delta x)^2\}\right\} = 0. \tag{5.1.72}$$

Rearranging and then dropping the higher-order terms gives the two-level implicit finite difference equation

$$-\frac{c}{4}\tau_{j-1}^{n+1} + \tau_j^{n+1} + \frac{c}{4}\tau_{j+1}^{n+1} = \frac{c}{4}\tau_{j-1}^{n} + \tau_j^{n} - \frac{c}{4}\tau_{j+1}^{n}. \qquad (5.1.73)$$

This can be used to solve the one-dimensional convection equation, with a truncation error of $O\{(\Delta t)^2,(\Delta x)^2\}$ and a contribution to the discretisation error of $O\{(\Delta t)^3,\Delta t(\Delta x)^2\}$ at each time step.

Knowing the values of τ_j^n at the nth time level and assuming cyclic boundary conditions so that $\tau_0^{n+1} = \tau_J^{n+1}$ and $\tau_{-1}^{n+1} = \tau_{J-1}^{n+1}$ at the (n+1)th time level, the values τ_j^{n+1}, $j=0(1)J-1$, can be computed from the set of Equations (5.1.73), $j=1(1)J$, using the algorithm described in Appendix 2.

Generally speaking, implicit methods are unconditionally stable. That is, round-off errors introduced at time level n are not magnified in modulus when values of the dependent variable are computed at time level (n+1). We will now determine the stability criteria for this finite difference equation using the von Neumann method.

If the initial error distribution has an mth Fourier mode of the form $\xi_j^0 = e^{i\beta j}$, $\beta = \pi m\Delta x$, then at the nth time level it will have the form

$$\xi_j^n = (G)^n e^{i\beta j},$$

where G is found by substituting ξ_j^n in the error equation which is analogous to (5.1.73). Division by $(G)^n e^{i\beta j}$ then gives

$$G\left(-\frac{c}{4}e^{-i\beta} + 1 + \frac{c}{4}e^{i\beta}\right) = \frac{c}{4}e^{-i\beta} + 1 - \frac{c}{4}e^{i\beta}, \qquad (5.1.74)$$

so that

$$G = \frac{1 - i\,(\tfrac{1}{2}c\,\sin\beta)}{1 + i\,(\tfrac{1}{2}c\,\sin\beta)}. \qquad (5.1.75)$$

Clearly, $|G| = 1$ for all β and c, so that (5.1.73) is stable for all values of c.

Previously, we saw that the value of G also gave some quantitative information about the accuracy of the finite difference method, besides qualitative information about the stability of the finite difference equation. We saw that,for two-level methods such as this, an infinite train of waves of length $\lambda = 2/m$, with initial form $\bar{\tau}_j^0 = e^{i\beta j}$, is propagated such that the amplitude of the wave is multiplied by $|G|$ with each step taken in time, and moves with a speed relative to the true speed given by $u_N/u = -\mathrm{Arg}\{G\}/(\beta c)$. Since $|G| = 1$ for all β and c, the amplitude of such a wave train is not changed by this method of solution. However, since $\arg\{G\} = -2\arctan\{\tfrac{1}{2}c\sin\beta\}$, then the relative speed of the wavetrain propagated by (5.1.73) to that propagated by the convection equation (5.1.2) is

$$u_N/u = 2\arctan\{\tfrac{1}{2}c\sin\beta\}/(\beta c).$$

If the wavelength λ is much greater than Δx, then $\beta \ll 1$ and $u_N/u \simeq 2(\tfrac{1}{2}c\beta)/(\beta c)=1$. Therefore the wave train found numerically has the same amplitude and very nearly the same speed as the true solution of the convection equation. Note that, in this method $u_N/u \le 1$, so the speed of propagation of the numerical wave is always less than the true wave speed.

Graphs of the relative wave speed

$$\mu = \frac{u_N}{u} = \frac{N_\lambda}{\pi c} \arctan\{\frac{c}{2} \sin (\frac{2\pi}{N_\lambda})\} \tag{5.1.76}$$

against the number of gridspacings per wavelength, N_λ, for various values of c, are shown in Figure 5.7. Clearly, for $N_\lambda > 50$, the speed of the numerical solution is close to the true wave speed.

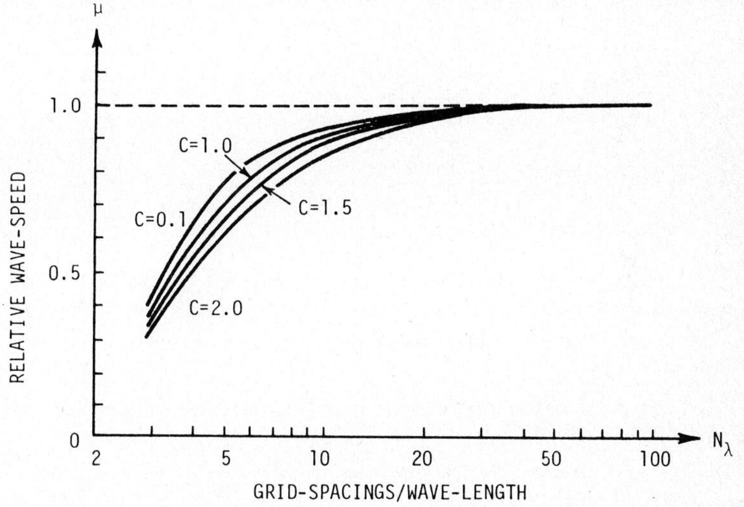

Figure 5.7 : Relative wave speed for the centred-time centred-space implicit method of solving the convection equation.

For this method, the contribution to the discretisation error at each time step is $O\{(\Delta t)^3, \Delta t(\Delta x)^2\}$, which is the same contribution as in the leap-frog method. Clearly, because the time derivative has been differenced about a point midway between two time levels to get an $O\{(\Delta t)^2\}$ truncation error in this case, this is a two-level method. Therefore the computational mode does not appear.

However, even though application of (5.1.73) does not amplify round-off errors so the finite difference equation is stable for all values of c, and the discretisation error is small so the method appears very accurate, there is a condition which must be placed on c in order to actually solve the system of linear algebraic equations using the cyclic algorithm. The linear algebraic system of equations can be solved without loss of accuracy because of round-off, if the coefficient matrix on the left side of (5.1.73) is diagonally dominant. This requires that

$$1 \geq |-\frac{c}{4}| + |\frac{c}{4}| = \frac{c}{2}, \qquad \text{or} \qquad c \leq 2.$$

For values of c > 2, the loss of accuracy in the solution of the algebraic equations *may* make the results from an application of this algorithm quite worthless.

A further point which bears examination, is whether this numerical method introduces artificial diffusion. For example, the upwind difference method introduces some diffusion but the leap-frog method does not.

Carrying out a formal consistency analysis, by substituting $\bar{\tau}$ for τ in (5.1.73) and expanding all terms in their Taylor series, gives the corresponding partial differential equation

$$\frac{\partial\bar{\tau}}{\partial t} + u\frac{\partial\bar{\tau}}{\partial x} = -\frac{\Delta t}{2}\frac{\partial^2\bar{\tau}}{\partial t^2} - \tfrac{1}{2}u\Delta t\,\frac{\partial^2\bar{\tau}}{\partial t\partial x} + O\{(\Delta t)^2,(\Delta x)^2\}, \qquad (5.1.77)$$

at the (j,n) grid-point. This equation is clearly consistent with the one-dimensional convection equation. Also, since $\bar{\tau}$ satisfies the convection equation then procedures similar to those used to derive (5.1.47), namely

$$\frac{\partial^2\bar{\tau}}{\partial t^2} = u^2\,\frac{\partial^2\bar{\tau}}{\partial x^2},$$

may be used to show that

$$\frac{\partial^2\bar{\tau}}{\partial t\partial x} = -u\,\frac{\partial^2\bar{\tau}}{\partial x^2}\;.$$

Substitution of these equations in (5.1.77) gives

$$\frac{\partial\bar{\tau}}{\partial t} + u\,\frac{\partial\bar{\tau}}{\partial x} = O\{(\Delta t)^2,(\Delta x)^2\}, \qquad (5.1.78)$$

as the terms involving $\partial^2\bar{\tau}/\partial x^2$ have cancelled one another. Therefore this numerical method introduces no artificial diffusion and has a truncation error of $O\{(\Delta t)^2,(\Delta x)^2\}$.

A semi-implicit method, similar to Saul'yev's method for solving the one-dimensional diffusion equation, has been developed by Roberts and Weiss (1966) for solving the convection equation. It is based on the following discretisation of the equation at the point $(j\Delta x, n\Delta t+\tfrac{1}{2}\Delta t)$, using only values of $\bar{\tau}_j^n$, $\bar{\tau}_{j+1}^n$, $\bar{\tau}_j^{n+1}$, $\bar{\tau}_{j-1}^{n+1}$.

Consider the equation (5.1.70), namely

$$\left.\frac{\partial\bar{\tau}}{\partial t}\right|_j^{n+\frac{1}{2}} + u\left.\frac{\partial\bar{\tau}}{\partial x}\right|_j^{n+\frac{1}{2}} = 0. \qquad (5.1.79)$$

Replacing the space derivatives as follows,

$$\left.\frac{\partial\bar{\tau}}{\partial t}\right|_j^{n+\frac{1}{2}} + \frac{u}{2}\left\{\left.\frac{\partial\bar{\tau}}{\partial x}\right|_j^{n+1} + \left.\frac{\partial\bar{\tau}}{\partial x}\right|_j^{n} + O\{(\Delta t)^2\}\right\} = 0,$$

followed by the discretisation

$$\frac{\bar{\tau}_j^{n+1} - \bar{\tau}_j^n}{\Delta t} + O\{(\Delta t)^2\} + \frac{u}{2}\left\{\frac{\bar{\tau}_j^{n+1} - \bar{\tau}_{j-1}^{n+1}}{\Delta x} + \frac{\bar{\tau}_{j+1}^n - \bar{\tau}_j^n}{\Delta x} + O\{(\Delta t)^2,\Delta x\}\right\} = 0, \qquad (5.1.80)$$

in which the backward difference form replaces $\partial\bar{\tau}/\partial x|_j^{n+1}$ and the forward difference form replaces $\partial\bar{\tau}/\partial x|_j^n$, gives

$$\tau_j^{n+1} = \tau_j^n + \frac{C}{2+C}(\tau_{j-1}^{n+1} - \tau_{j+1}^n), \qquad (5.1.81)$$

on dropping the terms of $O\{(\Delta t)^2,\Delta x)\}$, multiplying through by Δt and rearranging. Equation (5.1.81) is implicit, since it involves two values of τ from the (n+1)th time level, but can be calculated explicitly if the order of calculation is j=1(1)J-1, since τ_{j-1}^{n+1} is then known when τ_j^{n+1} is computed.

A von Neumann stability analysis of (5.1.81) yields the amplification factor

$$G = \frac{[2 + c(1 - \cos \beta)] - i\, c \sin \beta}{[2 + c(1 - \cos \beta)] + i\, c \sin \beta}. \tag{5.1.82}$$

Since $|G| = 1$ for all c and s, Equation (5.1.81) is stable for all c.

This is a particular case of a finite difference equation which is diagonally symmetrical; that is, a two level equation which includes values of τ at gridpoints symmetrical about $(j\Delta x, n\Delta t+\frac{1}{2}\Delta t)$ with equal coefficients when written as an implicit equation with terms at the (n+1)th time level on the left side and terms at the nth time level on the right side. The general form of such a finite difference equation is

$$\sum_{m=-M}^{M} a_m \tau_{j+m}^{n+1} = \sum_{m=-M}^{M} a_m \tau_{j-m}^{n} \tag{5.1.83}$$

(see Figure 5.8).

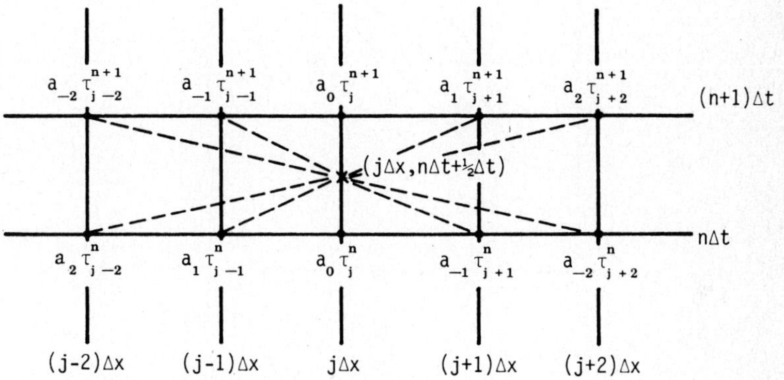

Figure 5.8 : The diagonally symmetrical finite difference form.

Diagonally symmetrical equations are always stable. This can be seen by finding the amplification factor G from a von Neumann analysis, namely

$$G = \sum_{m=-M}^{M} a_m e^{-im\beta} \Big/ \sum_{m=-M}^{M} a_m e^{im\beta}. \tag{5.1.84}$$

Since the coefficients a_m are real, then the numerator is the complete conjugate of the denominator, so both have the same modulus. Therefore,

$$|G| = 1, \text{ for all } \beta, \tag{5.1.85}$$

and (5.1.83) is always stable.

The finite difference equation (5.1.81) has, therefore, an amplitude ratio of $\gamma = 1$ for all c and N_λ; that is, the amplitude of a numerically computed infinite wave is the same as that of the true wave. The relative wave speed is given by (5.1.39) with

$$\text{Arg}\{G\} = -2\ \arctan\{\frac{c \sin\beta}{2 + 2c \sin^2(\beta/2)}\},$$

from Equation (5.1.82). Thus

$$\mu = \frac{N_\lambda}{\pi c} \arctan\{\frac{\frac{1}{2}c \sin(2\pi/N_\lambda)}{1 + c \sin^2(\pi/N_\lambda)}\}. \qquad (5.1.86)$$

Graphs of the relative wave speed against the number of grid spacings per wavelength, for various values of c, are shown in Figure 5.9. So long as $N_\lambda > 50$, the value of c is very close to 1 and the numerically computed infinite wave is almost identical to the true wave.

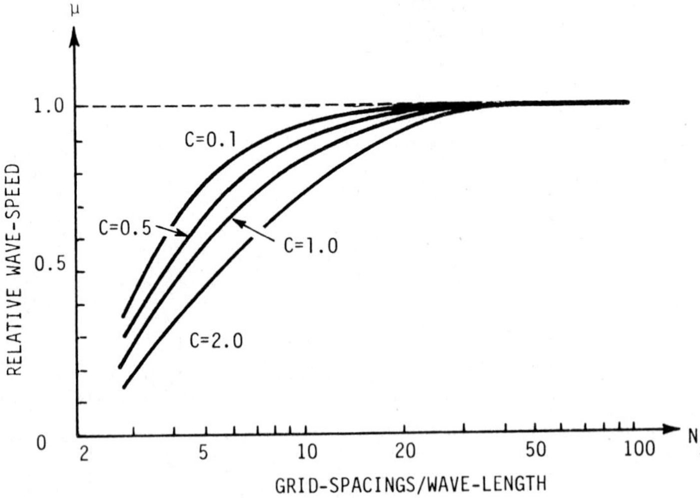

Figure 5.9 : Relative wave speed of the semi-implicit method of solving the convection equation.

Because $|G| = 1$, it could be expected that no numerical diffusion occurs with this finite difference equation. A consistency analysis of (5.1.81) gives the equivalent partial differential equation

$$\frac{\partial\bar{\tau}}{\partial t} + u\frac{\partial\bar{\tau}}{\partial x} = E(x,t) \qquad (5.1.87)$$

where

$$E(x,t) = -\tfrac{1}{2}\Delta t\,\frac{\partial^2\bar{\tau}}{\partial t^2} - \tfrac{1}{2}u\Delta t\,\frac{\partial^2\bar{\tau}}{\partial x\partial t} + O\{(\Delta x)^2, \Delta x\Delta t, (\Delta t)^2\}. \qquad (5.1.88)$$

Clearly $E(x,t) \to 0$ as $\Delta t \to 0$, $\Delta x \to 0$, so (5.1.81) is consistent with the convection equation. However, since both $\partial^2\bar{\tau}/\partial t^2$ and $\partial^2\bar{\tau}/\partial x\partial t$ can be transformed to terms involving $\partial^2\bar{\tau}/\partial x^2$, it would appear that some numerical diffusion may be introduced by the method. Substitution of

$$\frac{\partial^2\bar{\tau}}{\partial t^2} = u^2\frac{\partial^2\bar{\tau}}{\partial x^2}$$

and

$$\frac{\partial^2\bar{\tau}}{\partial x\partial t} = -u\frac{\partial^2\bar{\tau}}{\partial x^2}$$

into (5.1.88) gives

$$E(x,t) = -\tfrac{1}{2}u^2\Delta t\frac{\partial^2\bar{\tau}}{\partial x^2} + \tfrac{1}{2}u^2\Delta t\frac{\partial^2\bar{\tau}}{\partial x^2} + O\{(\Delta x)^2,\Delta x\Delta t,(\Delta t)^2\}, \qquad (5.1.89)$$

in which the terms in $\partial^2\bar{\tau}/\partial x^2$ cancel. The method introduces *no* numerical diffusion, which is consistent with the fact that $|G| = 1$ for all β and c.

Table 5.1 summarises information about some of the methods of solving the convection equation, including details about their accuracy, stability, and solvability.

TABLE 5.1

Finite difference techniques for solving the convection equation $\partial\bar{\tau}/\partial t + u\,\partial\bar{\tau}/\partial x = 0$, $u > 0$.

Method	Finite Difference Equation	Truncation Error E	Numerical Diffusion α'
FTCS Explicit 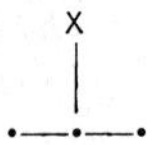	Eq. (5.1.9) $\tau_j^{n+1} = \tau_j^n - \frac{c}{2}(\tau_{j+1}^n - \tau_{j-1}^n)$	$O\{\Delta t, (\Delta x)^2\}$ $\equiv O\{\Delta x\}$ for fixed c	$-\frac{1}{2}u^2\Delta t$
Upwind Explicit 	Eq. (5.1.14) $\tau_j^{n+1} = (1-c)\tau_j^n + c\tau_{j-1}^n$	Eq. (5.1.46) $O\{\Delta t, \Delta x\}$ $\equiv O\{\Delta x\}$ for fixed c	Eq. (5.1.49) $\frac{1}{2}u\Delta x(1-c)$
Leapfrog Explicit 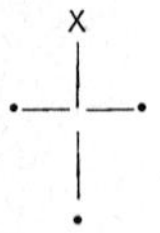	Eq. (5.1.52) $\tau_j^{n+1} = \tau_j^{n-1} + c(\tau_{j-1}^n - \tau_{j+1}^n)$	$O\{(\Delta t)^2, (\Delta x)^2\}$ $\equiv O\{(\Delta x)^2\}$ for fixed c	None
CTCS Implicit 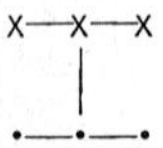	Eq. (5.1.73) $-\frac{c}{4}\tau_{j-1}^{n+1} + \tau_j^{n+1} + \frac{c}{4}\tau_{j+1}^{n+1}$ $= \frac{c}{4}\tau_{j-1}^n + \tau_j^n - \frac{c}{4}\tau_{j+1}^n$	$O\{(\Delta t)^2, (\Delta x)^2\}$ $\equiv O\{(\Delta x)^2\}$ for fixed c	None
Roberts & Weiss Semi-implicit 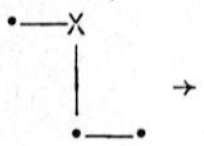	Eq. (5.1.81) $\tau_j^{n+1} = \tau_j^n +$ $+ \frac{c}{2+c}(\tau_{j-1}^{n+1} - \tau_{j+1}^n)$	Eq. (5.1.89) $O\{(\Delta t)^2, \Delta t\Delta x, (\Delta x)^2\}$ $\equiv O\{(\Delta x)^2\}$ for fixed c	None

TABLE 5.1. (Continued.)

Amplification Factor G	Stability Condition	Solvability Condition	Comments
Eq. (5.1.10) $1 - ic \sin \beta$	Unstable for all $c > 0$	None	Unstable
Eq. (5.1.16) $\{1 - 2c \sin^2(\frac{\beta}{2})\} - ic \sin \beta$	$c \leq 1$	None	Highly damped
Eq. (5.1.54) $\pm \sqrt{1 - c^2 \sin^2 \beta} - ic \sin \beta$	$c \leq 1$	None	Computational mode introduced
Eq. (5.1.75) $\dfrac{1 - i(\frac{c}{2} \sin \beta)}{1 + i(\frac{c}{2} \sin \beta)}$	None	$c \leq 2$	$c \leq 2$ required for solution of algebraic system
Eq. (5.1.82) $\dfrac{2 + 2c \sin^2(\frac{\beta}{2}) - ic \sin \beta}{2 + 2c \sin^2(\frac{\beta}{2}) + ic \sin \beta}$	None	None	Explicit

5.2 The Transport Equation

Consider the complete one dimensional transport equation (5.1.1) evaluated at the (j,n) grid-point, that is

$$\left.\frac{\partial\bar{\tau}}{\partial t}\right|_j^n + u\left.\frac{\partial\bar{\tau}}{\partial x}\right|_j^n = \alpha\left.\frac{\partial^2\bar{\tau}}{\partial x^2}\right|_j^n . \qquad (5.2.1)$$

If the central difference form (2.3.18) is used for the spatial derivative in the convection term with the forward difference form (2.3.7a) for the time derivative and the central difference form (2.3.19) for the second order derivatives in the diffusion term, this becomes

$$\frac{\bar{\tau}_j^{n+1} - \bar{\tau}_j^n}{\Delta t} + O\{\Delta t\} + u\{\frac{\bar{\tau}_{j+1}^n - \bar{\tau}_{j-1}^n}{2\Delta x} + O\{(\Delta x)^2\}\}$$

$$= \alpha\{\frac{\bar{\tau}_{j+1}^n - 2\bar{\tau}_j^n + \bar{\tau}_{j-1}^n}{(\Delta x)^2} + O\{(\Delta x)^2\}\}. \qquad (5.2.2)$$

Rearranging (5.2.2) so that $\bar{\tau}_j^{n+1}$ only appears on the left side, gives

$$\bar{\tau}_j^{n+1} = (s+\tfrac{1}{2}c)\bar{\tau}_{j-1}^n + (1-2s)\bar{\tau}_j^n + (s-\tfrac{1}{2}c)\bar{\tau}_{j+1}^n + O\{(\Delta t)^2,\Delta t(\Delta x)^2\}. \qquad (5.2.3)$$

By dropping the terms of $O\{(\Delta t)^2,\Delta t(\Delta x)^2\}$ from (5.2.3) an explicit finite difference equation is obtained which can be used to solve the one-dimensional transport equation. This equation is

$$\tau_j^{n+1} = (s+\tfrac{1}{2}c)\tau_{j-1}^n + (1-2s)\tau_j^n + (s-\tfrac{1}{2}c)\tau_{j+1}^n , \qquad (5.2.4)$$

which contributes an error of $O\{(\Delta t)^2,\Delta t(\Delta x)^2\}$ to the discretisation error at each time step.

Note that similar differencing produced a method which was always unstable in the absence of diffusion. We therefore firstly examine the stability of Equation (5.2.4). The amplification factor in the von Neumann stability analysis of this equation is found to be

$$G = \{1 - 2s(1 - \cos\beta)\} - i\{c \sin\beta\}. \qquad (5.2.5)$$

The diffusion term in Equation (5.1.1) is responsible for the second term in the real part of (5.2.5). Because the real part of G is now less than 1, the non zero value of s exercises a stabilising effect. The value of the gain $|G|$ is given by the expression

$$|G|^2 = (4s^2-c^2)\cos^2\beta - (8s^2-4s)\cos\beta + (4s^2+c^2-4s+1),$$

where the superscripts are powers.

We wish to find the region of the s-c plane for which $|G|^2 - 1 \le 0$ for all values of β in the closed interval $<0,2\pi>$. This is equivalent to finding the values of s and c for which the function $F(\chi) \le 0$ on the interval $-1 \le \chi \le 1$, where

$$F(\chi) = (4s^2-c^2)\chi^2 - (8s^2-4s)\chi + (4s^2+c^2-4s), \qquad (5.2.6)$$

and $\chi = \cos\beta$. Firstly, if $4s^2-c^2 > 0$, that is if

$$0 < c < 2s, \qquad (5.2.7a)$$

then $F(\chi)$ is a quadratic function with a minimum value. Since $F(1) = 0$, then $F(\chi) \leq 0$ on $-1 \leq \chi \leq 1$ if $F(-1) \leq 0$; that is, if

$$16s(2s-1) \leq 0$$

or

$$0 \leq 2s \leq 1. \tag{5.2.7b}$$

Combining the inequalities (5.2.7a,b) gives

$$0 < c < 2s \leq 1. \tag{5.2.8}$$

If $c = 2s$, then $F(\chi)$ is a linear function of χ, and as $F(1) = 0$, then we also require $F(-1) \leq 0$ as above in order to have $F(\chi) \leq 0$ on $\langle -1,1 \rangle$ Thus (5.2.8) can be extended to

$$0 < c \leq 2s \leq 1, \tag{5.2.9}$$

which is the region marked R1 in the s-c plane of Figure 5.10.

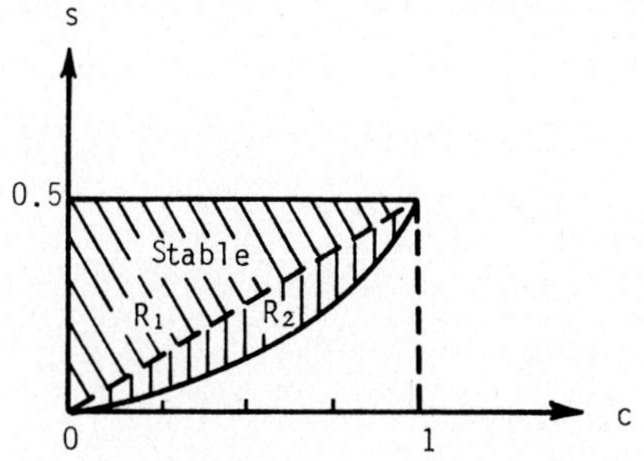

Figure 5.10: Stability region of the "centred space" finite difference method of solving the transport equation.

If $4s^2 - c^2 < 0$, that is if

$$0 < 2s < c, \tag{5.2.10a}$$

then $F(\chi)$ is a quadratic function with a maximum value. Since $F(1) = 0$, then $F(\chi) \leq 0$ on $-1 \leq \chi \leq 1$ if the product of the roots of $F(\chi) = 0$ is not less than 1; that is, if

$$\frac{4s^2+c^2-4s}{4s^2-c^2} \geq 1.$$

Multiplying both sides of this inequality by the negative quantity $4s^2-c^2$, gives

$$2s \geq c^2. \tag{5.2.10b}$$

Combining (5.2.10a,b) gives

$$0 < c^2 \leq 2s < c, \tag{5.2.11}$$

which is the region marked R2 in Figure 5.10. Combining the regions R1 and R2, it is seen that the finite difference equation (5.2.4) is stable if

$$c^2 \le 2s \le 1. \qquad (5.2.12)$$

This is a larger region than that defined by (5.2.9) which is the stability criterion usually given for this method (see for example, Isenberg and de Vahl Davis, 1975, and Roache, 1974).

By substituting $c = u\Delta t/\Delta x$ and $s = \alpha\Delta t/(\Delta x)^2$, the left inequality leads to the following condition applying to the time step Δt,

$$\Delta t \le 2\alpha/u^2, \qquad (5.2.13a)$$

and the right inequality leads to

$$\Delta t \le (\Delta x)^2/2\alpha. \qquad (5.2.13b)$$

The relations (5.2.13a,b) may be combined in the one condition on the time step, namely

$$\Delta t \le \min\{2\alpha/u^2,\ (\Delta x)^2/2\alpha\}. \qquad (5.2.14a)$$

For instance, with $\alpha = 10^{-3}$, $u = 10^{-1}$, $\Delta x = 10^{-2}$, the requirement is that

$$\Delta t \le \min\{2\times10^{-1},\ 5\times10^{-2}\},$$

so the time step should not exceed 5×10^{-2}.

Dividing (5.2.12) by $2c > 0$ gives

$$c/2 \le s/c \le 1/(2c),$$

which yields on inversion

$$2c \le c/s \le 2c^{-1}.$$

Substituting $c = u\Delta t/\Delta x$, $s = \alpha\Delta t/(\Delta x)^2$ in the central term gives

$$r = c/s = u\Delta x/\alpha,$$

which may be thought of as a Reynolds number associated with the grid size For stability, this number must satisfy the inequality

$$2c \le r \le 2c^{-1} \qquad (5.2.14b)$$

It is clear from the way that the finite difference equation (5.2.4) is derived from (5.2.2) that the truncation error for this method is usually $O\{\Delta t,(\Delta x)^2\}$. A formal consistency analysis, commencing with (5.2.4) and replacing the approximation τ by $\bar{\tau}$, which is the exact solution of (5.1.1), yields the partial differential equation

$$\begin{aligned}
&\frac{\partial\bar{\tau}}{\partial t} + \frac{\Delta t}{2}\frac{\partial^2\bar{\tau}}{\partial t^2} + \frac{(\Delta t)^2}{6}\frac{\partial^3\bar{\tau}}{\partial t^3} + \frac{(\Delta t)^3}{24}\frac{\partial^4\bar{\tau}}{\partial t^4} + \ldots \\
&\quad + u\frac{\partial\bar{\tau}}{\partial x} + u\frac{(\Delta x)^2}{6}\frac{\partial^3\bar{\tau}}{\partial x^3} + u\frac{(\Delta x)^4}{60}\frac{\partial^5\bar{\tau}}{\partial x^5} + \ldots \\
&\quad = \alpha\frac{\partial^2\bar{\tau}}{\partial x^2} + \alpha\frac{(\Delta x)^2}{12}\frac{\partial^4\bar{\tau}}{\partial x^4} + \ldots
\end{aligned} \qquad (5.2.15)$$

evaluated at the (j,n) gridpoint. Clearly (5.2.15) tends to the transport equation (5.1.1) in the limit as $\Delta x \to 0$, $\Delta t \to 0$. However, even though this implies that the difference equation (5.2.4) is consistent with the transport equation, in any practical situation (5.2.4) is solved with finite values of Δx and Δt. Therefore, in practice we actually are solving the partial differential equation (5.2.15).

Since $\bar{\tau}$ is the solution of the one-dimensional transport equation (5.1.1), it follows that

$$\frac{\partial\bar{\tau}}{\partial t} = [\frac{\partial}{\partial t}]\bar{\tau} = [-u\frac{\partial}{\partial x} + \alpha\frac{\partial^2}{\partial x^2}]\ \bar{\tau},$$

whence

$$\frac{\partial^2\bar{\tau}}{\partial t^2} = [\frac{\partial}{\partial t}]^2\bar{\tau} = [u^2\frac{\partial^2}{\partial x^2} - 2\alpha u\frac{\partial^3}{\partial x^3} + \alpha^2\frac{\partial^4}{\partial x^4}]\bar{\tau}, \tag{5.2.16a}$$

$$\frac{\partial^3\bar{\tau}}{\partial t^3} = [\frac{\partial}{\partial t}]^3\bar{\tau} = [-u^3\frac{\partial^3}{\partial x^3} + 3\alpha u^2\frac{\partial^4}{\partial x^4} - \ldots]\bar{\tau}, \tag{5.2.16b}$$

$$\frac{\partial^4\bar{\tau}}{\partial t^4} = [\frac{\partial}{\partial t}]^4\bar{\tau} = [u^4\frac{\partial^4}{\partial x^4} - \quad \ldots \quad]\bar{\tau}. \tag{5.2.16c}$$

Substitution in (5.2.15) yields the corresponding differential equation

$$\frac{\partial\bar{\tau}}{\partial t} + u\frac{\partial\bar{\tau}}{\partial x} = (\alpha-\alpha')\frac{\partial^2\bar{\tau}}{\partial x^2} + \frac{u(\Delta x)^2}{6}(6s+c^2-1)\frac{\partial^3\bar{\tau}}{\partial x^3} + \frac{\alpha(\Delta x)^2}{12}(1-6s-6c^2-\frac{c^4}{2s})\frac{\partial^4\bar{\tau}}{\partial x^4} + \ldots \tag{5.2.17}$$

where $\alpha' = u^2\Delta t/2$. Therefore, the solution of (5.2.4) is that of the transport equation, to $O\{\Delta t,(\Delta x)^2\}$, in which the diffusion coefficient α has been reduced by the amount α'. The numerical method has decreased the effect of the diffusion term $\alpha\ \partial^2\bar{\tau}/\partial x^2$ in (5.1.1) by an amount $\frac{1}{2}u^2\Delta t\ \partial^2\bar{\tau}/\partial x^2$, which can be large if both u and Δt are large. This is seen in the example shown in Figure 5.15, where the peak in the numerical solution is much larger than that in the true solution.

As seen in Section 5.1, the amplification factor G found in the von Neumann stability analysis not only gives a qualitative measure of the stability of a finite difference method but it also gives a quantitative measure of the change to the amplitude and the wave speed of an infinite travelling wave. With initial condition (5.1.22), namely

$$\bar{\tau}(x,0) = e^{i(\pi m x)}, \quad -\infty < x < \infty, \tag{5.2.18}$$

where m is real and the real part is intended, the numerical method produces the infinite wave solution (5.1.30) at time $n\Delta t$

$$\tau_j^n = |G|^n e^{i\pi m(j\Delta x + n\,\mathrm{Arg}\{G\}/\pi m)}. \tag{5.2.19}$$

The amplitude of the numerical solution is $|G|^n$ and it has a wave speed of

$$u_N = -\mathrm{Arg}\{G\}/\pi m\Delta t. \tag{5.2.20}$$

The exact solution of the one-dimensional transport equation with initial condition (5.2.18) is found by substituting the variables separable form

$$\bar{\tau}(x,t) = e^{\eta t}e^{i(\pi m x)}, \quad -\infty < x < \infty, \tag{5.2.21}$$

which clearly satisfies (5.2.18), into Equation (5.1.1). This gives

$$\eta = -\pi^2 m^2\alpha - i\pi m u. \tag{5.2.22}$$

The exact solution is therefore

$$\bar{\tau}(x,t) = e^{-\pi^2 m^2 \alpha t} e^{i\pi m(x-ut)}, \qquad (5.2.23)$$

or

$$\bar{\tau}_j^n = e^{-\pi^2 m^2 \alpha n\Delta t} e^{i\pi m(j\Delta x - un\Delta t)}. \qquad (5.2.24)$$

Like the numerical solution (5.2.19), the real part of this is an infinitely long wave travelling with speed u in the x direction, but it has an amplitude $\exp(-\pi^2 m^2 \alpha n\Delta t)$ which decreases as time passes because of the inclusion of the diffusion term $\alpha\, \partial^2\bar{\tau}/\partial x^2$.

The relative speed of the numerical wave to the true wave is

$$\mu = u_N/u = -\mathrm{Arg}\{G\}/\pi m u\Delta t, \qquad (5.2.25)$$

which is the same as (5.1.33) derived for the convection equation. This leads to

$$\mu = -\frac{N_\lambda}{2\pi c}\,\mathrm{Arg}\{G\}, \qquad (5.2.26)$$

the same as (5.1.39).

The ratio of the amplitude of the numerical wave to the true wave solution after time $n\Delta t$ is

$$|G|^n / e^{-\pi^2 m^2 \alpha n\Delta t} = \{|G| e^{\pi^2 m^2 \alpha \Delta t}\}^n. \qquad (5.2.27)$$

In one wave period, which is the time taken for the true wave to travel one wavelength, the number of time steps is given by (5.1.43), namely

$$n = N_P = N_\lambda/c. \qquad (5.2.28)$$

Therefore in one period the amplitude ratio (5.2.27) becomes

$$\gamma = |G|^{N_\lambda/c}\, e^{4\pi^2 s/cN_\lambda}, \qquad (5.2.29)$$

since from (5.1.23), (5.1.35) and (2.4.5)

$$\pi^2 m^2 \alpha \Delta t = 4\pi^2 s/(N_\lambda)^2. \qquad (5.2.30)$$

For the forward-time centred-space finite difference equation (5.2.4) the value of G, given by (5.2.5), may be written as a function of c, s and N_λ by replacing β by (5.1.38), giving

$$G(c,s,N_\lambda) = \{1-4s\sin^2(\pi/N_\lambda)\} - i\{c\sin(2\pi/N_\lambda)\}. \qquad (5.2.31)$$

In Figure 5.11, for $s = 0.25$ the relative wave speed μ given by (5.2.26) and the amplitude ratio γ given by (5.2.29) are graphed against the number of grid spacings N_λ, for various values of $c \le 1/\sqrt{2}$. The relative wave speed μ is close to 1 for all values of $N_\lambda > 20$. However, the amplitude ratio γ is markedly different from the ideal value of 1 even for relatively large numbers of grid spacings in a wavelength. In particular γ is larger than 1 for most values of c.

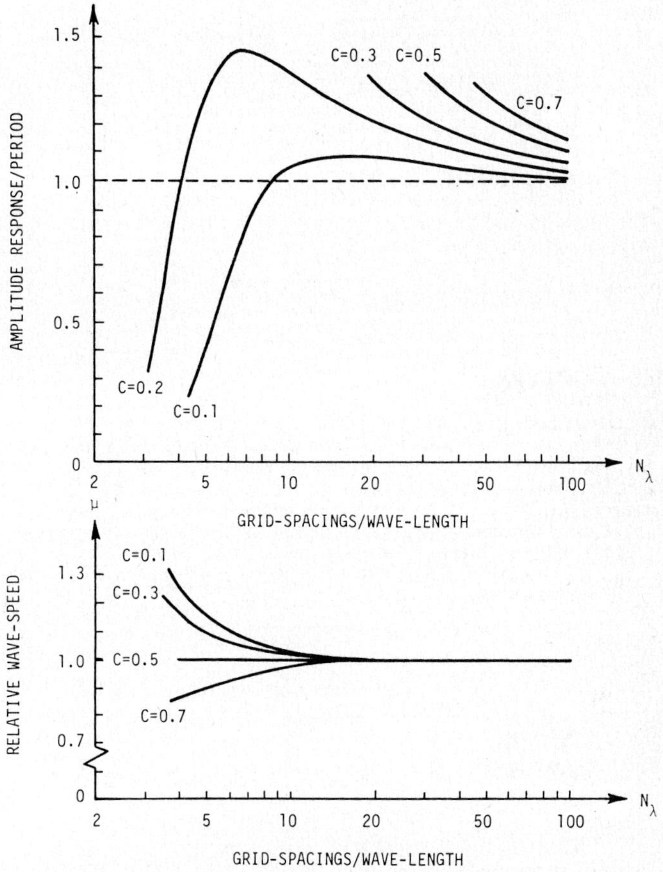

Figure 5.11 : Relative wave speed and amplitude response/period for the FTCS method of solving the transport equation, with s=0.25.

Figure 5.10indicates that the forward-time centred-space method is not stable if s is small, say s = 0.1, and c is relatively large, say 0.5. We will now investigate some other possible ways of solving (5.1.1) numerically.

In Section 4.1 it was seen that the major disadvantage of applying the Richardson method of using central time differencing with central space differencing to solve the one-dimensional diffusion equation, was that it gave a finite difference equation which was unstable for all s. However, the same procedure applied to the one-dimensional convection equation (5.1.2) in Section 5.1 gave the *leapfrog* difference equation which was stable for $c \leq 1$. When this method is applied to the complete one-dimensional transport equation (5.2.1), the following finite difference form is obtained:

$$\frac{\bar{\tau}_j^{n+1}-\bar{\tau}_j^{n-1}}{2\Delta t} + O\{(\Delta t)^2\} + u(\frac{\bar{\tau}_{j+1}^n-\bar{\tau}_{j-1}^n}{2\Delta x} + O\{(\Delta x)^2\})$$

$$= \alpha(\frac{\bar{\tau}_{j+1}^n-2\bar{\tau}_j^n+\bar{\tau}_{j-1}^n}{(\Delta x)^2} + O\{(\Delta x)^2\}). \tag{5.2.32}$$

Rearrangement of this equation gives

$$\bar{\tau}_j^{n+1} = \bar{\tau}_j^{n-1} + 2s(\bar{\tau}_{j-1}^n - 2\bar{\tau}_j^n + \bar{\tau}_{j+1}^n) + c(\bar{\tau}_{j-1}^n - \bar{\tau}_{j+1}^n)$$
$$+ O\{(\Delta t)^3, \Delta t(\Delta x)^2\}, \tag{5.2.33}$$

which yields the three-level explicit finite difference equation to solve (5.2.1),

$$\tau_j^{n+1} = (2s+c)\tau_{j-1}^n - 2s\tau_j^n + (2s-c)\tau_{j+1}^n + \tau_j^{n-1} \tag{5.2.34}$$

which contributes a term of $O\{(\Delta t)^3, \Delta t(\Delta x)^2\}$ to the discretisation error at each time step. Unfortunately, like the pure diffusion case, application of the von Neumann stability analysis shows that this method is also unstable for all values of $s \neq 0$. The method is stable for $s=0$, that is for the pure convection equation, provided that $c \leq 1$. This case was investigated in Section 5.1. However, application of the method of DuFort and Frankel (1953), produces a conditionally stable system. In this method the value $\bar{\tau}_j^n$ in the finite difference form used for the second-order space derivative $\partial^2\bar{\tau}/\partial x^2$ of the finite difference equation (5.2.32) is replaced by its average at time levels (n-1) and (n+1), giving

$$\frac{\bar{\tau}_j^{n+1}-\bar{\tau}_j^{n-1}}{2\Delta t} + O\{(\Delta t)^2\} + u(\frac{\bar{\tau}_{j+1}^n-\bar{\tau}_{j-1}^n}{2\Delta x} + O\{(\Delta x)^2\})$$

$$= \alpha\{\frac{\bar{\tau}_{j+1}^n-(\bar{\tau}_j^{n+1}+\bar{\tau}_j^{n-1})+\bar{\tau}_{j-1}^n+O\{(\Delta t)^2\}}{(\Delta x)^2} + O\{(\Delta x)^2\}\}. \tag{5.2.35}$$

Rearranging this equation so $\bar{\tau}_j^{n+1}$ is the subject, gives

$$\bar{\tau}_j^{n+1} = (\frac{1-2s}{1+2s})\bar{\tau}_j^{n-1} + (\frac{c+2s}{1+2s})\bar{\tau}_{j-1}^n - (\frac{c-2s}{1+2s})\bar{\tau}_{j+1}^n$$

$$+ O\{(\Delta t)^3, \Delta t(\Delta x)^2, (\Delta t)^3/(\Delta x)^2\}. \tag{5.2.36}$$

Therefore the three-level explicit finite difference equation

$$\tau_j^{n+1} = (\frac{1-2s}{1+2s})\tau_j^{n-1} + (\frac{c+2s}{1+2s})\tau_{j-1}^n - (\frac{c-2s}{1+2s})\tau_{j+1}^n, \tag{5.2.37}$$

may be used to get approximate answers to the one-dimensional transport problem with an $O\{(\Delta t)^3, \Delta t(\Delta x)^2, (\Delta t)^3/(\Delta x)^2\}$ contribution to the discretisation error.

Most references (for example, Isenberg and de Vahl Davis, 1975), state that the only stability restriction on Equation (5.2.37) is the requirement that $c \leq 1$ with no restriction on the value of s. However, application of this matrix method shows that the complete region of stability in the s-c plane is actually $c \leq 1$ for all s *and* $s \geq \frac{1}{2}\sqrt{c^2-1}$ for $c > 1$.

Using a Taylor series expansion about the (j,n) grid-point with the exact value $\bar{\tau}$ replacing τ in Equation (5.2.37), produces the differential equation

$$\alpha\left(\frac{\Delta t}{\Delta x}\right)^2 \frac{\partial^2\bar{\tau}}{\partial t^2} + \frac{\partial\bar{\tau}}{\partial t} + u\,\frac{\partial\bar{\tau}}{\partial x} = \alpha\frac{\partial^2\bar{\tau}}{\partial x^2} + O\{(\Delta t)^2,(\Delta x)^2,(\Delta t)^4/(\Delta x)^2\} \qquad (5.2.38)$$

at the (j,n) grid-point. The method is therefore consistent with the transport equation as $\Delta t \to 0$, $\Delta x \to 0$ only if Δt and Δx approach the limiting value of zero in such a way that $\Delta t/\Delta x \to 0$. This is an additional restriction on the size Δt can take, apart from the condition $\Delta t \le \Delta x/u$ which results from the stability requirement $c \le 1$.

DuFort and Frankel (1953) suggest that the stability of the method is due to the introduction of the first term in Equation (5.2.38), which makes the equation hyperbolic in nature.

It is clear that this scheme does have certain disadvantages, but it is a useful explicit conditionally stable method. For instance, Pearson (1965a) has shown that the DuFort-Frankel method is more accurate than the previously described forward-time centred-space method (5.2.4) when applied to certain problems involving the transport equation.

Using the forward difference form for the time derivative and the central difference form for the space derivative in the diffusion term of Equation (5.2.1), together with the upwind difference form for the convection term gives, for $u > 0$,

$$\frac{\bar{\tau}^{n+1}_j - \bar{\tau}^n_j}{\Delta t} + O\{\Delta t\} + u\left(\frac{\bar{\tau}^n_j - \bar{\tau}^n_{j-1}}{\Delta x} + O\{\Delta x\}\right)$$

$$= \alpha\left\{\frac{\bar{\tau}^n_{j+1} - 2\bar{\tau}^n_j + \bar{\tau}^n_{j-1}}{(\Delta x)^2} + O\{(\Delta x)^2\}\right\}, \qquad (5.2.39)$$

which can be rewritten

$$\bar{\tau}^{n+1}_j = (s+c)\bar{\tau}^n_{j-1} + (1-2s-c)\bar{\tau}^n_j + s\bar{\tau}^n_{j+1} + O\{(\Delta t)^2,\Delta t\Delta x\}. \qquad (5.2.40)$$

The explicit finite difference equation

$$\tau^{n+1}_j = (s+c)\tau^n_{j-1} + (1-2s-c)\tau^n_j + s\tau^n_{j+1}, \qquad (5.2.41)$$

can therefore be used to compute approximate solutions to the one-dimensional transport equation, with a truncation error of $O\{\Delta t,\Delta x\}$.

The amplification factor in the von Neumann stability analysis of Equation (5.2.41) is

$$G = \{1-(2s+c)(1-\cos\beta)\} - i\{c\sin\beta\}, \qquad (5.2.42)$$

so that the gain $|G|$ is given by the solution of

$$|G|^2 = (4s^2+4sc)\cos^2\beta - (8s^2+8sc+2c^2-4s-2c)\cos\beta$$

$$+ (1+4s^2+4sc+2c^2-4s-2c), \qquad (5.2.43)$$

where the indices represent powers. Consider the function $F(\chi)=|G|^2-1$, where $\chi=\cos\beta$. For stability we require $|G|\leq 1$. Therefore we wish to find the region of the s-c plane in which $F(\chi)\leq 0$ for all β, that is for all values of χ in $<-1,1>$. Now

$$F(\chi) = (4s^2+4sc)\chi^2 - (8s^2+8sc+2c^2-4s-2c)\chi + (4s^2+4sc+2c^2-4s-2c). \tag{5.2.44}$$

This is a quadratic function, with a minimum value because the coefficient $(4s^2+4sc)$ of χ^2 is positive. But $F(1) = 0$ and

$$F(-1) = 16s^2 + 16sc + 4c^2 - 8s - 4c$$

$$= 4(2s+c-1)(2s+c).$$

Therefore, so long as

$$2s + c \leq 1, \tag{5.2.45}$$

then $F(-1) \leq 0$, and $F(\chi) \leq 0$ for $-1 \leq \chi \leq 1$. Therefore, the inequality (5.2.45) is the required stability criterion. The region of the c-s plane for which (5.2.45) is true is shown in Figure 5.12. This condition implies that the time step must satisfy the inequality

$$\Delta t \leq 2\left\{\frac{\alpha}{(\Delta x)^2} + \frac{u}{\Delta x}\right\}^{-1}. \tag{5.2.46a}$$

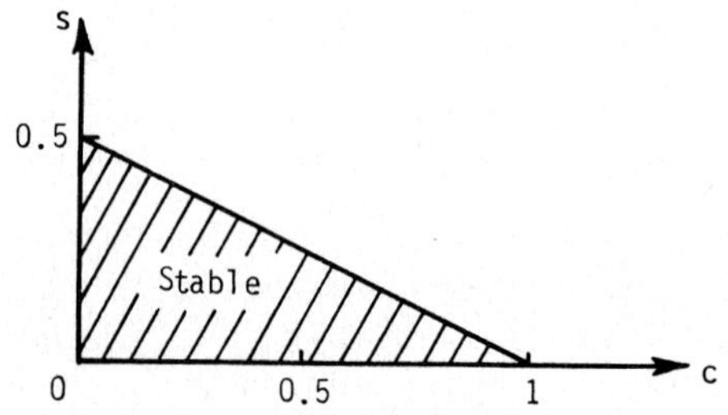

Figure 5.12: Stability region for the upwind difference method of solving the transport equation.

The condition (5.2.45), on division by $c = u\Delta t/\Delta x$ and substitution of $s = \alpha\Delta t/(\Delta x)^2$, can be written

$$r \geq 2c/(1-c) \tag{5.2.46b}$$

where $r = u\Delta x/\alpha$ is the Reynolds number based on the grid-spacing. This is less restrictive than the corresponding relation (5.2.14b) using the central space form (5.2.4) for the convective derivative (see Figure 5.13).

As with the forward-time centred-space equation for solving the transport equation, the amplification factor G obtained in the von Neumann stability analysis may be used to determine the amplitude ratio γ and the relative wave speed μ produced for an infinite wave. The form for G for the upwind equation is given by (5.2.42) which can be expressed as

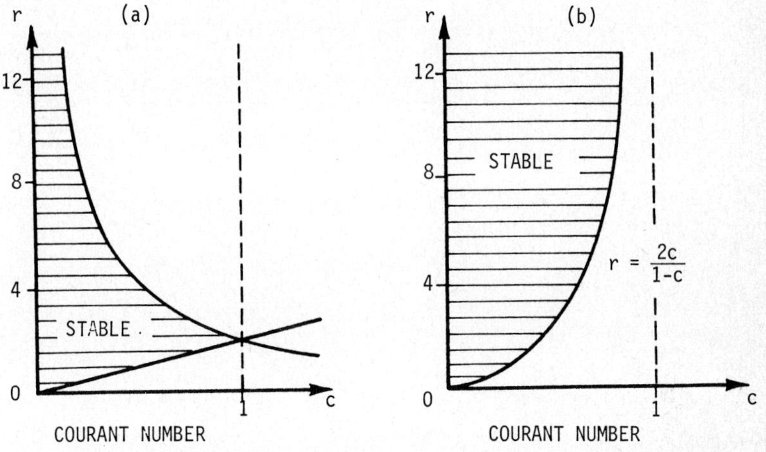

Figure 5.13 : Stability region of the r-c plane for (a) the FTCS and (b) upwind difference methods of solving the transport equation.

a function of c, s and N_λ, using (5.1.39):

$$G = \{1 - 2(2s+c)\sin^2(\pi/N_\lambda)\} - i\{c \sin(2\pi/N_\lambda)\}. \qquad (5.2.47)$$

This can be used with (5.2.29) and (5.2.26) to compute the amplitude ratio γ and the relative wave speed μ. Figure 5.14 shows graphs of γ and μ plotted against the number of grid spacings per wavelength, N_λ, for various values of c when s = 0.25. The relative wave speed μ is greater than 1 for all values of c and N_λ, and is very close to 1 for all c if $N_\lambda > 50$. The amplitude ratio γ is much less than 1 even for very large values of N_λ, which is reflected in the very poor results in Figure 5.15, obtained using this method.

A consistency analysis of the finite difference equation (5.2.41) gives the corresponding partial differential equation at the (j,n) gridpoint

$$\frac{\partial\bar{\tau}}{\partial t} + \frac{\Delta t}{2}\frac{\partial^2\bar{\tau}}{\partial t^2} + \frac{(\Delta t)^2}{6}\frac{\partial^3\bar{\tau}}{\partial t^3} + \frac{(\Delta t)^3}{24}\frac{\partial^4\bar{\tau}}{\partial t^4} + \ldots$$

$$+ u\frac{\partial\bar{\tau}}{\partial x} - u\frac{\Delta x}{2}\frac{\partial^2\bar{\tau}}{\partial x^2} + u\frac{(\Delta x)^2}{6}\frac{\partial^3\bar{\tau}}{\partial x^3} - u\frac{(\Delta x)^3}{24}\frac{\partial^4\bar{\tau}}{\partial x^4} + \ldots$$

$$= \alpha\frac{\partial^2\bar{\tau}}{\partial x^2} + \frac{(\Delta x)^2}{12}\frac{\partial^4\bar{\tau}}{\partial x^4} + \ldots \qquad (5.2.48)$$

Clearly this becomes equation (5.1.1) in the limit as $\Delta x \to 0$, $\Delta t \to 0$, so the upwind difference equation (4.2.41) is consistent with the one-dimensional transport equation. However, when the upwind difference equation is used both Δx and Δt are finite, and (5.2.48) is the partial

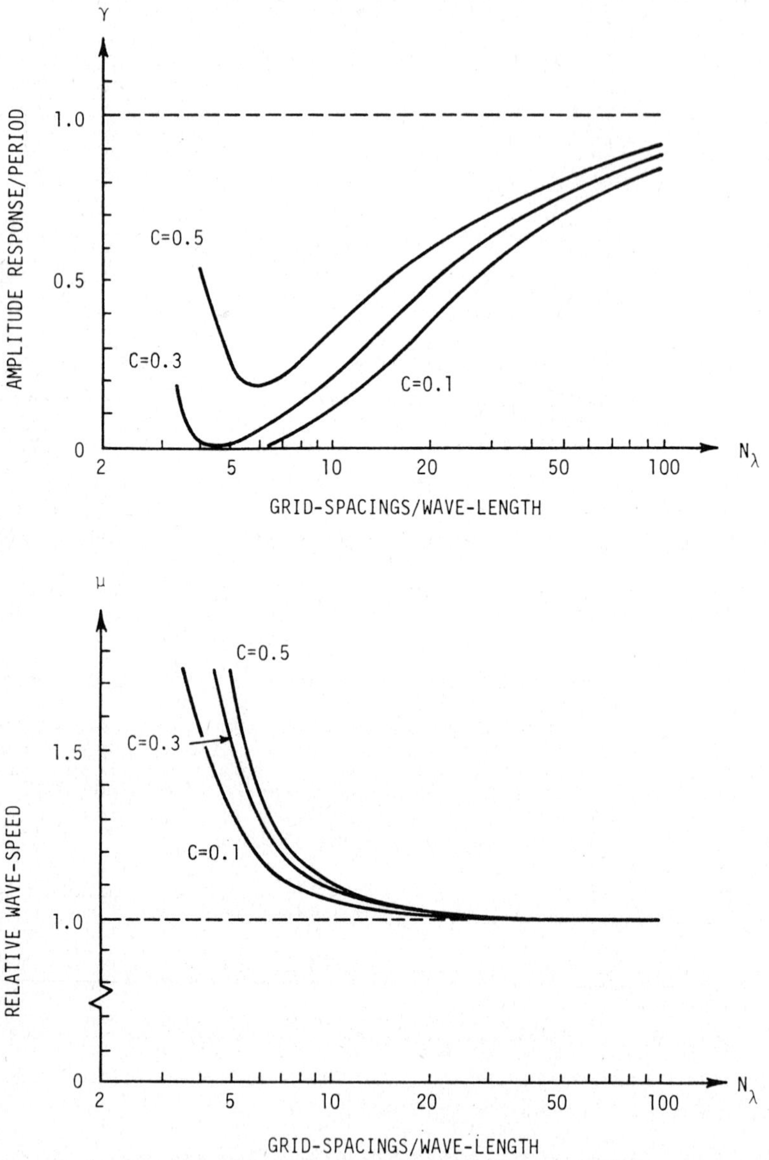

Figure 5.14 : The relative wave speed and the amplitude ratio/period for the upwind method of solving the transport equation, with s = 0.25.

differential equation actually solved. Also, because $\bar{\tau}$ is the solution of (5.1.1), $\partial^2\bar{\tau}/\partial t^2$ can be replaced by (5.2.16a), $\partial^3\bar{\tau}/\partial t^3$ by (5.2.16b) and $\partial^4\bar{\tau}/\partial t^4$ by (5.2.16c), giving the corresponding partial differential equation

$$\frac{\partial\bar{\tau}}{\partial t} + u\frac{\partial\bar{\tau}}{\partial x} = (\alpha+\alpha')\frac{\partial^2\bar{\tau}}{\partial x^2} + \frac{u(\Delta x)^2}{6}(6s+c^2-1)\frac{\partial^3\bar{\tau}}{\partial x^3}$$

$$+ \frac{(\Delta x)^4}{24\Delta t}(c+2s-12s^2-12sc^2-c^4)\frac{\partial^4\bar{\tau}}{\partial x^4} + \ldots \qquad (5.2.49)$$

where

$$\alpha' = \tfrac{1}{2}u\Delta x(1-c). \qquad (5.2.50)$$

Equation (5.2.50) indicates that the method introduces numerical diffusion of $O\{\Delta x\}$ which is minimized by taking c as close to 1 as possible. However, from (5.2.45), c must satisfy the relation

$$c \le 1 - 2s,$$

which implies that the maximum value c can take is 1-2s. Choosing this value gives the numerical diffusion coefficient

$$\alpha' = \tfrac{1}{2}u\Delta x(2s) = c\alpha.$$

Therefore, *in this case* the value of the diffusion coefficient in (5.2.48) is

$$\alpha + \alpha' = \alpha(1+c).$$

The relative error in α caused by the introduction of numerical diffusion is c. In general, accurate solutions are not possible unless $\alpha' \ll \alpha$ which, for a given value of $c < 1$, requires that $\tfrac{1}{2}u\Delta x \ll \alpha$ or $c \ll 1$.

The effect of the increase of the diffusion coefficient by the numerical diffusion of the upwind finite difference equation (5.2.41), is evident in Figure 5.15. The one-dimensional diffusion equation (2.1.1) was solved on $0 \le x \le 1$ with the initial condition

$$\bar{\tau}(x,0) = \exp\{-(x-0.2)^2/\alpha\},\ 0 \le x \le 1, \qquad (5.2.51a)$$

and boundary conditions

$$\bar{\tau}(0,t) = \frac{1}{\sqrt{4t+1}}\exp\{\frac{-(0.2+ut)^2}{\alpha(4t+1)}\},\ t \ge 0, \qquad (5.2.51b)$$

$$\bar{\tau}(1,t) = \frac{1}{\sqrt{4t+1}}\exp\{\frac{-(0.8-ut)^2}{\alpha(4t+1)}\},\ t \ge 0, \qquad (5.2.51c)$$

using $\alpha = 0.001$, $u = 0.2$, $s = 0.25$, $c = 0.5$, $\Delta x = 0.01$. When the numerical solution is compared with the exact solution

$$\bar{\tau}(x,t) = \frac{1}{\sqrt{4t+1}}\exp\{\frac{-(x-0.2-ut)^2}{\alpha(4t+1)}\},\ 0 \le x \le 1,\ t \ge 0, \qquad (5.2.52)$$

at $t = 2$, the peak value at $x = 0.6$ is clearly suppressed too much.

This is in distinct contrast with the results obtained using the forward-time centred-space finite difference equation (5.2.4), in which the diffusion coefficient is decreased by "negative" numerical diffusion. The peak at x = 0.6 in the numerical solution using the FTCS method is too high, as can be seen in Figure 5.15.

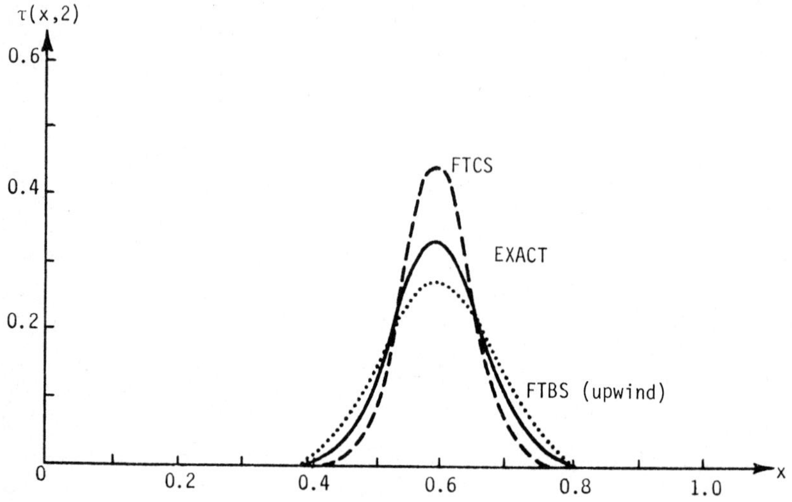

Figure 5.15 : A comparison of the FTCS and "upwind" numerical solutions with with true solution, for the transport equation.

It would seem that some form of weighted average of the backward and centred difference forms of the convective derivative $\partial\bar{\tau}/\partial x$ may produce a method with minimal numerical diffusion. Discretising the one-dimensional diffusion equation at the (j,n) gridpoint, see Equation (5.2.1), using the forward difference form for the time derivative, the centred difference form for the second order spatial derivative, and a weighted backward and centred difference form for the first order spatial derivative, namely

$$\left.\frac{\partial\bar{\tau}}{\partial x}\right|_j^n = \psi\left.\frac{\partial\bar{\tau}}{\partial x}\right|_j^n + (1-\psi)\left.\frac{\partial\bar{\tau}}{\partial x}\right|_j^n, \quad 0 \le \psi \le 1, \tag{5.2.53}$$

with the backward space form used for the part with coefficient ψ and the centred space form for the part associated with the weight $(1-\psi)$, gives

$$\begin{aligned}&\frac{\bar{\tau}_j^{n+1} - \bar{\tau}_j^n}{\Delta t} + O\{\Delta t\}\\ &\quad + u\{\psi(\frac{\bar{\tau}_j^n - \bar{\tau}_{j-1}^n}{\Delta x} + O\{\Delta x\}) + (1-\psi)(\frac{\bar{\tau}_{j+1}^n - \bar{\tau}_{j-1}^n}{2\Delta x} + O\{(\Delta x)^2\})\}\\ &\quad = \alpha\{\frac{\bar{\tau}_{j+1}^n - 2\bar{\tau}_j^n + \bar{\tau}_{j-1}^n}{(\Delta x)^2} + O\{(\Delta x)^2\}\}.\end{aligned} \tag{5.2.54}$$

Dropping terms of $O\{\Delta t, \psi\Delta x, (1-\psi)(\Delta x)^2\}$ gives, on rearrangement, the explicit finite difference equation

$$\tau_j^{n+1} = \{s + \tfrac{1}{2}(1+\psi)c\}\tau_{j-1}^n + (1-2s-\psi c)\tau_j^n + \{s - \tfrac{1}{2}(1-\psi)c\}\tau_{j+1}^n,$$

$$j=1(1)J-1. \qquad (5.2.55)$$

A formal consistency analysis of the finite difference equation (5.2.55) shows that it is consistent with the one-dimensional transport equation, with a truncation error

$$E(x,t) = \tfrac{1}{2}\psi u\Delta x\frac{\partial^2\bar{\tau}}{\partial x^2} - \tfrac{1}{2}\Delta t\frac{\partial^2\bar{\tau}}{\partial t^2} + O\{(\Delta x)^2,(\Delta t)^2\}. \qquad (5.2.56)$$

Rearranging the transport equation and differentiating as in (5.2.16a-c) then substituting in (5.2.56) gives

$$E(x,t) = \tfrac{1}{2}u(\psi\Delta x-u\Delta t)\frac{\partial^2\bar{\tau}}{\partial x^2} + O\{\Delta t,(\Delta x)^2\}. \qquad (5.2.57)$$

Clearly, $E(x,t) \to 0$ as $\Delta x \to 0$, $\Delta t \to 0$, so Equation (5.2.55) is consistent with the transport equation. However, since Δx, Δt are finite we are actually solving (5.1.1) with additional terms on the right side, namely

$$\frac{\partial\bar{\tau}}{\partial t} + u\frac{\partial\bar{\tau}}{\partial x} = (\alpha+\alpha')\frac{\partial^2\bar{\tau}}{\partial x^2} + O\{\Delta t,(\Delta x)^2\} \qquad (5.2.58)$$

where

$$\alpha' = \tfrac{1}{2}u(\psi\Delta x-u\Delta t), \qquad (5.2.59)$$

is the change in the diffusion coefficient α introduced by the use of (5.2.55).

A von Neumann stability analysis yields the amplification factor

$$G = \{1 - (2s+\psi c) + (2s+\psi c)\cos\beta\} - i\{c\sin\beta\}. \qquad (5.2.60)$$

Therefore

$$|G|^2 = \{1 - (2s+\psi c) + (2s+\psi c)\cos\beta\}^2 + \{c\sin\beta\}^2,$$

and

$$\begin{aligned}|G|^2 - 1 = &\ \{4s^2 + 4\psi sc + (\psi^2-1)c^2\}\cos^2\beta \\ + &\ \{-8s^2 - 8\psi sc - 2\psi^2c^2 + 4s + 2\psi c\}\cos\beta \\ + &\ \{4s^2 + 4\psi sc + (\psi^2+1)c^2 - 4s - 2\psi c\}. \qquad (5.2.61)\end{aligned}$$

We wish to find the values of s, c for which $|G|^2 - 1 \le 0$, for all β, which is the same as finding values of s, c for which $f(x) \le 0$, $-1\le x\le 1$, where $f(x)$ is the quadratic

$$\begin{aligned}f(x) = &\ \chi^2\{4s^2 + 4\psi sc + (\psi^2-1)c^2\} + \chi\{-8s^2 - 8\psi sc - 2\psi^2c^2 + 4s + 2\psi c\} \\ &+ \{4s^2 + 4\psi sc + (\psi^2+1)c^2 - 4s - 2\psi c\}. \qquad (5.2.62)\end{aligned}$$

Firstly, $f(1) = 0$, so $\chi = 1$ is a zero of $f(x)$. Therefore we require (see Figure 5.16)

$$f(-1) = 16s^2 + 16\psi sc + 4\psi^2c^2 + 8s + 4\psi c \le 0,$$

or

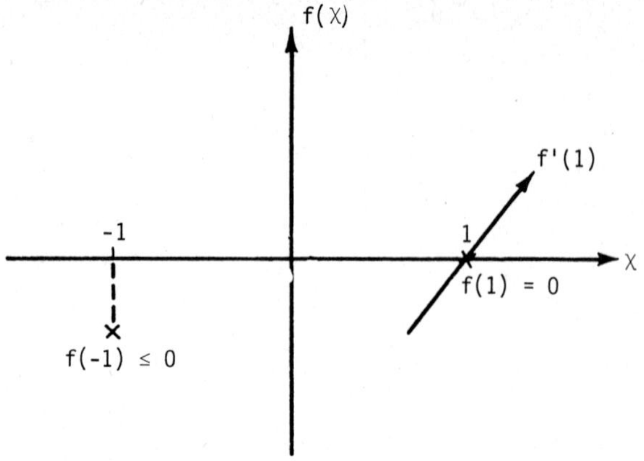

Figure 5.16: Conditions required to establish $f(\chi)\le 0$ for $-1\le\chi\le 1$.

$$4(2s+\psi c)(2s+\psi c-1) \le 0.$$

Since $2s + \psi c > 0$, we require

$$2s \le 1 - \psi c. \tag{5.2.63}$$

In addition we also require $f'(1) \ge 0$ (see Figure 5.16) namely

$$[\chi(8s^2+8\psi sc+2\psi^2c^2-2c^2) + (-8s^2-8\psi sc-2\psi^2c^2+4s+2\psi c)]_{\chi=1} \ge 0$$

which leads to the requirement that

$$2s \ge c^2 - \psi c. \tag{5.2.64}$$

The region of the s-c plane for which (5.2.63) and (5.2.64) both hold, in which (5.2.55) is stable, is therefore

$$c^2 - \psi c \le 2s \le 1 - \psi c. \tag{5.2.65}$$

This is shown in the Figure 5.17.

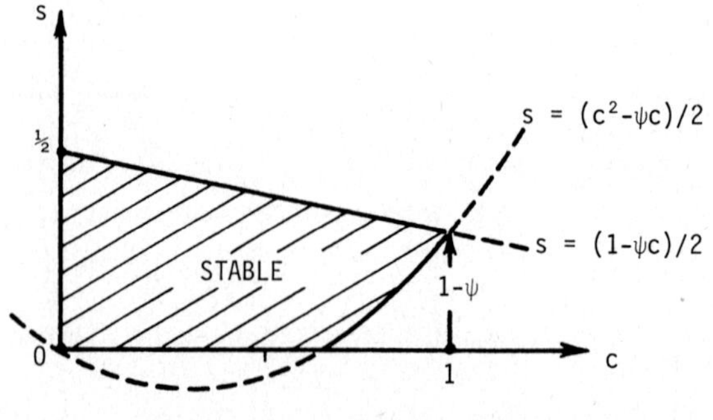

Figure 5.17 : Stability region for the space-weighted explicit method of solving the transport equation.

Note that if c = 0 Equation (5.2.55) becomes

$$\tau_j^{n+1} = s\tau_{j-1}^n + (1-2s)\tau_j^n + s\tau_{j+1}^n, \; j=1(1)J-1,$$

which is the FTCS explicit finite difference equation for solving the transport equation with u = 0, namely the diffusion equation. Also, putting c = 0 in (5.2.65) gives the stability requirement $0 < 2s \le 1$, which checks with the stability requirement given in Section 3.3 for the FTCS method.

Putting $\psi = 0$ in (5.2.55) gives the forward-time centred-space finite difference equation (5.2.4) for solving the transport equation, with the stability criterion obtained from (5.2.65) $c^2 \le 2s \le 1$, which is the same as (5.2.12). Putting $\psi = 1$ in (5.2.55) gives the upwind finite difference equation (5.2.41) which, from (5.2.65), is stable for $c^2 - c \le 2s \le 1 - c$. This is the same as $c + 2s \le 1$, the corresponding result for the upwind method.

Putting $\psi = c$ in (5.2.57) eliminates artificial diffusion, since then $\alpha' = 0$. The corresponding finite difference equation is (5.2.55), with $\psi = c$, namely

$$\tau_j^{n+1} = \{s + \tfrac{1}{2}(1+c)c\}\tau_{j-1}^n + \{1-2s-c^2\}\tau_j^n + \{s-\tfrac{1}{2}(1-c)c\}\tau_{j+1}^n . \tag{5.2.66}$$

This is stable for

$$0 < 2s \le 1 - c^2, \tag{5.2.67}$$

obtained by putting $\psi = c$ in (5.2.65). The stability region for this method is shown in Figure 5.18.

Figure 5.18 : Stability region for the optimal explicit method of solving the transport equation.

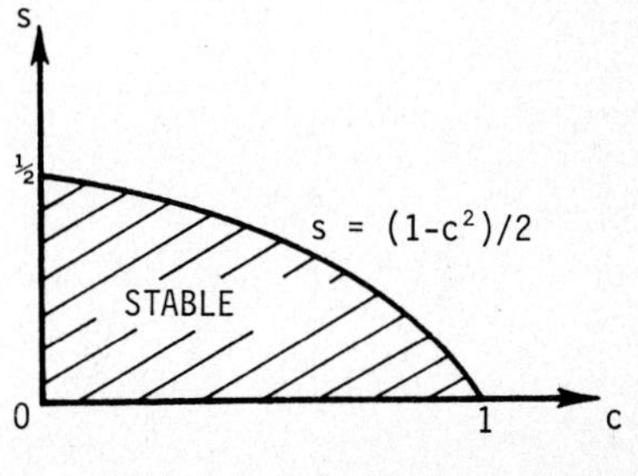

Since s = 0.25, c = 0.5 lies in the stability region of this method, it was used with these values to solve the diffusion problem with initial and boundary conditions (5.2.51a-c). The result was a numerical solution at t = 2 which was almost identical with the true solution shown in Figure 5.15.

The amplification factor G for this method is given by putting $\psi = c$ in Equation (5.2.60), which becomes

$$G = \{1 - 2(2s+c^2)\sin^2(\pi/N_\lambda)\} - i\{c \sin(2\pi/N_\lambda)\}, \tag{5.2.68}$$

on replacing β by $2\pi/N_\lambda$ (see Equation (5.1.38)). Substitution of (5.2.68) in Equation (5.2.29) gives the amplitude ratio γ, and in Equation (5.2.27) gives the relative wave speed μ, for Equation (5.2.66) used to solve the transport equation. Figure 5.19 shows, for s = 0.25, graphs of γ and μ plotted against N_λ, the number of grid spacings per wavelength, for various values of the Courant number c.

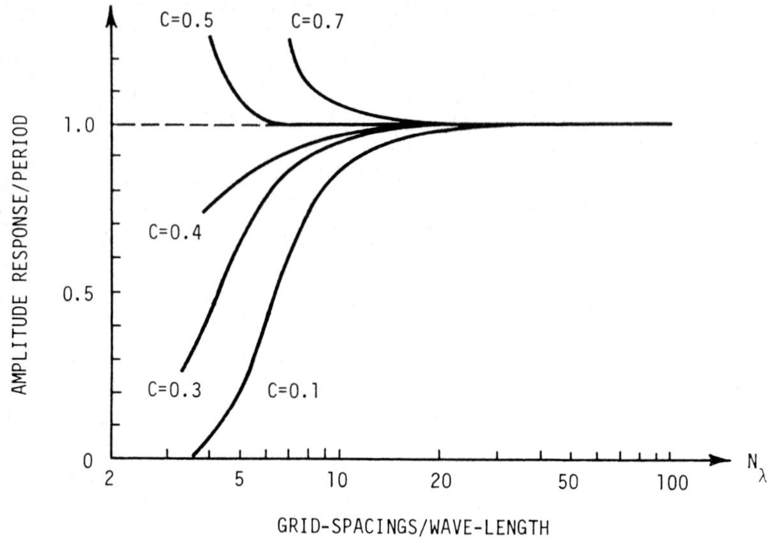

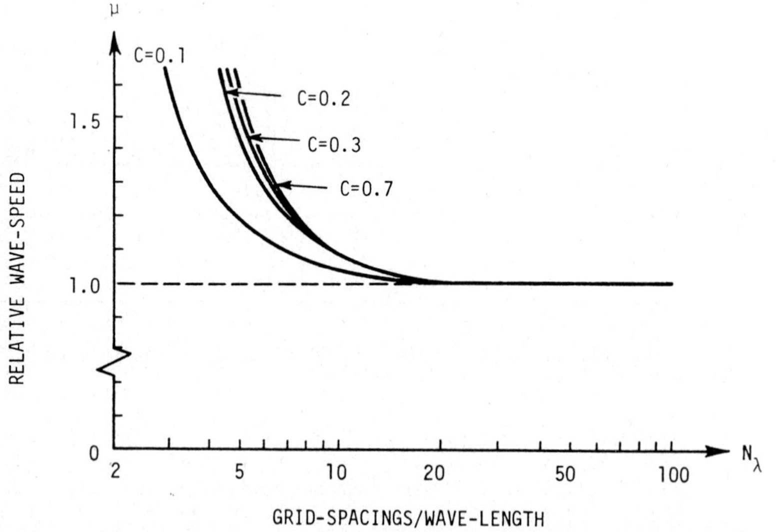

Figure 5.19 : The amplitude ratio/period and the relative wave speed for the optimal explicit method of solving the transport equation with $s = 0.25$. The method is then stable for $c \le 1/\sqrt{2}$.

Implicit methods may also be used to solve the one-dimensional transport equation. Consider the scheme, similar to the Crank-Nicolson type for solving the one-dimensional diffusion equation, in which the given partial differential equation at the point $(j\Delta x, n\Delta t+\frac{1}{2}\Delta t)$ in the solution domain is approximated in terms of values of the dependent variable at nearby grid-points. Thus we have

$$\left.\frac{\partial \bar{\tau}}{\partial t}\right|_j^{n+\frac{1}{2}} + u\left.\frac{\partial \bar{\tau}}{\partial x}\right|_j^{n+\frac{1}{2}} = \alpha\left.\frac{\partial^2 \bar{\tau}}{\partial x^2}\right|_j^{n+\frac{1}{2}}, \tag{5.2.69}$$

in which we write the two space derivatives in terms of the average of their values the time levels n and (n+1). Thus (5.2.69) becomes

$$\left.\frac{\partial \bar{\tau}}{\partial t}\right|_j^{n+\frac{1}{2}} + \frac{u}{2}\left\{\left.\frac{\partial \bar{\tau}}{\partial x}\right|_j^{n} + \left.\frac{\partial \bar{\tau}}{\partial x}\right|_j^{n+1} + O\{(\Delta t)^2\}\right\}$$

$$= \frac{\alpha}{2}\left\{\left.\frac{\partial^2 \bar{\tau}}{\partial x^2}\right|_j^{n} + \left.\frac{\partial^2 \bar{\tau}}{\partial x^2}\right|_j^{n+1} + O\{(\Delta t)^2\}\right\}. \tag{5.2.70}$$

Writing all the derivatives in (5.2.70) in their centred-difference forms gives

$$\frac{\bar{\tau}_j^{n+1} - \bar{\tau}_j^n}{\Delta t} + O\{(\Delta t)^2\} + \frac{u}{2}\left\{\frac{\bar{\tau}_{j+1}^n - \bar{\tau}_{j-1}^n}{2\Delta x} + \frac{\bar{\tau}_{j+1}^{n+1} - \bar{\tau}_{j-1}^{n+1}}{2\Delta x} + O\{(\Delta t)^2,(\Delta x)^2\}\right\}$$

$$= \frac{\alpha}{2}\left\{\frac{\bar{\tau}_{j+1}^n - 2\bar{\tau}_j^n + \bar{\tau}_{j-1}^n}{(\Delta x)^2} + \frac{\bar{\tau}_{j+1}^{n+1} - 2\bar{\tau}_j^{n+1} + \bar{\tau}_{j-1}^{n+1}}{(\Delta x)^2} + O\{(\Delta t)^2,(\Delta x)^2\}\right\}. \tag{5.2.71}$$

On rearranging this equation with the terms at the time level (n+1) on the left side and terms at the time level n on the right side, and dropping the terms of $O\{(\Delta t)^3, \Delta t(\Delta x)^2\}$, gives the implicit finite difference equation

$$\left(-\frac{s}{2} - \frac{c}{4}\right)\tau_{j-1}^{n+1} + (1+s)\tau_j^{n+1} + \left(-\frac{s}{2} + \frac{c}{4}\right)\tau_{j+1}^{n+1}$$

$$= \left(\frac{s}{2} + \frac{c}{4}\right)\tau_{j-1}^{n} + (1-s)\tau_j^{n} + \left(\frac{s}{2} - \frac{c}{4}\right)\tau_{j+1}^{n}, \quad j=1(1)J-1. \tag{5.2.72}$$

This equation may be used to find approximate solutions to the one-dimensional transport equation, with a truncation error of $O\{(\Delta t)^2,(\Delta x)^2\}$ and a contribution of $O\{(\Delta t)^3, \Delta t(\Delta x)^2\}$ to the discretisation error at every time step.

Note that, if u = 0 so that c = 0, this becomes the Crank-Nicolson equation (4.2.11) for solving the one-dimensional diffusion equation (2.1.1). Also, when $\alpha = 0$ so that s = 0, this becomes the corresponding equation (5.1.73) used to solve the one-dimensional convection equation (5.1.2). Application of the Thomas method then gives the solutions τ_j^{n+1}, j=1(1)J-1, of the tri-diagonal set of algebraic equations (5.2.72), j=1(1)J-1, if all values of τ_j^n at the previous time level are known together with the boundary values τ_0^{n+1}, τ_J^{n+1} (compare Equations (4.2.12) for the diffusion equation).

As seen previously, implicit finite difference equations are very stable. The stability of this method can be determined by applying the von Neumann analysis. The amplification factor is given by

$$G = \frac{(1-s+s\cos\beta) - i(\tfrac{1}{2}c\sin\beta)}{(1+s-s\cos\beta) + i(\tfrac{1}{2}c\sin\beta)}. \qquad (5.2.73)$$

Therefore

$$|G|^2 = \frac{(1-s+s\cos\beta)^2 + (\tfrac{1}{2}c\sin\beta)^2}{(1+s-s\cos\beta)^2 + (\tfrac{1}{2}c\sin\beta)^2}, \qquad (5.2.74)$$

and

$$|G|^2 - 1 = \frac{-4s(1-\cos\beta)}{(1+s-s\cos\beta)^2 + (\tfrac{1}{2}c\sin\beta)^2}. \qquad (5.2.75)$$

Since the factor $(1-\cos\beta) \geq 0$ for all β, and the denominator of (5.2.75) is always positive, then

$$|G|^2 - 1 \leq 0$$

and

$$|G| \leq 1,$$

for all c, s, β. The equation (5.2.72) is therefore *unconditionally stable.*

The set of equations (5.2.72), $j=1(1)J-1$, may be solved using the Thomas algorithm or any other elimination technique provided the system is diagonally dominant. This requires

$$|1+s| \geq |-\frac{s}{2} - \frac{c}{4}| + |-\frac{s}{2} + \frac{c}{4}|,$$

or

$$1 + s \geq \frac{s}{2} + \frac{c}{4} + |\frac{c}{4} - \frac{s}{2}|. \qquad (5.2.76)$$

If $s \leq \tfrac{1}{2}c$, this becomes $s \geq \tfrac{1}{2}c - 1$. The region of the s-c plane in which both these inequalities hold is labelled R1 in Figure 5.20. If $s \geq \tfrac{1}{2}c$, (5.2.76) becomes $1 + s \geq s$ which is always true. The latter inequalities hold in region R2 of Figure 5.20.

The system of equations (5.2.72) is therefore *solvable* in the shaded region of Figure 5.20, namely for

$$s \geq \tfrac{1}{2}c - 1. \qquad (5.2.77)$$

If $\alpha = 0$, then $s = 0$ and this becomes $c \leq 2$, which is the condition required for the corresponding equations to be solvable when this method is applied to the pure convection equation. If $u = 0$, then $c = 0$, and (5.2.77) is always satisfied, when this method is always solvable. This is the same result as that found when the Crank-Nicolson method for the diffusion equation was considered in Section 4.2.

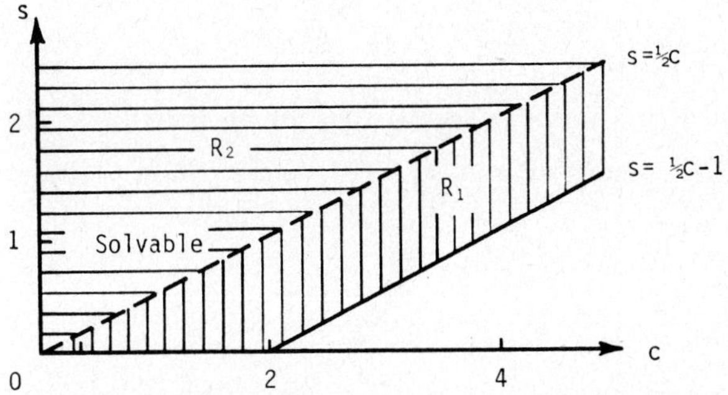

Figure 5.20: The region of the s-c plane in which Equations (5.2.38) are solvable.

Application of this method with $s = 0.25$, $c = 0.5$, $\Delta x = 0.01$, $u = 0.2$, $\alpha = 0.001$ to solve the transport equation with initial and boundary conditions (5.2.51a-c), gave a numerical solution almost the same as the true solution at $t = 2$ (see Figure 5.15).

Substitution of $2\pi/N_\lambda$ for β in Equation (5.2.73) gives the amplification factor as a function of c, s and N_λ, namely

$$G = \frac{\{1 - 2s\sin^2(\pi/N_\lambda)\} - i\{\tfrac{1}{2}c\sin(2\pi/N_\lambda)\}}{\{1 + 2s\sin^2(\pi/N_\lambda)\} + i\{\tfrac{1}{2}c\sin(2\pi/N_\lambda)\}}. \tag{5.2.78}$$

Equation (5.2.29) then gives the amplitude ratio γ, and Equation (5.2.27) the relative wave speed μ, as functions of c, s and N_λ. Figure 5.21 shows graphs of γ and μ, for $s = 0.25$, plotted against N_λ, the number of grid spacings per wavelength, for different values of c, the Courant number. The most notable feature of these results is the small value of the relative wave speed for small values of N_λ.

A *semi-implicit* method of solving the one-dimensional transport equation may be obtained by combining the differencing of the derivative $\partial\bar{\tau}/\partial x$ used by Roberts and Weiss (1966) for the convection equation and the differencing of the derivative $\partial^2\bar{\tau}/\partial x^2$ used by Saul'yev (1964) for the diffusion equation.

Discretising the transport equation (5.1.1) at the point $(j\Delta x, n\Delta t+\tfrac{1}{2}\Delta t)$, using the form (5.1.80) for the left hand side and (4.2.35) for $\partial^2\bar{\tau}/\partial x^2|_j^{n+\frac{1}{2}}$ on the right hand side, gives

$$\frac{\bar{\tau}_j^{n+1} - \bar{\tau}_j^{n}}{\Delta t} + O\{(\Delta t)^2\} + \frac{u}{2}\left\{\frac{\bar{\tau}_j^{n+1} - \bar{\tau}_{j-1}^{n+1}}{\Delta x} + \frac{\bar{\tau}_{j+1}^{n} - \bar{\tau}_j^{n}}{\Delta x} + O\{(\Delta t)^2, \Delta x\}\right\}$$

$$= \alpha\left\{\frac{1}{\Delta x}\left[\frac{\bar{\tau}_{j+1}^{n} - \bar{\tau}_j^{n}}{\Delta x} - \frac{\bar{\tau}_j^{n+1} - \bar{\tau}_{j-1}^{n+1}}{\Delta x} + O\{(\Delta x)^2\}\right]\right.$$

$$\left. + O\left\{\frac{\Delta t}{\Delta x}, (\Delta x)^2\right\}\right\}. \tag{5.2.79}$$

Dropping terms of $O\{\Delta x, \Delta t/\Delta x\}$, multiplying through by Δt, and rearranging, gives the semi-implicit equation

$$\tau_j^{n+1} = \frac{1}{1+s+\tfrac{1}{2}c}\{(s+\tfrac{1}{2}c)\tau_{j-1}^{n+1} + (1-s+\tfrac{1}{2}c)\tau_j^{n} + (s-\tfrac{1}{2}c)\tau_{j+1}^{n}\},\ j=1(1)J-1. \tag{5.2.80}$$

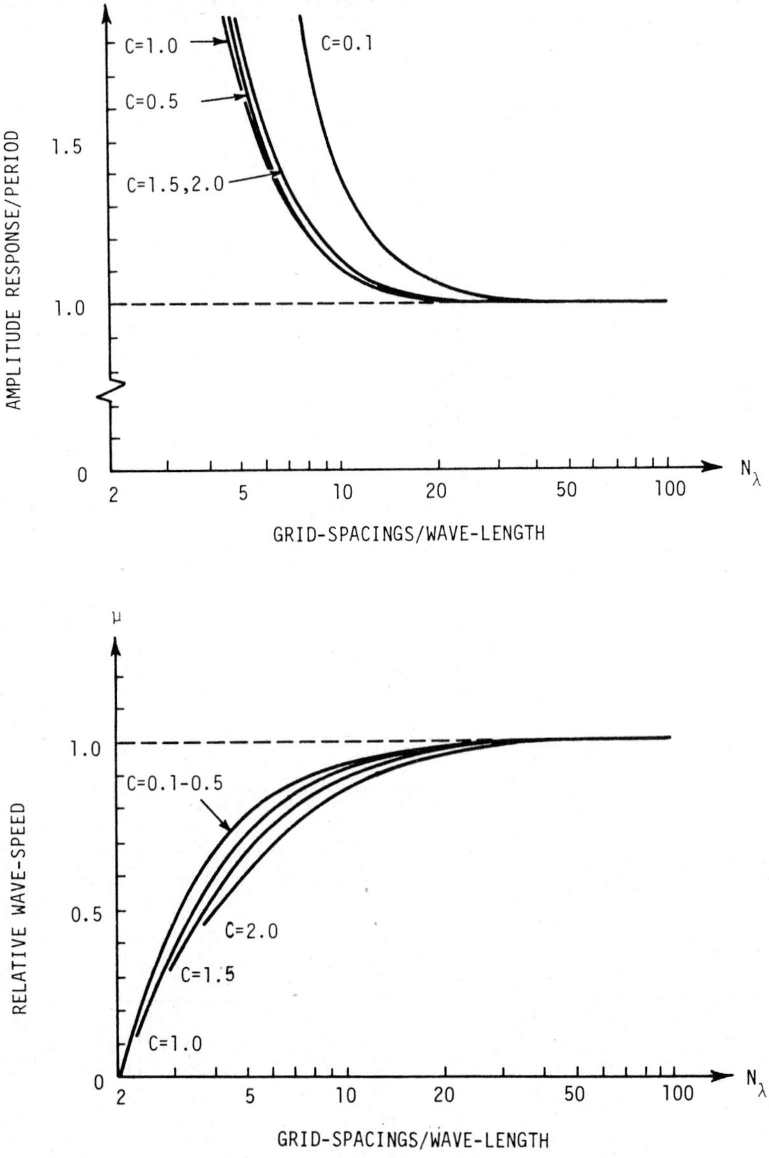

Figure 5.21 : The amplitude ratio/period and the relative wave speed for the implicit centred-time centred-space method, with $s = 0.25$. For this s the scheme is solvable for $c \leq 2.5$.

This equation can be used in a left to right sweep across the spatial grid, commencing with the known boundary value τ_0^{n+1} at the (n+1)th time level, in the manner illustrated in Figure 5.22.

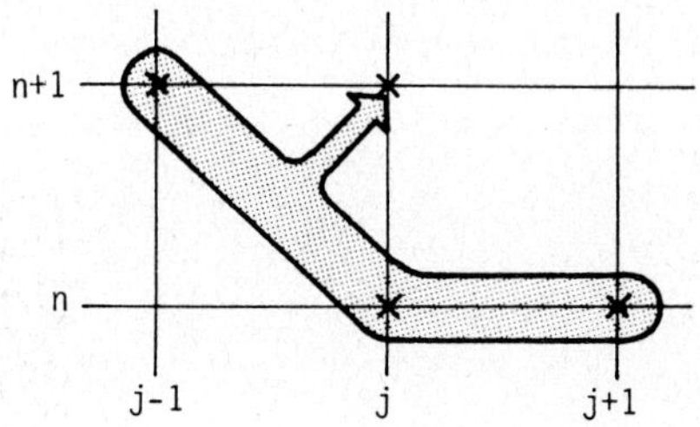

Figure 5.22 : A left to right sweep computes values at the (n+1)th time level from values at the nth time level, using Equation (5.2.80).

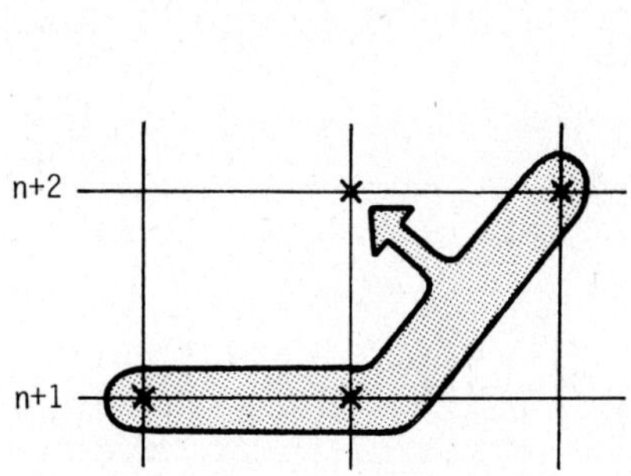

Figure 5.23 : A right to left sweep computes values at the (n+2)th time level from values at the (n+1)th time level, using Equation 5.2.81.

The transport equation may also be discretised at the point $(j\Delta x, n\Delta t+\frac{3}{2}\Delta t)$, giving the following semi-implicit equation which may be used in a right to left sweep commencing with the known bounday value τ_J^{n+2}:

$$\tau_j^{n+2} = \frac{1}{1+s-\frac{1}{2}c}\{(s-\tfrac{1}{2}c)\tau_{j+1}^{n+2} + (1-s-\tfrac{1}{2}c)\tau_j^{n+1} + (s+\tfrac{1}{2}c)\tau_{j-1}^{n+1}\},\ j=J-1(1)1. \tag{5.2.81}$$

The use of (5.2.80) followed by (5.2.81), for n = 0, 2, 4, 6, ... reduces the truncation errors of the individual equations from their individual value of $O\{\Delta x, \Delta t/\Delta x\}$ to the smaller value of $O\{(\Delta t)^2, (\Delta x)^2, (\Delta t)^2/\Delta x\}$. The application of the double sweep is illustrated in Figures 5.22 and 5.23. Application of a consistency analysis to a rearranged form of (5.2.81), which is used in the right to left sweep, with n - 2 replacing n, namely

$$\frac{1}{\Delta t}(\bar{\tau}_j^{n} - \bar{\tau}_j^{n-1}) + \frac{u}{2\Delta x}\{(\bar{\tau}_{j+1}^{n} - \bar{\tau}_j^{n}) + (\bar{\tau}_j^{n-1} - \bar{\tau}_{j-1}^{n-1})\}$$

$$= \frac{\alpha}{(\Delta x)^2}\{(\bar{\tau}_{j+1}^{n} - \bar{\tau}_j^{n}) - (\bar{\tau}_j^{n-1} - \bar{\tau}_{j-1}^{n})\}, \tag{5.2.82}$$

using the forms given in Appendix 3, yields

$$\left.\frac{\partial\bar{\tau}}{\partial t}\right|_j^n + u\left.\frac{\partial\bar{\tau}}{\partial x}\right|_j^n = \alpha\left.\frac{\partial^2\bar{\tau}}{\partial x^2}\right|_j^n + E^{(1)}\Big|_j^n \tag{5.2.83}$$

where

$$E^{(1)} = \frac{\Delta t}{2}\frac{\partial^2\bar{\tau}}{\partial t^2} - \frac{(\Delta t)^2}{6}\frac{\partial^3\bar{\tau}}{\partial t^3} - \frac{u(\Delta x)^2}{6}\frac{\partial^3\bar{\tau}}{\partial x^3} + \frac{u\Delta t}{2}\frac{\partial^2\bar{\tau}}{\partial x\partial t} - \frac{u\Delta x\Delta t}{4}\frac{\partial^3\bar{\tau}}{\partial x^2\partial t}$$

$$- \frac{u(\Delta t)^2}{4}\frac{\partial^3\bar{\tau}}{\partial x\partial t^2} + \frac{\alpha(\Delta x)^2}{12}\frac{\partial^4\bar{\tau}}{\partial x^4} + \frac{\alpha\Delta t}{\Delta x}\frac{\partial^2\bar{\tau}}{\partial x\partial t} - \frac{\alpha\Delta t}{2}\frac{\partial^3\bar{\tau}}{\partial x^3\partial t}$$

$$- \frac{\alpha(\Delta t)^2}{2\Delta x}\frac{\partial^3\bar{\tau}}{\partial x\partial t^2} + \frac{\alpha\Delta x\Delta t}{6}\frac{\partial^4\bar{\tau}}{\partial x^3\partial t} + \frac{\alpha(\Delta t)^2}{4}\frac{\partial^4\bar{\tau}}{\partial x^2\partial t^2} + \frac{\alpha(\Delta t)^3}{6\Delta x}\frac{\partial^4\bar{\tau}}{\partial x\partial t^3} + \ldots \tag{5.2.84}$$

Taken individually, this step of the double sweep method has a truncation error of $O\{(\Delta x)^2, \Delta t/\Delta x\}$, and it is consistent with the transport equation only if $\Delta t/\Delta x \to 0$ in addition to $\Delta t \to 0$, $\Delta x \to 0$.

Application of a consistency analysis to (5.2.80), which is used in the left to right sweep, gives

$$\left.\frac{\partial\bar{\tau}}{\partial t}\right|_j^n + u\left.\frac{\partial\bar{\tau}}{\partial x}\right|_j^n = \alpha\left.\frac{\partial^2\bar{\tau}}{\partial x^2}\right|_j^n + E^{(2)}\Big|_j^n \tag{5.2.85}$$

where

$$E^{(2)} = -\frac{\Delta t}{2}\frac{\partial^2\bar{\tau}}{\partial t^2} - \frac{(\Delta t)^2}{6}\frac{\partial^3\bar{\tau}}{\partial t^3} - \frac{u(\Delta x)^2}{6}\frac{\partial^3\bar{\tau}}{\partial x^3} - \frac{u\Delta t}{2}\frac{\partial^2\bar{\tau}}{\partial x\partial t} + \frac{u\Delta x\Delta t}{4}\frac{\partial^3\bar{\tau}}{\partial x^2\partial t}$$

$$- \frac{u(\Delta t)^2}{4}\frac{\partial^3\bar{\tau}}{\partial x\partial t^2} + \frac{\alpha(\Delta x)^2}{12}\frac{\partial^4\bar{\tau}}{\partial x^4} - \frac{\alpha\Delta t}{\Delta x}\frac{\partial^2\bar{\tau}}{\partial x\partial t} + \frac{\alpha\Delta t}{2}\frac{\partial^3\bar{\tau}}{\partial x^2\partial t}$$

$$- \frac{\alpha(\Delta t)^2}{2\Delta x}\frac{\partial^3\bar{\tau}}{\partial x\partial t^2} - \frac{\alpha\Delta x\Delta t}{6}\frac{\partial^4\bar{\tau}}{\partial x^3\partial t} + \frac{\alpha(\Delta t)^2}{4}\frac{\partial^4\bar{\tau}}{\partial x^2\partial t^2} - \frac{\alpha(\Delta t)^3}{6\Delta x}\frac{\partial^4\bar{\tau}}{\partial x\partial t^3} + \ldots \tag{5.2.86}$$

The truncation error is of the same order as that for the right to left sweep, and for consistency it is required that $\Delta t/\Delta x \to 0$ as well as $\Delta t \to 0$, $\Delta x \to 0$.

The truncation error for the double sweep is the average of $E^{(1)}$ and $E^{(2)}$, namely

$$E(x,t) = -\frac{(\Delta t)^2}{6}\frac{\partial^3\bar{\tau}}{\partial t^3} - \frac{u(\Delta x)^2}{6}\frac{\partial^3\bar{\tau}}{\partial x^3} - \frac{u(\Delta t)^2}{4}\frac{\partial^3\bar{\tau}}{\partial x\partial t^3} + \frac{\alpha(\Delta x)^2}{12}\frac{\partial^4\bar{\tau}}{\partial x^4}$$

$$- \frac{\alpha(\Delta t)^2}{2\Delta x}\frac{\partial^3\bar{\tau}}{\partial x\partial t^2} + \frac{\alpha(\Delta t)^2}{4}\frac{\partial^4\bar{\tau}}{\partial x^2\partial t^2} + \ldots \quad . \qquad (5.2.87)$$

This is smaller than both $E^{(1)}$ and $E^{(2)}$, because of the cancellation of a number of terms. Thus $E(x,t)$ is $O\{(\Delta x)^2, (\Delta t)^2/\Delta x\}$, and the method is consistent so long as Δt and Δx are of the same order of magnitude as they tend to zero.

Since

$$\frac{\partial^3\bar{\tau}}{\partial t^3} = [-u\frac{\partial}{\partial x} + \alpha\frac{\partial^2}{\partial x^2}]^3\bar{\tau} = -u^3\frac{\partial^3\bar{\tau}}{\partial x^3} + 3u^2\alpha\frac{\partial^4\bar{\tau}}{\partial x^4} + \ldots$$

$$\frac{\partial^3\bar{\tau}}{\partial x\partial t^2} = \frac{\partial}{\partial x}[-u\frac{\partial}{\partial x} + \alpha\frac{\partial^2}{\partial x^2}]\bar{\tau} = u^2\frac{\partial^3\bar{\tau}}{\partial x^3} - 2u\alpha\frac{\partial^4\bar{\tau}}{\partial x^4} + \ldots$$

$$\frac{\partial^4\bar{\tau}}{\partial x^2\partial t^2} = \frac{\partial}{\partial x}[\frac{\partial^3\bar{\tau}}{\partial x\partial t^2}] = u^2\frac{\partial^4\bar{\tau}}{\partial x^4} + \ldots,$$

Equation (5.2.87) may be written

$$E(x,t) = -\frac{u(\Delta x)^2}{6}\{1 + \tfrac{1}{2}c^2 + 3cs\}\frac{\partial^3\bar{\tau}}{\partial x^3} + \frac{\alpha(\Delta x)^2}{12}\{1 + 3c^2 + 12cs\}\frac{\partial^4\bar{\tau}}{\partial x^4} + \ldots \qquad (5.2.88)$$

from which it is clear that there is no possible choice of u, α, c or s which will eliminate the coefficients of $\partial^3\bar{\tau}/\partial x^3$ or $\partial^4\bar{\tau}/\partial x^4$.

The von Neumann stability analysis gives an amplification factor of

$$G^{(1)} = \frac{\{1 - (s-\tfrac{1}{2}c) + (s-\tfrac{1}{2}c)\cos\beta\} + i(s-\tfrac{1}{2}c)\sin\beta}{\{1 + (s+\tfrac{1}{2}c) - (s+\tfrac{1}{2}c)\cos\beta\} + i(s+\tfrac{1}{2}c)\sin\beta} \qquad (5.2.89)$$

for the propagation of errors from one time level to the next, using Equation (5.2.80). Therefore

$$|G^{(1)}|^2 = \frac{1 - 4\{(s-\tfrac{1}{2}c) - (s-\tfrac{1}{2}c)^2\}\sin^2(\beta/2)}{1 + 4\{(s+\tfrac{1}{2}c) + (s+\tfrac{1}{2}c)^2\}\sin^2(\beta/2)}, \qquad (5.2.90)$$

and

$$|G^{(1)}|^2 - 1 = \frac{-8s(1+c)\sin^2(\beta/2)}{1 + 4\{(s+\tfrac{1}{2}c) + (s+\tfrac{1}{2}c)^2\}\sin^2(\beta/2)} \le 0 \qquad (5.2.91)$$

for all β when $c > 0$ and $s > 0$. Therefore $|G^{(1)}| \le 1$ for all β in practical situations, and Equation (5.2.80) is unconditionally stable.

The amplification factor for Equation (5.2.81) is

$$G^{(2)} = \frac{\{1 - (s+\frac{1}{2}c) + (s+\frac{1}{2}c)\cos\beta\} - i(s+\frac{1}{2}c)\sin\beta}{\{1 + (s-\frac{1}{2}c) - (s-\frac{1}{2}c)\cos\beta\} - i(s-\frac{1}{2}c)\sin\beta}, \qquad (5.2.92)$$

whence

$$|G^{(2)}|^2 = \frac{1 - 4\{(s+\frac{1}{2}c) - (s+\frac{1}{2}c)^2\}\sin^2(\beta/2)}{1 + 4\{(s-\frac{1}{2}c) + (s-\frac{1}{2}c)^2\}\sin^2(\beta/2)} \qquad (5.2.93)$$

and

$$|G^{(2)}|^2 - 1 = \frac{-8s(1-c)\sin^2(\beta/2)}{1 + 4\{(s-\frac{1}{2}c) + (s-\frac{1}{2}c)^2\}\sin^2(\beta/2)}. \qquad (5.2.94)$$

Since the numerator of (5.2.94) is non-positive so long as $c \leq 1$, and the denominator, being the modulus squared of a complex number, is always positive, then

$$|G^{(2)}|^2 - 1 \leq 0$$

or

$$|G^{(2)}| \leq 1,$$

for all $s > 0$ and all β, so long as $c \leq 1$. Therefore the equation used in the right to left sweep is stable so long as $c \leq 1$.

When the two equations are used consecutively in the two-step process, the amplification factor for the double step is

$$G^* = G^{(1)} \cdot G^{(2)} \qquad (5.2.95)$$

where $G^{(1)}$ and $G^{(2)}$ are given by (5.2.89) and (5.2.92). Therefore

$$|G^*|^2 = |G^{(1)}|^2 \cdot |G^{(2)}|^2,$$

and substitution of (5.2.90) and (5.2.93) into this gives

$$|G^*|^2 = \frac{1 - 8(s-s^2-\frac{1}{4}c^2)\sin^2(\beta/2) + 16(s^2-\frac{1}{4}c^2)(1-2s+s^2-\frac{1}{4}c^2)\sin^4(\beta/2)}{1 - 8(-s-s^2-\frac{1}{4}c^2)\sin^2(\beta/2) + 16(s^2-\frac{1}{4}c^2)(1+2s+s^2-\frac{1}{4}c^2)\sin^4(\beta/2)}. \qquad (5.2.96)$$

Hence

$$|G^*|^2 - 1 = \frac{-16s\sin^2(\beta/2)\{1 + 4(s^2-\frac{1}{4}c^2)\sin^2(\beta/2)\}}{D}, \qquad (5.2.97)$$

where the denominator D is positive, since it is the square of the modulus of a complex number. The stability requirement $|G^*| \leq 1$ is therefore satisfied if

$$1 + 4(s^2-\tfrac{1}{4}c^2)\sin^2(\beta/2) \geq 0, \qquad (5.2.98)$$

which is certainly true if $s \geq \frac{1}{2}c$.

However, a less restrictive criterion can be found the following way. Consider the function

$$F(\chi) = 1 + 4(s^2-\tfrac{1}{4}c^2)\chi, \qquad (5.2.99)$$

where $\chi = \sin^2(\beta/2)$. We wish to find the values of $s \geq 0$, $c \geq 0$ for which $f(\chi) \geq 0$ for all χ in $<0,1>$. Clearly, $F(\chi) \geq 0$ for all $\chi \geq 0$ so long as $s \geq \frac{1}{2}c$. If

$$s \leq \tfrac{1}{2}c, \qquad (5.2.100)$$

then the straight line (5.2.99) in Figure 5.24 cuts the x-axis at

$$\chi = 1/(c^2-4s^2).$$

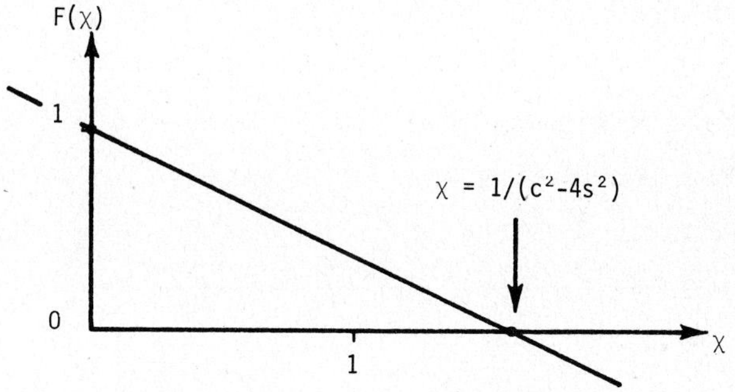

Figure 5.24 : The graph of $F(\chi) = 1 + 4(s^2-\frac{1}{4}c^2)\chi$.

We require this value of χ to be greater than, or equal, to 1; that is, we require

$$0 < (c^2-4s^2) \leq 1 \qquad (5.2.101)$$

or

$$\tfrac{1}{2}\sqrt{c^2-1} \leq s < \tfrac{1}{2}c. \qquad (5.2.102)$$

Condition (5.2.102) includes (5.2.100), and this, together with condition (5.2.98) gives the required stability criterion for the diagonal differenced method of solving the transport equation, namely

$$s \geq \tfrac{1}{2}\sqrt{c^2-1} \qquad (5.2.103)$$

which is the region shown in Figure 5.25.

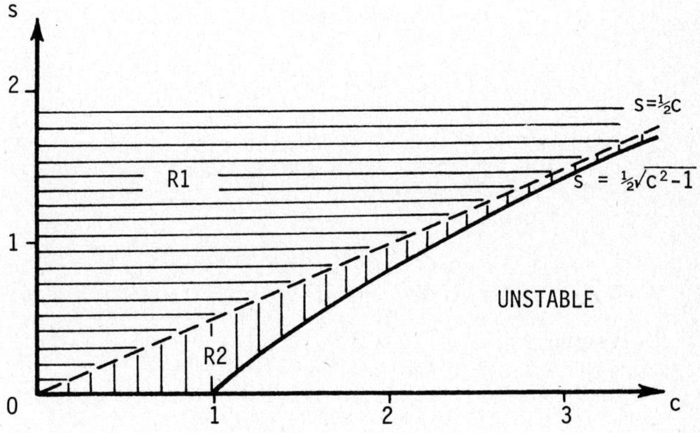

Figure 5.25 : Region R1 is that covered by the criterion $s \geq \frac{1}{2}c$.
The region R1 plus R2 is the complete region of stability for the semi-implicit method.

Since $s = 0.25$, $c = 0.5$ lies in the stability region of this method, it was used with these values to solve the one-dimensional transport equation with the initial and boundary conditions given in (5.2.51a-c). The solution at $t = 2$ was almost identical with the true solution shown in Figure 5.15, with an almost identical peak value and a very small phase error which produced a slight retardation of the peak.

The amplification factor G for time steps of Δt may be considered to be given by

$$(G)^2 = G^*, \qquad (5.2.104)$$

since G^* is the amplification factor for the double sweep.

Therefore, from (5.2.95), (5.2.89) and (5.2.92), with (5.1.38), G may be written as the following function of c, s and N_λ:

$$G = \{\frac{1-(2s-c)\sin^2(\pi/N_\lambda) + i(s-\frac{1}{2}c)\sin(2\pi/N_\lambda)}{1+(2s+c)\sin^2(\pi/N_\lambda) + i(s+\frac{1}{2}c)\sin(2\pi/N_\lambda)} \times$$

$$\times \frac{1-(2s+c)\sin^2(\pi/N_\lambda) - i(s+\frac{1}{2}c)\sin(2\pi/N_\lambda)}{1+(2s-c)\sin^2(\pi/N_\lambda) - i(s-\frac{1}{2}c)\sin(2\pi/N_\lambda)}\}^{\frac{1}{2}}. \qquad (5.2.105)$$

Substitution of this form for G in Equation (5.2.28) gives the amplitude ratio γ, and in Equation (5.2.26) the relative wave speed μ, for the use of the double sweep method to solve the transport equation. Figure 5.26 contains graphs of γ and μ, for $s = 0.25$, plotted against N_λ, the number of grid spacings per wavelength, for various values of c, the Courant number. The fact that the value of μ is noticeably less than 1 for most values of N_λ is the reason for the lag of the peak in the numerical solution relative to the true solution shown in Figure 5.15, reported earlier.

One feature of the amplitude response of the various finite difference methods of solving the transport equation deserves consideration. In some cases the value of γ increases greatly as N_λ, the number of grid-spacings per wave-length, becomes small and tends to 2. This occurs for some values of c and s for the FTCS explicit method (eg. $c = 0.3$ when $s = 0.25$, in Figure 5.11), for the optimal explicit method (eg. $c = 0.5$ when $s=0.25$, in Figure 5.19), for the implicit centred-time centred-space method (for all c for $s = 0.25$, in Figure 5.21) and for the semi-implicit method (for all c for $s = 0.25$, in Figure 5.26).

In such cases, components of short wave-length in the Fourier decomposition of the initial condition, will amplify very rapidly relative to their true value in the exact solution. They will then become evident in the numerical solution if they travel at a speed different from their true wave speed, and will be a major source of error. This occurs in all the cases mentioned above.

If the amplitude response is not 1 for small values of N_λ, it is preferable that short waves of small amplitude be attenuated rather than amplified. This occurs for some small values of c for the FTCS method (for example, $c = 0.1$, $s = 0.25$), for most values of c for the optimal explicit method ($c \leq 0.4$ for $s = 0.25$) and for all c for the upwind method. However, because the latter has such a poor amplitude response even for large N_λ, the optimal explicit method with small values of c is probably the best choice of those listed above.

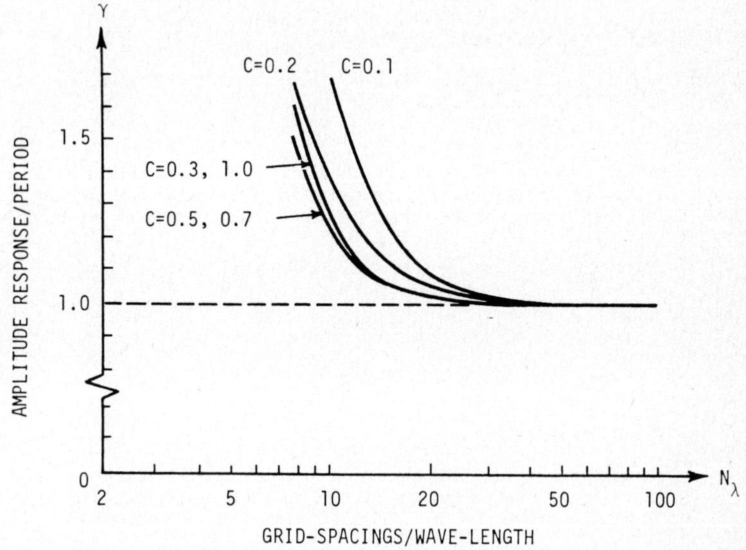

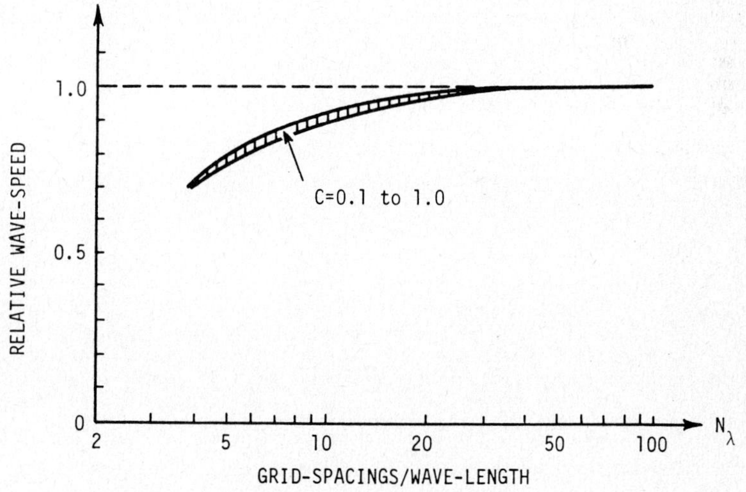

Figure 5.26 : The amplitude ratio/period and the relative wave speed for the diagonal differenced semi-implicit method for solving the transport equation, with $s = 0.25$. For this s, the method is stable for $c \leq \sqrt{5}/2 \simeq 1.12$.

Warming and Hyett (1974) have shown a connection between the amplitude ratio γ and the relative wave speed μ of a finite difference method of solving the transport equation and its corresponding differential equation. Consider a finite difference equation which has an equivalent partial differential equation

$$\frac{\partial\bar{\tau}}{\partial t} = \sum_{p=1}^{\infty} C_{2p-1} \frac{\partial^{2p-1}\bar{\tau}}{\partial x^{2p-1}} + \sum_{p=1}^{\infty} C_{2p} \frac{\partial^{2p}\bar{\tau}}{\partial x^{2p}}, \tag{5.2.106}$$

obtained by means of a consistency analysis. For instance, a consistency analysis of the explicit upstream finite difference equation (5.2.40) shows that it corresponds to the partial differential equation

$$\frac{\partial\bar{\tau}}{\partial t} + u\frac{\partial\bar{\tau}}{\partial x} = (\alpha+\alpha')\frac{\partial^2\bar{\tau}}{\partial x^2} + \frac{u(\Delta x)^2}{6}(6s+c^2-1)\frac{\partial^3\bar{\tau}}{\partial x^3} + \ldots \tag{5.2.107}$$

which is the same as (5.2.106) with

$$C_1 = -u,\ C_2 = \alpha + \alpha',\ C_3 = \frac{u(\Delta x)^2}{6}(6s+c^2-1),\ \ldots \tag{5.2.108}$$

where $\alpha' = \frac{1}{2}u\Delta x(1-c)$.

The solution of (5.2.106) with an initial condition which is the infinite wave train with wave number πm (5.2.18) may be written in the form (5.2.21), namely

$$\bar{\tau}(x,t) = e^{\eta t}e^{i(m\pi x)},\quad -\infty < x < \infty, \tag{5.2.109}$$

the real part intended, where

$$\eta = a + i\sigma, \tag{5.2.110}$$

a and σ being real.

A solution of this form may be written

$$\bar{\tau}(x,t) = e^{at}e^{i(m\pi x+\sigma t)}, \tag{5.2.111}$$

the real part of which is

$$\mathrm{Re}\{\bar{\tau}(x,t)\} = e^{at}\cos m\pi(x-bt) \tag{5.2.112}$$

where

$$b = -\sigma/m\pi. \tag{5.2.113}$$

Equation (5.2.111) represents a travelling wave of length $\lambda = 2/m$ moving with speed b in the positive x-direction. At time t, this wave has an amplitude $\exp(at)$, which is greater than 1 if $a > 0$, is equal to 1 if $a = 0$ and is less than 1 if $a < 0$. In particular, if a is negative the wave amplitude is increasingly attenuated as time passes.

Substitution of (5.2.109) into Equation (5.2.106) gives

$$\eta e^{\eta t}e^{i(m\pi x)} = \sum_{p=1}^{\infty} C_{2p-1}(im\pi)^{2p-1}e^{\eta t}e^{i(m\pi x)} + \sum_{p=1}^{\infty} C_{2p}(im\pi)^{2p}e^{\eta t}e^{i(m\pi x)}.$$

Division by $e^{\eta t}e^{i(m\pi x)}$ and substitution of (5.2.110) then gives

$$a + i\sigma = i\sum_{p=1}^{\infty} C_{2p-1}(-1)^{p-1}(m\pi)^{2p-1} + \sum_{p=1}^{\infty} C_{2p}(-1)^{p}(m\pi)^{2p}. \qquad (5.2.114)$$

Equating the real andimaginary parts of (5.2.114) yields the following expressions for a and σ:

$$a = \sum_{p=1}^{\infty} (-1)^{p}(m\pi)^{2p}C_{2p}, \qquad (5.2.115)$$

$$\sigma = \sum_{p=1}^{\infty} (-1)^{p-1}(m\pi)^{2p-1}C_{2p-1}. \qquad (5.2.116)$$

The solution of (5.2.106) is therefore an infinite wave travelling with speed

$$b = \sum_{p=1}^{\infty} (-1)^{p}(m\pi)^{2p-2}C_{2p-1}$$

$$= -C_1 + m^2\pi^2C_3 - m^4\pi^4C_5 + \ldots \qquad (5.2.117)$$

and have an amplitude

$$\exp(-at) = \exp\{(\sum_{p=1}^{\infty} (-1)^{p}(m\pi)^{2p}C_{2p})t\}. \qquad (5.2.118)$$

Therefore, in one wave period, $P = \lambda/u$, the amplitude of the wave is multiplied by the factor

$$f = \exp\{(-m^2\pi^2C_2 + m^4\pi^4C_4 - \ldots)\lambda/u\}. \qquad (5.2.119)$$

If the transport equation (5.1.1), which is the same as Equation (5.2.106) with $C_1 = -u$, $C_2 = \alpha$, $C_p = 0$ for $p = 3,4,5,\ldots$, is being solved subject to the initial condition (5.2.18), then the solution has wave speed

$$b = u, \qquad (5.2.120)$$

with a time dependent amplitude such that during one wave period the amplitude is multiplied by the factor

$$f_T = \exp\{(-m^2\pi^2\alpha)\lambda/u\}. \qquad (5.2.121)$$

The corresponding finite difference solution with equivalent partial differential equation (5.2.106), has wave speed $u_N = b$ given by (5.2.117) and a change of amplitude f_N during one period, given by (5.2.119). Therefore, the relative wave speed for the numerical method is

$$\mu = u_N/u = -C_1/u + m^2\pi^2C_3/u + \ldots, \qquad (5.2.122)$$

and the amplitude ratio per period is

$$\gamma = f_N/f_T = \exp\{[-(C_2-\alpha) + m^2\pi^2C_4 - \ldots]m^2\pi^2\ \lambda/u\}$$

$$= \exp\{[-(C_2-\alpha) + m^2\pi^2C_4 - \ldots]\ 4\pi^2/(u\Delta xN_\lambda)\}. \qquad (5.2.123)$$

Clearly, the relative wave speed will be close to 1 if $C_1 = -u$ and the amplitude ratio will be near to 1 if $C_2 = \alpha$. Any numerical diffusion α'

TABLE 5.2

Finite difference techniques for solving the one-dimensional transport equation dimensional transport equation $\frac{\partial\bar{\tau}}{\partial t} + u\frac{\partial\bar{\tau}}{\partial x} = \alpha\frac{\partial^2\bar{\tau}}{\partial x^2}$, $u > 0$, $\alpha > 0$.

Method	Finite Difference Equation	Truncation Error E	Numerical Diffusion α'
FTCS Explicit 	Eq. (5.2.4) $\tau_j^{n+1} = (s+\frac{1}{2}c)\tau_{j-1}^n + (1-2s)\tau_j^n$ $+ (s-\frac{1}{2}c)\tau_{j+1}^n$	Eq. (5.2.17) $O\{\Delta t,(\Delta x)^2\}$	$-u^2\Delta t/2$
Upwind Explicit 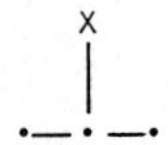	Eq. (5.2.41) $\tau_j^{n+1} = (s+c)\tau_{j-1}^n + (1-2s-c)\tau_j^n$ $+ s\tau_{j+1}^n$	Eq. (5.2.49) $O\{\Delta t,\Delta x\}$	$\frac{1}{2}u\Delta x(1-c)$
Optimal Explicit 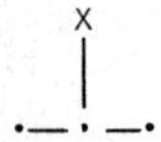	Eq. (5.2.66) $\tau_j^{n+1} = \{s + \frac{c}{2} + \frac{c^2}{2}\}\tau_{j-1}^n$ $+ \{1-2s-c^2\}\tau_j^n$ $+ \{s - \frac{c}{2} + \frac{c^2}{2}\}\tau_{j+1}^n$	Eq. (5.2.57) $O\{\Delta t,(\Delta x)^2\}$	None
DuFort-Frankel Explicit 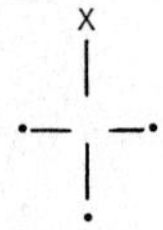	Eq. (5.2.37) $\tau_j^{n+1} = (\frac{1-2s}{1+2s})\tau_j^{n-1} + (\frac{c+2s}{1+2s})\tau_{j-1}^n$ $- (\frac{c-2s}{1+2s})\tau_{j+1}^n$	Eq. (5.2.38) $O\{(\frac{\Delta t}{\Delta x})^2,(\Delta x)^2\}$	$-\alpha c^2$
CTCS Implicit 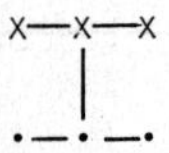	Eq. (5.2.72) $(-\frac{s}{2} - \frac{c}{4})\tau_{j-1}^{n+1} + (1+s)\tau_j^{n+1}$ $+ (-\frac{s}{2} + \frac{c}{4})\tau_{j+1}^{n+1} = (\frac{s}{2} + \frac{c}{4})\tau_{j-1}^n$ $+ (1-s)\tau_j^n + (\frac{s}{2} - \frac{c}{4})\tau_{j+1}^n$	$O\{(\Delta t)^2,(\Delta x)^2\}$	None
Diagonal-Diff. Semi-implicit 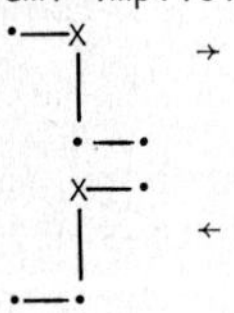	Eq. (5.2.80) $\tau_j^{n+1} = \frac{1}{1+s+\frac{1}{2}c}\{(s+\frac{1}{2}c)\tau_{j-1}^{n+1}$ $+ (1-s+\frac{1}{2}c)\tau_j^n+(s-\frac{1}{2}c)\tau_{j+1}^n\}$ Eq. (5.2.81) $\tau_j^{n+2} = \frac{1}{1+s-\frac{1}{2}c}\{(s-\frac{1}{2}c)\tau_{j+1}^{n+2}$ $+(1-s-\frac{1}{2}c)\tau_j^{n+1}+(s+\frac{1}{2}c)\tau_{j-1}^{n+1}\}$	Eq. (5.2.88) $O\{(\Delta x)^2\}$	None

Amplification Factor G	Stability Conditions	Solvability Conditions	Remarks
Eq. (5.2.5) $\{1 - 4s\sin^2(\frac{\beta}{2})\} - i\{c\sin\beta\}$	Eq. (5.2.12) $c^2 \le 2s \le 1$ *or* Eq. (5.2.14b) $2c \le r \le 2c^{-1}$	None	Accuracy requires $N_\lambda >> 100$
Eq. (5.2.42) $\{1 - 2(2s+c)\sin^2(\frac{\beta}{2})\} - i\{c\sin\beta\}$	Eq. (5.2.45) $2s + c \le 1$ *or* Eq. (5.2.46b) $r \ge 2c/(1-c)$	None	Accuracy requires $N_\lambda >> 100$
Eq. (5.2.60) with $\psi = c$ $\{1 - 2(2s+c^2)\sin^2(\frac{\beta}{2})\} - i\{c\sin\beta\}$	Eq. (5.2.67) $2s \le 1 - c^2$	None	Accurate for $N_\lambda > 50$
$\frac{\sqrt{(4s^2-c^2)}\cos\beta \pm \sqrt{(4s^2-c^2)\cos^2\beta + (1-4s^2)}}{(1+2s)}$	All s for $c \le 1$; $s \ge \frac{1}{2}\sqrt{c^2-1}$ for $c > 1$	None	Consistency requires $\Delta t << \Delta x$ *or* $c << 1$
Eq. (5.2.73) $\frac{1-2s\sin^2(\frac{\beta}{2}) - i\frac{c}{2}\sin\beta}{1+2s\sin^2(\frac{\beta}{2}) + i\frac{c}{2}\sin\beta}$	None	$s \ge \frac{c}{2} - 1$ or $r \le 2+2s^{-1}$	$r \le 2$ to prevent spatial oscillations
Eq. (5.2.95) with Eq. (5.2.89) and Eq. (5.2.92)	$s \ge \frac{1}{2}\sqrt{c^2-1}$	None	Accurate for $N_\lambda > 50$

which means $C_2 = \alpha + \alpha'$, will provide a large error in γ. A more accurate wave speed will be obtained if, in addition, $C_3 = 0$. This is a desirable correction which should be incorporated into methods which have an amplitude ratio close to 1 but a large wave speed error which shifts peaks in the numerical solution relative to their true position.

For the upwind (forward-time backward-space) equation (5.2.41) for solving the transport equation, consistency analysis shows that it is equivalent to the partial differential equation (5.2.106) with coefficients given by (5.2.108). The amplitude ratio per period for this method is therefore

$$\begin{aligned}\gamma &= \exp\{(-\alpha' + m^2\pi^2 C_4 - \ldots)4\pi^2/(u\Delta x N_\lambda)\}\\ &= \exp\{-2\pi^2(1-c)/N_\lambda + \ldots\},\end{aligned} \tag{5.2.124}$$

and the relative wave speed is

$$\begin{aligned}\mu &= 1 + \frac{1}{6}m^2\pi^2(\Delta x)^2(6s+c^2-1) + \ldots\\ &= 1 + \frac{2}{3}\pi^2(6s+c^2-1)/(N_\lambda)^2 + \ldots .\end{aligned} \tag{5.2.125}$$

Both γ and μ tend to 1 as N_λ increases, but for values of $c \neq 1$ they both differ markedly from 1 for values of N_λ up to 20. This is particularly true for the amplitude ratio γ which is smaller than 1. It is also clear that the relative wave speed μ will be larger than 1 for large values of s, c. These results are reflected in the graphs of γ and μ shown, for $s = 0.25$ and various c, in Figure 5.14.

For the FTCS (forward-time centred space) equation (5.2.4) of solving the transport equation, the equivalent partial differential equation found by consistency analysis is (5.2.17), which is the same as (5.2.106) with

$$C_1 = -u,\ C_2 = \alpha - \alpha',\ C_3 = \frac{u(\Delta x)^2}{6}(6s+c^2-1),\ \ldots \tag{5.2.126}$$

where $\alpha' = u^2\Delta t/2$. Substitution in (5.2.122) and (5.2.123) gives the amplitude ratio per period

$$\begin{aligned}\gamma &= \exp\{[\alpha' + m^2\pi^2 C_4 - \ldots]4\pi^2/(u\Delta x N_\lambda)\},\\ &= \exp\{\frac{2\pi^2 c}{N_\lambda} + \ldots\}\end{aligned} \tag{5.2.127}$$

which is larger than 1 for all values of c and N_λ, and the relative wave speed μ which is the same as (5.2.125) for the first two terms. These results are reflected in the nature of the graphs of γ and μ shown in Figure 5.11.

For the semi-implicit method using Equations (5.2.80) and (5.2.81), consistency analysis shows that the method actually solves the transport equation with truncation error (5.2.87), which is the same as (5.2.106) with

$$C_1 = -u,\ C_2 = \alpha,\ C_3 = -\frac{u(\Delta x)^2}{6}(1+\tfrac{1}{2}c^2+3cs),\ C_4 = \frac{\alpha(\Delta x)^2}{12}(1+3c^2+12cs),\ \ldots \tag{5.2.128}$$

Substitution of these values in (5.2.123) gives the amplitude ratio per period

$$\gamma = \exp\{[\frac{m^2\pi^2\alpha(\Delta x)^2}{12}(1+3c^2+12cs) - \ldots]4\pi^2/(u\Delta x N_\lambda)\}$$

$$= \exp\{\frac{4\pi^4 s}{3c(N_\lambda)^3}(1+3c^2+12cs) - \ldots\} \quad (5.2.129)$$

which tends to 1 very rapidly as N_λ increases. Therefore the amplitude of an infinite travelling wave is well represented by the method, which is clear from the graph in Figure 5.26. Substitution of the values of (5.2.128) in (5.2.122) gives the relative wave speed

$$\mu = 1 - \frac{2\pi^2}{3(N_\lambda)^2}(1+\tfrac{1}{2}c^2+3cs) + \ldots \quad (5.2.130)$$

The relative wave speed rapidly approaches the value 1 as the number of gridspacings per wavelength increases, and is less than 1; that is, the numerical wave is slower than the true wave (see Figure 5.26). For a given N_λ, the amplitude and phase errors produced by this method clearly become very large as c and s increase, particularly when N_λ is small. Table 5.2 summarises information about these methods of solving the one-dimensional transport equation.

5.3 Steady State Solution of the Transport Equation

There are two ways of obtaining the steady solution to the one-dimensional transport equation. In the first, the time varying solution of the transport equation (5.1.1) with appropriate boundary values is obtained and the steady solution is found in the limit as time $t \to \infty$. In the second method, the term $\partial\bar{\tau}/\partial t$ in (5.1.1) is set to zero, giving the steady state transport equation

$$u\frac{d\bar{\tau}}{dx} = \alpha\frac{d^2\bar{\tau}}{dx^2}, \quad 0 \le x \le 1. \quad (5.3.1)$$

This is then solved to obtain the steady solution $\bar{\tau}(x)$, subject to the given boundary conditions

$$\bar{\tau}(0) = g_0, \quad \bar{\tau}(1) = g_1. \quad (5.3.2)$$

The steady equation (5.3.1) may be solved by discretising it at the j-th gridpoint, when

$$u\frac{d\bar{\tau}}{dx}\bigg|_j = \alpha\frac{d^2\bar{\tau}}{dx^2}\bigg|_j, \quad j=1(1)J-1. \quad (5.3.3)$$

Substitution of centred difference forms for both space derivatives yields

$$u\{\frac{\bar{\tau}_{j+1} - \bar{\tau}_{j-1}}{2\Delta x} + O\{(\Delta x)^2\}\} = \alpha\{\frac{\bar{\tau}_{j+1} - 2\bar{\tau}_j + \bar{\tau}_{j-1}}{(\Delta x)^2} + O\{(\Delta x)^2\}\}, \quad (5.3.4)$$

which may be rearranged to give

$$r(\bar{\tau}_{j+1} - \bar{\tau}_{j-1}) = 2(\bar{\tau}_{j+1} - 2\bar{\tau}_j + \bar{\tau}_{j-1}) + O\{(\Delta x)^4\}, \quad (5.3.5)$$

where $r = u\Delta x/\alpha$ is the cell Reynolds number. Dropping terms of $O\{(\Delta x)^4\}$ yields the implicit finite difference equation

$$(2+r)\tau_{j-1} - 4\tau_j + (2-r)\tau_{j+1} = 0, \quad j=1(1)J-1. \quad (5.3.6)$$

This can be solved by direct methods such as the Thomas algorithm, by indirect methods such as Successive Over-Relaxation (see Section 4.3), or by the marching method (see Section 4.4). In each case the numerical stability of the computational process involved in solving (5.3.6) must be established. For instance, the above set of equations is *solvable* using the Thomas algorithm if they are diagonally dominant; that is, if

$$|-4| \geq |2+r| + |2-r|.$$

If $0 < r \leq 2$, the requirement is $4 \geq 2 + r + 2 - r = 4$, which is always true; if $2 < r$, it is $4 \geq 2 + r - 2 + r = 2r$, which is never true. Therefore the system is solvable if $0 < r \leq 2$.

Alternatively, the steady solution may be obtained by solving the time-dependent transport equation (5.1.1) using the FTCS finite difference equation (5.2.4), commencing with any suitable initial condition, and finding the limiting solution for very large values of t, that is, for a large number of time steps. However, it must be remembered that solving the FTCS finite difference equation is equivalent to solving the transport equation with a reduced diffusion coefficient $(\alpha-\alpha')$ because of the introduction of the negative numerical diffusion $\alpha' = \frac{1}{2}u^2\Delta t$. In fact, we are solving the equivalent partial differential equation (5.2.17). In the limit as $t \to \infty$, we actually solve the steady transport equation

$$u\frac{\partial\bar{\tau}}{\partial x} = (\alpha-\alpha')\frac{\partial^2\bar{\tau}}{\partial x^2} + E(x), \qquad (5.3.7)$$

where $E(x)$ is $O\{(\Delta x)^2\}$.

We now investigate whether the solution of the finite difference equation (5.3.6) also involves numerical diffusion. A formal consistency analysis of this equation yields the equivalent partial differential equation

$$u\frac{\partial\bar{\tau}}{\partial x} = \alpha\frac{\partial^2\bar{\tau}}{\partial x^2} + O\{(\Delta x)^2\}. \qquad (5.3.8)$$

Clearly, no numerical diffusion is introduced by this method of solution, which contrasts with the corresponding time dependent approach to obtaining the steady solution.

Ifthe time-dependent equation (5.1.1) is solved using the upwind method of solution (5.2.41), and the steady solution obtained in the limit as $t \to \infty$, the result is affected by the positive numerical diffusion

$$\alpha' = \tfrac{1}{2}u\Delta x(1-c), \quad c < 1, \qquad (5.3.9)$$

which is introduced. When the same differencing is used to solve (5.3.3) numerically, using the backward difference form for $\partial\bar{\tau}/\partial x$ and the centred difference form for $\partial^2\bar{\tau}/\partial x^2$, the resulting discretisation is

$$\bar{\tau}_{j-1} - (2-r)\bar{\tau}_j + (1-r)\bar{\tau}_{j+1} = O\{(\Delta x)^3\}. \qquad (5.3.10)$$

This yields the implicit finite difference equation

$$\tau_{j-1} - (2-r)\tau_j + (1-r)\tau_{j+1} = 0, \quad j=1(1)J-1. \qquad (5.3.11)$$

This is diagonally dominant, and therefore solvable using the Thomas algorithm, so long as $0 < r \leq 1$. A formal consistency analysis of (5.3.11) yields the equivalent partial differential equation

$$u\frac{\partial\bar{\tau}}{\partial x} = (\alpha+\alpha_s')\frac{\partial^2\bar{\tau}}{\partial x^2} + E(x), \qquad (5.3.12)$$

where $E(x)$ is $O\{(\Delta x)^2\}$ and

$$\alpha_s' = \tfrac{1}{2}u\Delta x. \qquad (5.3.13)$$

Both numerical methods based on upwind differencing of the first-order spatial derivative $\partial\bar{\tau}/\partial x$, introduce spurious numerical diffusion of different amounts, given by the coefficients (5.3.9) and (5.3.12).

The different results obtained using the same forms of spatial differencing when the steady solution to the transport equation is found by solving the steady state and time-dependent forms, is seen when they are compared with the exact solution to (5.3.1) with boundary conditions (5.3.2). This is

$$\bar{\tau}(x) = \{g_0(e^{u/\alpha} - e^{ux/\alpha}) + g_1(e^{ux/\alpha} - 1)\}/(e^{u/\alpha} - 1). \qquad (5.3.14)$$

Table 5.3 lists the exact solution and the errors produced by both the abovementioned methods of solving the steady transport equation with $u = 0.1$, $\alpha = 0.1$, $g_0 = 0$ and $g_1 = 1$. The exact solution in this case is

$$\bar{\tau}(x) = (e^x - 1)/(e - 1). \qquad (5.3.15)$$

The numerical diffusion of the upwind (UW) method (5.3.11) is evident in the large values of the error ($\bar{\tau}-\tau$) when compared to the errors in the centred space (CS) method (5.3.6). Results are given at every second gridpoint for the case when $\Delta x = 1/10$.

TABLE 5.3
Solution of the steady transport equation (5.3.1), $u = \alpha = 0.1$, with boundary values $\bar{\tau}(0) = 0$, $\bar{\tau}(1) = 1$, and $\Delta x = 1/10$, so $r = 1/10$.

	x	0	0.2	0.4	0.6	0.8	1.0
$\bar{\tau}$ (Exact)		0.000	0.129	0.286	0.478	0.713	1.000
Errors	CS	0	5×10^{-5}	9×10^{-5}	9×10^{-5}	8×10^{-5}	0
	UW	0	3×10^{-3}	6×10^{-3}	6×10^{-3}	5×10^{-3}	0

Table 5.4 compares the errors produced by finding the steady solution of the transport equation (5.1.1) by solving it by three different finite difference methods for as many time steps as needed to obtain numerical solutions the same to five decimal places at successive time levels. The errors, $\bar{\tau}-\tau$, for the forward-time centred-space (FTCS) method are an order of magnitude larger than the errors obtained using the centred space (CS) method of solving the steady equation (5.3.1), and took 228 time steps to reach the required accuracy. The forward-time upwind (FTUW) method produced errors an order of magnitude larger than the FTCS method, about the same as the upwind (UW) method of solving the steady equation, and took 218 time steps to reach steady state. Both these time-stepping methods produced errors larger than the centred-time centred-space (CTCS) method (5.2.72), though the FTCS method was only marginally worse in this case. The centred-space method (5.3.6) of solving the steady transport equation corresponds to the centred-time centred-space method (5.2.72) of solving the time-dependent transport equation, since in both methods no numerical diffusion is introduced.

Clearly, the centred space method (5.3.6) of solving the steady transport equation corresponds to solving the time-dependent transport equation (5.1.1) by centred-time and centred-space methods, such as the two-level implicit method (5.2.72) or the three-level explicit method (5.2.37). In all these methods, numerical diffusion is absent.

TABLE 5.4

Solution of the time-dependent transport equation (5.1.1), $u = \alpha = 0.1$ with boundary values $\bar{\tau}(0,t) = 0$, $\bar{\tau}(1,t) = 1$, $t \geq 0$, and initial condition $\bar{\tau}(x,0) = x$, $0 \leq x \leq 1$. $\Delta t = 1/40$, so $s=1/4$, $c = 1/40$.

	x	0	0.2	0.4	0.6	0.8	1.0
	$\bar{\tau}$ (Exact)	0.000	0.129	0.286	0.478	0.713	1.000
Errors	FTCS	0	-1×10^{-4}	-3×10^{-4}	-3×10^{-4}	-2×10^{-4}	0
	FTUW	0	-3×10^{-3}	-5×10^{-3}	-6×10^{-3}	-5×10^{-3}	0
	CTCS	0	-1×10^{-4}	-3×10^{-4}	-3×10^{-4}	-2×10^{-4}	0

Table 5.5 summarises the properties of the methods used in this Section, and explains the nature of the results obtained in Tables 5.3 and 5.4. The upwind (UW) equation (5.3.11) for solving the steady equation (5.3.1) introduced numerical diffusion α_s' which increased the diffusion coefficient α by 5%, and the forward-time upwind equation (5.2.41) for solving the time dependent equation (5.1.1) introduced artificial diffusion α' which increased α by 4.88%. Errors in both methods were of order 10^{-3}. The FTCS method, with artificial diffusion only 1/40th of that of the upwind methods, about 0.1% of α, produced solutions with errors of order 10^{-4}, similar to those of the CTCS method.

TABLE 5.5

Comparison of properties of various numerical methods of finding steady solution of the transport equation.

Solving steady equation (5.3.1)			Solving unsteady equation (5.1.1) $t \to \infty$.			
Method	Numerical Diffusion	Solvability Condition	Method	Numerical Diffusion	Stability Condition	Solvability Condition
Centred Space(CS) Eq. (5.3.6)	None	$0 < r \leq 2$	FTCS Eq.(5.2.4)	$-\frac{1}{2}cr\alpha$	$c^2 \leq 2s \leq 1$	None
			Upwind Eq.(5.2.41)	$\frac{1}{2}(1-c)r\alpha$	$c + 2s \leq 1$	None
Upwind(UW) Eq. (5.3.11)	$\frac{1}{2}r\alpha$	$0 < r \leq 1$	CTCS Eq.(5.2.72)	None	None	$c \leq 2s + 2$ or $r \leq 2 + 2s^{-1}$

Provided that the more accurate centred-space method (5.3.6) is used, it is better to solve the steady equation (5.3.1) than use the limiting solution of the time dependent equation (5.1.1), even if the most accurate method of solving the latter is used. This is more evident when the computational effort is assessed : to solve the centred-space equation (5.3.6) required only one application of the Thomas algorithm, whereas 231 applications were required to obtain the steady solution in Table 5.4 using the CTCS method (5.2.72) to solve the time dependent transport equation.

6. THE TRANSPORT EQUATION IN MULTI-DIMENSIONAL SPACE

6.1 Explicit Finite Difference Methods

Finite difference methods of solving the two- and three-dimensional transport equations are now considered. The transport equation in two-dimensional space is

$$\frac{\partial\bar{\tau}}{\partial t} + u\frac{\partial\bar{\tau}}{\partial x} + v\frac{\partial\bar{\tau}}{\partial y} = \alpha_x \frac{\partial^2\bar{\tau}}{\partial x^2} + \alpha_y \frac{\partial^2\bar{\tau}}{\partial y^2}, \tag{6.1.1}$$

while in three dimensions it is given by

$$\frac{\partial\bar{\tau}}{\partial t} + u\frac{\partial\bar{\tau}}{\partial x} + v\frac{\partial\bar{\tau}}{\partial y} + w\frac{\partial\bar{\tau}}{\partial z} = \alpha_x \frac{\partial^2\bar{\tau}}{\partial x^2} + \alpha_y \frac{\partial^2\bar{\tau}}{\partial y^2} + \alpha_z \frac{\partial^2\bar{\tau}}{\partial z^2}. \tag{6.1.2}$$

The Pure Diffusion Equation

The equation for the pure diffusion problem in two space dimensions is obtained from (6.1.1) by putting $u = v = 0$, giving

$$\frac{\partial\bar{\tau}}{\partial t} = \alpha_x \frac{\partial^2\bar{\tau}}{\partial x^2} + \alpha_y \frac{\partial^2\bar{\tau}}{\partial y^2}. \tag{6.1.3}$$

Since the FTCS (forward-time centred-space) finite difference approximation for the diffusion equation in one dimension was stable and convergent for $s \le \frac{1}{2}$, we will try this method of differencing Equation (6.1.3) at the (j,k,n) grid-point, that is at the point $(j\Delta x,k\Delta y,n\Delta t)$ in the solution domain in x-y-t space. Replacing the time derivative by the forward-difference form and the space derivatives by their centred-difference forms, gives

$$\begin{aligned}\frac{\bar{\tau}^{n+1}_{j,k}-\bar{\tau}^{n}_{j,k}}{\Delta t} + O\{\Delta t\} &= \alpha_x \{\frac{\bar{\tau}^{n}_{j+1,k}-2\bar{\tau}^{n}_{j,k}+\bar{\tau}^{n}_{j-1,k}}{(\Delta x)^2} + O\{(\Delta x)^2\}\} \\ &+ \alpha_y \{\frac{\bar{\tau}^{n}_{j,k+1}-2\bar{\tau}^{n}_{j,k}+\bar{\tau}^{n}_{j,k-1}}{(\Delta y)^2} + O\{(\Delta y)^2\}\},\end{aligned} \tag{6.1.4}$$

where $j=1(1)J-1$, $k=1(1)K-1$, and

$$\bar{\tau}^{n}_{j,k} = \bar{\tau}(j\Delta x,k\Delta y,n\Delta t).$$

Rearranging (6.1.4), and dropping the higher order terms, gives the finite difference approximation to the two-dimensional diffusion equation:

$$\begin{aligned}\tau^{n+1}_{j,k} &= s_y\tau^{n}_{j,k-1} + s_x\tau^{n}_{j-1,k} + (1-2s_x-2s_y)\tau^{n}_{j,k} \\ &+ s_x\tau^{n}_{j+1,k} + s_y\tau^{n}_{j,k+1}, \quad j=1(1)J-1, k=1(1)K-1,\end{aligned} \tag{6.1.5}$$

where $s_x = \alpha_x\Delta t/(\Delta x)^2$, $s_y = \alpha_y\Delta t/(\Delta y)^2$. In general, this equation has a truncation error of $O\{\Delta t,(\Delta x)^2,(\Delta y)^2\}$ and contributes an error of $O\{(\Delta t)^2,\Delta t(\Delta x)^2,\Delta t(\Delta y)^2\}$ to the discretisation error e^n_j at each time step.

The stability of Equation (6.1.5) may be determined by application of the von Neumann method in the following way. Each Fourier component of the error distribution $\xi^n_{j,k}$ at the nth time level may be written

$$\xi^{n}_{j,k} = (G)^n e^{i\pi(m_x j\Delta x+m_y k\Delta y)}, \tag{6.1.6}$$

where $(G)^n$ is the complex amplitude at time level n of the particular finite Fourier component whose wave numbers in the x and y directions are m_x and m_y, and where $i = \sqrt{-1}$. Defining x and y phase angles by $\beta_x = \pi m_x \Delta x$ and $\beta_y = \pi m_y \Delta y$, Equation (6.1.6) then becomes

$$\xi^n_{j,k} = (G)^n e^{i(j\beta_x + k\beta_y)}. \tag{6.1.7}$$

Substitution of (6.1.7) into the error equation corresponding to the finite difference equation (6.1.5) gives, on division by $(G)^n e^{i(j\beta_x + k\beta_y)}$,

$$G = 1 - 4s_x \sin^2(\beta_x/2) - 4s_y \sin^2(\beta_y/2) \tag{6.1.8}$$

where G is a function of the parameters s_x, s_y and the phase angles β_x, β_y.

Because G is real, the stability requirement $|G| \le 1$ becomes

$$-1 \le 1 - 4\{s_x \sin^2(\beta_x/2) + s_y \sin^2(\beta_y/2)\} \le 1.$$

The right hand inequality is true for all s_x, s_y, β_x, β_y. The left hand inequality holds if

$$s_x \sin^2(\beta_x/2) + s_y \sin^2(\beta_y/2) \le \tfrac{1}{2}.$$

Since the left side of this inequality has a maximum value of $(s_x + s_y)$ then $|G| \le 1$, and Equation (6.1.5) is stable, if

$$s_x + s_y \le \tfrac{1}{2}. \tag{6.1.9}$$

If the dimensionless diffusion coefficients α_x, α_y and the grid spacings Δx, Δy are such that $s_x = s_y = s$, then this condition becomes $s \le \tfrac{1}{4}$, a condition which is twice as restrictive on the time step Δt as in the one-dimensional case.

The three-dimensional diffusion equation is obtained from (6.1.2) by putting $u = v = w = 0$, giving

$$\frac{\partial \bar{\tau}}{\partial t} = \alpha_x \frac{\partial^2 \bar{\tau}}{\partial x^2} + \alpha_y \frac{\partial^2 \bar{\tau}}{\partial y^2} + \alpha_z \frac{\partial^2 \bar{\tau}}{\partial z^2}. \tag{6.1.10}$$

Using von Neumann's method, it can be shown that the corresponding FTCS finite difference equation is stable if

$$s_x + s_y + s_z \le \tfrac{1}{2}. \tag{6.1.11}$$

If the dimensionless diffusion coefficients α_x, α_y, α_z and the grid-spacings Δx, Δy, Δz are such that $s_x = s_y = s_z = s$, then (6.1.11) becomes $s \le 1/6$, which is more restrictive than either the one- or two-dimensional cases.

The Pure Convection Equation

The equation which describes the pure convection problem in two space dimensions is obtained from Equation (6.1.1) by taking $\alpha_x = \alpha_y = 0$; this gives

$$\frac{\partial \bar{\tau}}{\partial t} + u\frac{\partial \bar{\tau}}{\partial x} + v\frac{\partial \bar{\tau}}{\partial y} = 0. \tag{6.1.12}$$

Since application of forward-time and centred-space differencing of the derivatives in the one-dimensional convection equation gave an unstable explicit finite difference equation, extension of this method into two space dimensions would also give an unstable finite difference equation. However, as upwind differencing of the spatial derivative gave a stable equation in the one-dimensional case, we will try this procedure in the two-dimensional case.

If u and v are both positive, then backward differencing of the spatial derivatives and forward differencing of the time derivative, gives the finite difference form of (6.1.12) at the (j,k,n) grid-point:

$$\frac{\bar{\tau}^{n+1}_{j,k} - \bar{\tau}^{n}_{j,k}}{\Delta t} + O\{\Delta t\} + u\{\frac{\bar{\tau}^{n}_{j,k} - \bar{\tau}^{n}_{j-1,k}}{\Delta x} + O\{\Delta x\}\}$$
$$+ v\{\frac{\bar{\tau}^{n}_{j,k} - \bar{\tau}^{n}_{j,k-1}}{\Delta y} + O\{\Delta y\}\} = 0. \qquad (6.1.13)$$

Rearranging and dropping terms of $O\{(\Delta t)^2, \Delta t\Delta x, \Delta t\Delta y\}$ gives an explicit two level finite difference equation for finding an approximate solution to (6.1.12), namely

$$\tau^{n+1}_{j,k} = c_y \tau^{n}_{j,k-1} + c_x \tau^{n}_{j-1,k} + (1-c_x-c_y)\tau^{n}_{j,k}, \; j=1(1)J-1, k=1(1)K-1, \qquad (6.1.14)$$

where $c_x = u\Delta t/\Delta x$ and $c_y = v\Delta t/\Delta y$.

Application of the von Neumann stability analysis shows that the amplification factor for the (m_x,m_y) Fourier mode of the error distribution at any time level as it propagates to the next time level is

$$G = 1 - c_x(1-\cos\beta_x) - c_y(1-\cos\beta_y) - i(c_x \sin\beta_x + c_y \sin\beta_y).$$

The gain $|G|$ is therefore found by solving the equation

$$|G|^2 = (1-c_x+c_x\cos\beta_x-c_y+c_y\cos\beta_y)^2 + (c_x\sin\beta_x+c_y\sin\beta_y)^2. \qquad (6.1.15)$$

It can be shown that $|G|^2 \le 1$ for all values of β_x and β_y, and therefore (6.1.14) is stable, if

$$c_x + c_y \le 1. \qquad (6.1.16)$$

If u, v and Δx, Δy are such that $c_x = c_y = c$, Equation (6.1.16) becomes $c \le \frac{1}{2}$, a condition which is twice as restrictive as it is when the corresponding numerical method is used to solve the one-dimensional convection equation.

A consistency analysis of Equation (6.1.14) yields the equivalent partial differential equation

$$\frac{\partial\bar{\tau}}{\partial t} + u\frac{\partial\bar{\tau}}{\partial x} + v\frac{\partial\bar{\tau}}{\partial y} = \alpha'_x\frac{\partial^2\bar{\tau}}{\partial x^2} + \alpha'_y\frac{\partial^2\bar{\tau}}{\partial y^2} + O\{(\Delta t)^2,(\Delta x)^2,(\Delta y)^2\}, \qquad (6.1.17)$$

where

$$\alpha'_x = \tfrac{1}{2}u\Delta x(1-c_x), \qquad (6.1.18a)$$

$$\alpha'_y = \tfrac{1}{2}v\Delta y(1-c_y). \qquad (6.1.18b)$$

Since the stability condition (6.1.16) implies that both c_x, c_y must be less than 1, the use of upwind differencing has introduced numerical diffusion of $O\{\Delta t,\Delta x,\Delta y\}$ because of the positive coefficients α_x', α_y' of $\partial^2\bar{\tau}/\partial x^2$, $\partial^2\bar{\tau}/\partial y^2$.

The upstream differencing method can also be used to solve approximately the three-dimensional convection problem, subject to the stability condition

$$c_x + c_y + c_z \le 1, \tag{6.1.19}$$

or, if $c_x = c_y = c_z = c$, that $c \le 1/3$. Again, some artificial diffusion is introduced by this numerical scheme.

The Transport Equation

Using central difference forms for the spatial derivatives with forward differencing in time in Equation (6.1.1) at the (j,k,n) grid-point, namely

$$\left.\frac{\partial\bar{\tau}}{\partial t}\right|^n_{j,k} + u\left.\frac{\partial\bar{\tau}}{\partial x}\right|^n_{j,k} + v\left.\frac{\partial\bar{\tau}}{\partial y}\right|^n_{j,k} = \alpha_x \left.\frac{\partial^2\bar{\tau}}{\partial x^2}\right|^n_{j,k} + \alpha_y \left.\frac{\partial^2\bar{\tau}}{\partial y^2}\right|^n_{j,k} \tag{6.1.20}$$

gives

$$\begin{aligned}\frac{\bar{\tau}^{n+1}_{j,k}-\bar{\tau}^n_{j,k}}{\Delta t} &+ O\{\Delta t\} + u\{\frac{\bar{\tau}^n_{j+1,k}-\bar{\tau}^n_{j-1,k}}{2\Delta x} + O\{(\Delta x)^2\}\} \\ &+ v\{\frac{\bar{\tau}^n_{j,k+1}-\bar{\tau}^n_{j,k-1}}{2\Delta y} + O\{(\Delta y)^2\}\} \\ &= \alpha_x\{\frac{\bar{\tau}^n_{j+1,k}-2\bar{\tau}^n_{j,k}+\bar{\tau}^n_{j-1,k}}{(\Delta x)^2} + O\{(\Delta x)^2\} \\ &+ \alpha_y\{\frac{\bar{\tau}^n_{j,k+1}-2\bar{\tau}^n_{j,k}+\bar{\tau}^n_{j,k-1}}{(\Delta y)^2} + O\{(\Delta y)^2\}\}.\end{aligned} \tag{6.1.21}$$

Rearranging and neglecting terms of $O\{(\Delta t)^2,\Delta t(\Delta x)^2,\Delta t(\Delta y)^2\}$, then gives the explicit two-level finite difference equation

$$\begin{aligned}\tau^{n+1}_{j,k} &= (s_x+\tfrac{1}{2}c_x)\tau^n_{j-1,k} + (s_y+\tfrac{1}{2}c_y)\tau^n_{j,k-1} + (1-2s_x-2s_y)\tau^n_{j,k} \\ &+ (s_y-\tfrac{1}{2}c_y)\tau^n_{j,k+1} + (s_x-\tfrac{1}{2}c_x)\tau^n_{j+1,k},\ j=1(1)J-1, k=1(1)K-1.\end{aligned} \tag{6.1.22}$$

Substitution of Equation (6.1.7) for $\xi^n_{j,k}$ in the error equation corresponding to (6.1.22) gives the amplification factor G of the (m_x,m_y) Fourier component of the error distribution at any time level as it propagates to the next time level, namely

$$G = 1-2\{(s_x+s_y) - (s_x\cos\beta_x+s_y\cos\beta_y)\} - i(c_x\sin\beta_x + c_y\sin\beta_y). \tag{6.1.23}$$

The value of the gain $|G|$ is therefore given by the equation

$$\left.\begin{aligned}|G|^2 = 1 - 4\{(s_x+s_y) - (s_x\cos\beta_x+s_y\cos\beta_y)\}\\ + 4\{(s_x+s_y) - (s_x\cos\beta_x+s_y\cos\beta_y)\}^2\\ + (c_x\sin\beta_x)^2 + 2c_xc_y\sin\beta_x\sin\beta_y + (c_y\sin\beta_y)^2.\end{aligned}\right\} \quad (6.1.24)$$

Let $c_x \le 2s_x$ and $c_y \le 2s_y$, so that (6.1.24) becomes

$$\begin{aligned}|G|^2 \le 1 - 4\{(s_x+s_y) - (s_x\cos\beta_x+s_y\cos\beta_y)\}\\ + 4\{(s_x+s_y)^2 - 2(s_x+s_y)(s_x\cos\beta_x+s_y\cos\beta_y)\\ + (s_x)^2+2s_xs_y\cos(\beta_x-\beta_y) + (s_y)^2\}.\end{aligned} \quad (6.1.25)$$

Since the maximum value of $\cos(\beta_x-\beta_y)$ is 1, then

$$(s_x)^2 + 2s_xs_y\cos(\beta_x-\beta_y) + (s_y)^2 \le (s_x+s_y)^2,$$

and

$$|G|^2 \le 1 - 4\{s_x(1 - \cos\beta_x) + s_y(1 - \cos\beta_y)\}\{1 - 2(s_x+s_y)\}. \quad (6.1.26)$$

Furthermore, because $|\cos(\beta_x)| \le 1$ and $|\cos(\beta_y)| \le 1$, it follows that $|G|^2 \le 1$ so long as

$$s_x + s_y \le \tfrac{1}{2}. \quad (6.1.27a)$$

Together with

$$c_x \le 2s_x, \; c_y \le 2s_y, \quad (6.1.27b)$$

the relation (6.1.27a) is sufficient for the stability of the finite difference equation (6.1.22).

These stability conditions imply that the time step is limited by the inequality

$$\Delta t \le \tfrac{1}{2}\{\frac{\alpha_x}{(\Delta x)^2} + \frac{\alpha_y}{(\Delta y)^2}\}^{-1} \quad (6.1.28)$$

and the space steps by the inequalities

$$\Delta x \le 2\alpha_x/u, \quad \Delta y \le 2\alpha_y/v, \quad (6.1.29)$$

if $u > 0$, $v > 0$.

If $s_x = s_y = s$, condition (6.1.27a) becomes $s \le \frac{1}{4}$ which is twice as restrictive as the stability criterion for the FTCS finite difference method used to solve the one-dimensional transport equation. Also, if $c_x = c_y = c$, then we require $c \le 2s \le \frac{1}{2}$, which is again more restrictive than the corresponding one-dimensional case.

A consistency analysis of the difference equation (6.1.22) shows that it is equivalent to the partial differential equation

$$\frac{\partial\bar{\tau}}{\partial t} + u\frac{\partial\bar{\tau}}{\partial x} + v\frac{\partial\bar{\tau}}{\partial y} = (\alpha_x - \alpha_x')\frac{\partial^2\bar{\tau}}{\partial x^2} + (\alpha_y - \alpha_y')\frac{\partial^2\bar{\tau}}{\partial y^2} + O\{\Delta t, (\Delta x)^2, (\Delta y)^2\} \quad (6.1.30)$$

where

$$\alpha_x' = \tfrac{1}{2}u^2\Delta t > 0, \quad (6.1.31a)$$

$$\alpha_y' = \tfrac{1}{2}v^2\Delta t > 0, \quad (6.1.31b)$$

if $u > 0$, $v > 0$. Even though (6.1.30) is the same as the two-dimensional transport equation (6.1.20) in the limit as $\Delta x \to 0$, $\Delta y \to 0$, $\Delta t \to 0$, so the method is consistent, reduction of the diffusion coefficients α_x, α_y occurs for all finite Δt. This "negative" numerical diffusion causes this method to be unstable for the two-dimensional convection equation, when $\alpha_x = \alpha_y = 0$.

If u and v are positive in Equation (6.1.20), using the upwind difference forms for the spatial derivatives in the convective terms gives

$$\frac{\bar{\tau}^{n+1}_{j,k} - \bar{\tau}^n_{j,k}}{\Delta t} + O\{\Delta t\} + u\{\frac{\bar{\tau}^n_{j,k} - \bar{\tau}^n_{j-1,k}}{\Delta x} + O\{\Delta x\}\} + v\{\frac{\bar{\tau}^n_{j,k} - \bar{\tau}^n_{j,k-1}}{\Delta y} + O\{\Delta y\}\}$$

$$= \alpha_x\{\frac{\bar{\tau}^n_{j+1,k} - 2\bar{\tau}^n_{j,k} + \bar{\tau}^n_{j-1,k}}{(\Delta x)^2} + O\{(\Delta x)^2\}\}$$

$$+ \alpha_y\{\frac{\bar{\tau}^n_{j,k+1} - 2\bar{\tau}^n_{j,k} + \bar{\tau}^n_{j,k-1}}{(\Delta y)^2} + O\{(\Delta y)^2\}\}. \quad (6.1.32)$$

Rearranging and dropping terms of $O\{(\Delta t)^2, \Delta t\Delta x, \Delta t\Delta y\}$ gives the corresponding finite difference equation

$$\tau^{n+1}_{j,k} = (s_x + c_x)\tau^n_{j-1,k} + (s_y + c_y)\tau^n_{j,k-1} + (1 - 2s_x - 2s_y - c_x - c_y)\tau^n_{j,k}$$

$$+ s_x\tau^n_{j+1,k} + s_y\tau^n_{j,k+1}, \quad j=1(1)J-1, \ k=1(1)K-1. \quad (6.1.33)$$

Application of the von Neumann method shows that the stability requirement for (6.1.33) is

$$2s_x + 2s_y + c_x + c_y \le 1. \quad (6.1.34)$$

In particular, if $c_x = c_y = c$ and $s_x = s_y = s$, it follows that this upwind difference method is stable if

$$c + 2s \le \tfrac{1}{2}, \quad (6.1.35)$$

which again is twice as restrictive as in the one-dimensional case. In general, for any u and v the requirement on the time step is found to be

$$\Delta t \le \{\frac{2\alpha_x}{(\Delta x)^2} + \frac{2\alpha_y}{(\Delta y)^2} + \frac{|u|}{\Delta x} + \frac{|v|}{\Delta y}\}^{-1}. \quad (6.1.36)$$

Taylor series expansion of each term of Equation (6.1.33) about the (j,k,n) grid-point, with $\bar{\tau}$ substituted for τ, yields the corresponding differential equation at that grid-point,

$$\frac{\partial\bar{\tau}}{\partial t} + u\frac{\partial\bar{\tau}}{\partial x} + v\frac{\partial\bar{\tau}}{\partial y} = (\alpha_x + \alpha_x')\frac{\partial^2\bar{\tau}}{\partial x^2} + (\alpha_x + \alpha_y')\frac{\partial^2\bar{\tau}}{\partial y^2} + O\{\Delta t, (\Delta x)^2, (\Delta y)^2\} \qquad (6.1.37)$$

where

$$\left.\begin{aligned}\alpha_x' &= \tfrac{1}{2}u\Delta x(1-c_x),\\ \alpha_y' &= \tfrac{1}{2}v\Delta y(1-c_y).\end{aligned}\right\} \qquad (6.1.38)$$

As $\Delta x \to 0$, $\Delta y \to 0$, $\Delta t \to 0$ then $\alpha_x' \to 0$, $\alpha_y' \to 0$ and the higher order terms on the right side of Equation (6.1.37) also tend to zero. Therefore, the finite difference equation (6.1.33) is consistent with the two-dimensional transport equation.

Even though the upwind differencing method leads to convergence of the solution of (6.1.33) to the solution of the two-dimensional transport equation if $c_x + c_y + 2s_x + 2s_y \le 1$, the accuracy of the solution for a given Δx, Δy, Δt is greatly reduced by the false diffusion introduced by the terms $\alpha_x' \,\partial^2\bar{\tau}/\partial x^2$ and $\alpha_y' \,\partial^2\bar{\tau}/\partial y^2$. Note that the stability requirement (6.1.34) implies that $c_x < 1$, $c_y < 1$ which means that, for $u > 0$ and $v > 0$, the coefficients α_x' and α_y' are positive.

The appearance of the coefficients α_x' and α_y' is important because with small values of α_x and α_y, and large values of u and v, the false diffusion introduced will dominate the true diffusion unless Δx and Δy are sufficiently small. de Vahl Davis and Mallinson (1976) give examples from their work on forced convection in two directions which illustrate this point. They showed that the effect of false diffusion was greater in regions where the fluid was moving at 45 degrees to the two coordinate directions. Also, for a nominal Reynolds number of 1000 they found that the effective Reynolds number was as low as 240 in some regions, due to the effective increase in the diffusion coefficients α_x, α_y by the amounts α_x' and α_y'.

Although the use of upwind differences has a beneficial effect on the stability of the solution, the method only achieves this at the cost of reduced accuracy, which may be dramatic at large Reynolds numbers and may make the method almost useless. Whenever the upwind differencing method is used, the real diffusion coefficient α must be large enough to ensure that the numerical diffusion coefficient α' is negligible in comparison.

6.2 Implicit Finite Difference Methods

The stability restrictions which applied to the explicit equations described in the previous section for solving the transport equation in multi-dimensional space are overcome if implicit methods are used, thereby permitting the use of much larger time steps when using finite difference methods. For example, the Crank-Nicolson method for solving the two-dimensional *diffusion equation* (6.1.3) is based on evaluating the differential equation (6.1.3) at the point $(j\Delta x, k\Delta y, n\Delta t + \tfrac{1}{2}\Delta t)$, namely

$$\left.\frac{\partial\bar{\tau}}{\partial t}\right|_{j,k}^{n+\frac{1}{2}} = \alpha_x \left.\frac{\partial^2\bar{\tau}}{\partial x^2}\right|_{j,k}^{n+\frac{1}{2}} + \alpha_y \left.\frac{\partial^2\bar{\tau}}{\partial y^2}\right|_{j,k}^{n+\frac{1}{2}} . \qquad (6.2.1)$$

Replacing the space derivatives by the average of their values at the time levels n and (n+1), gives

$$\left.\frac{\partial\bar{\tau}}{\partial t}\right|_{j,k}^{n+\frac{1}{2}} = \frac{\alpha_x}{2}\left\{\left.\frac{\partial^2\bar{\tau}}{\partial x^2}\right|_{j,k}^{n} + \left.\frac{\partial^2\bar{\tau}}{\partial x^2}\right|_{j,k}^{n+1} + O\{(\Delta t)^2\}\right\}$$

$$+ \frac{\alpha_y}{2}\left\{\left.\frac{\partial^2\bar{\tau}}{\partial y^2}\right|_{j,k}^{n} + \left.\frac{\partial^2\bar{\tau}}{\partial y^2}\right|_{j,k}^{n+1} + O\{(\Delta t)^2\}\right\}. \qquad (6.2.2)$$

Replacing all derivatives by their centred difference forms, then rearranging as an implicit equation with $\bar{\tau}_{j-1,k}^{n+1}$, $\bar{\tau}_{j,k-1}^{n+1}$, $\bar{\tau}_{j,k}^{n+1}$, $\bar{\tau}_{j+1,k}^{n+1}$, $\bar{\tau}_{j,k+1}^{n+1}$ on the left side and neglecting terms of $O\{(\Delta t)^3, \Delta t(\Delta x)^2, \Delta t(\Delta y)^2\}$, gives

$$-\tfrac{1}{2}s_x\,\tau_{j+1,k}^{n+1} - \tfrac{1}{2}s_x\,\tau_{j-1,k}^{n+1} + (1+s_x+s_y)\tau_{j,k}^{n+1} - \tfrac{1}{2}s_y\,\tau_{j,k+1}^{n+1} - \tfrac{1}{2}s_y\,\tau_{j,k-1}^{n+1}$$

$$= \tfrac{1}{2}s_x\,\tau_{j+1,k}^{n} + \tfrac{1}{2}s_x\,\tau_{j-1,k}^{n} + (1-s_x-s_y)\tau_{j,k}^{n}$$

$$+ \tfrac{1}{2}s_y\,\tau_{j,k+1}^{n} + \tfrac{1}{2}s_y\,\tau_{j,k-1}^{n}. \qquad (6.2.3)$$

For j=1(1)J-1, k=1(1)K-1, the set of Equations (6.2.3) can be written with the unknown values of $\tau_{j,k}^{n+1}$ on the left side, and all known terms on the right side, namely the values of τ from the nth time level plus known values from the boundaries at the (n+1)th time level. By solving this system of linear algebraic equations the unknown values of τ at interior grid-points at the (n+1)th time level may be computed.

Application of von Neumann's stability analysis shows that the system of Equations (6.2.3) is unconditionally stable: that is, errors introduced at any time level due to round-off or other causes do not magnify exponentially at higher time levels. However, it does require the solution, at each time level, of a system of equations of order equal to the number of unknown values of τ, namely $(J-1)\times(K-1)$ in the case of the rectangular region in space with known boundary values. The major problem with this method is therefore the large amount of computational time needed to obtain a solution at each time level using conventional elimination methods of solution. Because this system is not tri-diagonal, the very rapid Thomas algorithm cannot be used. However, iterative methods or certain specially developed direct methods may be used to compute the values of τ in an economical fashion.

The set of equations (6.2.3) can be written with the unknown values of $\tau_{j,k}^{n+1}$ occurring in the $\{(k-1)\times(J-1)+j\}$th place and the equations written in order such that the one with the right side $d_{j,k}^{n}$ is the $\{(k-1)\times(J-1)+j\}$th equation. Then the term $(1+s_x+s_y)\tau_{j,k}^{n+1}$ lies on the leading diagonal of the resulting set of $(J-1)\times(K-1)$ linear equations. The terms $-\tfrac{1}{2}s_x\tau_{j-1}^{n+1}$ and $-\tfrac{1}{2}s_x\tau_{j+1,k}^{n+1}$ lie each side of the term $(1+s_x+s_y)\tau_{j,k}^{n}$ and the term $-\tfrac{1}{2}s_y\tau_{j,k-1}^{n+1}$ lies (J-1) places to the left and $-\tfrac{1}{2}s_y\tau_{j,k+1}^{n+1}$ lies (J-1) places to the right of the leading diagonal. Therefore the coefficient matrix has zero coefficients everywhere except on the leading diagonal, one diagonal each side of it, and two diagonals (J-1) places each side of the leading diagonal.

Written in this way the coefficient matrix formed from all equations of the form (6.2.3) is diagonally dominant for all s_x and s_y, because

$$|1+s_x+s_y| > |-\tfrac{1}{2}s_x| + |-\tfrac{1}{2}s_x| + |-\tfrac{1}{2}s_y| + |-\tfrac{1}{2}s_y|. \qquad (6.2.4)$$

Since diagonal dominance of the coefficient matrix of a set of linear equations implies unconditional stability in the solution of these equations, it follows that the implicit method defined by Equation (6.2.3) is always solvable. Some direct methods of solving such systems using elimination techniques are described in Mann (1981), and various iterative techniques are given in Colgan (1981).

When the full *transport equation* is considered, the form of the implicit finite difference equations change. Using the central difference approximation to the convection terms, the finite difference approximation to Equation (6.1.1) evaluated at the point $(j\Delta x, k\Delta y, n\Delta t+\frac{1}{2}\Delta t)$ is

$$\frac{\tau^{n+1}_{j,k} - \tau^{n}_{j,k}}{\Delta t} = \{P^{n+1}_{j,k} + P^{n}_{j,k}\}/2, \tag{6.2.5a}$$

where

$$P^{n}_{j,k} = -u\frac{(\tau^{n}_{j+1,k} - \tau^{n}_{j-1,k})}{2\Delta x} - v\frac{(\tau^{n}_{j,k+1} - \tau^{n}_{j,k-1})}{2\Delta y} + \alpha_x\frac{(\tau^{n}_{j+1,k} - 2\tau^{n}_{j,k} + \tau^{n}_{j-1,k})}{(\Delta x)^2} + \alpha_y\frac{(\tau^{n}_{j,k+1} - 2\tau^{n}_{j,k} + \tau^{n}_{j,k-1})}{(\Delta y)^2} \tag{6.2.5b}$$

and $P^{n+1}_{j,k}$ is a similar expression in terms of τ^{n+1}. Equation (6.2.5a) has a truncation error of $O\{(\Delta t)^2,(\Delta x)^2,(\Delta y)^2\}$. Rearranging it gives the implicit finite difference equation

$$(s_y+\tfrac{1}{2}c_y)\tau^{n+1}_{j,k-1} + (s_x+\tfrac{1}{2}c_x)\tau^{n+1}_{j-1,k} - 2(1+s_x+s_y)\tau^{n+1}_{j,k} + (s_x-\tfrac{1}{2}c_x)\tau^{n+1}_{j+1,k} + (s_y-\tfrac{1}{2}c_y)\tau^{n+1}_{j,k+1} = Q^{n}_{j,k} \quad j=1(1)J-1,\ k=1(1)K-1, \tag{6.2.6a}$$

where $Q^{n}_{j,k}$ contains all the terms at time level n, namely

$$Q^{n}_{j,k} = -(s_y+\tfrac{1}{2}c_y)\tau^{n}_{j,k-1} - (s_x+\tfrac{1}{2}c_x)\tau^{n}_{j-1,k} - 2(1-s_x-s_y)\tau^{n}_{j,k} - (s_x-\tfrac{1}{2}c_x)\tau^{n}_{j+1,k} - (s_y-\tfrac{1}{2}c_y)\tau^{n}_{j,k+1}. \tag{6.2.6b}$$

At each time step, the use of Equation (6.2.6a) contributes an error of $O\{(\Delta t)^3,\Delta t(\Delta x)^2,\Delta t(\Delta y)^2\}$ to the discretisation error.

Diagonal dominance and hence solvability of the set of linear algebraic equations (6.2.6) is assured if the set of unknowns $\tau^{n+1}_{j,k}$ are ordered as described previously, and

$$2(1+s_x+s_y) \geq (s_y+\tfrac{1}{2}c_y) + (s_x+\tfrac{1}{2}c_x) + |s_x-\tfrac{1}{2}c_x| + |s_y-\tfrac{1}{2}c_y|. \tag{6.2.7}$$

A number of cases need to be considered, depending on the sign of $s_x-\frac{1}{2}c_x$ and $s_y-\frac{1}{2}c_y$. For simplicity, assume that $s_x = s_y = s$ and $c_x = c_y = c$. Then the solvability requirement (6.2.7) is

$$1 + 2s \geq s+\tfrac{1}{2}c + |s-\tfrac{1}{2}c|.$$

For $c \leq 2s$, solvability of the system (6.2.6a) is assured. However, if $c > 2s$, the requirement becomes

$$c \leq 2s + 1. \tag{6.2.8}$$

Therefore, c may exceed 2s, but only by 1. Clearly, if the cell Reynolds number $r = c/s = u\Delta x/\alpha$ is less than $(2+s^{-1})$, then the above condition (6.2.8) is met.

The stability of (6.2.6a) can be investigated using the von Neumann method of analysis. Considering the simplified case with $s_x = s_y = s$, $c_x = c_y = c$, then the amplification factor G is found to be

$$G = \frac{2-4s+2s(\cos\beta_x+\cos\beta_y) - ic(\sin\beta_x+\sin\beta_y)}{2+4s-2s(\cos\beta_x+\cos\beta_y) + ic(\sin\beta_x+\sin\beta_y)}. \qquad (6.2.9)$$

Therefore

$$|G|^2 - 1 = \frac{-8s\{(1-\cos\beta_x) + (1-\cos\beta_y)\}}{\{2+4s-2s(\cos\beta_x+\cos\beta_y)\}^2 + c^2\{\sin\beta_x+\sin\beta_y\}^2}, \qquad (6.2.10)$$

which is non-positive for all values of β_x and β_y since s is positive. Therefore $|G|\le 1$ for all s and c, so the finite difference equation is unconditionally stable.

This finite difference method is therefore both stable and solvable if (6.2.8) is satisfied.

A formal consistency analysis of Equation (6.2.6) shows that this finite difference equation is consistent with the two-dimensional transport equation (6.1.20) with a truncation error of $O\{(\Delta t)^2,(\Delta x)^2,(\Delta y)^2\}$. There is no numerical diffusion introduced by this method.

If the convection terms in the transport equation evaluated at the point $(j\Delta x, k\Delta y, n\Delta t+\frac{1}{2}\Delta t)$ are approximated by upwind differences, the corresponding finite difference equation for $u > 0$, $v > 0$ is

$$(s_y+c_y)\tau^{n+1}_{j,k-1} + (s_x+c_x)\tau^{n+1}_{j-1,k} - 2(1+s_x+s_y+\tfrac{1}{2}c_x+\tfrac{1}{2}c_y)\tau^{n+1}_{j,k}$$

$$+ s_x\tau^{n+1}_{j+1,k} + s_y\tau^{n+1}_{j,k+1} = Q^n_{j,k}, \quad j=1(1)J-1,\ k=1(1)K-1, \qquad (6.2.11a)$$

where $Q^n_{j,k}$ contains all the terms in τ^n and is given by

$$Q^n_{j,k} = -(s_y+c_y)\tau^n_{j,k-1} - (s_x+c_x)\tau^n_{j-1,k} - 2(1-s_x-s_y-\tfrac{1}{2}c_x-\tfrac{1}{2}c_y)\tau^n_{j,k}$$

$$-s_x\tau^n_{j+1,k} - s_y\tau^n_{j,k+1}. \qquad (6.2.11b)$$

Solvability of (6.2.11a) is assured if there is diagonal dominance of the coefficient matrix; that is, with the unknowns $\tau^{n+1}_{j,k}$ ordered as described previously, the condition to be met is that

$$|-2(1+s_x+s_y+\tfrac{1}{2}c_x+\tfrac{1}{2}c_y)| \ge |s_y+c_y| + |s_x+c_x| + |s_x| + |s_y|. \qquad (6.2.12)$$

Since this is true for all positive values of s_x, s_y, c_x, c_y, the system of equations (6.2.11) is always solvable.

The stability of the finite difference equation (6.2.11a) may be determined by the von Neumann method. If $s_x = s_y = s$, $c_x = c_y = c$, then the amplification factor is

$$G = \frac{-(s+c)e^{-i\beta_y} - (s+c)e^{-i\beta_x} - 2(1-2s-c) - se^{i\beta_x} - se^{i\beta_y}}{(s+c)e^{-i\beta_y} + (s+c)e^{-i\beta_x} - 2(1+2s+c) + se^{i\beta_x} + se^{i\beta_y}}$$

$$= \frac{\{2-(2s+c)(2-\cos\beta_x - \cos\beta_y)\} - ic\{\sin\beta_x + \sin\beta_y\}}{\{2+(2s+c)(2-\cos\beta_x - \cos\beta_y)\} + ic\{\sin\beta_x + \sin\beta_y\}}. \qquad (6.2.13)$$

Therefore

$$|G|^2 - 1 = \frac{\{2-(2s+c)(2-\cos\beta_x - \cos\beta_y)\}^2 - \{2+(2s+c)(2-\cos\beta_x - \cos\beta_y)\}^2}{\{2+(2s+c)(2-\cos\beta_x - \cos\beta_y)\}^2 + c^2\{\sin\beta_x + \sin\beta_y\}^2}$$

$$= \frac{-8(2s+c)\{(1-\cos\beta_x) + (1-\cos\beta_y)\}}{\{2+(2s+c)(2-\cos\beta_x - \cos\beta_y)\}^2 + c^2\{\sin\beta_x + \sin\beta_y\}^2}. \qquad (6.2.14)$$

Since s, c are positive and $-1 \le \cos\beta_x \le 1$, $-1 \le \cos\beta_y \le 1$, then the right side of (6.2.14) is always less than, or equal to, zero. Therefore

$$|G| \le 1 \quad \text{for all } s, c, \beta_x, \beta_y,$$

and (6.2.11) is unconditionally stable.

A consistency analysis of Equation (6.2.11) indicates that it is consistent with the two-dimensional transport equation, with a truncation error of $O\{\Delta t, \Delta x, \Delta y\}$. It should be noted that, for finite Δx, Δy, Δt, numerical diffusion is introduced by this method, which increases the diffusion coefficients α_x, α_y by

$$\alpha_x' = \tfrac{1}{2}u\Delta x(1-c_x), \qquad \alpha_y' = \tfrac{1}{2}v\Delta y(1-c_y), \qquad (6.2.15)$$

if u, v are positive.

6.3 Time-Splitting Methods

Alternating Direction Implicit Methods

Using the Crank-Nicolson method to solve the two dimensional diffusion equation (6.1.3) gave the unconditionally stable implicit finite difference equation (6.2.3). The resulting set of algebraic equations is not tri-diagonal so the Thomas algorithm cannot be used in their solution.

The *alternating direction implicit* method, sometimes called the method of variable direction, was introduced by Peaceman and Rachford (1955) and Douglas (1955) in order to develop an implicit finite difference method which only required the inversion of tri-diagonal systems equations. This method requires two sets of calculations in order to take one step forward in time; a set of intermediate values τ^* are computed at the midpoint of the normal time interval Δt.

Consider the procedure of stepping from time level n to time level (n+1), with boundary values $\bar{\tau}$ given on x = 0,1 and y = 0,1. Firstly, values of $\tau^*_{j,k}$ are calculated at the space position $(j\Delta x, k\Delta y)$ at time level $(n+\frac{1}{2})\Delta t$ using central difference approximations for both $\partial^2\bar{\tau}/\partial x^2$ and $\partial^2\bar{\tau}/\partial y^2$, the former being written at the time level $(n+\frac{1}{2})\Delta t$ and the latter at the time level $n\Delta t$.

Evaluating Equation (6.1.3) at the point $(j\Delta x, k\Delta y, n\Delta t)$ gives

$$\left.\frac{\partial\bar{\tau}}{\partial t}\right|^n_{j,k} = \alpha_x \left.\frac{\partial^2\bar{\tau}}{\partial x^2}\right|^n_{j,k} + \alpha_y \left.\frac{\partial^2\bar{\tau}}{\partial y^2}\right|^n_{j,k}, \tag{6.3.1}$$

which may be rewritten

$$\left.\frac{\partial\bar{\tau}}{\partial t}\right|^n_{j,k} = \alpha_x \{\left.\frac{\partial^2\bar{\tau}}{\partial x^2}\right|^{n+\frac{1}{2}}_{j,k} + O\{\Delta t\}\} + \alpha_y \left.\frac{\partial^2\bar{\tau}}{\partial y^2}\right|^n_{j,k}. \tag{6.3.2}$$

Denoting values of $\bar{\tau}$ at the time $(n+\frac{1}{2})\Delta t$ by $\bar{\tau}^*$, and replacing the time derivative in Equation (6.3.2) by its forward-difference form and the space derivatives by their centred-difference forms, gives

$$\frac{\bar{\tau}^*_{j,k} - \bar{\tau}^n_{j,k}}{\frac{1}{2}\Delta t} + O\{\Delta t\} = \alpha_x \{\frac{\bar{\tau}^*_{j+1,k} - 2\bar{\tau}^*_{j,k} + \bar{\tau}^*_{j-1,k}}{(\Delta x)^2} + O\{(\Delta x)^2\}\}$$
$$+ \alpha_y \{\frac{\bar{\tau}^n_{j,k+1} - 2\bar{\tau}^n_{j,k} + \bar{\tau}^n_{j,k-1}}{(\Delta y)^2} + O\{(\Delta y)^2\}\} + O\{\Delta t\}. \tag{6.3.3}$$

Dropping terms of $O\{\Delta t,(\Delta x)^2,(\Delta y)^2\}$ yields the implicit finite difference equation

$$-\tfrac{1}{2}s_x \tau^*_{j-1,k} + (1+s_x)\tau^*_{j,k} - \tfrac{1}{2}s_x \tau^*_{j+1,k}$$
$$= \tfrac{1}{2}s_y \tau^n_{j,k-1} + (1-s_y)\tau^n_{j,k} + \tfrac{1}{2}s_y \tau^n_{j,k+1}, \quad j=1(1)J-1, \tag{6.3.4}$$

for each k=1(1)K-1.

The sytem of equations obtained for a given k by substituting j=1,2,3,.., J-1, in Equation (6.3.4) and using the known values $\tau^n_{j,k}$, j=0(1)J, $\tau^*_{0,k} = \bar{\tau}^{n+\frac{1}{2}}_{0,k}$ and $\tau^*_{J,k} = \bar{\tau}^{n+\frac{1}{2}}_{J,k}$ is tridiagonal, so it can be solved using the Thomas algorithm. On a rectangular grid where j=0(1)J, and k=0(1)K, then (K-1) applications of the Thomas algorithm gives the values $\tau^*_{j,k}$ for all interior grid-points.

Checking the stability of the finite difference equation (6.3.4) by the von Neumann method, the error distribution at the time level $n\Delta t$ is considered to have spatial Fourier components of the form $\xi^n_{j,k} = e^{i(j\beta_x + k\beta_y)}$ while it has the component $\xi^*_{j,k} = G'e^{i(j\beta_x+k\beta_y)}$ at the time level $(n+\frac{1}{2})\Delta t$. The amplification factor of this error when computing values at time level $(n+\frac{1}{2})\Delta t$ from those at time level $n\Delta t$ is

$$G' = \frac{1-2s_y \sin^2(\beta_y/2)}{1+2s_x \sin^2(\beta_x/2)}. \tag{6.3.5}$$

For $\beta_x \to 0$ and $\beta_y \to \pi$, $|G'| > 1$ for sufficiently large values of $s_y > s_x$. This step is therefore only conditionally stable.

At the second step, values of $\tau^{n+1}_{j,k}$ are computed from the values of $\tau^*_{j,k}$ found by the first step, by writing the forward difference approximation for $\partial\bar{\tau}/\partial t$ and the central difference approximation for $\partial^2\bar{\tau}/\partial x^2$ at the intermediate time level $(n+\frac{1}{2})\Delta t$ and the central difference approximation for $\partial^2\bar{\tau}/\partial y^2$ at the new time level $(n+1)\Delta t$. The resulting finite difference approximation for the two-dimensional diffusion equation has a truncation error of $O\{\Delta t,(\Delta x)^2,(\Delta y)^2\}$. It is

$$\frac{\tau^{n+1}_{j,k} - \tau^*_{j,k}}{\frac{1}{2}\Delta t} = \alpha_x\left\{\frac{\tau^*_{j+1,k} - 2\tau^*_{j,k} + \tau^*_{j-1,k}}{(\Delta x)^2}\right\} + \alpha_y\left\{\frac{\tau^{n+1}_{j,k+1} - 2\tau^{n+1}_{j,k} + \tau^{n+1}_{j,k-1}}{(\Delta y)^2}\right\}, \quad (6.3.6)$$

which can be rewritten in the implicit form for each $j=1(1)J-1$,

$$-\tfrac{1}{2}s_y\tau^{n+1}_{j,k-1} + (1+s_y)\tau^{n+1}_{j,k} - \tfrac{1}{2}s_y\tau^{n+1}_{j,k+1} = \tfrac{1}{2}s_x\tau^*_{j-1,k} + (1-s_x)\tau^*_{j,k} + \tfrac{1}{2}s_x\tau^*_{j+1,k}, \quad k=1(1)K-1. \quad (6.3.7)$$

In contrast to Equation (6.3.4), Equation (6.3.7) is implicit in the y-direction. Along each grid line $x = j\Delta x$, $j=1(1)J-1$, a tri-diagonal system of equations is obtained which can be solved by means of the Thomas algorithm. With $(J-1)$ applications of this algorithm, the values of τ can be found at all interior grid-points at the time level $(n+1)$.

Application of the von Neumann stability analysis shows that the second step (6.3.7) is also only conditionally stable, the amplification factor of the error propagation being

$$G'' = \frac{1-2s_x\sin^2(\beta_x/2)}{1+2s_y\sin^2(\beta_y/2)}. \quad (6.3.8)$$

However, the double process, in which Equation (6.3.4) is followed by Equation (6.3.7) at each time step, has an amplification factor of

$$G = G'G'' = \left\{\frac{1-2s_x\sin^2(\beta_x/2)}{1+2s_x\sin^2(\beta_x/2)}\right\}\cdot\left\{\frac{1-2s_y\sin^2(\beta_y/2)}{1+2s_y\sin^2(\beta_y/2)}\right\}, \quad (6.3.9)$$

and $|G| \le 1$ for all values of s_x, s_y, β_x, β_y. The two conditionally stable steps have combined to produce a method which is always stable.

Although it may not seem plausible that intermediate values computed half a time step forward by what may be an unstable method, can be used to obtain correct values another half time step later, it was seen in Section 4.2 that the Crank-Nicolson equation for the one-dimensional diffusion equation can be considered the result of such a process.

Even though each step has a truncation error of $O\{\Delta t,(\Delta x)^2,(\Delta y)^2\}$, the two-step procedure has a truncation error of $O\{(\Delta t)^2,(\Delta x)^2,(\Delta y)^2\}$. The second order accuracy is obvious when Equation (6.3.3), with the terms of $O\{\Delta t,(\Delta x)^2,(\Delta y)^2\}$ omitted, and Equation (6.3.6) are added to give

$$\frac{\tau^{n+1}_{j,k} - \tau^{n}_{j,k}}{\Delta t} = \alpha_x \left\{\frac{\tau^*_{j+1,k} - 2\tau^*_{j,k} + \tau^*_{j-1,k}}{(\Delta x)^2}\right\}$$

$$+ \tfrac{1}{2}\alpha_y \left\{\frac{\tau^{n}_{j,k+1} - 2\tau^{n}_{j,k} + \tau^{n}_{j,k-1}}{(\Delta y)^2} + \frac{\tau^{n+1}_{j,k+1} - 2\tau^{n+1}_{j,k} + \tau^{n+1}_{j,k-1}}{(\Delta y)^2}\right\}. \quad (6.3.10)$$

With the left side of this equation considered to be an approximation to $\partial\bar{\tau}/\partial t|^{n+\frac{1}{2}}_{j,k}$, the truncation error is of order $(\Delta t)^2$ because the central difference approximation was used about the point $(j\Delta x, k\Delta y, n\Delta t+\frac{1}{2}\Delta t)$. For the spatial derivative $\partial^2\bar{\tau}/\partial x^2|^{n+\frac{1}{2}}_{j,k}$ the error is also second order in Δx because the central difference approximation was used at the point $(j\Delta x, k\Delta y, n\Delta t+\frac{1}{2}\Delta t)$. The truncation error for the approximation to $\partial^2\bar{\tau}/\partial y^2$ at that point is $O\{(\Delta t)^2,(\Delta y)^2\}$, since it is the average of central difference approximations at the time levels n and (n+1). Clearly, the complete truncation error is $O\{(\Delta t)^2,(\Delta x)^2,(\Delta y)^2\}$, and the contribution to the discretisation error at each time step is $O\{(\Delta t)^3,\Delta t(\Delta x)^2,\Delta t(\Delta y)^2\}$.

In general, when time-splitting is used so that more than one step is used to move forward by Δt in time, the complete process involving all the steps needs to be examined in order to assess the accuracy of the method. For instance, errors of $O\{\Delta t\}$ in the separate steps of the previously described method cancel one another and the truncation error is actually $O\{(\Delta t)^2\}$.

The alternating direction implicit method can also be applied to solve the three dimensional diffusion problem. The most obvious method, described by Brian (1961), uses the same process as the two dimensional case; that is, the time interval Δt is divided up into three equal parts. Hence, three calculations are performed with two sets of intermediate values being calculated at $(n + 1/3)\Delta t$ and $(n + 2/3)\Delta t$. This method, however, has only first order accuracy in time and becomes unstable for $s > 3/2$.

Another method for the advancement of the diffusion equation (6.1.6) over the time interval Δt, which is second order accurate in time and is unconditionally stable, is described by Douglas (1962). This method is once again a three-step process but differs from Brian's method in that the two intermediate values are not calculated at the times $(n + 1/3)\Delta t$ and $(n + 2/3)\Delta t$ but rather they are used to find successive approximations for τ at the (n+1)th time level. That is, this process is similar to a three-step iterative procedure for calculating τ^{n+1} in which the formula for the iteration process changes at each step.

This method gives the following finite difference equations where $\tau^{(1)}_{j,k,\ell}$ and $\tau^{(2)}_{j,k,\ell}$ refer to the first and second approximations respectively for the value of $\tau^{n+1}_{j,k,\ell}$ at the grid-point $(x_j, y_k, z_\ell, t_{n+1})$:

$$\frac{\tau^{(1)}_{j,k,\ell} - \tau^{n}_{j,k,\ell}}{\Delta t} = \frac{\alpha_x}{2}\{\frac{\tau^{(1)}_{j+1,k,\ell} - 2\tau^{(1)}_{j,k,\ell} + \tau^{(1)}_{j-1,k,\ell}}{(\Delta x)^2}$$

$$+ \frac{\tau^{n}_{j+1,k,\ell} - 2\tau^{n}_{j,k,\ell} + \tau^{n}_{j-1,k,\ell}}{(\Delta x)^2}\}$$

$$+ \alpha_y\{\frac{\tau^{n}_{j,k+1,\ell} - 2\tau^{n}_{j,k,\ell} + \tau^{n}_{j,k-1,\ell}}{(\Delta y)^2}\}$$

$$+ \alpha_z\{\frac{\tau^{n}_{j,k,\ell+1} - 2\tau^{n}_{j,k,\ell} + \tau^{n}_{j,k,\ell-1}}{(\Delta z)^2}\}, \qquad (6.3.11a)$$

which is implicit in $\tau^{(1)}_{j-1,k,\ell}$, $\tau^{(1)}_{j,k,\ell}$ and $\tau^{(1)}_{j+1,k,\ell}$ for each value of k and ℓ;

$$\frac{\tau^{(2)}_{j,k,\ell} - \tau^{n}_{j,k,\ell}}{\Delta t} = \frac{\alpha_x}{2}\{\frac{\tau^{(1)}_{j+1,k,\ell} - 2\tau^{(1)}_{j.1.\ell} + \tau^{(1)}_{j-1,k,\ell}}{(\Delta x)^2}$$

$$+ \frac{\tau^{n}_{j+1,k,\ell} - 2\tau^{n}_{j,k,\ell} + \tau^{n}_{j-1,k,\ell}}{(\Delta x)^2}\}$$

$$+ \frac{\alpha_y}{2}\{\frac{\tau^{(2)}_{j,k+1,\ell} - 2\tau^{(2)}_{j,k,\ell} + \tau^{(2)}_{j,k-1,\ell}}{(\Delta y)^2}$$

$$+ \frac{\tau^{n}_{j,k+1,\ell} - 2\tau^{n}_{j,k,\ell} + \tau^{n}_{j,k-1,\ell}}{(\Delta y)^2}\}$$

$$+ \alpha_z\{\frac{\tau^{n}_{j,k,\ell+1} - 2\tau^{n}_{j,k,\ell} + \tau^{n}_{j,k,\ell-1}}{(\Delta z)^2}\}, \qquad (6.3.11b)$$

which is implicit in $\tau^{(2)}_{j,k-1,\ell}$, $\tau^{(2)}_{j,k,\ell}$ and $\tau^{(2)}_{j,k+1,\ell}$ for each (j,ℓ);

$$\frac{\tau^{n+1}_{j,k,\ell} - \tau^{n}_{j,k,\ell}}{\Delta t} = \frac{\alpha_x}{2}\{\frac{\tau^{(1)}_{j+1,k,\ell} - 2\tau^{(1)}_{j,k,\ell} + \tau^{(1)}_{j-1,k,\ell}}{(\Delta x)^2}$$

$$+ \frac{\tau^{n}_{j+1,k,\ell} - 2\tau^{n}_{j,k,\ell} + \tau^{n}_{j-1,k,\ell}}{(\Delta x)^2}\}$$

$$+ \frac{\alpha_y}{2}\{\frac{\tau^{(2)}_{j,k+1,\ell} - 2\tau^{(2)}_{j,k,\ell} + \tau^{(2)}_{j,k-1,\ell}}{(\Delta y)^2}$$

$$+ \frac{\tau^{n}_{j,k+1,\ell} - 2\tau^{n}_{j,k,\ell} + \tau^{n}_{j,k-1,\ell}}{(\Delta y)^2}\}$$

$$+ \frac{\alpha_z}{2}\{\frac{\tau^{n+1}_{j,k,\ell+1} - 2\tau^{n+1}_{j,k,\ell} + \tau^{n+1}_{j,k,\ell-1}}{(\Delta z)^2}$$

$$+ \frac{\tau^{n}_{j,k,\ell+1} - 2\tau^{n}_{j,k,\ell} + \tau^{n}_{j,k,\ell-1}}{(\Delta z)^2}\}, \qquad (6.3.11c)$$

which is implicit in $\tau^{n+1}_{j,k,\ell-1}$, $\tau^{n+1}_{j,k,\ell}$ and $\tau^{n+1}_{j,k,\ell+1}$. This method becomes only conditionally stable if in Equation (6.3.11c) the most recent approximation $\tau^{(2)}$ is used instead of $\tau^{(1)}$ in the finite difference approximation to the derivative $\partial^2\bar{\tau}/\partial x^2$ (see Richtmeyer and Morton, 1967). Also the Thomas algorithm can again be used to solve the sets of implicit equations which result from applying this method.

There can be a problem with storage with implicit time splitting methods such as this. For instance, with explicit methods it is possible to store the required values of $\tau_{j,k,\ell}$, $j=0(1)J$, $k=0(1)K$, $\ell=1(1)L$, some at time-level n and others at (n+1) in an array of size about $J\times K\times L$. The iterative method described here requires four times that amount of storage.

Alternating direction implicit methods are used widely for the solution of parabolic and hyperbolic partial differential equations in both two and three dimensions. However, Birkoff and Varga (1959) point out that the unconditional stability of the method applied to the diffusion equation (see Douglas 1955, 1957) does not necessarily apply to differential equations with variable coefficients nor to non-rectangular regions. Nevertheless, numerical experimentation indicates that they are applicable to a wider range of partial differential equations than those with constant coefficients prescribed on a rectangular domain.

When alternating direction implicit methods are applied to the complete transport equation, it is found (see Pearson, 1964) that the inclusion of the advection terms does not change the unconditional stability.

Locally One-Dimensional Methods

Recently, Russian numerical analysists such as D'Yakonov (1963), Samarskii (1964), Yanenko (1971) and Marchuk (1975) have developed a splitting of time dependent partial differential equations in two or more space variables, so that the resulting set of partial differential equations are one dimensional in space. These methods are called *locally one-dimensional.*

By this method the two-dimensional diffusion equation (6.1.3) is split into the pair of equations

$$\frac{1}{2}\frac{\partial\bar{\tau}}{\partial t} = \alpha_x\frac{\partial^2\bar{\tau}}{\partial x^2}, \tag{6.3.12a}$$

and

$$\frac{1}{2}\frac{\partial\bar{\tau}}{\partial t} = \alpha_y\frac{\partial^2\bar{\tau}}{\partial y^2}. \tag{6.3.12b}$$

In advancing a calculation from the time level n to time level (n+1) it is assumed that Equation (6.3.12a) holds from $t = n\Delta t$ to $t = (n+\frac{1}{2})\Delta t$ giving at each point in space a set of values τ^* which have no physical significance and must be discarded. Using these, the set of values of τ at $t = (n+1)\Delta t$ are computed using Equation (6.3.12b).

To give this procedure a physical interpretation we may consider that all the diffusion in the x-direction is lumped into the first half of the time step and all the diffusion in the y-direction is lumped into the second half of the time step.

The split equations (6.3.12) can be solved by either explicit or implicit methods. Application of the FTCS method to these two one-dimensional equations leads to the explicit forms

$$\tau^{*}_{j,k} = s_x\tau^n_{j-1,k} + (1-2s_x)\tau^n_{j,k} + s_x\tau^n_{j+1,k}, \quad j=1(1)J-1, k=1(1)K-1, \quad (6.3.13a)$$

and

$$\tau^{n+1}_{j,k} = s_y\tau^{*}_{j,k-1} + (1-2s_y)\tau^{*}_{j,k} + s_y\tau^{*}_{j,k+1}, \quad j=1(1)J-1, k=1(1)K-1. \quad (6.3.13b)$$

From our results for the FTCS method it is clear that Equation (6.3.13a) is stable if $s_x \le \frac{1}{2}$ and Equation (6.3.13b) is stable if $s_y \le \frac{1}{2}$. Alternatively, the stability condition for the pair of equations (6.3.13) can be determined by direct application of the von Neumann method. Inserting

$$\xi^n_{j,k} = e^{i(j\beta_x + k\beta_y)},$$

the Fourier component of the error distribution at the nth time level, into the corresponding error equations to (6.3.13a,b) with $\xi^{*}_{j,k} = G^{*}e^{i(j\beta_x+k\beta_y)}$ the error distribution produced by application of (6.3.13a) and $\xi^{n+1}_{j,k} = G\, e^{i(j\beta_x+k\beta_y)}$ the error distribution at time level (n+1), gives the amplification factor

$$G = \{1-4s_x\sin^2(\beta_x/2)\}\{1-4s_y\sin^2(\beta_y/2)\} \quad (6.3.14)$$

for the propagation from the time level n to (n+1). Therefore the stability requirement $|G| \le 1$ holds if

$$|1-4s_x\sin^2(\beta_x/2)||1-4s_y\sin^2(\beta_y/2)| \le 1. \quad (6.3.15)$$

This condition is satisfied if both $|1-4s_x\sin^2(\beta_x/2)| \le 1$ and $|1-4s_y\sin^2(\beta_y/2)| \le 1$. Using the results of Section 3.3 this is true if both $s_x \le \frac{1}{2}$ and $s_y \le \frac{1}{2}$.

When the non-dimensional diffusion coefficients α_x, α_y and the grid-spacings Δx, Δy are such that $s_x = s_y = s$, the stability requirement becomes $s \le \frac{1}{2}$. This is an improvement on the usual explicit restriction of $s \le \frac{1}{4}$ which holds for the one-step FTCS method of solving the two-dimensional diffusion equation. This is because Equations (6.3.13) relate the value of τ at one point at the time level (n+1) with values at nine points at the time level n, since elimination of τ^{*} from these finite difference equations gives

$$\left.\begin{aligned}\tau^{n+1}_{j,k} = {} & s_xs_y\tau^n_{j-1,k-1} + (1-2s_x)s_y\tau^n_{j,k-1} + s_xs_y\tau^n_{j+1,k-1} \\ & + s_x(1-s_y)\tau^n_{j-1,k} + (1-2s_x)(1-2s_y)\tau^n_{j,k} + s_x(1-s_y)\tau^n_{j+1,k} \\ & + s_xs_y\tau^n_{j-1,k+1} + (1-2s_x)s_y\tau^n_{j,k+1} + s_xs_y\tau^n_{j+1,k+1}.\end{aligned}\right\} \quad (6.3.16)$$

This contrasts with the one-step FTCS explicit method (6.1.5) which relates a value of τ at any time level with values of τ at only five points at the previous time level.

Two other features of this locally one-dimensional method are worth noting. Firstly, (6.3.16) approximates the two-dimensional diffusion equation (6.1.3) with a truncation error of $O\{\Delta t,(\Delta x)^2\}$, the same as the one-step FTCS method of solution. Secondly, the minimum number of operations required for a single time step of the one-step FTCS method (6.1.5) is about 3JK, which is less than for the time-split method which requires approximately 4JK multiplications or divisions, whether (6.3.13a,b) or (6.3.16) is used. But the more restrictive stability requirement of (6.1.5) means that twice as many time steps are required to reach the same time level if the largest possible time steps are taken with the same values of Δx and Δy, thereby doubling the effective number of applications of the one-step FTCS method. This method therefore requires fifty percent more computer time than the locally one-dimensional FTCS scheme.

Alternatively, implicit techniques may be used to solve (6.3.12a,b). Applying the Crank-Nicolson method gives the approximation to the one-dimensional equation (6.3.12a) at the point $(j\Delta x,k\Delta y,n\Delta t+\frac{1}{4}\Delta t)$ in the solution domain,

$$\frac{1}{2}\left\{\frac{\tau^*_{j,k}-\tau^n_{j,k}}{\frac{1}{2}\Delta t}\right\} = \frac{\alpha_x}{2}\left\{\frac{\tau^*_{j+1,k}-2\tau^*_{j,k}+\tau^*_{j-1,k}}{(\Delta x)^2}+\frac{\tau^n_{j+1,k}-2\tau^n_{j,k}+\tau^n_{j-1,k}}{(\Delta x)^2}\right\}. \tag{6.3.17}$$

Rearrangement yields, for each k=1(1)K-1, the set of implicit finite difference equations

$$-\tfrac{1}{2}s_x\tau^*_{j-1,k}+(1+s_x)\tau^*_{j,k}-\tfrac{1}{2}s_x\tau^*_{j+1,k}$$
$$=\tfrac{1}{2}s_x\tau^n_{j-1,k}+(1-s_x)\tau^n_{j,k}+\tfrac{1}{2}s_x\tau^n_{j+1,k},\quad j=1(1)J-1. \tag{6.3.18}$$

This one-dimensional finite difference equation is stable for all s_x.

Since $\tau^*_{0,k}=\tau^{\overline{n+1/2}}_{0,k}$, $\tau^*_{J,k}=\tau^{\overline{n+1/2}}_{J,k}$ are known boundary values for each k, then transferring these values to the right side of (6.3.18) with known values of $\tau^n_{j,k}$ gives a diagonally dominant tri-diagonal set of algebraic equations in the unknowns $\tau^*_{j,k}$, j=1(1)J-1. This set of equations may be solved using the Thomas algorithm.

Similarly, the Crank-Nicolson approximation to Equation (6.3.12b) yields, for each j=1(1)J-1, the set of unconditionally stable implicit finite difference equations

$$-\tfrac{1}{2}s_y\tau^{n+1}_{j,k-1}+(1+s_y)\tau^{n+1}_{j,k}-\tfrac{1}{2}s_y\tau^{n+1}_{j,k+1}$$
$$=\tfrac{1}{2}s_y\tau^*_{j,k-1}+(1-s_y)\tau^*_{j,k}+\tfrac{1}{2}s_y\tau^*_{j,k+1},\quad k=1(1)K-1. \tag{6.3.19}$$

For each j, use of the known values of $\tau^*_{j,k}$, k=0(1)K, with the boundary values $\bar{\tau}^{n+1}_{j,0}$ and $\bar{\tau}^{n+1}_{j,K}$ produces a diagonally dominant tri-diagonal set of algebraic equations, which may again be solved using the Thomas algorithm

Alternatively, the stability of the two-step process involving Equations (6.3.18) and (6.3.19) may be investigated using von Neumann's method for errors distributed in two dimensions. For an error distribution of the form $\xi^n_{j,k} = e^{i(j\beta_x + k\beta_y)}$ at the time level n, the amplification factor produced by application of the finite difference equation (6.3.18) is

$$G' = \frac{1-2s_x \sin^2(\beta_x/2)}{1+2s_x \sin^2(\beta_x/2)}.$$

Similarly, the amplification factor produced by application of (6.3.19) to step from $\tau^*_{j,k}$ at time $(n+\frac{1}{2})\Delta t$ to $\tau^{n+1}_{j,k}$ is

$$G'' = \frac{1-2s_y \sin^2(\beta_y/2)}{1+2s_y \sin^2(\beta_y/2)}.$$

The amplification factor for the complete step from time level n to (n+1) is

$$G = G'G'',$$

and as $|G'| \le 1$ for all s_x, β_x, $|G''| \le 1$ for all s_y, β_y, then $|G| \le 1$ for all s_x, s_y, and β_x, β_y. Therefore the combination of (6.3.18) with (6.3.19) is stable for all s_x, s_y.

Each of the implicit steps (6.3.18) and (6.3.19) is diagonally dominant for all s_x, s_y respectively. Therefore this locally one-dimensional method is solvable for all s_x, s_y.

In a similar way, the time-dependent diffusion equation (6.1.10) in three space dimensions can be split into three parts,

$$\frac{1}{3}\frac{\partial\bar{\tau}}{\partial t} = \alpha_x \frac{\partial^2\bar{\tau}}{\partial x^2}, \qquad (6.3.20a)$$

$$\frac{1}{3}\frac{\partial\bar{\tau}}{\partial t} = \alpha_y \frac{\partial^2\bar{\tau}}{\partial y^2}, \qquad (6.3.20b)$$

and

$$\frac{1}{3}\frac{\partial\bar{\tau}}{\partial t} = \alpha_z \frac{\partial^2\bar{\tau}}{\partial z^2}. \qquad (6.3.20c)$$

Again this gives a set of one-dimensional equations to be solved by either explicit or implicit methods for each time step of length $\Delta t/3$ to give values of τ^{n+1} at the end of the final step. Thus Equation (6.3.20a) is solved over the interval $t = n\Delta t$ to $t = (n + 1/3)\Delta t$ to give a fictitious set of values τ^* at the time $t = (n + 1/3)\Delta t$; Equation (6.3.20b) is then used with the values τ^* at the beginning of the time interval $(n + 1/3)\Delta t$ to obtain a fictitious set of values τ^{**} at the time $t = (n + 2/3)\Delta t$; these are used at the start of a further time step of $\Delta t/3$ with Equation (6.3.20c) to give an approximate value for $\bar{\tau}$ at the time $t = (n+1)\Delta t$.

If the corresponding FTCS equation is used at each stage of the time-splitting the following explicit finite difference method results:

$$\tau^*_{j,k,\ell} = s_x \tau^n_{j-1,k,\ell} + (1-2s_x)\tau^n_{j,k,\ell} + s_x \tau^n_{j+1,k,\ell},$$

$$j=1(1)J-1,\ k=1(1)K-1,\ \ell=1(1)L-1, \qquad (6.3.21a)$$

followed by

$$\tau^{**}_{j,k,\ell} = s_y\tau^{*}_{j,k-1,\ell} + (1+2s_y)\tau^{*}_{j,k,\ell} + s_y\tau^{*}_{j,k+1,\ell},$$
$$j=1(1)J-1,\ k=1(1)K-1,\ \ell=1(1)L-1, \qquad (6.3.21b)$$

and

$$\tau^{n+1}_{j,k,\ell} = s_z\tau^{**}_{j,k,\ell-1} + (1-2s_z)\tau^{**}_{j,k,\ell} + s_z\tau^{**}_{j,k,\ell+1},$$
$$j=1(1)J-1,\ k=1(1)K-1,\ \ell=1(1)L-1. \qquad (6.3.21c)$$

Like the two-dimensional case, this explicit method is only conditionally stable, requiring $s_x \le \frac{1}{2}$, $s_y \le \frac{1}{2}$ and $s_z \le \frac{1}{2}$.

If the corresponding Crank-Nicolson equation is used at each stage of the time-splitting an implicit finite difference method results, in which each set of linear algebraic equations which must be solve is tri-diagonal and diagonally dominant. The Thomas algorithm is therefore used for their solution. The resulting system is:

for each $k=1(1)K-1$, $\ell=1(1)L-1$, solve

$$-\tfrac{1}{2}s_x\tau^{*}_{j-1,k,\ell} + (1-s_x)\tau^{*}_{j,k,\ell} - \tfrac{1}{2}s_x\tau^{*}_{j+1,k,\ell}$$
$$= \tfrac{1}{2}s_x\tau^{n}_{j-1,k,\ell} + (1-s_x)\tau^{n}_{j,k,\ell} + \tfrac{1}{2}s_x\tau^{n}_{j+1,k,\ell},\ j=1(1)J-1; \quad (6.3.22a)$$

then, for each $j=1(1)J-1$, $\ell=1(1)L-1$, solve

$$-\tfrac{1}{2}s_y\tau^{**}_{j,k-1,\ell} + (1+s_y)\tau^{**}_{j,k,\ell} - \tfrac{1}{2}s_y\tau^{**}_{j,k+1,\ell}$$
$$= \tfrac{1}{2}s_y\tau^{*}_{j,k-1,\ell} + (1-s_y)\tau^{*}_{j,k,\ell} + \tfrac{1}{2}s_y\tau^{*}_{j,k+1,\ell},\ k=1(1)K-1; \quad (6.3.22b)$$

finally, for each $j=1(1)J-1$, $k=1(1)K-1$, solve

$$-\tfrac{1}{2}s_z\tau^{n+1}_{j,k,\ell-1} + (1+s_z)\tau^{n+1}_{j,k,\ell} - \tfrac{1}{2}s_z\tau^{n+1}_{j,k,\ell+1}$$
$$= \tfrac{1}{2}s_z\tau^{**}_{j,k,\ell-1} + (1-s_z)\tau^{**}_{j,k,\ell} + \tfrac{1}{2}s_z\tau^{**}_{j,k,\ell+1},\ \ell=1(1)L-1. \quad (6.3.22c)$$

This method is both *stable* and *solvable* for all s_x, s_y, s_z.

Because they are relatively new, not many applications of locally one-dimensional methods can be found in the literature, but their use is increasing, particularly because of their rather simple form when applied to three-dimensional problems.

6.4 An Implicit-Marching Method

Consider the time dependent diffusion equation (6.1.3) to be solved on the two-dimensional rectangular region $R: 0 \le x \le 1$, $0 \le y \le 1$ with $\bar{\tau}$, the dependent variable, defined initially on R and for all time t on the boundary of R.

The following marching method (Walsh & Noye, 1973, 1974) called the EVP (error vector propagation) method by Roache (1974), is based on the linearity of equation (6.1.3) and its corresponding finite difference approximation. It gives values of τ at the interior gridpoints of R, namely $(j\Delta x, k\Delta y)$, $j=1(1)J-1$, $k=1(1)K-1$, at the times $t = n\Delta t$, $n=1,2,3,..$ (see Figure 6.1).

In Section 6.2, application of the Crank-Nicolson method to Equation (6.1.3) gave the unconditionally stable implicit scheme (6.2.3). This equation may be rewritten in the form

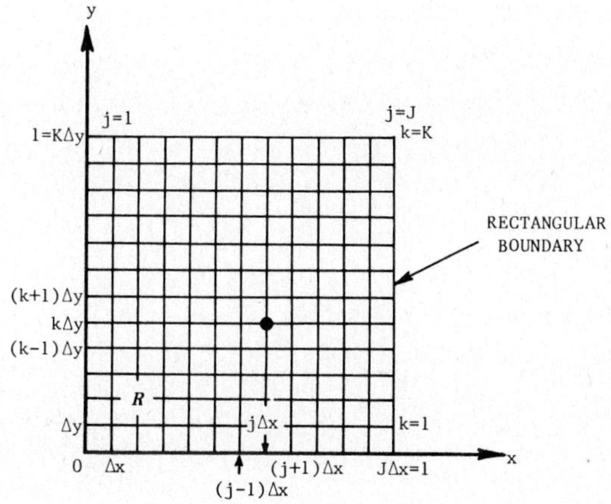

Figure 6.1 : Grid over space region R at time t = (n+1)Δt.

$$\tau_{j,k+1} = 2(1+s_x s_y^{-1}+s_y^{-1})\tau_{j,k} - s_x s_y^{-1}(\tau_{j+1,k}+\tau_{j-1,k})$$

$$- \tau_{j,k-1} + 2s_y^{-1}d^n_{j,k}, \quad j=1(1)J-1,\ k=0(1)K-1, \qquad (6.4.1)$$

where, for convenience, the superscripts (n+1) which indicate that the values of τ are those at the new time level have been omitted and the term $d^n_{j,k}$ contains the known terms from the previous time level n. This is a marching form for the computation of τ at time level (n+1) since, for j=1(1)J-1, Equation (6.4.1) defines values of τ explicitly at gridpoints along the line $y = (k+1)\Delta y$ in terms of values of τ along the two rows of gridpoints immediately below, namely along $y = k\Delta y$ and $y=(k-1)\Delta y$, as shown in Figure 6.1. Values of $\tau_{j,2}$ for j=1(1)J-1 can be calculated using Equation (6.4.1) with k=1, provided the values of $\tau_{j,1}$, j=1(1)J-1, are known in addition to the boundary values $\bar{\tau}_{j,0}$ for j=0(1)J, and $\bar{\tau}_{0,1}$, $\bar{\tau}_{J,1}$. Then taking k = 2(1)K-2 in order, the grid can be marched through to give $\tau_{j,k}$ at all interior gridpoints. Finally, with k = K-1 in Equation (6.4.1) the values of $\tau_{j,K}$ for j=1(1)J-1 can be computed.

The method described does not utilise the boundary conditions along the line $y = K\Delta y$. In fact, for each different set of starting values $\tau_{j,1}$ for j=1(1)J-1, different values of τ will be obtained along $y = K\Delta y$. Our problem is to determine the starting values $\tau_{j,1}$ so that the given boundary values are obtained along $y = K\Delta y$. If the vector $\underset{\sim}{e} = (e_1\ e_2\ ..\ e_{J-1})^T$ denotes the given boundary values along $y = K\Delta y$, then, with the correct starting values, the computed values

$$\underset{\sim}{\tau}_K = (\tau_{1,K}\,\tau_{2,K}\,\ldots\,\tau_{J-1,K})^T$$

must be such that

$$\underset{\sim}{\tau}_K = \underset{\sim}{e}.$$

Since the values of $\tau_{j,K}$ are linearly related to the values of $\tau_{j,1}$ used in the first application of (6.4.1), then

$$\underset{\sim}{\tau}_K = \underset{=}{A}\underset{\sim}{\tau}_1 + \underset{\sim}{b} \qquad (6.4.2)$$

where

$$\underset{\sim}{\tau}_1 = (\tau_{1,1}\ \tau_{2,1}\ \ldots\ \tau_{J-1,1})^T,$$

$\underset{\sim}{b}$ is the constant vector,

$$\underset{\sim}{b} = (b_1\ b_2\ \ldots\ b_{J-1})^T,$$

and $\underline{\underline{A}}$ is the (J-1) square matrix

$$\underline{\underline{A}} = [a_{i,j}], \qquad i,j = 1(1)J-1.$$

If $\underline{\underline{A}}$ and $\underset{\sim}{b}$ can be found, the required starting vector $\underset{\sim}{s}$ to be substituted for $\underset{\sim}{\tau}_1$ in order to obtain the correct boundary values along $y = K\Delta y$ is found by putting $\underset{\sim}{\tau}_K = \underset{\sim}{e}$ in Equation (6.4.2), giving

$$\underset{\sim}{s} = \underline{\underline{A}}^{-1}(\underset{\sim}{e}-\underset{\sim}{b}). \tag{6.4.3}$$

The vector $\underset{\sim}{b}$ and the columns $\underset{\sim}{a}_j$ of the matrix $\underline{\underline{A}}$ may be found in the following way. Firstly, use the marching process with $\underset{\sim}{\tau}_1 = \underset{\sim}{0}$ to obtain the end vector $\underset{\sim}{\tau}_K = \underset{\sim}{\varepsilon}^{(0)}$, where

$$\underset{\sim}{\varepsilon}^{(0)} = (\varepsilon_{1,0}\ \varepsilon_{2,0}\ \ldots\ \varepsilon_{J-1,0})^T.$$

From Equation (6.4.2) it is seen that $\underset{\sim}{b}$ is known, since substitution of $\underset{\sim}{\tau}_1 = \underset{\sim}{0}$ in (6.4.2) gives

$$\underset{\sim}{\varepsilon}^{(0)} = \underset{\sim}{b}.$$

Then progressively use the marching process with the vector $\underset{\sim}{\tau}_1$ consisting of elements of the Kronecker delta form; that is, start with $\underset{\sim}{\tau}_1 = \underset{\sim}{s}^{(\ell)}$, $\ell = 1(1)J-1$, where

$$\begin{aligned} s^{(\ell)} &= (\delta_{1,\ell}\ \delta_{2,\ell}\ \ldots\ \delta_{\ell,\ell}\ \ldots\ \delta_{J-1,\ell})^T \\ &= (0\quad 0\quad \ldots\ 1\quad \ldots\ 0\quad)^T. \end{aligned} \tag{6.4.4}$$

Let the corresponding end vectors $\underset{\sim}{\tau}_K = \underset{\sim}{\varepsilon}^{(\ell)}$, obtained on marching through the mesh with $\underset{\sim}{\tau}_1 = \underset{\sim}{s}^{(\ell)}$, be

$$\underset{\sim}{\varepsilon}^{(\ell)} = (\varepsilon_{1,\ell}\ \varepsilon_{2,\ell}\ \ldots\ \varepsilon_{J-1,\ell})^T$$

Substitution of the values $\underset{\sim}{\tau}_1 = \underset{\sim}{s}^{(\ell)}$ and $\underset{\sim}{\tau}_K = \underset{\sim}{\varepsilon}^{(\ell)}$ into Equation (6.4.2) gives

$$\underline{\underline{A}}\underset{\sim}{s}^{(\ell)} = \underset{\sim}{\varepsilon}^{(\ell)} - \underset{\sim}{\varepsilon}^{(0)}. \tag{6.4.5}$$

Since

$$\underline{\underline{A}}\underset{\sim}{s}^{(\ell)} = \begin{bmatrix} a_{1,1} & a_{1,2} & \cdots & a_{1,\ell} & \cdots & a_{1,J-1} \\ a_{2,1} & a_{2,2} & \cdots & a_{2,\ell} & \cdots & a_{2,J-1} \\ \vdots & \vdots & & \vdots & & \vdots \\ a_{J-1,1} & a_{J-1,2} & \cdots & a_{J-1,\ell} & \cdots & a_{J-1,J-1} \end{bmatrix} \begin{bmatrix} 0 \\ 0 \\ \vdots \\ 1 \\ \vdots \\ 0 \end{bmatrix} = \begin{bmatrix} a_{1,\ell} \\ a_{2,\ell} \\ \\ \\ a_{J-1,\ell} \end{bmatrix} = \underset{\sim}{a}^{(\ell)}$$

where $\underset{\sim}{a}^{(\ell)}$ is the ℓth column of the matrix $\underline{\underline{A}}$, it is seen from Equation (6.4.5) that

$$\underset{\sim}{a}^{(\ell)} = \underset{\sim}{\varepsilon}^{(\ell)} - \underset{\sim}{\varepsilon}^{(0)}. \tag{6.4.6}$$

Thus, from the end vectors $\underset{\sim}{\varepsilon}^{(\ell)}$ found by marching through the grid with starting vectors $\underset{\sim}{s}^{(\ell)}$, ℓ=0(1)J-1, the vector $\underset{\sim}{b}$ and the matrix $\underline{\underline{A}}$ can be constructed. Application of Equation (6.4.3) then gives the set of starting values $\underset{\sim}{\tau}_1 = \underset{\sim}{s}$ required to give the set of boundary values $\underset{\sim}{\tau}_K = \underset{\sim}{e}$. This procedure transforms the original problem requiring the solution of a large sparse linear system of (J-1)×(K-1) algebraic equations, to one requiring the solution of only (J-1) linear algebraic equations.

One more sweep through the grid, starting with $\underset{\sim}{\tau}_1 = \underset{\sim}{s}$ then gives the values of $\tau_{j,K}$, the solutions of the diffusion equation (6.1.3) in the rectangular region R at the time level (n+1), which satisfy the given boundary conditions.

Examination of the error propagation in the marching scheme (6.4.1) by means of discrete perturbation analysis indicates that an error $\xi_{j,k}$ in the computed value of $\tau_{j,k}$ produces an error of $2(1+s_x s_y^{-1}+s_y^{-1})\xi_{j,k}$ in the value of $\tau_{j,k+1}$. Therefore, in marching from one row of grid points to the next, the error is magnified by

$$|G| = 2(1 + s_x s_y^{-1} + s_y^{-1}) > 1,$$

so the scheme is unconditionally unstable.

However, the method is useful so long as the amplification of truncation errors using (6.4.1) does not alter the first four or five figures in the values of τ calculated along the row of gridpoints at y = KΔy. Progressing one grid spacing in the y direction produces a loss of one significant figure if the magnification of the truncation error is 10, there is a loss of two figures if the magnification is 10^2, and so on. The loss of significant figures is $\log_{10} 2(1+s_x s_y^{-1}+s_y^{-1})$ in marching one grid spacing in the y-direction using this method. Since Equation (6.4.1) must be applied (K-1) times before the top boundary is reached, there will be at most a loss of

$$L = (K-1)\log_{10} 2\{1+s_x s_y^{-1}+s_y^{-1}\} \tag{6.4.7}$$

significant figures on marching through the grid using this method.

Unfortunately, the final march through the grid with the correct starting values will produce a further loss of accuracy of the same amount, L. In order to avoid this, the linear nature of the problem can be exploited in the following way. The complete array of values $\tau_{j,k}^{(\ell)}$, j=1(1)J-1, k=1(1)K, is stored for each set of starting values of the form $\underset{\sim}{s}^{(\ell)}$ given by Equation (6.4.4). Then, for the set of correct starting values

$$\underset{\sim}{s} = (s_1\ s_2\ \dots\ s_{J-1})^T = s_1\,\underset{\sim}{s}^{(1)} + s_2\,\underset{\sim}{s}^{(2)} + \dots + s_{J-1}\,\underset{\sim}{s}^{(J-1)}$$

computed using Equation (6.4.3), the required array $\tau_{j,k}$ is given by

$$\tau_{j,k} = s_1\,\tau_{j,k}^{(1)} + s_2\,\tau_{j,k}^{(2)} + \dots + s_{J-1}\,\tau_{j,k}^{(J-1)} + \left(1 - \sum_{m=1}^{J-1} s_m\right)\tau_{j,k}^{(0)},$$

$$j=1(1)J-1,\ k=1(1)K-1. \tag{6.4.8}$$

An estimate of the loss of significant figures when using this method can be found from Equation (6.4.7). If K=11 and $s_x = \frac{1}{4}s_y = 2$, then the possible loss of significant figures would be of the order $10 \log_{10} 2.75 \simeq 4.4$. Clearly, the larger the ratio of s_x/s_y and the greater the value of s_y, the smaller is the loss of accuracy in this method. On a computer which stores 14 significant figures of each number, 7 figure accuracy would be retained. With K=46, the number of significant figures lost would be approximately $45 \log_{10} 2.75 \simeq 19.8$. Using double precision with storage of 28 significant figures of each number, at least 8 figure accuracy is achieved.

In practice, the accuracy of the results is indicated by a comparison of the computed values $\tau_{j,K}$ with the given boundary values $\bar{\tau}_{j,K} = e_j$. The effect of a discrete perturbation at the point $(\ell\Delta x, \Delta y)$ along the line k=1 is indicated by the values of $\varepsilon_{j,K}^{(\ell)} - \varepsilon_{j,K}^{(0)}$ for j=1(1)J-1, because the starting vector $s^{(\ell)}$ contains a unit perturbation in the value of $\tau_{\ell,1}$.

This method has been used to solve certain linearised problems involving wind induced circulation in lakes (for example, see Walsh and Noye, 1973, 1974, also Noye, 1977 and 1978).

6.5 Incorporating Boundary Conditions and Irregular Boundaries

When the two- or three-dimensional physical region to which the partial differential equation applies has an irregular boundary, that is, when segments of the boundary cannot be represented by a constant spatial coordinate, say x equals a constant in Figure 6.2, finite differences can still be employed to approximately solve the equation. However, care must be taken if the finite difference approximations to the boundary conditions are to be as accurate as the finite difference equation applied to the interior of the space region.

In the following, we suppose that the two-dimensional transport equation (6.1.1) is to be solved in a physical region R such as that shown in Figure 6.2, bounded by the curve R^+. As seen in Section 4.5, along different parts of R^+ the boundary conditions may be classified as one of three types. Either

(a) the dependent variable $\bar{\tau}$ is specified at each point (x,y) on the boundary of the space region R. As mentioned previously this type of boundary condition is said to be a *Dirichlet condition*;

(b) the normal derivative to the boundary is specified at each point on the boundary; that is $\partial\bar{\tau}/\partial\eta$ is known for all (x,y) in R^+, where η is measured normal to the boundary in a direction out from the region R. This type of boundary condition is referred to as the *Neumann condition*; or

(c) a combination of the Dirichlet and Neumann boundary conditions in which a linear function of $\bar{\tau}$ and $\partial\bar{\tau}/\partial\eta$ is specified at points on the boundary; that is

$$a\bar{\tau} + b\,\frac{\partial\bar{\tau}}{\partial\eta} = c$$

at points (x,y) on the boundary R^+, where $\bar{\tau}$, a, b, c may all be functions of x, y and t. This is called the *mixed boundary condition*.

In the process of finding a solution to Equation (6.1.1), these three types of boundary conditions must be treated separately. We first consider the Dirichlet condition.

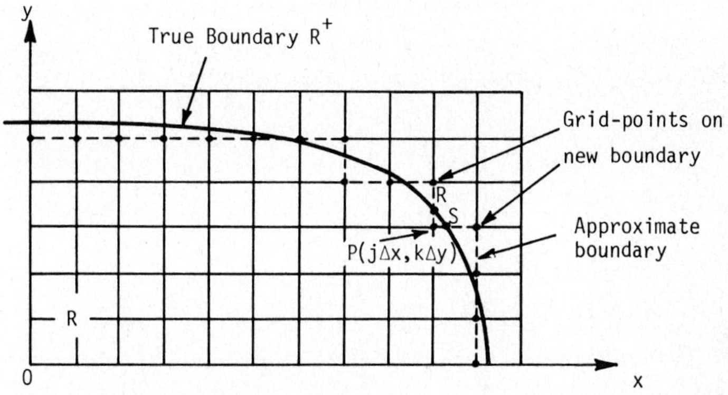

Figure 6.2: An irregular boundary in the rectangular cartesian coordinate system.

THE DEPENDENT VARIABLE SPECIFIED ON THE BOUNDARY - THE DIRICHLET CONDITION

Let the value of the dependent variable be given by

$$\bar{\tau}(x,y,t) = B(x,y,t)$$

for all points (x,y) on the boundary R^+.

Transfer of boundary values

The most obvious way to incorporate known values on a boundary into a finite difference scheme is to choose the interior gridpoint which lies closest to the point where the boundary intersects a grid-line, and at that gridpoint let the approximation τ take the boundary value. This has the effect of reshaping the boundary so that it conforms to the discrete points of the grid.

The dashed line in Figure 6.2 shows how a reshaped boundary could be drawn through gridpoints in the (x,y) plane if this method is used. Once the boundary is altered in this way the problem can be solved by any of the methods discussed previously.

The obvious disadvantage of treating the boundary condition in the above way is the low order of accuracy which results when the finite difference equations are applied at interior gridpoints near the boundary, since the shift of value introduces a contribution of either $O\{\Delta x\}$ or $O\{\Delta y\}$ to the discretisation error.

Taylor series method

This method generally results in a higher order of accuracy than reshaping the boundary so that it conforms with segments of the grid-lines. The extra work required is generally worthwhile because, in this case, the same sized truncation error at the boundary can be obtained using a coarse grid as that obtained using a very fine grid when boundary values are transferred to nearby gridpoints.

In this method, at any time level the gridpoints in the region R are divided into two disjoint sets. Regular interior gridpoints are those for which no neighbouring gridpoint in the direction of either coordinate axis lies outside the boundary. In Figure 6.3, these points are denoted by open circles. Interior gridpoints in R which have at least one neighbouring gridpoint in the direction of either coordinate axis lying outside the boundary are termed irregular points. These points are denoted by closed circles in Figure 6.3.

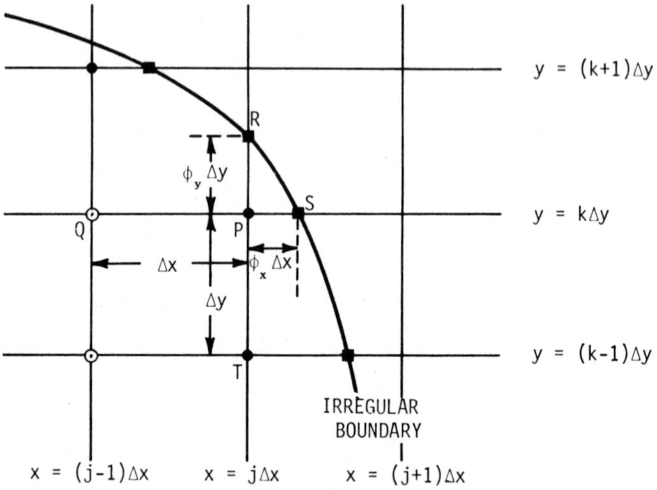

Figure 6.3 : Incorporating boundary values into finite difference equations

For regular points the usual finite difference forms for the derivatives apply, and equations such as those described in Sections 6.1 and 6.2 result. For irregular interior points the finite difference forms for the derivatives are altered in the following way. Since the values of $\bar{\tau}$ are known on the boundary, then they are known at points on the intersection of the boundary with each grid line. These points are shown as closed squares in Figure 6.3. Consider the point $P(j\Delta x, k\Delta y)$ in the figure. Let $R(j\Delta x, k\Delta y+\phi_y\Delta y)$ be the point on the boundary where it cuts the grid line between P and $(j\Delta x, k\Delta y+\Delta y)$, a distance $\phi_y\Delta y$ from the point P. Then at time $n\Delta t$ the value of $\bar{\tau}$ at R, $\bar{\tau}^n_{j,k+\phi_y}$, is given by the value of the function $B(j\Delta x, k\Delta y+\phi_y\Delta y, n\Delta t)$, which will be denoted $B^n_{j,k+\phi_y}$. Also denote by $B^n_{j+\phi_x,k}$ the known value of $\bar{\tau}$ at S on the boundary where it cuts the grid line $y = k\Delta y$ between $x = j\Delta x$ and $(j+1)\Delta x$, a distance $\phi_x\Delta x$ from the point P.

To illustrate the construction of the finite difference approximations for various spatial derivatives at irregular gridpoints at time $n\Delta t$, we will obtain $\partial\bar{\tau}/\partial x$, $\partial\bar{\tau}/\partial y$, $\partial^2\bar{\tau}/\partial x^2$, $\partial^2\bar{\tau}/\partial y^2$ simultaneously at P in terms of values of $\bar{\tau}$ at P and the four points R, S, T, Q surrounding P.

Clearly, the usual central difference approximations at P cannot be used since Q, T, R and S are not uniformly spaced about P along either grid line. Thus we have

$$\begin{aligned}
\bar{\tau}^n_{j-1,k} &= \bar{\tau}(j\Delta x-\Delta x,k\Delta y,n\Delta t)\\
\bar{\tau}^n_{j,k-1} &= \bar{\tau}(j\Delta x,k\Delta y-\Delta y,n\Delta t)\\
B^n_{j,k+\phi_y} &= \bar{\tau}(j\Delta x,k\Delta y+\phi_y\Delta y,n\Delta t)\\
B^n_{j+\phi_x,k} &= \bar{\tau}(j\Delta x+\phi_x\Delta x,k\Delta y,n\Delta t).
\end{aligned} \tag{6.5.1}$$

Expanding these functions of $\bar{\tau}$ in their Taylor series about $(j\Delta x,k\Delta y,n\Delta t)$ gives, up to terms of order $(\Delta x)^3$ and $(\Delta y)^3$, the four equations

$$\begin{aligned}
\bar{\tau}^n_{j-1,k} &= \bar{\tau}^n_{j,k} - \Delta x\left.\frac{\partial\bar{\tau}}{\partial x}\right|^n_{j,k} + \tfrac{1}{2}(\Delta x)^2\left.\frac{\partial^2\bar{\tau}}{\partial x^2}\right|^n_{j,k} + O\{(\Delta x)^3\},\\
\bar{\tau}^n_{j,k-1} &= \bar{\tau}^n_{j,k} - \Delta y\left.\frac{\partial\bar{\tau}}{\partial y}\right|^n_{j,k} + \tfrac{1}{2}(\Delta y)^2\left.\frac{\partial^2\bar{\tau}}{\partial y^2}\right|^n_{j,k} + O\{(\Delta y)^3\},\\
B^n_{j+\phi_x,k} &= \bar{\tau}^n_{j,k} + \phi_x\Delta x\left.\frac{\partial\bar{\tau}}{\partial x}\right|^n_{j,k} + \tfrac{1}{2}(\phi_x\Delta x)^2\left.\frac{\partial^2\bar{\tau}}{\partial x^2}\right|^n_{j,k} + O\{(\phi_x\Delta x)^3\},\\
B^n_{j,k+\phi_y} &= \bar{\tau}^n_{j,k} + \phi_y\Delta y\left.\frac{\partial\bar{\tau}}{\partial y}\right|^n_{j,k} + \tfrac{1}{2}(\phi_y\Delta y)^2\left.\frac{\partial^2\bar{\tau}}{\partial y^2}\right|^n_{j,k} + O\{(\phi_y\Delta y)^3\}.
\end{aligned} \tag{6.5.2}$$

These may be written in matrix form, as

$$\begin{bmatrix} -\Delta x & 0 & \tfrac{1}{2}(\Delta x)^2 & 0\\ 0 & -\Delta y & 0 & \tfrac{1}{2}(\Delta y)^2\\ \phi_x\Delta x & 0 & \tfrac{1}{2}(\phi_x\Delta x)^2 & 0\\ 0 & \phi_y\Delta y & 0 & \tfrac{1}{2}(\phi_y\Delta y)^2 \end{bmatrix} \begin{bmatrix} \left.\frac{\partial\bar{\tau}}{\partial x}\right|^n_{j,k}\\ \left.\frac{\partial\bar{\tau}}{\partial y}\right|^n_{j,k}\\ \left.\frac{\partial^2\bar{\tau}}{\partial x^2}\right|^n_{j,k}\\ \left.\frac{\partial^2\bar{\tau}}{\partial y^2}\right|^n_{j,k} \end{bmatrix} = \begin{bmatrix} \bar{\tau}^n_{j-1,k} - \bar{\tau}^n_{j,k} + O\{(\Delta x)^3\}\\ \bar{\tau}^n_{j,k-1} - \bar{\tau}^n_{j,k} + O\{(\Delta y)^3\}\\ B^n_{j+\phi_x,k} - \bar{\tau}^n_{j,k} + O\{(\phi_x\Delta x)^3\}\\ B^n_{j,k+\phi_y} - \bar{\tau}^n_{j,k} + O\{(\phi_y\Delta y)^3\} \end{bmatrix} \tag{6.5.3}$$

This equation contains four unknowns, the solutions for which are

$$\left.\frac{\partial\bar{\tau}}{\partial x}\right|^n_{j,k} = \frac{1}{\phi_x(1+\phi_x)\Delta x}\{B^n_{j+\phi_x,k} - [1-(\phi_x)^2]\bar{\tau}^n_{j,k} - (\phi_x)^2\bar{\tau}^n_{j-1,k}\} + O\{\phi_x(\Delta x)^2\},\quad 0<\phi_x\le 1, \tag{6.5.4a}$$

$$\left.\frac{\partial\bar{\tau}}{\partial y}\right|^n_{j,k} = \frac{1}{\phi_y(1+\phi_y)\Delta y}\{B^n_{j,k+\phi_y} - [1-(\phi_y)^2]\bar{\tau}^n_{j,k} - (\phi_y)^2\bar{\tau}^n_{j,k-1}\} + O\{\phi_y(\Delta y)^2\},\quad 0<\phi_y\le 1, \tag{6.5.4b}$$

$$\left.\frac{\partial^2\bar{\tau}}{\partial x^2}\right|^n_{j,k} = \frac{2}{\phi_x(1+\phi_x)(\Delta x)^2}\{B^n_{j+\phi_x,k} - (1+\phi_x)\bar{\tau}^n_{j,k} + \phi_x\bar{\tau}^n_{j-1,k}\} + O\{(1-\phi_x)\Delta x,(\Delta x)^2\},\ 0<\phi_x\le 1, \tag{6.5.4c}$$

$$\left.\frac{\partial^2\bar{\tau}}{\partial y^2}\right|_{j,k}^{n} = \frac{2}{\phi_y(1+\phi_y)(\Delta y)^2}\{B^n_{j,k+\phi_y} - (1+\phi_y)\bar{\tau}^n_{j,k} + \phi_y\bar{\tau}^n_{j,k-1}$$

$$+ O\{(1-\phi_y)\Delta y,(\Delta y)^2\},\ 0 < \phi_y \leq 1. \qquad (6.5.4d)$$

In this way, at irregular gridpoints the first and second order spatial derivatives in the partial differential equation being discretised can always be written in terms of values of the dependent variable at that gridpoint, at neighbouring gridpoints within R, and boundary values at points where the boundary R^+ intersects the gridlines leading to exterior gridpoints.

DERIVATIVE NORMAL TO THE BOUNDARY PRESCRIBED - THE NEUMANN CONDITION

Let the value of the normal derivative on the boundary R^+ be given by

$$\left.\frac{\partial\bar{\tau}}{\partial\eta}\right|_{(x,y,t)} = D(x,y,t).$$

First Method

Of the two methods available for handling this type of boundary condition, the simpler is that proposed by Fox (1944). The gridpoints inside the space region R are again divided into regular and irregular points. In both cases the centred finite difference forms for the spatial derivatives are of the same type as those given in (2.3.18), (2.3.19). However, in the case of the irregular points, standard centred finite difference forms, such as (2.3.18) and (2.3.19), of at least one of the x or y spatial derivatives will include a fictitious value of $\bar{\tau}$ from an exterior gridpoint. Thus, in Figure 6.4 the point $P(j\Delta x,k\Delta y)$ is an irregular point and the usual central difference forms for the first and second order derivatives $\partial\bar{\tau}/\partial y$ and $\partial^2\bar{\tau}/\partial y^2$ at that point include a value of $\bar{\tau}$ prescribed at the exterior point $E(j\Delta x,k\Delta y+\Delta y)$. An estimate for this value of $\bar{\tau}$ can be found in terms of the geometry of the boundary, the value of the normal derivative on the boundary and two interior gridpoints, in the following way.

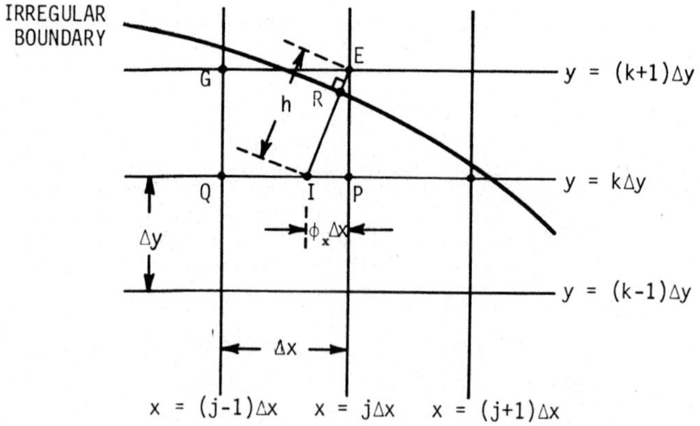

Figure 6.4 : Approximating derivative boundary conditions

From the exterior point E a line is drawn normal to the boundary to cut it at R, and extended until at I it intersects a grid line joining two adjacent interior gridpoints (see Figure 6.4). An estimate of the value of $\bar{\tau}$ at I can be obtained by linearly interpolating between the values of $\bar{\tau}$ at the two gridpoints P and Q, giving

$$\bar{\tau}_I^n = \phi_x \bar{\tau}_Q + (1-\phi_x)\bar{\tau}_P + O\{(\Delta x)^2\}, \quad 0 < \phi_x < 1, \qquad (6.5.5)$$

where the length of PI is $\overline{PI} = \phi_x \Delta x$. The fictitious value $\bar{\tau}_{j,k+1}^n$ is then allotted to the exterior point E by relating the values of $\bar{\tau}$ at E and I through the known derivative boundary condition at R. If, at time $t_n = n\Delta t$, $\partial\bar{\tau}/\partial\eta$ takes the value D_R^n at the point R, and if $\overline{EI}$ is of length h, then, unless R is the mid-point of EI,

$$\left.\frac{\partial\bar{\tau}}{\partial\eta}\right|_R^n = D_R^n = \frac{\bar{\tau}_E^n - \bar{\tau}_I^n}{h} + O\{h\}. \qquad (6.5.6)$$

However, if R is the midpoint of EI, the truncation error is $O\{h^2\}$. Combining (6.5.6) with (6.5.5) leads to the relation

$$\bar{\tau}_{j,k+1}^n = \phi_x \bar{\tau}_{j-1,k}^n + (1-\phi_x)\bar{\tau}_{j,k}^n + D_R^n h + O\{h^2\}. \qquad (6.5.7)$$

This value of $\bar{\tau}_{j,k+1}^n$ is then used in the central difference forms for the derivatives $\partial^2\bar{\tau}/\partial y^2$ and $\partial\bar{\tau}/\partial y$ so they can be written entirely in terms of values of $\bar{\tau}$ at interior gridpoints and the value of the normal derivative at R on the boundary. Thus we find, for example, that

$$\left.\frac{\partial\bar{\tau}}{\partial y}\right|_{j,k}^n = \frac{\phi_x \bar{\tau}_{j-1,k}^n + (1-\phi_x)\bar{\tau}_{j,k}^n - \bar{\tau}_{j,k-1}^n + D_R^n h}{2\Delta y} + O\{\Delta y\}, \qquad (6.5.8)$$

since h is proportional to Δy.

The finite difference equations which result on substituting these approximations for the derivatives in the transport equation evaluated at irregular points $(j\Delta x, k\Delta y)$ and time $n\Delta t$, can be combined with the set of finite difference equations for regular gridpoints, in order to find $\bar{\tau}$ at all interior gridpoints.

This method can be generalized to account for irregular points of the type P in Figure 6.3, where P is adjacent to boundary gridpoints in both x and y directions. Some variations in the way of approximating $\bar{\tau}_P^n$ occur in the literature, although they are essentially the same (see, for example, Ames (1965), and Fox (1944)).

Fox also gave a method which improves the accuracy of the approximation to the value of the normal derivative, but is more complicated. To illustrate his method, consider Figure 6.4. Instead of extending the normal through E to the first grid-line QP, he only extended it as far as the diagonal GP, where G, P were both interior points. He then used linear interpolation to obtain the value of $\bar{\tau}_H$ at the point of intersection H of the diagonal and the normal, in terms of values of $\bar{\tau}$ at G and P; the value of $\bar{\tau}_H$ written in terms of $\bar{\tau}_{j-1,k+1}^n$, $\bar{\tau}_{j,k}^n$, was then substituted into the value of the derivative $\partial\bar{\tau}/\partial\eta|_R^n$ expressed in terms of $\bar{\tau}_E$ and $\bar{\tau}_H$ to obtain $\bar{\tau}_{j,k+1}$ in terms of $\bar{\tau}_{j-1,k+1}^n$, $\bar{\tau}_{j,k}^n$.

The advantage of this method is that the normal line from E to H is shorter than the line $\overline{EI}$, so the approximation to the derivative $\partial\bar{\tau}/\partial\eta|_R^n$ is more accurate when using the analogous equation to (6.5.6). Note that, if $G(j\Delta x-\Delta x, k\Delta y+\Delta y)$ is an external gridpoint, this alternative method is not appropriate and the normal has to be extended to I as in Figure 6.4.

Taylor Series Method
In this method a governing difference equation is developed for every internal point. However, at an irregular point, modifications to the forms at regular points take place by taking into account the derivative boundary condition and the non-standard grid-spacings.

The procedure for treating those boundary conditions involving the normal derivative hinge on the expression for the directional derivative,

$$\frac{\partial\bar{\tau}}{\partial\eta} = \frac{\partial\bar{\tau}}{\partial x}\cos\gamma + \frac{\partial\bar{\tau}}{\partial y}\sin\gamma, \tag{6.5.9}$$

where all derivatives are evaluated at the boundary point in question and the angle γ specifies the direction relative to the x-axis of the normal to the boundary at that point. In Figure 6.5 $\partial\bar{\tau}/\partial\eta$ is known along the unit normals $\underset{\sim}{\eta}_R$, $\underset{\sim}{\eta}_N$, $\underset{\sim}{\eta}_S$.

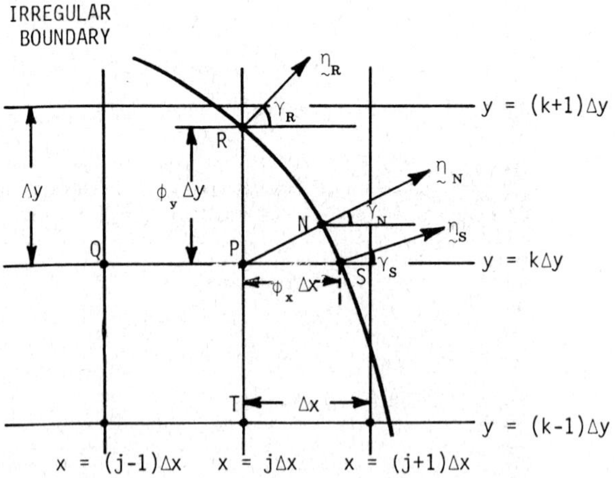

Figure 6.5 : Geometry of an irregular point near the boundary

Now, as for the Dirichlet problem, consider Taylor series expansions about the point P(jΔx,kΔy) for $\bar{\tau}_R^n$, $\bar{\tau}_S^n$, $\bar{\tau}_N^n$. Thus

$$\begin{aligned}\bar{\tau}_R^n &= \bar{\tau}(j\Delta x, k\Delta y + \phi_y\Delta y, n\Delta t)\\ &= \bar{\tau}_P^n + \phi_y\Delta y \left.\frac{\partial\bar{\tau}}{\partial y}\right|_P^n + \tfrac{1}{2}(\phi_y\Delta y)^2 \left.\frac{\partial^2\bar{\tau}}{\partial y^2}\right|_P^n + \ldots\end{aligned}$$

Replacing $\bar{\tau}$ by $\partial\bar{\tau}/\partial x$, then by $\partial\bar{\tau}/\partial y$, yields the pair of equations

$$\left.\frac{\partial\bar{\tau}}{\partial x}\right|_R^n = \left.\frac{\partial\bar{\tau}}{\partial x}\right|_P^n + \phi_y\Delta y \left.\frac{\partial^2\bar{\tau}}{\partial y\partial x}\right|_P^n + O\{(\phi_y\Delta y)^2\}, \tag{6.5.10a}$$

$$\left.\frac{\partial\bar{\tau}}{\partial y}\right|_R^n = \left.\frac{\partial\bar{\tau}}{\partial y}\right|_P^n + \phi_y\Delta y \left.\frac{\partial^2\bar{\tau}}{\partial y^2}\right|_P^n + O\{(\phi_y\Delta y)^2\}. \tag{6.5.10b}$$

Since R is a boundary point we may substitute Equations (6.5.10a,b) into the right side of Equation (6.5.9), giving

$$\left.\frac{\partial\bar{\tau}}{\partial\eta}\right|_R^n = \left[\left.\frac{\partial\bar{\tau}}{\partial x}\right|_P^n + \phi_y\Delta y\left.\frac{\partial^2\bar{\tau}}{\partial x\partial y}\right|_P^n\right]\cos\gamma_R$$

$$+ \left[\left.\frac{\partial\bar{\tau}}{\partial y}\right|_P^n + \phi_y\Delta y\left.\frac{\partial^2\bar{\tau}}{\partial y^2}\right|_P^n\right]\sin\gamma_R + O\{(\phi_y\Delta y)^2\}. \qquad (6.5.11)$$

For the boundary points N and S, two analogous equations are obtained, containing $\partial\bar{\tau}/\partial x|_P^n$, $\partial\bar{\tau}/\partial y|_P^n$, $\partial^2\bar{\tau}/\partial x^2|_P^n$, $\partial^2\bar{\tau}/\partial x\partial y|_P^n$ and $\partial^2\bar{\tau}/\partial y^2|_P^n$. Adding to these three equations the two Taylor series expansions about P for $\bar{\tau}_Q^n$ and $\bar{\tau}_T^n$ gives a total of five linear algebraic equations containing the five derivatives listed above.

$$\begin{bmatrix} \Delta x & 0 & -(\Delta x)^2/2 & 0 & 0 \\ \cos\gamma_S & \sin\gamma_S & \phi_x\Delta x\cos\gamma_S & \phi_x\Delta x\sin\gamma_S & 0 \\ \cos\gamma_N & \sin\gamma_N & \phi_x\Delta x\cos\gamma_N & \left\{\begin{matrix}\phi_y\Delta y\cos\gamma_N \\ +\phi_x\Delta x\sin\gamma_N\end{matrix}\right\} & \phi_y\Delta y\sin\gamma_N \\ \cos\gamma_R & \sin\gamma_R & 0 & \phi_y\Delta y\cos\gamma_R & \phi_y\Delta y\sin\gamma_R \\ 0 & \Delta y & 0 & 0 & -(\Delta y)^2/2 \end{bmatrix} \begin{bmatrix} \left.\frac{\partial\bar{\tau}}{\partial x}\right|_P^n \\ \left.\frac{\partial\bar{\tau}}{\partial y}\right|_P^n \\ \left.\frac{\partial^2\bar{\tau}}{\partial x^2}\right|_P^n \\ \left.\frac{\partial^2\bar{\tau}}{\partial x\partial y}\right|_P^n \\ \left.\frac{\partial^2\bar{\tau}}{\partial y^2}\right|_P^n \end{bmatrix}$$

$$= \begin{bmatrix} \bar{\tau}_Q^n - \bar{\tau}_P^n \\ \left.\frac{\partial\bar{\tau}}{\partial\eta}\right|_S^n \\ \left.\frac{\partial\bar{\tau}}{\partial\eta}\right|_N^n \\ \left.\frac{\partial\bar{\tau}}{\partial\eta}\right|_R^n \\ \bar{\tau}_T^n - \bar{\tau}_P^n \end{bmatrix} + O\{(\Delta x)^2, \Delta x\Delta y, (\Delta y)^2\}. \qquad (6.5.12)$$

After dropping terms of $O\{(\Delta x)^2, \Delta x\Delta y, (\Delta y)^2\}$, the solution of this system of equations gives expressions for the various first and second order spatial derivatives at the irregular point P in terms of the known values $\partial\bar{\tau}/\partial\eta|_R^n = D_R^n$, $\partial\bar{\tau}/\partial\eta|_N^n = D_N^n$, $\partial\bar{\tau}/\partial\eta|_S^n = D_S^n$, and the quantities τ_Q^n, τ_T^n, τ_P^n.

THE MIXED BOUNDARY CONDITION

In this case $\bar{\tau}$ and $\partial\bar{\tau}/\partial\eta$ are linearly related at all points on the boundary. As seen earlier, this may be represented by the relation (6.5.3), namely,

$$a(x,y,t)\bar{\tau} + b(x,y,t)\frac{\partial\bar{\tau}}{\partial\eta} = c(x,y,t), \qquad (6.5.13)$$

for all points $B(x,y)$ on R^+. If $b(x,y,t) \equiv 0$ for all t this degenerates to the Dirichlet boundary condition and if $a(x,y,t) \equiv 0$, the Neumann boundary condition is obtained. At the point $R(x,y)$ where the boundary cuts the grid-line $x = j\Delta x$, the value of $\bar{\tau}_R^n$ is not known explicitly if $b(x,y,t) \neq 0$. Therefore, a linear relation is sought which involves $\bar{\tau}_R^n$ and values of $\bar{\tau}$ at neighbouring interior gridpoints and points where the boundary cuts the grid-lines - see Figure 6.6.

As for the Neumann problem treated earlier, an approximation to $\partial\bar{\tau}/\partial\eta|_R^n$ may be found by extending the normal at R to intersect a grid line. By

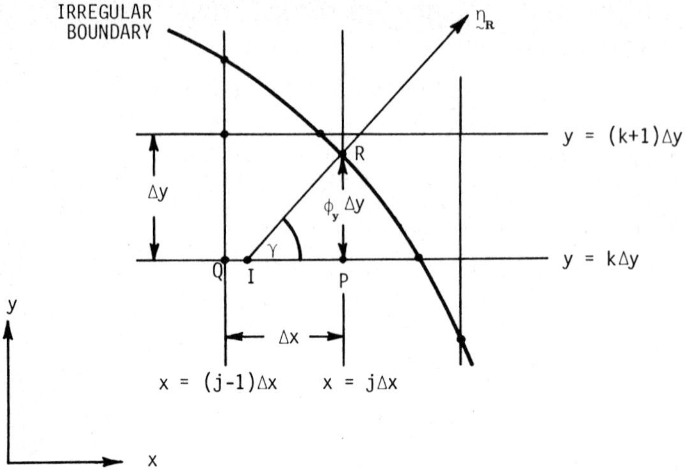

Figure 6.6 : Determination of the normal derivative

analogy with the manner in which $\partial\bar{\tau}/\partial\eta|_N^n$ is approximated in the Taylor series method of treating the Neumann problem, it is found, in general, that

$$\left.\frac{\partial\bar{\tau}}{\partial\eta}\right|_R^n = \frac{\bar{\tau}_R^n - \bar{\tau}_I^n}{\overline{IR}} + O\{\overline{IR}\}. \tag{6.5.14}$$

The value of $\bar{\tau}_I^n$ may be found by interpolation between $\bar{\tau}_Q^n$, $\bar{\tau}_P^n$, as before, giving

$$\bar{\tau}_I^n = \frac{1}{\Delta x}\{\overline{IP}\ \bar{\tau}_Q^n + \overline{QI}\ \bar{\tau}_P^n\} + O\{(\Delta x)^2\}$$

and, as $\overline{IP} = \overline{RP}\cot\gamma = \phi_y\Delta y\cot\gamma$, and $\overline{QI} = \Delta x - \overline{IP}$, then

$$\bar{\tau}_I^n = \phi_y\frac{\Delta y}{\Delta x}\cot\gamma\ \bar{\tau}_Q^n + \left(1 - \phi_y\frac{\Delta y}{\Delta x}\cot\gamma\right)\bar{\tau}_P^n + O\{(\Delta x)^2\}. \tag{6.5.15}$$

Substitution of (6.5.15) and $\overline{IR} = \phi_y\Delta y\ \mathrm{cosec}\ \gamma$ in (6.5.14), then gives

$$\left.\frac{\partial\bar{\tau}}{\partial\eta}\right|_R^n = \frac{\sin\gamma}{\phi_y\Delta y}(\bar{\tau}_R^n - \bar{\tau}_P^n) - \frac{\cos\gamma}{\Delta x}(\bar{\tau}_Q^n - \bar{\tau}_P^n) + O\{(\Delta x)^2/\Delta y\}. \tag{6.5.16}$$

Substituting (6.5.16) into the given boundary condition (6.5.13)

$$a_R^n\bar{\tau}_R^n + b_R^n\left.\frac{\partial\bar{\tau}}{\partial\eta}\right|_R^n = c_R^n \tag{6.5.17}$$

gives a relation for $\bar{\tau}_P^n$ in terms of $\bar{\tau}_Q^n$, $\bar{\tau}_R^n$, since the values a_R^n, b_R^n, c_R^n are known for the point $R(j\Delta x, k\Delta y+\phi_y\Delta y)$ at time $t = n\Delta t$. This relation is

$$\bar{\tau}_P^n = \frac{1}{b_R^n(\Delta x\sin\gamma - \phi_y\Delta y\cos\gamma)}\{\Delta x\phi_y\Delta y\ a_R^n + b_R^n\sin\gamma)\bar{\tau}_R^n$$

$$- \phi_y\Delta y\ b_R^n\cos\gamma\ \bar{\tau}_Q^n - \phi_y\Delta y\Delta x c_R^n\} + O\{(\Delta x)^2, (\Delta x)^3/\Delta y\}. \tag{6.5.18}$$

In conjunction with the finite difference equation applied at regular interior gridpoints, (6.5.18) permits approximations to $\bar{\tau}$ to be calculated at irregular gridpoints.

In general, problems involving mixed boundary conditions on irregular boundaries are difficult to solve.

When considering boundary conditions on irregular boundaries in three-dimensional space, little is different from the two-dimensional case. The techniques described in this section may be extended to the three-dimensional case, but the methods are much more complicated. In general, results are usually less accurate because of the further approximations which must be made due to the incorporation of derivatives involving the extra space dimension.

7. THE WAVE EQUATION

7.1 Introduction

The one-dimensional wave equation may be written

$$\frac{\partial^2\bar{\tau}}{\partial t^2} = u^2\,\frac{\partial^2\bar{\tau}}{\partial x^2}, \qquad 0 \le x \le 1,\ t \ge 0. \tag{7.1.1}$$

Associated with this equation are *two* initial conditions, such as

$$\bar{\tau}(x,0) = f(x),\ \partial\bar{\tau}/\partial t\Big|_{(x,0)} = h(x), \tag{7.1.2a}$$

and two boundary conditions, for example

$$\bar{\tau}(0,t) = g_0(t),\ \bar{\tau}(1,t) = g_1(t),\ t \ge 0. \tag{7.1.2b}$$

If u is *constant*, an exact solution of Equation (7.1.1) subject to initial conditions (7.2.1a), on the infinite interval $-\infty < x < \infty$, may be found using D'Alembert's method. Using the transformation $\mu = x+ut$, $\nu = x-ut$ in Equation 7.1.1 yields

$$\frac{\partial^2\bar{\tau}(\mu,\nu)}{\partial\mu\partial\nu} = 0, \tag{7.1.3}$$

which has the general solution

$$\bar{\tau}(\mu,\nu) = F(\mu) + G(\nu), \tag{7.1.4}$$

where F and G are arbitrary functions. Thus,

$$\bar{\tau}(x,t) = F(x+ut) + G(x-ut). \tag{7.1.5}$$

Clearly, this solutions consists of two parts, initially F(x) and G(x), both moving with speed u, the first in the negative x-direction and the second in the positive x-direction. The functions of F and G must satisfy the initial and boundary conditions of a particular problem. For instance, with initial condition (7.1.2a) it can be shown that

$$\bar{\tau}(x,t) = \tfrac{1}{2}\{f(x+ut) + f(x-ut)\} + \frac{1}{2u}\int_{x-ut}^{x+ut} h(\chi)d\chi,\ -\infty < x < \infty. \tag{7.1.6}$$

7.2 An Explicit Finite Difference Method

One way to obtain a finite difference approximation to Equation (7.1.1) is to use central difference forms for the time and space derivatives at the (j,n) gridpoint. In this case, if u is constant,

$$\frac{\partial^2\bar{\tau}}{\partial t^2}\bigg|_j^n = u^2\,\frac{\partial^2\bar{\tau}}{\partial x^2}\bigg|_j^n \tag{7.2.1}$$

may be written

$$\frac{\bar{\tau}_j^{n+1} - 2\bar{\tau}_j^n + \bar{\tau}_j^{n-1}}{(\Delta t)^2} + O\{(\Delta t)^2\} = u^2\{\frac{\bar{\tau}_{j+1}^n - 2\bar{\tau}_j^n + \bar{\tau}_{j-1}^n}{(\Delta x)^2} + O\{(\Delta x)^2\}\}. \tag{7.2.2}$$

A suitable finite difference approximation to Equation (7.2.2) with truncation error $O\{(\Delta t)^2,(\Delta x)^2\}$ is therefore

$$\frac{\tau_j^{n+1} - 2\tau_j^n + \tau_j^{n-1}}{(\Delta t)^2} = u^2\{\frac{\tau_{j+1}^n - 2\tau_j^n + \tau_{j-1}^n}{(\Delta x)^2}\}. \quad (7.2.3)$$

Rearranging (7.2.3) gives the three-level explicit finite difference equation

$$\tau_j^{n+1} = c^2(\tau_{j-1}^n + \tau_{j+1}^n) + 2(1-c^2)\tau_j^n - \tau_j^{n-1}, \quad (7.2.4)$$

where, as before $c = u\Delta t/\Delta x$. At each time step the contribution to the discretisation error in this numerical method is $O\{(\Delta t)^4,(\Delta t)^2(\Delta x)^2\}$. Since this is a linear initial value problem, and the finite difference equation is consistent because the truncation error tends to zero as both Δx and Δt tend to zero, then convergence of the solution of this equation to the solution of the wave equation is ensured if (7.2.4) is stable.

If the stability is investigated by means of the von Neumann method, the amplification factor is given by the solution of

$$G^2 - 2\gamma G + 1 = 0, \quad (7.2.5)$$

where $\gamma = 1 - 2c^2 \sin^2(\beta/2)$. The solutions of (7.2.5) are

$$G_1,G_2 = \gamma \pm \sqrt{\gamma^2-1}. \quad (7.2.6)$$

If $c \le 1$, then $|\gamma| \le 1$ and $\sqrt{1-\gamma^2}$ is real for all β. In this case,

$$G_1,G_2 = \gamma \pm i\sqrt{1-\gamma^2},$$

so that

$$|G_1| = |G_2| = \sqrt{\gamma^2 + (1-\gamma^2)} = 1,$$

for all β and (7.2.4) is stable. If $c \ge 1$, then for some β we have $|\gamma| > 1$ and $\sqrt{\gamma^2-1}$ in (7.2.6) is real. It follows that one of G_1, G_2 has a modulus which is larger than 1 and equation (7.2.4) is unstable. For instance, if $\gamma < -1$ then $\gamma - \sqrt{\gamma^2-1} < -1$. So one of G_1, G_2 is smaller than -1 and therefore has a modulus which is larger than 1. Similarly, if $\gamma > 1$ then $\gamma + \sqrt{\gamma^2-1} > 1$.

Equation (7.2.4) defines a three-level finite difference scheme. Therefor a special starting procedure is required to obtain the set of values of τ at the first time level from the set of initial values τ_j^0, $j=0(1)J$. This involves the use of the initial derivative condition. Given that

$$\frac{\partial\bar{\tau}(x,0)}{\partial t} = h(x),$$

the derivative may be replaced by a central difference form at the grid-point (j,0), giving

$$\left.\frac{\partial\bar{\tau}}{\partial t}\right|_j^0 = \frac{\bar{\tau}_j^1 - \bar{\tau}_j^{-1}}{2\Delta t} + O\{(\Delta t)^2\} = h_j\,,\ j=1(1)J, \quad (7.2.7)$$

where $h_j = h(j\Delta x)$ and $\bar{\tau}_j^{-1} = \bar{\tau}(j\Delta x,-\Delta t)$. A finite difference approximation to the initial derivative condition, with truncation error $O\{(\Delta t)^2\}$, is therefore

$$\frac{\tau_j^1 - \tau_j^{\circ 1}}{2\Delta t} = h_j \text{ , } j=0(1)J, \tag{7.2.8}$$

where $\tau_j^{\circ 1}$ is a fictitious value of τ assigned to the position $j\Delta x$ at a time Δt before the initial instant. Therefore,

$$\tau_j^{\circ 1} = \tau_j^1 - 2\Delta t\, h_j \text{ , } j=0(1)J, \tag{7.2.9}$$

with a contribution of $O\{(\Delta t)^3\}$ to the discretisation error in τ_j^n. Substitution of this value of $\tau_j^{\circ 1}$ in (7.2.4) with n=0, gives

$$\tau_j^1 = \tfrac{1}{2}c^2(f_{j-1} + f_{j+1}) + (1-c^2)f_j + \Delta t\, h_j \text{ , } j=1(1)J-1, \tag{7.2.10}$$

where $f_j = f(j\Delta x)$.

By means of (7.2.10) the values of τ at the first time level are computed from the known initial values $\bar{\tau}_j^0 = f_j$, j=0(1)J and derivative initial conditions $\partial\bar{\tau}/\partial t|_j^0 = h_j$, j=1(1)J-1. Equation (7.2.9) contributes a term of $O\{(\Delta t)^3\}$ to the discretisation error while at each subsequent time step (7.2.4) contributes an error of $O\{(\Delta t)^4,(\Delta t)\ (\Delta x)^2\}$, so that in practical terms this method has a discretisation error of $O\{N(\Delta t)^4,N(\Delta t)^2\ (\Delta x)^2\}$ in the value of τ_j^N if N is large.

In the special case with $h(x) \equiv 0$, so $h_j = 0$ for all j=0(1)J, and with c = 1, then (7.2.10) becomes

$$\tau_j^1 = \tfrac{1}{2}(f_{j-1} + f_{j+1}). \tag{7.2.11}$$

As a test (7.2.11) was used in conjunction with (7.2.4) to solve a wave-equation problem with a zero derivative initial condition and the results compared with the exact solution obtained using D'Alembert's formula (7.1.6). The initial condition chosen was $f(x) = \sin(\pi x)$, $0 \le x \le 1$, with boundary values $\bar{\tau}(0,t) = \bar{\tau}(1,t) = 0$ for $t \ge 0$. With u = 1, $\Delta t = 0.1$, $\Delta x = 0.1$, then c = 1, and the analytic and numerical solutions for $\bar{\tau}(j\Delta x,n\Delta t)$, j=1(1)9, n=0(1)100 gave the same answers to 10 significant figures on a DEC-VAX 11 computer.

An alternative and more accurate way of calculating the values at the first time level takes advantage of D'Alembert's solution (7.1.6). In discrete form his formula applied for $x = j\Delta x$, j=1(1)J-1, and $t = \Delta t$, is

$$\begin{aligned}\bar{\tau}_j^1 &= \bar{\tau}(j\Delta x,n\Delta t)\\ &= \tfrac{1}{2}[f(j\Delta x+u\Delta t) + f(j\Delta x-u\Delta t)] + \frac{1}{2u}\int_{j\Delta x-u\Delta t}^{j\Delta x+u\Delta t} h(\chi)\,d\chi.\end{aligned}$$

With c = 1, so $\Delta x = u\Delta t$, it follows that

$$\begin{aligned}\bar{\tau}_j^1 &= \tfrac{1}{2}[f(j\Delta x+\Delta x) + f(j\Delta x-\Delta x)] + \frac{1}{2u}\int_{j\Delta x-\Delta x}^{j\Delta x+\Delta x} h(\chi)\,d\chi\\ &= \tfrac{1}{2}[f_{j-1} + f_{j+1}] + \frac{1}{2u}\int_{X_{j-1}}^{X_{j+1}} h(\chi)\,d\chi,\end{aligned} \tag{7.2.12}$$

where the integral can be calculated using Simpson's Rule, namely

$$\int_{x_{j-1}}^{x_{j+1}} h(\chi)d\chi = \frac{\Delta x}{3}\{h_{j-1} + 4h_j + h_{j+1}\} + O\{(\Delta x)^4\}.$$

Therefore

$$\tau_j^1 = \tfrac{1}{2}[f_{j-1} + f_{j+1}] + \frac{\Delta t}{6}[h_{j-1} + 4h_j + h_{j+1}], \qquad (7.2.13)$$

with a contribution to the discretisation error at later time levels, of $O\{(\Delta x)^4\}$. Note the similarity between this result and (7.2.10) with c=1, which contributes a term of $O\{(\Delta t)^3\} \equiv O\{(\Delta x)^3\}$ to the discretisation error. If $h(x) \equiv 0$, again this results in τ_j^1 being computed by means of (7.2.11), which from (7.2.12) is clearly the exact value $\bar{\tau}_j^1$.

7.3 Implicit Methods

If, in Equation (7.2.1), $\partial^2\bar{\tau}/\partial x^2|_j^n$ is replaced by the average of its values at time levels (n-1) and (n+1), then

$$\left.\frac{\partial^2\bar{\tau}}{\partial t^2}\right|_j^n = \frac{u^2}{2}\left\{\left.\frac{\partial^2\bar{\tau}}{\partial x^2}\right|_j^{n-1} + \left.\frac{\partial^2\bar{\tau}}{\partial x^2}\right|_j^{n+1} + O\{(\Delta t)^2\}\right\}. \qquad (7.3.1)$$

Replacing all the derivatives in (7.3.1) by their central difference forms gives

$$\frac{\tau_j^{n+1} - 2\bar{\tau}_j^n + \bar{\tau}_j^{n-1}}{(\Delta t)^2} + O\{(\Delta t)^2\} = \frac{u^2}{2}\left\{\frac{\bar{\tau}_{j+1}^{n-1} - 2\bar{\tau}_j^{n-1} + \bar{\tau}_{j-1}^{n-1}}{(\Delta x)^2}\right.$$

$$\left. + \frac{\bar{\tau}_{j+1}^{n+1} - 2\bar{\tau}_j^{n+1} + \bar{\tau}_{j-1}^{n+1}}{(\Delta x)^2} + O\{(\Delta t)^2,(\Delta x)^2\}\right\}. \qquad (7.3.2)$$

Multiplying through by $(\Delta t)^2$, then rearranging terms so the values at the time level (n+1) are on the left side and dropping terms of $O\{(\Delta t)^4,(\Delta t)^2(\Delta x)^2\}$, the following implicit equation is obtained for j=1(1)J-1:

$$-\tfrac{1}{2}c^2\tau_{j-1}^{n+1} + (1+c^2)\tau_j^{n+1} - \tfrac{1}{2}c^2\tau_{j+1}^{n+1}$$

$$= 2\tau_j^n + \tfrac{1}{2}c^2\tau_{j-1}^{n-1} - (1+c^2)\tau_j^{n-1} + \tfrac{1}{2}c^2\tau_{j+1}^{n-1}. \qquad (7.3.3)$$

If the values of $\bar{\tau}$ are known on the boundary, say $\tau_0^{n+1} = g_0(t_{n+1})$, $\tau_J^{n+1} = g_1(t_{n+1})$, then the set of simultaneous algebraic equations (7.3.3) in the unknowns τ_j^{n+1}, j=1(1)J-1, is tridiagonal, and as

$$|1+c^2| > |-\tfrac{1}{2}c^2| + |-\tfrac{1}{2}c^2|$$

for all c, then the system is also diagonally dominant. The set of linear algebraic equations can therefore be solved economically using the Thomas algorithm.

In order to start the solution process, substitution of (7.2.9) into (7.3.3) with n = 0 gives, for j=1(0)J-1,

$$-c^2\tau_{j-1}^1 + 2(1+c^2)\tau_j^1 - c^2\tau_{j+1}^1$$

$$= 2f_j - c^2\Delta t\, h_{j-1} + \tfrac{1}{2}(1+c^2)\Delta t\, h_j - c^2\Delta t\, h_{j+1}, \qquad (7.3.4)$$

which may also be solved using the Thomas algorithm.

The stability of Equation (7.3.3) may be investigated by the von Neumann method. The amplification factor G of the m-th Fourier component of the error distribution as it propagates from one time level to the next is given by the solution of the quadratic equation

$$G^2 - 2\gamma G + 1 = 0, \qquad (7.3.5)$$

where $\gamma = \{1 + 2c^2 \sin^2(\beta/2)\}^{-1} \leq 1$ for all c and β. It follows that the values of G are

$$G_1, G_2 = \gamma \pm i\sqrt{1-\gamma^2}, \qquad (7.3.6)$$

where $\sqrt{1-\gamma^2}$ is real. Hence

$$|G_1| = |G_2| = \sqrt{\gamma^2 + (1-\gamma^2)} = 1 \qquad (7.3.7)$$

for all c and β, and the implicit scheme given by (7.3.3) is always *stable*. Investigation of the stability of the starting equation (7.3.4) is not necessary, since it is only applied to obtain values at the first time level.

The above implicit method is a special case of a general form involving a weighting factor θ in time. Consider

$$\bar{\tau}_j^n = \theta\bar{\tau}_j^{n+1} + (1-2\theta)\bar{\tau}_j^n + \theta\bar{\tau}_j^{n-1} + O\{\theta(\Delta t)^2\},\ 0 < \theta \leq 1,$$

This is exact if $\theta = 0$. It follows that the wave equation at the (j,n) gridpoint, namely Equation (7.2.1) can be written

$$\left.\frac{\partial^2\bar{\tau}}{\partial t^2}\right|_j^n = u^2\{\theta \left.\frac{\partial^2\bar{\tau}}{\partial x^2}\right|_j^{n+1} + (1-2\theta)\left.\frac{\partial^2\bar{\tau}}{\partial x^2}\right|_j^n + \theta \left.\frac{\partial^2\bar{\tau}}{\partial x^2}\right|_j^{n-1} + O\{\theta(\Delta t)^2\},$$

$$0 < \theta \leq 1. \qquad (7.3.8)$$

On substitution of the central difference forms for the derivatives in (7.3.8), then dropping terms of $O\{(\Delta t)^2, (\Delta x)^2\}$, and rearranging, the following implicit finite difference equation is obtained for j=1(1)J-1:

$$-\theta c^2\tau_{j-1}^{n+1} + (1+2\theta c^2)\tau_j^{n+1} - \theta c^2\tau_{j+1}^{n+1}$$

$$= (1-2\theta)c^2\tau_{j-1}^n + 2[1 - (1-2\theta)c^2]\tau_j^n + (1-2\theta)c^2\tau_{j+1}^n$$

$$+ \theta c^2\tau_{j-1}^{n-1} - (1+2\theta c^2)\tau_j^{n-1} + \theta c^2\tau_j^{n-1}, \qquad (7.3.9)$$

for n=1(1)N. Equation (7.3.9) has a truncation error of $O\{(\Delta t)^2, (\Delta x)^2\}$ and contributes terms of $O\{(\Delta t)^4, (\Delta t)^2(\Delta x)^2\}$ to the discretisation error at each time step.

The equation used for n = 0 is found by substitution of (7.2.9) into (7.3.9) with n = 0, giving

$$-2\theta c^2\tau_{j-1}^1 + 2(1+2c^2\theta)\tau_j^1 - 2\theta c^2\tau_{j+1}^1$$

$$= (1-2\theta)c^2 f_{j-1} + 2[1-(1-2\theta)c^2]f_j + (1-2\theta)c^2 f_{j+1}$$

$$- 2\theta c^2 \Delta t\, h_{j-1} + 2(1+2\theta c^2)\Delta t\, h_j - 2\theta c^2 \Delta t\, h_{j+1}, \qquad (7.3.10)$$

for $j=1(1)J-1$.

Not that $\theta = 0$ gives the explicit equation (7.2.4) considered in Section 7.2 and $\theta = \frac{1}{2}$ gives the implicit equation (7.3.3) previously described in this section. The von Neumann stability analysis shows that this method is stable for all c if $\theta \geq \frac{1}{4}$, and is stable for $c \leq (1-4\theta)^{-\frac{1}{2}}$ if $\theta < \frac{1}{4}$. Since the set of tridiagonal equations (7.3.9) is diagonally dominant for all θ and c, the solutions may be computed with the aid of the Thomas algorithm.

Another implicit method (see Wachspress, 1966) uses central differences to approximate $\partial^2\bar{\tau}/\partial x^2$ at the $(j,n+1)$ gridpoint, namely

$$\left.\frac{\partial^2\bar{\tau}}{\partial x^2}\right|_j^{n+1} = \frac{\bar{\tau}_{j-1}^{n+1} - 2\bar{\tau}_j^{n+1} + \bar{\tau}_{j+1}^{n+1}}{(\Delta x)^2} + O\{(\Delta x)^2\}, \qquad (7.3.11)$$

with a backward difference form for $\partial^2\bar{\tau}/\partial t^2$ at the same gridpoint, that is

$$\left.\frac{\partial^2\bar{\tau}}{\partial t^2}\right|_j^{n+1} = \frac{\bar{\tau}_j^{n+1} - 2\bar{\tau}_j^{n} + \bar{\tau}_j^{n-1}}{(\Delta t)^2} + O\{\Delta t\}. \qquad (7.3.12)$$

Substitution of (7.3.11) and (7.3.12) in the wave equation applied at the $(j,n+1)$ gridpoint gives the finite difference equation used for $n=1,2,3,\ldots$,

$$c^2\tau_{j-1}^{n+1} - (1+2c^2)\tau_j^{n+1} + c^2\tau_{j+1}^{n+1} = -2\tau_j^{n} + \tau_j^{n-1}, \; j=1(1)J-1, \qquad (7.3.13)$$

with a truncation error of $O\{\Delta t, (\Delta x)^2\}$ and a contribution of $O\{(\Delta t)^3, (\Delta t)^2(\Delta x)^2\}$ to the discretisation error at each time step. These errors are of the same order as (7.3.9) with $\theta \neq \frac{1}{2}$, but (7.3.12) is less accurate than (7.3.9) with $\theta = \frac{1}{2}$, that is than (7.3.3). Again, the tridiagonal set of algebraic equations (7.3.12) is diagonally dominant, and the Thomas algorithm may be used to solve the system.

The finite difference equation used for $n = 0$ is obtained by substituting τ_j^{n-1} from Equation (7.2.9) into Equation (7.3.13) with $n = 0$, giving

$$c^2\tau_{j-1}^{1} - 2(1+c^2)\tau_j^{1} + c^2\tau_{j+1}^{1} = -2f_j - 2\Delta t\, h_j, \; j=1(1)J-1. \qquad (7.3.14)$$

Application of the von Neumann method shows (7.3.12) is stable for all values of c. Again it should be noted that backward differences in times are used to guarantee stability for all values of c in many finite difference schemes.

7.4 Conversion to Two First Order Partial Differential Equations

If we define $\bar{p}(x,t) = u\partial\bar{\tau}/\partial x$ and $\bar{q}(x,t) = \partial\bar{\tau}/\partial t$, the wave equation (7.1.1) with constant u can be expressed as two simultaneous first order partial differential equations,

$$\frac{\partial\bar{p}}{\partial t} = u\frac{\partial\bar{q}}{\partial x} \qquad (7.4.1a)$$

and

TABLE 7.1

Finite difference techniques for solving the one-dimensional wave equation $\partial^2\bar{\tau}/\partial t^2 = u^2\ \partial^2\bar{\tau}/\partial x^2,\ u > 0$

Method	Finite Difference Equation	Truncation Error E	Amplification Factor G	Stability Conditions	Solvability Conditions
CTCS Explicit	Eqn. (7.2.4) $\tau_j^{n+1} = c^2(\tau_{j-1}^n + \tau_{j+1}^n) + 2(1-c^2)\tau_j^n - \tau_j^{n-1}$	$O\{(\Delta t)^2, (\Delta x)^2\}$	$\gamma \pm i\sqrt{1-\gamma^2}$, $\gamma = 1-2c^2\sin^2(\frac{\beta}{2})$	$c \leq 1$	None
CTCS Implicit	Eqn. (7.3.3) $-\frac{c^2}{2}\tau_{j-1}^{n+1} + (1+c^2)\tau_j^{n+1} - \frac{c^2}{2}\tau_{j+1}^{n+1} = 2\tau_j^n + \frac{c^2}{2}\tau_{j-1}^{n-1} - (1+c^2)\tau_j^{n-1} + \frac{c^2}{2}\tau_{j+1}^{n-1}$	$O\{(\Delta t)^2, (\Delta x)^2\}$	$\gamma \pm i\sqrt{1-\gamma^2}$, $\gamma = \{1+2c^2\sin^2(\frac{\beta}{2})\}^{-1}$	None	None
Time-Weighted $0 < \theta \leq 1$	Eqn. (7.3.9) $-\theta c^2\tau_{j-1}^{n+1} + (1+2\theta c^2)\tau_j^{n+1} - \theta c^2\tau_{j+1}^{n+1} = (1-2\theta)c^2\tau_{j-1}^n + 2[1-(1-2\theta)c^2]\tau_j^n + (1-2\theta)c^2\tau_{j+1}^n + \theta c^2\tau_{j-1}^{n-1} - (1+2\theta)c^2\tau_j^{n-1} + \theta c^2\tau_{j+1}^{n-1}$	$O\{(\Delta t)^2, (\Delta x)^2\}$	$\gamma \pm i\sqrt{1-\gamma^2}$, $\gamma = \frac{1-2(1-2\theta)c^2\sin^2(\beta/2)}{1+4\theta c^2\sin^2(\beta/2)}$	For $\theta \geq \frac{1}{4}$, none. For $0 \leq \theta < \frac{1}{4}$ $c \leq (1-4\theta)^{-\frac{1}{2}}$	None
Wachspress Implicit	Eqn. (7.3.13) $c^2\tau_{j-1}^{n+1} - (1+2c^2)\tau_j^{n+1} + c^2\tau_{j+1}^{n+1} = -2\tau_j^n + \tau_j^{n-1}$	$O\{\Delta t, (\Delta x)^2\}$	$\frac{1 \pm i\{2c\sin(\beta/2)\}}{1 + 4c^2\sin^2(\beta/2)}$	None	None

$$\frac{\partial \bar{q}}{\partial t} = u \frac{\partial \bar{p}}{\partial x}. \tag{7.4.1b}$$

The obvious finite difference equations with which one could try to solve this pair of simultaneous partial differential equations would be based on forward difference forms for $\partial\bar{p}/\partial t$ and $\partial\bar{q}/\partial t$ with central difference forms for $\partial\bar{p}/\partial x$ and $\partial\bar{q}/\partial x$.

Thus, at the (j,n) gridpoint, the equation

$$\left.\frac{\partial \bar{p}}{\partial t}\right|_j^n = u \left.\frac{\partial \bar{q}}{\partial x}\right|_j^n \tag{7.4.2a}$$

may be written

$$\frac{\bar{p}_j^{n+1} - \bar{p}_j^n}{\Delta t} + O\{\Delta t\} = u\{\frac{\bar{q}_{j+1}^n - \bar{q}_{j-1}^n}{2\Delta x} + O\{(\Delta x)^2\}\},$$

which becomes, on dropping terms of $O\{\Delta t,(\Delta x)^2\}$ and rearranging,

$$p_j^{n+1} = p_j^n + \tfrac{1}{2}c(q_{j+1}^n - q_{j-1}^n), \quad j=1(1)J-1. \tag{7.4.2b}$$

Similarly,

$$\left.\frac{\partial \bar{q}}{\partial t}\right|_j^n = u \left.\frac{\partial \bar{p}}{\partial x}\right|_j^n \tag{7.4.3a}$$

may be written

$$\frac{\bar{q}_j^{n+1} - \bar{q}_j^n}{\Delta t} + O\{\Delta t\} = u\{\frac{\bar{p}_{j+1}^n - \bar{p}_{j-1}^n}{2\Delta x} + O\{(\Delta x)^2\}\},$$

which yields the finite difference equation

$$q_j^{n+1} = q_j^n + \tfrac{1}{2}c(p_{j+1}^n - p_{j-1}^n), \quad j=1(1)J-1, \tag{7.4.3b}$$

with a truncation error of $O\{\Delta t,(\Delta x)^2\}$. At each time step, the contribution to the discretisation error of p_j^n and q_j^n is $O\{(\Delta t)^2,\Delta t(\Delta x)^2\}$.

But checking the stability of the simultaneous pair of equations (7.4.2b), (7.4.3b) it is found that the system is unstable for all values of $c > 0$. This result is similar to that obtained when solving the convection equation

$$\frac{\partial \bar{\tau}}{\partial t} = -u \frac{\partial \bar{\tau}}{\partial x}$$

using the Forward-Time Central-Space method : the method is therefore of no practical use.

One variation of this method uses forward difference forms for $\partial\bar{p}/\partial t$ and $\partial\bar{q}/\partial x$ in Equation (7.4.1a) discretised at the (j,n) gridpoint. Then backward difference forms for the derivatives are used to discretise Equation (7.4.1b) at the gridpoint (j,n+1). This gives, on dropping terms of $O\{\Delta t,\Delta x\}$ and rearranging, the pair of explicit difference equations for j=1(1)J-1,

$$p_j^{n+1} = p_j^n + c(q_{j+1}^n - q_j^n), \qquad\qquad n = 0,1,2,3,\ldots \qquad (7.4.4a)$$

$$q_j^{n+1} = q_j^n + c(p_j^{n+1} - p_{j-1}^{n+1}). \qquad (7.4.4b)$$

These equations are similar to (7.4.2b), (7.4.3b), except that the latest values of p are used in (7.4.4b). In general, these equations contribute errors of $O\{(\Delta t)^2, \Delta t \Delta x\}$ to the discretisation errors of p_j^{n+1}, q_j^{n+1} at each time step.

By adapting the von Neumann stability analysis it can be applied to simultaneous partial differential equations, and may be used to find the values of c for which the system of equations (7.4.4) is stable.

The errors ζ_j^n in p_j^n and η_j^n in q_j^n propagate in a way which is analogous to the calculation of p and q using Equations (7.4.4a,b). If the m-th Fourier component of the distribution of errors in p_j^n, j=0(1)J, at time level n is given by

$$\zeta_j^n = \delta_p (G)^n e^{i\beta j}, \qquad (7.4.5a)$$

and the corresponding component of the distribution of errors in q_j^n, j=0(1)J, is given by

$$\eta_j^n = \delta_q (G)^n e^{i\beta j}, \qquad (7.4.5b)$$

where $\beta = m\pi\Delta x$, then substitution of these forms into the corresponding error equations to (7.4.4a,b) gives, on division by $e^{i\beta j}$,

$$\delta_p (G)^{n+1} = \delta_p (G)^n + c\delta_q (G)^n (e^{i\beta}-1), \qquad (7.4.6a)$$

$$\delta_q (G)^{n+1} = \delta_q (G)^n + c\delta_p (G)^{n+1} (1-e^{-i\beta}). \qquad (7.4.6b)$$

Rearrangement of equations (7.4.6) gives

$$(1-G)\delta_p + c(e^{i\beta}-1)\delta_q = 0, \qquad (7.4.7a)$$

$$cG(1-e^{-i\beta})\delta_p + (1-G)\delta_q = 0. \qquad (7.4.7b)$$

Solutions for δ_p, δ_q can be found if, and only if,

$$\begin{vmatrix} (1-G) & c(e^{i\beta}-1) \\ cG(1-e^{-i\beta}) & (1-G) \end{vmatrix} = 0, \qquad (7.4.8)$$

which gives the following equation for G:

$$G^2 - 2\gamma G + 1 = 0, \qquad (7.4.9)$$

where $\gamma = 1 - 2c^2 \sin^2(\beta/2)$. The roots of (7.4.9) are

$$G_1, G_2 = \gamma \pm \sqrt{\gamma^2 - 1}.$$

If $|\gamma| > 1$, then G_1, G_2 are real and one of $|G_1|$, $|G_2|$ is larger than 1. If $|\gamma| \le 1$, then G_1 and G_2 are complex conjugates given by

$$G_1, G_2 = \gamma \pm i\sqrt{1-\gamma^2}.$$

In this case

$$|G_1| = |G_2| = \sqrt{\gamma^2+(1-\gamma^2)} = 1.$$

The set of equations (7.4.4a,b) is therefore stable if $|\gamma| \leq 1$, which is true if

$$-1 \leq 1 - 2c^2\sin^2(\beta/2) \leq 1,$$

or

$$0 \leq c^2\sin^2(\beta/2) \leq 1.$$

This is true for all β if $c \leq 1$. Therefore the system of Equations (7.4.4a,b) is stable if $c \leq 1$ and may be used to compute p_j^n, q_j^n, $j=1(1)J-1$, $n=1,2,3,\ldots$. From these values we must compute the values of τ_j^n, $j=1(1)J-1$, $n=1,2,3,\ldots$.

Consider the case in which the boundary values $\bar{\tau}_0^n$, $\bar{\tau}_J^n$, J odd, are given for all $n=0,1,2,\ldots$ Once the p_j^n are known for $j=1(1)J-1$, then application of

$$\bar{p}_j^n = u\left.\frac{\partial\bar{\tau}}{\partial x}\right|_j^n, \tag{7.4.10}$$

using central differencing for the space derivative on the right side, gives

$$\bar{p}_j^n = u\left\{\frac{\bar{\tau}_{j+1}^n - \bar{\tau}_{j-1}^n}{2\Delta x} + O\{(\Delta x)^2\}\right\},\ j=1(1)J-1. \tag{7.4.11}$$

Rearranging, and dropping terms of $O\{(\Delta x)^3\}$, gives the two forms

$$\tau_{j+1}^n = \tau_{j-1}^n + 2\Delta x\ p_j^n/u, \tag{7.4.12a}$$

$$j=1(1)J-1,$$

$$\tau_{j-1}^n = \tau_{j+1}^n - 2\Delta x\ p_j^n/u, \tag{7.4.12b}$$

from which τ_j^n at all interior gridpoints may be found as follows.

We commence application of the system of finite difference equations (7.4.4a,b) with n=0. Since the initial values $\bar{\tau}_j^0 = f_j$, $j=0(1)J$, are known, using n=0 in (7.4.11) and dropping terms of $O\{(\Delta x)^2\}$ gives

$$p_j^0 = \frac{u}{2\Delta x}\{f_{j+1} - f_{j-1}\},\ j=1(1)J-1. \tag{7.4.13}$$

In order to obtain p_j^0, j=0, with the same accuracy as (7.4.13) we proceed as follows. Substituting

$$\left.\frac{\partial\bar{\tau}}{\partial x}\right|_j^n = \frac{-3\bar{\tau}_j^n + 4\bar{\tau}_{j+1}^n - \bar{\tau}_{j+2}^n}{2\Delta x} + O\{(\Delta x)^2\}$$

into (7.4.10) then putting j=0 gives

$$p_0^n = \frac{u}{2\Delta x}\{-3\tau_0^n + 4\tau_1^n - \tau_2^n\}, \tag{7.4.14}$$

on dropping terms of $O\{(\Delta x)^2\}$. With n=0, we have

$$p_0^0 = \frac{u}{2\Delta x}\{-3f_0 + 4f_1 - f_2\}. \tag{7.4.15}$$

Also, from the given initial derivative condition

$$\bar{q}_j^0 = \left.\frac{\partial\bar{\tau}}{\partial t}\right|_j^0 = h_j\ ,\ j=0(1)J, \tag{7.4.16}$$

is known. Hence p_j^1, $j=0(1)J-1$ may be found by applying Equation (7.4.4a) with $n=0$. Then use of Equation (7.4.12a) with $n=1$, $j=1(2)J-2$, commencing with the known value $\bar{\tau}_1^1 = g_0^1$, followed by Equation (7.4.12b) with $n=1$, $j=J(-2)1$ starting with the boundary value $\bar{\tau}_J^1 = g_J^1$, gives τ_j^1, $j=1(1)J-1$. Application of Equation (7.4.4b), with $n=0$, then gives q_j^1, $j=1(1)J-1$.

Repeated application of Equations (7.4.4a,b) with $n=1,2,3,...,$ gives τ_j^{n+1}, $j=1(1)J-1$. The only additional requirements are that q_0^n, $n=1,2,3,..$ must be known whenever Equation (7.4.4a) is used with $j=0$ and q_J^n, $n=1,2,3,..$ must be known whenever that equation is used with $j=J-1$. Since

$$\bar{q}_j^n = \left.\frac{\partial\bar{\tau}}{\partial t}\right|_j^n \tag{7.4.17}$$

we may write

$$\bar{q}_j^n = \frac{\bar{\tau}_j^{n+1} - \bar{\tau}_j^{n-1}}{2\Delta t} + O\{(\Delta t)^2\}.$$

Therefore, putting $j=0$ and dropping terms of $O\{(\Delta t)^2\}$, we obtain

$$q_0^n = \frac{1}{2\Delta t}\{g_0^{n+1} - g_0^{n+1}\},\ n=1,2,3,..., \tag{7.4.18}$$

and similarly, putting $j=J$ gives

$$q_J^n = \frac{1}{2\Delta t}\{g_J^{n+1} - g_0^{n+1}\},\ n=1,2,3,... \tag{7.4.19}$$

A less accurate method, but one which applies for values of J which are either odd or even, and which can readily incorporate derivative boundary conditions is now described. The method is the same as the previous one until p_j^1, $j=0(1)J-1$ is found by applying Equation (7.4.4a) with $n=0$. Then τ_j^1, $j=1(1)J-1$, is found in the following manner.

Substituting the forward difference form for $\partial\bar{\tau}/\partial x|_j^n$ in (7.4.10) gives

$$\bar{p}_j^n = u\{\frac{\bar{\tau}_{j+1}^n - \bar{\tau}_j^n}{\Delta x} + O\{\Delta x\}\},\ j=0(1)J-1. \tag{7.4.20}$$

Rearranging and neglecting terms of $O\{(\Delta x)^2\}$, gives

$$\tau_{j+1}^n = \tau_j^n + \Delta x\ p_j^n/u\ ,\ j=0(1)J-2, \tag{7.4.21}$$

from which values of τ_j^n, $n=1$, at all interior gridpoints may be found, starting with the boundary value $\bar{\tau}_0^1 = g_0^1$. Application of Equation (7.4.4b), with $n=0$, then gives q_j^1, $j=1(1)J-1$.

Repeated application of Equations (7.4.4a,b) with $n=1,2,3,...,$ gives τ_j^{n+1}, $j=1(1)J-1$, with the additional requirements that q_0^n, q_J^n, $n=1,2,3,..$ are found using (7.4.18) and (7.4.19).

Since the use of (7.4.21) uses only the boundary value at $x=0$, and never uses the given values at x=1, alternate use of (7.4.21) with the rearranged form,

$$\tau_j^n = \tau_{j+1}^n - \Delta x\ p_j^n/u,\ j=J-1(-1)1, \tag{7.4.22}$$

starting with $\bar{\tau}_J^n = g_J^n$, will prevent accumulation of errors at x=1 caused by not using the boundary value there to compute τ_j^n, j=0(1)J-1 from p_j^n, j=0(1)J-1.

If derivative boundary conditions are given, say $\partial\bar{\tau}/\partial x|_J^n = c_J^n$ is known for all n, as well as $\bar{\tau}_0^n$, then only (7.4.21) can be used as $\bar{\tau}_J^n$, required in the use of (7.4.22), is not known. In order to find an approximation to τ_J^n, we may use the backward difference form for the derivative in the boundary condition, giving

$$\frac{\bar{\tau}_J^n - \bar{\tau}_{J-1}^n}{\Delta x} + O\{\Delta x\} = c_J^n .$$

Neglecting terms of O{Δx} and smaller, gives

$$\tau_J^n = \tau_{J-1}^n + \Delta x\ c_J^n . \tag{7.4.23}$$

These values of τ_J^n may then be used to find q_J^n, n=1,2,3,..., which are required in the use of Equation (7.4.4a) with j=J-1, by using the backward difference form of $\partial\bar{\tau}/\partial t|_j^n$ in the equation (7.4.17), then substituting j=J, giving

$$q_J^n = \frac{1}{\Delta t}\{\tau_J^n - \tau_J^{n-1}\} \tag{7.4.24}$$

with an error of O{Δt}.

There are other forms of the finite difference approximations to (7.4.2a), (7.4.3a) which may be used. For example, after the first time step in which Equtions (7.4.2b), (7.4.3b) may be used, central difference forms can be used for the time derivatives in Equations (7.4.2a, 7.4.3a). This gives, for j=1(1)J-1,

$$p_j^{n+1} = p_j^n + c(q_{j+1}^n - q_{j-1}^n) \tag{7.4.25a}$$

$$n=1,2,3,...$$

$$q_j^{n+1} = q_j^{n-1} + c(p_{j+1}^n - p_{j-1}^n) \tag{7.4.25b}$$

The truncation errors for these equations are of $O\{(\Delta t)^2,(\Delta x)^2\}$. Use of the von Neumann stability analysis shows that this method is stable for $c \le 1$.

Another variation is to replace p_j^n in (7.4.2b) by $\frac{1}{2}(p_{j+1}^n + p_{j-1}^n)$ and q_j^n in (7.4.3b) by a similar expression, which yields the following set of difference equations for j=1(1)J-1,

$$p_j^{n+1} = \tfrac{1}{2}(p_{j+1}^n + p_{j-1}^n) + \tfrac{1}{2}c(q_{j+1}^n - q_{j-1}^n) \tag{7.4.26a}$$

$$n=0,1,2,3,...$$

$$q_j^{n+1} = \tfrac{1}{2}(q_{j+1}^n + q_{j-1}^n) + \tfrac{1}{2}c(p_{j+1}^n - p_{j-1}^n) \tag{7.4.26b}$$

which is stable for $c \le 1$. The truncation errors for these equations are of $O\{\Delta t,(\Delta x)^2/\Delta t\}$, and at each time step, the contribution to the discretisation errors of p_j^n, q_j^n are $O\{(\Delta t)^2,(\Delta x)^2\}$.

Yet another system of finite difference equations used to solve the pair of equations (7.4.1) is to replace the approximation to the derivative $\partial\bar{p}/\partial x|_j^n$ in equation (7.4.3a) by $\partial\bar{p}/\partial x|_j^{n+1}$ which introduces an error of $O\{\Delta t\}$. With (7.4.2a) this gives, for $j=1(1)J-1$,

$$\left.\begin{aligned} p_j^{n+1} &= p_j^n + \tfrac{1}{2}c(q_{j+1}^n - q_{j-1}^n) \\ q_j^{n+1} &= q_j^n + \tfrac{1}{2}c(p_{j+1}^{n+1} - p_{j-1}^{n+1}) \end{aligned}\right\} \quad n=0,1,2,3,\ldots \qquad (7.4.27a)$$

with truncation errors of $O\{\Delta t,(\Delta x)^2\}$. The stability requirement for this scheme is $c \le 2$.

7.5 Shock Waves

When dealing with shocks or discontinuities we may not know where the shock is, so we are unable to put the usual Eulerian grid over the space we are considering and still consider the shock; that is, the computation would go on as though the schock were not there.

The best method of handling the shock wave is to make one of the grid lines coincide with the shock at all time steps and calculate the effect of the shock on the neighbouring spatial grid points by using the Rankine-Hugoniot shock conditions. This method can be applied if the Lagrangian coordinate system, which follows the motion of the particles, is used.

Another method for treating shocks is to introduce a pseudo viscosity term. The idea behind this is to add an artificial viscosity term into the equation we are looking at, in such a way that there results a smooth shock transition extending over a small number of spatial grid spacings (see Richtmeyer & Morton, 1967).

8. DIFFERENT GRID AND COORDINATE SYSTEMS

In the previous sections the transport and wave equations were differenced on a uniformly spaced grid in the rectangular cartesian coordinate system. However, this may not be the most appropriate system to use.

While a finite difference approximation developed on a uniform grid may be the simplest and formally the most accurate scheme to use, it can prove deficient in some respects. For example, when computing the flow around solid bodies or solving other boundary layer problems, there are quite large velocity gradients in certain regions of the flow near the body. If there are not enough grid points across the boundary layer to resolve the velocity, the numerical solution may be a very poor approximation to the true solution in this layer. To increase the resolution in this region we could still use a uniform grid but with a much smaller grid spacing over the complete area of flow. However, this may lead to unacceptably large computational times.

To avoid excessive numbers of calculations, while still obtaining a high degree of resolution in particular regions, a non-uniform grid system may be used with a fine grid in the area of interest and a coarse grid over the rest of the flow field. Different types of non-uniform grids may be used. For instance, it is possible to have a coarse grid which changes abruptly to a fine grid in the region of interest, as described in Section 8.1, or it is possible to have a grid in which there is a gradual reduction from a large to a small grid size. This case is dealt with in Section 8.2. Grids need not be rectangular or even regular; specially shaped grids of variable size may be used in problems involving irregular boundaries (for example, see Thoman and Szewczyk, 1969).

While there are considerable benefits to be gained by having a non-uniform grid, these may be offset by other problems introduced by the change in grid size. There may be wave distortion due to a phase change at the interface of two grids of different size or shape or a wave reflection off the interface. It may also be difficult to match the finite difference approximations at the interface of different grids. The introduction of a change in grid spacing also may adversely affect the formal truncation error and the stability of the system.

Because of these difficulties, a coordinate transformation along one or more of the coordinate directions may be preferred to a change in grid shape or size in order to obtain higher resolution in a particular area of flow. It is often desirable to make a change of the complete coordinate system if a problem can be handled more easily in another system. By aligning a coordinate surface along a boundary, it is easier to apply boundary conditions than to use the methods described in Section 6.5. For example, if a plane region is circular in shape, then the use of polar coordinates is a better choice than rectangular cartesian coordinates. However, changing the coordinate system or making a coordinate transformation can also lead to complications which are outlined later.

8.1 Abrupt Changes in Grid Spacing

The simplest non-uniform grid scheme has a large constant grid spacing over most of the flow field with an abrupt change to a smaller constant grid spacing to cover the region of particular interest. In making this change of grid size it is desirable that the second order spatial accuracy, which occurs with central differences on a uniform grid, should be retained; a decrease in computational accuracy due to the change in grid size may more than offset the gain in accuracy due to the use of a finer grid.

Crowder and Dalton (1971) considered the case where the grid spacing is decreased from a constant value $\Delta x_{J-1} = x_J - x_{J-1}$ to the smaller value $\Delta x_J = x_{J+1} - x_J$ at a point J, as in Figure 8.1, in order to see how the order of the truncation error of the approximations to the spatial derivatives $\partial\bar{\tau}/\partial x$ and $\partial^2\bar{\tau}/\partial x^2$ of a function $\bar{\tau}$ was affected by this change. Expanding the function $\bar{\tau}$ in a Taylor series forward and backward from the J-th grid point, at a particular time level, the superscript for which is omitted in the following, gives

$$\bar{\tau}_{J+1} = \bar{\tau}_J + \Delta x_J \left.\frac{\partial\bar{\tau}}{\partial x}\right|_J + \frac{(\Delta x_J)^2}{2!}\left.\frac{\partial^2\bar{\tau}}{\partial x^2}\right|_J + \frac{(\Delta x_J)^3}{3!}\left.\frac{\partial^3\bar{\tau}}{\partial x^3}\right|_J + O\{(\Delta x_J)^4\}, \tag{8.1.1}$$

$$\bar{\tau}_{J-1} = \bar{\tau}_J - \Delta x_{J-1}\left.\frac{\partial\bar{\tau}}{\partial x}\right|_J + \frac{(\Delta x_{J-1})^2}{2!}\left.\frac{\partial^2\bar{\tau}}{\partial x^2}\right|_J - \frac{(\Delta x_{J-1})^3}{3!}\left.\frac{\partial^3\bar{\tau}}{\partial x^3}\right|_J + O\{(\Delta x_{J-1})^4\}. \tag{8.1.2}$$

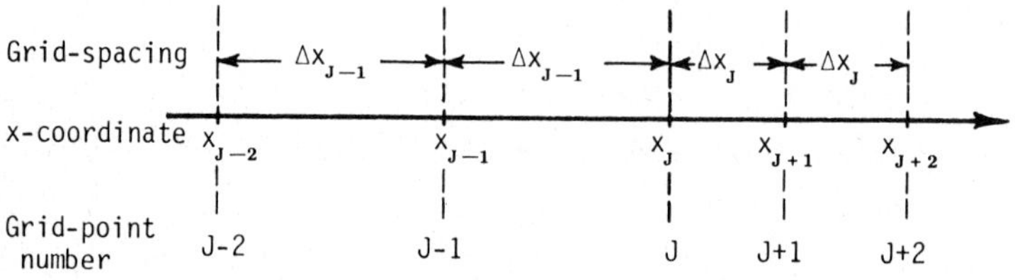

Figure 8.1: A sudden change in grid-spacing at the J-th gridpoint.

Eliminating the terms containing $\partial^2\bar{\tau}/\partial x^2|_J$, by multiplying Equation (8.1.1) by $(\Delta x_{J-1})^2$ and subtracting Equation (8.1.2) multiplied by $(\Delta x_J)^2$, yields

$$\left.\frac{\partial\bar{\tau}}{\partial x}\right|_J = \frac{\bar{\tau}_{J+1} + [(\nu)^2 - 1]\bar{\tau}_J - (\nu)^2\bar{\tau}_{J-1}}{\Delta x_J(\nu + 1)} - \frac{\Delta x_{J-1}\Delta x_J}{6}\left.\frac{\partial^3\bar{\tau}}{\partial x^3}\right|_J + O\{(\Delta x_{J-1})^3\}, \tag{8.1.3}$$

where $\nu = \Delta x_J/\Delta x_{J-1}$. The quasi-central difference approximation to the first derivative at the point where the grid spacing changes is given by the first term on the right hand side of (8.1.3), and has a truncation error of $O\{\Delta x_{J-1}\Delta x_J, (\Delta x_{J-1})^3\}$.

At all other gridpoints the standard central finite difference form (2.3.18), with $\Delta x = \Delta x_{J-1}$ if $j < J$ or $\Delta x = \Delta x_J$ if $j > J$, is used.

Similarly, by eliminating the terms with $\partial\bar{\tau}/\partial x|_J$ from Equations (8.1.1) and (8.1.2) gives

$$\left.\frac{\partial^2\bar{\tau}}{\partial x^2}\right|_J = \frac{\bar{\tau}_{J+1} - (\nu + 1)\bar{\tau}_J + \nu\,\bar{\tau}_{J-1}}{\frac{1}{2}\Delta x_{J-1}\Delta x_J(\nu + 1)} + \frac{\Delta x_{J-1}}{3}(1-\nu)\left.\frac{\partial^3\bar{\tau}}{\partial x^3}\right|_J + O\{(\Delta x_{J-1})^2\}. \tag{8.1.4}$$

Therefore the quasi-central difference approximation to the second derivative at x_J is given by the first term on the right side of (8.1.4), and has a truncation error of $O\{\frac{1}{3}\Delta x_{J-1}(1-\nu), (\Delta x_{J-1})^2\}$. At the other gridpoints the standard finite difference form (2.3.19) is used.

Note that the choice $\Delta x_{J-1} = \Delta x_J = \Delta x$ in (8.1.3) and (8.1.4) yield the usual central difference forms (2.3.18) and (2.3.19) for a uniform grid spacing.

Because the finite difference approximation based on Equation (8.1.4) has a truncation error of $O\{\frac{1}{3}\Delta x_{J-1}(1-\nu),(\Delta x_{J-1})^2\}$, any finite difference methods for second order partial differential equations based upon this approximation will involve a truncation error of $O\{(\Delta x_{J-1})^2\}$ at the interface where the grid size changes, only if

$$\Delta x_{J-1}(1-\nu) = \kappa(\Delta x_{J-1})^2, \quad \kappa = O\{1\}, \tag{8.1.5}$$

so that

$$\Delta x_J = \Delta x_{J-1}(1-\kappa\Delta x_{J-1}). \tag{8.1.6}$$

Therefore, the truncation error is worse than that with a uniform grid, rather than an improvement, unless the change in grid spacing is small. Crowder and Dalton (1971) compared results obtained using six different grids and concluded that a uniform square mesh was the best grid to use for the Poiseuille flow problem they were studying. This was contrary to their original hypothesis that using non-uniform, non-square grids would give a minimum overall error with their finite difference methods. Commenting on these findings, Blottner and Roache (1971) pointed out that such results are not generally true and that they probably occurred because change of grid spacing from 0.05 to 0.10 by Crowder and Dalton did not satisfy condition (8.1.5). Some applications where variable grids give better results than a constant grid are described by Moretti and Salas (1970) and Anthes (1970). MacCormack (1971) and Chavez and Richards (1970) also indicate that the loss of accuracy of the solution in a finite difference approximation in which grid size varies is not usually as bad as may be indicated by the formal truncation error, particularly if an isolated change in mesh size is used.

When a variable grid spacing is used, one of the main problems is to obtain consistent difference equations at interfaces in the grid without introducing instability. For explicit finite difference approximations, in a one-dimensional problem, stability conditions such as

$$s = \alpha\Delta t/(\Delta x)^2 \le \tfrac{1}{2}$$

for the diffusion equation, and the Courant-Friedrichs-Lewy condition

$$c = u\Delta t/\Delta x \le 1$$

for the convection equation, are needed. The first of these requires

$$\Delta t \le (\Delta x)^2/2\alpha, \tag{8.1.7}$$

while the second needs

$$\Delta t \le \Delta x/u. \tag{8.1.8}$$

From Equations (8.1.7) and (8.1.8) we can see that if the grid size Δx is reduced, then the time step Δt may also have to be reduced if the method is to remain stable. If Δx_J is the smallest grid-spacing used then either $\Delta t \le (\Delta x_J)^2/2\alpha$ or $\Delta t \le \Delta x_J/u$ are required. Consequently, the number of computations required may increase significantly. Since such restrictive conditions do not apply to implicit methods, these are more useful in such cases, although it must be remembered that truncation errors do increase as Δt increases, thereby reducing accuracy even with implicit methods.

Osher (1970) considered implicit difference equations appropriate for the solution of the diffusion equation (2.1.1) on a grid with an abrupt change in grid size and showed that they were stable over the entire spatial domain, especially at the interface. Ciment (1971) studied the solution of the convection equation (5.1.2) across a change in grid size and showed that certain difference schemes, such as the Lax-Wendroff scheme (Lax and Wendroff, 1960, 1964) are stable approximations to this equation over the whole domain. Similar stability considerations are discussed by Venit (1973) and Ciment and Sweet (1973).

A further complication arising from the use of a non-uniform grid is the possibility of wave reflection off the interface between the regions of different grid size. Browning et al (1973) state this need not be a problem but they point out another phenomenon, causing wave distortion, which cannot be avoided. If a wave is already well represented in the coarse grid it should propagate through the interface without any difficulty. But any wave which is poorly represented in the coarser grid will change phase speed when it passes through the interface into a finer grid region. If this wave passes back into the coarse grid at some other place there can be a serious interaction of the part which propagated through the fine grid with the part of the wave which remained in the coarse grid.

8.2 Variable Grid Spacing

One way of avoiding some of the problems caused by a sudden change in grid size is to use a gradual change of grid spacing in which the intervals are varied continuously from a coarse grid covering most of the flow field to a fine grid in the region of special interest. One advantage of this gradual reduction in grid size is the decrease in both the computer memory necessary and computational time required to obtain improved solutions over the same area as using a small grid spacing everywhere.

In order to have the same formal accuracy as using centred differences on a uniform grid, the change in grid-spacing at each gridpoint must be small. Therefore, in order to change the spacing by a large amount it is necessary to use the procedure defined by Equations (8.1.3), (8.1.4) and (8.1.6) at *each* gridpoint.

Consider the interval $0 \le x \le 1$ divided into intervals of decreasing length. The gridpoints are numbered $j=0,1,2,3,..$, the value of x at the j-th gridpoint being denoted x_j and the interval between x_j and x_{j+1} denoted Δx_j. Now consider $\bar{\tau}(x)$ on the domain $\langle 0,1\rangle$ with $\bar{\tau}_j = \bar{\tau}(x_j)$. Then by (8.1.3) and (8.1.4) we have:

$$\left.\frac{\partial\bar{\tau}}{\partial x}\right|_j = \frac{\bar{\tau}_{j+1}+[(\nu_j)^2-1]\bar{\tau}_j-(\nu_j)^2\bar{\tau}_{j-1}}{\Delta x_j(\nu_j+1)} - \frac{1}{6}\nu_j(\Delta x_{j-1})^2\left.\frac{\partial^3\bar{\tau}}{\partial x^3}\right|_j + $$

$$+\frac{1}{24}\nu_j(1-\nu_j)(\Delta x_{j-1})^3\left.\frac{\partial^4\bar{\tau}}{\partial x^4}\right|_j + \ldots \qquad (8.2.1)$$

$$\left.\frac{\partial^2\bar{\tau}}{\partial x^2}\right|_j = \frac{\bar{\tau}_{j+1}-(\nu_j+1)\bar{\tau}_j+\nu_j\bar{\tau}_{j-1}}{\frac{1}{2}\Delta x_j\,\Delta x_{j-1}(\nu_j+1)} + \frac{1}{3}(1-\nu_j)\Delta x_{j-1}\left.\frac{\partial^3\bar{\tau}}{\partial x^3}\right|_j -$$

$$-\frac{1}{12}(1-\nu_j+\nu_j^2)(\Delta x_{j-1})^2\left.\frac{\partial^4\bar{\tau}}{\partial x^4}\right|_j + \ldots \qquad (8.2.2)$$

where $\nu_j = \Delta x_j / \Delta x_{j-1}$. By choosing $\Delta x_{j-1} - \Delta x_j = O\{(\Delta x_{j-1})^2\}$ the first terms on the right hand side of Equations (8.2.1) and (8.2.2) approximate the derivatives on the left hand side with a truncation error of order $O\{(\Delta x_{j-1})^2\}$ - compare with Equation (8.1.6).

In this way Sundqvist and Veronis (1970) solved the boundary layer problem proposed by Stommel (1948), using a non-uniform grid with spacings

$$\Delta x_0 = h, \; \Delta x_j = \Delta x_{j-1}(1-\kappa\Delta x_{j-1}), \tag{8.2.3}$$

where κ is a constant of $O\{1\}$ - compare Equation (8.1.6). They compared their results with those obtained using a uniform grid, for which the solution exhibited large variations from the exact solution with an oscillatory behaviour near the boundary. It was found that the non-uniform grid solution was much more accurate near the boundary and no oscillations were evident. While this method of choosing the grid improves resolution near the boundary, overall it still requires a large number of grid points to significantly reduce the grid spacing there if κ is small.

In their analysis of the formation of a shock in a viscous fluid Moretti and Salas (1970) used both constant and variable size grid spacings. With the constant grid size their solutions exhibited oscillations while with a variable grid these oscillations were no longer present and the solution was more accurate.

There are many geophysical situations in which variable grids are used in order to get a higher resolution in certain areas. For example, ocean models may require higher resolution over continental boundary regions than over other areas of the ocean because of the rapidly changing shore line and associated shallow waters. Variable grids are also useful in modelling atmospheric phenomena. Numerical experiments by Anthes (1970) and Harrison (1973) indicate that a variable grid may give superior results to a constant grid when forecasting the movement and development of tropical storms.

We now consider the application of variable spaced grids to finding finite difference approximations to the solution of the one dimensional diffusion equation (2.1.1).

VARIABLE GRID PRODUCED BY THE κ-METHOD

As seen in (8.1.5), in order to set up a grid in which there is a gradual change in grid size, we use the relation

$$\Delta x_j = \Delta x_{j-1}(1+\kappa\Delta x_{j-1}). \tag{8.2.4}$$

By varying the value of κ, different types of non-uniform grids are obtained. If $\kappa > 0$, a non-uniform grid which starts with a small spacing and gradually changes to a large spacing is obtained. If $\kappa < 0$, the reverse is obtained - the grid gradually changes from coarse to fine. Figure 8.2 shows the different grid systems obtained by choosing $\kappa = 0$ (uniform), 0.5, 1, 3, 5 with the number of gridpoints J=10.

$\kappa = 0.5$ does not give a much better resolution near $x = 0$ than a uniform grid. However, for $\kappa = 5$, good resolution is obtained with x close to 0. A very fine grid near $x = 0$ is achieved as the number of gridpoints is increased.

Consider the one-dimensional diffusion equation (2.1.1) with $\alpha = 0.01$, with the initial conditions

$$\bar{\tau}(x,0) = 0, \; 0 \leq x \leq 1,$$

and the boundary conditions

$$\left.\begin{aligned} \bar{\tau}(0,t) &= 1.0, \; t > 0, \\ \bar{\tau}(1,t) &= \operatorname{erfc}(1/(2\sqrt{\alpha t})), \; t > 0. \end{aligned}\right\} \tag{8.2.5}$$

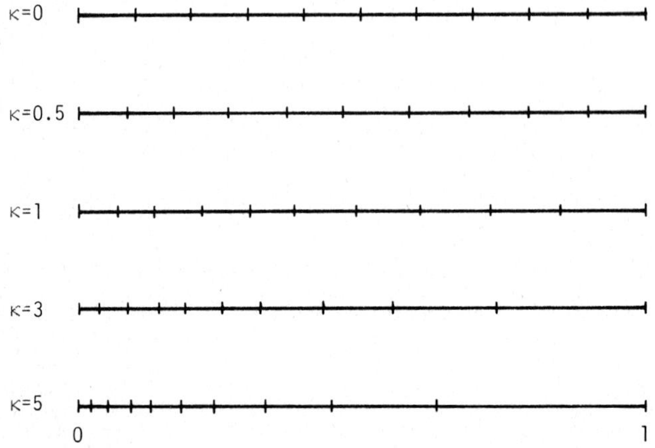

Figure 8.2 : The variable grid obtained using various values of κ with J=10.

The exact solution to this problem is

$$\bar{\tau}(x,t) = \operatorname{erfc}(x/(2\sqrt{\alpha t})), \; t > 0. \tag{8.2.6}$$

Evaluated at the (j,n) gridpoint, the one-dimensional diffusion equation becomes

$$\left.\frac{\partial \bar{\tau}}{\partial t}\right|_j^n = \alpha \left.\frac{\partial^2 \tau}{\partial x^2}\right|_j^n . \tag{8.2.7}$$

Taking the finite difference approximation based on a non-uniform grid for the second space derivative, and using the forward-time finite difference approximation for the time derivative, yields

$$\frac{\bar{\tau}_j^{n+1} - \bar{\tau}_j^n}{\Delta t} + O\{(\Delta t)\} = \alpha \left\{ \frac{\bar{\tau}_{j+1}^n - (\nu_j + 1)\bar{\tau}_j^n + \nu_j \bar{\tau}_{j-1}^n}{\frac{1}{2}\Delta x_{j-1} \Delta x_j (\nu_j + 1)} + \frac{\Delta x_{j-1}(1-\nu_j)}{3} \left.\frac{\partial^3 \bar{\tau}}{\partial x^3}\right|_j + O\{(\Delta x_{j-1})^2\} \right\}. \tag{8.2.8}$$

After dropping the terms of $O\{\Delta t, \Delta x_{j-1}(1-\nu_j), (\Delta x_{j-1})^2\}$ in the above equation and rearranging the terms, we obtain the finite difference equation

$$\tau_j^{n+1} = s_j \nu_j \tau_{j-1}^n + \{1-s_j(\nu_j+1)\}\tau_j^n + s_j \tau_{j+1}^n, \tag{8.2.9}$$

where

$$s_j = \frac{\alpha\Delta t}{\frac{1}{2}(1+\nu_j)\Delta x_{j-1}\Delta x_j}. \tag{8.2.10}$$

With the given initial and boundary conditions (8.2.5) this gives approximate values τ_j^n to the true solution $\bar{\tau}(x_j, t_n)$ of the one-dimensional diffusion equation (2.1.1). Obviously, the use of Equation (8.2.9) involves a truncation error of $O\{\Delta t, \Delta x_{j-1}(1-\nu_j), (\Delta x_{j-1})^2\}$ and a contribution to the discretisation error of $O\{(\Delta t)^2, \Delta t\Delta x_{j-1}(1-\nu_j), \Delta t(\Delta x_{j-1})^2\}$ at each time step. For convenience, we shall refer to the use of Equation (8.2.9) as the Forward-Time Variable-Space (FTVS) method. This reduces to the FTCS (Forward-Time Central-Space) equation when $\Delta x_j = \Delta x_{j-1} = \Delta x$, since then $\nu_j = 1$ for all j.

When using Equation (8.2.9) care must be taken to ensure that Δt is small enough for the equation to be stable. Based on the fact that the FTCS method is stable if and only if $s \le \frac{1}{2}$, where $s = \alpha\Delta t/(\Delta x)^2$, it follows that the FTVS method is certainly stable if $\Delta t \le (\Delta x_{min})^2/2\alpha$. For instance, with $\alpha = 0.01$ and $\Delta x_{min} = 0.01$, choice of $\Delta t = 0.005$ ensures stability of the FTVS method.

The non-uniform grid most appropriate to this problem must now be determined. Since the finite difference equation (8.2.9) involves a truncation error

$$E_j = \frac{1}{3}\Delta x_{j-1}(1-\nu_j)\left.\frac{\partial^3\bar{\tau}}{\partial x^3}\right|_j + O\{\Delta t, (\Delta x_{j-1})^2\}, \tag{8.2.11}$$

in order to improve accuracy either the coefficient $\frac{1}{3}\Delta x_{j-1}(1-\nu_j)$ or the derivative $\partial^3\bar{\tau}/\partial x^3|_j$ need to be small. If $\partial^3\bar{\tau}/\partial x^3$ is large, the coefficient must be small in order to reduce the error introduced by this method. On the other hand, if $\partial^3\bar{\tau}/\partial x^3$ is small the coefficient $\frac{1}{3}\Delta x_{j-1}(1-\nu_j)$ need not be small. Investigating the value of $\partial^3\bar{\tau}/\partial x^3$ in our problem, we have

$$\text{at } x = 0.1,\ t = 1,\ \frac{\partial^3\bar{\tau}}{\partial x^3} = 1.10 \times 10^2,$$

$$\text{at } x = 0.9,\ t = 1,\ \frac{\partial^3\bar{\tau}}{\partial x^3} = 1.79 \times 10^{-5}.$$

It is obvious from the above that to reduce the error, $\frac{1}{3}\Delta x_{j-1}(1-\nu_j)$ has to be small if x is near zero.

This means that a very fine grid size is required at x near to 0, with a coarse grid used near x = 1 for consistent accuracy.

To compare the accuracy of the methods on this non-uniform grid, a point at which the gradient was steep is chosen for comparison, for example at x = 0.08. Since the non-uniform grid used may not have a grid point exactly at x = 0.08, interpolation must be used to find the approximate value of τ at this point. Exponential interpolation involving values from the three closest grid point to x = 0.08 was used.

TABLE 8.1

Errors* in the use of the FTVS method used to solve problem (8.2.5) at time t = 1.

κ	J=20, Δt=0.01		J=40, Δt=0.001	
	Average error	Error at x=0.08	Average error	Error at x=0.08
0	8.69	-28.72	2.02	-0.92
0.4	6.55	-12.89	1.46	-1.46
0.6	5.65	-7.40	1.23	-0.82
0.8	5.05	-3.79	1.04	-0.12**
1.0	4.44	-1.94**	0.88	-1.23
2.0	2.98	-3.77	0.38	-0.64
2.5	2.85+	-6.60	0.32+	-0.46
3.0	2.94	-4.28	0.36	-0.35
4.0	3.64	-4.75	0.57	-0.28
5.0	4.71	-5.39	0.84	-0.20

*The errors have all been multiplied by 10^4. Thus the actual error at x=0.08 when κ = 1, J = 20, Δt = 0.01 is -1.94×10^{-4}.

The errors are defined by

$$\text{Average error} = \{\sum_{j=1}^{J-1} |\bar{\tau}_j - \tau_j|\}/(J-1)$$

$$\text{Error (at } x = 0.8) = \bar{\tau}_j - \tau_j .$$

Note that the average error is always smaller for the variable grid with $0 < \kappa < 5$. At x = 0.08, the error for J = 20 is always smaller for the non-uniform grid; for J = 40, the error is generally smaller for the non-uniform grid.

**For the point x = 0.08, when J = 20 the minimum error of -1.94×10^{-4} was found at κ = 1, and when J = 40 the minimum error of -1.2×10^{-5} for J = 40 with κ = 2.5.

+The average error at t = 1 was found to be a minimum of 2.85×10^{-4} for J = 20 with κ = 2.5, and a minimum of 3.2×10^{-5} for J = 40 with κ = 2.5.

VARIABLE GRID PRODUCED BY A MAPPING FUNCTION

In the previous problem the solution had a steep gradient and relatively large values of the spatial derivative at only one end of the interval $0 \leq x \leq 1$. We now consider a one-dimensional initial value problem which has a solution with a steep gradient at each end of the unit interval: this is the one-dimensional diffusion equation (2.1.1) with initial and boundary conditions (2.1.3) described in Section 2.1. The exact solution for this heated rod problem is given by Equation (2.1.4).

The κ-method can be used to set up a suitable grid by dividing the spatial domain into two half regions, $0 \leq x \leq \frac{1}{2}$ and $\frac{1}{2} \leq x \leq 1$. A variable grid is set up on the first half region in the manner described previously, and a mirror image of this is taken for the grid on the second half.

However, there is an alternative way of producing a suitable grid. The required grid is obtained by mapping a uniform grid in some variable, η, onto a non-uniform grid in the x-variable, using a suitable "mapping function"

$$x = F(\eta).$$

One such function is

$$x = \sin^2(\pi\eta/2). \tag{8.2.12}$$

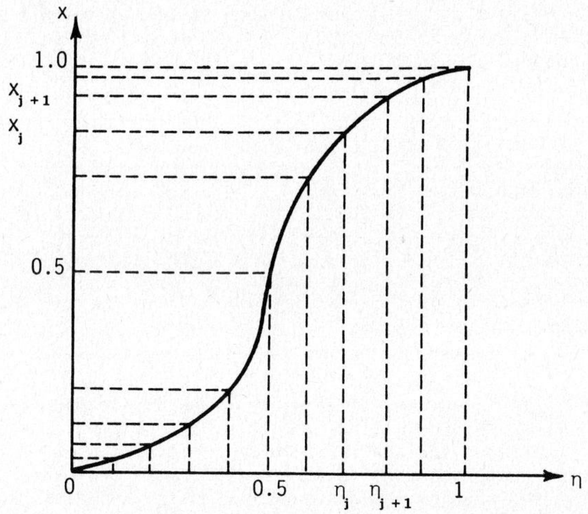

Figure 8.3 : The mapping of $\eta_j = j/J$ onto x_j using the mapping function $x = sin^2(\pi\eta/2)$.

Figure 8.3 shows how the variable grid is obtained for the x-coordinate, for J=10. Clearly

$$\Delta x_j = x_{j+1} - x_j, \; j=0(1)J-1, \tag{8.2.13a}$$

where

$$x_j = \sin^2(\pi j/2J). \tag{8.2.13b}$$

It follows that

$$\Delta x_j = \sin(\pi/J).\sin\{(2j+1)\pi/J\}, \; j=0(1)J-1. \tag{8.2.14}$$

Figure 8.4 shows the variable grids obtained using (8.2.14) with J=10 and J=20.

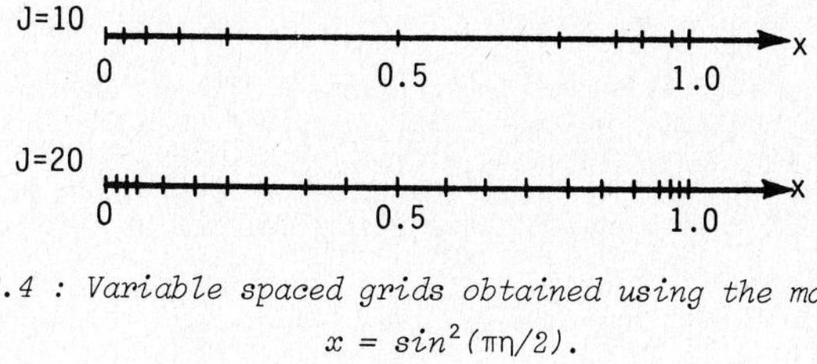

Figure 8.4 : Variable spaced grids obtained using the mapping $x = sin^2(\pi\eta/2)$.

The finite difference approximation to $\bar{\tau}(x_j, t_{n+1})$ may then be found using the FTVS equation (8.2.9). Table 8.2 shows the results obtained at t=1,2,5 for this method with the variable grid defined by (8.2.13b) and those obtained using a uniform grid and the FTCS equation.

TABLE 8.2

Results obtained using the FTVS method to the heated rod problem (2.1.2)-(2.1.4) using the mapping (8.2.12) compared to the FTCS method, with J=20, Δt=0.01. The asterisk indicates the more accurate result.

	Uniform grid		Variable grid	
t	Average error	Error at x=0.08	Average error	Error at x=0.08
1	0.0017389	0.0028722*	0.0014473*	0.0031201
2	0.0012161*	0.0012375	0.0010359	0.0009701*
3	0.0004895	0.0002119*	0.0001655*	0.0002255

These results for the variable grid are generally better across one time level than those for the uniform grid, but at the point x = 0.08 the error is generally greater for the variable grid. This may be due to the fact that interpolation had to be used to find the value of τ at x=0.08, and methods of interpolation produce more accurate results when equispaced data are used than when the original data is not equally spaced. It may also be due to the fact that the mapping was equivalent to using a variable value of κ ranging from 0.5 to 50, thereby violating the criterion that κ should be O{1}.

8.3 Stretched Coordinates

Increasing resolution in certain areas may be achieved with a coordinate stretching transformation, such as an exponential stretch. For example, Pao and Daugherty (1969) transformed the cartesian coordinates (x,y) to the coordinates (χ,γ) by means of the relations

$$\left.\begin{aligned} x &= \chi \\ y &= b\{\exp(a\gamma)-1\}, \end{aligned}\right\} \qquad (8.3.1)$$

where a and b are arbitrary constants.

While such stretching transformations have the same purpose as that of an expanding grid, in that both reduce the number of grid points in certain regions, the two methods are fundamentally different. When the original equations are differenced on an expanding grid, there is a deterioration in the formal truncation error, whereas when the transformed equations are differenced on a uniform grid there is no such loss in formal accuracy.

However, while the formal truncation error is not degraded by coordinate stretching, other problems do arise. Stability and convergence may be affected by the introduction of new terms in the transformed equations which can lead to the computation time per step being increased. There is also the possibility of wave distortion due to phase changes and damping.

Kalnay de Rivas (1972) investigated coordinate stretching and found the following.

Consider a function $\bar{\tau}(x)$ defined on a non-uniform grid in the x-direction as in Figure 8.5. Suppose grid intervals are varied by defining a stretched coordinate χ where $x = f(\chi)$ in such a way that grid intervals $\Delta\chi$ are constant. If the function being studied is defined on the region $0 \le x \le 1$ with a fine grid required near $x = 0$, then $f(\chi)$ should have the properties:

(i) $df/d\chi$ should be finite over the whole interval - if it becomes infinite at some point, then there is poor resolution near that point;

(ii) $df/d\chi$ must be smaller at $x = 0$ than elsewhere in $0 < x \le 1$, which ensures high resolution near the point $x = 0$, but $df/d\chi$ should be non-zero at $x = 0$.

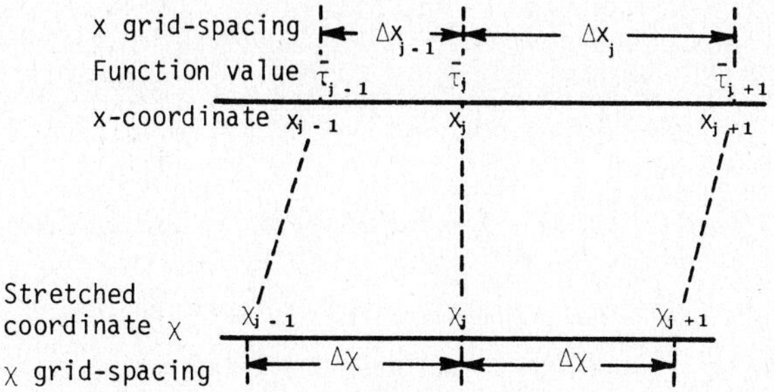

Figure 8.5 : Non-uniform grid defined through the use of a stretched coordinate

By taking Taylor series expansions about the point x_j to obtain the approximations to the first and second order derivatives, it was found that the truncation error is second order in $\Delta\chi$ even near the boundary $x=0$, where the errors may be first order with respect to Δx_j.

This may be seen as follows. Since

$$\begin{aligned}\Delta x_j &= x_{j+1} - x_j \\ &= f(j\Delta\chi+\Delta\chi) - f(j\Delta\chi),\end{aligned} \tag{8.3.2}$$

then application of Taylor series expansion of the first function about $x_j = f(j\Delta\chi)$ gives

$$\Delta x_j = \Delta\chi\{\frac{df}{d\chi}\Big|_j + \frac{\Delta\chi}{2}\frac{d^2f}{d\chi^2}\Big|_j + \frac{(\Delta\chi)^2}{6}\frac{d^3f}{d\chi^3}\Big|_j + \frac{(\Delta\chi)^3}{24}\frac{d^4f}{d\chi^4}\Big|_j + \ldots\}. \tag{8.3.3}$$

Similarly we find

$$\Delta x_{j-1} = \Delta\chi\{\frac{df}{d\chi}\Big|_j - \frac{\Delta\chi}{2}\frac{d^2f}{d\chi^2}\Big|_j + \frac{(\Delta\chi)^2}{6}\frac{d^3f}{d\chi^3} - \frac{(\Delta\chi)^3}{24}\frac{d^4f}{d\chi^4}\Big|_j + \ldots\}. \tag{8.3.4}$$

Substitution into Equation (8.2.2), written in terms of Δx_{j-1} and Δx_j, namely

$$\left.\frac{\partial^2\bar{\tau}}{\partial x^2}\right|_j = \frac{\bar{\tau}_{j+1}-\bar{\tau}_j}{\frac{1}{2}\Delta x_j(\Delta x_j+\Delta x_{j-1})} - \frac{\bar{\tau}_j-\bar{\tau}_{j-1}}{\frac{1}{2}\Delta x_{j-1}(\Delta x_j+\Delta x_{j-1})}$$

$$-\frac{\Delta x_j-\Delta x_{j-1}}{3}\left.\frac{\partial^3\tau}{\Delta x^3}\right|_j - \frac{1}{12}[(\Delta x_j)^2-\Delta x_j\Delta x_{j-1}+(\Delta x_{j-1})^2]\left.\frac{\partial^4\tau}{\partial x^4}\right|_j + \ldots \tag{8.3.5}$$

gives

$$\left.\frac{\partial^2\bar{\tau}}{\partial x^2}\right|_j = \frac{(\bar{\tau}_{j+1}-\bar{\tau}_j)/\left.\frac{df}{d\chi}\right|_j - (\bar{\tau}_j-\bar{\tau}_{j-1})/\left.\frac{df}{d\chi}\right|_{j-1}}{(\Delta\chi)^2[\left.\frac{df}{d\chi}\right|_j + \frac{5}{24}(\Delta\chi)^2\left.\frac{d^3f}{d\chi^3}\right|_j]}$$

$$-\frac{(\Delta\chi)^2/12}{1+\frac{5}{24}(\Delta\chi)^2\left.\frac{d^3f}{d\chi^3}\right|_j/\left.\frac{df}{d\chi}\right|_j}\left\{\begin{matrix}4\left.\frac{\partial^3\bar{\tau}}{\partial x^3}\right|_j[\left.\frac{d^2f}{d\chi^2}\right|_j + \frac{(\Delta\chi)^2}{12}\{\left.\frac{d^4f}{d\chi^4}\right|_j + \frac{5}{2}\left.\frac{d^3f}{d\chi^3}\right|_j\left.\frac{d^2f}{d\chi^2}\right|_j/\left.\frac{df}{d\chi}\right|_j\}] \\ +\left.\frac{\partial^4\bar{\tau}}{\partial x^4}\right|_j[\left.\frac{df}{d\chi}\right|_j + \frac{(\Delta\chi)^2}{4}\{3\left.\frac{d^2f}{d\chi^2}\right|_j + \frac{13}{6}\left.\frac{d^3f}{d\chi^3}\right|_j\left.\frac{df}{d\chi}\right|_j\}]\end{matrix}\right\} + O\{(\Delta\chi)^3\}. \tag{8.3.6}$$

A similar result is obtained for the first spatial derivative, $\partial\bar{\tau}/\partial x|_j$.

This shows that any smooth function $f(\chi)$ which satisfies conditions (i) and (ii) will give an approximation to the first and second spatial derivatives with second order accuracy, because the truncation errors due to the non-uniformity of the grid are of second order in $\Delta\chi$.

Consider the relation

$$x = \sinh(\chi)/\sinh(1), \tag{8.3.7}$$

which satisfies properties (i) and (ii). Transforming the one-dimensional diffusion equation (2.1.1) using (8.3.7) gives

$$\frac{\partial\bar{\tau}}{\partial t} + \frac{\alpha(k_1)^2\sinh\chi}{\cosh^3\chi}\cdot\frac{\partial\bar{\tau}}{\partial x} = \frac{\alpha(k_1)^2}{\cosh^2\chi}\cdot\frac{\partial^2\bar{\tau}}{\partial x^2}, \tag{8.3.8}$$

where $k_1 = \sinh(1)$. Discretising this equation using forward difference in time and centred differences in space, (8.3.8) becomes

$$\bar{\tau}_j^{n+1} = (s_j+c_j)\bar{\tau}_{j-1}^n + (1-2s_j)\bar{\tau}_j^n + (s_j-c_j)\bar{\tau}_{j+1}^n + O\{(\Delta t)^2,\Delta t(\Delta\chi)^2\} \tag{8.3.9}$$

where

$$s_j = \frac{\alpha(k_1)^2\Delta t}{(\Delta\chi)^2\cosh^2(j\Delta\chi)}, \quad c_j = \frac{\alpha(k_1)^2\Delta t\sinh(j\Delta\chi)}{2\Delta\chi\cosh^3(j\Delta\chi)}. \tag{8.3.10}$$

This can be approximated by the explicit finite difference equation

$$\tau_j^{n+1} = (s_j+c_j)\tau_{j-1}^n + (1-2s_j)\tau_j^n + (s_j-c_j)\tau_{j+1}^n \tag{8.3.11}$$

with a contribution of $O\{(\Delta t)^2,\Delta t(\Delta\chi)^2\}$ to the discretisation error, at each time step.

Similarly, the relation

$$x = (e^{\chi}-1)/(e-1), \tag{8.3.12}$$

which also satisfies properties (i) and (ii), transforms the one-dimensional diffusion equation to

$$\frac{\partial\bar{\tau}}{\partial t}+\alpha(k_2)^2\exp(-2\chi)\frac{\partial\bar{\tau}}{\partial\chi}=\alpha(k_2)^2\exp(-2\chi)\frac{\partial^2\bar{\tau}}{\partial\chi^2}, \tag{8.3.13}$$

where $k_2 = e - 1$. Using the forward-time central-space discretisation of (8.3.13) gives the finite difference equation (8.3.11), with

$$s_j = \frac{\alpha(k_2)^2\Delta t}{(\Delta\chi)^2\exp(2j\Delta\chi)},\ c_j = \frac{\alpha(k_2)^2\Delta t}{2\Delta\chi\exp(2j\Delta\chi)}\ . \tag{8.3.14}$$

TABLE 8.3

Comparison of solutions for the one-dimensional diffusion equation (2.1.1) with $\alpha = 0.01$ and initial and boundary conditions (8.2.5). Parameters used were J=20 and Δt=0.001. The exact solution at x=0.1, t=5 is $\bar{\tau}$ = 0.7518297.

	Uniform grid	Transformation $x = \frac{e^{\chi}-1}{e-1}$	Transformation $x = \frac{\sinh(\chi)}{\sinh(1)}$
Numerical solution x=0.1, t=5	0.7516206	0.7517336	0.7518342
Error at x=0.1, t=5	2.1×10^{-4}	-1.0×10^{-4}	0.03×10^{-4}
Average error at t=5	3.5×10^{-4}	1.9×10^{-4}	2.3×10^{-4}

Table 8.3 compares the results obtained using a uniform grid and the transformations (8.3.7) and (8.3.12) to solve the one-dimensional diffusion equation with initial and boundary conditions defined by Equations (8.2.5). Because the gridpoints are different in each case, x = 0.1 was chosen as the point at which comparisons would be made, using exponential interpolation of values at the three nearest gridpoints to x = 0.1. It is clear that the stretching coordinate method with $x = \sinh(\chi)/\sinh(1)$ and $x = (e^{\chi}-1)/(e-1)$ as the transformations, results in lower average errors across the grid at the time t=5 and more accurate approximations at x=0.1 than for the uniform grid.

An interesting comparison between the mapping method to obtain a variable grid described in Section 8.2 and the coordinate stretching technique described in this section, is given in Table 8.4. For the same functional forms, the two methods gave comparable results and both were more accurate than using a uniform grid.

TABLE 8.4

Comparison of the mapping method and coordinate stretching to solve the one-dimensional diffusion equation (2.1.1) with $\alpha = 0.01$ and initial and boundary conditions (8.2.5). Parameters used were J=20, Δt=0.01. Forward time and "centred" space differencing were used in both cases.

MAPPING METHOD			COORDINATE STRETCHING		
Map	Error at x=0.1,t=4	Av.error at t=4	Stretch	Error at x=0.1,t=4	Av.error at t=4
$x=\frac{\sinh(\eta)}{\sinh(1)}$	2.7×10^{-5}	22.7×10^{-5}	$x=\frac{\sinh(\chi)}{\sinh(1)}$	2.7×10^{-5}	23.3×10^{-5}
$x=\frac{e^{\eta}-1}{e-1}$	13.2×10^{-5}	21.8×10^{-5}	$x=\frac{e^{\chi}-1}{e-1}$	-9.8×10^{-5}	19.0×10^{-5}

Another form which has been suggested as a stretching transformation to give improved resolution near $x = 0$ is

$$x = \chi^2. \tag{8.3.15}$$

If a fine grid is required at both $x = 0$ and $x = 1$, a suitable choice of stretched coordinate is

$$x = \sin^2(\pi\chi/2). \tag{8.3.16}$$

Both types of stretched coordinates defined by (8.3.15) and (8.3.16) have been successfully used in two-dimensional numerical models of the atmosphere of Venus (Kalnay de Rivas, 1971). Comparing the percentage errors introduced by the Sundqvist-Veronis grid for Stommel's problem with those introduced by using the stretched coordinate $x = \chi^2$, Kalnay de Rivas found that the latter is more accurate and that there is no tendency for relative errors to grow as $x \to 0$. However, tests carried out by the author indicate that the stretchings (8.3.7) and (8.3.12) were superior to $x = \chi^2$ for all cases tested.

Blottner and Roache (1971) noted that, for the problem considered by Crowder and Dalton (1971), a large improvement in accuracy of the solution is achieved by transforming the independent variables, rather than by changing the grid spacing as in Sections 8.1 and 8.2. In fact, the exact solution to their problem can be obtained by means of a parabolic transformation.

8.4 Change of Coordinate System

While ad hoc grid refinements have proved useful for improving the solution of a flow in special regions such as boundary layers, a coordinate transformation is generally preferred for increasing resolution in a particular area. However. tests using grid refinements based on the κ-method have consistently given better results than using variable grid spacing based on the mapping method or using uniform grids afer applying coordinate stretching. Choice of other mappings or transformations could change this finding. Similarly the preferred method for treating non-rectangular boundaries is to express the governing partial differential equation in a non-rectangular coordinate system in which boundaries are coordinate surfaces. For example, flow over a cylinder of elliptic cross-section is naturally calculated in elliptic cylinder coordinates. Roache (1974) describes a number of examples in which different coordinate systems are used for specific problems.

Coordinate transformations may also be used to align coordinate surfaces along physical boundaries and so simplify the modelling of such boundaries. Thus, problems involving circular boundaries are usually solved more easily in polar coordinates because they avoid the use of complicated finite difference formulae near the curved boundary. Consider the two-dimensional diffusion equation (6.1.3) with $\alpha_x = \alpha_y = \alpha$; in polar coordinates in two-dimensional space this becomes

$$\frac{\partial\bar{\tau}}{\partial t} = \alpha\{\frac{\partial^2\bar{\tau}}{\partial r^2} + \frac{1}{r}\frac{\partial\bar{\tau}}{\partial r} + \frac{1}{r^2}\frac{\partial^2\tau}{\partial\theta^2}\}. \tag{8.4.1}$$

If Equation (8.4.1) is differenced on a uniform rectangular grid in the r-θ plane this corresponds to differencing (6.1.3) on a non-uniform grid consisting of radial and circular lines in the x-y plane.

However, the changing coordinate system may pose problems. Thus, the origin of the polar coordinate system, r=0, becomes a line r=0 in the r-θ plane. Also, a singularity may be introduced at the origin because of the nature of the second and third terms in (8.4.1).

The numerical solution of certain equations, such as the Navier-Stokes equations, is often hindered by the fact that with the cartesian coordinate system and the standard finite difference approximation, we have to deal with complicated boundaries and so it is difficult to use the correct boundary conditions. Gal-Chen (1975) used a special coordinate transformation to change a domain with an irregular boundary into a regular domain; the transformed system had flat boundaries and homogeneous boundary conditions. Again, while there are considerable advantages to be obtained by this transformation, difficulties may arise due to extra terms being obtained in the transformed governing equations. A more serious problem can occur if the transformation causes new singularities to appear in the transformed equations. Since these singularities are not part of the true solution they may considerably affect numerical stability. Nevertheless, the transformation approach to flow over irregular bodies, such as atmospheric flows over the earth's surface, seems preferable to that of retaining the rectangular cartesian system and using special techniques to cope with the lower flow boundary.

Recently there has been a great deal of emphasis on producing boundary-conforming curvilinear coordinate systems for use with finite difference methods when solving partial differential equations on arbitrarily shaped regions. Such systems have a coordinate surface corresponding to each segment of the boundary. Once the partial differential equations are transformed onto such a coordinate system, they can be discretised without the need of interpolation whatever the original boundary shape. Since all computations are carried out on a rectangular grid in the transformed region, general computer programs can be written for the numerical solution of partial differential equations on any physical region.

Boundary-conforming coordinate systems may be generated numerically by determining the coordinates from information about the boundary. One way of doing this is to interpolate algebraically into the interior of the physical region from coordinates on the boundary. There are three ways of doing this: the method of transfinite interpolation is described in Gordon and Hall (1973) and Eriksson (1980), the multisurface method is described by Eiseman (1979) and the two-boundary technique is described by Smith (1981). In general, grid spacing with algebraic generation systems is achieved by means of stretching functions incorporated in the interpolation formulae. This can be achieved interactively by means of computer graphics (see Smith, 1982). Roberts (1982) also describes an automatic method of algebraic coordinate generation.

A second method of generating boundary fitted coordinate systems is to solve a set of partial differential equations with associated boundary conditions which describe the boundaries of the physical region. This technique includes conformal mapping, as described in Ives (1976), Caughey (1978), Davis (1979) and Anderson et.al. (1982), as well as formation of the grid from the solution of general elliptic, parabolic and hyperbolic systems. Warsi (1982) discusses, in general, the construction of a grid by means of differential systems. In particular, elliptic systems are considered by Thompson (1982), Sorenson (1980) and Thomas (1981). A hyperbolic grid generation method is described for airfoil computations by Steger and Chaussee (1980), and grids generated by parabolic systems for fuselage-wing flow calculations are described by Nakamura (1982).

The coordinate system generated in physical space should have grid-lines concentrated in regions where the physical solution is expected to vary rapidly, but the grid-spacing should change only gradually. The spacing of grid-lines, and their orientation, may be controlled by certain adjustable terms in the partial differential equations in the second method described. This is discussed generally by Warsi (1982), while Thompson (1982) describes ways of governing grid-line spacing with elliptic generating systems.

Orthogonality is desirable but not necessary. For instance, non-orthogonal transformations may produce additional cross-derivative terms in the transformed partial differential equations. The successful application of non-orthogonal transformations to the solution of a three-dimensional tidal problem is described in Noye et.al. (1981, 1982). The requirement of orthogonality places certain constraints on the distribution of grid-points. These are considered in Eiseman (1982) who describes the generation of orthogonal coordinate systems by both algebraic methods and the use of hyperbolic partial differential equations. Visbal and Knight (1982) and Christov(1982) consider how control functions can be determined in elliptic systems, in order to achieve orthogonality.

Once a grid has been constructed on the physical region, there are two ways of discretising the partial differential equations to be solved. One method involves the transformation of the equations so the dependent variables are written as functions of the new independent variables defined on a rectangular region (see Noye et.al., 1981, 1982). The differencing is then straightforward, although additional terms and singularities may have to be considered: this approach was considered in the one-dimensional case in Section 8.3 Alternatively, the equations may be differenced in terms of the original independent variables at the grid-points generated in the physical space: the one-dimensional application of this approach was considered in Sections 8.1 and 8.2. The finite difference forms of derivatives on general grids is discussed in detail by Hyman and Larrouturou (1982), who have written a FORTRAN package, called DERMOD, for calculating numerical approximations for the spatial derivates of a function defined only on an arbitrary discrete set of grid-points.

8.5 Non-Rectangular Grids

In the previous sections only finite difference schemes on rectangular grids have been considered. However, other geometrical shapes may have advantages when the boundaries are irregular in shape. In particular, triangular meshes have been sucessfully used in solving differential equations on a planar region, by Winslow (1966), Williamson (1969) and

Thacker (1977). A hexagonal grid has been used for a spherical surface by Sadourny and Morel (1969).

There are only three basic two-dimensional lattices that can be formed from regular polygons, namely, those constructed of squares, equilateral triangles and regular hexagons. Other lattices can be obtained from these by conformal mapping.

In order to show there are no other regular polygons which can be used as the elements of a grid, let θ be the interior angle formed by any two adjacent sides of a regular polygon. In order for a regular polygon to fit with its neighbours to form a grid,

$$N\theta = 360^0, \tag{8.5.1}$$

where N is the number of sides which meet at one vertex, $N \geq 3$. Since equilateral triangles have the smallest interior angle of such polygons, then $\theta \geq 60^0$, so that $N \leq 6$. Considering the cases N = 3, 4, 5, 6, we find

$N = 3 \rightarrow \theta = 120^0 \rightarrow$ hexagonal grid,

$N = 4 \rightarrow \theta = 90^0 \rightarrow$ square grid,

$N = 5 \rightarrow \theta = 72^0 \rightarrow$ *no* regular polygon,

$N = 6 \rightarrow \theta = 60^0 \rightarrow$ triangular grid.

The use of such non-rectangular grids in spatial discretisations will be illustrated by considering the Laplacian $\nabla^2\bar{\tau}$ in the two-dimensional diffusion equation

$$\frac{\partial\bar{\tau}}{\partial t} = \alpha\nabla^2\bar{\tau}, \tag{8.5.2}$$

where

$$\nabla^2\bar{\tau} = \frac{\partial^2\bar{\tau}}{\partial x^2} + \frac{\partial^2\bar{\tau}}{\partial y^2}\,. \tag{8.5.3}$$

First, consider the triangular grid with spacing h, shown in Figure 8.6. The six nearest gridpoints to $(j\Delta x,k\Delta y)$ are $(j\Delta x\pm\Delta x,k\Delta y)$ and $(j\Delta x\pm\Delta x/2,k\Delta y+\Delta y)$. Consider the sum of the values of the dependent variable $\bar{\tau}$ at these six grid points, namely

$$\begin{aligned}[\textstyle\sum\bar{\tau}]_{j,k} = {} & \bar{\tau}_{j+1,k} + \bar{\tau}_{j+\frac{1}{2},k+1} + \bar{\tau}_{j-\frac{1}{2},k+1} + \bar{\tau}_{j-1,k} \\ & + \bar{\tau}_{j-\frac{1}{2},k-1} + \bar{\tau}_{j+\frac{1}{2},k-1}\,.\end{aligned} \tag{8.5.4}$$

Expanding each term in the right hand side of (8.5.4) as Taylor series about the (j,k) gridpoint, gives

$$\begin{aligned}[\textstyle\sum\bar{\tau}]_{j,k} = {} & \bar{\tau}_{j,k} + \Delta x\left.\frac{\partial\bar{\tau}}{\partial x}\right|_{j,k} + \frac{(\Delta x)^2}{2}\left.\frac{\partial^2\bar{\tau}}{\partial x^2}\right|_{j,k} + \frac{(\Delta x)^3}{6}\left.\frac{\partial^3\bar{\tau}}{\partial x^3}\right|_{j,k} + \ldots \\ & + \bar{\tau}_{j,k} + [\frac{\Delta x}{2}\frac{\partial}{\partial x} + \Delta y\frac{\partial}{\partial y}]\bar{\tau}|_{j,k} + \tfrac{1}{2}[\frac{\Delta x}{2}\frac{\partial}{\partial x} + \Delta y\frac{\partial}{\partial y}]^2\bar{\tau}|_{j,k} + \ldots \\ & + \ldots \\ & + \bar{\tau}_{j,k} + [\frac{\Delta x}{2}\frac{\partial}{\partial x} - \Delta y\frac{\partial}{\partial y}]\bar{\tau}|_{j,k} + \tfrac{1}{2}[\frac{\Delta x}{2}\frac{\partial}{\partial x} - \Delta y\frac{\partial}{\partial y}]^2\bar{\tau}|_{j,k} + \ldots\end{aligned} \tag{8.5.5}$$

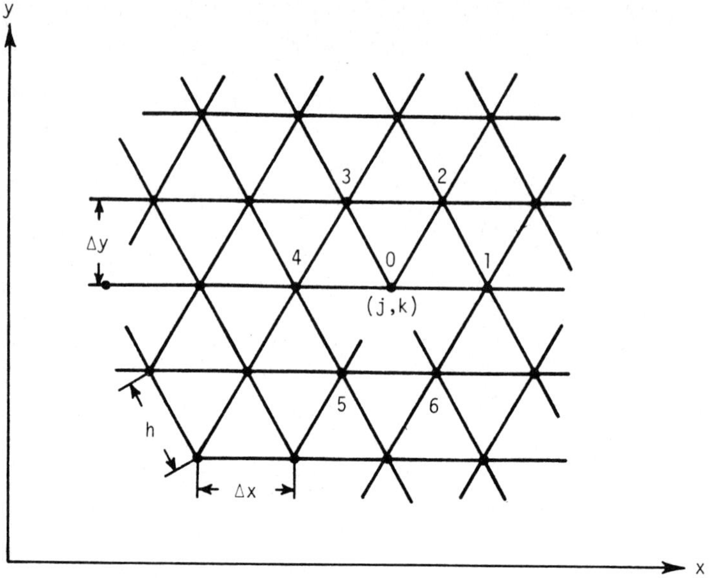

Figure 8.6 : The triangular grid in x-y space

Since $\Delta x = h$, $\Delta y = \sqrt{3}h/2$, simplification of (8.5.5) yields

$$[\sum\bar{\tau}]_{j,k} = 6\bar{\tau}_{j,k} + \frac{3(h)^2}{2}[\nabla^2\bar{\tau}]_{j,k} + \frac{3(h)^2}{32}[\nabla^2\bar{\tau}]_{j,k} + O\{(h)^6\}. \qquad (8.5.6)$$

Rearrangement gives

$$[\nabla^2\bar{\tau}]_{j,k} = \frac{[\sum\bar{\tau}]_{j,k} - 6\bar{\tau}_{j,k}}{3(h)^2} - \frac{(h)^2}{16}[\nabla^4\tau]_{j,k} + O\{(h)^4\}, \qquad (8.5.7)$$

or, using the notation in Figure 8.6,

$$[\nabla^2\bar{\tau}]_0 = \frac{2(\bar{\tau}_1 + \bar{\tau}_2 + \bar{\tau}_3 + \bar{\tau}_4 + \bar{\tau}_5 + \bar{\tau}_6 - 6\bar{\tau}_0)}{3(h)^2} + O\{(h)^2\}. \qquad (8.5.8)$$

This may be used with (say) the forward difference form for the time derivative in (8.5.2) to discretise this equation at the (j,k) gridpoint, giving an alternative FTCS explicit equation to solve the two-dimensional diffusion equation.

Now, consider the hexagonal grid shown in Figure 8.7. Then we may write the Laplacian at the (j,k) gridpoint in terms of values of the dependent variable $\bar{\tau}$ at the points $(j\Delta x+\Delta x, k\Delta y)$, $(j\Delta x-\Delta x/2, k\Delta y\pm\Delta y)$ in the x-y plane. These values are denoted $\bar{\tau}_1$, $\bar{\tau}_2$, $\bar{\tau}_3$. Expanding these terms in Taylor series about the (j,k) gridpoint gives

$$\bar{\tau}_1 = \bar{\tau}_{j+1,k}$$

$$= \bar{\tau}_{j,k} + \Delta x\frac{\partial\bar{\tau}}{\partial x}\bigg|_{j,k} + \frac{(\Delta x)^2}{2}\frac{\partial^2\bar{\tau}}{\partial x^2}\bigg|_{j,k} + \frac{(\Delta x)^3}{6}\frac{\partial^3\bar{\tau}}{\partial x^3}\bigg|_{j,k} + \ldots$$

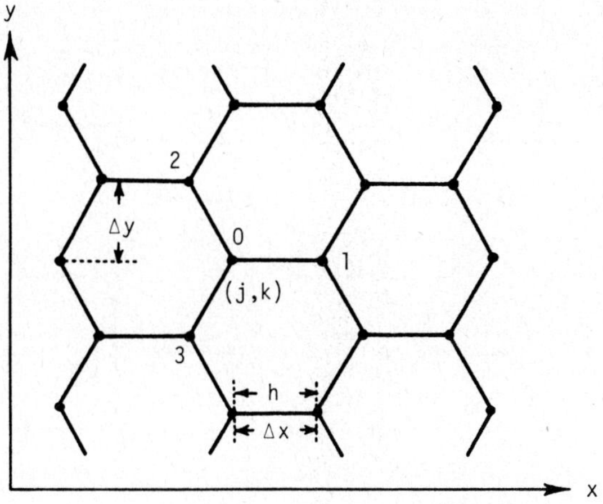

Figure 8.7 : The hexagonal grid in the x-y plane.

$$\bar{\tau}_{2,3} = \bar{\tau}_{j-\frac{1}{2},k\pm 1}$$

$$= \bar{\tau}_{j,k} + [-\frac{\Delta x}{2}\frac{\partial}{\partial x} \pm \Delta y\frac{\partial}{\partial y}]\bar{\tau}|_{j,k} +$$

$$+ \tfrac{1}{2}[\ \frac{(\Delta x)^2}{4}\frac{\partial^2}{\partial x^2} \mp \Delta x\Delta y\frac{\partial^2}{\partial x\partial y} + (\Delta y)^2\frac{\partial^2}{\partial y^2}]\bar{\tau}|_{j,k}$$

$$+ \frac{1}{6}[-\ \frac{(\Delta x)^3}{8}\frac{\partial^3}{\partial x^3} \pm \frac{3(\Delta x)^2\Delta y}{4}\ \frac{\partial^3}{\partial x^2\partial y} - \frac{3\Delta x(\Delta y)^2}{2}\ \frac{\partial^3}{\partial x\partial y^2} \pm$$

$$\pm (\Delta y)^3\frac{\partial^3}{\partial y^3}]\bar{\tau}|_{j,k} + \ldots$$

Therefore, as $\bar{\tau}_{j,k} = \bar{\tau}_0$, and $\Delta x = h$, $\Delta y = \sqrt{3}h/2$,

$$\bar{\tau}_1 + \bar{\tau}_2 + \bar{\tau}_3 = 3\bar{\tau}_0 + \frac{3(h)^2}{4}[\nabla^2\bar{\tau}]_{j,k} + O\{(h)^3\}, \qquad (8.5.9)$$

or

$$[\nabla^2\bar{\tau}]_0 \quad = \frac{4(\bar{\tau}_1 + \bar{\tau}_2 + \bar{\tau}_3 - 3\bar{\tau}_0)}{3(h)^2} + O\{h\}. \qquad (8.5.10)$$

Since the truncation error is O{h}, this is not as good an approximation as (8.5.8) was for the triangular grid.

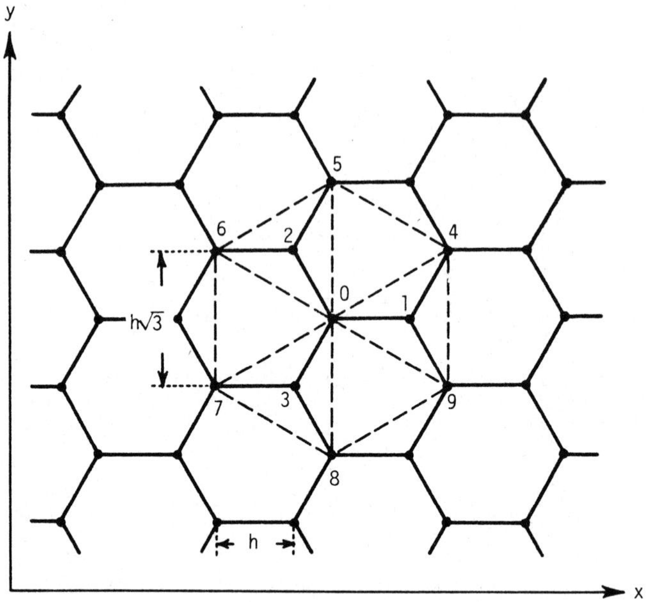

Figure 8.8 : The connection between the hexagonal grid (solid line) and the triangular grid (dashed line)

There is an equivalent approximation to (8.5.8) for the hexagonal grid. This is seen in Figure 8.8, in which the gridpoints labelled 4, 5, 6, 7, 8, 9 in the hexagonal scheme are seen to be equivalent to those labelled 1, 2, 3, 4, 5, 6 in the triangular scheme of Figure 8.6. The only difference is that the h of the triangular scheme must be replaced by $h\sqrt{3}$ in the hexagonal scheme. Therefore the equivalent equation to (8.5.8) is

$$[\nabla^2\tau]_0 = \frac{2(\bar{\tau}_4 + \bar{\tau}_5 + \bar{\tau}_6 + \bar{\tau}_7 + \bar{\tau}_8 + \bar{\tau}_9 - 6\bar{\tau}_0)}{9h^2} + O\{(h)^2\}. \qquad (8.5.11)$$

Because they are more complicated to use than rectangular grids, triangular and hexagonal grids are rarely used.

9. IMPROVING THE ACCURACY OF SOLUTIONS

A number of techniques for improving the accuracy of solutions when using finite difference methods have already been considered. They involved reducing the truncation error by using central differences, using methods which minimise numerical diffusion, damping and dispersion, and using more accurate derivative approximations at boundaries. Other methods include:
(1) reducing the size of the grid spacing,
(2) using higher order difference approximations,
(3) using predictor-corrector methods, and
(4) reducing the discretisation error by extrapolation.

9.1 Reducing the Grid Spacing

No very useful results which connect the magnitude of the discretisation error to the size of the grid spacing have yet been found. However, it is clear that the errors decrease as the grid spacing is reduced (for example, see Table 3.1). One therefore expects that using smaller and smaller grid spacings will eventually produce successive finite difference solutions which differ from the true solution by less than some required amount.

Also, use of a fine grid permits a more accurate approximation to the irregular shape of a physical boundary. It also enables derivative boundary conditions to be included more accurately.

However, this approach can be very uneconomical. For instance, when solving the one-dimensional diffusion equation using the FTCS method with a fixed value of the parameter s, a reduction of the grid spacing from $\Delta x = 0.1$ (the second row in Table 3.1) to $\Delta x = 0.01$ (the last row) increased the required number of calculations 1000-fold in order to compute τ at a given position and time. In general, for a given s, the number of arithmetic operations in this case is proportional to $1/(\Delta x)^3$. Solving the time dependent diffusion equation in two and three space dimensions with the same decrease in size of grid spacing is even more uneconomical. If the FTCS method with fixed s is used in the three-dimensional case, then the number of arithmetic operations is multiplied by a factor of r^5 when the grid intervals in each coordinate direction in the physical space are divided by r. Therefore, reduction of Δx, Δy and Δz to one-tenth their original values, keeping s_x, s_y, s_z constant, increases 100,000-fold the required number of calculations to reach a fixed time. A one second computer run increases to more than one day in these circumstances.

9.2 Higher Order Finite Difference Approximations

More accurate solutions to a problem can be found by representing the derivatives in the given partial differential equation by finite difference approximations of a higher order of accuracy. In this way, the truncation error of the finite difference approximation to the partial differential equation is reduced. For example, in Section 2.3 it was seen that the three-point central difference approximation to a spatial derivative on a grid with spacing Δx has a truncation error of $O\{(\Delta x)^2\}$, whereas the two-point forward and backward difference approximations have a truncation error of $O\{\Delta x\}$. From the point of view of accuracy, it is preferable to use central difference approximations for spatial derivatives.

Finite difference equations of higher order of accuracy usually contain a larger number of different values of the dependent variable τ, but in many cases the increase is not large.

Consider the spatial derivative in the one-dimensional diffusion equation (2.1.1). At the (j,n) gridpoint, the three-point centred-space finite difference form is given by Equation (2.3.19), namely

$$\left.\frac{\partial^2\bar{\tau}}{\partial x^2}\right|_j^n = \frac{\bar{\tau}_{j+1}^n - 2\bar{\tau}_j^n + \bar{\tau}_{j-1}^n}{(\Delta x)^2} + O\{(\Delta x)^2\}. \tag{9.2.1}$$

A higher order accuracy for this spatial derivative can be obtained using $\bar{\tau}$ values two gridpoints either side of the point $(j\Delta x, n\Delta t)$, along the nth time level. By means of Taylor series expansions of $\bar{\tau}_{j+2}^n$, $\bar{\tau}_{j+1}^n$, $\bar{\tau}_{j-1}^n$, $\bar{\tau}_{j-2}^n$ about values at $(j\Delta x, n\Delta t)$, it can be shown that $\partial^2\bar{\tau}/\partial x^2|_j^n$ can be written in the form

$$\left.\frac{\partial^2\bar{\tau}}{\partial x^2}\right|_j^n = \frac{-\bar{\tau}_{j+2}^n + 16\bar{\tau}_{j+1}^n - 30\bar{\tau}_j^n + 16\bar{\tau}_{j-1}^n - \bar{\tau}_{j-2}^n}{12(\Delta x)^2} + O\{(\Delta x)^4\}. \tag{9.2.2}$$

The right side of this equation is called the 5-point centred space form for $\partial^2\bar{\tau}/\partial x^2|_j^n$.

Substitution of (9.2.2) for the space-derivative of the one-dimensional diffusion equation evaluated at the (j,n) gridpoint, together with the forward finite difference form (2.3.7a) for the time derivative, yields

$$\frac{\bar{\tau}_j^{n+1} - \bar{\tau}_j^n}{\Delta t} + O\{\Delta t\} =$$

$$\frac{-\bar{\tau}_{j+2}^n + 16\bar{\tau}_{j+1}^n - 30\bar{\tau}_j^n + 16\bar{\tau}_{j-1}^n - \bar{\tau}_{j-2}^n}{12(\Delta x)^2} + O\{(\Delta x)^4\}. \tag{9.2.3}$$

Dropping terms of $O\{\Delta t, (\Delta x)^4\}$ gives, on rearrangement, the explicit finite difference equation

$$\tau_j^{n+1} = \frac{1}{12}\{-s\tau_{j-2}^n + 16s\tau_{j-1}^n + (12-30s)\tau_j^n + 16s\tau_{j+1}^n - s\tau_{j+2}^n\}. \tag{9.2.4}$$

However, when methods of improved accuracy are developed for gridpoints in the interior of the solution domain, it should be kept in mind that the formulation of the boundary conditions should be of comparable accuracy. There is little point in using finite difference approximations to the space derivatives in a partial differential equation with truncation error of fourth order, if the finite difference approximations to the space derivatives in derivative boundary conditions have truncation errors of first or second order (see Section 4.4).

It should also be noted that even when the boundary values are known, special treatment is required at interior gridpoints next to the boundary if the finite difference equation used there is to have a fourth order spatial truncation error. When the boundary values $\tau_{0,k}$, $\tau_{J,k}$, $k=0(1)K$ are known, the form (9.2.4) can be used to compute values of τ_j^n, $j=2(1)J-2$, with fourth order spatial truncation errors. However, with $j=1$, the use of (9.2.4) requires knowledge about $\tau_{-1,k}$, and with $j=J-1$, $\tau_{J+1,k}$ is needed; both are values at gridpoints exterior to the region considered. Equation (9.2.1) could be used since, with $j=1$ for example, it gives

$$\left.\frac{\partial^2\bar{\tau}}{\partial x^2}\right|_1^n = \frac{\bar{\tau}_2^n - 2\bar{\tau}_1^n + \bar{\tau}_0^n}{(\Delta x)^2} + O\{(\Delta x)^2\}. \tag{9.2.5}$$

Use of (9.2.5) with a forward difference form for the time derivative in the one-dimensional diffusion equation discretised as the (1,n) gridpoint yields the previously derived three-point FTCS equation (2.4.4) with j=1, namely

$$\tau_1^{n+1} = s\tau_0^n + (1-2s)\tau_1^n + s\tau_2^n. \qquad (9.2.6)$$

However, this has a truncation error which is only second order in the space variable. This is much less accurate than the fourth order spatial truncation error involved in the use of (9.2.4) for j=2(1)J-2.

The truncation error of $\partial^2\bar{\tau}/\partial x^2|_1^n$ can be made $O\{(\Delta x)^4\}$ if the derivative is expressed as a linear combination of $\bar{\tau}_j^n$, j=0(1)5. Use of Taylor series about the gridpoint (1,n) then gives the form

$$\left.\frac{\partial^2\bar{\tau}}{\partial x^2}\right|_1^n = \frac{10\bar{\tau}_0^n - 15\bar{\tau}_1^n - 4\bar{\tau}_2^n + 14\bar{\tau}_3^n - 6\bar{\tau}_4^n + \bar{\tau}_5^n}{12(\Delta x)^2} + O\{(\Delta x)^4\}. \qquad (9.2.7)$$

Use of this form in place of (9.2.2) in the discretisation of the one-dimensional diffusion equation at the (1,n) gridpoint yields the explicit finite difference equation

$$\tau_1^{n+1} = \frac{1}{12}\{10s\tau_0^n + (12-15s)\tau_1^n - 4s\tau_2^n + 14s\tau_3^n - 6s\tau_4^n + s\tau_5^n\}, \qquad (9.2.8)$$

with a truncation error of the same order as that for (9.2.4). For j=J-1, use of the analogous expression to (9.2.7), namely

$$\left.\frac{\partial^2\bar{\tau}}{\partial x^2}\right|_{J-1}^n = \frac{\bar{\tau}_{J-5}^n - 6\bar{\tau}_{J-4}^n + 14\bar{\tau}_{J-3}^n - 4\bar{\tau}_{J-2}^n - 15\bar{\tau}_{J-1}^n + 10\bar{\tau}_J^n}{12(\Delta x)^2} + O\{(\Delta x)^4\}, \qquad (9.2.9)$$

in the discretisation of the diffusion equation at the (J-1,n) gridpoint, gives the equation

$$\tau_{J-1}^{n+1} = \frac{1}{12}\{s\tau_{J-5}^n - 6s\tau_{J-4}^n + 14s\tau_{J-3}^n - 4s\tau_{J-2}^n + (12-15s)\tau_{J-1}^n + 10s\tau_J^n\}. \qquad (9.2.10)$$

Use of (9.2.8), followed by (9.2.4) with j=2(1)J-2, then (9.2.10) gives values of τ_j^{n+1}, j=1(1)J-1, in terms of τ_j^n, j=0(1)J, with a consistent truncation error of $O\{\Delta t,(\Delta x)^4\}$.

However, the use of finite difference equations based on derivative approximations of higher order accuracy does not necessarily mean that the finite difference solution obtained will be more accurate. Cheng (1970), for example, indicates that second order methods are generally optimal.

Examples given by Hamming (1962), Cyrus and Fulton (1967) and Cheng (1968) clearly show that some lower order methods produce results at least as accurate as corresponding methods of higher order. The lack of success with the latter may have been due to poor treatment of the boundary conditions; often the accuracy of the solution is restricted by the order of the boundary value approximation. The order of the discretisation errors of values computed at the boundary and in the interior of the space region concerned should be consistent.

One of the reasons for the lack of improvement when higher order methods are used to solve partial differential equations is that the order of the method is significant only as $\Delta x \to 0$, $\Delta y \to 0$, $\Delta z \to 0$, $\Delta t \to 0$, because the order of the truncation error is only that of the leading term which has been omitted. This order is more significant to ordinary differential equations where much smaller grid spacings, Δx say, can be used than the Δx, Δy, Δz, Δt for partial differential equations. This is because finite difference solutions in the former case only involve the storage of arrays of dimension J. For partial differential equations in three space dimensions, difficulty is often encountered because arrays of dimension $J \times K \times L$, the product of the number of grid points in the x direction with the numbers in the y and z directions, must be stored. Halving the grid spacing in each direction increases the necessary storage eight-fold; a further halving of grid spacing increases the required storage 64-fold.

However, the possibility of improved accuracy is not the only fact to be considered when higher order methods are evaluated. In practice, stability, rate of convergence of the solution and special procedures required near the boundaries must also be considered. For instance, stability is important, since conditional stability requirements restrict the size of the time step which may be used with time dependent problems, and this can greatly increase the number of calculations required. Even with unconditionally stable methods, such as many of the implicit methods, the rate of convergence of the solutions found (say) by iterative methods become important.

For instance, consider the stability of the finite difference equation (9.2.4) for j=2(1)J-2. Application of the von Neumann stability analysis yields the amplification factor

$$G = (1 - \tfrac{7}{3}s) + \tfrac{8}{3}s \cos\beta - \tfrac{1}{3}s \cos^2\beta, \quad \beta = m\pi\Delta x, \qquad (9.2.11)$$

which applies to each Fourier component, m=1,2,..., of the error distribution as it is transmitted from one time level to the next. The requirement for stability, namely $|G| \le 1$, is therefore

$$-1 \le (1 - \tfrac{7}{3}s) + \tfrac{8}{3}s \cos\beta - \tfrac{1}{3}s \cos^2\beta \le 1,$$

since G is real. Substitution of $\chi = \cos\beta$ and rearranging yields the stability requirement that

$$0 \le \chi^2 - 8\chi + 7 \le 6s^{-1} \qquad (9.2.12)$$

for all χ in <-1,1>. From the graph of $F(\chi) = \chi^2 - 8\chi + 7$, shown in Figure 9.1, it is clear that (9.2.12) is satisfied for $-1 \le \chi \le 1$ so long as

$$16 \le 6s^{-1},$$

or

$$s \le \frac{3}{8} = 0.375. \qquad (9.2.13a)$$

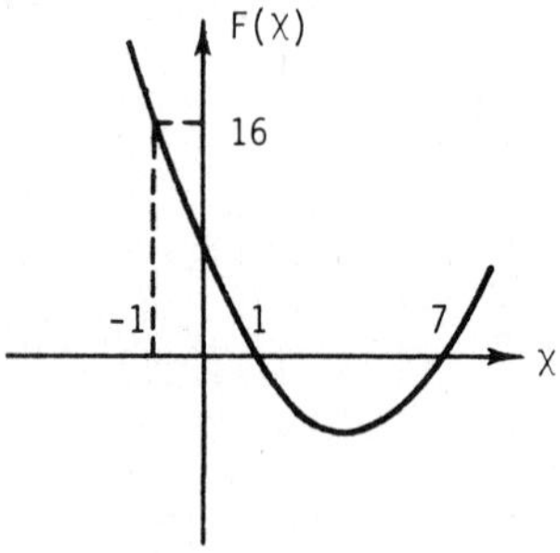

Figure 9.1

Since Equation (9.2.4), j=2(1)J-2, is used with (9.2.8) for j=1 and (9.2.10) for j=J-1, we must also examine the stability of the latter two equations. The amplification factor G for Equation (9.2.8) is

$$G(s,\beta) = \frac{5s}{6} e^{-i\beta} + \frac{1}{4}(3-5s) - \frac{s}{3} e^{i\beta} + \frac{7s}{6} e^{i2\beta} - \frac{s}{2} e^{i3\beta} + \frac{s}{12} e^{i4\beta}.$$

Testing whether $|G| \leq 1$ for β in $<0,2\pi>$ for various values of s, using complex arithmetic on a computer, it was found that (9.2.8) is stable so long as

$$s \leq 0.351. \tag{9.2.13b}$$

The same criterion guaranteed stability for Equation (9.2.10).

It follows that the FTCS method, using a 5 point centred difference for the space derivative, is stable for values of s which satisfy both (9.2.13a,b). The method is therefore stable for $s \leq 0.351$. This is a more restrictive condition on the size of Δt for a given Δx, than the FTCS method which used a 3 point centred difference for the space derivative since the stability criterion in the latter case is $s \leq \frac{1}{2}$. The use of the 5 point equation with its truncation error of $O\{\Delta t,(\Delta x)^4\}$ is certainly not justified when it is considered that a choice of $s = 1/6$ with the 3 point equation produces a more accurate method with a truncation error of $O\{(\Delta t)^2,(\Delta x)^4\}$.

9.3 Predictor-Corrector Methods

The following technique of solving the one-dimensional diffusion equation is essentially time and space centred, but involves two explicit steps, thereby avoiding the necessity of solving the set of linear algebraic equations which is required in the Crank-Nicolson method. The method is based on the use of Equation (4.2.10), with the terms of $O\{(\Delta t)^2,(\Delta x)^2\}$ omitted, which yields the corresponding finite difference equation

$$\frac{\tau_j^{n+1} - \tau_j^n}{\Delta t} = \frac{\alpha}{2}\left\{\frac{\tau_{j+1}^{n+1} - 2\tau_j^{n+1} + \tau_{j-1}^{n+1}}{(\Delta x)^2} + \frac{\tau_{j+1}^n - 2\tau_j^n + \tau_{j-1}^n}{(\Delta x)^2}\right\}. \tag{9.3.1}$$

First estimates of the values of τ_{j-1}^{n+1}, τ_j^{n+1}, τ_{j+1}^{n+1} on the right-hand side of (9.3.1) are obtained by the FTCS equation (2.4.4). These values are then substituted in (9.3.1) to give a corrected estimate of τ_j^{n+1}.

The method therefore consists of two steps, a *predictor*, which gives the first estimate $\hat{\tau}_j^{n+1}$ of the values at the (n+1)th time level:

$$\hat{\tau}_j^{n+1} = s\,\tau_{j-1}^n + (1-2s)\tau_j^n + s\,\tau_{j+1}^n, \qquad j=1(1)J-1, \tag{9.3.2a}$$

followed by a second step, which uses the first predictions to compute the final estimate τ_j^{n+1} from

$$\frac{\tau_j^{n+1} - \tau_j^n}{\Delta t} = \frac{\alpha}{2}\left\{\frac{\hat{\tau}_{j+1}^{n+1} - 2\hat{\tau}_j^{n+1} + \hat{\tau}_{j-1}^{n+1}}{(\Delta x)^2} + \frac{\tau_{j+1}^n - 2\tau_j^n + \tau_{j-1}^n}{(\Delta x)^2}\right\},$$

which becomes, on rearranging,

$$\tau_j^{n+1} = \tau_j^n + \frac{s}{2}\{\hat{\tau}_{j-1}^{n+1} - 2\hat{\tau}_j^{n+1} + \hat{\tau}_{j+1}^{n+1} + \tau_{j-1}^n - 2\tau_j^n + \tau_{j+1}^n\},$$
$$j=1(1)J-1. \tag{9.3.2b}$$

This step is sometimes called a *corrector*, since it can be written in the form

$$\tau_j^{n+1} = \hat{\tau}_j^{n+1} + \frac{s}{2}\{\tau_{j-1}^n + 2(s^{-1}-1)\tau_j^n + \tau_{j+1}^n - \hat{\tau}_{j-1}^{n+1} - 2(s^{-1}-1)\hat{\tau}_j^{n+1} + \hat{\tau}_{j+1}^{n+1}\}, \tag{9.3.3}$$

where the term added to $\hat{\tau}_j^{n+1}$ may be considered a correction.

Both the predictor step (9.3.2a) and the corrector step (9.3.2b) are explicit.

In order to investigate this two-step method, we express the right side of (9.3.2b) in terms of τ_i^n, j=1(1)J-1, and boundary values $\bar{\tau}_0^n$, $\bar{\tau}_J^n$, $\bar{\tau}_0^{n+1}$, $\bar{\tau}_J^{n+1}$, by substituting (9.3.2a) into (9.3.2b). This gives

$$\tau_j^{n+1} = \tfrac{1}{2}s^2\tau_{j-2}^n + (s-2s^2)\tau_{j-1}^n + (1-2s+3s^2)\tau_j^n + (s-2s^2)\tau_{j+1}^n + \tfrac{1}{2}s^2\tau_{j+2}^n, \quad j=2(1)J-2, \tag{9.3.4a}$$

$$\tau_1^{n+1} = (1-2s+\tfrac{5}{2}s^2)\tau_1^n + (s-2s^2)\tau_2^n + \tfrac{1}{2}s^2\tau_3^n + (\tfrac{1}{2}s-s^2)\bar{\tau}_0^n + \tfrac{1}{2}s\bar{\tau}_0^{n+1}, \tag{9.3.4b}$$

$$\tau_{J-1}^{n+1} = \tfrac{1}{2}s^2\tau_{J-3}^n + (s-2s^2)\tau_{J-2}^n + (1-2s+\tfrac{5}{2}s^2)\tau_{J-1}^n + (\tfrac{1}{2}s-s^2)\bar{\tau}_J^n + \tfrac{1}{2}s\bar{\tau}_J^{n+1}. \tag{9.3.4c}$$

The accuracy and stability of the predictor-corrector method (9.3.2a,b) may be determined by an analysis of the equivalent Equations (9.3.4a-c).

Application of the von Neumann stability analysis to Equation (9.3.4a) gives the amplification factor

$$G = 1 - 2s + 3s^2 + (2s-4s^2)\cos\beta + s^2\cos 2\beta, \tag{9.3.5}$$

or

$$G(\chi) = 1 - 4s\chi + 8s^2\chi^2, \tag{9.3.6}$$

where $\chi = \sin^2(\beta/2)$. Since G is real, we require

$$-1 \le 1 - 4s\chi + 8s^2\chi^2 \le 1$$

for all χ in <0,1>.

The right-hand inequality may be written

$$F(\chi) = 8s^2\chi(\chi - 1/2s) \le 0,$$

which is true for all $0 \le \chi \le 1/2s$. Clearly, so long as $s \le \tfrac{1}{2}$, $F(\chi) \le 0$ for all χ in <0,1>. The left-hand inequality may be written

$$8s^2\chi^2 - 4s\chi + 2 \ge 0$$

or

$$(4s\chi-1)^2 \ge -3$$

which is true for all s and χ. Thus the finite difference equation (9.3.4a) is *stable* so long as $s \le \tfrac{1}{2}$.

A von Neumann analysis of Equations (9.3.4 b,c) indicates they are stable for $s \le 2/3$. Therefore the system (9.3.4) is stable so long as

$$s \le 1/2. \tag{9.3.7}$$

This is the same as the stability condition for the FTCS method used in the predictor step.

Consistency analyses of Equations (9.3.4a,b,c) give the corresponding partial differential equations

$$\left.\frac{\partial\bar{\tau}}{\partial t}\right|_j^n = \alpha \left.\frac{\partial^2\bar{\tau}}{\partial x^2}\right|_j^n + E_j^n, \tag{9.3.8}$$

where the truncation error is

$$E_j^n = \frac{\alpha}{12}(\Delta x)^2 \left.\frac{\partial^4\bar{\tau}}{\partial x^4}\right|_j^n + \frac{1}{6}(\Delta t)^2 \left.\frac{\partial^3\bar{\tau}}{\partial t^3}\right|_j^n + O\{(\Delta t)^3, \Delta t(\Delta x)^2, (\Delta x)^4\} \tag{9.3.9}$$

for (9.3.4a), for example. In all cases, E_j^n is $O\{(\Delta t)^2,(\Delta x)^2\}$. Clearly, the use of Equations (9.3.2a,b) results in a second order truncation error in both Δt and Δx, which is equivalent to time and space centreing of the finite difference approximation of the given partial differential equation.

While predictor-corrector methods have been used for some time in the numerical solution of ordinary differential equations under labels such as Adams-Bashforth methods (see May and Noye, 1983), they have seldom been used for the solution of partial differential equations. As has been seen in this section, previously computed results up to time level n are first used to predict results at time level n+1 using an explicit formula generally involving forward time differencing. These predictions are then improved by using them in more accurate centred time methods which require information from time level n+1. This forms the corrector step. If necessary, the corrector may be used repeatedly to obtain even more accurate final results.

Because the very efficient Thomas algorithm can be used to solve the Crank-Nicolson equation (4.2.11), there is little to gain in using predictor-corrector methods to solve the one-dimensional diffusion equation. However, for the two-dimensional diffusion equation (6.1.3) the use of such a method permits time-centreing of the finite difference equation while avoiding the problems associated with the inversion of a large sparse system of linear algebraic equations (see Section 6.2).

For the two-dimensional diffusion equation (6.1.3), the predictor step consists of the FTCS equation (6.1.5), namely

$$\begin{aligned}\hat{\tau}_{j,k}^{n+1} = s_y\tau_{j,k-1}^n + s_x\tau_{j-1,k}^n + (1-2s_x-2s_y)\tau_{j,k}^n \\ + s_x\tau_{j+1,k}^n + s_y\tau_{j,k+1}^n, \quad j=1(1)J-1,\ k=1(1)K-1, \end{aligned} \tag{9.3.10a}$$

followed by the equivalent of the Crank-Nicolson equation (6.2.3) in which the values of $\tau_{j,k}^{n+1}$ used in the discretisation of the diffusion terms $\partial^2\bar{\tau}/\partial x^2$ and $\partial^2\bar{\tau}/\partial y^2$ are approximated by the values $\hat{\tau}_{j,k}^{n+1}$. The corrector step is therefore

$$\begin{aligned}\tau_{j,k}^{n+1} &= \tau_{j,k}^n \\ &+ \tfrac{1}{2}s_x\{\tau_{j-1,k}^n - 2\tau_{j,k}^n + \tau_{j+1,k}^n + \hat{\tau}_{j-1,k}^{n+1} - 2\hat{\tau}_{j,k}^{n+1} + \hat{\tau}_{j+1,k}^{n+1}\} \\ &+ \tfrac{1}{2}s_y\{\tau_{j,k-1}^n - 2\tau_{j,k}^n + \tau_{j,k+1}^n + \hat{\tau}_{j,k-1}^{n+1} - 2\hat{\tau}_{j,k}^{n+1} + \hat{\tau}_{j,k+1}^{n+1}\} \\ &\qquad j=1(1)J-1,\ k=1(1)K-1. \end{aligned} \tag{9.3.10b}$$

The use of Equations (9.3.10a,b) results in a time centred explicit method, with conditional stability governed by the predictor step.

An alternative predictor-corrector method of solving the one-dimensional diffusion Equation (2.1.1) is obtained by predicting values of $\hat{\tau}_j^{n+1/2}$, for j=1(1)J-1, using the FTCS method with a time step of $\frac{1}{2}\Delta t$, and then substituting these values in the discretised form of the diffusion term $\partial^2\bar{\tau}/\partial x^2|_j^{n+1/2}$ of Equation (4.2.9). This results in the predictor

$$\hat{\tau}_j^{n+1/2} = \tfrac{1}{2}s\tau_{j-1}^n + (1-s)\tau_j^n + \tfrac{1}{2}s\tau_{j+1}^n, \; j=1(1)J-1, \qquad (9.3.11a)$$

with the corrector based on the time and space centred difference equation

$$\frac{\tau_j^{n+1} - \tau_j^n}{\Delta t} = \alpha\{\frac{\hat{\tau}_{j+1}^{n+1/2} - 2\hat{\tau}_j^{n+1/2} + \hat{\tau}_{j-1}^{n+1/2}}{(\Delta x)^2}\},$$

which may be written in the form

$$\tau_j^{n+1} = \tau_j^n + s\{\hat{\tau}_{j-1}^{n+1/2} - 2\hat{\tau}_j^{n+1/2} + \hat{\tau}_{j+1}^{n+1/2}\}, \; j=1(1)J-1. \qquad (9.3.11b)$$

Again, both the predictor and corrector steps are explicit.

The accuracy and stability of this method is determined by substituting the values $\hat{\tau}_j^{n+1/2}$ from (9.3.11a) into Equation (9.3.11b), so the right side of the latter is written only in terms of values τ_j^n, j=1(1)J-1 and the boundary values $\bar{\tau}_0^n$, $\bar{\tau}_J^n$ and $\bar{\tau}_0^{n+1/2}$, $\bar{\tau}_J^{n+1/2}$. The resulting set of equations is identical to (9.3.4a,b,c) if the boundary values $\bar{\tau}_0^{n+1/2}$, $\bar{\tau}_J^{n+1/2}$ are replaced by their averages at time levels n and n+1. Therefore it may be considered to be the same as the predictor-corrector method (9.3.2a,b).

9.4 Reduction of the Discretisation Error by Extrapolation

This method, introduced by Richardson (1910) and called by him the *deferred approach to the limit*, may lead to considerable improvements of numerical results when solving partial differential equations by finite difference methods. Applications of this method to equations solved on rectangular regions in space are described in Richardson and Gaunt (1927) and Salvadori (1951).

The reduction of the error depends upon there being a reliable estimate of the discretisation error as a function of grid spacing. That is, m_1, m_2, m_3, m_4 are known, where

$$e = O\{(\Delta x)^{m_1},(\Delta y)^{m_2},(\Delta z)^{m_3},(\Delta t)^{m_4}\}.$$

Since

$$e = \bar{\tau} - \tau,$$

then the exact solution $\bar{\tau}$ may be written in terms of the numerical solution τ as

$$\bar{\tau} = \tau + k_{0,1}(\Delta x)^{m_1} + k_{0,2}(\Delta y)^{m_2} + k_{0,3}(\Delta z)^{m_3} + k_{0,4}(\Delta t)^{m_4}$$
$$+ k_{1,1}(\Delta x)^{m_1+1} + k_{1,2}(\Delta y)^{m_2+1} + \ldots \qquad (9.4.1)$$

where the k's are constants at a given point in the solution domain.

Consider a grid-scheme in which Δx, Δy, Δz, Δt are proportional to the power of some spacing h. Then

$$\Delta x = C_1(h)^{P_1}, \; \Delta y = C_2(h)^{P_2}, \; \Delta z = C_3(h)^{P_3}, \; \Delta t = C_4(h)^{P_4},$$

where p_1, p_2, p_3 and p_4 are known numbers. Substitution in (9.4.1) gives

$$\bar{\tau} = \tau + K_0(h)^m + K_1(h)^{m+1} + \ldots \qquad (9.4.2)$$

where

$$m = \mathrm{Min}\{m_1p_1, m_2p_2, m_3p_3, m_4p_4\}. \qquad (9.4.3)$$

Since the m_j's and p_j's are known, m is known.

Let $\tau^{(1)}$ and $\tau^{(2)}$ be the finite difference solutions obtained at a particular point in the solution domain, computed using spacings h_1 and h_2, respectively. Substitution in (9.4.2) gives

$$\bar{\tau} = \tau^{(1)} + K_0(h_1)^m + K_1(h_1)^{m+1} + \ldots \qquad (9.4.4)$$

and

$$\bar{\tau} = \tau^{(2)} + K_0(h_2)^m + K_1(h_2)^{m+1} + \ldots . \qquad (9.4.5)$$

Multiplying Equation (9.4.4) by $(h_2)^m$ and subtracting from this the product of Equation (9.4.5) with $(h_1)^m$, eliminates the terms of the form $K_0(h)^m$, giving

$$\bar{\tau} = \frac{(h_2)^m\tau^{(1)} - (h_1)^m\tau^{(2)}}{(h_2)^m - (h_1)^m} - K_1\Big[\frac{(h_1/h_2) - 1}{1 - (h_2/h_1)^m}\Big](h_2)^{m+1} + \ldots . \qquad (9.4.6)$$

Clearly,

$$\tau_E = \frac{(h_2)^m\tau^{(1)} - (h_1)^m\tau^{(2)}}{(h_2)^m - (h_1)^m} \qquad (9.4.7)$$

is an approximation with a discretisation error of $O\{(h_2)^{m+1}\}$. Using $\tau^{(1)}$ and $\tau^{(2)}$ in this manner, a more accurate value for $\bar{\tau}$ is obtained, provided a reliable estimate of m is available.

For instance, the FTCS equation (2.4.4) used to solve the one-dimensional diffusion equation (2.1.1), with initial and boundary conditions given in Section 2.1, contributes a term of $O\{(\Delta t)^2, \Delta t(\Delta x)^2\}$ to the discretisation error at each time step if $s \neq 1/6$. Consider the approximations to the value of $\bar{\tau}(0.4,8)$ if $s = \frac{1}{2}$. As the number of time steps required to reach $t_N = N\Delta t = 8$ is inversely proportional to Δt, then the total discretisation error in the approximations to $\bar{\tau}(0.4,8)$ are

$$e = O\{N(\Delta t)^2, N\Delta t(\Delta x)^2\} = O\{\Delta t, (\Delta x)^2\}.$$

Also, as Δt is proportional to $(\Delta x)^2$ for a fixed value of s, then the total discretisation error may be written as

$$e = O\{(\Delta x)^2\}.$$

Here Δx is equivalent to h in (9.4.2), so we may take $m = 2$ in this example as long as s is held constant.

From Table 3.1, it is seen that with $J = 10$, the approximation to $\bar{\tau}(0.4,8) = 0.4503963$ is $\tau^{(1)} = 0.4754791$, and with $J = 20$, $\tau^{(2)} = 0.4566436$. Substitution of $\tau^{(1)}$ and $\tau^{(2)}$ into (9.4.7) with $h_1 = 0.1$, $h_2 = 0.05$ gives the improved approximation $\tau_E = 0.4503651$. Note that, while $\tau^{(1)}$

is correct to only one significant figure and $\tau^{(2)}$ to only two figures, the extrapolated value τ_E is correct to four figures.

An alternate form of (9.4.6) is obtained by writing

$$r = h_1/h_2, \tag{9.4.8}$$

where we may consider $r > 1$, that is, $h_1 > h_2$. Then

$$\tau_E = \frac{(r)^m\tau^{(2)} - \tau^{(1)}}{(r)^m - 1}, \tag{9.4.9}$$

and

$$\bar{\tau} = \frac{(r)^m\tau^{(2)} - \tau^{(1)}}{(r)^m - 1} + K_1^*(h_2)^{m+1} + O\{(h_2)^{m+2}\}, \tag{9.4.10}$$

where

$$K_1^* = \left(\frac{1-r}{1-r^{-m}}\right)K_1. \tag{9.4.11}$$

Generally, r is chosen to be order 1, say $r = 2$, when the factor $r-1/1-r^{-m}$ is $O\{1\}$, and K_1 and K_1^* are of the same order of magnitude.

If $\tau^{(3)}$ is the numerical solution obtained using $h = h_3$ where $h_3 = h_2/r$, application of (9.4.10) gives

$$\bar{\tau} = \frac{(r)^m\tau^{(3)} - \tau^{(2)}}{(r)^m - 1} + K_1^*\left(\frac{h_2}{r}\right)^{m+1} + O\left\{\left(\frac{h_2}{r}\right)^{m+2}\right\}. \tag{9.4.12}$$

Multiplying (9.4.12) by r^{m+1} and subtracting (9.4.10) then gives

$$\bar{\tau} = \frac{(r)^{2m+1}\tau^{(3)} - (r)^m(r+1)\tau^{(2)} + \tau^{(1)}}{(r)^{2m+1} - (r)^m(r+1) + 1} + O\{(h_3)^{m+2}\}. \tag{9.4.13}$$

Clearly,

$$\tau_E^* = \frac{(r)^{2m+1}\tau^{(3)} - (r)^m(r+1)\tau^{(2)} + \tau^{(1)}}{(r)^{2m+1} - (r)^m(r+1) + 1}, \tag{9.4.14}$$

is a more accurate extrapolation than (9.4.9) as it has a discretisation error of order $(m+2)$.

This method of reducing the discretisation error has been applied to the one-dimensional diffusion problem presented in Section 2.1, which was solved using the FTCS explicit method in Section 2.4. From the results presented in Table 3.1 it is seen that, when $s = \frac{1}{2}$ and $J=10,20,40$ the approximations to $\bar{\tau}(0.4,8)$ are $\tau^{(1)} = 0.4754791$, $\tau^{(2)} = 0.4566436$, $\tau^{(3)} = 0.4519564$, respectively. The first two approximations to the exact value $\bar{\tau} = 0.4503963$ have relative errors of 5.6% and 1.4% respectively, and the third has a relative error of 0.35%. Substitution of these values into Equation (9.4.14) gives the value $\tau_E^* = 0.4503981$ which has a relative error of 0.0004%. While $\tau^{(1)}$ is correct to only one figure, $\tau^{(2)}$ to two figures and $\tau^{(3)}$ to three figures, the extrapolated value τ_E^* is correct to five figures.

Use of these extrapolation formulae increases the accuracy of the result at a given point in the solution domain without having to excessively increase the number of calculations required. To obtain the result τ_E^* with $s = \frac{1}{2}$ required 16 + 64 + 256 = 336 applications of the explicit finite difference equation (2.4.4). This result is much more accurate than that obtained in Table 3.1 using the FTCS method with J=100 and s = 1/6, which required 4800 applications of the Equation (2.4.4). However, note that the more accurate values are only obtained on the coarsest grid used, that is at x = j/10, j=1(1)9, in the above example.

If the value of m is not known for a finite difference method which has discretisation errors of consistent order for all approximations used at grid points in the interior and on the boundary of the solution domain, it may be estimated in the following manner. Consider three finite difference solutions $\tau^{(1)}$, $\tau^{(2)}$, $\tau^{(3)}$ obtained at the one point in the solution domain using grid spacings which have been successively reduced by dividing by some factor r, so that

$$h_3 = h_2/r = h_1/(r)^2. \tag{9.4.15}$$

Subtraction of (9.4.5) from (9.4.4) gives

$$\tau^{(1)} - \tau^{(2)} = K_0[(r)^m - 1](h_2)^m + O\{(h_2)^{m+1}\}. \tag{9.4.16}$$

Substitution of h_3 for h_2 and h_2 for h_1 in (9.4.16) gives

$$\tau^{(2)} - \tau^{(3)} = K_0[(r)^m-1](h_3)^m + O\{(h_3)^{m+1}\}. \tag{9.4.17}$$

Dividing (9.4.16) by (9.4.17), ignoring the terms of $O\{(h)^{m+1}\}$ on the right side, and substituting $h_2 = rh_3$, gives

$$\frac{\tau^{(1)} - \tau^{(2)}}{\tau^{(2)} - \tau^{(3)}} \simeq (r)^m. \tag{9.4.18}$$

Therefore

$$m \simeq \ln\{\frac{\tau^{(1)} - \tau^{(2)}}{\tau^{(2)} - \tau^{(3)}}\}/\ln r. \tag{9.4.19}$$

If h has been chosen so that (9.4.3) must give an integer value for m, then the approximation obtained by (9.4.19) can be rounded off to give the required value of m.

For the results $\tau^{(1)} = 0.4754791$, $\tau^{(2)} = 0.4566436$, $\tau^{(3)} = 0.4519564$ obtained with the FTCS method with r = 2, it is found that $m \simeq 2.006$. Clearly, m has the value 2, which was established in a theoretical discussion earlier in this section.

This extrapolation method may be of uncertain value near boundaries which are not straight since boundary value interpolation is then required, near corners with interior angles greater than 180^0, and near boundaries on which the given values of the dependent variable or its normal derivative are not spatially smooth. For details about these restrictions see Wasow (1955) and Forsythe and Wasow (1960).

10. SUMMARY

Analytical methods of solving partial differential equations are usually restricted to linear cases with simple geometries and boundary conditions. The increasing availability of more and more powerful digital computers has made more common the use of numerical methods for solving such equations, in addition to non-linear equations with more complicated boundaries and boundary conditions.

In the previous sections, one particular numerical method, the method of finite differences, has been described. This method is based on the representation of the continuously defined function $\bar{\tau}(x,y,z,t)$ and its derivatives in terms of values of an approximation τ defined at particular, discrete points called gridpoints. These are generally the points of intersection of a grid of straight lines $x = x_j$, $y = y_k$, $z = z_\ell$, $t = t_n$, $j,k,\ell,n=1,2,3...$, that cover the domain of $\bar{\tau}$. From the appropriate Taylor's series expansions of $\bar{\tau}$ about such gridpoints, forward, backward and central difference approximations to derivatives of $\bar{\tau}$ can be developed to convert the given partial differential equation and its initial and boundary conditions to a set of linear algebraic equations linking the approximations τ defined at the gridpoints.

The resulting numerical model is composed of the finite difference analogues (the difference equations) of the mathematical model (the partial differential equation and the initial and boundary conditions) that describes the original problem. While the mathematical model represents the problem in terms of the dependent variable $\bar{\tau}$ defined continuously in the domain of the independent variables x, y, z and t, the finite difference numerical model defines the dependent variable pointwise at the discrete gridpoints (x_j, y_k, z_ℓ, t_n) within the domain of the mathematical model. The resulting numerical solution of the finite difference equation is a table of numbers, each one being associated with one point in the grid. Distinct finite difference numerical models of the same mathematical model are obtained by using different grids and different finite difference approximations to the continuous derivatives involved in the mathematical model.

The finite difference method can be used to solve even the most complicated mathematical models where they consist of partial differential equations and their associated auxiliary conditions. The basic techniques used have been described in this article and are illustrated by considering a number of simple models based on the transport equation, which is a mathematical model of the variation in space and time of continuum properties such as temperature and vorticity. More information about these techniques may be found in books by Ames (1969), Forsythe and Wasow (1960), Fox (1962), Richtmeyer and Morton (1967) and Roache (1974), as well as in scientific and engineering journals such as the Journal of Computational Physics.

The formulation of a well posed mathematical model of a continuum problem requires that the problem be reduced to a set of differential equations and associated boundary conditions that accurately describe the problem. The formulation of a well posed numerical model requires both the derivation of a set of difference equations based on the mathematical model and a demonstration of the adequacy of these difference equations. The latter involves an analysis of the consistency, convergence and stability of the numerical model.

The difference equations and auxiliary conditions of the numerical model must be *consistent* with the differential equations and initial and boundary conditions of the mathematical model. The numerical model is consistent if the truncation error, that is, the discrepancy between the finite difference approximations and the continuous derivatives, tends to zero as the grid spacings get smaller and smaller. Second, the solution to the finite difference equation must *converge* to the solution of the differential equation as the grid spacings get smaller and smaller. That is, the difference between the exact solutions of the numerical and mathematical models should vanish as the grid spacings tend to zero. Finally, it must be possible to solve the finite difference equation on a computer. As computers can store only a finite number of digits to represent each number (between 7 and 28 decimal digits) at each step of any calculation, round-off errors may occur. The computation is *stable* if the growth of these errors is reasonable or controlled.

A numerical model with consistent equations, convergent solutions and stable error propagation forms a computational scheme which gives results which closely approximate the exact solution of the mathematical model. This was illustrated in Sections 2, 4, 5, 6 and 7, in which finite difference solutions obtained from various numerical models were compared with exact solutions of the original mathematical models.

Convergence of the numerical solution of particular finite difference schemes is often difficult to prove. However, for mathematical models which are well posed, linear, initial-value boundary-value problems, Lax's Equivalence Theorem can be used. This theorem states that the solution of a numerical model which is consistent with such a mathematical model, converges if and only if the finite difference equation is stable.

In principle, the method of finite differences can be used on non-linear mathematical models, but consistency, stability and convergence are more difficult to prove. For instance, the matrix method and von Neumann's method of stability analysis cannot be used other than locally, since they only apply to linear finite difference equations. In many cases, numerical experimentation, such as solving the finite difference equation using progressively smaller grid spacings and examining the behaviour of the sequence of values of τ obtained at a given point (x,y,z,t), is the only method available with which to assess the numerical model.

The various methods of obtaining a finite difference numerical model corresponding to a particular mathematical model may result in either explicit or implicit finite difference equations. The former may be solved directly with a minimum of programming effort and computation time in order to advance from one time level to another in an initial-value problem, but are generally only conditionally stable. It was found that the FTCS explicit method for solving the one dimensional diffusion equation is stable only if $s = \alpha\Delta t/(\Delta x)^2 \le \frac{1}{2}$ and then only with given boundary values. Inclusion of derivative boundary conditions further restricted s in some cases. This limits the size of the time increments possible with an initial-value problem of fixed grid size in space.

However, implicit finite difference schemes are often unconditionally stable. Thus, the Crank-Nicolson method for solving the one dimensional diffusion equation was found to be stable for all values of s. This implies computational stability for any size of the time increment. However, the size of Δt is still limited by the accuracy required in the answers; use of very large values of the time step results in poor answers because of unacceptably large truncation errors produced.

Implicit finite difference schemes depend on the solution of a large number of linear algebraic equations. These may be solved by elimination processes, such as the Thomas algorithm used for the tri-diagonal system of linear algebraic equations obtained when the Crank-Nicolson finite difference method is used to solve the one-dimensional diffusion equation. Alternatively, iterative methods such as successive over-relaxation may be used. Iterative methods may be used at each time level when solving numerical models of multi-dimensional equations in space. Recently, alternating direction implicit methods and locally one-dimensional methods have been increasingly used to solve initial-value, boundary-value problems involving more than one space dimension.

A similar numerical technique in vogue for solving equilibrium type problems is the finite element method (see Zienkiewicz, 1967). This method is not classified as a finite difference technique since it differs from finite difference methods in two ways. First, instead of involving a set of grid points at the intersection of a set of parallel lines, it employs a series of "finite elements" connected to discrete points on the boundary of the space region and covering the region. Second, the left side of the partial differential equation, rearranged so it is equated to zero, is replaced over the region by its equivalent integral form, the value of which is minimized using the calculus of variations. The resulting set of linear algebraic equations connecting values of the dependent variable at various points over the space region involved, must then be solved. Unlike the well-structured sets of equations which result from implicit finite differencing, which are usually strongly banded and can be solved using special algorithms, the form of the equations resulting from the finite element method are more random in nature. They usually require reordering to develop any inherent banding. For time-dependent initial-value problems in two and three dimensions the finite element method in space is usually linked with finite differencing in time. The finite element method is seldom extended to include the independent variable, time. Recent articles describing this method include those by Fletcher (1978, 1982) and Tomas (1982).

ACKNOWLEDGEMENT

I thank the many students who have attended my postgraduate courses on the numerical solutions of partial differential equations, for their interest and enthusiasm which lead to the writing of this article. In particular, the following assisted in the calculation of some of the tabulated numerical results : Robert Arnold, David Beard, Grant Bigg, Malcolm Stevens and Stella Suhana.

APPENDIX 1 : The Thomas Algorithm

The tri-diagonal system of linear algebraic equations

$$
\begin{aligned}
b_1\tau_1 + c_1\tau_2 &= d_1 \\
a_2\tau_1 + b_2\tau_2 + c_2\tau_3 &= d_2 \\
a_3\tau_2 + b_3\tau_3 + c_3\tau_4 &= d_3 \\
\vdots \qquad &\quad \vdots \\
a_j\tau_{j-1} + b_j\tau_j + c_j\tau_{j+1} &= d_j \\
\vdots \qquad &\quad \vdots \\
a_{M-1}\tau_{M-2} + b_{M-1}\tau_{M-1} + c_{M-1}\tau_M &= d_{M-1} \\
a_M\tau_{M-1} + b_M\tau_M &= d_M
\end{aligned}
\qquad \text{(A1.1)}
$$

in which the a's, b's, c's and d's are known, is usually solved for the unknown values τ_j , j=1,2,...,M, by an elimination procedure attributed to Thomas (1949). The first equation listed in (A1.1) is used to eliminate τ_1 from the second equation, the new second equation is used to eliminate τ_2 from the third equation and so on until finally, the new second-last equation is used to eliminate τ_{M-1} from the last equation giving one equation with only one unknown, τ_M. The unknowns $\tau_{M-1}, \tau_{M-2}, \ldots, \tau_1, \tau_2$ are then found in turn by back-substitution. This method is therefore a particular example of the Gauss elimination method with back substitution.

Suppose τ_{j-2} has been eliminated from the (j-1)th equation in the set (A1.1) which has become

$$
\begin{aligned}
\beta_1\tau_1 + \gamma_1\tau_2 &= \delta_1 \\
\beta_2\tau_2 + \gamma_2\tau_3 &= \delta_2 \\
\beta_3\tau_3 + \gamma_3\tau_4 &= \delta_3 \\
\vdots \qquad &\quad \vdots \\
\beta_{j-1}\tau_{j-1} + \gamma_{j-1}\tau_j &= \delta_{j-1} \\
a_j\tau_{j-1} + b_j\tau_j + c_j\tau_{j+1} &= d_j \\
\vdots \qquad &\quad \vdots \\
a_M\tau_{M-1} + b_M\tau_M &= d_M
\end{aligned}
\qquad \text{(A1.2)}
$$

where $\beta_1 = b_1$, $\gamma_1 = c_1$, $\delta_1 = d_1$. Using the new (j-1)th equation to eliminate τ_{j-1} from the jth equation in (A1.2) gives

$$
\left(b_j - \frac{a_j\gamma_{j-1}}{\beta_{j-1}}\right)\tau_j + c_j\tau_{j+1} = d_j - \frac{a_j\delta_{j-1}}{\beta_{j-1}}, \qquad \text{(A1.3a)}
$$

or

$$
\beta_j\tau_j + \gamma_j\tau_{j+1} = \delta_j \qquad \text{(A1.3b)}
$$

A comparison of Equations (A1.3a) and (A1.3b) shows that the coefficients β_j, γ_j, δ_j are given in terms of β_{j-1}, γ_{j-1}, δ_{j-1} and a_j, b_j, c_j, d_j by the relations

$$\varepsilon_j = a_j/\beta_{j-1}, \tag{A1.4a}$$

$$\beta_j = b_j - \varepsilon_j \gamma_{j-1}, \tag{A1.4b}$$

$$\gamma_j = c_j, \tag{A1.4c}$$

$$\delta_j = d_j - \varepsilon_j \delta_{j-1}. \tag{A1.4d}$$

Commencing with $\beta_1 = b_1$, $\gamma_1 = c_1$ and $\delta_1 = d_1$, Equations (A1.4) provide a method of finding ε_j, β_j, γ_j, δ_j for j=2(1)M.

When j=M, we obtain

$$\beta_M \tau_M = \delta_M,$$

since $\gamma_M = c_M = 0$. Thus

$$\tau_M = \delta_M/\beta_M$$

and the remaining values of τ are found by substitution into a rearranged form of Equation (A1.3b), namely

$$\tau_j = (\delta_j - \gamma_j \tau_{j+1})/\beta_j, \quad j=M-1(-1)1. \tag{A1.5}$$

The complete algorithm, with c_j substituted for γ_j in the previous results, is:

Set $\beta_1 = b_1$, $\delta_1 = d_1$.

For j=2(1)M *compute*

$$\varepsilon_j = a_j/\beta_{j-1}$$

$$\beta_j = b_j - \varepsilon_j c_{j-1}$$

$$\delta_j = d_j - \varepsilon_j \delta_{j-1}.$$

Then

$$\tau_M = \delta_M/\beta_M.$$

For j=M-1(-1)1 *compute*

$$\tau_j = (\delta_j - c_j \tau_{j+1})/\beta_j.$$

This method is stable with regard to propagation of round-off error as long as $|b_j| \geq |a_j| + |c_j|$ for all j, with strict inequality for one value of j.

The normal Gauss elimination methods require about $M^3/3$ multiplications and divisions (about 9,000 if M=30) to solve a system of M linear algebraic equations in M unknowns. The Thomas method, which takes advantage of the tridiagonal nature of the system, requires only approximately 5M multiplications and divisions (150 if M=30) since the algorithm eliminates all multiplications by zero coefficients of τ which take place when the general Gauss elimination method is applied.

If there are N such systems as (A1.1) to be solved, the quantities ε_j and β_j need be computed only once because the values of a_j , b_j , c_j are fixed. Equations (A1.4d) and (A1.5) must each be computed for N systems, for a total of approximately 3MN arithmetic operations.

Even greater economy can be achieved if the values of the a_j , b_j , c_j are the same for each j in most of the equations (A1.1). This is the case for the equations (4.2.3) of the classical implicit method and the equations (4.2.12) of the Crank-Nicolson method for solving the diffusion problem.

APPENDIX 2 : Solving a cyclic tri-diagonal linear algebraic system

The system

$$
\begin{array}{lllll}
b_1\tau_1 + c_1\tau_2 & & & + a_1\tau_N & = d_1 \\
a_2\tau_1 + b_2\tau_2 + c_2\tau_3 & & & & = d_2 \\
\quad a_3\tau_2 + b_3\tau_3 + c_3\tau_4 & & & & = d_3 \\
\quad\quad \vdots & & & & \quad\vdots \\
\quad\quad a_j\tau_{j-1} + b_j\tau_j + c_j\tau_{j+1} & & & & = d_j \\
\quad\quad\quad \vdots & & & & \quad\vdots \\
& a_{N-1}\tau_{N-2} & + b_{N-1}\tau_{N-1} & + c_{N-1}\tau_N & = d_{N-1} \\
c_N\tau_1 + & & a_N\tau_{N-1} & + b_N\tau_N & = d_N
\end{array}
\qquad (A2.1)
$$

in which the a's, b's, c's and d's are known, is solved for the unknown values τ_j, $j=1,2,\ldots,N$ by an elimination algorithm described by Evans and Atkinson (1970).

The first equation listed in (A2.1) is normalised by dividing through by b_1, and then used to eliminate τ_1 from the second equation. This new second equation is also normalized and then used to eliminate τ_2 from the third equation and so on until finally the new normalized third last equation is used to eliminate τ_{N-2} from the second last equation, which is then normalized.

Suppose τ_{j-2} has been eliminated from the (j-1)th equation in the set (A2.1) which has become

$$
\begin{array}{lllll}
\tau_1 - g_1\tau_2 & & & - h_1\tau_N & = f_1 \\
\quad \tau_2 - g_2\tau_3 & & & - h_2\tau_N & = f_2 \\
\quad\quad \tau_3 - g_3\tau_4 & & & - h_3\tau_N & = f_3 \\
\quad\quad\quad \vdots & & & & \quad\vdots \\
\quad\quad\quad \tau_{j-1} - g_{j-1}\tau_j & & & - h_{j-1}\tau_N & = f_{j-1} \\
\quad\quad\quad a_j\tau_{j-1} + b_j\tau_j + c_j\tau_{j+1} & & & & = d_j \\
& & \vdots & & \quad\vdots \\
c_N\tau_1 + & & a_N\tau_{N-1} & + b_N\tau_N & = d_N
\end{array}
\qquad (A2.2)
$$

where $g_1 = -c_1/b_1$, $h_1 = -a_1/b_1$, $f_1 = d_1/b_1$. Using the new (j-1)th equation to eliminate τ_{j-1} from the ith equation in (A2.2) gives

$$(b_j + a_j g_{j-1})\tau_j + c_j\tau_{j+1} + a_j h_{j-1}\tau_N = d_j - a_j f_{j-1} .$$

Normalizing gives

$$\tau_j + \frac{c_j}{b_j + a_j g_{j-1}}\tau_{j+1} + \frac{a_j h_{j-1}}{b_j + a_j g_{j-1}}\tau_N = \frac{d_j - a_j f_{j-1}}{b_j + a_j g_{j-1}} \qquad (A2.3a)$$

or

$$\tau_j - g_j \tau_{j+1} - h_j \tau_N = f_j \, . \tag{A2.3b}$$

A comparison of the equations (A2.3a) and (A2.3b) shows that the coefficients g_j, h_j, f_j are given in terms of a_j, b_j, c_j, d_j and g_{j-1}, h_{j-1}, f_{j-1} by the relations

$$t_j = b_j + a_j g_{j-1} \tag{A2.4a}$$

$$g_j = - c_j / t_j \tag{A2.4b}$$

$$h_j = - a_j h_{j-1} / t_j \tag{A2.4c}$$

$$f_j = (d_j - a_j f_{j-1}) / t_j \tag{A2.4d}$$

Commencing with $g_1 = - c_1/b_1$, $h_1 = - a_1/b_1$, $f_1 = - d_1/b_1$, Equations (A2.4) provide a method of finding t_j, g_j, h_j, f_j for j=2(1)N-1.

The sytem may now be written

$$\begin{array}{llllll}
\tau_1 - g_1\tau_2 & & & & - h_1\tau_N & = f_1 \\
 & \tau_2 - g_2\tau_3 & & & - h_2\tau_N & = f_2 \\
 & & \tau_3 - g_3\tau_3 & & - h_3\tau_N & = f_3 \\
 & & & \vdots & & \vdots \\
 & & & \tau_{N-1} & - (g_{N-1} + h_{N-1})\tau_N & = f_{N-1} \\
c_N \tau_1 & & & + a_N \tau_{N-1} & + \quad b_N \tau_N & = d_N
\end{array} \tag{A2.5}$$

where the g's, h's, f's can be derived from a's, b's, c's, d's and a_N, b_N and c_N are known. In the last equation τ_1 is eliminated by use of the first equation but produces a τ_2 term. The τ_2 in this new equation is then eliminated by use of the second equation and so on across the row until only a τ_N term remains. τ_N is therefore found and back-substitution is used to solve for τ_{N-1}, τ_{N-2}, ..., $\tau_3 \tau_1$.

Suppose τ_{j-2} has been eliminated from the last equation and (A2.5) has become

$$\begin{array}{lllll}
\tau_1 - g_1\tau_2 & & & - h_1\tau_N & = f_1 \\
 & \tau_2 - g_2\tau_2 & & - h_2\tau_N & = f_2 \\
 & \vdots & & & \vdots \\
 & & \tau_{j-1} - g_{j-1}\tau_j & - h_{j+1}\tau_N & = f_{j-1} \\
 & & \vdots & & \vdots \\
 & -p_{j-1}\tau_{j-1} & & a_N\tau_{N-1} + q_{j-1}\tau_N & = r_{j-1}
\end{array} \tag{A2.6}$$

where $p_1 = - c_N$, $q_1 = b_N$, $r_1 = d_N$. Using the (j-1)th equation in (A2.6) to eliminate τ_{j-1} from the present last equation gives

$$- p_{j-1} g_{j-1} \tau_j + a_N \tau_{N-1} + (q_{j-1} - p_{j-1} h_{j-1}) \tau_N = r_{j-1} + p_{j-1} f_{j-1} \tag{A2.7a}$$

or

$$- p_j \tau_j + a_N \tau_{N-1} + q_j \tau_N = r_j \, . \tag{A2.7b}$$

A comparison of Equations (A2.7a) and (A2.7b) shows that the coefficients p_j, q_j, r_j are given in terms of p_{j-1}, q_{j-1}, r_{j-1}, g_{j-1}, h_{j-1}, f_{j-1} by the relations

$$p_j = p_{j-1} g_{j-1} \tag{A2.8a}$$

$$q_j = q_{j-1} - p_{j-1} h_{j-1} \tag{A2.8b}$$

$$r_j = r_{j-1} + p_{j-1} f_{j-1}. \tag{A2.8c}$$

Commencing with $p_1 = -c_N$, $q_1 = b_N$, $r_1 = d_N$, Equations (A2.8) provide a method of finding p_j, q_j, r_j for j=2(1)N-1.

The system now is

$$\begin{array}{l} \tau_1 - g_1\tau_2 \qquad\qquad\qquad - h_1\tau_N = f_1 \\ \qquad \tau_2 - g_2\tau_3 \qquad\qquad - h_2\tau_N = f_2 \\ \qquad\qquad \vdots \qquad\qquad\qquad\qquad \vdots \\ \qquad\qquad\qquad \tau_{N-1} - (h_{N-1}+g_{N-1})\tau_N = f_{N-1} \\ \qquad\qquad (a_N-p_{N-1})\tau_{N-1} + \qquad q_{N-1}\tau_N = r_{N-1}. \end{array} \tag{A2.9}$$

Eliminating τ_{N-1} from last equation in (A2.9) gives

$$[q_{N-1} + (a_N-p_{N-1})(h_{N-1}+g_{N-1})]\tau_N = r_{N-1} - (a_N-p_{N-1})f_{N-1} \tag{A2.10a}$$

or

$$q_N\tau_N = r_N. \tag{A2.10b}$$

A comparison of Equations (A2.10a) and (A2.10b) shows that

$$q_N = q_{N-1} + (a_N-p_{N-1})(h_{N-1}+g_{N-1})$$

$$r_N = r_{N-1} - (a_N-p_{N-1})f_{N-1}.$$

Finally the system becomes

$$\begin{array}{l} \tau_1 - g_1\tau_2 \qquad\qquad\qquad - h_1\tau_N = f_1 \\ \qquad \tau_2 - g_2\tau_3 \qquad\qquad - h_2\tau_N = f_2 \\ \qquad\qquad \tau_3 - g_3\tau_4 \qquad - h_3\tau_N = f_3 \\ \qquad\qquad\qquad \vdots \qquad\qquad\qquad \vdots \\ \qquad\qquad \tau_{N-1} - (g_{N-1}+h_{N-1})\tau_N = f_{N-1} \\ \qquad\qquad\qquad\qquad\qquad q_N\tau_N = r_N \end{array} \tag{A2.11}$$

Hence,

$$\tau_N = r_N/q_N. \tag{A2.12}$$

Backsubstitution yields the values for τ_{N-1}, ..., τ_2, τ_1 using a rearranged form of (A2.3b),

$$\tau_j - g_j\tau_{j+1} - h_j\tau_N = f_j.$$

Therefore, for j=N-1(-1)1

$$\tau_j = f_j + g_j \tau_{j+1} + h_j \tau_N. \tag{A2.13}$$

The complete algorithm combining the above steps in one process is

$$g_1 = -c_1/b_1, \; h_1 = -a_1/b_1, \; f_1 = d_1/b_1$$

and

$$p_1 = -c_N, \; q_1 = b_N, \; r_1 = d_N.$$

For j=2(1)N-1 compute

$$t_j = b_j + a_j g_{j-1}$$

$$g_j = -c_j/t_j$$

$$h_j = -a_j h_{j-1}/t_j$$

$$f_j = (d_j - a_j f_{j-1})/t_j$$

$$p_j = p_{j-1} g_{j-1}$$

$$q_j = g_{j-1} - p_{j-1} h_{j-1}$$

$$r_j = r_{j-1} + p_{j-1} f_{j-1}$$

and for j=N, compute

$$q_N = q_{N-1} + (a_N - p_{N-1})(g_{N-1} + h_{N-1})$$

$$r_N = r_{N-1} - (a_N - p_{N-1}) f_{N-1}.$$

Then

$$\tau_N = r_N/q_N,$$

and for j=N-1(-1)1

$$\tau_j = p_j + g_j \tau_{j+1} + h_j \tau_N.$$

This method is stable with regard to propagation of round-off errors as long as $|b_j| \geq |a_j| + |c_j|$ for all j with strict inequality for one value of j, otherwise it is possible that a value of t_j will be small or zero, which causes problems. For example, if $b_1 = 3$, $c_1 = 2$, $a_2 = 6$, $b_2 = 4$, then for j=2 in the above

$$t_2 = b_2 + a_2 g_1 = 4 + 6g_1$$

where $g_1 = -c_1/b_1 = -2/3$. Therefore

$$t_2 = 0,$$

so that g_2, h_2 and f_2 cannot be calculated.

This method, which takes advantage of the large number of zeros in the given set of linear algebraic equations, requires only approximately 12N multiplications and divisions, which gives 360 operations if M = 30 compared with approximately 9,000 for normal Gauss elimination.

If there are M similar systems with d_j's different, only the f_j's and r_N (r_j's) need to be recalculated and then the back substitution carried out. After one system is solved, each new system will require approximately a further 3N multiplications and divisions (90 if M = 30).

Greater economy of effort can be obtained if the a_j, b_j, c_j are the same for all values of j.

References

Evans, D.J. and Atkinson, L.V. (1970). *An algorithm for the solution of general three term linear systems.* pp.323-324.

APPENDIX 3 : Taylor series expansions about $(j\Delta x, n\Delta t)$ for various finite difference forms

$$\bar{\tau}^n_{j+1} - \bar{\tau}^n_j = \Delta x \left.\frac{\partial\bar{\tau}}{\partial x}\right|^n_j + \frac{(\Delta x)^2}{2}\left.\frac{\partial^2\bar{\tau}}{\partial x^2}\right|^n_j + \frac{(\Delta x)^3}{6}\left.\frac{\partial^3\bar{\tau}}{\partial x^3}\right|^n_j + O\{(\Delta x)^4\} \qquad (A3.1)$$

$$\bar{\tau}^n_j - \bar{\tau}^n_{j-1} = \Delta x \left.\frac{\partial\bar{\tau}}{\partial x}\right|^n_j - \frac{(\Delta x)^2}{2}\left.\frac{\partial^2\bar{\tau}}{\partial x^2}\right|^n_j + \frac{(\Delta x)^3}{6}\left.\frac{\partial^3\bar{\tau}}{\partial x^3}\right|^n_j + O\{(\Delta x)^4\} \qquad (A3.2)$$

$$\bar{\tau}^n_{j+1} - \bar{\tau}^n_{j-1} = 2\Delta x \left.\frac{\partial\bar{\tau}}{\partial x}\right|^n_j + \frac{(\Delta x)^3}{3}\left.\frac{\partial^3\bar{\tau}}{\partial x^3}\right|^n_j + O\{(\Delta x)^5\} \qquad (A3.3)$$

$$\bar{\tau}^{n\pm1}_{j+1} - \bar{\tau}^{n\pm1}_j = \Delta x \left.\frac{\partial\bar{\tau}}{\partial x}\right|^n_j + \frac{(\Delta x)^2}{2}\left.\frac{\partial^2\bar{\tau}}{\partial x^2}\right|^n_j + \frac{(\Delta x)^3}{6}\left.\frac{\partial^3\bar{\tau}}{\partial x^3}\right|^n_j \pm \Delta x\ \Delta t \left.\frac{\partial^2\bar{\tau}}{\partial x\partial t}\right|^n_j$$

$$\pm \tfrac{1}{2}(\Delta x)^2\Delta t \left.\frac{\partial^3\bar{\tau}}{\partial x^2\partial t}\right|^n_j + \tfrac{1}{2}\Delta x\ (\Delta t)^2 \left.\frac{\partial^3\bar{\tau}}{\partial x\partial t^2}\right|^n_j$$

$$+ O\{(\Delta x)^4, (\Delta x)^3\Delta t, (\Delta x)^2(\Delta t)^2, \Delta x\ (\Delta t)^3\} \qquad (A3.4)$$

$$\bar{\tau}^{n\pm1}_j - \bar{\tau}^{n\pm1}_{j-1} = \Delta x \left.\frac{\partial\bar{\tau}}{\partial x}\right|^n_j - \frac{(\Delta x)^2}{2}\left.\frac{\partial^2\bar{\tau}}{\partial x^2}\right|^n_j + \frac{(\Delta x)^3}{6}\left.\frac{\partial^3\bar{\tau}}{\partial x^3}\right|^n_j \pm \Delta x\Delta t \left.\frac{\partial^2\bar{\tau}}{\partial x\partial t}\right|^n_j$$

$$\mp \frac{(\Delta x)^2\Delta t}{2}\left.\frac{\partial^3\bar{\tau}}{\partial x^2\partial t}\right|^n_j + \tfrac{1}{2}\Delta x(\Delta t)^2 \left.\frac{\partial^3\bar{\tau}}{\partial x\partial t^2}\right|^n_j$$

$$+ O\{(\Delta x)^4, (\Delta x)^3\Delta t, (\Delta x)^2(\Delta t)^2, \Delta x\ (\Delta t)^3\} \qquad (A3.5)$$

$$\bar{\tau}^{n\pm1}_{j+1} - \bar{\tau}^{n\pm1}_{j-1} = 2\Delta x \left.\frac{\partial\bar{\tau}}{\partial x}\right|^n_j + \frac{(\Delta x)^3}{3}\left.\frac{\partial^3\bar{\tau}}{\partial x^3}\right|^n_j \pm 2\ \Delta x\ \Delta t \left.\frac{\partial^2\bar{\tau}}{\partial x\partial t}\right|^n_j$$

$$+ \Delta x(\Delta t)^2 \left.\frac{\partial^3\bar{\tau}}{\partial x\partial t^2}\right|^n_j + O\{(\Delta x)^5, (\Delta x)^3\Delta t, \Delta x(\Delta t)^3\} \qquad (A3.6)$$

$$\bar{\tau}^{n+1}_j - \bar{\tau}^n_j = \Delta t \left.\frac{\partial\bar{\tau}}{\partial t}\right|^n_j + \frac{(\Delta t)^2}{2}\left.\frac{\partial^2\bar{\tau}}{\partial t^2}\right|^n_j + \frac{(\Delta t)^3}{6}\left.\frac{\partial^3\bar{\tau}}{\partial t^3}\right|^n_j + O\{(\Delta t)^4\} \qquad (A3.7)$$

$$\bar{\tau}^n_j - \bar{\tau}^{n-1}_j = \Delta t \left.\frac{\partial\bar{\tau}}{\partial t}\right|^n_j - \frac{(\Delta t)^2}{2}\left.\frac{\partial^2\bar{\tau}}{\partial t^2}\right|^n_j + \frac{(\Delta t)^3}{6}\left.\frac{\partial^3\bar{\tau}}{\partial t^3}\right|^n_j + O\{(\Delta t)^4\} \qquad (A3.8)$$

$$\bar{\tau}^{n+1}_j - \bar{\tau}^{n-1}_j = 2\Delta t \left.\frac{\partial\bar{\tau}}{\partial t}\right|^n_j + \frac{(\Delta t)^3}{3}\left.\frac{\partial^3\bar{\tau}}{\partial t^3}\right|^n_j + O\{(\Delta t)^5\} \qquad (A3.9)$$

$$\bar{\tau}_j^{n+1} - 2\bar{\tau}_j^{n} + \bar{\tau}_j^{n-1} = (\Delta t)^2 \left.\frac{\partial^2 \bar{\tau}}{\partial t^2}\right|_j^n + \frac{(\Delta t)^4}{12} \left.\frac{\partial^4 \bar{\tau}}{\partial t^4}\right|_j^n + O\{(\Delta t)^6\} \quad \text{(A3.10)}$$

$$\bar{\tau}_j^{n} - 2\bar{\tau}_j^{n-1} + \bar{\tau}_j^{n-2} = (\Delta t)^2 \left.\frac{\partial^2 \bar{\tau}}{\partial t^2}\right|_j^n - (\Delta t)^3 \left.\frac{\partial^3 \bar{\tau}}{\partial t^3}\right|_j^n + O\{(\Delta t)^4\} \quad \text{(A3.11)}$$

$$\bar{\tau}_{j+1}^{n} - 2\bar{\tau}_j^{n} + \bar{\tau}_{j-1}^{n} = (\Delta x)^2 \left.\frac{\partial^2 \bar{\tau}}{\partial x^2}\right|_j^n + \frac{(\Delta x)^4}{12} \left.\frac{\partial^4 \bar{\tau}}{\partial x^4}\right|_j^n + O\{(\Delta x)^6\} \quad \text{(A3.12)}$$

$$\begin{aligned} \bar{\tau}_{j+1}^{n\pm 1} - 2\bar{\tau}_j^{n\pm 1} + \bar{\tau}_{j-1}^{n\pm 1} &= (\Delta x)^2 \left.\frac{\partial^2 \bar{\tau}}{\partial x^2}\right|_j^n \pm (\Delta x)^2 \Delta t \left.\frac{\partial^3 \bar{\tau}}{\partial t\, \partial x^2}\right|_j^n \\ &+ \frac{(\Delta x)^2 (\Delta t)^2}{2} \left.\frac{\partial^4 \bar{\tau}}{\partial x^2 \partial t^2}\right|_j^n + \frac{(\Delta x)^4}{12} \left.\frac{\partial^4 \bar{\tau}}{\partial x^4}\right|_j^n \\ &+ O\{(\Delta x)^6, (\Delta x)^4 \Delta t, (\Delta x)^2 (\Delta t)^3\} \end{aligned} \quad \text{(A3.13)}$$

APPENDIX 4 : Glossary of symbols

English letters - Lower case

a — coefficient $\kappa/\rho C_p$ in heat diffusion equation - see Eq.(1.2.1a),
or real constant,
or function of space and time in mixed boundary conditions - see Section (4.5).

b — real constant,
or function of space and time in mixed boundary conditions.

$\underset{\sim}{b}$ — constant vector.

c — real constant,
or function of space and time in mixed boundary conditions,
or Courant number $u\Delta t/\Delta x$ (non-dimensional).

c_x, c_y — Courant numbers $u\Delta t/\Delta x$, $v\Delta t/\Delta y$.

d — indicates ordinary derivative, e.g. $d\bar{\tau}(x)/dx$.

$\underset{\sim}{d}$ — constant vector.

d_m — constant on right side of m-th equation in a set of linear algebraic equations.

d_j^n — collection of known values at n-th time level and boundary values at (n+1)th time level, in implicit finite difference equation at (j,n) gridpoint.

e — discretisation error at (x,t).

$\underset{\sim}{e}$ — vector of end values in marching method, Section 6.4.

e_j^n — discretisation error at (j,n) gridpoint.

f(x) — initial values for partial differential equation given as a function of x.

f_j — initial value at $x = x_j$.

$f(\chi)$ — function of $\chi = \sin^2(\beta/2)$.

g(t) — boundary values given as a function of time t e.g. $g_0(t)$ is boundary value at x = 0.

g^n — boundary value at n-th time level.

h — a distance used in approximating the Neumann boundary conditions in Section 4.5.

h(x) — initial derivative value for partial differential equation given as a function of x.

h_j — initial derivative value at $x = x_j$ - see Eq. (7.2.7).

i — $\sqrt{-1}$.

j — subscript indicating grid coordinate in the x-direction, e.g. $x_j = j\Delta x$.

k — subscript indicating grid coordinate in the y-direction.

k_1 — constant, sinh(1).

k_2 — constant, exp(1)-1.

ℓ — subscript indicating grid coordinate in the z-direction.

m — subscript,
or running index in a summation, e.g. $\sum_m C_m \nu_m$,
or order of leading term in truncation error - see Eq.(9.4.2).

n superscript indicating time level, e.g. $\bar{\tau}_j^n = \bar{\tau}(j\Delta x, n\Delta t)$,
or running index in a summation, e.g. $\sum_n$,
or superscript indicating power, e.g. $(G)^n$.

o subscript indicating value at x = 0, e.g. g_0 is boundary value at x = 0.

p running index in a summation, e.g. $\sum_p$
or superscript indicating iteration number, e.g. $\tau^{(p)}$ is p-th iterate,
or function used in solving the wave equation, namely $p = u\,\partial\bar{\tau}/\partial x$.

q function used in solving the wave equation, namely $q = \partial\bar{\tau}/\partial t$.

r grid (or cell) Reynolds number, $r = c/s = u\Delta x/\alpha$,
or radial distance in polar coordinates,
or ratio/factor.

s non-dimensional constant, $\alpha\Delta t/(\Delta x)^2$.

s_x, s_y, s_z non-dimensional constants in x,y,z directions, namely
$s_x = \alpha_x\Delta t/(\Delta x)^2$, $s_y = \alpha_y\Delta t/(\Delta y)^2$, $s_z = \alpha_z\Delta t/(\Delta z)^2$.

s_j parameter $\alpha\Delta t/(\Delta x_j)^2$ used with variable grid-spacing.

$\underset{\sim}{s}$ vector of starting values in marching method, Section 6.4.

t time variable.

u velocity component in x-direction.

v velocity component in y-direction.

w velocity component in z-direction.

x cartesian coordinate in x-direction,
or subscript indicating phase-angle (β_x), Courant number (c_x), or diffusion coefficient (α_x) in x-direction.

y cartesian coordinate in y-direction,
or subscript indicating y-direction, e.g. phase-angle $\beta_y = \pi m_y \Delta_y$

z cartesian coordinate in z-direction,
or subscript indicating y-direction.

English letters - Upper case

$\underline{\underline{A}}$ square matrix.

B measure of breadth,
or functional form B(x,y,t) of boundary values.

$\underline{\underline{B}}$ tri-diagonal square matrix.

C_m constants.

C_p specific heat at constant pressure.

D domain of solution of partial differential equation, in general a region of x-y-z-t space,
or functional value D(x,y,t) of derivative boundary condition on planar boundary.

D_j^n accumulation of known values at n-th time level, on right side of an implicit finite difference equation at (j,n) gridpoint.

E value of truncation error at point (x,t),
or an exterior gridpoint.

E_j^n value of truncation error at (x_j, t_n).

$F(\chi)$ function of $\chi = \cos\beta$.

G amplification factor, in general complex, of von Neumann stability analysis
or a gridpoint

H measure of height,
or a function.

I a gridpoint.

$\underline{\underline{I}}$, $\underline{\underline{I}}^*$ identity matrix.

J upper limit of range of integral values of j.

K constant, such as coefficients K_0, K_1, .. in truncation error (9.4.2),
or upper limit of range of integral values of k.

L measure of length,
or greatest lower bound,
or number of lost significant figures in marching method, Section (6.4).

M subscript indicating medium surrounding physical region,
or maximum value.

N upper limit of rangeof integral values of n,
or point on a boundary surface.

N_p number of gridspacings, Δt, in a wave period, P.

N_λ number of gridspacings, Δx, in a wave length, λ.

0 origin of polar coordinate system.

O{ } order of magnitude e.g. $O\{(\Delta x)^2\}$, $\Delta x \to 0$.

P typical interval of time, such as the period of an oscillation,
or a gridpoint.

Q a gridpoint.

R a region in space,
or a gridpoint,
or a point on the boundary where it is cut by a gridline.

R^+ boundary (surface) of a region in space.

S subscript indicating a solid,
or a gridpoint,
or a point on the boundary where it is cut by a gridline.

T temperature,
or a gridpoint,
or a superscript indicating the transpose of a matrix.

Greek letters - Lower case

α non-dimensional diffusion coefficient.

$\alpha_x, \alpha_y, \alpha_z$ diffusion coefficients in x,y,z directions.

β phase angle $m\pi\Delta x$ - see Eq.(3.3.19).

$\beta_x, \beta_y, \beta_z$ phase angles $m_x\pi\Delta x$, $m_y\pi\Delta y$, $m_z\pi\Delta z$ in x,y,z directions.

γ amplitude response of finite difference equation with infinite wave initial condition,
or angle of normal direction to boundary relative to x-axis,
or a function of $\sin^2(\beta/2)$, e.g. Eq. (4.1.8).

δ a constant.

$\delta_{j,k}$ Kroneker delta, zero if $j \neq k$, 1 if $j = k$.

ε a small number,
or an error.

$\underset{\sim}{\varepsilon}$ error vector.

$\underset{\sim}{\zeta}$ vector of errors occurring along two successive time levels.

η measure of distance along a vector normal to a boundary,
or variable used in the mapping function $x = x(\eta)$.

$\underset{\sim}{\eta}$ unit vector normal to a boundary surface.

θ parameter used to form weighted average of different time levels,
or angle used in polar coordinates.

κ coefficient of thermal conductivity,
or wave number of progressive (or stationary) wave,
or constant used in the determination of a variable grid.

λ wavelength of a progressive (or stationary) wave,
or an eigenvalue, e.g. λ_A is an eigenvalue of matrix $\underset{=}{A}$.

μ spectral radius,
or relative wave speed of computed to true wave for infinite initial wave.

ν ratio of change in grid-size at step change of grid-spacing, viz. $\Delta x_J/\Delta x_{J-1}$.

ν_j ratio of change in grid-size at x_j in continuously variable grid.

$\underset{\sim}{\nu}$ an eigenvector.

ξ round-off or truncation error in computing the solution of a finite difference equation.

$\underset{\sim}{\xi}$ vector of round-off/truncation errors.

π 3.14159265... .

ρ density of a fluid or solid.

σ circular frequency, $2\pi/P$.

τ computed approximation to some non-dimensional scalar property, such as temperature, concentration or vorticity.

$\bar{\tau}$ exact value of some scalar property, such as temperature.

$\underset{\sim}{\tau}$ vector of computed approximations τ.

ϕ fraction of a gridspacing in a spatial grid.

ϕ_x, ϕ_y fraction of x,y gridspacing in a spatial grid.

χ the variable $\cos\beta$, when $-1 \le \chi \le 1$,
or the variate sin (/2), when $0 \le \chi \le 1$,
or a variable used in the stretching transformation $x = x(\chi)$.

ψ negative of the argument of the amplification factor G,
or parameter used in spatial weighting.

ω relaxation factor in the successive over-relaxation method.

ω_0 optimal relaxation factor.

Greek letters - Upper case

Δ stepped increments along a coordinate direction e.g. Δx.

$\sum_p$ sum over a given range of the index p.

Other symbols

∂ indicates partial derivative, such as $\partial\bar{\tau}(x,t)/\partial t$.

∇^2	Laplacian operator, $\partial^2/\partial x^2 + \partial^2/\partial y^2 + \partial^2/\partial z^2$.
$\hat{}$	over-symbol indicating first approximation in a predictor-corrector method.
$\bar{}$	over-symbol indicating an exact value.
*	pre-superscript indicating numerically computed solution of a finite difference equation.
*	superscript indicating first approximation in a time-splitting method.
**	superscript indicating second approximation in a time-splitting method.
~	under-symbol indicating a vector, e.g. $\underset{\sim}{b}$.
$=$	under-symbol indicating a square matrix, e.g. $\underline{\underline{I}}$.

REFERENCES

Ames, W.F. (1965), *Nonlinear Partial Differential Equations in Engineering*, Academic Press, New York.

Ames, W.F. (1969), *Numerical Methods for Partial Differential Equations*, Barnes and Noble, Inc., New York, New York.

Amsden, A.A., and Harlow, F.H. (1964), "Slip Instability", *Physics of Fluids*, Vol. 7, pp. 327-334.

Anderson, O.L., Davis, R.T., Hankins, G.B., and Edwards, D.E. (1982), "Solution Of Viscous Internal Flows On Curvilinear Grids Generated By The Schwarz-Christoffel Transformation", *Numerical Grid Generation*, edited J.F. Thompson, Elsevier Science Publishing Coy., pp. 507-524.

Anthes, R.A. (1970), "Numerical Experiments with a Two Dimensional Horizontal Variable Grid", *Monthly Weather Review*, Vol. 98, No. 11, pp. 810-822.

Birkhoff, G., and Varga, R.S. (1959), "Implicit Alternating Direction Methods", *Transactions American Mathematical Society*, Vol. 92, pp. 13-24.

Blottner, F.G., and Roache, P.J. (1971), "Nonuniform Mesh Systems", *Journal of Computational Physics*, Vol. 8, pp. 498-499.

Brian, P.L.T. (1961), "A Finite-Difference Methods of High-Order Accuracy for the Solution of Three-Dimensional Transient Heat Conduction Problems", *American Institute of Chemical Engineering Journal*, Vol.7, No.3, pp.367-370.

Browning, G., Kreiss, H.O., and Oliger, J. (1973) "Mesh Refinement", *Mathematics of computation*, Vol.27, No. 121, pp. 29-39.

Carnahan, B., Luther, H.A., and Wilkes, J.O. (1969), *Applied Numerical Methods*, John Wiley, New York.

Carre, B.A. (1961), "The Determination of the Optimum Accelerating Factor for Successive Over-Relaxation", *Computer Journal*, Vol.4, No.1,pp. 73-78.

Carslaw, H.S., and Jaeger, J.C. (1962), *Conduction of Heat in Solids*, Oxford University Press, London.

Caughey, D. (1978) "A Systematic Procedure for Generating Useful Conformal Mappings", *International Journal of Numerical Methods in Engineering*, Vol. 12, p. 1651.

Chavez, S.P., and Richards, C.G. (1970), *A Numerical Study of the Coanda Effect*, 70-FIcs-12, The American Society of Mechanical Engineers, United Engineering Center, New York, New York.

Cheng, S.I. (1968), *Accuracy of Difference Formulation of Navier-Stokes Equations*, A.M.S. Department, Princeton University, Princeton, New Jersey.

Cheng, S.I. (1970), "Numerical Integration of Navier-Stokes Equations", *A.I.A.A. Journal*, Vol. 8, No. 12, pp. 2115-2122.

Christov, C.I. (1982), "Orthogonal Coordinate Meshes With Manageable Jacobian", *Numerical Grid Generation*, edited J.F. Thompson, Elsevier Science Publishing Coy., pp. 885-894.

Ciment, M. (1971), "Stable Difference Schemes with Uneven Mesh Spacings", *Mathematics of Computation*, Vol. 25, No. 114, pp. 219-227.

Ciment, M., and Sweet, R.A. (1973), "Mesh Refinements for Parabolic Equations", *Journal of Computational Physics*, Vol.12, pp. 513-525.

Colgan, L.H. (1981), "Iterative Methods for Solving Large Sparse Linear Systems", *Numerical Solution of Partial Differential Equations*, editor John Noye, North-Holland Publishing Coy., Amsterdam, pp. 367-396.

Courant, R., Isaacson, E., and Rees, M. (1952), "On the Solution of Nonlinear Hyperbolic Differential Equations by Finite Differences", *Communications on Pure and Applied Mathematics,* Vol. 5, pp. 243-255.

Crandall, S.H. (1955), "An Optimum Implicit Recurrence Formula for the Heat Conduction Equation", *Quarterly of Applied Mathematics*, Vol. 13, No. 3, pp. 318-320.

Crank, J. (1975), *The Mathematics of Diffusion*, 2nd Edition, Oxford University Press, London.

Crank, J., and Nicolson, P. (1947), "A Practical Method for Numerical Evaluation of Solutions of Partial Differential Equations of the Heat-Conduction Type", *Proceedings of the Cambridge Philosophical Society*, Vol. 43, No. 50, pp. 50-67.

Crowder, H.J., and Dalton, C. (1971), "Errors in the Use of Nonuniform Mesh Systems", *Journal of Computational Physics*, Vol. 7, pp. 32-45.

Cyrus, N.J., and Fulton, R.E. (1967), *Accuracy Study of Finite Difference Methods*, NASA TN D-4372, National Aeronautics and Space Administration, Langley Research Center, Langley Station, Hampton, Virginia.

Davis, R.T. (1979), "Numerical Methods for Coordinate Generation Based on Schwartz-Christoffel Transformation", *Proceedings AIAA 4th Computational Fluid Dynamics Conference*, Williamsburg, Virginia, Paper No. 79-1463.

de Vahl Davis, G., and Mallinson, G.D. (1976), "An Evaluation of Upwind and Central Difference Approximations by a Study of Recirculating Flow", *Computers and Fluids*, Vol. 4, pp. 29-43.

Douglas, J., Jr. (1955), "On the Numerical Integration of $\partial^2u/\partial x^2 + \partial^2u/\partial y^2 = \partial u/\partial t$ by Implicit Methods", *Journal Society of Industrial Applied Mathematics*, Vol. 3, No. 1, pp. 42-65.

Douglas, J., Jr. (1957), "A note on the Alternating Direction Implicit Method for Numerical Solution of Heat Flow Problems", *Proceedings of the American Mathematical Society*, Vol. 8, pp. 409-412.

Douglas, J., Jr. (1962), "Alternating Direction Methods for Three Space Variables", *Numerische Mathematik*, Vol. 4, pp. 41-63.

DuFort, E.C., and Frankel, S.P. (1953), "Stability Conditions in the Numerical Treatment of Parabolic Differential Equations", *Mathematical Tables and Other Aids to Computation*, Vol. 7, pp. 135-152.

D'Yakonov, E.G. (1963), "Difference Schemes with Split Operators for Multi-Dimensional Unsteady Problems", *U.S.S.R. Computational Mathematics*, Vol. 4, No. 2, pp. 92-110.

Eddy, E.P. (1949), *Stability in the Numerical Solution of Initial Value Problems in Partial Differential Equations*, NOLM 10232, Naval Ordinance Laboratory, White Oak, Silver Spring, Maryland.

Eiseman, P.R. (1979), "A Multi-Surface Method of Coordinate Generation", *Journal of Computational Physics*, Vol. 33(1), pp. 118.

Eiseman, P.R. (1982), "Orthogonal Grid Generation", *Numerical Grid Generation* edited J.F. Thompson, Elsevier Science Publishing Coy., pp. 193-234.

Eriksson, L-E. (1980), "Three-Dimensional Spline-Generated Coordinate Transformations for Grids Around Wing-Body Configurations". *Numerical Grid Generation Techniques*, NASA Conference Publication No. 2166.

Fletcher, C.A.J. (1978), "The Galerkin Method : An Introduction", *Numerical Simulation of Fluid Motion*, editor John Noye, North-Holland Publishing Coy., Amsterdam, pp. 113-170.

Fletcher, C.A.J. (1982), "Burgers' Equation : A Model for all Reasons", *Numerical Solutions of Partial Differential Equations*, editor John Noye, North-Holland Publishing Coy., Amsterdam, pp. 139-226.

Forsythe, G.E., and Wasow, W. (1960), *Finite Difference Methods for Partial Differential Equations*, Wiley, New York.

Fox, L. (1944), "Solution by Relaxation Methods of Plane Potential Problems with Mixed Boundary Conditions", *Quarterly of Applied Mathematics*, Vol. 2, pp. 251-257.

Fox, L. (1962), *Numerical Solution of Ordinary and Partial Differential Equations*, Addison-Wesley Publishing Company, Inc., Reading, Massachusetts.

Frankel, S.P. (1950), "Convergence Rates of Iterative Treatments of Partial Differential Equations", *Mathematical Tables and Other Aids to Computation*, Vol. 4, pp. 65-75.

Frankel, S.P. (1956), "Some Qualitative Comments on Stability Considerations in Partial Differential Equations", *Proceedings Sixth Symposia in Applied Mathematics, AMS, Vol. 6: Numerical Analysis*, pp. 73-75.

Fromm, J.E., and Harlow, F.H. (1963), "Numerical Solution of the Problem of Vortex Sheet Development", *Physics of Fluids*, Vol. 6, No. 7, pp. 975-982.

Gal-Chen, T. (1975), "On the Use of a Coordinate Transformation for the Solution of the Navier Stokes Equations", *Journal of Computational Physics*, Vol. 17, pp. 209-228.

Gordon, W.J. and Hall, C.A. (1973), "Construction of Curvilinear Coordinate Systems and Applications to Mesh Generation", *International Journal for Numerical Methods in Engineering*, Vol. 7, pp. 461-477.

Hahn, S.G. (1958), "Stability Criteria for Difference Schemes", *Communications on Pure and Applied Mathematics*, Vol. 11, pp. 243-255.

Hamming, R.W. (1962), *Numerical Methods for Scientists and Engineers*, McGraw-Hill Book Company, Inc., New York.

Harrison, E.J. (1973), "Three-Dimensional Numerical Simulations of Tropical Systems Utilizing Nested Finite Grids", *Journal of the Atmospheric Sciences*, Vol. 30, pp. 1528-1543.

Hirt, C.W. (1968), "Heuristic Stability Theory for Finite-Difference Equations", *Journal of Computational Physics*, Vol. 2, pp. 339-355.

Hung, T.K., and Macagno, E.O. (1966), "Laminar Eddies in a Two-Dimensional Conduit Expansion", *La Houille Blanche*, Vol. 21, No. 4, pp. 391-400.

Hyman, J.M., and Larrouturou, B. (1982), "The Numerical Differentiation of Discrete Functions Using Polynomial Interpolation Methods", *Numerical Grid Generation*, edited J.F. Thompson, Elsevier Science Publishing Coy., pp.487-506.

Isenberg, J., and de Vahl Davis, G. (1975), "Finite Difference Methods in Heat and Mass Transfer", in *Topics in Transport Phenomena*, ed. C. Gutfinger, Hemisphere Publishing Company, Washington, D.C.

Ives, D.C. (1976), "A Modern Look at Conformal Mapping, Including Multiply Connected Regions", *AIAA Journal*, Vol.14, pp. 1006-1011.

John, F. (1952), "On the Integration of Parabolic Equations by Difference Methods", *Communications on Pure and Applied Mathematics*, Vol. 5, pp. 155-211.

Kalnay de Rivas, E. (1971), *The Circulation of the Atmosphere of Venus*, Ph.D. Thesis, Department of Meteorology, M.I.T., Massachusetts, U.S.A.

Kalnay de Rivas, E. (1972), "On the Use of Non-Uniform Grids in Finite-Difference Equations", *Journal of Computational Physics*, Vol. 10, pp. 202-210.

Kreyszig, E. (1979), *Advanced Engineering Mathematics*, 4th Edition, John Wiley & Sons, Inc., New York.

Kusic, G.L., and Lavi, A. (1968), "Stability of Difference Methods for Initial-Value Type Partial Differential Equations", *Journal of Computational Physics*, Vol. 3, pp. 358-378.

Kusic, G.L. (1969), "On Stability of Numerical Methods for Systems of Initial-Value Partial Differential Equations", *Journal of Computational Physics*, Vol. 4, pp. 272-275.

Laasonen, P. (1949), "Uber eine Methode zur Losung der Warmeleitungsgleichung", *Acta Mathematica*, Vol. 81, p. 309.

Lax, P.D., and Richtmyer, R.D. (1956), "Survey of the Stability of Linear Finite Difference Equations", *Communications on Pure and Applied Mathematics*, Vol. 9, pp. 267-293.

Lax, P.D., and Wendroff, B. (1960), "Systems of Conservation Laws", *Communications on Pure and Applied Mathematics*, Vol. 13, pp. 217-237.

Lax, P.D., and Wendroff, B. (1964), "Difference Schemes with High Order of Accuracy for Solving Hyperbolic Equations", *Communications on Pure and Applied Mathematics*, Vol. 17, pp. 381-398.

Levy, H., and Lessman, F. (1959), *Finite Difference Equations*, Sir Isaac Pitman & Sons, Ltd., London.

MacCormack, R.W. (1971), "Numerical Solution of the Interaction of a Shock Wave with a Laminar Boundary Layer", *Proceedings of Second International Conference on Numerical Methods in Fluid Dynamics*, ed. M. Holt, Lecture Notes in Physics, Vol. 8, Springer-Verlag, New York.

Mann, K.J. (1981), "Inversion of Large Sparse Matrices - Direct Methods", *Numerical Solution of Partial Differential Equations*, editor John Noye, North-Holland Publishing Coy., Amsterdam, pp. 311-355.

Marchuk, G.I. (1975), *Methods of Numerical Mathematics*, Springer-Verlag, New York.

May, R.L., and Noye, B.J. (1983), The Numerical Solution of Ordinary Differential Equations: Initial Value Problem, *Computational Techniques for Differential Equations*, editor John Noye, North-Holland Publishing Coy., Amsterdam, pp. 1-94.

Moore, A.H., Kaplan, B., and Mitchell, D.B. (1975), "A Comparison of Crandall and Crank-Nicolson Methods for Solving a Transient Heat Conduction Problem", *International Journal for Numerical Methods in Engineering*, Vol. 9, pp. 938-943.

Moretti, G., and Salas, M.D. (1970), "Numerical Analysis of Viscous One-Dimensional Flows", *Journal of Computational Physics*, Vol. 5, pp. 487-506.

Myers, G.E. (1971), *Analytical Methods in Conduction Heat Transfer*, McGraw-Hill Book Company.

Nakamura, S. (1982), "Marching Grid Generation Using Parabolic Partial Differential Equations", *Numerical Grid Generation*, edited J.F. Thompson, Elsevier Science Publishing Coy., pp. 775-786.

Noye, B.J. (1970), *The Physical Limnology of Shallow Lakes and The Theory of Tide Wells*, Ph.D. Thesis, University of Adelaide, South Australia.

Noye, B.J. (1977), "Wind-Induced Circulation and Water Level Changes in Lakes", *Proceedings of International Conference on Applied Numerical Modelling*, University of Southampton, 11-15 July, 1977, Pentech Press, pp. 135-145.

Noye, B.J. (1978), "Finite Difference Schemes for the Solution of Linearised Wind Effect Equations", *Numerical Simulation of Fluid Motion*, editor John Noye, North-Holland Publishing Coy., Amsterdam, pp. 435-451.

Noye, B.J., May, R.L., and Teubner, M.D. (1981), "Three-Dimensional Numerical Model of Tides in Spencer Gulf", *Ocean Management*, Vol.6, pp.137-148.

Noye, B.J., May, R.L., and Teubner, M.D. (1982), "A Three-Dimensional Tidal Model for a Shallow Gulf", *Numerical Solutions of Partial Differential Equations*, editor John Noye, North-Holland Publishing Coy., pp. 417-436.

O'Brien, G.G., Hyman, M.A., and Kaplan, S. (1950), "A study of the Numerical Solution of Partial Differential Equations", *Journal of Mathematics and Physics*, Vol. 29, pp. 223-251.

Osher, S.J. (1970), "Mesh Refinements for the Heat Equation", *S.I.A.M. Journal of Numerical Analysis*, Vol. 7, pp. 219-227.

Pao, Y-H., and Daugherty, R.J. (1969), *Time-Dependent Viscous Incompressible Flow Past a Finite Flat Plate*, Boeing Scientific Research Laboratories, D1-82-0822, January, 1969.

Payne, R.B. (1958), "Calculation of Unsteady Viscous Flow Past a Circular Cylinder", *Journal of Fluid Mechanics*, Vol. 4, p. 81.

Peaceman, D.W., and Rachford, H.H., Jr. (1955), "The Numerical Solution of Parabolic and Elliptical Differential Equations", *Journal Society Indust. Applied Mathematics*, Vol. 1, No. 1, March 1955, pp. 28-41.

Pearson, C.E. (1964), *A Computational Method for Time-Dependent Two-Dimensional Incompressible Viscous Flow Problems,* Sperry Rand Research Centre Report, SRRC-RR-64-17, February, 1964.

Pearson, C.E. (1965a), "A Computational Method for Viscous Flow Problems", *Journal of Fluid Mechanics*, Vol. 21, Part 4, pp. 611-622.

Pearson, C.E. (1965b), "Numerical Solutions for the Time-Dependent Viscous Flow Between Two Rotating Coaxial Disks", *Journal of Fluid Mechanics*, Vol. 21, Part 4, pp. 623-633.

Richardson, L.F. (1910), "The Approximate Arithmetical Solution by Finite Differences of Physical Problems Involving Differential Equations, with an Application to the Stresses in a Masonry Dam", *Transactions of the Royal Society of London*, Series A., Vol. 210, pp. 307-357.

Richardson, L.F., and Gaunt, J.A. (1927), "The Deferred Approach to the Limit", *Philosohical Transactions of the Royal Society, London*, Series A, Vol. 226, pp. 299-361.

Richtmyer, R.D. (1957), *Difference Methods for Initial-Value Problems*, Interscience Publishers, Inc., New York, New York.

Richtmyer, R.D., and Morton, K.W. (1967), *Difference Methods for Initial-Value Problems*, Second Edition, Interscience Publishers, J. Wiley and Sons, New York, New York.

Roache, P.J. (1974), *Computational Fluid Dynamics*, Hermosa Publishers, Albuquerque.

Roberts, A. (1982), "Automatic Topology Generation and Generalised B-Spline Mapping", *Numerical Grid Generation*, edited J.F. Thompson, Elsevier Science Publishing Coy., pp. 41-78.

Roberts, K.V., and Weiss, N.O. (1966), "Convective Difference Schemes", *Mathematics of Computation*, Vol. 20 (94), pp. 272-329.

Sadourny, R., and Morel, P. (1969), "A Finite-Difference Approximation of the Primitive Equations for a Hexagonal Grid on a Plane", *U.S. Monthly Weather Review,* Vol. 97, No. 6, pp. 439-445.

Salvadori, M.G. (1951), "Extrapolation Formulas in Linear Difference Operators", *Proceedings 1st Congress Applied Mechanics*, New York, pp. 15-81.

Samarskii, A.A. (1964), "Local One-Dimensional Difference Schemes for Multi-Dimensional Hyperbolic Equations in an Arbitrary Region", *Z. Vycisl. Mat. i Mat. Fiz.*, Vol. 4, pp. 21-35.

Saul'yev, V.K. (1964), *Integration of Equations of Parabolic Type by the Methods of Nets*. Translated G.J.Tee, Pergamon Press.

Smith, G.D. (1969), *Numerical Solution of Partial Differential Equations*, Oxford University Press, London.

Smith, R.E. (1981), "Two-Boundary Grid Generation for the Solution of the Three-Dimensional Navier-Stokes Equations", *NASA Report TM-83123, May 1981*.

Smith, R.E. (1982), "Algebraic Grid Generation", *Numerical Grid Generation*, edited J.F. Thompson, Elsevier Science Publishing Coy., pp. 137-170.

Sorenson, R.L. (1980), "A Computer Program to Generate Two-Dimensional Grids About Airfoils and Other Shapes by the Use of Poisson's Equation", *NASA Report TM-81198, 1980*.

Steger, J.S., and Chaussee, D.S. (1980), "Generation of Body Fitted Coordinates Using Hyperbolic Partial Differential Equations", *FSI Report 80-1, Flow Simulation Inc.*, Sunnyvale, California, Jan. 1980.

Stommel, H. (1948), "The Westward Intensification of Wind-Driven Ocean Currents", *Transactions of American Geophysical Union*, Vol. 29, pp. 202-206.

Sundqvist, H., and Veronis, G. (1970), "A Simple Finite-Difference Grid with Non-Constant Intervals", *Tellus*, Vol. 22, pp. 26-31.

Thacker, W.C. (1977), "Irregular Grid Finite-Difference Techniques: Simulation in Shallow Circular Basins", *Journal of Physical Oceanography*, Vol. 7(2), pp. 284-292.

Thoman, D., and Szewczyk, A.A. (1969), "Time Dependent Viscous Flow over a Circular Cylinder", *The Physics of Fluids Supplement II*, pp. 76-87.

Thomas, G.B. (1968), *Calculus and Analytic Geometry*, 4th Edition, Addison-Wesley Publishing Co., Reading, Massachusetts.

Thomas, L.H. (1949), *Elliptic Problems in Linear Difference Equations over a Network*, Watson Scientific Computing Laboratory, Columbia University, New York.

Thomas, P.D. (1981), "Construction of Composite Three Dimensional Grids from Subregion Grids Generated by Elliptic Systems", *Proceedings AIAA 5th Computational Fluid Dynamics Conference*, Palo Alto, pp.24-32.

Thompson, J.F. (1982), "Elliptic Grid Generation", *Numerical Grid Generation*, edited J.F. Thompson, Elsevier Science Publishing Coy., pp. 79-106.

Tomas, J.A. (1982), "The Finite Element Method in Engineering Practice and Education", *Numerical Solutions of Partial Differential Equations*, editor John Noye, North Holland Publishing Coy., Amsterdam, pp. 227-288.

Torrance, K.E. (1968), "Comparison of Finite-Difference Computations of Natural Convection", *Journal of Research of the National Bureau of Standards*, Vol. 72B, No. 4, pp. 281-301.

Trapp, J.A., and Ramshaw, J.D. (1976), "A simple Heuristic Method for Analyzing the Effect of Boundary Conditions on Numerical Stability", *Journal of Computational Physics*, Vol. 20, pp. 238-242.

Venit, S. (1973), "Mesh Refinements for Parabolic Equations of Second Order", *Mathematics of Computation*, Vol. 27, No. 124, pp. 745-754.

Visbal, M., and Knight, D. (1982), "Generation of Orthogonal and Nearly Orthogonal Coordinates with Grid Control Near Boundaries", *AIAA Journal*, Vol. 20, pp. 305-306.

Wachspress, E.L. (1966), *Iterative Solution of Elliptic Systems and Applications to the Neutron Diffusion Equations of Reactor Physics*, Prentice-Hall, Englewood Cliffs, New Jersey.

Wade, T.L. (1951), *The Algebra of Vectors and Matrices*, Addison-Wesley Publishing Company Inc., Reading, Massachusetts, U.S.A.

Walsh, P.J., and Noye, B.J., (1973), "Modelling Long-Wave Wind Effects on the Lakes of the Murray Mouth", *Proceedings 2nd. South Australian Regional Conference on Physical Oceanography*, editor B.J. Noye, University of Adelaide, pp. 187-215.

Walsh, P.J., and Noye, B.J., (1974), "A Numerical Model of Wind-Induced Circulation in the Murray Mouth Lakes, South Australia", *Proceedings 5th. Australasian Conference on Hydraulics and Fluid Mechanics*, Christchurch, New Zealand, pp. 284-293.

Warming, R.F., and Hyett, B.J., (1974) "The Modified Equation Approach to the Stability and Accuracy Analysis of Finite-Difference Methods", *Journal of Computational Physics*, Vol. 14, pp. 159-179.

Warsi, Z.U.A. (1982), "Basic Differential Models for Coordinate Generation", *Numerical Grid Generation*, edited J.F. Thompson, Elsevier Science Publishing Coy., pp. 41-78.

Wasow, W. (1955), "Discrete Approximations to Elliptic Differential Equations", *Zeitschrift Angewandte Mathematical Physics*, Vol. 6, pp. 81-97.

Williamson, D. (1969), "Numerical Integration of Fluid Flow Over Triangular Grids", *Monthly Weather Review*, Vol. 97, No. 12, pp. 885-895.

Winslow, A.M. (1966), "Numerical Solution of the Quasilinear Poisson Equation in a Nonuniform Triangle Mesh", *Journal of Computational Physics*, Vol. 1, No. 2, p. 149.

Yanenko, N.N. (1971), *The Method of Fractional Steps: The Solution of Problems of Mathematical Physics in Several Variables*, English Translation edited M. Holt, Springer-Verlag, New York.

Young, D. (1954), "Iterative Methods for Solving Partial Differential Equations of Elliptic Type", *Transactions American Mathematical Society*, Vol. 76, pp. 92-111.

Zienkiewicz, O.C. (1967), *The Finite Element Method in Structural and Continuum Mechanics*, McGraw-Hill, London.

Computational Techniques for Differential Equations
J. Noye (Editor)

THE GALERKIN METHOD AND BURGERS' EQUATION

CLIVE FLETCHER
University of Sydney, New South Wales, Australia

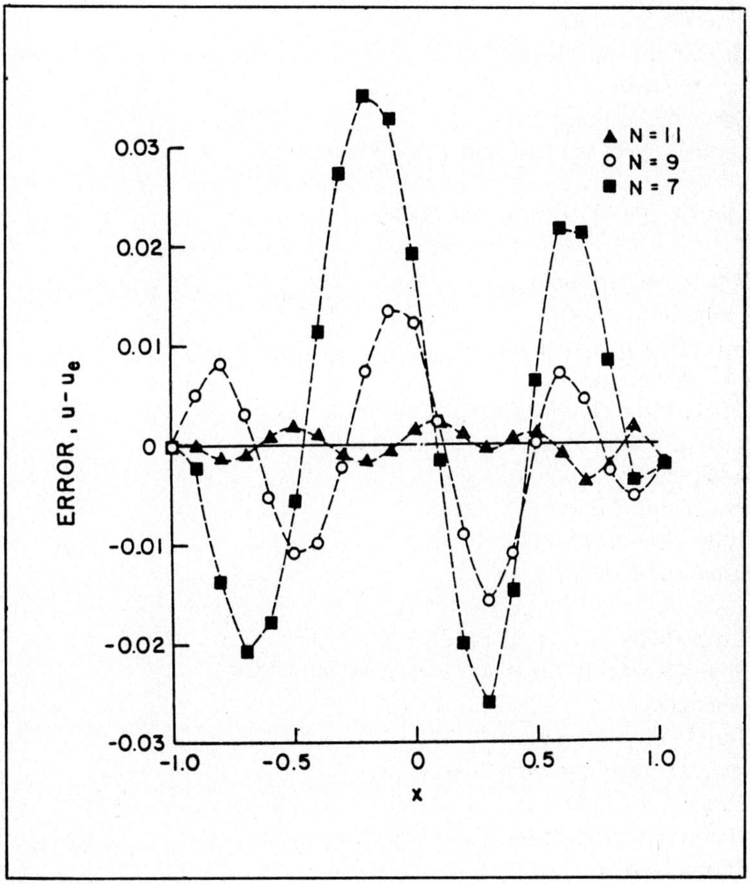

CONTENTS

1. INTRODUCTION

In this chapter we describe the Galerkin method in its three major disguises: the traditional Galerkin method, the Galerkin spectral method and the Galerkin finite element method. Simple examples will be used to indicate the mechanics of applying the Galerkin methods and to demonstrate particular features of the individual methods.

All the methods will be applied to Burgers' equation. Burgers' equation is a nonlinear partial differential equation that closely models the balance of nonlinear convection and (viscous) dissipation that is an important feature of most fluid flows.

Galerkin formulations have been used to solve ordinary differential equation, partial differential equations and integral equations in such diverse areas as structural mechanics, dynamics, fluid flow, acoustics and microwave theory. The Galerkin formulation is equally appropriate to steady, unsteady or eigenvalue problems. Broadly any physical problem for which governing equations are available is a candidate for a Galerkin method.

The original method, which may well have been in informal use for many years, is attributed to Galerkin (1915). In this paper Galerkin was concerned with the elastic equilibrium of rods and thin plates. Galerkin was a Russian engineer and applied mechanician who graduated from the St. Petersburg (Leningrad) Technological Institute in 1896 and spent most of his research career in Leningrad at the St. Petersburg Polytechnical Institute. It appears possible (Anon. (1941)) that Galerkin was formulating the method while in prison during 1906-07 for his anti-Tsarist political views.

The Galerkin method was widely used in Russia for a number of years. It first attracted attention in the Western literature through the work of Duncan (1937, 1938) in relation to the dynamics of aeronautical structures. Bickley (1941) applied the method to the unsteady heat conduction equation and compared the solutions obtained using the Galerkin method with those obtained using the collocation method and method of least-squares. The use of the method has increased steadily since the 1940's and shows no sign of diminishing at present.

A particular feature of the method, that contributed to its early popularity, is the ability to obtain relatively accurate solutions with *relatively few unknown parameters*. This attribute is clearly in evidence for the examples we solve with the traditional Galerkin method in Section 2.

The advent of almost unlimited computing power (compared with that available to Galerkin) has seen the Galerkin method develop in two distinct directions. Firstly the trial functions, that are used to represent the behaviour of the dependent variables, are chosen from global, orthogonal functions. This feature preserves the high accuracy of the traditional Galerkin method and leads directly to *Galerkin spectral methods* that are described in Section 3.

The second direction has seen the trial functions being deliberately restricted to a local domain. By also recasting the unknown parameters in the problem as nodal unknowns one level of computation is eliminated and the unknowns acquire direct physical relevance. This second direction leads to *Galerkin finite element methods* which retain much of the accuracy of the traditional Galerkin method while providing a very flexible and economical solution technique. Galerkin finite element methods are developed in Section 4.

To illustrate the various Galerkin methods we will make continual use of Burgers' equation. This is not inappropriate since Burgers' equation has a ubiquity that begins to rival that of the Galerkin method.

Burgers' equation is one of the few nonlinear partial differential equations for which *exact solutions* can be readily obtained. Depending on the magnitude of various terms in Burgers' equation it behaves as an elliptic, parabolic or hyperbolic partial differential equation. The general properties of Burgers' equation, including the *Cole-Hopf transformation* and some of the more interesting exact solutions, are provided in Section 5.

The ready availability of exact solutions and the ability to control the type of partial differential equation has made Burgers' equation, particularly in one and two dimensions, a popular model equation for testing and comparing computational techniques. These aspects are illustrated in Section 6.

Burgers' equation is very closely related to the momentum equations that govern, in part, viscous flow. In addition Burgers' equation happens to be the governing equation for such diverse physical phenomenon as acoustic transmission, traffic flow and the supersonic flow about aerofoils. Some of the physical problems governed by Burgers' equation are described in Section 7.

A list of the symbols used in the text is provided in the appendix.

2. TRADITIONAL GALERKIN METHOD

We begin this section by laying out the essential features that form the foundation of all Galerkin methods. We will see subsequently, through specific examples and various classifications, how these essential features affect the accuracy, economy and generality of the Galerkin method.

In order to appreciate where the Galerkin method fits into the overall framework of computational techniques it is useful to introduce and describe briefly the method of weighted residuals. To demonstrate the procedure for applying the traditional Galerkin we will examine two problems at length. Firstly an ordinary differential equation will be solved. This problem was used by Duncan (1937) to introduce the Galerkin method. This example will be used here to indicate the effect of non-optimal choices for the weight (test) function and the role of boundary conditions. The second example is Burgers' equation which will illustrate the additional complexity that arises in handling nonlinear terms.

The traditional Galerkin method has, over the years, been applied to problems of considerable diversity. Here we will pick a few examples to reflect this diversity. The examples will reflect the fact that the traditional Galerkin method is a pre-computer method. Attempts to extend the method by direct computer solution reveal a number of inherent limitations. These limitations are examined as a prelude to considering the spectral and finite element methods in Sections 2 and 3 respectively.

2.1 The Galerkin Concept

The essential features of the Galerkin method can be stated quite concisely. Suppose that a two-dimensional problem is governed by a linear differential equation

$$L(u) = 0 \tag{2.1}$$

in a domain, $D(x,y)$, with boundary conditions

$$S(u) = 0 \tag{2.2}$$

on ∂D the boundary of D. The Galerkin method assumes that u can be accurately represented by an *approximate* (trial) solution,

$$u_a = u_o(x,y) + \sum_{j=1}^{N} a_j\phi_j(x,y) \tag{2.3}$$

where the ϕ_j's are known *analytic functions*, u_o is introduced to satisfy the boundary conditions, and the a_j's are coefficients to be determined. Substitution of eq. (2.3) into eq. (2.1) produces a non-zero *residual*, R, given by

$$R(a_o, a_1 \ldots a_N, x,y) = L(u_a) = L(u_o) + \sum_{j=1}^{N} a_j L(\phi_j). \tag{2.4}$$

It is convenient to define an *inner product* (f,g), in the following manner

$$(f,g) = \iint_D f\, g\, dx\, dy\ . \tag{2.5}$$

In the Galerkin method the unknown coefficients, a_j, in eq. (2.3) are obtained by solving the following system of equations

$$(\phi_k, R) = 0 \quad , \quad k = 1, \quad \ldots \quad N. \tag{2.6}$$

In eq. (2.6), R is just the equation residual, eq. (2.4), and the ϕ_k's are the *same* analytic functions that appear in eq. (2.3). Since this example is based on a linear differential equation, eq. (2.6) can be written directly as a matrix equation for the coefficients, a_j, as

$$\sum_{j=1}^{N} a_j(L(\phi_j), \phi_k) = -(L(u_o), \phi_k) \ , \quad k = 1, \ldots N. \tag{2.7}$$

Substitution of the a_j's resulting from the solution of eq. (2.7), into eq. (2.3) gives the required approximate solution, u_a.

2.2 Relation to Other Methods

A modern viewpoint would treat the Galerkin method as an example of the broader class of methods of weighted residuals (MWR). This class of methods is explored in Section 2.2.1. In some situations the Galerkin method produces identical equations and solutions to other methods. Some of these other methods are indicated in Section 2.2.2.

2.2.1 Method of weighted residuals (MWR)

The name, methods of weighted residuals, was first used by Crandall (1956). MWR are discussed by Finlayson and Scriven (1966), Finlayson (1972) and Ames (1972). Collatz (1960) arrived at a similar classification by considering different *"error distribution principles"*.

The method of weighted residuals (MWR) can be described in the following way. In obtaining the solution of a differential equation

$$L(u) = 0 \ , \tag{2.8}$$

with initial condition, $I(u) = 0$, and boundary conditions, $S(u) = 0$, an approximate (trial) solution, u_a, is sought which satisfies both $L(u_a) = R$ and $S(u_a) = R_b$. One would like the equation residual, R, and the boundary residual, R_b, to be as small as possible.

A trial solution, of the following form is assumed,

$$u_a(\overline{x},t) = u_o(\overline{x},t) + \sum_{j=1}^{N} a_j(t)\phi_j(\overline{x}) \ , \tag{2.9}$$

in which $\phi_j(\overline{x})$ are known analytic functions and the a_j's are to be determined as the solution develops (in time dependent problems) or once and for all (in steady problems).

By forcing the form of the analytic solution on the approximate solution, the approximate solution is constrained. In a sense making N larger relaxes the constraints. However, for computational efficiency the error between u and u_a should be acceptably small with N as small as possible. The Galerkin method, as used by Galerkin and Duncan, etc., was a pre-computer method for which a typical value of N would be N = 5.

The form of u_0 (in eq. (2.9)) is chosen to satisfy the boundary conditions and the initial conditions exactly if possible. Ames (1965) calls a method of weighted residuals (MWR) which satisfies the boundary conditions exactly and the governing equation approximately an *interior method*. It is also possible to formulate a method (Shuleshko,(1959)) in which both the partial differential equation and the boundary conditions are satisfied approximately.

Clearly substitution of the trial solution, eq. (2.9), into the governing equation, (2.8), will cause a non-zero residual for all choices of a_j unless the trial solution contains the exact solution, which is unlikely to occur in real problems. The equation residual, R, is typically continuous in the spatial domain ($\overline{x}$) and with increasing N a choice of the a_j's is possible for which $\|R\|$ becomes smaller and smaller. For a given N the a_j's are chosen by requiring that an integration of the weighted residual (and hence the name) over the domain is zero. Thus

$$(W_k, R) = 0 \ . \qquad (2.10)$$

By letting k = 1,...N a system of equations involving only the a_j's is obtained. For unsteady problems this would be a system of ordinary differential equations; for steady problems a system of algebraic equations is obtained. *Different choices of W_k give rise to the different methods within the class.* Some of these methods are:

i) Subdomain method.

The domain is split up into N subdomains, D_k, which may overlap, and

$$\begin{aligned} W_k &= 1 \quad \text{in } D_k \\ &= 0 \quad \text{outside } D_k \ . \end{aligned} \qquad (2.11)$$

An example of this method is the solution of the blunt body problem in supersonic inviscid flow (Belotserkovskii and Chushkin (1965)).

ii) Collocation method

$$W_k(\overline{x}) = \delta(\overline{x}-\overline{x}_k) \qquad (2.12)$$

where δ is the Dirac delta function. Thus the collocation method sets R = 0 at $\overline{x}=\overline{x}_k$. Typical finite difference methods are collocation methods with $\overline{x}_k$ the nodes of a uniform grid. Villadsen and Stewart (1967) choose $\overline{x}_k$ as the zeros of Jacobi polynomials which gives a method of orthogonal collocation. They have applied this method to viscous flow in a channel.

iii) Least squares

$$W_k(\overline{x}) = \partial R/\partial a_k \ . \qquad (2.13)$$

Then eq. (2.13) is equivalent to the requirement that

$$\int R^2 \, dx \text{ is a minimum .}$$

The method of least-squares has been used by Narasimha and Deshpande (1969) to examine shock structure via the Boltzmann equation. In conjunction with the finite element method, Lynn (1974) has applied a least-squares formulation to the boundary layer equations, Fletcher (1979) and Chattot et al. (1981) have computed solutions to compressible, inviscid flows and Steven and Milthorpe (1978) have considered viscous, incompressible flows.

iv) Galerkin method

$$W_k(\bar{x}) = \phi_k(\bar{x}) \qquad (2.14)$$

i.e., the weighting functions are chosen from the same family as the trial functions in eq. (2.9). Then eq. (2.10) coincides with eq. (2.6). If the trial functions form a *complete set*, eq. (2.10) indicates that the residual must be orthogonal to every member of the complete set. As N tends to infinity the approximate solution, u_a, will converge to the exact solution, u.

An interesting feature of the method (WRM) explored by Finlayson (1972) is that evaluation of $\|R\|$, which is straightforward, will give some indication of the solution error $\|u-u_a\|$. Thus for nonlinear problems where an iterative procedure is required monitoring $\|R\|$ or $R(\bar{x})$ will indicate the progress towards convergence and the regions of the domain where convergence is worst.

The various methods of weighted residuals have been compared by Fletcher (1983a). The conclusions are summarised in Table 1. In addition Fletcher makes the following observation, "the Galerkin method produces results of consistently high accuracy and has a breadth of application as wide as any method of weighted residuals".

Table 1. Subjective comparison of different methods of weighted residuals

MWR	Galerkin	Least-squares	Subdomain	Collocation
Accuracy	very high	very high (when applicable)	high	moderate
Ease of formulation	moderate	poor	good	very good
Additional remarks	Equivalent to Ritz method where applicable	Not suited to eigenvalue or evolutionary problems	Equivalent to finite volume method; suited to conservation formulation	Orthogonal collocation gives high accuracy

The Galerkin method is also closely related to the *tau* method (Lanczos (1956)). Conceptually the tau method perturbs the given problem until an exact solution can be obtained. If the perturbation is small the exact solution of the perturbed problem is an accurate approximate solution of the given problem.

In practice the tau method uses some of the unknown coefficients in the trial solution to satisfy the boundary conditions exactly. This relaxes the strict interpretation of the Galerkin method that each trial function should satisfy the boundary conditions (Fletcher (1983a)) exactly, and produces trial functions that are simpler in form and computationally more economical to evaluate. The relative accuracy of the two methods is problem dependent (Fletcher (1983a)).

2.2.2 Equivalence to other methods

Here we are interested in situations where the Galerkin method is equivalent to other methods in the sense of leading to the same approximate equations and solutions.

The first situation is relatively trivial. Suppose a given problem can be solved by the method of separation of variables. If the same problem is solved by the Galerkin method with the trial solution based on the eigenfunctions of the problem (e.g. as obtained by the separation of variables approach) then the same solution will be obtained. Thus, if the trial solution contains the exact solution the Galerkin method will capture it.

There is an equivalence between the Galerkin method and the *orthogonal collocation* metnod (Villadsen and Stewart (1967)) if Gauss quadrature is used to evaluate the inner products in equations like (2.6) and certain other conditions are met (Fletcher (1983a)).

The most important link is between the Galerkin method and variational methods, in particular the *Rayleigh-Ritz method*. Under many circumstances both methods produce the same approximate equations and solutions. The importance is that if a variational formulation for the problem exists tnen the approximate solution will be an upper or a lower bound on the exact solution. Clearly the Galerkin solution will also be the same bound on the exact solution. Of course the Galerkin method has wider application in that it can be used when a variational formulation does not exist.

In two dimensions it is assumed that a variational formulation can be written,

$$I(u) = \iint F(x,y,u_x',u_y')\,dx\,dy \qquad (2.15)$$

with appropriate boundary conditions, and that the exact solution, $u = u_e(x,y)$ corresponds to $I(u_e)$ having a minimum. The Rayleigh-Ritz method assumes a trial solution for u e.g.

$$u_a = \sum_{j=1}^{N} a_j \phi_j(x,y) \quad . \qquad (2.16)$$

It is expected that $I(u_a)$ will be close to $I(u_e)$ in some sense if N is large. The coefficients a_j can be obtained by making $I(u_a)$ a minimum, i.e. by setting

$$\frac{\partial I(u_a)}{\partial a_k} = 0 \quad , \quad k = 1,2,\ldots N \quad . \qquad (2.17)$$

The equivalence with the Galerkin method can be illustrated by considering the Poisson equation,

$$u_{xx} + u_{yy} = f \quad . \qquad (2.18)$$

with boundary condition u = 0 on the boundary. The equivalent variational problem is that

$$I(u) \equiv \iint (u_x^2 + u_y^2 + 2fu)dx\,dy \qquad (2.19)$$

has a minimum corresponding to the exact solution, with the boundary condition u = 0 on the boundary. Substitution of eq. (2.16) into eq. (2.19) and evaluation of eq. (2.17) gives

$$\frac{\partial I(u_a)}{\partial a_k} = 2 \iint \{\frac{\partial \phi_k}{\partial x} [\sum_{j=1}^{N} a_j \frac{\partial \phi_j}{\partial x}] + \frac{\partial \phi_k}{\partial y} [\sum_{j=1}^{N} a_j \frac{\partial \phi_j}{\partial y}] + f\phi_k\}dx\,dy=0. \qquad (2.20)$$

Application of the Galerkin method to eq. (2.18) gives

$$\iint \phi_k [\frac{\partial^2 u_a}{\partial x^2} + \frac{\partial^2 u_a}{\partial y^2} - f]\,dx\,dy = 0 \quad . \qquad (2.21)$$

Application of Green's theorem and noting the homogeneous boundary condition gives

$$-\iint \{\frac{\partial \phi_k}{\partial x}\frac{\partial u_a}{\partial x} + \frac{\partial \phi_k}{\partial y}\frac{\partial u_a}{\partial y} + f\phi_k\} = 0 \quad . \qquad (2.22)$$

Substitution of eq. (2.16) into (2.22) gives eq. (2.20) demonstrating that the two methods are equivalent for this particular problem.

2.3 Examples

The mechanics and some of the features of the Galerkin method are demonstrated by considering two simple examples. The first example, of an ordinary differential equation, was first given by Duncan (1937). This example is particularly effective in demonstrating the need to choose the weight functions from the lowest members of a complete set of functions.

Burgers' equation is the second example and here illustrates the reduction of a parabolic partial differential equation to a system of ordinary differential equations in time. In addition this example indicates the added complication that nonlinear terms introduce.

2.3.1 An ordinary differential equation

Consider the ordinary differential equation,

$$\frac{dy}{dx} - y = 0 \quad , \qquad (2.23)$$

with boundary condition y = 1 at x = 0. The exact solution is y = exp(x). A Galerkin method starts by assuming an appropriate trial solution, e.g., let

$$y_a = 1 + \sum_{j=1}^{N} a_j x^j \quad . \qquad (2.24)$$

The form of eq. (2.24) satisfies the boundary condition *exactly*. Substitution of eq. (2.24) into eq. (2.23) produces the residual

$$R = -1 + \sum_{j=1}^{N} a_j \{jx^{j-1} - x^j\} \quad . \tag{2.25}$$

For the Galerkin method,

$$W_k(x) = x^{k-1} \quad , \quad k = 1,2,\ldots N \quad , \tag{2.26}$$

and evaluation of eq. (2.10) gives

$$\underline{MA} = \underline{B} \tag{2.27}$$

where $\underline{A}$ is the vector of unknowns a_j. An element of $\underline{M}$ is evaluated from

$$m_{kj} = \int_0^1 (jx^{j-1} - x^j)x^{k-1} \, dx \quad , \tag{2.28}$$

and an element of $\underline{B}$ from

$$b_k = \int_0^1 1 \; x^{k-1} \, dx \quad ,$$

For N = 3 the approximate solution, eq. (2.24), is

$$y_a = 1 + 1.0141 \; x + 0.4225 \; x^2 + 0.2817 \; x^3 \quad . \tag{2.29}$$

Solutions of increasing order are compared with the exact solution in Table 2. It is clear that both the solution error, $\|y-y_a\|_{rms}$, and the equation residual, $\|R\|_{rms}$, diminish rapidly *with increasing order*. To achieve high accuracy with few unknowns is an important property for a hand method. It can also be seen, from Table 2 that $\|R\|_{rms}$ is a reasonable indicator of $\|y-y_a\|_{rms}$. The importance of choosing $W_k(x)$ from a complete set of functions can be appreciated from the results shown in Table 3. Each approximate solution is a cubic but progressively higher-order weight functions have been used for successive solutions to the right. It can be seen, by examining $\|y-y_a\|_{rms}$ and $\|R\|_{rms}$, that the *solution accuracy progressively deteriorates as the order of weight function is increased.*

Table 2. Traditional Galerkin solutions for Duncan's example

x	Approximate solution, y_a			Exact solution $y = \exp(x)$
	linear	quadratic	cubic	
0	1	1	1	1
0.2	1.4	1.2057	1.2220	1.2214
0.4	1.8	1.4800	1.4913	1.4918
0.6	2.2	1.8229	1.8214	1.8221
0.8	2.6	2.2349	2.2259	2.2251
1.0	3.0	2.7143	2.7183	2.7183
$\|y-y_a\|_{rms}$	0.3129	0.0097	0.0006	
$\|R\|_{rms}$	0.5774	0.0639	0.0053	

Table 3. Different weight functions for Duncan's example

x	Cubic solution; with weight function, x^k			exact $y = \exp(x)$
	k = 0,1,2	k = 1,2,3	k = 2,3,4	
0	1	1	1	1
0.2	1.2220	1.2248	1.2282	1.2214
0.4	1.4913	1.4952	1.5004	1.4918
0.6	1.8214	1.8255	1.8317	1.8221
0.8	2.2259	2.2303	2.2375	2.2251
1.0	2.7183	2.7241	2.7330	2.7183
$\lVert y-y_a \rVert_{rms}$	0.0006	0.0039	0.0096	
$\lVert R \rVert_{rms}$	0.0053	0.0086	0.0150	

For this example it is possible to obtain *more accurate* solutions (Fletcher (1983a)) by imposing the additional boundary condition $dy/dx = 1$ at $x = 0$. This is obtained by combining eq. (2.23) with the given boundary condition $y = 1$ at $x = 0$.

2.3.2 Burgers' equation

Applying the traditional Galerkin method to Burgers' equation is more complicated than for the ordinary differential equation considered in Section 2.3.1. Firstly Burgers' equation is a partial differential equation so that application of the traditional Galerkin method produces a system of ordinary differential equations which have to be integrated in time. In principle it would be possible to introduce a trial function that was a function of both independent variables and hence obtain a system of algebraic equations. However, for a hand calculation, this would require a greater effort to solve. A second complication is that Burgers' equation is nonlinear. If the system of ordinary equations, for the $a_j(t)$'s, is integrated explicitly this complication manifests itself as the evaluation of a double summation. The complication is greater if the partial differential equation is reduced to a system of algebraic equations. Then an iterative method, like Newton's method, will be required.

Consider Burgers' equation in the form

$$u_t + uu_x - \frac{1}{Re} u_{xx} = 0 \ , \tag{2.30}$$

with boundary conditions

$$u(-1,t) = 1 \ , \quad u(1,t) = 0 \ , \tag{2.31}$$

and initial conditions

$$u(x,0) = 1 \ , \quad -1 \leqslant x \leqslant 0$$

$$u(x,0) = 0 \ , \quad 0 < x \leqslant 1 \ . \tag{2.32}$$

Equations (2.30) to (2.32) govern a physical situation in which an initially discontinuous shock wave propagates to the right and whose profile is slowly smoothed.

Although it would be possible to introduce a trial solution similar to eq. (2.24) it is preferable to develop the trial solution as an ascending series of *Chebyshev polynomials*.

The reason for introducing Chebyshev polynomials is that by interpolating the solution more accurately, particularly close to the boundaries $x = \pm 1$, the resulting equations are better conditioned than if polynomial trial functions (as in eq. (2.24)) were used. Chebyshev polynomials, $T_j(x)$, are orthogonal over the interval (-1,1) with respect to the weight, $(1-x^2)^{-1/2}$, i.e.,

if
$$(f,g) \equiv \int_{-1}^{1} f(x)\, g(x)\, (1-x^2)^{-1/2}\, dx \tag{2.33}$$

then

$$\begin{aligned}(T_i,T_j) &= 0 && \text{if} \quad i \neq j\\ &= \pi/2 && \text{if} \quad i = j \neq 0\\ &= \pi && \text{if} \quad i = j = 0\end{aligned} \tag{2.34}$$

It can be seen, from the nature of weight function, that the use of Chebyshev polynomials gives more significance to points adjacent to $x = \pm 1$. In contrast the use of conventional polynomials for interpolation can be appreciated by considering Runge's example (Forsythe, Malcolm and Moler, 1977), in which the function $f(x) = 1/(1+25x^2)$ is fitted with a high order polynomial over the interval (-1,1). It is found that although the interpolating polynomial matches f(x) at the interpolation points (equally spaced) the interpolating functions display gross oscillations between the interpolation points near the boundary of the region.

The relevance of considering the interpolatory nature of the trial solution, is that the initial values of the a_j's are determined by *fitting the initial data* and the subsequent growth of the a_j's, according to the Galerkin equation, is such as to keep the trial solution *a reasonable interpolatory fit* of the exact solution.

Returning to Burgers' equation, the following trial solution is assumed

$$u(x,t) = \sum_{j=0}^{N} a_j(t)\, T_j(x) \quad . \tag{2.35}$$

A strict interpretation of the Galerkin method would require the substitution of eq. (2.35) into eq. (2.31) and the elimination of two of the a_j coefficients. The result would be test functions that were combinations of the Chebyshev functions, $T_j(x)$, rather than individual Chebyshev functions. Here we obtain N-1 coefficients, $a_j(t)$, by solving eq. (2.10) and the other two coefficients, $a_{N-1}(t)$ and $a_N(t)$, by satisfying eq. (2.31). This is the tau method which is closely related to the Galerkin method (Fletcher (1983a)). Substituting eq. (2.35) into eq. (2.30) produces the residual

$$R = \sum_j \dot{a}_j\, T_j(x) + \sum_j a_j \sum_i a_i\, T_j(x)\, \frac{dT_i(x)}{dx} - \frac{1}{Re} \sum_j a_j \frac{d^2 T_j(x)}{dx^2} \quad . \tag{2.36}$$

Application of the Galerkin method gives

$$\int_{-1}^{1} R\, T_k(x)dx = 0 \quad , \quad k = 0,\ldots N \ . \tag{2.37}$$

So that only first derivatives of T_j will appear it is necessary to apply Green's theorem to the viscous term. The overall result is

$$\underline{M}\,\underline{\dot{A}} + (\underline{B}+\underline{C})\underline{A} = 0 \quad , \tag{2.38}$$

where an element of $\underline{\dot{A}}$ is $\dot{a}_j$, an element of $\underline{M}$ is given by

$$m_{k,j} = \int_{-1}^{1} T_j(x)\ T_k(x)dx \quad , \tag{2.39}$$

an element of $\underline{B}$ is given by

$$b_{k,j} = \sum_i a_i \int_{-1}^{1} \frac{dT_i(x)}{dx}\, T_j(x)\ T_k(x)dx \tag{2.40}$$

and an element of $\underline{C}$ by

$$c_{k,j} = \frac{1}{Re}\int_{-1}^{1} \frac{dT_j(x)}{dx}\frac{dT_k(x)}{dx}\, dx \ . \tag{2.41}$$

The initial profile (2.32) gives the initial values for the a_j's, after solving the system of algebraic equations

$$\underline{M}\ \underline{A} = \underline{D} \tag{2.42}$$

where an element of $\underline{D}$ is given by

$$d_k = \int_{-1}^{0} T_k(x)\ 1\ dx \ . \tag{2.43}$$

The system of equations (2.38) is integrated using a fourth-order Runge-Kutta scheme. The scheme uses a variable stepsize in time that is adjusted to be as large as possible without causing instability. After a fixed time, typically $\Delta t = 0.92$, the solution is compared with that of the exact solution, eq. (5.21). Results for Re = 10 for various orders of representation are shown in Table 4 and Figure 1.

For the solutions shown in Figure 1 the last two unknown coefficients (a_j's) in the trial solution have been determined as functions of the other unknown coefficients to satisfy the boundary conditions at $x = -1$ and 1. At the Reynolds number considered (Re = 10) and at the time of the comparison, the boundary conditions at $x = -1$ and 1 are no longer 1 and 0 precisely. For the solutions presented the values of u at $x = \pm 1$ have been set equal to the exact solution there every five steps. It can be seen that the largest errors in the approximate solutions occur at the point in the flowfield where the exact solution is changing *most rapidly*. This, of course, occurs where the shock is situated. A consideration of $\|u-u_e\|$ and Figure 1 indicates that the increase in accuracy with increasing order is *substantial*.

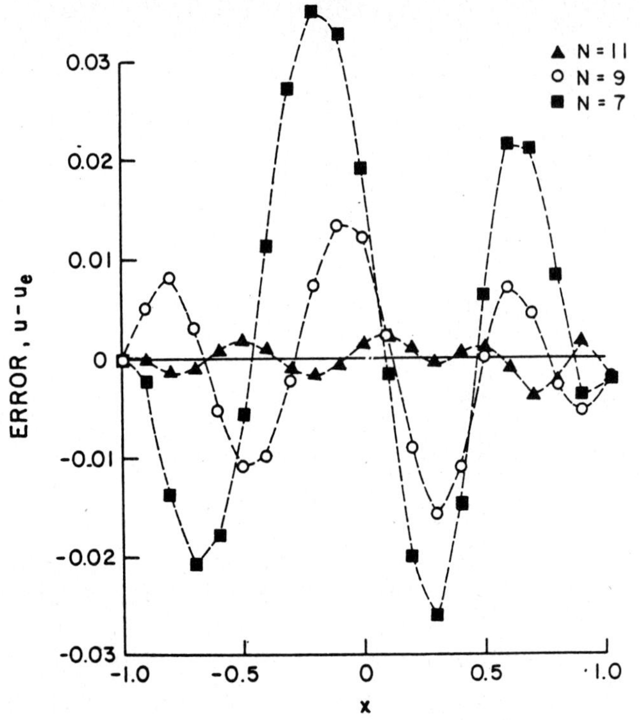

Figure 1. Solution of Burgers' equation by traditional Galerkin method at Re = 10

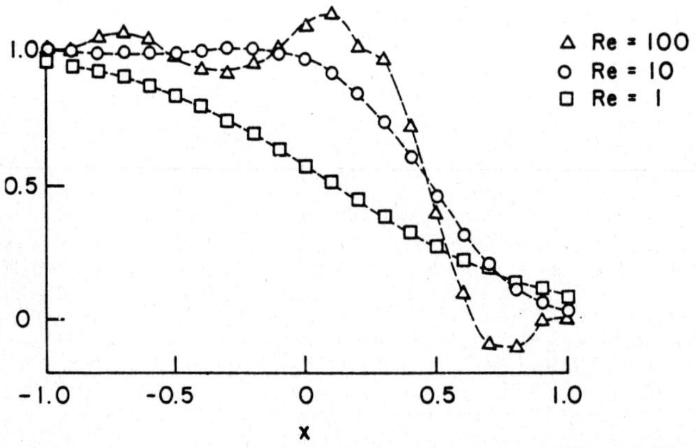

Figure 2. Solution of Burger's equation for increasing Reynolds number

Table 4. Solution of Burgers' equation by traditional Galerkin method. t = 0.92, Re = 10.

x	Approximate Solution			Exact Solution
	N = 7	N = 9	N = 11	
-1.0	1.0000	1.0000	1.0000	1.0000
-0.9	0.9978	1.0054	1.0003	1.0000
-0.8	0.9863	1.0082	0.9987	0.9999
-0.7	0.9792	1.0029	0.9987	0.9998
-0.6	0.9818	0.9942	1.0002	0.9996
-0.5	0.9932	0.9881	1.0007	0.9990
-0.4	1.0088	0.9879	0.9985	0.9977
-0.3	1.0216	0.9924	0.9969	0.9948
-0.2	1.0241	0.9960	0.9869	0.9888
-0.1	1.0096	0.9903	0.9759	0.9768
0	0.9730	0.9661	0.9551	0.9539
0.1	0.9115	0.9159	0.9151	0.9131
0.2	0.8251	0.8359	0.8461	0.8451
0.3	0.7169	0.7272	0.7426	0.7428
0.4	0.5927	0.5962	0.6077	0.6074
0.5	0.4606	0.4543	0.4554	0.4545
0.6	0.3306	0.3162	0.3080	0.3092
0.7	0.2134	0.1969	0.1885	0.1926
0.8	0.1192	0.1084	0.1088	0.1110
0.9	0.0562	0.0550	0.0618	0.0602
1.0	0.0288	0.0288	0.0288	0.0309
$\|u-u_e\|_{rms}$	0.0189	0.0083	0.0016	

For the same order of trial function (N = 9) a comparison of exact and approximate solutions are presented for various Reynolds numbers in Table 5 and Figure 2. Each solution has been obtained at a different time. The effect of increasing the Reynolds number is to increase the sharpness of the shock and this can be seen, most readily, by comparing the exact solutions. At a high Reynolds number the approximate solution demonstrates its inability to follow the exact profile by developing substantial oscillations in space adjacent to the location of the shock (Figure 2). In turn this produces a substantial inaccuracy as is apparent by considering $\|u-u_e\|_{rms}$. In principle the oscillations could be reduced, and the accuracy increased, by increasing N. However this would increase the evaluation time and the influence on computational efficiency is not obvious.

No comment is made on the computational efficiency of the traditional Galerkin method applied to Burgers' equation. A high-order time integrator, e.g., the fourth-order Runge-Kutta scheme, has been used with a step-size small enough to ensure stability. Also the error associated with the temporal integration is negligible compared with the error in spatial representation. This has been done to facilitate comparison with the spectral and Galerkin finite element methods but is not optimal for computational efficiency.

Table 5. Solution of Burgers' equation by traditional method, N = 9, various Reynolds numbers.

x	Re = 1.0,	t = 0.23	Re = 10,	t = 0.92	Re = 100,	t = 0.92
	Approx	Exact	Approx	Exact	Approx	Exact
-1.0	0.9609	0.9599	1.0000	1.0000	1.0000	1.0000
-0.9	0.9443	0.9447	1.0054	1.0000	0.9956	1.0000
-0.8	0.9263	0.9253	1.0082	0.9999	1.0456	1.0000
-0.7	0.9029	0.9008	1.0029	0.9998	1.0672	1.0000
-0.6	0.8731	0.8709	0.9942	0.9996	1.0402	1.0000
-0.5	0.8369	0.8350	0.9881	0.9990	0.9831	1.0000
-0.4	0.7947	0.7931	0.9879	0.9977	0.9303	1.0000
-0.3	0.7469	0.7452	0.9924	0.9948	0.9128	1.0000
-0.2	0.6938	0.6919	0.9960	0.9889	0.9444	1.0000
-0.1	0.6359	0.6340	0.9903	0.9768	1.0159	1.0000
0	0.5742	0.5728	0.9661	0.9539	1.0963	1.0000
0.1	0.5099	0.5097	0.9159	0.9131	1.1411	1.0000
0.2	0.4449	0.4464	0.8359	0.8451	1.1057	1.0000
0.3	0.3814	0.3844	0.7272	0.7428	0.9613	0.9998
0.4	0.3213	0.3254	0.5962	0.6074	0.7099	0.9714
0.5	0.2665	0.2706	0.4543	0.4545	0.3933	0.1861
0.6	0.2179	0.2210	0.3162	0.3092	0.0905	0.0015
0.7	0.1756	0.1772	0.1969	0.1925	-0.1017	0.0000
0.8	0.1388	0.1395	0.1084	0.1110	-0.1154	0.0000
0.9	0.1066	0.1077	0.0550	0.0601	0.0091	0.0000
1.0	0.0793	0.0817	0.0288	0.0309	0.0000	0.0000
$\|u-u_e\|_{rms}$	0.0022		0.0083		0.1049	

2.4 Typical Applications

The traditional Galerkin method has been widely used in the past. Many of the applications prior to 1972 have been reviewed by Ames (1972) and by Finlayson (1972). More recently the traditional Galerkin method has been substantially replaced by the Galerkin finite element method and to a lesser extent by the Galerkin spectral method. However applications of the traditional method do still appear occasionally.

2.4.1 Natural convection

Steady two-dimensional natural convection in a rectangular slot, which is oriented arbitrarily with respect to the gravity vector, provides an interesting temperature/velocity interaction. The basic geometry is shown in Figure 3

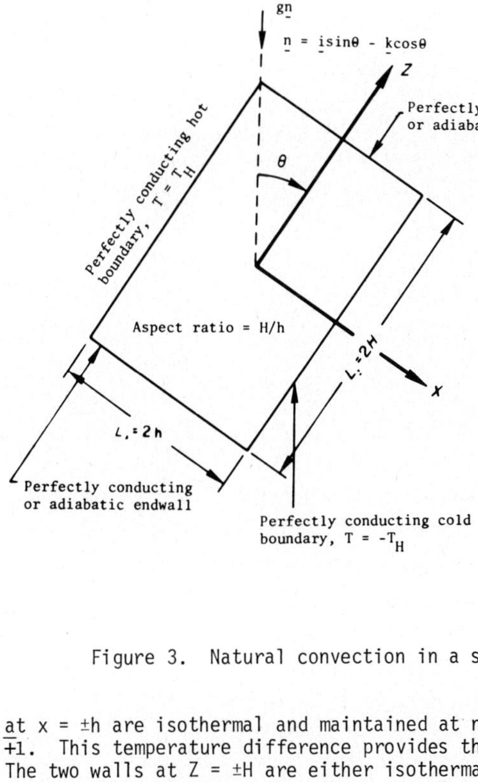

Figure 3. Natural convection in a slot

The walls at x = ±h are isothermal and maintained at nondimensional temperatures $T(\pm 1, Z) = \mp 1$. This temperature difference provides the driving mechanism for the problem. The two walls at Z = ±H are either isothermal or adiabatic. By varying θ the influence of slot inclination on the convection characteristics can be considered. The aspect ratio, A = H/h, can also be varied and is found to have a significant effect on the temperature and velocity profiles within the slot.

The equations governing the temperature and velocity behaviour can be written in nondimensional form as

$$\nabla \cdot \underline{v} = 0 \tag{2.44}$$

$$G^*(\underline{v} \cdot \nabla)\underline{v} = -\nabla p + \nabla^2 \underline{v} - T\underline{n} \tag{2.45}$$

and $$G^*(\underline{v} \cdot \nabla)T = \frac{1}{Pr} \nabla^2 T \, , \tag{2.46}$$

where G* is the Grashof number and Pr is the Prandtl number. The two-dimensional velocity vector $\underline{v}$ is just

$$\underline{v} = \underline{i}u + \underline{k}w \tag{2.47}$$

and $\underline{i}$ and $\underline{k}$ are unit vectors in the x and z directions. Boundary conditions for this problem are, for the velocity,

$$u(\pm 1,z) = w(\pm 1,z) = 0$$
$$u(x,\pm A) = w(x,\pm A) = 0 \tag{2.48}$$

For the isothermal side walls (at $x = \pm 1$) the boundary conditions are

$$T(\pm 1,z) = \mp 1 \quad . \tag{2.49}$$

For the end walls the boundary conditions are

$$T(x,\pm A) = -x \tag{2.50}$$

or

$$\partial T/\partial z(x,\pm A) = 0 \tag{2.51}$$

Equation (2.50) is an isothermal boundary condition that is compatible with eq. (2.49). Equation (2.51) is an adiabatic boundary condition.

Trial solutions are introduced for the velocity and temperature distributions as follows

$$u(x,z) = \sum_{j=1}^{N} B_j u_j(x,z) \tag{2.52}$$

$$w(x,z) = \sum_{j=1}^{N} B_j w_j(x,z) \tag{2.53}$$

and

$$T(x,z) = \left(\sum_{j=1}^{N} C_j f_j(x,z)\right) - x \quad . \tag{2.54}$$

The trial functions, $u_j(x,z)$ and $w_j(x,z)$, are chosen so that the continuity equation, (2.44), is automatically satisfied. This is brought about by defining a stream function, ψ_j, such that

$$u_j = -\partial\psi_j/\partial z \quad \text{and} \quad w_j = \partial\psi_j/\partial x \tag{2.55}$$

and choosing a trial function for the stream function that will satisfy the boundary conditions, eq. (2.48). Because of the geometric symmetry it is convenient to define odd and even stream functions

$$\psi_j^{(e)} = C_{M_j}(x) \quad C_{L_j}(z/A)$$

and (2.56)

$$\psi_j^{(o)} = S_{M_j}(x) \quad S_{L_j}(z/A)$$

where

$$C_{M_j}(x) = \cosh[\lambda_{M_j} x]/\cosh \lambda_{M_j} - \cos[\lambda_{M_j} x]/\cos \lambda_{M_j} .$$

and (2.57)

$$S_{M_j}(x) = \sinh[\mu_{M_j} x]/\sinh \mu_{M_j} - \sin[\mu_{M_j} x]/\sin \mu_{M_j} .$$

The parameters λ_m, μ_m must be chosen to satisfy the boundary conditions on u_j and v_j. The technique for doing this and the corresponding functions C_{M_j} and S_{M_j} are described in Chandrasekhar (1961).

The trial functions, f_j, in eq. (2.54) are

$$f_j^{(e)}(x,j) = \sin[M_j \pi x] \cos[2L_m - 1 - \alpha) \, 0.5\pi z/A]$$

and (2.58)

$$f_j^{(o)}(x,j) = \cos[(2M_j - 1) \, 0.5\pi x] \sin[(2L_m - \alpha) \, 0.5\pi_z/A],$$

where $\alpha = 0$ corresponds to the isothermal boundary condition, eq. (2.50)

and $\alpha = 1$ corresponds to the adiabatic boundary condition, eq. (2.51).

Introduction of eqs. (2.52) to (2.54) into eqs. (2.45) and (2.46) produces equation residuals. Evaluating inner products of the residuals with the vector velocity trial function, $\underline{v_j} = \underline{i}\, u_j + \underline{k}\, w_j$ and the temperature trial function, f_j, produces the following system of matrix equations for the unknown coefficients B_j and C_j in eqs. (2.52) to(2.54)

$$\underline{L}^{(1)} \underline{B} + \underline{L}^{(2)} \underline{C} + \frac{R^*}{Pr} \underline{L}^{(6)} \underline{B} = \underline{L}^{(3)} \qquad (2.59)$$

and
$$R^* \underline{L}^{(4)} \underline{B} + \underline{L}^{(5)} \underline{C} - R^* \underline{L}^{(7)} \underline{C} = 0 . \qquad (2.60)$$

In eqs. (2.59) and (2.60) the Rayleigh number, $R^* = G^*Pr$. The elements of the matrices $\underline{L}^{(1)}$ etc. are given by Catton et al. (1974). The matrices $L^{(6)}$ and $\underline{L}^{(7)}$ depend on the unknown B_j coefficients so that the equations are nonlinear.

However in the limit of $Pr \to \infty$, eq. (2.59) is *linear* and it is convenient to use this equation to eliminate $\underline{B}$ from eq. (2.60). The term dropped from eq. (2.59) implies that the convection induced velocities will be small. The resulting equation for $\underline{C}$ is nonlinear and can be solved efficiently using the Newton-Raphson technique.

Catton et al. have obtained solutions for Rayleigh numbers up to 2×10^6, aspect ratios varying from 0.2 to 20 and tilt angles, θ, varying from -30° to 75°. Galerkin solutions have been obtained with $N \leqslant 32$ in the trial solutions. For a vertical configuration, $\theta = 0°$, results for Nusselt number variation with Rayleigh number are shown in Fig. 4. Good agreement with the data of Poots (1958) and De Vahl Davis (1968) can be seen. The Nusselt number is a nondimensional measure of the heat transfer. Here it is defined by

$$Nu = -\frac{1}{2A}\int_{-A}^{+A} \left.\frac{\partial T}{\partial x}\right|_{x=1} dz \quad . \tag{2.61}$$

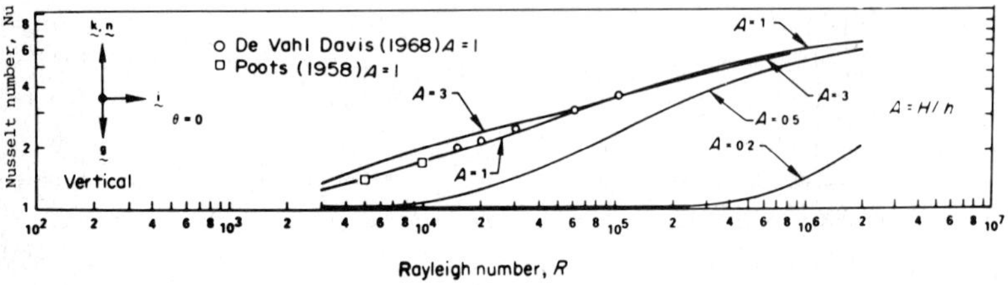

Figure 4. Heat transfer with isothermal endwalls

2.4.2 Motion of a mooring cable

Williams (1975) has used a traditional Galerkin method to investigate the motion of a mooring cable which is given a harmonic displacement at its upper end. Williams considered a two-dimensional configuration in which the applied force was in the plane of the cable and tangential to the cable at the upper end. The cable was assumed to be elastic and under the external loading of its own weight and a hydrodynamic drag force.

Williams develops the general equations for the tension and instantaneous displacement by considering the small perturbations from the catenary shape that is the appropriate static orientation of the cable in the absence of hydrodynamic drag. In addition, the equations are simplified by assuming small strain and rotation. If the displacement of the disturbing force is made small compared with the length of the cable and if the motion away from the ends of the cables is considered, the following system of partial differential equations can be obtained,

$$u_{ss} - \frac{N_F}{\overline{E}} u_{tt} = \left(\frac{\ell}{R} v\right)_s \tag{2.62}$$

and

$$\frac{N_F}{\overline{E}} v_{tt} + \frac{C_D N_D}{2\overline{E}} v_t \ |v_t| + \left(\frac{\ell}{R}\right)^2 v = \frac{\ell}{R} u_s \quad , \tag{2.63}$$

where u and v are the displacements due to the dynamic motion tangential to and normal to the cable. s is the nondimensional distance along the cable. ℓ is the length and R the radius of curvature of the cable. $N_F = Aw^2/g$ where A is the amplitude of the disturbing force and w is the frequency of the disturbing force. $\bar{E} = A/\ell\varepsilon_1$ where $\varepsilon_1 = 4\mu g\ell/\pi d^2 E$ and μ is the mass per unit length, d the cable diameter and E the Young's modulus for the cable material. C_D is the drag coefficient and $N = \rho_w A^2 w^2 d/\mu g$ where ρ_w is the density of the fluid surrounding the cable. The parameters in the problem can be rearranged to give

$$N_d = \frac{4}{\pi} \left(\frac{\rho_w}{\rho_s}\right) \left(\frac{A}{d}\right) N_f \quad , \tag{2.64}$$

where ρ_s is the density of the cable. A considerable simplification can be made to eqs. (2.62) and (2.63) if $N_F \to 0$ as N_D remains fixed. This implies a very low frequency disturbing force whose amplitude is large compared with the cable diameter. Then one obtains

$$v = \left(\frac{\ell}{R}\right)^{\frac{1}{2}} F(t) \tag{2.65}$$

$$T = 0.5\ C_D\ N_D\ F_t\ |F_t| \tag{2.66}$$

and

$$u = F \int_0^s \left(\frac{\ell}{R}\right)^{3/2} dx + 0.5\ \frac{C_D\ N_D}{\bar{E}}\ F_t\ |F_t|\ s\ . \tag{2.67}$$

F(t) is a function to be determined and T is the nondimensional tension associated with the dynamic motion. F is obtained from the boundary condition

$$u(1,t) = \lambda F + 0.5\ \frac{C_D\ N_D}{\bar{E}}\ F_t\ |F_t| = \cos t \quad , \tag{2.68}$$

where λ is related to the cable configuration. Equation (2.68) is solved by the Galerkin method by assuming a trial solution for F of the form,

$$F = A \sin t + B \cos t \quad . \tag{2.69}$$

Imposing the orthogonality condition (weighted integral of the residual) allows eq. (2.69) to be written

$$F = \frac{\cos \alpha_o}{\lambda} \cos(t - \alpha_o) \tag{2.70}$$

where

$$\cos \alpha_o = \left[(1 + 4C_o^2)^{\frac{1}{2}} - 1\right]^{\frac{1}{2}}/(\sqrt{2}C_o) \tag{2.71}$$

and

$$C_o = 4C_D\ N_D/3\pi\lambda^2\bar{E} \quad .$$

The corresponding maximum dynamic tension is given by

$$T_{dyn,max} = 0.5\ C_D\ (\cos \alpha_o/\lambda)^2\ \rho_w \ell\ d\ (Aw)^2 \quad . \tag{2.72}$$

It is also possible to obtain F by direct numerical integration of eq. (2.68) but this does not indicate the nature of the dependence of F on t, as does Eq. (2.70). Some indication of the accuracy of the Galerkin formulation can be obtained from Table 6 (after Williams, 1975).

Table 6. Galerkin and numerical solutions of eq. (2.68)

λ	N_D	α_0 (deg.)	$F(t)_{max}$	
			Galerkin	Numerical
0.5431	0.2137	8.95	1.819	1.839
	0.4808	18.8	1.743	1.800
	0.8547	29.2	1.607	1.689
0.1664	0.2137	48.5	3.979	4.305
	0.4808	61.5	2.872	3.149
	0.8547	68.4	2.216	2.431

2.4.3 Aerofoil flow

For the inviscid, incompressible flow around a lifting aerofoil shown in Fig. 5 it is possible to write down a Fredholm integral

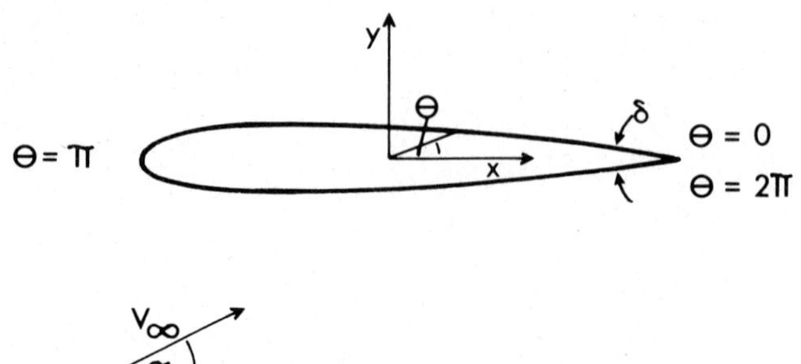

Figure 5. Flowfield geometry for inclined aerofoil

equation of the second kind for the tangential velocity, V, at the surface,

$$g(\theta) - \frac{1}{2\pi}\int_0^{2\pi} K(\theta,\phi)\ g(\phi)d\phi = f(\theta) \ , \qquad (2.73)$$

where

$$g(\theta) = (\dot{x} + \dot{y})^{\frac{1}{2}}\ V(\theta)/V_\infty$$

$$K(\theta,\phi) = 2\,\{\dot{y}(x-\xi) - \dot{x}(y-\eta)\}/\{(x-\xi)^2 + (y-\eta)^2\} \tag{2.74}$$

$$f(\theta) = -2(\dot{x}\cos\alpha + \dot{y}\sin\alpha)$$

$$\dot{x} = dx/d\theta, \quad \dot{y} = dy/d\theta \quad .$$

θ is a surface coordinate defined by $x = 0.5\cos\theta$. Then $y = y(x(\theta))$ is the ordinate. (ξ,η) are dummy variables related to ϕ as (x,y) are to θ.

For aerofoils with a non-zero trailing edge angle, δ, an appropriate trial solution for $g(\theta)$ is (after Nugmanov, (1975))

$$g(\theta) = \left[\sum_{j=1}^{N} A_j^* \frac{(1 - \cos j\theta)}{\sqrt{3\pi}} - \sum_{j=1}^{N} B_j \frac{\sin j\theta}{\sqrt{\pi}}\right]\lambda(\theta) \quad . \tag{2.75}$$

The function $\lambda(\theta)$ is introduced to provide the Kutta condition when $\delta \neq 0$ and is defined as follows

$$\lambda(\theta) = [(1 - \cos\theta)/(1 - \cos\theta^*)]^{\delta/2\pi}, \quad -\theta^* \leqslant \theta \leqslant \theta^* \tag{2.76}$$

$$= 1 \text{ otherwise} \quad .$$

θ^* is typically given by $\cos\theta^* = 1 - \nu\tau^2$ where τ is the thickness chord ratio and ν is a free parameter. The coefficients A_j^* and B_j can be found iteratively by applying the Galerkin method at each step. At the first step, $\lambda^{(0)}(\theta) = 1$. Mikhlin (1964) establishes that Galerkin solutions of the present problem will converge *in the mean* to the exact solution of eq. (2.73) if $K(\theta,\phi)$ and $f(\theta)$ are continuous functions of ϕ and θ and if the trial functions are orthonormal and members of a complete set.

An orthonormal trial solution based on eq. (2.75) is

$$g(\theta) = \left[\sum_{j=1}^{N} A_j^* \omega_j(\theta) + \sum_{j=1}^{N} B_j \sigma_j(\theta)\right]\lambda(\theta) \quad , \tag{2.77}$$

where $\quad \omega_1(\theta) = \dfrac{1 - \cos\theta}{\sqrt{3\pi}}$

and $\quad \omega_j(\theta) = M_j\left[\dfrac{1 - \cos j\theta}{\sqrt{3\pi}} - \dfrac{2}{2j-1}\left\{\dfrac{1-\cos(j-1)\theta}{\sqrt{3\pi}} \ldots + \dfrac{1 - \cos\theta}{\sqrt{3\pi}}\right\}\right]$

with $\quad M_j = [3(2j - 1)/(2j + 1)]^{\frac{1}{2}}$

$$\sigma_j = (\sin j\theta)/\sqrt{\pi} \quad .$$

Substitution of eq. (2.77) into eq. (2.73) and application of the Galerkin method produces the following system of algebraic equations

$$\underline{D}\,\underline{A} + \underline{C}\,\underline{B} = \underline{E} \quad \text{and} \quad \underline{Q}\,\underline{A} + \underline{P}\,\underline{B} = \underline{F} \quad . \tag{2.78}$$

Elements of the various matrices in eq. (2.78) are given by

$$d_{kj} = -\frac{1}{2\pi}\int_o^{2\pi}\int_o^{2\pi} K(\theta,\phi)\ \omega_j(\phi)\ \sigma_k(\theta)\ d\phi d\theta$$

$$c_{kj} = \delta_{kj} - \frac{1}{2\pi}\int_o^{2\pi}\int_o^{2\pi} K(\theta,\phi)\ \sigma_j(\phi)\ \sigma_k(\theta)\ d\phi d\theta$$

$$q_{kj} = \delta_{kj} - \frac{1}{2\pi}\int_o^{2\pi}\int_o^{2\pi} K(\theta,\phi)\ \omega_j(\phi)\ \omega_k(\theta)\ d\phi d\theta \qquad (2.79)$$

$$p_{kj} = -\frac{1}{2\pi}\int_o^{2\pi}\int_o^{2\pi} K(\theta,\phi)\ \sigma_j(\phi)\ \omega_k(\theta)\ d\phi d\theta$$

$$e_k = e^* - 2\sin\alpha\int_o^{2\pi} \dot{y}(\theta)\ \sigma_k(\theta)\ d\theta$$

and $$f_k = -2\sin\alpha\int_o^{2\pi} \dot{y}(\theta)\ \omega_k(\theta)\ d\theta\ .$$

In eq. (2.79), $\delta_{kj} = 1$ if $k = j$, $\delta_{kj} = o$ otherwise and $e^* = \sqrt{\pi}\cos\alpha$ for $k = 1$, $e^* = o$ otherwise.

If the solutions of eq. (2.78) are labelled $\underline{A}^{(o)}$ and $\underline{B}^{(o)}$ then

$$g^{(o)}(\theta) = \left[\sum_{j=1}^{N} A_j^{(o)}\ \omega_j(\theta) + \sum_{j=1}^{N} B_j^{(o)}\ \sigma_j(\theta)\right]\lambda(\theta)\ . \qquad (2.80)$$

Because of the introduction of $\lambda(\theta)$ into eq. (2.80), $g^{(o)}(\theta)$ will no longer satisfy eq. (2.73). Therefore iterative solutions are sought as

$$\underline{A}^{(\nu)} = \underline{A}^{(o)} + \underline{\Delta A}^{(1)} \ldots \underline{\Delta A}^{(\nu)}$$

and $$\underline{B}^{(\nu)} = \underline{B}^{(o)} + \underline{\Delta B}^{(1)} \ldots \underline{\Delta B}^{(\nu)} \qquad (2.81)$$

where $\underline{\Delta A}^{(\nu)}$ and $\underline{\Delta B}^{(\nu)}$ can be obtained by solving

$$\Delta g^{(\nu)}(\theta) - \frac{1}{2\pi}\int_o^{2\pi} K(\theta,\phi)\ \Delta g^{(\nu)}(\phi)\ d\phi = f^{(\nu)}(\theta)$$

$$(2.82)$$

and $$f^{(\nu)}(\theta) = f(\theta) - g^{(\nu-1)}(\theta) + \frac{1}{2\pi}\int_o^{2\pi} K(\theta,\phi)\ g^{(\nu-1)}(\phi)\ d\phi\ .$$

Convergence of the iterative scheme corresponds to $\|\Delta A^{(\nu)}\| < \varepsilon_1$
$\|\Delta B^{(\nu)}\| < \varepsilon_2$

The lift coefficient, C_L, is given by

$$C_L = 2\int_0^{2\pi} g(\theta)\,d\theta = \frac{4\pi}{N}\sum_{j=1}^{N-1} g(\theta_i) \tag{2.83}$$

and the pressure coefficient, C_p, is given by

$$C_p(\theta_i) = 1 - \frac{g^2(\theta_i)}{(\dot{x}_i^2 + \dot{y}_i^2)} \tag{2.84}$$

where $\theta_i = 2\pi i/N, \quad i = 1, \quad 2 \ldots N-1$.

For a 20% circular arc aerofoil at $\alpha = 0°$, the pressure coefficient variation with x is shown in Fig. 6. It was found (Nugmanov) that the

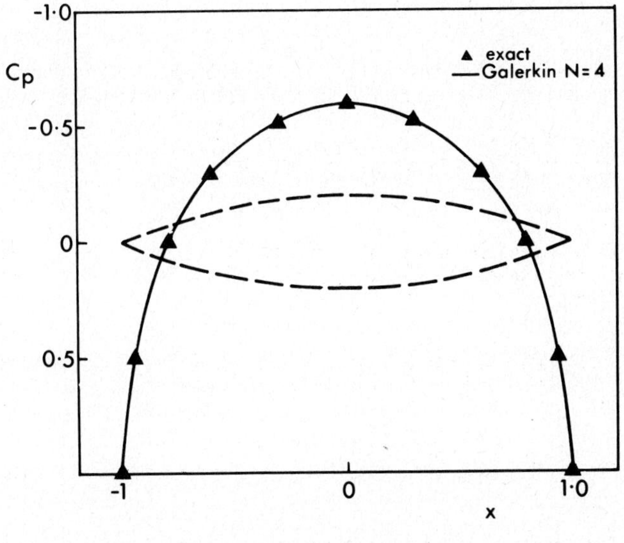

Figure 6. Pressure distribution for 20% circular-arc aerofoil

solution agreed with the exact for $N \geqslant 4$ and that for this example, A_j and B_j were insensitive to $\lambda(\theta)$. Thus the solution could be taken as eq. (2.80) without the need for further iteration.

2.4.4. Other applications

The early work of Galerkin (1917) and Duncan (1937,1938) has already been mentioned. Bickley (1941) applied the Galerkin method to the partial differential equation,

$$\phi_t - a\phi_{xx} = 0 \quad , \tag{2.85}$$

to investigate problems of electrical circuit design. Bickley also presented some comparisons with the methods of collocation and least-squares.

An early application in fluid dynamics is due to Kawaguti (1955) who studied low speed viscous flow past a sphere to determine the critical Reynolds above which a previously steady flow would become unsteady. This was done by perturbing an initially steady solution. The Galerkin method was used to provide the steady solution. The governing equation for steady, axisymmetric viscous, incompressible flow is

$$\psi_r \{\frac{1}{r^2 \sin^2\theta} D^2\psi\}_\theta - \psi_\theta \{\frac{1}{r^2 \sin^2\theta} D^2\psi\} - \frac{2}{Re \sin\theta} D^4\psi = 0 \tag{2.86}$$

where

$$D^2 \equiv \frac{\partial^2}{\partial r^2} + \frac{\sin\theta}{r^2} \frac{\partial}{\partial\theta} (\frac{1}{\sin\theta} \frac{\partial}{\partial\theta})$$

In eq. (2.86) ψ is the stream function, (r,θ,ϕ) are spherical polar coordinates and velocity components (u,v,w) are related to the stream function by

$$u = \frac{1}{r^2 \sin\theta} \frac{\partial\psi}{\partial\theta} , \quad v = \frac{-1}{r \sin\theta} \frac{\partial\psi}{\partial r} \quad \text{and} \quad w = 0 . \tag{2.87}$$

In eq. (2.86) Re is the Reynolds number based on the diameter of the sphere. Boundary conditions are required at the surface of the sphere and far from the sphere. At the surface,

$$\psi = \frac{\partial\psi}{\partial r} = 0 \quad \text{at} \quad r = 1 . \tag{2.88}$$

In the farfield,

$$\psi = \frac{r^2}{2} \sin^2\theta \text{ at } r \to \infty \tag{2.89}$$

Kawaguti introduced a trial solution in r and θ. For larger Reynolds numbers this was

$$\psi = \{\frac{r^2}{2} + \frac{a_1}{r} + \frac{a_2}{r^2} + \frac{a_3}{r^3} + \frac{a_4}{r^4}\} \sin^2\theta$$

$$+ \{\frac{b_1}{r} + \frac{b_2}{r^2} + \frac{b_3}{r^3} + \frac{b_4}{r^4}\} \sin^2\theta \cos\theta \quad . \tag{2.90}$$

The form of eq. (2.90) automatically satisfies the farfield boundary condition, eq. (2.89). To satisfy the surface boundary conditions for all θ, four of the unknowns in eq. (2.90) can be eliminated. If eq. (2.90) is substituted into eq. (2.86) some residual, R, results. By forcing the residual to be zero on the surface of the body, i.e. by demanding that eq. (2.86) is satisfied exactly, Kawaguti obtains two additional conditions for the four remaining unknown coefficients in eq. (2.90). The remaining two equations are obtained by imposing the following weighted residual relationships.

$$\int_1^\infty \int_{-1}^1 \frac{1}{r}\{P_0(z) - P_2(z)\}\, R\, dz\, dr = 0 \tag{2.91}$$

and

$$\int_1^\infty \int_{-1}^1 \frac{1}{r^2}\{P_1(z) - P_3(z)\}\, R\, dz\, dr = 0 \ . \tag{2.92}$$

In eqs. (2.91) and (2.92), $P_n(z)$ is a Legendre function of order n and $z = \cos\theta$.

The results, that Kawaguti obtained, predicted the drag coefficient quite closely but were less accurate for the pressure distribution and vorticity distribution at the surface.

Snyder and Stewart (1966) have considered an important problem in chemical engineering; I.e. very slow viscous flow through a densely packed bed of spheres. For very slow (creeping) flow the convective (inertial) terms can be ignored, in comparison with the viscous and pressure terms; this renders the equations linear. Imposition of symmetry provides many of the boundary conditions. After applying the traditional Galerkin formulation it was found that not enough equations were available, therefore additional orthogonal relations had to be imposed. The evaluations of the integrals, like eq. (2.41), was undertaken with Gaussian quadrature. An indication of the accuracy can be obtained from the fact that the predicted flow rate was within 5% of the measured flow rate with 28 unknowns in the expansions for the three velocity components and pressure.

Durvasula (1971) has examined the flutter characteristics of clamped skew panels in supersonic flow using a 16-term beam function expansion for the trial solution. Chopra and Durvasula (1971) have considered the vibration of trapezoidal plates with up to a six-term Fourier sine series to represent the deflected surface. Prabhu and Durvasula (1976) have analysed the post-buckling characteristics of clamped skew plates when subjected to various thermal and mechanical loads.

2.5 Limitations of the Traditional Galerkin Method

For the two model problems given in Section 2.3, high accuracy is obtained with a relatively low-order trial solution (and hence small effort). However the ready availability of computers has fostered a greater interest in more complex problems and the achievement of higher accuracy.

A direct increase in the order of the analytic representation in the Galerkin method creates an immediate problem. That is, there is *little difference*, typically, between the nth and $(n-1)^{th}$ weight functions. Thus, the algebraic equations that result from the application of the traditional Galerkin method would be almost linearly dependent for large n.

If Duncan's example is considered with the weight function, x^{99} and x^{100}, the resulting equations are

$$0.000099\ a_1 + 0.009998\ a_2 + 0.019703\ a_3 \ldots 0.494975\ a_{99} + 0.497513\ a_{100} = 0.01 \tag{2.93}$$

and

$$0.000097\ a_1 + 0.009899\ a_2 + 0.019511\ a_3 \ldots 0.492487\ a_{99} + 0.495025\ a_{100} = 0.0099\ . \tag{2.94}$$

Clearly the resulting matrix of equations will be very *ill-conditioned*. In addition, the matrix will be dense and consequently will have a *high operation count*. In particular if any of the matrices have to be factorised, e.g. the matrix M in eq. (2.38), the operation count will be of $O(N^3)$ where N is the order of the matrix.

A second disadvantage is that imposing the weighted residual integral has required that the limits of integration *coincide with coordinate lines*. In principle this problem can be avoided for distorted but simple geometries by mapping the distorted physical region into a regular region. However this not only increases the algebraic complexity but one would also expect the accuracy to be degraded since the grid is being severely distorted in the physical plane.

A third disadvantage concerning products of series arise with nonlinear equations, e.g. the convective term in eq. (2.30). In typical nonlinear problems perhaps three or more series will be multiplied together and, since an interative solution will probably be required, the coefficients in the series will need to be upgraded at every step of the iteration. Clearly this will impose a *severe operational overhead* if the number of terms in each series is large.

Improvements, associated with computer applications, have gone in two directions. Firstly the problem of matrix ill-conditioning can be avoided by using *orthogonal* basis functions so that a matrix like M in eq. (2.27) has only diagonal entries and is trivial to invert. *Spectral* methods are of this type and will be considered in Section 3.

The second direction is to give the unknowns a *direct* physical significance by making them nodal unknowns of the problem. The related basis functions then become *interpolatory* in nature. In addition the basis functions are deliberately restricted to *low order* and *local support*. Linear and quadratic polynomials are typical. As a consequence matrices like M, B or C in Section 2.3.2, although not diagonal, are *very sparse* and the non-zero entries in the matrix occur close to the main diagonal. If the trial solutions are introduced in discrete subregions or '*finite elements*' the treatment of arbitrary domain boundaries is greatly facilitated. The Galerkin finite element method incorporates the above features and it will be described in Section 4.

3. GALERKIN SPECTRAL METHOD

Like the traditional Galerkin method, spectral methods are *global* in the sense that the trial solution spans the whole region. Thus altering a coefficient in the trial solution immediately *influences the whole region*. The essential differences from the traditional Galerkin method are in the use of *orthogonal* functions and the special treatment given to nonlinear terms. The use of orthogonal functions overcomes the problem of matrix illconditioning (Section 2.5) associated with the use of almost linearly-dependent test functions in the traditional Galerkin method. For large values of N, the number of unknowns in the trial solution, special procedures are needed to handle nonlinear terms if reasonable economy is to be achieved. The relative merits of spectral methods, in relation to finite difference and finite element methods, are considered by Fletcher (1983a).

3.1 Orthogonal Functions

The test and trial functions, ϕ_k, are orthogonal if

$$(\phi_k,\phi_j) = 0 \quad \text{with} \quad k \neq j$$
$$\neq 0 \quad \text{with} \quad k = j \quad . \tag{3.1}$$

The result of using orthogonal functions with the examples considered in Section 2.3 is that the "mass" matrix, $\underline{M}$, in eqs. (2.27) and (2.38), only has non-zero entries on the diagonal. Consequently the computationally expensive, $O(N^3)$, factorisation of $\underline{M}$ is avoided and the ill-conditioning problem (eqs. (2.93) and (2.94)) does not arise. In addition there is no loss of accuracy in using orthogonal functions.

Typical choices for test and trial functions for spectral methods are shown in Table 7

Table 7. Heirarchy of trial functions

Trial function	Comments
Eigenfunctions	Suggested by a related problem
Fourier series	Periodic boundary conditions; infinitely differentiable
Legendre polynomials	Good wavelength resolution; nonperiodic
Chebyshev polynomials	Very robust; nonperiodic; minimax

If *eigenfunctions* of a closely related problem are available, accurate results are obtained with only a few terms, N, in the trial solution. Bourke et al.

(1977) use surface spherical harmonics as trial functions in global weather prediction. Surface spherical harmonics are eigenfunctions of the Laplace equation written in spherical co-ordinates.

For eigenfunctions to yield accurate results for small values of N it is necessary, typically, for the exact solution to be infinitely differentiable and for particular, often homogeneous, boundary conditions to apply. If the various conditions do not hold the convergence rate with N will be much lower.

For problems with periodic boundary conditions *Fourier series* are an appropriate choice for the orthogonal trial solution. If the exact solution is infinitely differentiable and if sufficient homogeneous boundary conditions on higher-order derivatives are available a convergence rate *greater than any power of* 1/N as $N \to \infty$ is possible (Orszag (1980)).

If a Fourier series is used as the trial solution for a problem that has nonperiodic boundary conditions or if a *discontinuity* occurs in the exact solution in the interior of the domain the convergence rate is typically $O(N^{-1})$ or $O(N^{-2})$ (Fletcher 1983a)). The problem of nonperiodic boundary conditions could be overcome by adding a few terms of a polynomial to the Fourier series trial solution. However then the convergence rate is controlled by the number of polynomial terms added (Gottlieb and Orszag (1977)).

A better strategy for problems with nonperiodic boundary conditions is to use orthogonal polynomials, such as *Legendre* or *Chebyshev* polynomials, as test and trial functions.

Legendre polynomials are orthogonal over the range $-1 \leqslant x \leqslant 1$. Gottlieb and Orszag (1977) indicate that a very rapid rate of convergence with N is achieved if *at least* Π *terms are included* in the trial solution for every complete wave expected in the exact solution. If discontinuities occur in the exact solution the use of Legendre polynomial trial functions will, typically, produce a convergence rate of only $O(N^{-\frac{1}{2}})$ near the boundaries of the domain. This is a weakness of Legendre polynomials.

Chebyshev polynomials have often been used with spectral methods. Chebyshev polynomials do not obey eq. (3.1) but instead obey the *weighted* orthogonal relationship,

$$\left(T_j(x), \frac{T_k(x)}{(1-x^2)^{\frac{1}{2}}} \right) = 0 \text{ with } k \neq j$$
$$\neq 0 \text{ with } k = j \qquad (3.2)$$

To make use of the weighted orthogonal property, eq. (3.2), requires a generalisation of the Galerkin procedure (Fletcher (1983a)) where Chebyshev polynomials, $T_j(x)$, are used as trial functions but the groups, $T_k(x)/(1 - x^2)^{\frac{1}{2}}$, are used as test functions. If the exact solution is sufficiently smooth the use of Chebyshev polynomials will produce a convergence rate comparable to that of a Fourier series. If an internal discontinuity occurs the convergence rate is like $O(N^{-1})$, both in the interior and adjacent to the boundaries.

The orthogonal functions shown in Table 7 have been placed in a particular order. The higher an orthogonal function is on the list in Table 7 the more restrictions on differentiability and boundary conditions are required. However if the restrictions are met the convergence rate will be higher for the same number of terms in the trial solution. Conversely the lower an item is on the list the less sensitive to boundary condition specification and solution differentiability will be the accuracy and convergence properties of the spectral solution.

However if discontinuities in the exact solutions occur convergence rates *no better than* $O(N^{-1})$ or $O(N^{-2})$ are obtained irrespective of the choice of orthogonal functions. Such a convergence rate is equivalent to $O(\Delta x)$ or $O(\Delta x^2)$ for a finite element or finite difference method.

3.2 Examples

Here we will consider two examples to demonstrate the application of the spectral method. The first is the linear heat conduction equation which is often chosen as the archetypical parabolic partial differential equation. Burgers' equation is included as the second example. This illustrates how the spectral method handles nonlinear problems. For both examples the spectral method will be applied to the spatial terms, only.

3.2.1 Heat conduction equation

This is a linear problem to which the spectral method will be applied to reduce the original partial differential equation to an ordinary differential equation. The governing equation is

$$\psi_x - \psi_{yy} = 0 \ , \tag{3.3}$$

with boundary conditions

$$\psi(0,y) = \sin \pi y + y \ , \quad \psi(x,0) = 0 \ , \quad \psi(x,1) = 1 \tag{3.4}$$

In this problem ψ is a nondimensional velocity of the viscous flow between two parallel plates (Fletcher, 1978a). A trial solution can be introduced that incorporates the boundary condition at $x = 0$

$$\psi_a = \sin \pi y + y + \sum_{j=1}^{N} a_j(x) \sin j\pi y \ . \tag{3.5}$$

Application of the spectral method produces the following ordinary differential equation,

$$\frac{da_k}{dx} + (k\pi)^2 a_k + r_k = 0 \ , \quad k = 1,\ldots N \ , \tag{3.6}$$

where

$$\begin{aligned} r_k &= \pi^2 \quad \text{if} \quad k = 1 \\ &= 0 \quad \text{if} \quad k \neq 1 \ . \end{aligned} \tag{3.7}$$

The solution of eqs. (3.6), (3.7) can be written down directly as

$$a_1 = \exp(-\pi^2 x) - 1$$

and

$$a_k = 0 \ , \quad k = 2,\ldots N \tag{3.8}$$

so that eq. (3.5) becomes

$$\psi_a = \sin \pi y \exp(-\pi^2 x) + y \quad , \tag{3.9}$$

which is, in fact, the exact solution of eqs. (3.3) and (3.4). This has happened because the trial solution, fortuitously, *includes* the exact solution. One would expect this to be a rather rare occurrence. A more realistic solution can be achieved if the entry velocity distribution at $x = 0$ is given by

$$\psi(o,y) = 5y - 4y^2 \quad . \tag{3.10}$$

The development is similar to the previous cases except that r_k in eq. (3.7) is replaced by

$$r_k = 8 \int_0^1 \sin k\pi y \ \ dy \quad . \tag{3.11}$$

For this case eq. (3.6) has been integrated with an Euler scheme and $\Delta x = 0.001$. The results for different N, in eq. (3.5), are shown in Table 8. Also shown in Table 8 are the exact solution and a solution ψ_b which only has errors associated with the numerical integration scheme. The exact solution, shown in Table 8, can be obtained by the separation of variables technique as

$$\psi = y + 4 \sum_{j=1}^{\infty} \left[\frac{2}{(2j-1)\pi}\right]^3 \sin(2j - 1)\pi y \exp\left[-\{(2j - 1)\pi\}^2_x\right] . \tag{3.12}$$

The approximate solution, ψ_b, is obtained from

$$\psi_b = y + 4 \sum_{j=1}^{\infty} \left[\frac{2}{(2j-1)\pi}\right]^3 \sin(2j-1)\pi y \ X_j(x) \tag{3.13}$$

where $X_j(x)$ is obtained from the numerical solution, using the same Euler scheme as for ψ_a, of

$$\frac{\partial X_j}{\partial x} + \{(2j-1)\pi\}^2 X_j = 0 \quad . \tag{3.14}$$

Examination of Table 8 indicates that better agreement is achieved with increasing N. The error between ψ_a and the exact solution, ψ, comes partly from the numerical integration error and partly from the approximation inherent in the Galerkin formulation. The difference between ψ_a and ψ_b is independent of x and arises from the Galerkin formulation; in fact, from the fitting of the initial data (on $x = 0$) in terms of the trial functions.

When the traditional Galerkin method is applied to this problem (Fletcher (1983a)) comparable accuracy to that shown in Table 8 with N is achieved. However for the spectral method the expensive factorisation of $\underline{M}$, e.g. as in eq. (2.38), is avoided.

Table 8. Solutions of eq. (3.3)

x (at y=0.5)	Approximate solution, ψ_a N = 1	N = 3	N = 5	Exact Solution, ψ	Approximate Solution, ψ_b
0	1.5000	1.5000	1.5000	1.5000	1.5000
0.02	1.3143	1.3466	1.3384	1.3408	1.3404
0.04	1.1620	1.1993	1.1911	1.1943	1.1931
0.06	1.0371	1.0752	1.0670	1.0707	1.0690
0.08	0.9347	0.9729	1.9647	0.9686	0.9667
0.10	0.8507	0.8889	0.8807	0.8847	0.8828
0.12	0.7819	0.8201	0.8118	0.8158	0.8139
0.14	0.7254	0.7636	0.7553	0.7592	0.7574
0.16	0.6791	0.7173	0.7090	0.7182	0.7111
0.18	0.6411	0.6793	0.6710	0.6747	0.6731
0.20	0.6099	0.6481	0.6399	0.6434	0.6420

3.2.2 Burgers' equation

The governing equation is

$$u_t + uu_x - \frac{1}{Re} u_{xx} = 0 \quad , \tag{3.15}$$

with boundary conditions

$$u(-1,t) = 1 \quad , \quad u(1,t) = 0$$

and

$$u(x,0) = 1 \quad , \quad x \leqslant 0 \tag{3.16}$$

$$u(x,0) = 0 \quad , \quad x > 0 \quad .$$

This problem is that of a propagating "shock" which was also considered in conjunction with the traditional Galerkin method in Section 2.3.2. The main complication compared with the heat conduction equation is the appearance of the nonlinear term. However, the method of solution develops in essentially the same way. A trial solution is introduced by

$$u(x,t) = \sum_{j=0}^{N} a_j(t)\, P_j(x) \quad , \tag{3.17}$$

where $P_j(x)$ are Legendre functions (Dahlquist et al., 1974) which are orthogonal polynomials over the interval $-1 \leqslant x \leqslant 1$. Following the same procedure as in Section 2.3.2, one obtains a set of algebraic equations which can be written,

$$\underline{M}\,\dot{\underline{A}} + (\underline{B} + \underline{C})\underline{A} = 0 \tag{3.18}$$

where an element of $\underline{M}$ is given by

$$m_{k,k} = 2/(2k+1)$$

$$m_{j,k} = 0 \quad , \quad \text{if } j \neq k \quad . \tag{3.19}$$

An element of $\underline{B}$ is given by

$$b_{k,j} = \sum_i a_i \int_{-1}^{1} \frac{dP_i(x)}{dx} P_j(x)\, P_k(x) dx \quad , \tag{3.20}$$

and an element of $\underline{C}$ by

$$c_{k,j} = \frac{1}{Re} \int_{-1}^{1} \frac{dP_j(x)}{dx} \frac{dP_k(x)}{dx} dx \quad . \tag{3.21}$$

Table 9. Solution of Burgers' equation by spectral method. T = 0.80 and Re = 10

x	Approximate Solution			Exact Solution
	N = 7	N = 9	N = 11	
-1.0	1.0000	1.0000	1.0000	1.0000
-0.9	0.9974	1.0054	0.9986	1.0000
-0.8	0.9856	1.0082	0.9997	0.9999
-0.7	0.9799	1.0020	1.0017	0.9998
-0.6	0.9855	0.9926	1.0012	0.9996
-0.5	1.0000	0.9870	0.9978	0.9988
-0.4	1.0176	0.9882	0.9937	0.9972
-0.3	1.0290	0.9934	0.9904	0.9934
-0.2	1.0258	0.9947	0.9857	0.9852
-0.1	1.0007	0.9816	0.9730	0.9687
0	0.9491	0.9442	0.9422	0.9372
0.1	0.8696	0.8758	0.8829	0.8816
0.2	0.7643	0.7748	0.7887	0.7920
0.3	0.6390	0.6463	0.6611	0.6648
0.4	0.5025	0.5011	0.5109	0.5104
0.5	0.3657	0.3546	0.3573	0.3542
0.6	0.2405	0.2233	0.2224	0.2225
0.7	0.1377	0.1213	0.1236	0.1280
0.8	0.0653	0.0561	0.0646	0.0684
0.9	0.0259	0.0241	0.0330	0.0344
1.0	0.0136	0.0078	0.0102	0.0164
$\Vert u-u_e \Vert_{rms}$	0.0200	0.0086	0.0031	

The initial condition, eq. (3.16), is converted into initial values of the a_j's by solving the system,

$$\underline{M}\,\underline{A} = \underline{D} \tag{3.22}$$

where an element of $\underline{D}$ is given by

$$d_k = \int_{-1}^{0} P_k(x)\ 1\ dx\quad . \tag{3.23}$$

The system of equations, eq. (3.18), is integrated in time with a fourth-order Runge-Kutta schemes and a variable step-size (temporal). After a certain time, $\Delta t \,\underline{\Omega}\, 0.80$, the solution is compared with the exact solution. For Re = 10, solutions with different orders of the representation, N in eq. (3.17), are presented in Table 9 and Figure 7. The last two coefficients in eq. (3.17) have been determined, in terms of the other coefficients in eq. (3.17) so that the boundary conditions at x = ±1 are satisfied.

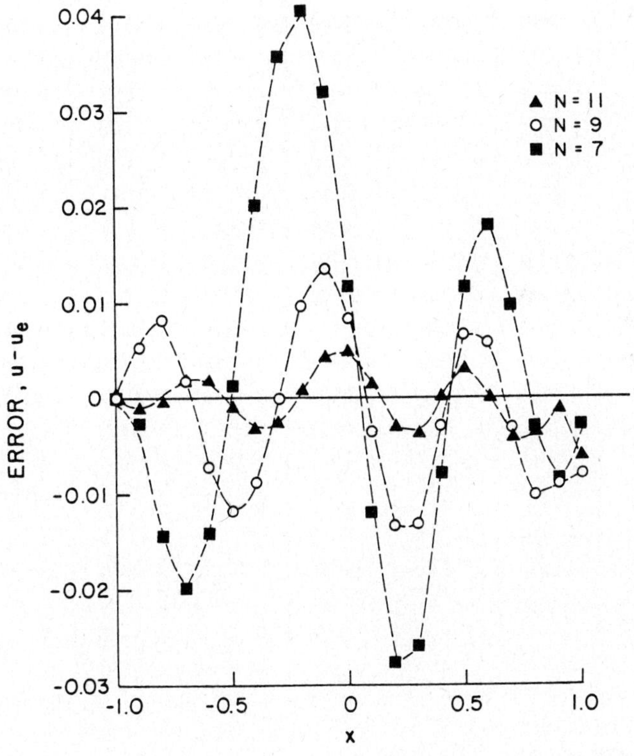

Figure 7. Distribution of error for the solution of Burgers' equation, Re = 10

For the solutions presented in Tables 9 and 10 the boundary conditions at x = ±1 have been set equal to the exact solution every five steps of the time integration. The results presented in Table 9 indicate that as the order of the representation is increased *the accuracy improves rapidly*. Comparing the values of $\|u-u_e\|_{rms}$ in both Tables 4 and 9 it can be seen that the accuracy of the spectral and traditional Galerkin methods is comparable although less computational work is required with the spectral method.

Table 10. Solution of Burgers' equation by spectral method, N = 9, various Reynolds number.

	Re = 1.0, t = 0.19		Re = 10, t = 0.80		Re = 100, t = 0.66	
x	Approx.	Exact	Approx.	Exact	Approx.	Exact
-1.0	0.9718	0.9706	1.0000	1.0000	1.0000	1.0000
-0.9	0.9593	0.9574	1.0054	1.0000	1.0052	1.0000
-0.8	0.9426	0.9395	1.0082	0.9999	1.0830	1.0000
-0.7	0.9198	0.9162	1.0020	0.9998	1.0646	1.0000
-0.6	0.8898	0.8865	0.9926	0.9996	0.9760	1.0000
-0.5	0.8521	0.8499	0.9870	0.9988	0.8968	1.0000
-0.4	0.8068	0.8058	0.9882	0.9971	0.8853	1.0000
-0.3	0.7544	0.7545	0.9934	0.9933	0.9515	1.0000
-0.2	0.6957	0.6965	0.9947	0.9849	1.0603	1.0000
-0.1	0.6321	0.6328	0.9816	0.9681	1.1509	1.0000
0	0.5648	0.5651	0.9442	0.9360	1.1634	1.0000
0.1	0.4956	0.4954	0.8758	0.8793	1.0609	1.0000
0.2	0.4262	0.4258	0.7748	0.7882	0.8436	0.9986
0.3	0.3585	0.3585	0.6463	0.6514	0.5508	0.8218
0.4	0.2943	0.2955	0.5011	0.5040	0.2503	0.0297
0.5	0.2355	0.2383	0.3546	0.3479	0.0165	0.0002
0.6	0.1834	0.1879	0.2233	0.2174	-0.0967	0.0000
0.7	0.1393	0.1449	0.1213	0.1244	-0.0813	0.0000
0.8	0.1037	0.1092	0.0561	0.0662	0.0095	0.0000
0.9	0.0762	0.0804	0.0241	0.0331	0.0718	0.0000
1.0	0.0555	0.0578	0.0078	0.0157	0.0000	0.0000
$\|u-u_e\|_{rms}$	0.0029		0.0086		0.1175	

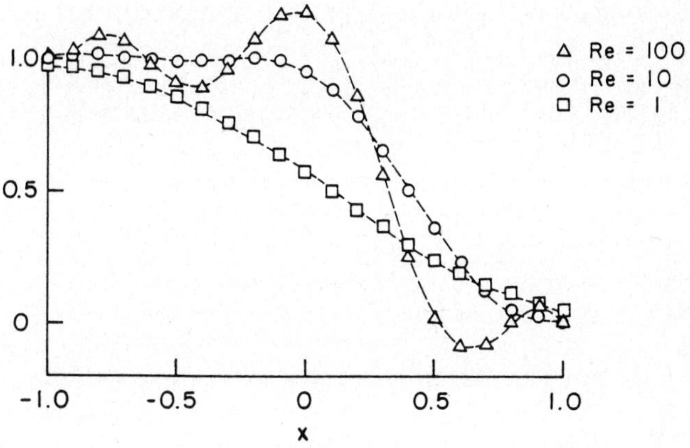

Figure 8. Solution of Burgers' equation for increasing Reynolds number

Solutions at different Reynolds number are presented in Table 10 and Figure 8. The main problem in obtaining accurate solutions is related to the difficulty of the approximate solution in representing the exact solution in the region where the shock is situated. In this region u is changing most rapidly. This problem becomes more difficult as the shock becomes steeper, i.e. as the Reynolds number increases. This can be seen most easily by examining the values of the rms norm, $\|u-u_e\|_{rms}$. The solutions can be compared with those for the traditional Galerkin method, which are presented in Table 5. It can be observed that the solutions are of comparable accuracy, as might be expected.

Gottlieb and Orszag (1977) have formulated a general Chebyshev spectral approach to Burgers' equation and have considered in more detail a linearised Burgers' equation, i.e. the convective term is replaced by $u_o u_x$ where u_o is a constant. They found that a Legendre Galerkin spectral approximation was stable and convergent for the linearised Burgers' equation used with boundary conditions

$$u(-1,0) = 0 \quad , \quad u(1,0) = 0 \tag{3.24}$$

and initial conditions

$$u(x,0) = g(x) \quad . \tag{3.25}$$

However for a Chebyshev Galerkin spectral formulation the solution was not convergent unless $N > N_{crit}$ where N_{crit} depended on the value of Re u_o. For Re u_o = 100, N_{crit} = 15, for Re u_o = 2500, N_{crit} = 101. It turns out that for the boundary conditions considered a narrow boundary layer develops at the outflow (x = 1) for large Re u_o and satisfying the outflow boundary condition is the cause of the instability.

3.3 Special Procedure for Nonlinear Terms

The evaluation of quadratically nonlinear terms, in the direct manner, eq. (3.20), is an $O(N^3)$ process. This feature essentially limited the use of the spectral method to small values of N until Orszag (1969) introduced the *transform* technique.

More recently, Orszag (1980) has expressed the spirit of transform techniques in the following way: "Transform freely between the physical and spectral representations, evaluating each term in whatever representation that term is most accurately, and simply, evaluated."

Implicit in the above remarks is the requirement that a typical transformation, e.g.

$$u(x_\ell) = \sum_{j=1}^{N} a_j \, \phi_j(x_\ell), \quad \ell = 1, \dots N \qquad (3.26)$$

and inverse transformation (assuming ϕ_j are orthonormal functions),

$$a_k = \int_R u \, \phi_k(x) \, dx, \quad k = 1, \dots N \qquad (3.27)$$

and can be evaluated *very economically*. A direct evaluation of eqs. (3.26), or (3.27) using numerical quadrature, would require an $O(N^2)$ process. However if the trial and test functions, ϕ_j, are members of a Fourier series, the *fast Fourier transform* will require only $O(N \log_2 N)$ operations.

The use of the fast Fourier transform with the Galerkin method is described by Orszag (1971c). However Orszag (1980) notes that comparable fast transformations are possible for other orthogonal trial functions. We will use the expression fast transform to imply a technique that will evaluate systems of equations like (3.26) or (3.27) in $O(N \log N)$ operations rather than $O(N^2)$ operations.

Orszag (1980) points out that the main advantage of transform methods comes from their ability *to split up multidimensional transforms into a sequence of one-dimensional transforms*. An example is given of solving the Navier-Stokes equations for three dimensional incompressible flow with periodic boundary conditions. A trial solution with 128 unknown coefficients in each direction is used. The evaluation of all the nonlinear terms in spectral space requires about 5×10^5 s per time-step on a CRAY-1 computer. Using a fast transform to physical space permits an evaluation in 20 s per time-step. However the fast transform provides a speed-up by a factor of 2 and the conversion to a sequence of one-dimensional transformations provides the rest of the speed-up, $O(10^4)$!

We will now describe, conceptually, the integration of one time-step of the spectral formulation of Burgers' equation using some of the above ideas.

We assume that a_j^n are known at time level n. The following sequence is required:

1) evaluate $u_a^n(x_\ell) = \sum_j a_j^n \, \phi_j(x_\ell)$, $\ell = 1, \dots 2N$, F.T.: $O(2N \log 2N)$

2) evaluate $b_j^{(1)n}$ from a_j^n by recurrence, : $O(2N)$

3) evaluate $\partial u_a^n(x_\ell)/\partial x = \sum_j b_j^{(1)n} \phi_j(x_\ell)$, $\ell = 1...2N$, F.T.: $0(2N \log 2N)$

4) evaluate $w^n(x_\ell) = u_a^n(x_\ell)\ \partial u_a^n(x_\ell)/\partial x$, $\ell = 1...2N$, : $0(2N)$

5) evaluate $d_k^n = \int_R w^n \phi_k \, dx$, F.T.: $0(2N \log 2N)$

6) evaluate $s_k^n = \sum_j \left(\phi_k, \dfrac{\partial^2\phi_j}{\partial x^2}\right) a_j^n$ by recurrence, : $0(N)$

7) evaluate $da_k^{n+1}/dt = d_k^n - s_k^n/Re$, : $0(N)$

8) evaluate $a_k^{n+1} = a_k^n + f\left(da_k^{n+1}/dt\right)$. : $0(N)$

In the above, d_k^n is evaluated with a quadrature scheme using the values of $w(x_\ell)$ evaluated at step 4. For large values of N it is clear that the above sequence is considerably more economical than the more direct procedure described in Section 3.2.2.

It is necessary to evaluate $w(x_\ell)$ at 2N points to avoid aliasing errors. *Aliasing* is the phenomenon where high frequencies of the solution on a discrete grid appears as low frequencies. In the present situation high frequencies are generated by the products that occur in the nonlinear terms.

Aliasing is often cited as the underlying cause of the nonlinear instabilities that can occur in the long-term integration of equations governing global atmospheric circulation. However for many problems there is sufficient natural dissipation to prevent aliasing from destabilising the time integration (Orszag, 1972). For such problems $w(x_\ell)$, in step 4, need only be evaluated at N points.

3.4 Spectral Method for Boundary Layer Equations

In this section we will start with the method of integral relations, which is a member of the class of methods of weighted residuals, and show that it can be upgraded into a Galerkin spectral method, i.e. *Dorodnitsyn spectral method*. Like the traditional Galerkin method, the method of integral relations endeavours to use special test and trial functions so that an accurate solution can be obtained with relatively few unknown coefficients. The essential feature of the Dorodnitsyn spectral method is to take the special test functions and construct orthonormal functions from them with respect to a weight function suggested by the problem at hand.

The method of integral relations was applied to the boundary layer equations,

$$u_x + u_y = 0 \tag{3.28}$$

$$uu_x + vu_y = u_e\, u_{e_x} + \nu u_{yy} \ , \tag{3.29}$$

by Dorodnitsyn (1960). The method consists of eliminating v from eqs. (3.28) and (3.29) by forming the product, $f_k \times$ eq. (3.28) + $df_k/du \times$ eq. (3.29) and integrating the resulting equation across the boundary layer with respect to y. Subsequently the independent variables are changed from (x,y) to (x,u). The result is an ordinary differential equation,

$$\frac{d}{dx}\int_0^1 \theta\, u\, f_k\, du = \frac{u_{e_x}}{u_e}\int_0^1 \frac{df_k}{du}\,\theta\,(1-u^2)du - \left[\frac{df_k}{du}\,\tau\right]_{wall} - \int_0^1 \frac{d^2 f_k}{du^2}\,\tau\, du\,. \quad (3.30)$$

In eq. (3.30), $\theta = 1/\tau$ and τ is the shear stress, $f_k(u)$ is a weight function to be prescribed and u_e is the known velocity at the outer edge of the boundary layer. The method proceeds by prescribing the dependence of θ on u as

$$\theta = \frac{1}{1-u}\left(a_o + \sum_j a_j\, u^j\right) \quad . \qquad (3.31)$$

The factor (1-u) is required so that τ goes to zero at the outer edge of the boundary layer (u = 1). For the weight function, f_k, Dorodnitsyn used

$$f_k(u) = (1 - u)^k \quad . \qquad (3.32)$$

Substitution of eqs. (3.31) and (3.32) into eq. (3.30) and evaluation of the integrals produces a system of ordinary differential equations for the a_j's that can be written,

$$\underline{B}\ \ [da_j/dx] = \underline{C}\ , \qquad (3.33)$$

where $\underline{B}$ and $\underline{C}$ are matrices that must be re-evaluated at every step since they depend on the solution. It is apparent, from the form of f_k in eq. (3.32), that for large k the difference between f_k and f_{k-1} is not very great and the resulting equations in the system (3.33) will be almost linearly dependent. Consequently the equations $\underline{B}$ and $\underline{C}$ become progressively more ill-conditioned, and in the case of $\underline{B}$, difficult to invert accurately.

This problem is overcome in the Dorodnitsyn spectral method by generating a set of *orthonormal functions*, g_j, from the Dorodnitsyn weighting functions, f_k. The functions g_j replace f_k in eq. (3.30). In addition, the analytic representation for θ, eq. (3.31), is replaced by an analytical representation in terms of the g_j's. Thus a Galerkin spectral formulation is obtained in which matrix $\underline{B}$ is diagonal, and hence trivial to invert, and matrix $\underline{C}$ is well-conditioned. This interpretation of the orthonormal method of integral relations as a Galerkin spectral method is described in more detail in Fletcher (1978b).

The orthonormal functions, g_j, are generated as follows. Firstly, a general dependence on f_k is introduced by

$$g_j = \sum_{k=1}^{j} e_{kj}\, f_k \quad , \qquad (3.34)$$

where the coefficients e_{kj} are evaluated using the *Gram-Schmidt* orthonormalisation process (e.g. Isaacson and Keller, 1966). To obtain a diagonal form for B, the functions g_j are made orthonormal with respect to a given weighting function, w(u), according to the following inner product

$$(g_i,g_j) = \int_0^1 g_i(u)\ g_j(u)\ w(u)du \tag{3.35}$$

and

$$(g_i,g_j) = 0 \quad \text{if} \quad i \neq j$$
$$= 1 \quad \text{if} \quad i = j \ .$$

The following analytic representation for θ is assumed (to replace eq. (3.31)),

$$\theta = \frac{1}{1-u}\left(b_o + \sum_j b_j\ g_j\right) \ . \tag{3.36}$$

The non-orthogonal leading term, b_o, is retained so that θ will behave correctly at $u = 1$. Substitution of eq. (3.36) into eq. (3.30) and replacement of f_k with g_k gives

$$\frac{d}{dx}\int_0^1 \left(b_o + \sum_j b_j g_j\right) g_k \frac{u}{1-u}\, du = C_k' \quad , \quad k = 1,\ldots N \tag{3.37}$$

where

$$C_k' = \frac{u_{e_x}}{u_e}\int_0^1 \theta\,\frac{dg_k}{du}(1-u^2)du - \left[\frac{dg_k}{du}\,\tau\right]_{wall} - \int_0^1 \frac{d^2 g_k}{du^2}\,\tau\, du \ . \tag{3.38}$$

Comparing eqs. (3.35) and (3.38), it is clear that if B is to be diagonal then

$$w(u) = u/(1-u) \ . \tag{3.39}$$

Thus eq. (3.37) can be written

$$\frac{db_o}{dx} V_k + \frac{db_k}{dx} = C_k' \quad , \quad k = 1,\ldots N-1 \tag{3.40}$$

where

$$V_k = \int_0^1 g_k\ w(u)\ du; \tag{3.41}$$

when $k = N$,

$$\frac{db_o}{dx} = C_N'/V_N \ . \tag{3.42}$$

Thus eq. (3.40) can be replaced by

$$\frac{db_k}{dx} = C_k' - C_N'\ V_k/V_N \ . \tag{3.43}$$

From the form of eq. (3.41), V_k can be evaluated once and for all; however $C'(k)$ must be re-evaluated at every step. Results are presented by Fletcher and Holt (1975) indicating that the Dorodnitsyn spectral method is considerably more accurate and much more economical, for the same value of N, than the method of integral relations. Fletcher and Holt (1976) have used the Dorodnitsyn spectral method to study the boundary layer development on inclined cones in supersonic flow.

The Dorodnitsyn spectral method has been compared with the Dorodnitsyn finite element method by Fleet and Fletcher (1982) and with a typical finite difference method for laminar (Fletcher and Fleet, (1983a)) and turbulent (Yeung and Yang (1981), Fletcher and Fleet (1983b)) boundary layer flow.

3.5 Applications

Spectral methods have been applied in many areas with particular success in two: weather prediction and turbulence simulation. Most applications have been to time-dependent mixed initial-boundary value problems with finite difference schemes to provide time differencing and integration.

3.5.1. Weather prediction

The first application of a spectral method to a meteorological flow was due to Silberman (1954) who considered the vorticity equation in a spherical coordinate system. Lorenz (1960) established that a truncated spectral representation of nondivergent barotropic flow conserved mean square kinetic energy and mean square vorticity. Platzman (1960) showed that this property would prevent nonlinear instability from developing.

In the early spectral methods all nonlinear terms were evaluated in spectral space via *interaction coefficients*. Necessarily the order N of the trial solution was limited. Orszag (1970) demonstrated the transform technique on the vorticity equation and showed that the $O(N^5)$ operations per time-step associated with the interaction coefficient formulation could be reduced to $O(N^3)$ operations per time-step if a transform method was used.

Subsequently Bourke (1972) applied a transform Galerkin spectral method to the divergent barotropic vorticity equations in two dimensions (latitude/longitude), and made detailed comparisons of the transform and interaction coefficient techniques.

Bourke et al. (1977) subsequently described an operational model based on a wave number truncation of N = 15. However to suit a cut-off of N = 15, Bourke et al. found it necessary to smooth the topology of the earth's surface.

They used a semi-implicit time integration scheme of a leapfrog type. For a typical equation, the vorticity equation, this scheme is written

$$\zeta^{n+1} = \zeta^{n-1} + 2\Delta t\ a + 2\Delta t\ V\ \zeta^{n+1} \quad . \tag{3.44}$$

a contains the nonlinear terms which are evaluated at time level n (the convection terms) and time level n-1 (horizontal diffusion terms). V represents the linearised internal vertical diffusion. The semi-implicit scheme requires the solution of full matrices by Gaussian elimination at each time-step, but permits a maximum time-step increase from 10 mins to 1 hr with only a 3% computational overhead at N = 15. Bourke et al. also discussed the possibility of time-splitting.

Current operational models (Puri, 1981) use nine vertical levels and wave number cut-offs in the range N = 21 to 31. Projected operational models (Puri, 1982) are expected to approximately double the wave number cut-off. The main limitation on the useful prediction period appears to be the quality and resolution of the initial data. Currently, reliable four day and two day predictions are possible for Northern and Southern Hemispheres respectively.

3.5.2. Turbulence simulation

In contrast to the empirical modelling of the effects of turbulence it is possible to solve the incompressible unsteady Navier-Stokes equations, if the details of the flow can be resolved sufficiently. The governing equations for unsteady incompressible flow are

$$\frac{\partial v_\ell}{\partial t} + v_j \frac{\partial v_\ell}{\partial x_j} = \frac{-1}{\rho}\frac{\partial p}{\partial x_\ell} + \nu\left\{\frac{\partial^2 v_\ell}{\partial x_j \partial x_j}\right\} \tag{3.45}$$

and

$$\frac{\partial v_\ell}{\partial x_\ell} = 0 \tag{3.46}$$

where the indices ℓ, $j = 1, 2, 3$ in three dimensions. $v_\ell(x_j, t)$ are the velocity components and $p(x_j, t)$ is the pressure. The flow becomes turbulent if the kinematic viscosity is sufficiently small.

A spectral formulation for the direct simulation of turbulence was given by Orszag and Kruskal (1968). However it was the introduction of the transform technique to handle nonlinear terms (Section 3.3) by Orszag (1969) that permitted computational results to be obtained at a reasonable cost.

Early applications of the method have included the simulation of three-dimensional homogeneous isotropic turbulence at moderate Reynolds number (Orszag and Patterson, 1972) and the use of two-dimensional turbulence simulations to provide test data to assess analytic theories of turbulence (Herring et al., 1974).

Homogeneous turbulence can be simulated by solving eqs. (3.45) and (3.46) with periodic spatial boundary conditions applied to the velocity components. As indicated in Section 3.1, a Fourier series is an appropriate trial solution if periodic boundary conditions apply. Thus the velocity components, v_ℓ, are represented as follows,

$$v_\ell(\underline{x}, t) = \sum_{|k|<N} u_\ell(\underline{k}, t) \exp(i\underline{k}\underline{x}) \; . \tag{3.47}$$

A similar trial solution can be introduced for the pressure, p. However p is subsequently eliminated (see Orszag and Kruskal for details). Application of the spectral method to eqs. (3.45) and (3.46) gives

$$\left(\frac{\partial}{\partial t} + \nu k^2\right) u_\ell(\underline{k}, t) = - ik_m(\delta_{\ell j} - k_\ell k_j/k^2) \sum_{\underline{p}+\underline{q}=\underline{k}} u_j(\underline{p}, t)\, u_m(\underline{q}, t). \tag{3.48}$$

In eq. (3.48) $k = |\underline{k}|$ and $\delta_{\ell j}$ is the Kronecker delta.

A simulation of a turbulent free shear layer has been employed to assess a traditional mixing length hypothesis, which relates the Reynolds stress, $\overline{uw}$, to the mean velocity, $\overline{U}(z)$, in the following way,

$$\overline{uw} = -L^2 \frac{\partial \overline{U}}{\partial z} \left| \frac{\partial \overline{U}}{\partial z} \right| \quad . \tag{3.49}$$

The shear layer is directed along the x-axis with the z axis normal to the shear layer. This comparison was obtained from a two-dimensional direct simulation (Patera and Orszag (1981)).

The testing of simpler, semi-empirical turbulence models is an important function of direct turbulence simulation by spectral techniques (Orszag, 1977).

3.5.3. Other applications

Moin and Kim (1980) discuss the difficulties of providing boundary conditions for a spectral formulation of channel flow when a Poisson equation is to be solved for pressure. They avoid this problem by making explicit use of the continuity equation. A pseudospectral (Fletcher (1983a)) method is used with trial solutions for velocity components and pressure in terms of a Fourier series in the horizontal direction (with periodic boundary conditions), and Chebyshev polynomials in the normal direction.

Orszag (1971b) has solved the Orr-Sommerfeld equation to predict transition in a two-dimensional channel flow. Orszag introduced a Chebyshev trial solution for the disturbance field and predicted a critical Reynolds number of 5772.22.

However it is well known that transition is very susceptible to three dimensional disturbances and that transition has been observed with a Reynolds number as low as 1000. Orszag and Kells (1980) applied a composite pseudospectral, spectral-tau fractional step method to plane Couette and Poiseuille flow. The method used, as a trial solution, Fourier series in the plane of the flow and Chebyshev polynomials normal to the flow. It was confirmed computationally that transition can take place at Reynolds numbers of order 1000 if three dimensional disturbances are introduced.

Taylor and Murdock (1981) have applied a novel pseudospectral method to two dimensional laminar flow over a flat plate to test the stability of the flow to disturbances in the inflow boundary conditions. An unsteady primitive variable formulation (u,v,p) employs a parabolised x-momentum equation to obtain the longitudinal velocity component, u, the continuity equation to obtain the normal velocity component, v, and a Poisson equation to obtain the pressure, p. Each variable (u,v,p) is represented by a series of Chebyshev polynomials in the longitudinal (x) and normal (y) directions.

Orszag (1971a) has applied a Galerkin spectral method to the inviscid convection of a 'cone' and used the results to compare Galerkin methods with finite difference methods.

McCrory and Orszag (1980) have investigated the feasibility of applying spectral methods to very distorted regions for diffusion-dominated problems. In this case a pseudospectral method, based on a mixed Fourier cosine, Chebyshev polynomial trial solution, was applied to the heat conduction equation.

Haidvogel et al. (1980) have applied a pseudospectral method to the inviscid vorticity equation that models certain classes of ocean flows. They have used this as a test problem to compare spectral, finite element and finite difference methods.

4. GALERKIN FINITE ELEMENT METHOD

Both the traditional Galerkin method and the spectral Galerkin method are global in character. This follows directly from the use of global test and trial functions. In contrast the finite element method uses *local* test and trial functions.

4.1 Improvements Over the Traditional Galerkin Method

Firstly we introduce a trial solution *directly for the nodal unknowns*,

$$u(x,y) = \sum_{j=1}^{N} \bar{u}_j \, \phi_j(x,y) \quad , \tag{4.1}$$

where $\bar{u}_j$ are the nodal values of u. Clearly $\bar{u}_j$ have an immediate physical significance. The trial functions, $\phi_j(x,y)$, are *interpolatory*. That is $\phi_j = 1$ at node j and $\phi_j = 0$ at all other nodes. Between node j and adjacent nodes ϕ_j is non-zero.

The form of the trial solution given by eq. (4.1) can be related to the form used with the traditional Galerkin method, i.e.

$$u = \sum_{\ell=1}^{N} a_\ell \, \psi_\ell(x,y) \; . \tag{4.2}$$

If eq. (4.2) is evaluated at the nodal locations, (x_j, y_j), the following matrix equation is obtained,

$$\underline{\psi} \; \underline{A} = \underline{u} \tag{4.3}$$

where an element of $\underline{\psi}$ is $\psi_\ell(x_j, y_j)$, an element of $\underline{A}$ is a_ℓ and an element of $\underline{u}$ is $\bar{u}_j$. The coefficients, $\underline{A}$, can be obtained from the nodal unknowns, $\underline{u}$, by

$$\underline{A} = \underline{\psi}^{-1} \, \underline{u} \; . \tag{4.4}$$

Consequently eq. (4.2) can be written

$$u = \sum_{\ell=1}^{N} \sum_{j=1}^{N} \psi_{\ell j}^{-1} \, \bar{u}_j \, \psi_\ell(x,y) \tag{4.5}$$

or

$$u = \sum_{j=1}^{N} \bar{u}_j \left\{ \sum_{\ell=1}^{N} \psi_{\ell j}^{-1} \, \psi_\ell(x,y) \right\} . \tag{4.6}$$

Thus the trial functions, ϕ_j in eq. (4.1), could be evaluated from

$$\phi_j(x,y) = \sum_{\ell=1}^{N} \psi_{\ell j}^{-1} \, \psi_\ell(x,y) \quad . \tag{4.7}$$

The use of eq. (4.7) to obtain ϕ_j would require evaluation of $\underline{\psi}^{-1}$ which is an $O(N^3)$ process. Fortunately it is usually straightforward to obtain ϕ_j directly.

In one dimension the lowest-order representation for ϕ_j is shown in Fig. 9. The trial function, ϕ_j, takes the value $\phi_j = 1$ at $x = x_j$ and $\phi_j = 0$ at $x \leqslant x_{j-1}$ and $x \geqslant x_{j+1}$. Between nodes j-1 and j and j and j+1 ϕ_j varies linearly. Higher order interpolating functions, in one and more than one dimension, are discussed by Fletcher (1983a).

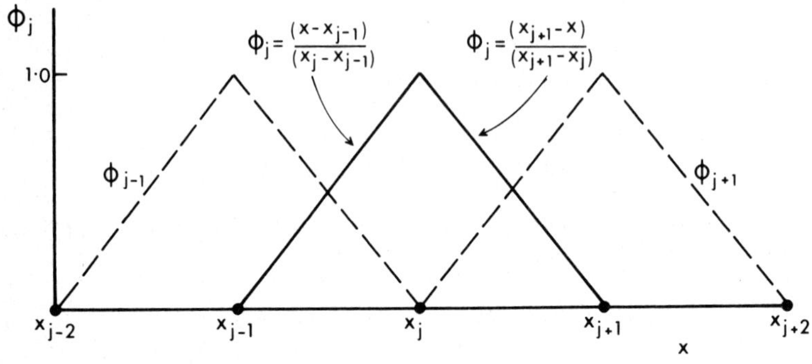

Figure 9. Finite element interpolation using linear shape functions

Application of a Galerkin method with low-order test and trial functions leads to low-order integrands in evaluating the inner product, eq. (2.6). For complex problems the inner product has to be evaluated numerically. A lower-order integrand, arising from the use of lower-order test and trial functions, can be evaluated more economically since a lower-order quadrature formula can be used.

It may be recalled, Section 2.5, that the matrix equation generated by the traditional Galerkin method becomes progressively more ill-conditioned as the order, N, of the matrix equation increases. However when ϕ_k (see Fig. 9) is used as a test function it is clear that *linear independence* is maintained and that increasing N, i.e. a mesh refinement, does not cause an ill-conditioned matrix equation to occur. Fletcher (1983a) gives a numerical example of this for the ordinary differential equation considered in Section 2.3.1.

The local nature of the test and trial functions gives a very sparse form to the resulting algebraic equation. Thus applying the Galerkin finite element method, with the linear test and trial functions shown in Figure 9, to eq. (2.23) in section 2.3.1, produces the algebraic formula,

$$\frac{\bar{u}_{k+1} - \bar{u}_{k-1}}{2\,\Delta x} - \left\{\frac{1}{6}\bar{u}_{k+1} + \frac{2}{3}\bar{u}_k + \frac{1}{6}\bar{u}_{k-1}\right\} = 0 . \qquad (4.8)$$

The corresponding matrix, $\underline{M}$, will be tridiagonal and can be factorised in 5N - 4 operations. This may be compared with the $O(N^3)$ operation count if $\underline{M}$ is full

Nonlinear terms, such as occur in Burgers' equation, can be evaluated *very econo-*

mically when local test and trial functions are used in the Galerkin method. The greater economy is accompanied by some loss of accuracy. Comparison by Fletcher (1983a) indicate that the solution with local test and trial functions is less accurate than when the same number of global test and trial functions are used.

The use of local, low-order interpolating functions in more than one dimension is most easily handled via the introduction of finite elements. A typical situation is shown in Figure 10. Here the (i,j)th node is surrounded

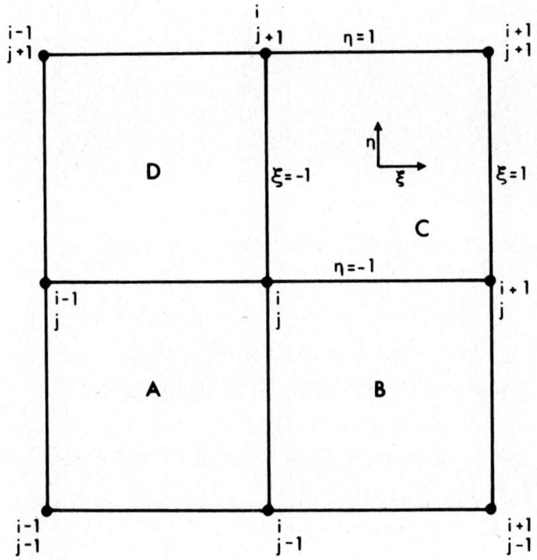

Figure 10. Element coordinates for a linear rectangular element

by four elements. Test and trial functions are defined for each element in turn by introducing an *element-based* coordinate system (ξ,η). Thus in element C u is interpolated by

$$u = \sum_{j=1}^{4} \phi_j(\xi,\eta)\,\bar{u}_j \quad , \tag{4.9}$$

and $\phi_j(\xi,\eta)$ are referred to as bilinear shape functions. The form of ϕ_j is

$$\phi_j(\xi,\eta) = \frac{1}{4}\,(1 + \xi_j\xi)(1 + \eta_j\eta) \tag{4.10}$$

i.e. $\phi_j = 1$ at node j and $\phi_j = 0$ at all other nodes. It is clear that u is continuous at element boundaries but derivatives of u are not continuous. The form of ϕ_j in eq. (4.10) implies that contributions to the global equation come only from the four elements, A, B, C and D.

A particular strength of the finite element method is the relative ease with which complex domain shapes are handled. Consider the situation shown in Figure 11,

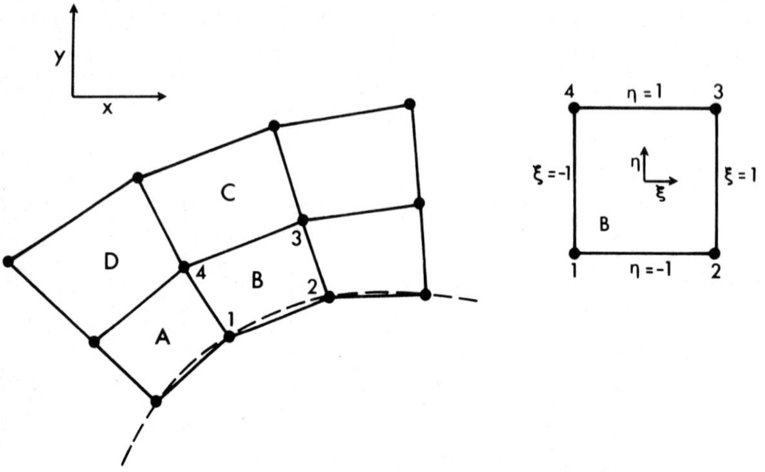

Figure 11. Isoparametric mapping at a boundary

which requires the use of distorted elements to follow the local boundary geometry. By using the element based co-ordinate system, (ξ,η), a distorted element in physical space (x,y) can be mapped to a uniform element in element space (ξ,η). The mapping is called an *isoparametric* transformation. For element B this is defined by

$$x = \sum_{\ell=1}^{4} \phi_\ell(\xi,\eta)\, \bar{x}_\ell$$

and (4.11)

$$y = \sum_{\ell=1}^{4} \phi_\ell(\xi,\eta)\, \bar{y}_\ell \quad .$$

In eq. (4.11) $\bar{x}_\ell$, $\bar{y}_\ell$ are the coordinates of the ℓth corner of element B in physical space. $\phi_\ell(\xi,\eta)$ are the same functions as in eqs. (4.9) and (4.10).

The isoparametric transformation is used in the evaluation of the inner products. A typical contribution to the algebraic equations is

$$I = \iint_D \frac{\partial\phi_j}{\partial x} \frac{\partial\phi_k}{\partial x}\, dx\, dy \quad . \tag{4.12}$$

After application of the isoparametric transformation, the contribution to eq.

(4.12) from each element is

$$Ie = \int_{-1}^{1}\int_{-1}^{1} \frac{1}{\det J}\left\{\frac{\partial y}{\partial \xi}\frac{\partial \phi_j}{\partial \eta} - \frac{\partial y}{\partial \eta}\frac{\partial \phi_j}{\partial \xi}\right\}\left\{\frac{\partial y}{\partial \xi}\frac{\partial \phi_k}{\partial \eta} - \frac{\partial y}{\partial \eta}\frac{\partial \phi_k}{\partial \xi}\right\} d\xi d\eta \quad . \tag{4.13}$$

The details are given by Fletcher (1983a). All the terms in eq. (4.13) are functions of (ξ,η) so that Ie can be readily evaluated. Typically this requires the use of Gauss quadrature.

4.2 Examples

Here we present some simple examples to demonstrate the Galerkin finite element method. The first examples utilises a one-dimensional ordinary differential equation. The second example is two-dimensional inviscid, incompressible flow and introduces the use of the isoparametric formulation. The third example is Burgers' equation which is nonlinear. This example permits a direct comparison with the traditional and spectral Galerkin methods.

4.2.1 Sturm-Liouville equation

We consider the simplified Sturm-Liouville equation,

$$\frac{d^2y}{dx^2} + y = f = -\sum_{\ell=1}^{M} a_\ell \sin(\ell-0.5)\pi x \tag{4.14}$$

with boundary conditions

$$y(0) = 0 \quad \text{and} \quad dy/dx\,(1) = 0 \quad . \tag{4.15}$$

The form of f allows an exact solution to be written

$$y = \sum_{\ell=1}^{M} [a_\ell/\{1 - ((\ell-0.5)\pi)^2\}] \sin[(\ell-0.5)\pi x] \quad . \tag{4.16}$$

A trial solution based on linear elements is

$$y_a = \sum_{j=1}^{N} \bar{y}_{aj} N_j(x) \tag{4.17}$$

where $N_j(x)$ is conveniently expressed in an element coordinate system, $x(\xi)$, as

$$\begin{aligned} &\text{in element } j\ , \quad N_j(\xi) = 0.5\,(1 + \xi) \\ &\text{in element } j{+}1\ , \quad N_j(\xi) = 0.5\,(1 - \xi) \quad . \end{aligned} \tag{4.18}$$

Application of the Galerkin finite element method produces the equations

$$(d^2y_a/dx^2, N_k) + (y_a\ , N_k) - (f, N_k) = 0. \tag{4.19}$$

The use of function values only, as nodal unknowns, requires that the highest derivative appearing in eq. (4.19) is the first. Consequently the first term in eq. (4.18) is integrated by parts,

$$(d^2y_a/dx^2, N_k) \equiv \int_o^1 \frac{d^2y_a}{dx^2} N_k \, dx = \left[\frac{dy_a}{dx} N_k\right]_o^1 - \int_o^1 \frac{dy_a}{dx}\frac{dN_k}{dx} \, dx \quad . \qquad (4.20)$$

For the boundary conditions used, eq. (4.15), there is no contribution from the first term. Thus

$$(d^2y_a/dx \, , N_k) = - (dy_a/dx, dN_k/dx) \qquad (4.21)$$

Substitution of eq. (4.21) into (4.19) and rearrangement gives

$$\underline{B} \, \underline{Ya} = \underline{G} \, , \qquad (4.22)$$

where b_{kj} is an element of $\underline{B}$ and

$$b_{kj} = - \left(\frac{dN_j}{dx} , \frac{dN_k}{dx}\right) + \left(N_j , N_k\right) \quad . \qquad (4.23)$$

With linear elements, only three of the b_{kj} coefficients are nonzero (for internal nodes)

$$b_{k,k-1} = \frac{1}{\Delta x_j} + \frac{\Delta x_j}{6}$$

$$b_{k,k} = - \left(\frac{1}{\Delta x_j} + \frac{1}{\Delta x_{j+1}}\right) + \frac{1}{3}\left(\Delta x_j + \Delta x_{j+1}\right) \qquad (4.24)$$

and $$b_{k,k+1} = \frac{1}{\Delta x_{j+1}} + \frac{\Delta x_{j+1}}{6} \quad .$$

An element of $\underline{G}$ is given by

$$g_k = (f, N_k) \quad . \qquad (4.25)$$

In principle eq. (4.25) could be evaluated exactly. However it is convenient to interpolate f in the same manner as y_a i.e.

$$f = \sum_{j=1}^{N} \bar{f}_j \, N_j(x) \qquad (4.26)$$

where the nodal values, $\bar{f}_j$, are given by eq. (4.14). Consequently eq. (4.25) becomes

$$g_k = \sum_{j=1}^{N} \bar{f}_j \, (N_j, N_k) \quad . \tag{4.27}$$

This is an example of a *group* finite element formulation (Fletcher (1983b)). Equation (4.27) is evaluated, for linear elements, as

$$g_k = \frac{\Delta x_k}{6} \bar{f}_{k-1} + \frac{(\Delta x_k + \Delta x_{k+1})}{3} \bar{f}_k + \frac{\Delta x_{k+1}}{6} \bar{f}_{k+1} \quad . \tag{4.28}$$

Clearly eq. (4.22) is a tridiagonal system and can be evaluated efficiently using the Thomas algorithm.

Solutions have been obtained with

$$a_1 = 1.0,\ a_2 = -0.5,\ a_3 = 0.3,\ a_4 = -0.2 \text{ and } a_5 = 0.1 \ .$$

For equally-spaced points the error in the discrete L_2 norm for various values of N is shown in Table 11. The discrete L_2 norm is defined by

$$\|y_a - y\|_{2,d} = \left\{ \sum_{i=1}^{N} (y_{ai} - y_i)^2 \right\}^{\frac{1}{2}} \quad .$$

Table 11. Error in the discrete L_2 norm for simplified Sturm-Liouville problem (p=1, q=1)

N	Δx	$\|y_a - y\|_{2,d}$
6	0.2	19.180×10^{-3}
11	0.1	6.169×10^{-3}
21	0.05	2.048×10^{-3}
41	0.025	0.698×10^{-3}
81	0.0125	0.243×10^{-3}

The reduction in error with reducing Δx shown in Table 11 is proportional to Δx^2. This is the theoretically expected result for linear elements.

4.2.2 Inviscid, incompressible flow

This example is considered to illustrate the use of the isoparametric formulation and the need to introduce numerical integration. This example also demonstrates the ability of achieving significantly higher accuracy by a small modification to the Galerkin method. As with the Sturm-Liouville equation this problem has an exact solution.

The flow geometry for the flow past a circular cylinder is shown in Figure 12. Because of the symmetry only a quarter-plane need be considered. Also shown in Figure 12 are the distribution of elements used, in this case eight-noded

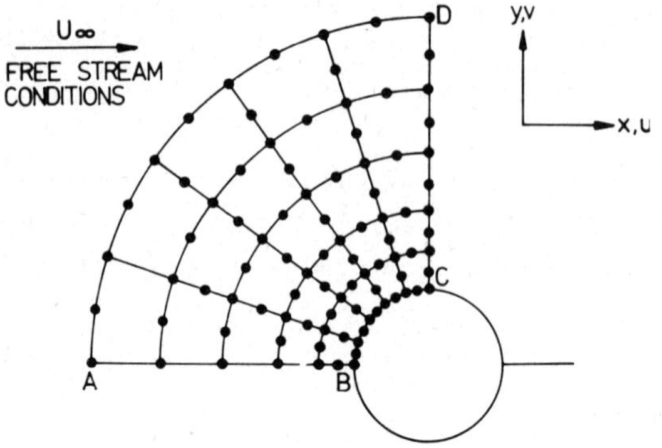

Figure 12. Element distribution for flow past a circular cylinder

quadratic (Serendipity) rectangular elements. The governing equations are

$$\frac{\partial u}{\partial x} + \frac{\partial v}{\partial y} = 0 \tag{4.29}$$

$$\frac{\partial u}{\partial y} - \frac{\partial v}{\partial x} = 0 \quad , \tag{4.30}$$

with boundary conditions

$$u = 1, v = 0 \text{ at the farfield boundary}$$

and (4.31)

$$u \cos \theta - v \sin \theta = 0 \text{ at the body surface} \quad .$$

Trial solutions for u and v are introduced by

$$u = \sum_j \phi\,(x,y)\bar{u}_{aj} \tag{4.32}$$

and

$$v = \sum_j \phi_j(x,y)\bar{v}_{aj} \quad , \tag{4.33}$$

where the ϕ_j's are quadratic, rectangular shape functions. Execution of the Galerkin formulation produces the following algebraic equations,

$$\sum_j a_{kj}\,\bar{u}_{aj} + \sum_j b_{kj}\,\bar{v}_{aj} = 0 \tag{4.34}$$

and

$$\sum_j b_{kj}\,\bar{u}_{aj} - \sum_j a_{kj}\,\bar{v}_{aj} = 0 \tag{4.35}$$

where

$$a_{kj} = \sum_e \iint \{\phi_k\;\partial\phi_j/\partial x\}dx\;dy \tag{4.36}$$

and

$$b_{kj} = \sum_e \iint \{\phi_k\;\partial\phi_j/\partial y\}dx\;dy \quad . \tag{4.37}$$

$\sum_e$ denotes the summation over the elements adjacent to the kth node, i.e., those elements for which ϕ_k is non-zero. Since the elements are defined on a polar grid it is difficult to evaluate eqs. (4.36) and (4.37) in the physical plane. However it is straightforward to introduce an isoparametric transformation which transforms a distorted element in the physical (x,y) plane into a regular shape in the (ξ,η) plane. This is accomplished by defining

$$x = \sum_i \phi_i(\xi,\eta)\bar{x}_i \tag{4.38}$$

$$y = \sum_i \phi_i(\xi,\eta)\bar{y}_i \quad , \tag{4.39}$$

where the ϕ_i's are just the shape functions defined in the (ξ,η) plane and $(\bar{x}_i,\bar{y}_i)$ are the physical coordinates of the ith node. After some algebra, eqs. (4.36) and (4.37) can be replaced by

$$a_{kj} = \sum_i c_{kji}\;\bar{y}_i \tag{4.40}$$

and

$$b_{kj} = \sum_i c_{kji}\;\bar{x}_i \tag{4.41}$$

where

$$c_{kji} = \int_{-1}^{1}\int_{-1}^{1} \phi_k\left\{\frac{\partial\phi_j}{\partial\xi}\,\frac{\partial\phi_i}{\partial\eta} - \frac{\partial\phi_i}{\partial\xi}\,\frac{\partial\phi_j}{\partial\eta}\right\} d\xi\;d\eta \quad . \tag{4.42}$$

It may be noted that c_{kji} is independent of the physical plane and may be evaluated, once and for all, in the (ξ,η) plane. That this is possible is a direct result of applying the Galerkin formulation to a *first derivative*. Thus the Jacobian, which is a function of the physical location, associated with the derivative cancels with the Jacobian associated with the integration. Clearly this situation produces a considerable economy. Typically eq. (4.42) is sufficiently complex that numerical integration is required. With quadratic elements a 3 x 3 Gauss quadrature scheme produces exact integration.

Solutions have been obtained with 25 elements and 149 nodal unknowns in the quarter-plane. ABCD (Figure 12). To facilitate a comparison with the exact solution, the rms error between the finite element solution, q_{Ta}, and the exact solution, q_{Te}, for the tangential velocity at the body surface has been computed by

$$\phi = \{ \sum_{i=1}^{N} (q_{Ta} - q_{Te})^2 \}/N\}^{\frac{1}{2}} \quad . \qquad (4.43)$$

Table 12. Modified Galerkin finite element formulations

Case	Comments	σ
1	exact numerical integration, Eq. (4.42)	0.034
2	exact numerical integration, Eq. (4.46)	0.017
3	reduced numerical integration, Eq. (4.42)	0.005

A result for exact numerical integration is shown in Table 12.

Fletcher (1978c,1979,1980a) has considered a modification to the Galerkin method in which the weighted integration of the residual is written as

$$\iint \phi_k (R)_{\ell s} \, dx \, dy = 0 \quad , \qquad (4.44)$$

where $(R)_{\ell s}$ is a *least-squares fit* of the residual over an element. The idea here is that eq. (4.44) will constrain the approximate solution less than the conventional Galerkin formulation. If one evaluates eq. (4.42) using a 2 x 2 Guass integration formula (reduced integration), this is equivalent to replacing the weighted integral of the residual with

$$\iint (\phi_k)_{\ell s} (R)_{\ell s} \, dx \, dy = 0 \quad . \qquad (4.45)$$

Results for σ using eqs. (4.44) and (4.45) are also shown in Table 12. It can be seen that a significant improvement in the accuracy of the solution is possible. In applying eq. (4.44), eq. (4.42) is written

$$c_{kji} = \int_{-1}^{1}\int_{-1}^{1} \phi_k \{(\frac{\partial\phi_j}{\partial\xi})_{\ell s} \frac{\partial\phi_i}{\partial\eta} - (\frac{\partial\phi_j}{\partial\eta})_{\ell s} \frac{\partial\phi_i}{\partial\xi} \} d\xi \, d\eta \qquad (4.46)$$

and $(\frac{\partial\phi_j}{\partial\xi})$ etc. are represented by

$$(\partial\phi_j/\partial\xi)_{\ell s} = a_0 + a_1\xi + a_2\eta + a_3\xi\eta \quad . \qquad (4.47)$$

It was found by Fletcher (1980a) that these advantages were *restricted to rectangular elements*. For cases 2 and 3 with triangular elements inferior results were obtained.

4.2.3 Burgers' equation

Here we re-examine the propagating shock problem previously considered in Sections 2.3.2 and 3.2.2 in conjunction with the traditional and spectral Galerkin methods. Burgers' equation is

$$u_t + uu_x - \frac{1}{Re} u_{xx} = 0 \quad . \tag{4.48}$$

The appearance of the nonlinear convective term in eq. (4.48) creates severe gradients in the flow if Re is large. Equation (4.48) will be solved with boundary conditions

$$u(x_L,t) = 1 \quad , \quad u(x_R,t) = 0 \tag{4.49}$$

and initial conditions

$$\begin{aligned} u(x,0) &= 1 \quad , \quad x_L \leqslant x \leqslant 0 \\ u(x,0) &= 0 \quad , \quad 0 < x \leqslant x_R. \end{aligned} \tag{4.50}$$

x_L and x_R are chosen to be sufficiently large so that eq. (4.49) is true at the time and for the value of the Reynolds number at which a comparison is made with the exact solution, eq. (5.21). Typically $1 \leqslant |x_L|, x_R \leqslant 2$. Equations (4.48) to (4.50) describe the viscous propagation of a 'shock wave'.

A trial solution, with the same form as eq. (4.1), is introduced as

$$u_a = \sum_j N_j(x)\, \bar{u}_{a_j}(t) \tag{4.51}$$

Application of the Galerkin method produces a system of ordinary differential equations

$$\underline{M}\ \dot{\underline{U}}_a + (\underline{B} + \underline{C})\underline{U}_a = 0 \quad . \tag{4.52}$$

The similarity to eq. (2.38) may be noted. An element of $\underline{U}_a$ is a nodal unknown $u_{a_j}(t)$. An element of the mass matrix, $\underline{M}$, is

$$m_{kj} = \int_{x_L}^{x_R} \phi_k(x)\, \phi_j(x)\, dx \quad . \tag{4.53}$$

Since only one space dimension is present, only two elements will contribute to eq. (4.53) at interior nodes. An element of $\underline{B}$ is

$$b_{kj} = \sum_i \bar{u}_{a_i} \int_{x_L}^{x_R} \phi_k(x)\, \phi_i(x) \frac{d\phi_j(x)}{dx} dx \quad . \tag{4.54}$$

It can be seen that $\underline{B}$ is a function of the solution. After applying Green's theorem an element of $\underline{C}$ is given by

$$c_{kj} = \frac{1}{Re}\int_{x_L}^{x_R} \frac{d\phi_k}{dx}\frac{d\phi_j}{dx}\,dx \quad . \tag{4.55}$$

Since u at $x = x_L$ and $x = x_R$ is given there is no contribution from the boundaries.

The structure of eq. (4.52) can be readily appreciated for the special case of linear elements on a uniform grid. Then eq. (4.52) becomes

$$\frac{1}{6}\dot{\bar{u}}_{a_{k-1}} + \frac{2}{3}\dot{\bar{u}}_{a_k} + \frac{1}{6}\dot{\bar{u}}_{a_{k+1}} + \frac{(\bar{u}_{a_{k-1}} + \bar{u}_{a_k} + \bar{u}_{a_{k+1}})}{3}\frac{(\bar{u}_{a_{k+1}} - \bar{u}_{a_{k-1}})}{2\Delta x}$$

$$- \frac{1}{Re}\frac{(\bar{u}_{a_{k-1}} - 2\bar{u}_{a_k} + \bar{u}_{a_{k+1}})}{\Delta^2 x} = 0 \quad . \tag{4.56}$$

This form of eq. (4.56) is very similar to that of a centred finite difference scheme. The only difference being the distribution of u and $\dot{u}$ over adjacent nodes in the first two terms.

To obtain solutions to the system of equations like (4.56) a fourth-order Runge-Kutta scheme has been utilized to integrate the equations in time from t = 0 up to t = 0.80, typically. The integration scheme is the same as for the traditional Galerkin and spectral methods. At the final time the solution is compared with the exact solution. Some typical results for Re = 10 are shown in Table 13 and Figure 13. Results for a three-point finite difference scheme,

$$\dot{u}_k + u_k\frac{(u_{k+1} - u_{k-1})}{2\Delta x} - \frac{1}{Re}\frac{(u_{k-1} - 2u_k + u_{k+1})}{\Delta^2 x} = 0 \tag{4.57}$$

are shown in the second column for comparison. However, it should be noted that, due to the step-size control on the Runge-Kutta scheme, the solutions correspond to different times. Thus $\|u_a - u_e\|_{rms}$ gives a better means of comparing the various results; this has been computed from the exact solution corresponding to the approximate solution at the specified time rather than from the exact solution shown in the final column which is only applicable to t = 0.50.

It is apparent that the linear finite element scheme, eq. (4.56), is producing results of considerably greater accuracy than the comparable finite difference scheme, eq. (4.57). Also an increase in the number of nodal points for the finite element scheme causes a substantial increase in the accuracy as does an increase in the order of the representation.

The results for a linear element, N = 11, can be compared with the results with N = 11 for the traditional Galerkin method, Table 4. It can be seen that for the same number of unknowns, nine in both situations, *the traditional Galerkin method gives answers of higher accuracy*. Since the trial functions are of a higher order for the traditional Galerkin formulation this higher accuracy is to be

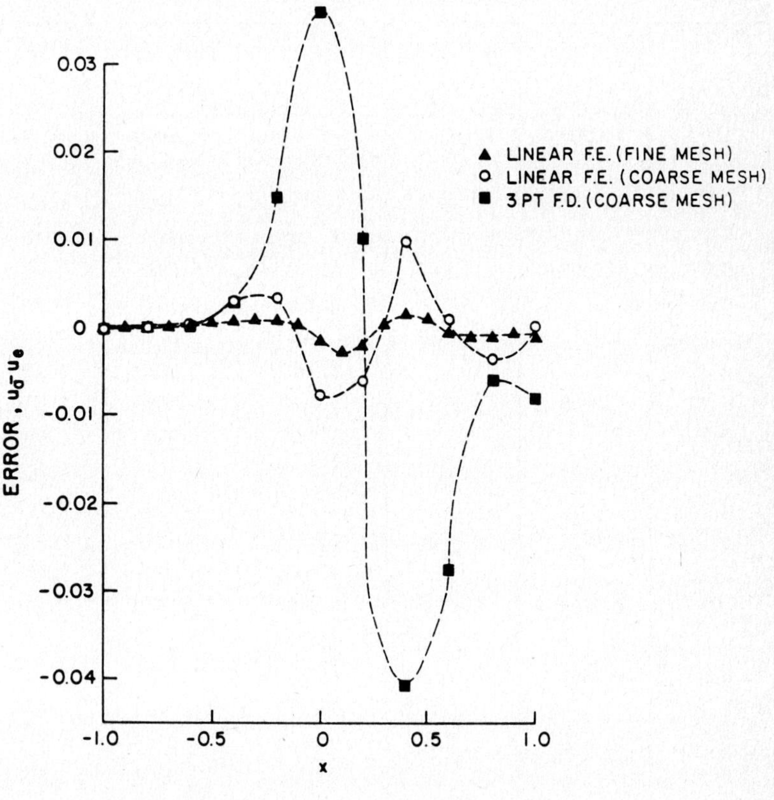

Figure 13. Distribution of error for the solution of Burgers' equation, Re = 10

expected. However if the comparison were made on the basis of computational efficiency, i.e. the achieved accuracy per unit of execution time, one would expect the finite element method to be more efficient.

Table 13. Solution of Burgers' equation by Galerkin finite element method, Re = 10

X	3PT F.D. N = 11 t = 0.81	Lin. F.E. N = 11 t = 0.47	Lin. F.E. N = 21 t = 0.50	Quad. F.E. N = 21 t = 50	Exact t = 0.50
-1.0	1.0000	1.0000	1.0000	1.0000	1.0000
-0.9			1.0000	1.0000	1.0000
-0.8	1.0000	1.0000	1.0000	1.0000	1.0000
-0.7			1.0000	0.9999	0.9999
-0.6	1.0000	1.0003	0.9998	0.9995	0.9996
-0.5			0.9991	0.9987	0.9987
-0.4	1.0000	0.9989	0.9967	0.9959	0.9960
-0.3			0.9907	0.9893	0.9891
-0.2	1.0000	0.9746	0.9734	0.9724	0.9725
-0.1			0.9367	0.9370	0.9366
0	0.9731	0.8514	0.8657	0.8671	0.8672
0.1			0.7482	0.7501	0.7509
0.2	0.8021	0.5608	0.5864	0.5882	0.5884
0.3			0.4065	0.4053	0.4062
0.4	0.4696	0.2339	0.2463	0.2440	0.2448
0.5			0.1307	0.1293	0.1301
0.6	0.1948	0.0540	0.0613	0.0611	0.0619
0.7			0.0256	0.0263	0.0268
0.8	0.0622	0.0049	0.0094	0.0103	0.0106
0.9			0.0029	0.0036	0.0039
1.0	0.0082	0.0005	0.0000	0.0011	0.0013
$\Vert u_a - u_e \Vert_{rms}$	0.0215	0.0050	0.0011	0.0005	

For one finite element scheme, a linear element with N = 21, comparative results for increasing Reynolds number are shown in Table 14 and Figure 14. As with the traditional Galerkin method and the spectral method, the difficulty of following the rapid change in u with increasing Reynolds number produces a corresponding reduction in accuracy. However with the finite element method it is relatively straightforward to introduce a variable grid for which more points are grouped where the solution is changing most rapidly. Thus higher accuracy is possible with little or no additional computation. Such a situation is not available, conveniently, to the traditional Galerkin method.

Table 14. Solution of Burgers' equation by Galerkin finite element method, linear element, N = 21, various Reynolds number.

Re = 1.0, t = 0.49			Re = 10, t = 0.50			Re = 100, t = 0.32		
x	Approx	Exact	x	Approx	Exact	x	Approx	Exact
-1.0	0.9179	0.9174	-1.0	1.0000	1.0000	-1.00	1.0000	1.0000
-0.9	0.8998	0.8996	-0.9	1.0000	1.0000	-0.762	1.0000	1.0000
-0.8	0.8792	0.8789	-0.8	1.0000	1.0000	-0.567	1.0000	1.0000
-0.7	0.8556	0.8554	-0.7	1.0000	0.9999	-0.412	1.0000	1.0000
-0.6	0.8289	0.8288	-0.6	0.9998	0.9996	-0.290	1.0000	1.0000
-0.5	0.7991	0.7991	-0.5	0.9991	0.9987	-0.197	1.0000	1.0000
-0.4	0.7664	0.7664	-0.4	0.9967	0.9960	-0.128	1.0000	1.0000
-0.3	0.7307	0.7308	-0.3	0.9901	0.9891	-0.079	1.0000	1.0000
-0.2	0.6923	0.6926	-0.2	0.9734	0.9725	-0.045	1.0000	1.0000
-0.1	0.6517	0.6520	-0.1	0.9367	0.9366	-0.020	1.0000	1.0000
0	0.6091	0.6095	0	0.8657	0.8672	0	1.0000	0.9998
0.1	0.5651	0.5656	0.1	0.7482	0.7509	0.020	1.0003	0.9994
0.2	0.5202	0.5207	0.2	0.5864	0.5884	0.045	0.9924	0.9976
0.3	0.4750	0.4757	0.3	0.4065	0.4062	0.079	1.0195	0.9845
0.4	0.4302	0.4309	0.4	0.2463	0.2448	0.128	0.8190	0.8281
0.5	0.3862	0.3871	0.5	0.1307	0.1301	0.197	0.0942	0.1236
0.6	0.3438	0.3447	0.6	0.0613	0.0619	0.290	-0.0212	0.0009
0.7	0.3032	0.3043	0.7	0.0256	0.0268	0.412	0.0045	0.
0.8	0.2651	0.2662	0.8	0.0095	0.0106	0.567	-0.0010	0.
0.9	0.2299	0.2309	0.9	0.0029	0.0039	0.762	0.0002	0.
1.0	0.1969	0.1984	1.0	0.0000	0.0013	1.000	0.	0.
$\|u_a-u_e\|_{rms}$	0.0007			0.0011			0.0119	

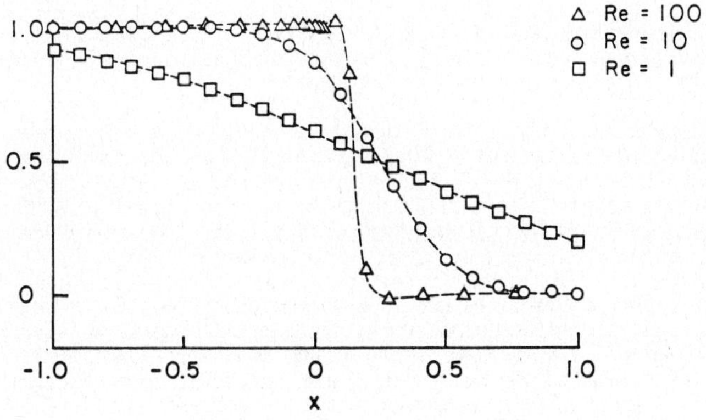

Figure 14. Solution of Burgers' equation for increasing Reynolds number

4.3 Theoretical Properties

The theoretical properties of the Galerkin finite element method have received considerable attention. Much of this work is discussed in Strang and Fix (1973), Oden and Reddy (1976) and Mitchell and Wait (1978).

The theory of the finite element method has developed from two important foundations. Firstly the trial solutions make use of *polynomial interpolation* between the nodal values. Polynomial interpolation is a well-established discipline e.g. Birkhoff and de Boor (1965). Secondly, where problems can be given an *equivalent variational interpretation*, the theoretical properties of the Ritz method (e.g. Mikhlin, 1964) carry over to the Galerkin finite element method.

Here we examine briefly three aspects. Firstly we look at the requirements for convergence of the finite element solution to the exact solution. Secondly we consider the a priori estimation of the errors in the solution in relation to element size and the order of the interpolation used. Thirdly we indicate the results of numerical experiments to measure the convergence rate.

4.3.1 Convergence

The Galerkin finite element method permits the highest derivative that appears in the weak form, i.e.

$$(L(u), \phi_k) = 0 \qquad (4.58)$$

to be minimised by shifting derivatives onto the test function, ϕ_k. If the highest derivative appearing in L(u) is of order 2m, application of Green's theorem m times reduces the highest derivative to order m. The result can be written

$$(L^m(u), \phi_k^m) = 0 \ . \qquad (4.59)$$

As the elements become smaller and smaller a derivative of order m in eq. (4.59) will assume a constant value over an element. *For convergence to the exact solution it is necessary that the trial functionsare capable of representing a constant value of the highest derivative,* m, *exactly*. With polynomial trial functions this requires a polynomial that is complete to order m. As an example, if second derivatives appear in eq. (4.59) then at least quadratic interpolating (shape) functions are required for convergence.

Secondly there is a requirement that elements be *conforming*. This means that, if m^{th} order derivatives appear in eq. (4.59), the trial solution and its first (m-1) derivatives should be continuous across element boundaries. This implies that if second derivatives appear in eq. (4.59) either hermitian elements (Mitchell and Wait (1978)) will be required or auxiliary variables must be introduced for the first derivatives of u.

The use of conforming elements ensure inter-element continuity. However convergence is often possible with nonconforming elements. To test the convergence of nonconforming elements, Irons (see Irons and Razzague (1972)) developed the *patch test*. The patch test, applied to finite element approximation, is equivalent to the idea that a finite difference approximation to a partial differential equation should be *consistent* with the partial differential equation. These similarities are developed further by Fletcher (1983a).

4.3.2 Error estimates

We will call the difference between the Galerkin solution and the exact solution the approximation error. As the number of nodes increases we expect the approximation error to reduce; eventually we expect convergence to the exact solution.

A major contribution to the approximation error comes from the interpolation error. The interpolation error depends on the nature and order of the trial functions used within each element. Additional contributions to the approximation error can arise from using numerical quadrature to evaluate the inner products, from interpolating the boundary conditions, from the failure of the finite element boundaries to match the actual boundaries and from round-off in the computer.

The approximation error can be related to the interpolation error in the following way. Consider the equation

$$A(u) = f$$

which satisfies the ellipticity condition

$$\sigma \|u\|^2_{H^m_{,R}} \leqslant (A(u),u) \leqslant K \|u\|^2_{H^m_{,R}} \quad . \tag{4.60}$$

Then the approximation error, e, can be written

$$\|e\|_{H^m_{,R}} \equiv \|u-u_a\|_{H^m_{,R}} \leqslant (1 + \frac{K}{\sigma}) \|u-u_I\|_{H^m_{,R}} \tag{4.61}$$

where $\|u-u_I\|_{H^m_{,R}}$ is the H^m norm of the interpolation error and the weak form of eq. (4.59) contains derivatives up to order m. The H^m norm is defined by

$$\|u\|^2_{H^m_{,R}} = \sum_{|\alpha| \leqslant M} \int_R \left| \frac{\partial^{|\alpha|} u(x_1, x_2 \ldots x_\eta)}{\partial x_1^{\alpha_1} \ldots x_\eta^{\alpha_\eta}} \right|^2 dx \quad . \tag{4.62}$$

If the exact solution, u, is sufficiently smooth the approximate error in the finite element solution, u_a, can be estimated from

$$\|u-u_a\|_{H^s_{,R}} \leqslant C\, h^{k+1-s} \|u\|_{H^{k+1}_{,R}} \quad \text{if} \quad s \geqslant 2m - k - 1 \tag{4.63}$$

$$\|u-u_a\|_{H^s_{,R}} \leqslant C\, h^{2(k+1-m)} \|u\|_{H^{k+1}_{,R}} \quad \text{if} \quad s \leqslant 2m - k - 1 \tag{4.64}$$

where C is a positive constant, k is the order of the interpolating function and h is the maximum element dimension. Strang and Fix (1973) note that eq. (4.63) is more likely to be limiting; it comes from the interpolation error. In contrast eq. (4.64) represents an upper limit.

If u is not as smooth as the interpolation, i.e. $u \in H^r(R)$ and $r \leqslant k+1$, eq. (4.63) is replaced by

$$\|u-u_a\|_{H,R}^{s} \leqslant C\, h^{r-s} \, \|u\|_{H,R}^{r} \; . \tag{4.65}$$

Element distortion causes a reduction in the accuracy (see Fletcher 1983a)).

For problems that satisfy the ellipticity condition, eq. (4.60), the Galerkin finite element method produces solutions that are *optimal* (Mitchell and Wait, 1978) in the Sobolev norm, eq. (4.62). For parabolic problems comparable error estimates can be obtained *if the initial data is sufficiently smooth.* However for hyperbolic problems the standard Galerkin solutions are sub-optimal (Fairweather (1978)).

The above error estimates have all been global estimates. However at specific points higher accuracy *(super convergence)* is possible. Such points might be the corner nodes of higher-order elements or the Gauss points used for numerical integration. This is explored further by Fletcher (1983a).

4.3.3 Numerical Convergence

Fletcher (1982a) has considered the convergence properties of the Galerkin finite element method applied to the modified Burgers' equation

$$\frac{\partial u}{\partial t} + (u - \alpha)\frac{\partial u}{\partial x} - \frac{1}{Re}\frac{\partial^2 u}{\partial x^2} = 0 \tag{4.66}$$

with linear, quadratic and cubic Lagrangian elements in x.

By using a value of $\alpha = 0.5$ and the boundary conditions given by eq. (4.49) the convection of the shock-like disturbance to the right *does not occur.*

The time-step has been kept sufficiently small that errors in the solution after a finite time are due to the spatial finite element representation, only. Steady-state solutions have been obtained by integrating to large t. By plotting the error in the L_2 norm, i.e.

$$\|e\|_2 = \|u_a - u_e\|_2 = \left[\int_{x_L}^{x_R} (u_a - u_e)^2 dx\right]^{\frac{1}{2}} \; , \tag{4.67}$$

against step-size, Δx, a spatial convergence rate can be obtained, in the form,

$$\|e\|_2 = k\Delta^{\alpha} x. \tag{4.68}$$

For $R_e = 10$ and a range of step-size, $0.002 < \Delta x < 0.2$, average values of α are shown in Table 15. The corresponding theoretical values of α are shown in Table 16.

Table 15. Measured spatial convergence rates

FINITE ELEMENT METHOD (INTERPOLATION)	SPATIAL CONVERGENCE RATE		FINITE DIFFERENCE METHOD	SPATIAL CONVERGENCE RATE
	CONVENTIONAL F.E.M.	GROUP F.E.M.		
Linear	$\alpha = 2.00$	$\alpha = 2.04$	three-point	$\alpha = 2.00$
Quadratic	$\alpha = 2.98$	$\alpha = 2.90$	five-point	$\alpha = 3.87$
Cubic	$\alpha = 3.90$	$\alpha = 3.94$	seven point	$\alpha = 5.58$

The conventional finite element results are based on eq. (4.66). For linear elements this corresponds to eq. (4.56) with the addition of the α term. Also shown in Table 15 are convergence rates for the group finite element formulation (Fletcher (1983c)). In the group formulation $0.5\partial(u^2)/\partial x$ replaces $u\partial u/\partial x$ and a separate trial solution is introduced for u^2. It is apparent that both formulations produce convergence rates in good agreement with the theoretical predictions (Table 16).

Table 16. Theoretical spatial convergence rates

Finite element method	Spatial convergence rate, $\Delta^{\alpha}x$	Finite difference method	Spatial convergence rate. $\Delta^{\alpha}x$
linear element	$\alpha = 2$	three-point	$\alpha = 2$
quadratic element	$\alpha = 3$	five-point	$\alpha = 4$
cubic element	$\alpha = 4$	seven-point	$\alpha = 6$

Spatial convergence rates have also been obtained for three-point, five-point and seven-point finite difference representations of eq. (4.66). The errors are measured in the rms norm i.e.

$$\|u_a - u_e\|_{rms} = \left[\sum_j (u_{a_j} - u_{e_j})^2/N\right]^{\frac{1}{2}} . \tag{4.69}$$

As with the finite element results, the finite difference convergence rates, Table 15 , substantially achieve the theoretical convergence rates (based on truncation error) shown in Table 16. It was found (Fletcher, 1982b) that both the finite element and the finite difference convergence rates improved with grid refinement.

Spatial convergence rates have also been determined for small values of time i.e. a parabolic situation. However it was found that the initial data had a significant influence (Fletcher, 1982a). If the discontinuous initial conditions, eq. (4.50), were used the convergence rate for the cubic finite element and five- and seven-point finite difference schemes was *reduced to second order*.

However if the exact solution at $t = 0.01$ was used for the initial conditions, convergence rates comparable to those shown in Table 15 were obtained for all methods. Similar convergence rates, for continuous initial conditions, were also obtained with Re = 1 and 100.

4.4 Special Trial Functions

In the traditional Galerkin method it is an important strategy to choose the nature of the trial functions so that an accurate solution can be obtained with as few trial functions as possible. In contrast almost all Galerkin finite element formulations use piecewise polynomial trial functions. The use of special functions with the finite element method is unusual and often difficult to implement efficiently. Here we indicate two situations where special functions have been used successfully within a finite element framework.

4.4.1 Dorodnitsyn finite element formulation

Our starting point is the Dorodnitsyn boundary layer equation,

$$\frac{\partial}{\partial x}\int_o^1 u f_k\ \theta^c du = \frac{u_{ex}}{u_e}\int_o^1 \frac{df_k}{du}(1-u^2)\ \theta^c du + u_e \int_o^1 \frac{df_k}{du}\frac{\partial T}{\partial u} du = 0 \qquad (4.70)$$

The derivation of this equation is given by Fletcher and Fleet (1983a). In eq. (4.70) x and u are the independent variables and θ^c and T are the dependent variables. θ^c and T are related to each other in the following manner

$$1/\theta^c = T = \partial u/\partial \eta \qquad (4.71)$$

where η is a non-dimensional coordinate measured across the boundary layer. Trial solutions are introduced for θ^c and T in the following way,

$$\theta^c = \sum_{j=1}^{M} N_j(u)/(1-u)\ \theta_j(x)$$

$$\text{and } T = \sum_{j=1}^{M} (1-u)N_j(u)\ \tau_j(x) \quad . \qquad (4.72)$$

In eqs. (4.72) $N_j(u)$ are one-dimensional shape functions, either linear or quadratic. The additional factor $(1-u)$ ensures that θ^c and T behave properly at the outer edge of the boundary layer. Equations (4.72) do not satisfy eq. (4.71) except at the nodes where $\theta_j = 1/\tau_j$.

The test function, $f_k(u)$, is given by

$$f_k(u) = (1-u)N_k(u) \quad . \qquad (4.73)$$

The substitution of eqs. (4.72) and (4.73) into eq. (4.70) indicates that a *modified Galerkin formulation* (Fletcher, (1983a)) is produced. Evaluation of the various inner products produces a system of ordinary differential equations for

θ_j and τ_j ,

$$\sum_j CC_{kj} \frac{d\theta_j}{dx} = \frac{u_{ex}}{u_e} \sum_j EF_{kj}\, \theta_j + u_e \sum_j AA_{kj}\, \tau_j \tag{4.74}$$

$$\text{where } CC_{kj} = \int_0^1 N_k N_j\, u\, du\ , \quad EF_{kj} = \int_0^1 N_j \left\{ \frac{dN_k}{du} (1-u) - N_k \right\} (1+u)\, du$$

$$\text{and } AA_{kj} = \int_0^1 \left\{ \frac{dN_j}{du} (1-u) - N_j \right\} \left\{ \frac{dN_k}{du} (1-u) - N_k \right\} du \ . \tag{4.75}$$

Equation (4.74) can be solved very efficiently using an implicit algorithm to march in the x - direction. The details are given by Fletcher and Fleet (1983a).

The use of this algorithm on a number of representative laminar boundary layer flows (Fletcher and Fleet (1983a)) indicates that quadratic elements are computationally *more efficient* than linear elements. However both elements produce solutions of high accuracy *on a relatively coarse grid*.

Essentially the same formulation has been applied to turbulent boundary layer flows (Fletcher and Fleet (1983b)) and compared with a representative finite difference computer package, STAN5. A comparison is indicated in Table 17. Both methods produced results of comparable accuracy. The skin friction variation for the turbulent boundary layer flow over a flat plate is shown in Fig. 15. However the finite element solutions are typically achieved with a *tenth of the execution time* required by STAN5.

Table 17. Comparison of DOROD-FEM and STAN5

Case (pressure gradient)	Computational method	Grid points across b.l.	$\Delta x/L$	No. of steps, $\Delta x/L$	Relative execution time
zero	DOROD-FEM	11	0.0001-0.070	205	1
(flat plate)	STAN5	33-39	0.0004-0.031	401	8.99
favourable	DOROD-FEM	11	0.0001-0.044	118	0.64
	STAN5	32-35	0.0017-0.0126	301	6.83
adverse	DOROD-FEM	11	0.0001-0.035	243	1.29
	STAN5	47-48	0.0002-0.0039	660	17.70
adverse	DOROD-FEM	11	0.002-0.023	188	0.92
(separating)	STAN5	33-36	0.001-0.031	238	5.71

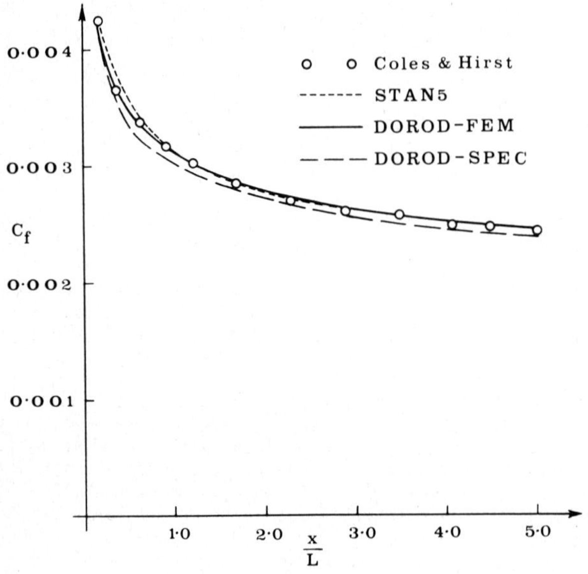

Fig. 15. Skin friction variation for a zero pressure gradient

4.4.2 Logarithmic elements in turbulent flow

For turbulent flows in pipes the longitudinal velocity increases very rapidly, over a small radial distance, from zero at the wall to an almost constant value across the radius. A large number of conventional elements would be required, close to the wall, to accurately represent the velocity profile.

Taylor et al. (1977) avoid this problem by introducing a *logarithmic element*, in place of a quadratic Serendipity element, adjacent to the wall. The turbulent velocity profile close to the wall is well represented by the function,

$$S_k(\eta) = \log_{M_k}\left[M_k - 0.5(M_k - 1)(1 + \eta)\right] \qquad (4.76)$$

which satisfies the conditions (Fig. 16).

$$S_k = 0 \quad \text{at } \eta = +1$$

and $$S_k = 1 \quad \text{at } \eta = -1 \ .$$

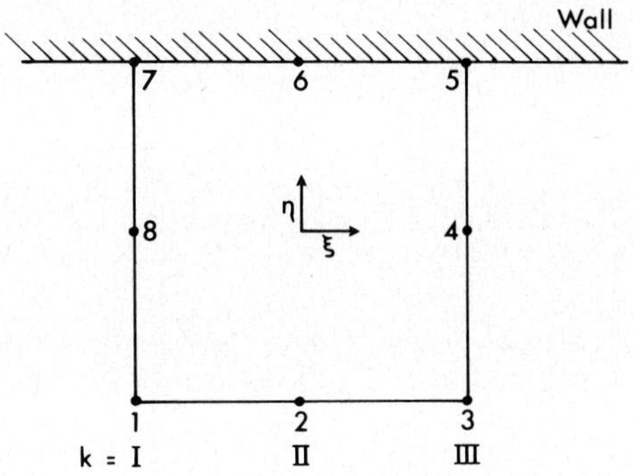

Figure 16. Near-wall element

The parameter M_k is solution dependent and can be obtained as follows. Firstly $S_k(0)$ is related to the nodal velocities (Fig. 16) by

$$S_I(0) = L_I = u_8/u_1$$

$$S_{III}(0) = L_{III} = u_4/u_3 \tag{4.77}$$

and $$S_{II}(0) = L_{II} = 0.5(L_I + L_{III}) .$$

Then M_k is chosen to satisfy

$$2M_k^{L_k} - M_k - 1 = 0 , \quad k = I, II, III . \tag{4.78}$$

The above functions have been used to generate the logarithmic interpolation. On a coarse grid, logarithmic interpolation is considerably more accurate than quadratic polynomial interpolation, when the underlying behaviour is logarithmic.

The logarithmic shape functions are expressed in the product form

$$J^k(\eta)N_i(\xi,\eta)$$

where $N_i(\xi,\eta)$ is a conventional quadratic Serendipity shape function (Fletcher 1983a)) and $J^k(\eta)$ is given by

$$J^k(\eta) = \frac{\log_{10}[M_k - 0.5(M_k-1)(1+\eta)]}{[(1-\eta^2)L_k - 0.5(\eta-\eta^2)]\log_{10}M_k} \quad (4.79)$$

and $k = I \quad i = 1,8$

$= II \quad i = 2$

$= III \quad i = 3,4$.

Due to the complex nature of the shape functions, eq. (4.79), Gauss quadrature is required to evaluate the usual Galerkin inner products. Due to the boundary conditions on u, no shape functions associated with nodes 5, 6 and 7 in Fig. 16 are required.

Typical results, using a coarse mesh, are shown in Fig. 17. The results with one logarithmic element adjacent to the wall show excellent agreement with the fine mesh results.

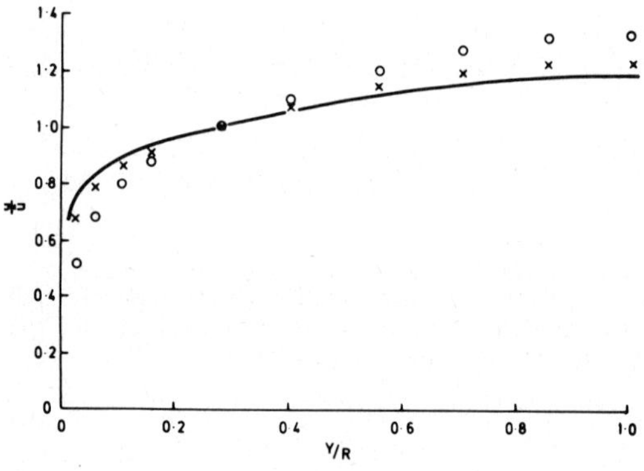

Figure 17. Velocity profile for developing pipe-flow — fine mesh; x logarithmic coarse grid; o quadratic coarse grid

4.5 Applications

Galerkin finite element methods have been applied to many problems. Early catalogues are given by Whiteman (1975) and Norrie and de Vries (1976). Here we pick out a few specific examples to indicate the power of the method and its flexibility.

4.5.1 Thermal entry problem

The first example is chosen to illustrate the possibility of combining the Galerkin finite element method with techniques for the efficient solution of the resulting algebraic equations. In this case the *alternating directions implicit* (ADI) technique for finding efficient solutions in two or more directions for problems posed in an unsteady formulation. Here the unsteady formulation is used as a convenient vehicle for obtaining the steady-state solution.

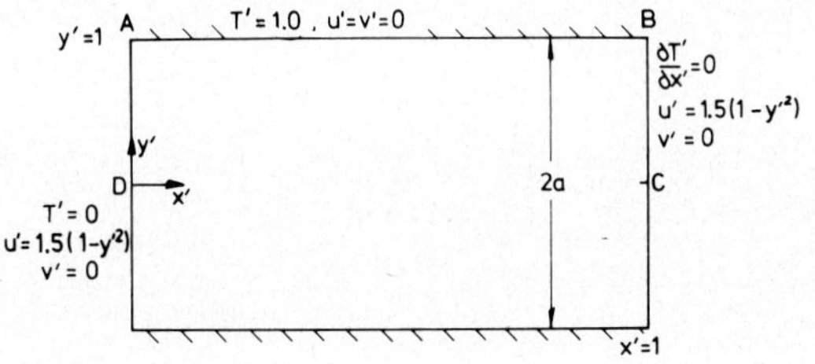

Figure 18. Thermal entry problem

The first example is that of the temperature distribution in a two-dimensional duct (Figure 18). The governing equation can be written,

$$\frac{\partial T}{\partial t} + \frac{\partial}{\partial x}(uT) + \frac{\partial}{\partial y}(vT) = \alpha_x \frac{\partial^2 T}{\partial x^2} + \alpha_y \frac{\partial^2 T}{\partial y^2} \ . \tag{4.80}$$

It is assumed that the velocity field (u,v) is already available from the solution of the hydrodynamic entry problem. Thus the problem is linear in the temperature, T. The temperature (and velocity components) have been nondimensionalised in such a way that the nondimensional temperature will have reached the fully developed value of T = 1 at the boundary, BC.

To obtain algebraic equations linear rectangular elements on a uniform mesh have been used to represent T and the groups (uT) and (vT). Representing groups in this way produces an important economy in the number of algebraic terms that must be manipulated. This idea has been used previously with a least-squares finite element formulation for subsonic compressible flow (Fletcher, 1979) and with the orthonormal method of integral relations (Fletcher and Holt, 1976). The present status of this group formulation is given by Fletcher (1983c).

Applying the Galerkin formulation as in Section 4.2.3 produces the following equations,

$$\sum_{k,\ell} [a_{k,\ell} \dot{T}_{k,\ell} + f_{k,\ell} (uT)_{k,\ell} + g_{k,\ell} (vT)_{k,\ell} - r_{k,\ell} T_{k,\ell} - t_{k,\ell} T_{k,\ell}] = 0 ,$$

$$k = i-1, i, i+1, \quad \ell = j-1, j, j+1 \tag{4.81}$$

where

$$a \equiv \frac{1}{36}\begin{bmatrix} 1 & 4 & 1 \\ 4 & 16 & 4 \\ 1 & 4 & 1 \end{bmatrix}$$

$$f \equiv \frac{1}{12\Delta x}\begin{bmatrix} -1 & 0 & 1 \\ -4 & 0 & 4 \\ -1 & 0 & 1 \end{bmatrix}, \quad g \equiv \frac{1}{12\Delta y}\begin{bmatrix} 1 & 4 & 1 \\ 0 & 0 & 0 \\ -1 & -4 & -1 \end{bmatrix} \qquad (4.82)$$

$$r \equiv \frac{1}{6\Delta^2 x}\begin{bmatrix} 1 & -2 & 1 \\ 4 & -8 & 4 \\ 1 & -2 & 1 \end{bmatrix}, \quad t \equiv \frac{1}{6\Delta^2 y}\begin{bmatrix} 1 & 4 & 1 \\ -2 & -8 & -2 \\ 1 & 4 & 1 \end{bmatrix}$$

It can be seen that eq. (4.81) is a nine-point scheme centered at the (i,j) node. Before introduction of the ADI formulation the $\dot{T}$ terms are replaced by first-order finite difference expressions. The form of the coefficients, eq. (4.82), can be seen to be similar to those associated with the use of centered second-order finite difference expression for the spatial derivatives in eq. (4.80). E.g.

$$f_{f.d.} \equiv \frac{1}{12\Delta x}\begin{bmatrix} 0 & 0 & 0 \\ -6 & 0 & 6 \\ 0 & 0 & 0 \end{bmatrix}, \quad t_{f.d.} \equiv \frac{1}{6\Delta^2 y}\begin{bmatrix} 0 & 6 & 0 \\ 0 & -12 & 0 \\ 0 & 6 & 0 \end{bmatrix}.$$

Thus the use of linear rectangular elements produces the same symmetry as the second-order finite difference expression except that some weight is given to adjacent grid lines.

The ADI concept is described in Mitchell and Griffiths, 1980. The basic idea is to consider all x-derivatives implicitly for the first half time-step and all y-derivatives implicitly for the second half time-step. For the finite difference formulae shown above this produces tridiagonal systems of equations along each grid-line which can be solved very efficiently by the Thomas algorithm.

For the finite element method a difficulty arises that part of the x-derivatives (see f in eq. (4.82)) comes from the $(i-1)^{th}$ and $(i+1)^{th}$ grid lines. To retain the availabilities of the tridiagonal algorithm it is necessary to treat these contributions explicitly. Although this degrades the accuracy of the transient solution it does not affect the accuracy of the steady-state solution. However it might be expected that the stability of the scheme will be reduced. A von Neumann stability analysis or direct calculation indicates that the scheme is, indeed, subject to a restriction on the time-step if steady-state solutions are to be obtained. The maximum time step for the scheme (1) is shown in Table 18. The main cause of this lack of stability is due to the coefficient, a, multiplying the time derivative term. In particular the weight associated with the (i,j) node is only 16/36 (eq. (4.82)) compared with 1 for the finite-difference method.

Since the time derivative term has no effect on the steady-state solution it is convenient to lump all the contributions from $a_{k,\ell}$ onto the $(i,j)^{th}$ node so that $a_{i,j}=1$. This is scheme 2 in Table 18. As can be seen, this scheme is an improvement over scheme 1 but still subject to a time-step restriction. It is apparent that certain terms in t (eq. (4.82)) could be treated implicitly during the first half-time-step and similarly certain terms in r during the second half-time-step. These have been treated implicitly in scheme 3 (Table 18) and the result is a scheme with no restriction on the time-step. Scheme 4 is the conventional ADI finite difference method. The rms difference shown in Table 18 is based on the difference between the computed steady-state solution and a more 'exact' solution on the centre line (DC in Figure 18). The solutions were obtained with a uniform 21 x 21 grid.

Table 18. Comparison of schemes for thermal entry length problem

Scheme	Description	Δt_{max}	Number of time steps	Relative execution time	rms difference
1	ADIFEM	0.0003	1250	7.73	0.0502
2	ADIFEM, fully lumped	0.002	247	1.49	0.0502
3	ADIFEM, fully lumped, fully implicit	—	40	0.21	0.0502
4	ADI finite difference	—	332	1.00	0.0601

It can be seen that the progressive reduction in the restriction on the time-step means less time-steps to reach the steady-state and consequently a shorter execution time. Although the finite difference scheme had no stability restriction it was found that too large a time-step slowed down the rate of convergence to the steady-state. Since the finite element schemes only differ in their transient behaviour the steady-state accuracy is the same for all three schemes. The entry-length problem and solution is described in more detail in Fletcher (1980b).

4.5.2 Viscous compressible flow

This example is a natural extension of the thermal entry problem. However a more efficient time-splitting algorithm is introduced and heavy reliance is placed on the group finite element formulation (Fletcher (1983c)). In addition for some of the problems considered on eddy viscosity formulation is introduced to represent the turbulent nature of the flow.

The system of nondimensional equations governing viscous compressible flow in two dimensions can be written

$$\frac{\partial \bar{q}}{\partial t} + \frac{\partial \bar{F}}{\partial x} + \frac{\partial \bar{G}}{\partial y} = \frac{\partial^2 \bar{R}}{\partial x^2} + \frac{\partial^2 \bar{S}}{\partial x \partial y} + \frac{\partial^2 \bar{T}}{\partial y^2} \tag{4.83}$$

where

$$\bar{q}^t = \{\rho u, \rho v, \rho\}$$

$$\bar{F}^t = \{\rho u, p + \rho u^2 - \sigma_x, \rho uv - \tau_{xy}\}$$

$$\overline{G}^t = \{\rho v,\ \rho uv - \tau_{xy},\ P + \rho v^2 - \sigma_y\}$$

$$\overline{R}^t = \{\theta_\rho^d\ \rho,\ \frac{4}{3Re}\ u,\ \frac{1}{Re}\ v\} \qquad (4.84)$$

$$\overline{S}^t = \{0,\ \frac{v}{3Re}\ ,\ \frac{u}{3Re}\}$$

$$\overline{T}^t = \{\theta_\rho^d\ \rho,\ \frac{1u}{Re}\ ,\ \frac{4}{3}\frac{v}{Re}\}$$

In the above equations ρ is the density, p is the pressure, u and v are the velocity components and Re is the Reynolds number. σ_x, σ_y and τ_{xy} are the Reynolds stresses associated with turbulent flow. The pressure, p, can be eliminated from the above equations by using

$$1 + \gamma M_\infty^2 P = \rho\{1 + 0.5(\gamma-1)\ M_\infty^2(1 - u^2 - v^2)\} \qquad (4.85)$$

where M_∞ is the freestream Mach number and γ is the specific heat ratio.

The Galerkin finite element method with linear, Lagrange rectangular elements (Fletcher 1983a)) are applied to eqs. (4.83). Trial solutions are introduced for the groups of terms, e.g. ρuv, that appear in eq. (4.84). Estimates given by Fletcher (1983c) indicate that the group formulation is approximately *seventeen times* more economical than the conventional finite element method applied to the nonlinear terms. In three dimensions the group formulation is estimated to be close to one hundred times more economical. This greater economy of the group formulation is also accompanied by *an improvement in accuracy* (Fletcher (1983b)).

After application of the Galerkin finite element formulat,on a system of ordinary differential equation is obtained which can be written,

$$M_x \otimes M_y\ \dot{\overline{q}} + M_y \otimes \delta_x \overline{F} + M_x \otimes \delta_y \overline{G} = M_y \otimes \delta_x^2\ \overline{R} + \delta_{xy}\overline{S} + M_x \otimes \delta_y^2\ \overline{T} \qquad (4.86)$$

where $\otimes$ denotes the *tensor product* and M_x, M_y, δ_x, δ_y, $\delta_x^{\ 2}$, δ_{xy} and δ_y^2 are operators that connect together the various nodal values i.e.

$$M_x \equiv \frac{1}{6}\ \{1,\ 2(1+r_x),\ r_x\}\ ,\qquad M_y^T \equiv \frac{1}{6}\ \{r_y,\ 2(1+r_y),\ 1\}$$

$$\delta_x \equiv \frac{1}{2\Delta x}\ \{-1,\ 0,\ 1\},\ \delta_y^T \equiv \frac{1}{2\Delta y}\ \{1,\ 0,\ -1\} \qquad (4.87)$$

$$\delta_x^2 \equiv \frac{1}{\Delta x^2}\{1,\ -2(1 + \frac{1}{r_x}),\ \frac{1}{r_x}\}\ ,\ (\delta_y^2\ )^T \equiv \frac{1}{\Delta y^2}\{\frac{1}{r_y},\ -2(1 + \frac{1}{r_y})\ ,\ 1\}$$

In the above r_x and r_y are the ratio of neighbouring elements sizes surrounding the node of interest (Fig. 19)

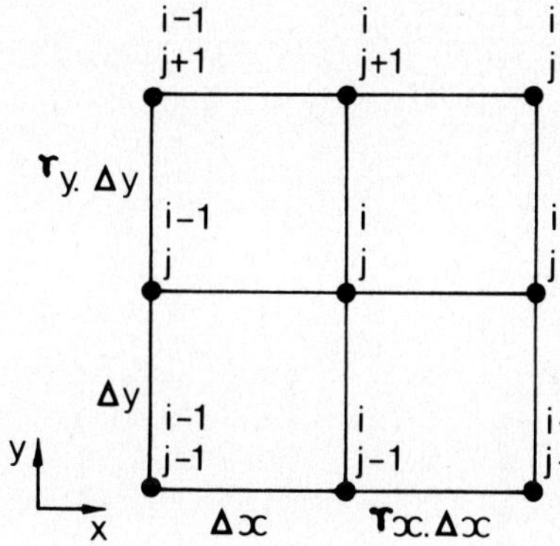

Figure 19. Global node orientation

The operator δ_{xy} is defined by

$$\delta_{xy}\,\overline{S} = \frac{0.25}{\Delta x\Delta y}\left\{(\overline{S}_{\substack{i+1\\j+1}} - \overline{S}_{\substack{i-1\\j+1}}) - (\overline{S}_{\substack{i+1\\j-1}} - \overline{S}_{\substack{i-1\\j-1}})\right\} \tag{4.88}$$

The above splitting is possible with Lagrange shape functions but not with Serendipity shape functions (Fletcher (1983a)). The time derivative term can be evaluated as follows

$$M_x \otimes M_y \left\{\alpha \frac{\Delta q^{n+1}}{\Delta t} + (1-\alpha)\frac{\Delta q^{n}}{\Delta t}\right\} = \beta\ RHS^{n+1} + (1-\beta)\ RHS^{n} \tag{4.89}$$

where $RHS = M_y \otimes \delta_x^2\,\overline{R} + \delta xy\overline{S} + M_x \otimes \delta_y^2\,\overline{T} - M_y \otimes \delta_x\,\overline{F} - M_x \otimes \delta_y\,\overline{G}$

and $\Delta q^{n+1} = q^{n+1} - q^n$

The above formulation is marched in the direction of increasing time by solving a block tridiagonal system of equations at each time-step. To obtain this system it is necessary to linearise the term RHS^{n+1} about the nth time level. This gives

$$RHS^{n+1} = RHS^n + \frac{\partial(RHS)}{\partial q}\Delta q^{n+1}$$

Then eq. (4.89) can be written

$$\alpha(M_x \otimes M_y - \Delta t\,\frac{\beta}{\alpha}\,\frac{\partial(RHS)}{\partial q})\ \Delta q^{n+1} = \Delta t\ RHS^n - (1-\alpha)\ M_x \otimes M_y\ \Delta q^n \tag{4.90}$$

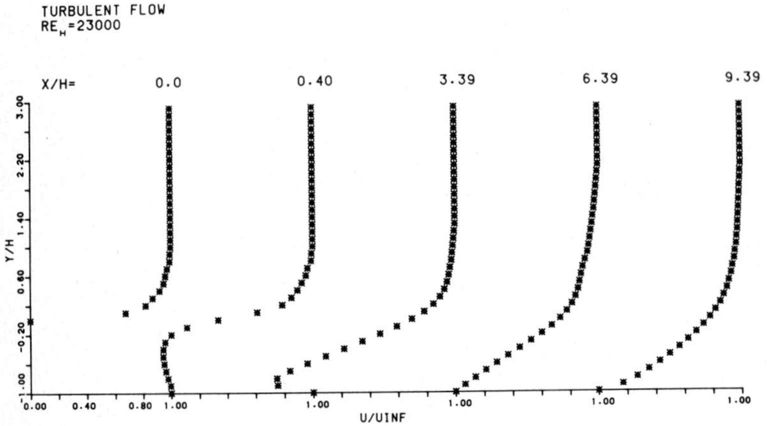

Figure 20. Mean velocity profiles behind a backward-facing step

4.5.3 Jetflap flow

This problem is relevant to the use of two-dimensional jets exhausting over the upper surface of aircraft wings to increase the lift without requiring undue structural complexity (flap retraction mechanisms etc.)

Computationally this problem introduces the technique of *parabolising* the Navier-Stokes equations (Davis and Rubin, 1980). Namely the deletion of the second derivative terms in one particular coordinate direction on an order-of-magnitude basis. The solution can then be obtained by marching in that direction. For the present problem this is the downstream direction.

Secondly the flow is turbulent and to properly account for the noise generated by the jet it is necessary to supplement the Navier-Stokes equations with transport equations for the turbulent kinetic energy, k, and the dissipation function, ε (Baker and Manhardt, 1978).

An idealised solution domain is shown in Fig. 21. The solution domain can be conveniently split into two parts. From the origin at the exit of the slot nozzle to the end of the flap the problem is solved by marching downstream, simultaneously considering the region adjacent to the flap (2DBL) and the region far from the flap (2DPNS). 2DBL stands for two-dimensional boundary layer equations and 2DPNS stands for two-dimensional parabolised Navier-Stokes equations. Downstream of the flap trailing edge (wake solution) the 2DPNS equations are solved across the whole region (in the direction x_2) as one marches in the direction x_1.

To facilitate the solution of this equation the left hand side is split in the following way

$$\alpha\{M_x + \frac{\beta}{\alpha}\Delta t\,(\delta_x \frac{\partial\bar{F}}{\partial\bar{q}} - \delta_x^2 \frac{\partial\bar{R}}{\partial\bar{q}})\}\otimes\{M_y + \frac{\beta}{\alpha}\Delta t\,(\delta_y \frac{\partial\bar{G}}{\partial\bar{q}} - \delta_y^2 \frac{\partial\bar{T}}{\partial\bar{q}})\}\ \Delta q^{n+1} = \Delta t\ RHS^n - (1-\alpha) M_x \otimes M_y \Delta q^n \tag{4.91}$$

The splitting is approximate in that it ignores the contribution from the cross-derivative term $\partial\bar{S}/\partial\bar{q}$ and introduces additional terms of $O(\Delta^2 t)$. However the splitting does permit a very efficient two-stage algorithm. In the first stage the following equation is solved

$$\alpha\left[M_x + \frac{\beta}{\alpha}\Delta t\,(\delta_x \frac{\partial\bar{F}}{\partial\bar{q}} - \delta_x^2 \frac{\partial\bar{R}}{\partial\bar{q}})\right]\Delta q^* = \Delta t\ RHS^n - (1-\alpha)\ M_x \otimes M_y\ \Delta q^n \tag{4.92}$$

and in the second stage

$$\left[M_y + \frac{\beta}{\alpha}\Delta t\,(\delta_y \frac{\partial\bar{G}}{\partial\bar{q}} - \delta_y^2 \frac{\partial\bar{T}}{\partial\bar{q}})\right]\Delta q^{n+1} = \Delta q^* \tag{4.93}$$

where $\Delta q^* = q^* - q^n$ and $\Delta q^{n+1} = q^{n+1} - q^n$.

The importance of the original splitting into the mass operators M_x, M_y and differential operators, δx etc., is now clear. At each stage of the splitting the block tridiagonal system of equations (4.92) and (4.93), are solved for Δq^* and Δq^{n+1}.

In the current situation only the steady-state solution is of interest. This corresponds to $RHS^n = 0$ and $\Delta q^* = \Delta q^{n+1} = 0$. This also implies that the factors on the left-hand-side could be changed without altering the steady-state solution. This strategy has been used to improve the stability by Fletcher (1982c).

In eqs. (4.92) and (4.93) two second-order marching schemes are of interest. These are

(a) $\alpha = 1$, $\beta = 0.5$

(b) $\alpha = 1.5$, $\beta = 1.0$

Scheme (a) is the Crank-Nicolson scheme which has the advantage of being a two-level scheme. However it is characterised by a relatively slow convergence to the steady-state. Scheme (b) is a three-level scheme but is generally more robust than scheme (a) and gives *faster convergence to the steady state*.

The above formulation has been applied to laminar and turbulent flows past rectangular obstacles, flat plates and backward-facing steps by Fletcher (1982c) and Srinivas and Fletcher (1983). Typical mean velocity profiles downstream of a backward-facing step are shown in Fig. 20.

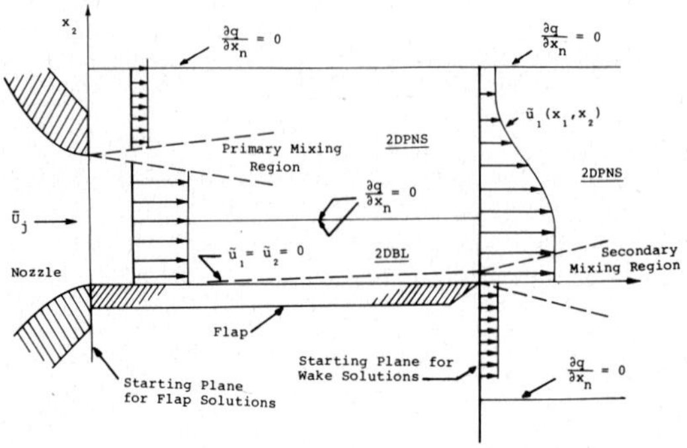

Figure 21. Idealised jet-flap flowfield geometry

For this problem the governing equations can be written (see Baker and Manhardt for the detailed derivation)

$$L_c(\rho) = \frac{\partial}{\partial x_i}(\rho u_i) = 0 \tag{4.94}$$

$$L_m(u_i) = \frac{\partial}{\partial x_j}(\rho u_i u_j) - \frac{\partial}{\partial x_\ell}\left[\mu^e\left(\frac{\partial u_i}{\partial x_\ell} + \frac{\partial u_\ell}{\partial x_i}\right)\right] + \frac{\partial p}{\partial x_i} = 0 \tag{4.95}$$

$$L_k(k) = \frac{\partial}{\partial x_j}(\rho u_j k) - \frac{\partial}{\partial x_\ell}\left[C_k \mu^e \frac{\partial k}{\partial x_\ell}\right] - \mu^e \frac{\partial u_1}{\partial x_\ell}\frac{\partial u_1}{\partial x_\ell} + \rho\varepsilon = 0 \tag{4.96}$$

$$L_\varepsilon(\varepsilon) = \frac{\partial}{\partial x_j}(\rho u_j \varepsilon) - \frac{\partial}{\partial x_\ell}\left[C_\varepsilon \mu^e \frac{\partial \varepsilon}{\partial x_\ell}\right] - C_\varepsilon^1 \frac{\varepsilon}{k}\mu^e \frac{\partial u_1}{\partial x_\ell}\frac{\partial u_1}{\partial x_\ell} + C_\varepsilon^2 \rho \frac{\varepsilon^2}{k} = 0 \tag{4.97}$$

where $\mu_e = \mu/Re + \rho\nu_T$

and $\nu_T = C_\nu k^2/\varepsilon$ (4.98)

For three-dimensional flow $1 \leq i,j \leq 3$ and $2 \leq \ell \leq 3$. Since $\ell \neq 1$ the solution can be obtained by marching in the x_1 direction.

The Galerkin finite element procedure is applied by introducing trial solutions for ρ, u_i, k and ε. E.g.

$$\rho = \sum_j N_j^{\rho} (x_2, x_3)\bar{\rho}_j(x_1) \qquad (4.99)$$

The form of eq. (4.99) is typical of that for time-dependent problems. In the present problem the coordinate x_1 has a time-like role. Thus the Galerkin finite element method is only applied over the x_2, x_3 domain.

Substitution of equations like eq. (4.99) into eqs. (4.94) to (4.97) produce residuals, $L_c(\bar{\rho}_j, \bar{u}_{ij})$ etc. Evaluation of the inner products, like

$$(L_c, N_k) = 0 \quad , \qquad (4.100)$$

produce a system of ordinary differential equations. A finite difference representation is introduced for the x_1 derivatives allowing a marching algorithm to be constructed. The 2DBL formulation is obtained from the above 3DPNS formulation by discarding eq. (4.95) when j = 2 and only letting $\ell = 2$, in the other equations.

Baker and Manhardt present data for the turbulent properties of the downstream jet development. A typical longitudinal velocity profile downstream of the jet flap is shown in Fig. 22. Excellent agreement with the experimental data of Schreker and Maus (1974) is indicated.

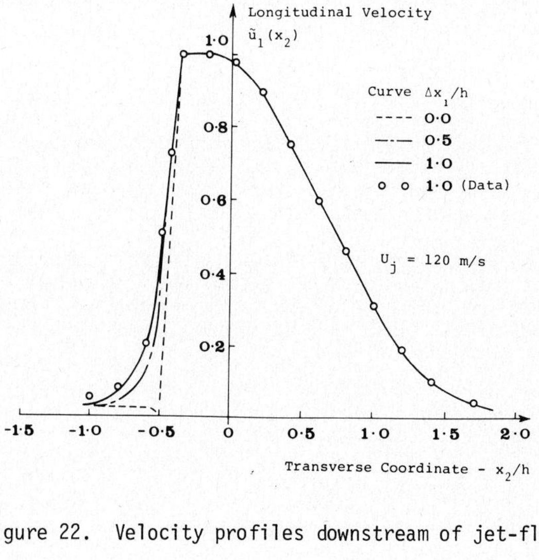

Figure 22. Velocity profiles downstream of jet-flap

4.5.4 Other applications

The general transient, convective heat-transfer problem, in which both velocity and temperature fields are solved for, has been investigated with a conventional Galerkin formulation by Gartling (1977). Astley and Eversman (1981) have used the Galerkin finite element method to study acoustic transmission in non-uniform, compressible duct flow. For this particular problem they found that the Galerkin formulation gave considerably more accurate results than a least-squares formulation.

The shallow-water equations which govern tidal flows have been solved, using the Galerkin finite element method, by Taylor and Davis (1975) and Kawahara et al. (1978) amongst others. A recent survey is provided by Taylor (1980).

The Galerkin finite element method has been used for weather forecasting by Staniforth and Daley (1979) and by Cullen and Hall (1979) and to study environmental pollution problems in rivers by Baker and Zelazny (1975).

It is well known (Roache, 1972) that when solutions for convection diffusion problems are obtained using a centered finite difference scheme for the convective terms the steady-state solution will show large non-physical oscillations if the Reynolds number based on a cell is greater than 2. The same phenomenon affects the conventional Galerkin finite element method. Historically, for the finite difference method, this stability problem has been avoided by using first-order differencing (often called *upwind differencing*) for the convective terms. However, such a scheme is only first-order accurate. If it is interpreted as a second order scheme it is equivalent to solving the governing partial differential equation(s) with additional viscous terms added. In many situations if the grid is locally refined to achieve acceptable accuracy the cell Reynolds number will be less than two and the conventional centered difference or standard Galerkin formulae give accurate solutions.

Even so, considerable effort, in the finite element area, has been expended to generalise the Galerkin method to give the *Petrov-Galerkin* method (Fletcher (1983a)). The thrust of this work has been to obtain the upwind feature of suppressing non-physical oscillations without the introduction of excessive artificial viscosity.

The earlier references, Christie et al (1976), Heinrich et al. (1977), Hughes (1978), were mainly concerned with simple convection/diffusion equations predominantly in one dimension. More recently, Griffiths and Mitchell (1979) and Hughes and Brooks (1979), have sought to minimise spurious cross wind diffusion in more than one dimension by aligning the weight function with the local flow direction. This approach, called *streamline upwind Petrov-Galerkin* by Hughes, has been further developed by Brooks and Hughes (1982) and by Hughes et al. (1982).

An alternative Petrov-Galerkin formulation based on extending the optimal nature of Galerkin methods for pure elliptic problems (typically no convection) to *convection dominated* problems has been developed by Barrett and Morton (1980, 1981) and by Morton and Parrott (1980) and Morton and Stokes (1981).

5. BURGERS' EQUATION

In this section we will look more closely at Burgers' equation

$$u_t + uu_x - \nu u_{xx} = 0 , \tag{5.1}$$

to explore its physical significance and to see under what circumstances an *exact* solution can be obtained. Subsequently, Sections 6 and 7 will demonstrate the use of Burgers' equation to test computational algorithms and to model various physical phenomenon respectively.

In a typical application u is a velocity-like dependent variable and ν is a viscosity-like parameter. The subscript t indicates a time derivative and the subscript x indicates a spatial derivative. If eq. (5.1) is interpreted as a non-dimensional equation $1/\nu$ is replaced by the Reynolds number, Re.

5.1 The Role of the Convective and Dissipative Terms

As it stands, eq. (5.1) is a quasi-linear parabolic partial differential equation that describes the evolution of some dependent variable, u, with time. The equation is linear in the second x, and t derivatives, of u but *nonlinear* in u. An important feature of the equation is that it is a prototype equation for the balance between the nonlinear convective term, uu_x, and the diffusive term, νu_{xx}. The ability to handle this balance efficiently is probably the single most difficult aspect of computing fluid dynamic problems.

The interaction of these two effects can be appreciated by simplifying eq. (5.1). Firstly, if the diffusive term is dropped, eq. (5.1) becomes

$$u_t + uu_x = 0 . \tag{5.2}$$

Equation (5.2) is hyperbolic and is a model for the convection of disturbances in inviscid flow. A typical situation is shown in Figure 23. A wave is convecting from left to right and solutions for successive times, t = 0, t_1, t_2 are illustrated. Points on the wave with larger u convect faster and consequently overtake parts of the wave convecting with smaller u. At t = t_2 u has three values at a

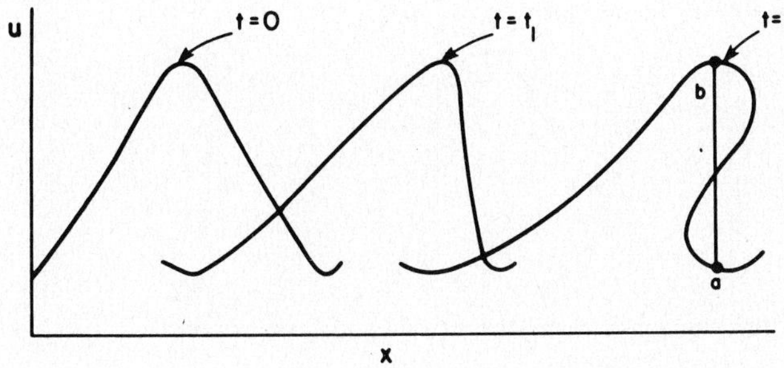

Figure 23. Formation of a multivalued solution of $u_t + uu_x = 0$

given x. For eq. (5.2) to have a unique solution (and a physically sensible result) it is necessary to postulate a *shock* (ab in Figure 23) across which u changes discontinuously.

If in eq. (5.1) the nonlinear convective term is dropped, the result is

$$u_t = \nu u_{xx} \tag{5.3}$$

which is the classical diffusion or heat conduction equation. If u is interpreted as the temperature of a bar of length, L, and ν as the thermal conductivity, the solution would be represented schematically as in Figure 24. An initial temperature profile is shown at t = 0. Due to heat transfer from the ends of the bar the 'temperature' falls everywhere and eventually, at t = ∞, the temperature will be constant at all points along the bar.

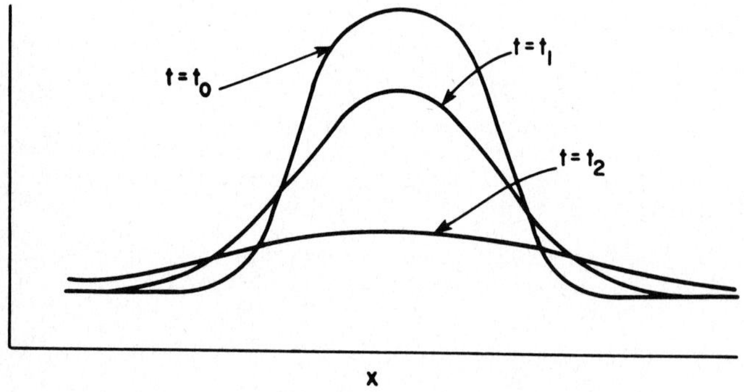

Figure 24. Evolution of the solution $u_t = u_{xx}$

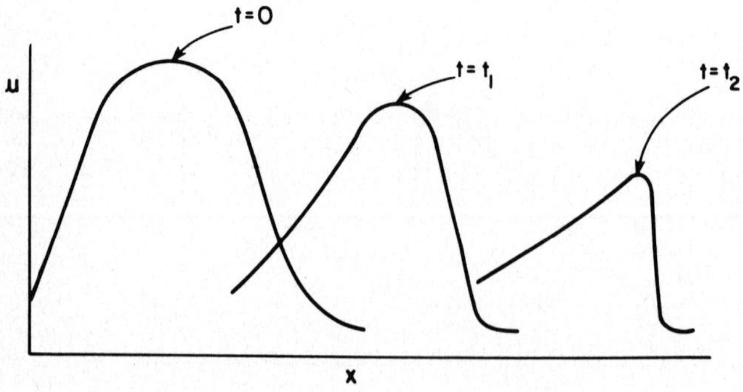

Figure 25. Evolution of the solution $u_t + uu_x = \nu u_{xx}$

If the two processes of convection and diffusion are considered together, i.e. eq. (5.1), the evolution of the solution can be shown schematically as in Figure 25. Clearly the maximum amplitude of u is diminishing and the leading profile (right hand side) becomes steeper. However, the dissipative term, νu_{xx}, always prevents a multivalued solution from developing, since its contribution becomes larger where the steepening occurs (shock formation).

Burgers' equation can also represent the balance between convection and dissipation in the steady-state sense. If u_t is dropped from eq. (5.1), the result is

$$uu_x = \nu u_{xx} \quad . \tag{5.4}$$

Equation (5.4) would be an elliptic partial differential equation in more than one spatial direction. If $u = u_1$ at $x = -\infty$ and $u = u_2$ at $x = +\infty$ the solution of (5.4) be shown schematically in Figure 26. In the limit $\nu \to 0$ the solution changes discontinuously. The solution, is

$$u = 0.5 \left[(u_1 + u_2) - (u_1 - u_2) \tan h\,[x(u_1 - u_2)/4\nu]\right] \tag{5.4a}$$

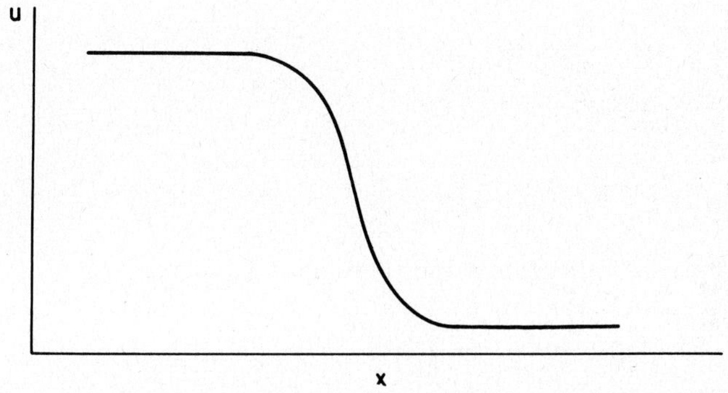

Figure 26. Typical solution of $uu_x = \nu u_{xx}$

Thus it is apparent that, for different conditions, Burgers' equation can be classified as an elliptic, hyperbolic or parabolic partial differential equation and that it is the simplest nonlinear model equation for considering convection diffusion interactions.

5.2 The Cole-Hopf Transformation

As a medium for testing new computational schemes Burgers' equation has one other very important attribute. For a wide range of initial and boundary conditions *exact solutions exist and can be obtained easily*. Independently Hopf (1950) and Cole (1951) discovered a transformation that would reduce eq. (5.1) to the linear diffusion equation, (5.3). The relevance of this transformation is that the large number of known exact solutions of eq. (5.3), e.g. see Carslaw and Jaegar

(1959), become, after transformation, solutions of eq. (5.1).

The Cole-Hopf transformation, after Whitham (1974), can be introduced in two stages. First let

$$u = \psi_x \tag{5.5}$$

so that eq. (5.1) becomes, after integration,

$$\psi_t + \tfrac{1}{2}\psi_x^2 = \nu\psi_{xx} \ . \tag{5.6}$$

Then if

$$\psi = -2\nu \log \phi \quad , \tag{5.7}$$

eq. (5.6) becomes

$$\phi_t = \nu\phi_{xx} \quad , \tag{5.8}$$

which is just eq. (5.3). The Cole-Hopf transformation can be generalised (Ames, 1965) as follows. Starting from the linear heat conduction equation (5.8) let

$$\phi = F(\psi) \quad , \tag{5.9}$$

so that eq. (5.8) becomes

$$\psi_t - \nu(\psi_x)^2 \frac{F''}{F'} = \nu\psi_{xx} \ . \tag{5.10}$$

Differentiating with respect to x and setting $u = \psi_x$ gives

$$u_t - 2\nu\, uu_x \frac{F''}{F'} - \nu u^3 \left\{\frac{F''}{F'} - \left(\frac{F''}{F'}\right)^2\right\} = \nu u_{xx} \quad , \tag{5.11}$$

and Burgers' equation is the special case

$$F(\psi) = \exp(-\psi/2\nu) \ . \tag{5.12}$$

Initial conditions can be readily transformed. However, boundary conditions are not handled so easily and many exact solutions are for the interval, $-\infty < x < \infty$.

As noted by Cole (1951) and Ames (1965), the Cole-Hopf transformation generalises to *multi-dimensions*. Thus, in two dimensions, the transformations

$$u = -2\nu \frac{\partial}{\partial x}(\ln \phi) = -2\nu\, \phi_x/\phi \tag{5.13}$$

$$v = -2\nu \frac{\partial}{\partial y}(\ln \phi) = -2\nu\, \phi_y/\phi \quad , \tag{5.14}$$

transform the two-dimensional heat conduction equation,

$$\phi_t = \phi_{xx} + \phi_{yy} \quad , \tag{5.15}$$

into

$$u_t + uu_x + vu_y = \nu(u_{xx} + u_{yy}) \tag{5.16}$$

and

$$v_t + uv_x + vv_y = \nu(v_{xx} + v_{yy}) \tag{5.17}$$

which are the two-dimensional Burgers' equations.

5.3 Exact Solutions in One Dimension

The known exact solutions of eq. (5.1) have been tabulated by Benton and Platzman (1972).

The Cole-Hopf transformation, eqs. (5.5) and (5.7), can be applied to the initial conditions for eq. (5.1), i.e.

$$u(x,0) = u_0(x) \quad , \tag{5.18}$$

to give

$$\phi(x,0) = \phi_0(x) = \exp\left[-\left\{\int_0^x u_0(\xi)d\xi\right\}/2\nu\right] \quad . \tag{5.19}$$

For the infinite domain ($-\infty < x < \infty$) the solution of eq. (5.8) subject to the initial condition, eq. (5.19), is particularly concise,

$$\phi(x,t) = \frac{1}{\sqrt{4\pi\nu t}} \int_{-\infty}^{\infty} \exp\left[\frac{-(x-\xi)^2}{4\nu t}\right] \phi_0(\xi)d\xi \quad . \tag{5.20}$$

Consequently the corresponding solution to Burgers' equation, (5.1), is, through the Cole-Hopf transformation,

$$u(x,t) = \int_{-\infty}^{\infty} \left\{\frac{x-\xi}{t}\right\} \exp\{-F/2\nu\}d\xi \Big/ \int_{-\infty}^{\infty} \exp\{-F/2\nu\}d\xi \tag{5.21}$$

where

$$F(\xi;x,t) = \int_0^{\xi} u_0(\xi')d\xi' + (x-\xi)^2/2t. \tag{5.22}$$

As soon as a boundary condition must be applied at a finite x location, a difficulty arises. Thus consider the half-plane problem with initial and boundary conditions

$$u(x,0) = u_0(x) \quad , \quad 0 \leqslant x \leqslant \infty \tag{5.23}$$

and

$$u(0,t) = g(t) \quad , \quad t > 0 \quad .$$

To solve eq. (5.1) subject to eq. (5.23) requires the solution of an integral equation for h(t).

$$g(t)\,h(t) = 2\nu \int_0^t \frac{h'(t)}{\sqrt{\pi(t-\tau)}}\,d\tau + P(t) \quad , \tag{5.24}$$

where P(t) is a function of $u_0(x)$. After applying the Cole-Hopf transformation, h(t) becomes the boundary condition for ϕ,

$$\phi(0,t) = h(t) \quad . \tag{5.25}$$

Rodin (1970a) concludes that the appearance of eq. (5.24) precludes the possibility of obtaining an exact solution and that it is more efficient to postulate a series form of solution from the start. Rodin considers the motion induced by a piston in the half-plane, $0 \leqslant x < \infty$, with initial motion of the piston, H(t). Thus the boundary condition at x = 0 becomes

$$u(H(t),t) = H'(t) \quad . \tag{5.26}$$

Expanding as a Taylor series and assuming that the displacement and velocity of the piston are large, he obtains

$$\begin{aligned} u_x(0,t) &= H'(t)/H(t) \; , \\ u(0,t) &= 0 \quad . \end{aligned} \tag{5.27}$$

Rodin then obtains a series solution of the form

$$u(x,t) = -2\nu \sum_{n=0}^{\infty} [H(t)^{-0.5}]^{(n+1)} \frac{x^{2n+1}}{\nu(2n+1)!} \Bigg/ \sum_{n=0}^{\infty} [H(t)^{-0.5}]^{(n)} \frac{x^{2n}}{\nu(2n)!} \; . \tag{5.28}$$

For the bounded region, $0 \leqslant x \leqslant a$, a separation of variables solution can be obtained (Dennemeyer, 1968). Thus let

$$\phi = X(x)\,T(t) \quad . \tag{5.29}$$

For homogeneous boundary conditions an eigenvalue problem results so that eq. (5.29) must be replaced by a series expansion which takes the particular form,

$$\phi\,(x,t) = \sum_{n=1}^{\infty} A_n \exp[-(\tfrac{n\pi}{a})^2 \nu t] \sin \tfrac{n\pi}{a} x \tag{5.30}$$

where

$$A_n = \frac{2}{a} \int_0^a \sin(\tfrac{n\pi}{a} x)\, \phi_0(x)\,dx \tag{5.31}$$

and

$$u(x,t) = \frac{\sum_{n=1}^{\infty} \frac{n\pi}{a} A_n \exp[-(\frac{n\pi}{a})^2 \nu t] \cos \frac{n\pi}{a} x}{\sum_{n=1}^{\infty} \frac{n\pi}{a} A_n \exp\left[-(\frac{n\pi}{a})^2 \nu t\right] \sin \frac{n\pi}{a} x} . \tag{5.32}$$

Boldrighini (1977) considers Burgers' equation with stationary non-zero boundary conditions on a finite domain and obtains a solution via a separation of variables approach. Boldrighini also demonstrates that the corresponding steady-state solution is unique.

Equation (5.1) can be manipulated to resemble a *Riccati* equation

$$v_x = f(x) + g(x)\, v + h(x)\, v^2 . \tag{5.33}$$

As with the Cole-Hopf transformation, the first step is to introduce

$$u = \psi_x . \tag{5.34}$$

A similarity variable, $z = x^2/t$, is introduced and if

$$\psi(x,t) = f(z) , \tag{5.35}$$

then eq. (5.1) becomes

$$f_{zz} = -[(z + 2\nu)/4\nu z]\, f_z + (f_z)^2/2\nu , \tag{5.36}$$

i.e., a Riccati equation for f_z. Rodin (1970b) uses this formulation to obtain some exact solutions to the inhomogeneous Burgers' equation

$$u_t + uu_x - \nu u_{xx} = F(x,t) \tag{5.37}$$

for particular choices of F.

An interesting problem is whether the solution of Burgers' equation as $\nu \to 0$ converges to the solution of the inviscid Burgers' equation, (5.2). In fact, it does and this has been established for the infinite x domain by Hopf (1950) and for the finite x domain by Ton (1975). Hopf established that Burgers' equation, as an initial value problem, is well posed and that solutions to eq. (5.1) are unique if $\mu > 0$. Hopf found that when $\mu = 0$ discontinuous solutions are possible and that it is necessary to consider the equivalent integral formulation to establish uniqueness. Thus instead of eq. (5.1) the following is considered,

$$\iint f(u_t + uu_x - \nu u_{xx})\, dx\, dt = 0 , \tag{5.38}$$

where $f(x,t)$ is any function of class C^2 in $t > 0$ and vanishes outside of some circle contained in $t > 0$. Then

$$\iint \left(uf_t + \frac{u^2}{2} f_x + \nu u f_{xx}\right) dx\, dt = 0 \tag{5.39}$$

will have a unique solution for u with $\nu \geq 0$. In order to model certain aspects of turbulence Hopf (1950) demonstrates that the double limit should be applied in the order $t \to \infty$, $\mu \to 0$. It is shown that, reversing the limit, produces a different solution.

5.4 Exact Solutions in Two Dimensions

For pure initial value problems (i.e. $-\infty \leq x \leq \infty$, $-\infty < y < \infty$) with

$$\phi(x,y,o) = \phi_0(x,y) \tag{5.40}$$

the general solution eq. (5.20), for the one-dimensional Burgers' equation can be extended to multi-dimensions. In two dimensions the solution is

$$\phi(x,y,t) = \frac{1}{4\pi\nu t} \int_{-\infty}^{\infty}\int_{-\infty}^{\infty} \exp\left[-\{(x-\xi)^2 + (y-\eta)^2\}/4\nu t\right] \phi_0(\xi,\eta)\, d\xi d\eta\,. \tag{5.41}$$

The initial conditions u_0, v_0 can be obtained from eq. (5.40) by applying eqs. (5.13) and (5.14), and similarly the solutions for u and v are

$$u(x,y,t) = \frac{\int_{-\infty}^{\infty}\int_{-\infty}^{\infty} \frac{(x-\xi)}{t} \exp\left[-\{(x-\xi)^2 + (y-\eta)^2\}/4\nu t\right] \phi_0(\xi,\eta)\, d\xi d\eta}{\int_{-\infty}^{\infty}\int_{-\infty}^{\infty} \exp\left[-\{(x-\xi)^2 + (y-\eta)^2\}/4\nu t\right] \phi_0(\xi,\eta)\, d\xi d\eta} \tag{5.42}$$

and

$$\frac{\int_{-\infty}^{\infty}\int_{-\infty}^{\infty} \frac{(y-\eta)}{t} \exp\left[-\{(x-\xi)^2 + (y-\eta)^2\}/4\nu t\right] \phi_0(\xi,\eta)\, d\xi d\eta}{\int_{-\infty}^{\infty}\int_{-\infty}^{\infty} \exp\left[-\{(x-\xi)^2 + (y-\eta)^2\}/4\nu t\right] \phi_0(\xi,\eta)\, d\xi d\eta}\,. \tag{5.43}$$

For two-dimensional *steady* solutions of Burgers' equations it is convenient to construct solutions of

$$\phi_{xx} + \phi_{yy} = 0\,, \tag{5.44}$$

i.e. the steady part of eq. (5.15). Subsequently the application of eqs. (5.13) and (5.14) give exact solutions of the two-dimensional Burgers' equations (Fletcher (1983d)).

To generate an exact solution with a shock-like structure the following general solution of eq. (5.44) suggests itself,

$$\phi = a_o + a_1x + a_2y + a_3xy + a_4\{\exp(k(x-x_o)) + \exp(-k(x-x_o))\} \cos ky \quad ,$$

where a_o, a_1, a_2, a_3, a_4, k and x_o can be chosen to give specific features to the flow. The application of eqs. (5.13) and (5.14) produces the following solutions of the Burgers' equations, (5.16) and (5.17),

$$u = \frac{-2\{a_1 + a_3y + ka_4[\exp(k(x-x_o)) - \exp(-k(x-x_o))] \cos ky\}}{\text{Re}\ \{a_o + a_1x + a_2y + a_3xy + a_4[\exp(k(x-x_o)) + \exp(-k(x-x_o))] \cos ky} \tag{5.45}$$

and

$$v = \frac{-2\{a_2 + a_3x - ka_4[\exp(k(x-x_o)) + \exp(-k(x-x_o))] \sin ky\}}{\text{Re}\{a_o + a_1x + a_2y + a_3xy + a_4[\exp(k(x-x_o)) + \exp(-k(x-x_o))] \cos ky\}} \ . \tag{5.46}$$

A typical solution is shown in Fig. 27

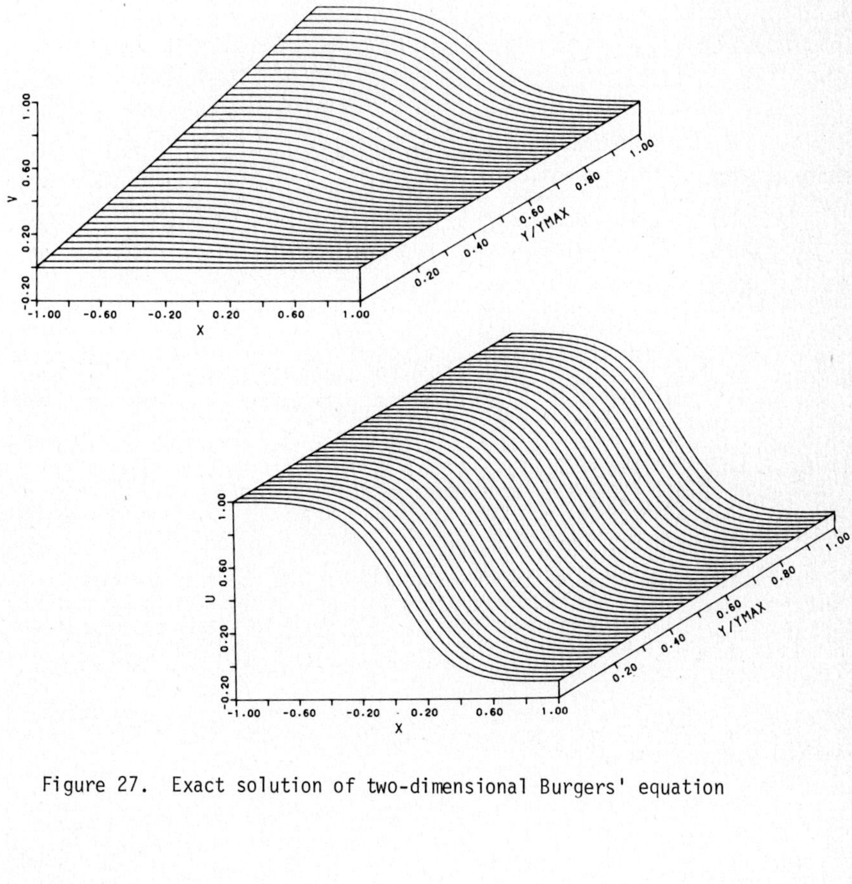

Figure 27. Exact solution of two-dimensional Burgers' equation

6. BURGERS' EQUATION AS A COMPUTATIONAL PROTOTYPE

As noted in Section 5 Burgers' equation is suitable for modelling the convective dissipative interactions that characterise many flows. The ready availability of exact solutions permits a direct measure of the accuracy of computational solutions of Burgers' equation. Consequently Burgers' equation is useful as a *non-linear test problem* for comparing computational algorithms. Three examples are provided where Burgers' equation has been used in this way.

6.1 One-Dimensional Comparison of the Finite Difference and Finite Element Methods

Here we use the modified Burgers' equation, eq. (4.66), to obtain solutions for the propagating shock problem, Figure 26. By choosing $\alpha = 0.5$ in eq. (4.66) the propagation of the shock is frozen.

Linear, quadratic and cubic elements have been compared with three, five and seven point finite difference schemes. In addition explicit (second-order Runge-Kutta; 2R-K) and implicit (C-N) time integration schemes have been compared (Fletcher (1982a)). Typical execution times per time-step are shown in Table 19.

Table 19. CPU time comparison for various schemes.

Case	CPU time per time-step	
	explicit (2R-K)	implicit (C-N)
linear finite element	0.30 s	0.55 s
quadr. finite element	0.39 s	0.61 s
cubic finite element	0.43 s	0.68 s
3 pt. finite difference	0.26 s	0.59 s
5 pt. finite difference	0.34 s	0.69 s
7 pt. finite difference	0.42 s	0.85 s

Explicit finite difference schemes are typically twice as economical as implicit schemes. Explicit finite element schemes are not quite so economical because of a once-only factorisation of $\underline{M}$ in eq. (4.52). Implicit schemes require factorisation of a spatially-augmented mass matrix, $\underline{M}^{sa}$, at each time step. The generalized Thomas algorithm, that is used to execute the factorisation, is capable of taking advantage of the narrower bandwidth associated with midside nodes of the higher-order finite elements. Consequently the execution time for higher-order finite elements schemes is less than for higher-order finite difference schemes. This is true *in one dimension only* (Fletcher (1983b))

The computational efficiency can be interpreted as the accuracy achieved per unit execution time. For a 121 node variable grid (refined close to the shock) solutions to the modified Burgers' equation, (4.66), have been obtained by marching variable time-step schemes from $t = 0.01$ to $t = 8.00$ at $Re = 100$. The solution at $t = 8.00$ has reached the steady state. The results are shown in Table 20.

The explicit schemes, the second and fourth-order Runge-Kutta schemes, have a maximum time-step limitation imposed by stability and are clearly not competitive with the implicit schemes.

Table 20. Summary of results to assess computational efficiency.

	Integration scheme	Δt_{max}	Number of time-steps	CPU time	$\|u-u_{ex}\|_{2,d}$
linear fin. elem.	2R-K	0.004	1900	570 s	1.04×10^{-2}
linear fin. elem.	4R-K	0.004	1900	988 s	1.04×10^{-2}
linear fin. elem.	C-N	0.32	32	18 s	1.03×10^{-2}
quadr. fin. elem.	C-N	0.32	32	20 s	0.55×10^{-2}
cubic fin. elem.	C-N	0.16	56	38 s	0.35×10^{-2}
3 pt. fin. diff.	C-N	0.32	32	19 s	1.01×10^{-2}
5 pt. fin. diff.	C-N	0.32	32	22 s	0.33×10^{-2}
7 pt. fin. diff.	C-N	0.32	32	27 s	0.35×10^{-2}

For this particular problem the quadratic finite element and five-point finite difference schemes are the *most efficient* schemes. Lower-order schemes are less accurate, higher-order schemes are less economical.

A related transient problem is the propagating sine-wave shown in Figure 28 with Re = 48. As time increases the top part of the sine-wave is convected downstream

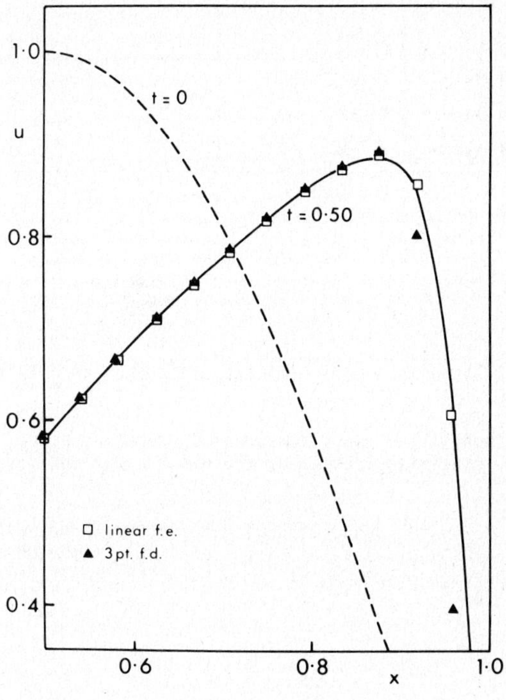

Figure 28. Propagating sine-wave at various times

and the amplitude of the sine-wave reduces due to the dissipative term, u_{xx}/Re. At t = 0.5 a well-defined boundary layer develops adjacent to x = 1.0. At $t = \infty$, u = 0 everywhere.

The governing equation is again Burgers' equation, (4.48), with initial and boundary conditions,

$$u(x,0) = \sin \pi x \quad , \quad 0 \leqslant x \leqslant 1 \tag{6.1}$$

and

$$u(0,t) = u(1,t) = 0 \tag{6.2}$$

Solutions on a uniform grid after a Crank-Nicolson integration from t = 0 to 0.5 at Re = 48 are shown in Table 21

Table 21. Computational efficiency of the finite element and finite difference methods for the propagating sine-wave problem

Case	Δ_{max}	No. of points	No. of time-steps	CPU time	$\lVert u - u_{ex} \rVert_{2,d}$
linear finite element	0.016	49	35	7.5s	1.00×10^{-2}
quadratic finite element	0.016	41	35	7.3s	0.84×10^{-2}
cubic finite element	0.016	37	35	7.3s	0.59×10^{-2}
3 pt. finite difference	0.016	151	35	26s	0.89×10^{-2}
5 pt. finite difference	0.004	57	127	41s	0.89×10^{-2}
7 pt. finite difference	0.004	37	127	33s	1.00×10^{-2}

For each scheme Δt and Δx have been chosen to give the smallest CPU time at a nominal one per-cent error level. Here *the finite element methods are more efficient* than low-order schemes. Higher-order finite difference schemes were rather inaccurate for this problem particularly near the downstream boundary, x = 1.0. This problem is discussed more fully by Fletcher (1982b).

6.2 Two-Dimensional Comparison of the Finite Difference and Finite Element Methods

We seek steady solutions of the two-dimensional Burgers' equation, (5.16) and (5.17), with boundary conditions chosen by the exact solutions, eqs. (5.45) and (5.46).

Because of the greater economy associated with the group formulation (Fletcher (1983c)) it is desirable to utilise the two-dimensional inhomogeneous Burgers' equations that also have the exact solution, eqs. (5.45) and (5.46). These equations can be written

$$u_t + (u^2)_x + (uv)_y - \frac{1}{Re}(u_{xx} + u_{yy}) = SX \tag{6.3}$$

and

$$v_t + (uv)_x + (v^2)_y - \frac{1}{Re}(v_{xx} + v_{yy}) = SY \quad , \qquad (6.4)$$

where the source terms are given by

$$SX = 0.5Re\ u(u^2 + v^2) \quad \text{and} \quad SY = 0.5Re\ v(u^2 + v^2) \quad .$$

These equations have been used to compare (Fletcher(1983b)) the first and second-order finite difference and finite element methods shown in Table 22. All methods have been integrated using a time-split algorithm, similar to that described in Section 4.5.2, until the steady state is reached.

Table 22. Schemes for solving the two-dimensional Burgers' equations

Scheme	Description
3 - FD	Three-point finite difference formulae
LFE(C)	Conventional linear finite element representation
LFE(G)	Linear group f.e. representation for u^2, uv etc.
QFE(G)	Quadratic group f.e. representation for u^2, uv etc.
5 - FD	Five-point finite difference formulae

The execution time per time-step for the various methods is shown in Table 23. It is clear that in two dimensions, finite element schemes are *less economical* than finite difference schemes and the disparity is greater in comparing quadratic finite element methods with five-point finite difference schemes. Although the linear group finite element formulation is less economical than the three-point finite difference scheme it is more economical than the conventional linear finite element formulation.

Table 23. Comparison of relative execution times per time-step

Mesh	Δx	3-FD	LFE(C)	LFE(G)	QFE(G)	5 - FD
6 x 6	0.4	1.0	4.9	2.2		2.9
7 x 7	0.3333				7.9	
11 x 11	0.2	4.0	25.3	10.2	25.9	13.0
21 x 21	0.1	16.1	111.2	45.3	110.7	53.9
41 x 41	0.05	67.8	478.1	189.8	464.1	224.7

The error in the solution for various mesh sizes are shown in Figure 29. These solutions have been obtained on a uniform grid with $\Delta x = \Delta y$. The exact solution has a moderate internal gradient orientated in the x direction. The exact solution has required the following parameter values: $a_o = a_1 = 110.13$, $a_2 = a_3 = 0$, $a_4 = 1.0$, $k = 5$, $x_o = 1$, $Re = 10$.

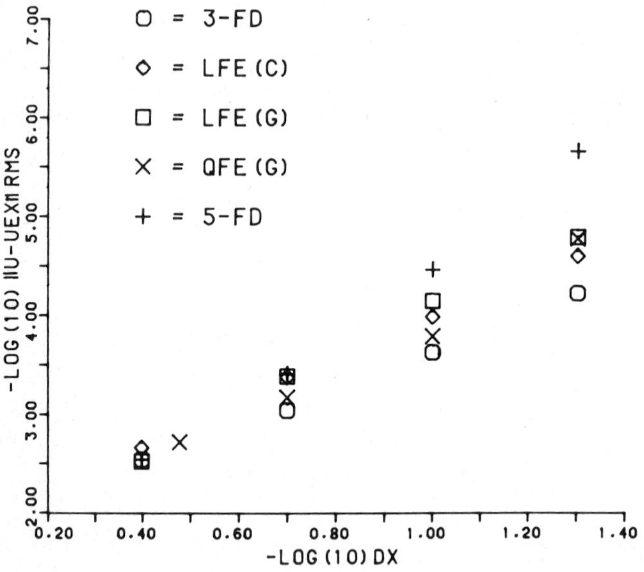

Figure 29. Spatial convergence properties for a moderate internal gradients; Re = 10.

The results indicate that the 3-FD scheme is least accurate and the 5-FD scheme is most accurate on a refined grid. The group linear finite element scheme, LFE (G), is more accurate than the conventional linear finite element scheme, LFE(C). After consideration of a number of gradient cases Fletcher (1983b) concludes that on a coarse grid, or where sharp gradients occur the 3-FD or LFE(G) schemes are *the most efficient*. On a refined mesh the 5-FD scheme is most efficient.

6.3 Compact Implicit Methods

Although the centered difference representation of u_x is second order accurate, a fourth-order accurate representation is possible from the scheme

$$\frac{1}{6}(u_x)_{i-1} + \frac{2}{3}(u_x)_i + \frac{1}{6}(u_x)_{i+1} = \frac{u_{i+1} - u_{i-1}}{2\Delta x} . \tag{6.5}$$

However, the derivatives now appear implicitly and, typically, auxiliary variables $(u_x)_i$, etc., need to be introduced to take advantage of the higher accuracy.

Thus if second derivatives occur the solution of a 3 x 3 block tridiagonal system of equations will be required (Hirsch, 1975). However, Adam (1975,1977) showed how to treat the second derivative implicitly so that only a 2 x 2 block tridiagonal system need be solved. Ciment et al. (1978) demonstrated that, by considering the complete spatial operator rather than individual terms, it is possible to generate a compact implicit formulation that only requires a scalar tridiagonal system of equations to be solved. However, the algebraic complexity is somewhat greater.

Burgers' equation, with Re = 1, was represented by Adam (1975) as

$$H\left\{\frac{u_i^{n+1} - u_i^n}{\Delta t}\right\} + \frac{H}{2}\{u_i^{n+1}\,u_{xi}^n\} + \frac{H}{2}\{u_i^n u_{xi}^{n+1}\} - \frac{\delta^2}{2\Delta x^2}\{u_i^{n+1} + u_i^n\} = 0 , \qquad (6.6)$$

where H and δ^2 are operators defined by

$$Hu_i = u_{i+1} + 10u_i + u_{i-1} ,$$

and

$$\delta^2 u_i = u_{i+1} - 2u_i + u_{i-1} . \qquad (6.7)$$

The second equation is

$$\frac{u_{xi+1}^{n+1}}{6} + \frac{2u_{xi}^{n+1}}{3} + \frac{u_{xi-1}^{n+1}}{6} = \frac{u_{i+1} - u_{i-1}}{2\Delta x} . \qquad (6.8)$$

Solutions to Burgers' equations have been obtained for initial conditions

$$u(x,0) = R \sin x , \qquad (6.9)$$

and boundary conditions, u=0 at $x = -\pi$ and 0. Results for the above scheme (eqs. (6.6) to (6.8)) are compared with a conventional second-order Crank-Nicolson scheme for various times in Table 24. ε is the absolute average error and m is the

Table 24. Comparison of compact implicit and Crank-Nicolson methods

t	0.1	0.2	0.3	0.4	0.5	1.0	m	Δt
ε/u_{max} for compact implicit	0.0049	0.0020	0.0012	0.0006	0.0004	0.7×10^{-4}	21	0.01
ε/u_{max} for Crank-Nicolson	0.083	0.066	0.037	0.024	0.019	0.010	21	0.01
ε/u_{max} for Crank-Nicolson (refined)	0.0029	0.0017	0.0011	0.0008	0.0006	0.0004	101	0.005

number of spatial nodal points. According to Adam (1975) the compact implicit method is 2.3 times more expensive than the Crank-Nicolson method per step but for comparable accuracy after the same time the Crank-Nicolson method is four times more expensive. Adam also suggested the loss of accuracy with a variable grid is *far less* for the compact implicit method than for conventional methods.

Ciment et al. (1978) have sought solutions of a modified Burgers' equation written in the form

$$u_t + (u-\alpha)u_x - \nu u_{xx} = 0 \quad , \qquad (6.10)$$

using a compact implicit scheme in which only the function values, u, need be solved for explicitly. Equation (6.10) was solved in the region $-5 \leqslant x \leqslant 5$ subject to the initial conditions $u = 1$ for $x < 0$, $u = 0.5$, for $x = 0$ and $u = 0$ for $x > 0$. Equation (6.10) was integrated until the steady state was reached using a Crank-Nicolson time integrator. The rms error is compared with corresponding errors for a second-order Crank-Nicolson scheme in Table 25.

Table 25. Comparison of compact implicit and second-order Crank-Nicolson for steady-state Burgers' equation, $\alpha = 0.5$

Scheme	N	Δx	ν	rms error
compact implicit	60	0.167	0.031	0.0346
(after Ciment et al., (1978))	100	0.10	0.031	0.0039
second-order	100	0.10	0.031	0.0334
Crank-Nicolson	200	0.05	0.031	0.0067

Clearly the compact implicit scheme is *an order of magnitude* more accurate for the same grid resolution. Ciment et al. indicated that the following cell Reynolds number restriction is necessary to achieve stable results

$$Re = (u-\alpha)\frac{\Delta x}{\nu} < 2.55 \ . \qquad (6.11)$$

Rubin and Graves (1975) also solved the propagating shock problem, but using a cubic spline formulation. For low to moderate Reynolds number (i.e., relatively diffuse 'shocks') the technique gives solutions of high accuracy. The formulation appears to be comparable in accuracy to the compact implicit schemes considered above. For higher Reynolds number the solution is oscillatory adjacent to the shock unless local mesh refinement is introduced. An interesting result, obtained by Rubin and Graves, is that the non-divergence form, uu_x, generally gives more accurate results than the divergence form, $0.5(u_x)^2$; this is opposite to the situation for finite difference schemes.

Rubin and Khosla (1976) developed comparable schemes to Adam by starting from high-order Hermitian splines and obtained solutions to Burgers' equation.

7. PHYSICAL PHENOMENA MODELLED BY BURGERS' EQUATION

In the preceding sections Burgers' equation has been presented as a model equation for the interaction between convection and diffusion (usually due to viscous dissipation). However a number of problems can be modelled, approximately or exactly, by Burgers' equation. Here we give three interesting examples.

7.1 Continuum Traffic Flow

A flux or current, J, and a related density or concentration, n, can, in general, be connected through a continuity equation (after Whitham, 1974). For a one-dimensional problem, we have

$$n_t + J_x = 0 \tag{7.1}$$

If another relationship can be prescribed between J and n, then closure may be expected.

For the traffic flow problem J may be interpreted as the volume flow rate of cars at a particular location and time. n may be interpreted as the number of cars per unit length of highway. Although a pedestrian view of the traffic flow problem might focus on the discrete or microscopic aspects, here we are constructing a *continuum or macroscopic model*. If v is the local mean velocity, a first approximation to the dependence of J on n might be

$$J = nv \quad . \tag{7.2}$$

Clearly in the limit $n \to 0$, $J \to 0$ however fast individual cars travel. As the concentration reaches saturation, i.e., the cars become nose-to-tail, we would expect v to fall to zero. A simple representation of this would be

$$v = v_o(1 - n/n_s) \quad , \tag{7.3}$$

i.e. v_o is the speed when the concentration is zero and n_s is the saturation concentration, at which $v = 0$. Combining (7.2) and (7.3) produces the relationship shown in Figure 30.

In practice drivers may be expected to look ahead and to modify their speed accordingly. To represent this, eq. (7.2) is replaced by

$$J = nv - Dn_x \quad , \tag{7.4}$$

where D is a diffusion coefficient given by (Musha and Higuchi (1978))

$$D = \tau\, v_r^2 \quad , \tag{7.5}$$

where v_r is a random velocity and τ is the mean collision time for the cars. Combining equations (7.1), (7.3) and (7.4) gives

$$(n_t + v_o n_x) - 2\left(\frac{n}{n_s}\right) v_o n_x - Dn_{xx} = 0 \quad . \tag{7.6}$$

It is convenient to introduce a moving reference frame by

$$\xi = -x + v_o t \quad . \tag{7.7}$$

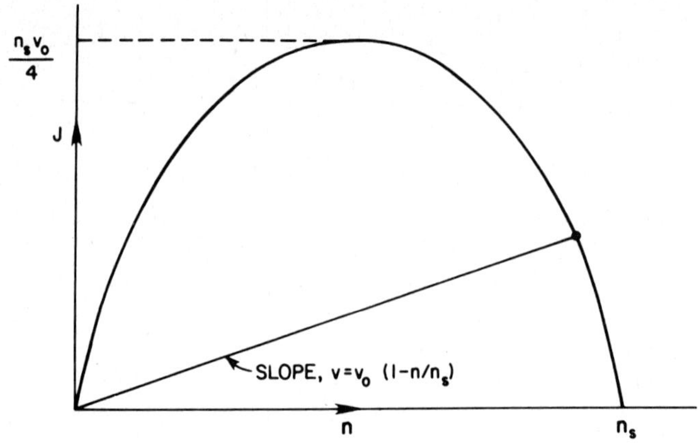

Figure 30. Idealised flux density relationship

If, in addition, n is nondimensionalised with $n_s/2$ and the time by t_0 the result is

$$n_t + nn_\xi - \frac{1}{Re} n_{\xi\xi} = 0 \quad , \tag{7.8}$$

i.e. *a Burgers' equation for the car concentration.* Re is a dimensionless constant, analogous to the Reynolds number, defined by

$$Re = \left(\frac{v_o}{v_r}\right)^2 \frac{t_o}{\tau} \quad , \tag{7.9}$$

where the characteristic time, t_0, is the mean time between successive cars passing a stationary observer.

In order to use eq. (7.8) as a model for real traffic flows it is necessary to consider the power spectrum of Burgers' flow. For transient Burgers' flow assuming an initial Gaussian concentration distribution it is found that the power spectral density depends on the normalised wave number K, like K^{-q}, where initially $q \approx 4$ and asymptotically $q \approx 1.0$.

To obtain the power spectrum for steady Burgers' flow a random force f is added to eq. (7.8). Taking the ensemble average and requiring that $\langle n \rangle$ is constant and $\langle f \rangle = 0$ gives

$$\tfrac{1}{2} \langle n^2 \rangle_x = 0 \quad . \tag{7.10}$$

If Burgers' equation plus the random force, f, are multiplied by n and an ensemble average taken, the result is, if $\langle n^2 \rangle$ is constant in space (from eq. (7.10)),

$$\tfrac{1}{2} \langle n^2 \rangle_t + \frac{1}{Re} \langle (n_x)^2 \rangle = \langle nf \rangle \quad . \tag{7.11}$$

But for steady flow, $\langle n^2 \rangle_t = 0$, so that eq. (7.11) gives the requirement that a random force will lead to a steady Burgers' equation. It is found that the power spectral density for the steady Burgers' equation leads to a dependence on wave number like $K^{-1.4}$ which is not very different from the transient case. For the traffic flow problem it is convenient to express the power spectral density, S, as a function of the frequency, f. Then

$$S(f) = \frac{10^{-2}}{(2\pi)^{0.4}} (n_s v_o)^2 \frac{1}{t_o^{0.4}} \frac{1}{f^{1.4}} + 2(nv_o) \quad . \tag{7.12}$$

A comparison of this equation with experimental observations is shown in Figure 31. Musha and Higuchi (1978) consider the agreement for the steady Burgers flow satisfactory.

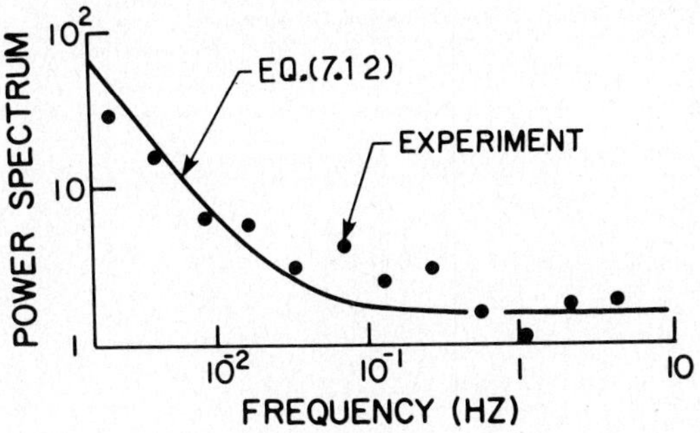

Figure 31. Comparison of eq. (7.12) with experimental power spectrum

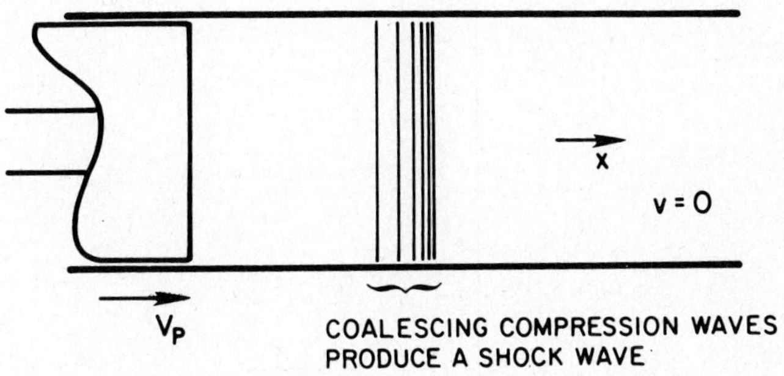

Figure 32. Incipient shock formation

7.2 Weak Shock Waves

Figure 32 shows an impulsively-started piston moving at a steady velocity, V_p, into a tube containing a compressible fluid, e.g. air, initially at rest. The equations that govern the one-dimensional unsteady motion of the fluid in the tube will be reduced to Burgers' equation.

The motion of the piston creates compression waves that propagate to the right at a velocity greater than V_p. The compression waves eventually coalesce, due to the nonlinear nature of the convection, to form a single *shock wave*. Ahead (to the right) of the shock wave the fluid is undisturbed, v = 0. Behind the shock (i.e., between the shock wave and the piston) the fluid moves with the speed of the piston, V_p. As the velocity profile, associated with the coalescence of the compression waves, steepens due to the nonlinear convective effect, viscous forces come into play tending to smooth the velocity profile through the shock wave and, by implication, spread the shock thickness over a finite width. Thus the velocity undergoes a rapid increase through the shock to reach a steady value adjacent to the piston. However, the shock location is a function of time, since the shock travels at a speed greater than V_p.

This problem is governed by the continuity equation

$$\rho_t + \rho v_x + v\rho_x = 0 , \tag{7.13}$$

where ρ is the density and v is the velocity. The x-momentum equation can be written

$$v_t + vv_x + p_x/\rho = \delta v_{xx} , \tag{7.14}$$

where δ is the "diffusivity of sound" (after Lighthill (1956)). It is assumed that entropy changes are small. It is convenient to replace the density by the sound speed, a, via

$$a/a_o = (\rho/\rho_o)^{0.5(\gamma-1)} , \tag{7.15}$$

where γ is the specific heat ratio and the subscript o refers to the undisturbed (constant) values. Equations (7.13) and (7.14) become

$$a_t + va_x + \frac{(\gamma-1)}{2} a\, v_x = 0 \tag{7.16}$$

and

$$v_t + vv_x + \frac{2}{\gamma-1} aa_x = \delta v_{xx} , \tag{7.17}$$

where δ, the "diffusivity of sound," is a function of the undisturbed (to the right of the shock) viscosity, density, specific heat and thermal conductivity of the medium.

Equations (7.16) and (7.17) can be simplified by introducing the Riemann invariants ,

$$r = a/(\gamma-1) + v/2$$

and (7.18)

$$s = a/(\gamma-1) - v/2 \quad ,$$

to give

$$r_t + (a+v)r_x = \tfrac{1}{2}\delta(r_{xx}-s_{xx}) \tag{7.19}$$

$$s_t - (a-v)s_x = \tfrac{1}{2}\delta(s_{xx}-r_{xx}) \quad . \tag{7.20}$$

For the particular problem under consideration, the propagation of a disturbance into an initially undisturbed region, $s = s_0$ where $s_0 = a_0/(\gamma-1)$ (from eq. (7.18)). The problem is governed by eq. (7.19). But from eq. (7.18),

$$(a+v) = \frac{\gamma+1}{2} r + \tfrac{1}{2}(\gamma-3)s_0 \quad , \tag{7.21}$$

so that eq. (7.19) becomes

$$r_t + \{0.5(\gamma+1)r + 0.5(\gamma-3)s_0\}r_x = \frac{\delta}{2} r_{xx} \quad . \tag{7.22}$$

As the final step we introduce

$$u = 0.5(\gamma+1)(r-r_0) = 0.5(\gamma+1)r + 0.5(\gamma-3)s_0 - a_0 \tag{7.23}$$

$$\xi = x - a_0 t \tag{7.24}$$

to give *Burgers' equation*

$$u_t + uu_\xi = \frac{\delta}{2} u_{\xi\xi} \quad . \tag{7.25}$$

From eq. (7.23) $u \approx a + v - a_0$, i.e., an excess wavelet velocity. The coordinate ξ is measured relative to a frame of reference moving with the undisturbed speed of sound, a_0.

The general solution to (7.25) is given by eq. (5.21). For the specific initial conditions

$$u = u_1 = 0.5(\gamma+1)V_p \quad \text{for} \quad \xi < 0,\ t = 0$$

and (7.26)

$$u = 0 \quad \text{for} \quad \xi \geqslant 0,\ t = 0 \quad ,$$

the solution is

$$U(\xi,t)/u_1 = I_A/(I_A + I_B \exp\{u_1(\xi - 0.5u_1 t)/\delta\}) \tag{7.27}$$

where

$$I_A = \int_{\xi-u_1t}^{\infty} \exp\{-\zeta^2/2\delta t\}d\zeta \tag{7.28}$$

and

$$I_B = \int_{-\xi}^{\infty} \exp\{-\zeta^2/2\delta t\}d\zeta \quad . \tag{7.29}$$

Velocity solutions for increasing time and diffusivity are shown in Figures 33 and 34. The effective propagation velocity in the (ξ,t) plane is $0.5u_1$.

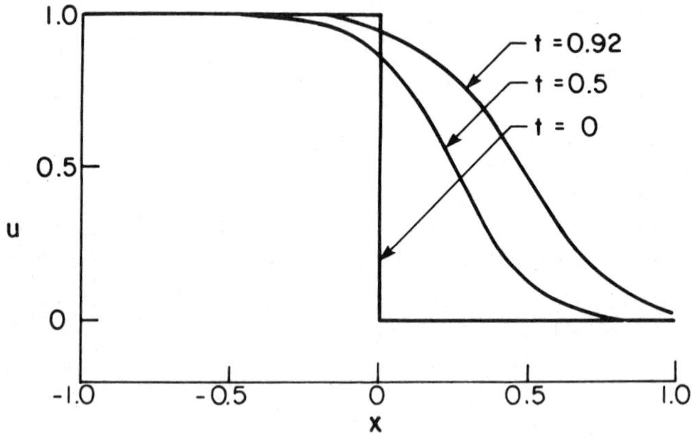

Figure 33. Velocity distributions through shock for various times, $\delta = 0.2$

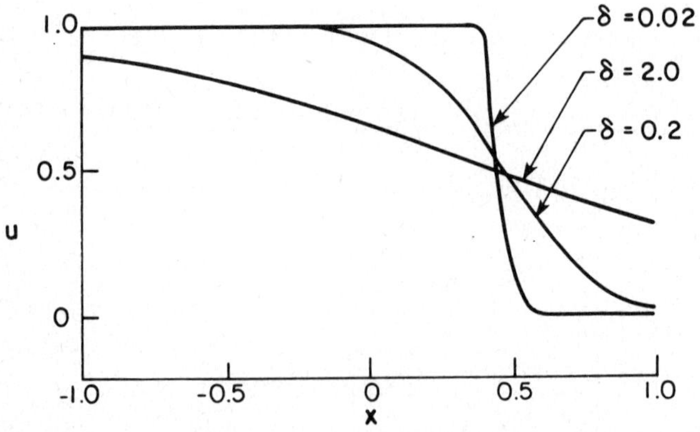

Figure 34. Velocity distributions through shock for different diffusivities, $t = 0.92$

The use of Burgers' equation is also appropriate to other gas dynamic phenomena. For instance, the convection and decay of a *compression pulse*, i.e., a Dirac delta function at zero time, is governed by Burgers' equation. For this problem a Reynolds number (expressing the balance between convection and viscous diffusion) can be defined by

$$Re = \frac{1}{\delta}\int_{-\infty}^{\infty} u \, d\xi \ . \tag{7.30}$$

For this situation, $u(t/\delta)^{\frac{1}{2}}$ is a function of the similarity variable, $\eta = \xi/(\delta t)^{\frac{1}{2}}$, only and the asymptotic solution is given by

$$u = (\delta/t)^{\frac{1}{2}} \exp\{-0.5\eta^2\}/[P + I_C] \tag{7.31}$$

where

$$P = \sqrt{2\pi}/(\exp(Re) - 1)$$

and

$$I_C = \int_{\eta}^{\infty} \exp(-0.5\zeta^2)d\zeta \ . \tag{7.32}$$

Some typical results are presented in Figure 35. At small Reynolds number the compression wave diffuses rapidly before significant convection can take place (in the ξ,t plane). However, at large Reynolds number, the compression wave develops a steep leading face and convects over a considerable range of ξ values before dissipation is effective.

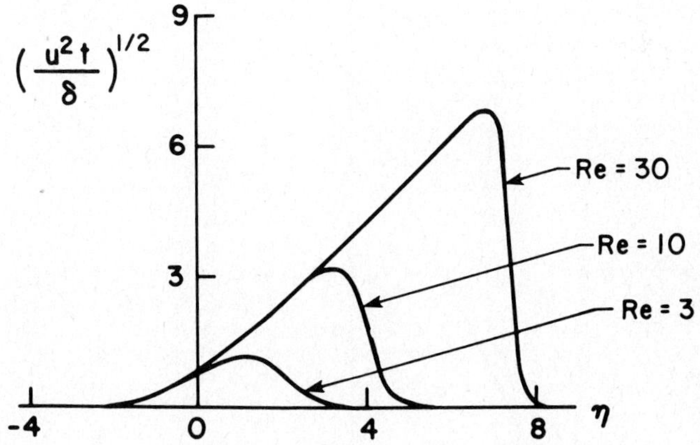

Figure 35. Compression pulse solutions for various Reynolds numbers

If the just described compression wave is followed by an equal (and opposite) expansion wave, the result is the N-*wave*. In this case a Reynolds number can be defined by considering half of the wave, i.e.,

$$Re = \frac{1}{\delta} \int_{\xi_u = 0}^{\infty} u \, d\zeta \quad , \tag{7.33}$$

and the solution is

$$u = \xi/t/\{1 + \exp(0.5\eta^2)/(\exp(Re) - 1.)\} \quad . \tag{7.34}$$

Typical solutions at various Reynolds numbers are shown in Figure 36. The steepening of the wave form with increasing Reynolds number is apparent. The corresponding pressure solution provides a useful model for sonic boom phenomena (Seebass (1969)).

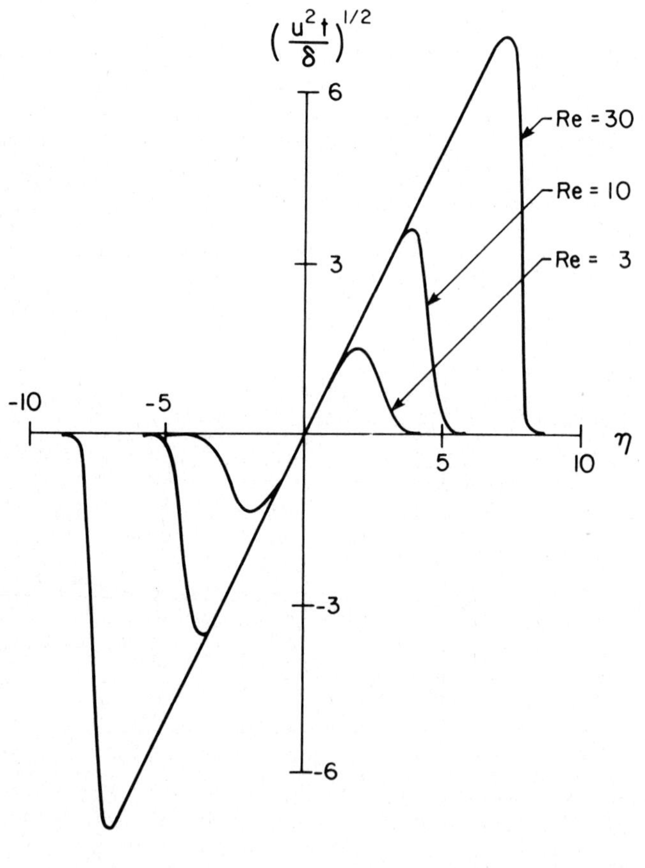

Figure 36. N-wave solutions for various Reynolds numbers

Sachdev and Seebass (1973) have looked at the more general N-wave problem which is governed by the equation

$$u_t + uu_\xi + 0.5\alpha\, u/t = 0.5\delta\, u_{xx} \tag{7.35}$$

where α = 0,1 and 2 for plane, cylindrical or spherical symmetry, respectively. For α = 1 or 2 the Cole-Hopf transformation does not give the linear heat conduction equation and no exact solutions are available. Therefore Sachdev and Seebass obtained numerical solutions using an implicit predictor-corrector finite difference scheme. The accuracy of the scheme was established for the plane case ($\alpha = 0$).

It was found by Sachdev and Seebass that, for Re (defined by eq. (7.33)) > 0.5, the expression for the velocity profile in a plane shock (after Taylor (1910)) i.e.

$$u = u_{max}/[1 + \exp\{u_{max}(\xi-\xi_s)/\delta\}] \tag{7.36}$$

is also accurate at the shock for the cylindrical and spherical cases. Solutions for cylindrical and spherical N-waves are relevant to the study of sonic booms from aircraft and explosions.

Lighthill (1956) also presents solutions for the problem of one shock overtaking and coalescing with another. In addition, as with N-waves, the formation and decay of shocks in nonplanar geometry are also shown to be governed by Burgers' equation.

The derivation of Burgers' equation (7.25) as the governing equation for many gas dynamic phenomenon has followed Lighthill (1956). However, using the method of matched asymptotic expansions it is also possible to derive Burgers' equation for any time after the initiation of the motion (e.g. Moran and Shen (1966)).

7.3 Supersonic Aerofoil Flow

For slender wings or aerofoils subjected to a supersonic inviscid flow, it is convenient to think of the wing as a perturbation of an, otherwise, uniform free stream. If the governing partial differential equation is expressed as a velocity potential (ϕ) as a function of position (x,y) it is possible to develop a hierarchy of solutions, the lowest order of which is linear.

In fact the first-order problem can be thought of as a body which communicates its disturbing effect on the flow by passing information along straight characteristics (or Mach lines) (the governing partial differential equation is hyperbolic) which emanate from the body. The governing equation is

$$(1 - M_\infty^2)\phi_{xx} + \phi_{yy} = 0 \quad , \tag{7.37}$$

where M_∞ is the Mach number of the freestream. Although solutions to eq. (7.37) give an accurate description of the flow at an intermediate distance from the body, in the farfield a nonlinear second-order correction must be added to account for the merging of the characteristics. In this case, eq. (7.37) is replaced by

$$(1 - M_\infty^2 - (\gamma+1)M_\infty^4\phi_x)\phi_{xx} + \phi_{yy} = 0 \quad . \tag{7.38}$$

However, when characteristics merge the velocity potential description is inappropriate locally, since a shock wave will occur. One of the current computational research problems is to use a potential description of the flow, like eq. (7.39), that strictly excludes the appearance of shocks, to describe physical situations, like the transonic flow around an aerofoil or through a nozzle where shocks may occur. The attraction of equations like (7.38) is that through use of the Prandtl-Glauert transformation, perturbations of the Laplace or wave equation can be obtained.

One way to permit shocks to appear in the solution is to add a diffusive term to eq. (7.38). This idea has been used in transonic flow by Sichel (1963) and for supersonic flow by Chin (1978). In this case eq. (7.38) is replaced by

$$\delta\phi_{xxx} + (1 - M_\infty^2 - (\gamma+1)M_\infty^4\phi_x)\phi_{xx} + \phi_{yy} = 0 \quad , \tag{7.39}$$

where δ is a small parameter. The term $\delta\phi_{xxx}$ will only contribute to the solution in regions where shocks occur. In the limit $\delta \to 0$ the inviscid solution is recovered. The effect of including the term $\delta\phi_{xxx}$ permits a continuous solution to be obtained in the vicinity of the shock so that the gross effect of the shock on, say, the pressure distribution on the aerofoil can be obtained. Typically, boundary conditions to be used with eq. (7.39) are

$$\phi_y(x,0) = \varepsilon T'(x) \quad , \quad \phi = 0 \quad \text{at} \quad x = -\infty \text{ (upstream)} \tag{7.40}$$

where ε is the thickness ratio of the body and T' is the local, normalised body slope.

The relevance of eq. (7.39) is that, with a suitable transformation (Chin (1978)), *Burgers' equation can be obtained and a closed form solution follows*. Physically the problem has two obvious co-ordinate directions, the freestream and the characteristics. This suggests the transformation

$$\xi = x - (M_\infty^2 - 1)^{\frac{1}{2}}y \quad \text{and} \quad \eta = (M_\infty^2 - 1)^{\frac{1}{2}}y \tag{7.41}$$

Substitution into eq. (7.39) generates certain second-order terms in $\phi_\eta, \phi_{\eta\eta}$ which can be dropped (Hayes (1954)). Additional transformations are introduced by

$$\zeta = (2(M_\infty^2 - 1))^{\frac{1}{2}}\xi$$

and (7.42)

$$u = \phi_x(\gamma+1)M_\infty^4/(2(M_\infty^2 - 1))^{\frac{1}{2}} \quad ,$$

where u is a modified velocity component in the freestream direction, Equations (7.39) and (7.40) become

$$u_\eta + uu_\zeta = \delta u_{\zeta\zeta} \tag{7.43}$$

and

$$u = -\varepsilon \frac{\gamma+1}{\sqrt{2}} \frac{M_\infty^4}{M_\infty^2-1} T'\xi \quad \text{at} \quad \eta = 0 \quad , \tag{7.44}$$

$$u = 0 \quad \text{at} \quad \zeta = -\infty \text{ (upstream)} \quad .$$

The solution to these equations is just eq. (5.4a).

All techniques for solving the transonic full potential equation are iterative. The solutions to eqs. (7.43) and (7.44) would form an appropriate starting point for the iterative scheme. To the extent that the solution to eqs. (7.43) and (7.44) are close to the desired solution in some sense one would expect the subsequent iteration to converge rapidly. To the author's knowledge, the solution of Burgers' equation for this problem has never been used in such a way.

8. CLOSURE

In this chapter we have examined the Galerkin method in depth. Firstly the traditional Galerkin method, essentially a pre-computer method, has been described and its relation to other methods of weighted residuals established. The mechanics of the method have been clarified through consideration of simple examples. A number of typical applications of the traditional Galerkin method were presented and considerable diversity was indicated.

The traditional Galerkin method is not well-suited to problems requiring a large number of unknowns to achieve acceptable accuracy, and by implication the execution of a computer program to obtain results. Modern developments of the traditional Galerkin method have gone in two directions. Firstly *orthogonal functions* have been introduced as test and trial functions. This leads to *spectral methods*, which like the traditional methods, are global and characterised by *high accuracy per degree of freedom* when the underlying solution is smooth enough. Spectral methods require special procedures to handle nonlinear terms but have proved to be very effective for simulating turbulence and in modelling global meteorological flows.

The second modern development of the traditional Galerkin method has produced *the Galerkin finite element method.* Finite element methods are characterised by the use of *piecewise polynomials of low order* for the test and trial functions and by making the unknown parameters the *nodal values* of the solution being sought. Consequently the resultant formulae bear a resemblance, particularly in one dimension, to finite difference formulae. Although the finite element method is less accurate per unknown (in the trial solution) than the spectral method it is *more economical* and *easier to code.* By introducing separate trial solutions in each finite element and by making use of the isoparametric mapping it is possible to handle complicated domain shapes *without difficulty.* The relative merits of the spectral, finite element and finite difference methods are considered by Fletcher (1983a).

To provide some basis of comparison the various Galerkin methods have been applied to *Burgers' equation* which governs the propagating shock problem. Burgers' equation is of interest in its own right as a model *nonlinear* equation that possesses an *exact solution* for many combinations of boundary and initial conditions. A number of interesting physical phenomena are represented quite accurately by Burgers' equation; a few of these have been described.

ACKNOWLEDGEMENT

The author is particularly grateful to Susan Lotho who typed this chapter both skillfully and cheerfully.

APPENDIX: List of Symbols

a	sound speed
a_o, a_1, a_2	unknown parameters in a global trial solution
b_o, b_1, b_2	unknown parameters in a Dorodnitsyn trial solution
C_D	drag coefficient
C_L	lift coefficient
C_P	pressure coefficient
D	diffusion coefficient
e	approximation error
f	trial (test) function; Dorodnitsyn function; frequency
g	surface velocity function; orthonormal Dorodnitsyn function
G*	Grashof number
h	maximum element dimension
H	Sobolev norm
J	Jacobian; logarithmic shape function; flux
k	wave number
ℓ	cable length
L	mixing length
M_x,M_y	mass operators
M	mass matrix; Mach number
n	(car) concentration
N	shape function; number of unknown coefficients in trial solution
Nu	Nusselt number
p	pressure
P_n	Legendre function of order n
P_r	Prandtl number
q_T	tangential velocity
r	radius

r_x	stepsize ratio in the x direction
r_y	stepsize ratio in the y direction
r,s	Reimann invariants
R	residual
Re	Reynolds number
R*	Rayleigh number
S	power spectral density
S_k	logarithmic function
t	time
T	Chebychev function;
u	dependent variable; excess wavelet velocity
u_a	approximate (trial) solution
u_o	auxiliary solution in global Galerkin trial solution
u,v,w	velocity components; spatial displacements
u_e	velocity at boundary layer outer edge
u_e, u_{ex}	exact solution
$\underline{v}$	velocity vector
W_k	weight (test) function
x,y,z	spatial co-ordinates
y_a	approximate (trial) solution
γ	specific heat ratio
δ	Dirac delta; Kronecker delta; diffusivity of sound
$\delta_x, \delta_x, \delta_{x^2}$ etc.	differential operators
Δ	stepsize
ζ	vorticity
θ	boundary layer function; surface co-ordinate; angular co-ordinate
Θ	continuous Dorodnitsyn stress function
μ	viscosity
ν	kinematic viscosity

ξ, η	local element co-ordinates
ρ	density
σ	rms error
τ	shear stress; aerofoil thickness
T	continuous Dorodnitsyn shear stress
ϕ	trial (test) function; velocity potential
ψ	dependent variable; trial function; streamfunction

REFERENCES

Adam, Y. (1975), "A Hermitian finite difference method for the solution of parabolic equations," Comp. & Maths. with Applics., 1, 393-406.

Adam, Y. (1977), "Highly accurate compact implicit methods and boundary conditions," Journal of Computational Physics, 24, 10-22.

Ames, W.F. (1965), "Nonlinear partial differential equations in engineering," Academic Press, New York, 23.

Anderssen, R.S. and Mitchell, A.R. (1979), "Analysis of generalised Galerkin methods in the numerical solution of elliptic equations," Math. Meth. Appl. Sci., 1, 3-15.

Anon. (1941), Prik. Mat. Mekh, 5, 337-341.

Arminjon, P. and Beauchamp, C. (1978), "A finite element method for Burgers' equation in hydrodynamics," Int. Journal for Numerical Methods in Engineering, 12, 415-428.

Arminjon, P. and Beauchamp, C. (1979), "Numerical solution of Burgers' equation in two space dimensions," Comp. Maths. Appl. Mech. Eng., 19, 351-365.

Arminjon, P. and Beauchamp, C. (1981), "Continuous and discontinuous finite element methods for Burgers' equation," Comp. Maths. Appl. Mech. Eng., 25, 65-84.

Astley, R.J. and Eversman, W. (1981), "Acoustic transmission in non-uniform ducts with mean flow, part II: the finite element method," J. Sound Vibration, 74, 103-121.

Babuska, I., Katz, N and Szabo, B.A. (1981), "The p-version of the finite element method," SIAM J. Num. Anal., 18, 515-545.

Baker, A.J. and Manhardt, P.D. (1978), AIAA J., 16, 807-814.

Baker, A.J. and Zelazny, S.W. (1975), "Predictions of environmental hydrodynamics using the finite element method - applications," AIAA J., 13, 43-46.

Barrett, J.W. and Morton, K.W. (1980), Int. J. Num. Meth. Eng., 15, 1457-1474.

Barrett, J.W. and Morton, K.W. (1981), "Optimal finite element approximation for diffusion-convection problems", Proc. Conf. Mathematics of Finite Elements and Their Applications (ed. J. Whiteman), Brunel University, U.K.

Bateman, H. (1915), "Some recent researches on the motion of fluids," Mon. Weather Rev., 43, 163-170.

Bellman, R., Azen, S.P. and Richardson, J.M. (1965), "On new and direct computational approaches to some mathematical models of turbulence," Quart. Appl. Math., 23, 55-67.

Belotserkovskii, O.M. and Chushkin, P.I. (1965), "The numerical solution of problems in gas dynamics," in "Basic developments in fluid dynamics," M. Holt (ed.), Academic Press, New York, 1-126.

Benton, E.R. and Platzman, G.W. (1972), "A table of solutions of the one-dimensional Burgers' equation," Quart. Appl. Math., 30, 195-212.

Bickley, W.E. (1941), "Experiments in approximating to solutions of a partial differential equation," Phil. Mag., 32, 50-66.

Birkhoff, G. and de Boor, C.P. (1965) in "Approximations of Functions", H.L. Garabedian (ed.), Elsevier, 164-190.

Boldrighini, C. (1977), "Absence of turbulence in a uni-dimensional model of fluid motion (Burgers model)," Meccanica, 12, 15-18.

Bourke, W. (1972), "An efficient, one-level, primitive equation spectral model," Mon. Weather Rev., 100, 683-689.

Bourke, W., McAvaney, B., Puri, K. and Thurling, R. (1977), "Global modeling of atmospheric flow by spectral methods," Methods in Comp. Phys., 17, 267-325.

Brooks, A.N. and Hughes, T.J.R. (1982), "Streamline upwind/Petrov-Galerkin formulations for convection dominated flows with particular emphasis on the incompressible Navier-Stokes equations," Comp. Meth. Appl. Mech Eng, 30.

Burgers, J.M. (1939), "Mathematical examples illustrating relations occurring in the theory of turbulent fluid motion," Trans. Roy. Neth. Acad. Sci., Amsterdam, 17, 1-53.

Burgers, J.M. (1948), "A mathematical model illustrating the theory of turbulence" in Adv. in App. Mech., 1, 171-199.

Burgers, J.M. (1964), "Statistical problems connected with the solution of a nonlinear partial differential equation," in Nonlinear problems in engineering (ed. W.F. Ames), Academic Press, New York, 123-137.

Burgers, J.M. (1972), "Statistical problems connected with asymptotic solution of one-dimensional nonlinear diffusion equation," in Statistical models and turbulence (eds. M. Rosenblatt and C. van Atta), Springer, Berlin, 41-60.

Burgers, J.M. (1974), "The nonlinear diffusion equation," Reidel, Holland.

Carslaw, H.S. and Jaeger, J.C. (1959), "Conduction of heat in solids," O.U.P., Clarendon, 2nd edition.

Cary, B.B. (1973), "An exact shock wave solution to Burgers' equation for parametric excitation of the boundary," Journal of Sound and Vibration, 30, 454-464.

Cary, B.B. (1975), "Asymptotic Fourier analysis of a 'sawtooth-like' wave for dual frequency source excitation," Journal of Sound and Vibration, 42, 235-241.

Catton, I., Ayyaswamy, P.S. and Clerer, R.M. (1974), Int. J. Heat Mass Trans., 17, 173-184.

Chandrasekhar, S. (1961), "Hydrodynamic and Hydromagnetic Stability", O.U.P., U.K.

Chattot, J.J., Guiu-Roux, J. and Laminie, J. (1981), "Finite element calculation of steady transonic flow in nozzles using primary variables," Lecture Notes in Physics, 141, Springer-Verlag, 107-112.

Cheng, S.I. and Shubin, G. (1978), "Computational accuracy and mesh Reynolds number," Journal of Computational Physics, 28, 315-326.

Chin, W.C. (1978), "Pseudo-transonic equation with a diffusion term," AIAA Journal, 16, 87-88.

Christie, I., Griffiths, D., Mitchell, A.R. and Zienkiewicz, O.C. (1976), "Finite element methods for second order differential equations with significant first derivatives," Int. J. Num. Meth. Engng., 10, 1389-1396.

Ciment, M., Leventhal, S.H. and Weinberg, B.C. (1978), "The operator compact implicit method for parabolic equations," Journal of Computational Physics, 28, 135-166.

Cole, J.D. (1951), "On a quasi-linear parabolic equation occurring in aerodynamics," Quart. Appl. Math., 9, 225-236.

Collatz, L. (1960), "The numerical treatment of differential equations," Springer-Verlag, Berlin, 28.

Crandall, S.H. (1956), "Engineering analysis," McGraw-Hill, New York, 151.

Cullen, M.J.P. and Hall, C.D. (1979), Quart. J.R. Met. Soc., 105, 571-592.

Dahlquist, G., Bjorek, A. and Anderson, N. (1974), "Numerical methods," Prentice-Hall, N.J., 104.

Davidson, G.A. (1975), "Sound propagation in fogs," Journal of the Atmospheric Sciences, 32, 2201-2205.

Davidson, G.A. (1976), "A Burgers' equation for finite amplitude acoustics in fogs," Journal of Sound and Vibration, 45 (4), 473-485.

Davies, A.M. (1977), "A numerical investigation of errors arising in applying the Galerkin method of the solution of nonlinear partial differential equations," Comp. Meth. Appl. Mech. Eng., 11, 341-350.

Davies, A.M. and Owen, A. (1979), "Three-dimensional numerical sea model using the Galerkin method with a polynomial basis set," Appl. Math. Mod., 3, 421-428.

Davis, R.T. and Rubin, S.G. (1980), Comp. and Fluids, 8, 101-132.

Dennemeyer, R. (1968), "Introduction to partial differential equations and boundary value problems," McGraw-Hill, New York, 295.

de Vahl Davis, G. (1968), Int. J. Heat Mass Trans., 11, 1675-1693.

Douglas, J. and Dupont, T. (1973), "Superconvergence for Galerkin methods for the two-point boundary problem via local projections," Numer. Math., 21, 270-278.

Duncan, W.J. (1937), "Galerkin's method in mechanics and differential equations," ARC R & M 1978.

Duncan, W.J. (1938), "The principles of Galerkin's method," ARC R & M 1848.

Durvasula, S. (1969), "Natural frequencies and modes of clamped skew plates," AIAA J., 7, 1164-1167.

Durvasula, S. (1971), "Flutter of clamped skew panels in supersonic flow," 8, 461-466.

Fairweather, G. (1978), "Finite element Galerkin methods for differential equations," Lecture notes in pure and applied mathematics, 34, Dekker, New York.

Finlayson, B.A. (1972), "The method of weighted residuals and variational principles," Academic Press, New York.

Finlayson, B.A. and Scriven, L.E. (1966), "The method of weighted residuals - a review," Applied Mechanics Reviews, 19, 735-748.

Fleet, R.W. and Fletcher, C.A.J. (1982), "A Comparison of the finite element and spectral methods for the Dorodnitsyn boundary layer formulation" in Finite Element Methods in Engineering (eds. P.J. Hoadley & L.K. Stevens), Melbourne Univ. Press, Australia, 59-63.

Fletcher, C.A.J. (1978a), "The Galerkin method: an introduction," in "Numerical simulation of fluid motion," B.J. Noye (ed.), North-Holland, 113-170.

Fletcher, C.A.J. (1978b), "Application of an improved method of integral relations to the supersonic boundary layer flow about cones at large angles of attack," in "Numerical simulation of fluid motion," J. Noye (ed.), North-Holland, 537-550.

Fletcher, C.A.J. (1978c), "An improved finite element formulation derived from the method of weighted residuals," Comp. Math. Appl. Mech. Eng., 15, 207-222.

Fletcher, C.A.J. (1979a), "A primitive variable finite element formulation for inviscid, compressible flow," J. Comp. Phys., 33, 301-312.

Fletcher, C.A.J. (1979b), "On the application of an improved finite element formulation with isoparametric elements," Proc. 3rd Int. Conf. in Australia on Finite Element Methods, Sydney, 1979, 671-681.

Fletcher, C.A.J. (1980a), "On the application of a least-squares residual fitting finite element formulation to fluid flow problems," Comp. Meth. App. Math. Eng., 24, 251-268.

Fletcher, C.A.J. (1980b), "On the application of alternating direction implicit finite element methods to flow problems," Proc. 3rd Int. Conf. Finite Elements in Flow Problems, Calgary, Canada.

Fletcher, C.A.J. (1981a), "An alternating direction implicit finite element method for compressible, viscous flows," Lecture Notes in Physics, 141, Springer-Verlag, 182-187.

Fletcher, C.A.J. (1981b), "On an alternating direction implicit finite element method for flow problems," Comp. Meth. Appl. Mech. Eng. (to appear).

Fletcher, C.A.J. (1982a), "A Comparison of the Finite Element and Finite Difference Methods for Computational Fluid Dynamics" in Finite Element Flow Analysis (ed. T. Kawai), Univ. of Tokyo Press, 1003-1010.

Fletcher, C.A.J. (1982b), "Finite element solutions of Burgers' equation" in Finite Element Methods in Engineering (eds P.J. Hoadley and L.K. Stevens), Melbourne Univ. Press, Australia, 49-53.

Fletcher, C.A.J. (1982c) Comp. Meth Appl. Mech. Eng., 30, 307-322.

Fletcher, C.A.J. (1983a), "Computational Galerkin Methods", Springer Verlag, Heidelberg.

Fletcher, C.A.J. (1983b), "A comparison of finite element and finite difference solutions of the one and two-dimensional Burgers' equations," J. Comp. Phys., (to appear).

Fletcher, C.A.J. (1983c), "The group finite element formulation," Comp. Meth. Appl. Mech. Eng. (to appear)

Fletcher, C.A.J. (1983d), "Generating exact solutions of the two-dimensional Burgers' equation," Int. J. Num. Meth. in Fluids (to appear).

Fletcher, C.A.J. and Fleet, R.W. (1983a), "A Dorodnitsyn finite element formulation for laminar boundary layer flow," Int. J. Num. Meth. Fluids (submitted)

Fletcher, C.A.J. and Fleet, R.W. (1983b), "A Dorodnitsyn finite element formulation for turbulent boundary layers," Comp. & Fluids (submitted).

Fletcher, C.A.J. and Holt, M. (1975), "An improvement to the method of integral relations," J. Comp. Phys., 18, 154-164.

Fletcher, C.A.J. and Holt, M. (1976), "Supersonic flow over cones at large angles of attack," J. Fluid Mechanics, 74, 561-591.

Forsythe, G.E., Malcolm, M.A. and Moler, C.B. (1977), "Computer methods for mathematical computations," Prentice-Hall, N.J.

Fortin, M. and Thomasset, F. (1979), "Mixed finite element methods for incompressible flow problems," Journal of Computational Physics, 31, 113-145.

Galerkin, B.G. (1915), "Rods and plates. Series occurring in some problems of elastic equilibrium of rods and plates," Vestnik Inzhenerov. Tech., 19, 897-908.

Gartling, D.K. (1977), "Convective heat transfer by the finite element method," Comp. Meth. App. Mech. Eng., 12, 365-382.

Gazdag, J. (1973), "Numerical convective schemes based on accurate computation of space derivatives," J. Comp. Phys., 13, 100-113.

Gear, C.W. (1971), "Numerical initial value problems in ordinary differential equations," Prentice-Hall, N.J.

Gey, F.C. and Lesser, M.B. (1969), "Computer generation of series and rational function solutions to partial differential initial value problems," Proc. 24th Nat. Conf. of A.C.M., Publ. P-69, 559-572.

Gottlieb, D. and Orszag, S.A. (1977), "Numerical analysis of spectral methods: theory and applications," SIAM Publications, Philadelphia.

Griffiths, D.F. and Mitchell, A.R. (1979), in "Finite Elements for Convection Dominated Flows" (ed. T.J.R. Hughes), AMD 34, ASME, New York, 91-104.

Gurbatov, S.N., Dubkov, A.A. and Lobachevskii, N.I. (1977), "Parametric excitation of low-frequency noise waves in a nonlinear medium," Soviet Physics of Acoustics, 23, 2, 146-148.

Haidvogel, D.B., Robinson, A.R. and Schulman, E.E. (1980), "The accuracy, efficiency and stability of three numerical models with application to open ocean problems," J. Comp. Phys., 37, 1-53.

Hayes, W.D. (1954), "Pseudotransonic similitude and first-order wave structure," Journal of the Aeronautical Sciences, 21, 721-730.

Heinrich, J.G., Huyakorn, P., Zienkiewicz, O.C. and Mitchell, A. (1977), "An upwind finite element scheme for two dimensional convective transport equation," Int. J. Num. Meth. Engng., 11, 131-143.

Herring, J.R., Orszag, S.A., Kraichman, R.H. and Fox, D.G. (1974), J. Fluid Mech., 66, 417-444.

Hirsh, R.S. (1975), "Higher order accurate difference solutions of fluid mechanics problems by a compact differencing technique," Journal of Computational Physics, 19, 90-109.

Hopf, E. (1950), "The partial differential equation $u_t + uu_x = \mu u_{xx}$." Comm. Pure Appl. Math., 3, 201-230.

Hsu, C.C. (1975), "A Galerkin method for a class of steady, two-dimensional, incompressible, laminar boundary-layer flows," J. Fluid Mechanics, 69, 783-802.

Hughes, T.J.R. (1978), Int. J. Num. Meth. Eng., 12, 1359-1365.

Hughes, T.J.R. (1979), "A simple scheme for developing upwind finite elements," Int. J. Num. Meth. Engng., 12, 1359-1365.

Hughes, T.J.R. and Brooks, A.N. (1979) in "Finite Elements for Convection Dominated Flows" (ed. T.J.R. Hughes), AMD 34, ASME, New York, 19-35.

Hughes, T.J.R., Tezduyar, T.E. and Brooks, A.N. (1982), "A Petrov-Galerkin finite element formulation for systems of conservation laws with special reference to the compressible Euler equations," Proc. IMA Conf. Numerical Methods in Fluid Dynamics, Univ of Reading, U.K., Mar 1982.

Irons, B.M. and Razzaque A.(1972) in "The Mathematical Foundations of the Finite Element Method with Applications to Partial Differential Equations" A.K. Aziz (ed.), Academic Press, 557-587.

Isaacson, E. and Keller, H.B. (1966), "Analysis of numerical methods," Wiley, New York, 199.

Jain, P.C. and Holla, D.N. (1978), "Numerical solution of coupled Burgers' equations," Int. J. Nonlinear Mech., 13, 213-222.

Jeffrey, A. and Kakutani, T. (1972), "Weak nonlinear dispersive waves: A discussion centered around the Korteweg-deVries equation," SIAM Review, 14, 4, 582-643.

Kawaguti, M. (1955), "The critical Reynolds number for the flow past a sphere," J. Phys. Soc. of Japan, 10, 694-699.

Kawahara, M. Takencki, N. and Yoshida, T. (1978), "Two step explicit finite element method for tsunami wave propagation analysis," Int. J. Num. Meth. Engng., 12, 331-351.

Kellogg, R.B., Shubin, G.R. and Stephens, A.B. (1980), "Uniqueness and the cell Reynolds number," SIAM J. Num. Anal., 17, 733-739.

Kelly, D.W., Nakazawa, S., Zienkiewicz, O.C. and Heinrich, J.C. (1980), "A note on upwinding and anisotropic balancing dissipation in finite element approximations to convective diffusion problems," Int. J. Num. Meth. Engng., 15, 1705-1711.

Khosla, P.K. and Rubin, S.G. (1979), "Filtering of nonlinear instabilities," J. Eng. Math., 13, 127-141.

Kida, S. (1979), "Asymptotic properties of Burgers' turbulence," Journal of Fluid Mechanics, 93, 337-377.

Kuo, Y. and Tanner, R.I. (1971), "Burgers-type model of turbulent decay in a non-Newtonian fluid," ASME paper 71-W/APM-11.

Kuznetsov, V.V., Nakoryakov, V.E., Pokusaev, B.G. and Shreiber, I.R. (1977), "Propagation of disturbances in a gas-liquid mixture," Soviety Physics of Acoustics, 23, 2, 153-156.

Lanczos, C. (1956), "Applied Analysis," Prentice-Hall, Englewood Cliffs, N.J.

Lerner, A.M. and Fridman, V.E. (1978), "Model equations for the nonlinear acoustics of media with a high thermal conductivity," Soviety Physics for Acoustics, 24, 2, 128-133.

Lick, W. (1967), "Wave propagation in real gases," Advanced in Applied Mechanics, 10, 1-72.

Lick, W. (1970), "Nonlinear wave propagation in fluids," Annual Review of Fluid Mechanics, 2, 113-136.

Lighthill, M.J. (1956), "Viscosity effects in sound waves of finite amplitude," in Surveys in Mechanics (eds. G.K. Batchelor and R.M. Davies), C.U.P., Cambridge, 250-351.

Lorenz, E.N. (1960), Tellus, 12, 243-254.

Love, M.D. (1980), "Subgrid modelling studies with Burgers equation," Journal of Fluid Mechanics, 100, 87-110.

Lynn, P.P. (1974), "Least-squares finite element analysis of laminar boundary layer flows," Int. J. Num. Meth. Eng., 8, 865-876.

Marshall, R.H., Churches, A.E. and Reizes, J.A. (1979), "A hybrid integration scheme for the solution of viscous compressible flows," Appl. Math. Modelling, 3, 459-465.

McCrory, R.L. and Orszag, S.A. (1980), J. Comp. Phys, 37, 93-112.

Mikhlin, S.G. (1964), "Variational Methods in Mathematical Physics" Pergamon, Oxford.

Mitchell, A.R. and Griffiths, D.F. (1980), "The finite difference method in partial differential equations," Wiley, New York.

Mitchell, A.R. and Wait, R. (1977), "The Finite Element Method in Partial Differential Equations," Wiley, London.

Miura, R.M. (1976), "The Korteweg-deVries Equation: A Survey of Results," SIAM Review, 18, 3, 412-459.

Moin, P. and Kim, J. (1980), J. Comp. Phys., 35, 381-392.

Morchoisne, Y. (1979), "Resolution of Navier-Stokes equations by a space-time pseudo-spectral method," La Recherche Aerospatiale, 1979-5, 11-31.

Moran, J.P. and Shen, S.F. (1966), "On the formation of weak plane shock waves by impulsive motion of a piston," Journal of Fluid Mechanics, 25, part 4, 705-718.

Morton, K.W. and Parrott, A.K. (1980), "Generalised Galerkin methods for first-order hyperbolic equations," J. Comp. Phys., 36, 249-270.

Morton, K.W. and Stokes, A. (1981), "Generalised Galerkin Methods of hyperbolic equations," Proc. Conf. Mathematics of Finite Elements and Their Applications (ed. J. Whiteman), Brunel University, U.K.

Murdock, J.W. (1979), "A numerical study of nonlinear effects on boundary layer stability," AIAA J., 15, 1167-1173.

Musha, T. and Higuchi, H. (1978), "Traffic current fluctuation and the Burgers' equation," Japanese Journal of Applied Physics, 17, 5, 811-816.

Narasimha, R. and Deshpande, S.M. (1969), "Minimum-error solutions of the Boltzmann equation for shock structure," J. Fluid Mech., 36, 555-570.

Nickell, R.E., Gartling, D. and Strang, G. (1979), "Spectral decomposition in advection-diffusion analysis by finite element method," Comp. Meths. Appl. Mech. Eng., 17/18, 561-580.

Norrie, D. and de Vries, G. (1976), Finite Element Bibliography, Plenum Press, New York.

Nugmanov, Z.kh. (1975), Izv. Vuz. Avia. Tekh., 18, 78-83.

Oden, J.T. and Reddy, J.N. (1976) "An Introduction to the Mathematical Theory of Finite Elements," Wiley, New York.

Orszag, S.A. (1969), Physics of Fluids, 12, Supplement II, 250-257.

Orszag, S.A. (1970), J. Atmos. Sci., 27, 890-895.

Orszag, S.A. (1971a), "Numerical simulation of incompressible flows with simple boundaries; accuracy," J. Fluid Mech., 49, 75-42.

Orszag, S.A. (1971b), "Accurate solution of the Orr-Sommerfeld equation," J. Fluid Mech., 50, 375-405.

Orszag, S.A. (1971c), Stud. Appl. Math., 50, 293-327

Orszag, S.A. (1972), "Comparison of pseudospectral and spectral approximation," Studies in App. Math., 51, 253-259.

Orszag, S.A. (1980), "Spectral methods for problems in complex geometries," J. Comp. Phys., 37, 93-112.

Orszag, S.A. and Kells, L.C. (1980), J. Fluid Mech., 96, 159-205.

Orszag, S.A. and Kruskal, M.D. (1968), "Formulation of the theory of turbulence," Phys. Fluids, Supplement II, 12, 250-257.

Orszag, S.A. and Patterson, G.S. (1972), "Numerical simulation of three-dimensional homogeneous, isotropic turbulence," Phys. Rev. Lett., 28, 76-79.

Patera, A.T. and Orszag, S.A. (1981), "Transition in planar channel flows," Proc. 7th Int. Conf. Num. Meth. in Fluid Dynamics, Stanford, publ. as Lecture

Notes in Physics, 141, Springer-Verlag, Berlin, 329-335.

Piacsek, S.A. and Williams, G.P. (1970), "Conservation properties of convection difference schemes," Journal of Computational Physics, 6, 392-405.

Pierson, B.L. and Kutler, P. (1980), "Optimal nodal point distribution for improved accuracy computational fluid dynamics," AIAA J., 18, 49-54.

Platzman, G.W. (1960), "The spectral form of the vorticity equation," J. Meteor., 17, 635-644.

Platzman, G.W. (1962), "The analytical dynamics of the spectral vorticity equation," J. Meteor., 19, 313-328.

Poots, G. (1968), Quart. J. Mech. Appl. Math, 11, 257-267.

Puri, K. (1981), Mon. Weath. Rev., 109, 286-305.

Puri, K. (1982), private communication.

Rizun, V.I. and Engel'brekht, Iu.K. (1975), "Application of the Burgers equation with a variable coefficient to the study of nonplanar wave transients," Applied Mathematics and Mechanics, 39, 3, 524-528.

Roache, P.J. (1972), "Computational Fluid Dynamics," Hermosa, Albuquerque.

Rodin, E.Y. (1970a), "On some approximate and exact solutions of boundary value problems for Burgers' equation," J. Math. Anal. Appl., 30, 401-414.

Rodin, E.Y. (1970b), "A Ricatti solution for Burgers' equation," Quart. Appl. Math., 27, 541-545.

Rubin, S.G. and Graves, R.A. (1975), "Viscous flow solutions with a cubic spline approximation," Computers and Fluids, 3, 1-36.

Rubin, S.G. and Khosla, P.K. (1976), "Higher-order numerical methods derived from three-point polynomial interpolation," NASA CR-2735.

Sachdev, R.L. and Seebass, R. (1973), "Propagation of spherical and cylindrical N-waves," Journal of Fluid Mech., 58, part 1, 197-205.

Schamel, H. and Elsasser, K. (1976), "The application of the spectral method to nonlinear wave propagation," Journal of Computational Physics, 22, 501-516.

Schreker, G.O. and Maus, J.R. (1974), "Noise characteristics of jet-flap type exhaust flows," NASA CR-2342.

Seebass, R. (1969), "Sonic boom theory," Journal of Aircraft, 6, 177-184.

Shuleshko, P. (1959), "A new method of solving boundary-value problems of mathematical physics," Australian J. Appl. Sci., 10, 1-16.

Silberman, I.S. (1954), Meteorology, 11, 27-34.

Snyder, L.J. and Stewart, W.E. (1966), "Velocity and pressure profiles for Newtonian creeping flow in regular packed beds of spheres," Am. Inst. Chem. Eng. J., 12, 167-173.

Srinivas, K. and Fletcher, C.A.J. (1983), "Finite element solutions for laminar and turbulent compressible flow", Int. J. Num. Meth. Fluids (submitted).

Staniforth, A.N. and Daley, R.W. (1979), Mon. Weath. Rev., 107, 107-121.

Steven, G.P. and Milthorpe, J. (1978), "On a least-squares approach to the integration of the Navier-Stokes equations," in Finite Elements in Fluids, 3, Wiley, 89-110.

Strang, G. and Fix, G.J. (1973), "The Analysis of the Finite Element Method," Prentice-Hall, Englewood Cliffs, N.J.

Tatsumi, T. (1980), "Theory of homogeneous turbulence," Advances in Applied Mechanics, 20, 39-133.

Tatsumi, T. and Kida, S. (1972), "Statistical mechanics of the Burgers model of turbulence," Journal of Fluid Mechanics, 55, 659-675.

Tatsumi, T. and Tokunaga, H. (1974), "One dimensional shock turbulence in a compressible fluid," Journal of Fluid Mechanics, 65, 581-601.

Taylor, G.I. (1910), "The conditions necessary for discontinuous motions in gases," Proc. Royal Society, Series A, 84, p. 371-377.

Taylor, C. (1980), "The utilisation of the F.E.M. in the solution of some free surface problems", 3rd Finite Element in Flow Problems Conference, Bauff, Canada, 54-81.

Taylor, C. and Davis, J. (1975), Comp. and Fluids, 3, 125-148.

Taylor, C., Hughes, T.G. and Morgan, K. (1977), Comp. and Fluids, 5, 191-204.

Taylor, T.D. and Murdock, J.W. (1981), Comp. and Fluids, 9, 255-263.

Taylor, T.D., Ndefo, E. and Masson, B.S. (1972), "A study of numerical methods for solving viscous and inviscid flow problems," Journal of Computational Physics, 9, 99-119.

Thompson, J.F., Thames, F.C. and Martin, C.M. (1974), "Automatic numerical generation of body-fitted curvilinear coordinate system for field containing any number of arbitrary two-dimensional bodies," J. Comp. Phys., 15, 299-319.

Ton, B.A. (1975), "On the behaviour of the solution of the Burgers equation as the viscosity goes to zero," J. Math. Anal. Appl., 49, 713-720.

Villadsen, J.V. and Stewart, W.E. (1967), "Solution of boundary-value problems by orthogonal collocation," Chem. Eng. Sci., 22, 1483-1501.

Whiteman, J.R. (1975), A Bibliography for Finite Elements, Academic Press, U.K.

Whitham, G.B. (1974), "Linear and nonlinear waves," Wiley, New York, 97.

Williams, H.E. (1975), "Motion of a cable used as a mooring," J. Hydronautics, 9, 107-118.

Yee, H.C., Beam, R.M. and Warming, R.F. (1981), "Stable boundary approximations for a class of implicit schemes for the one-dimensional inviscid equations of gas dynamics," AIAA paper 81-1009, AIAA 5th Computational Fluid Dynamics Conference, Palo Alto, June 22-23, 1981.

Yeung, W.S. and Yang, R.I. (1981), "Application of the method of integral relations to the calculation of two-dimensional incompressible turbulent boundary layers," J. Appl. Mech, 48, 701-706.

Computational Techniques for Differential Equations
J. Noye (Editor)

THE FINITE ELEMENT METHOD IN ENGINEERING APPLICATION

JOSEF TOMAS
Royal Melbourne Institute of Technology, Victoria, Australia

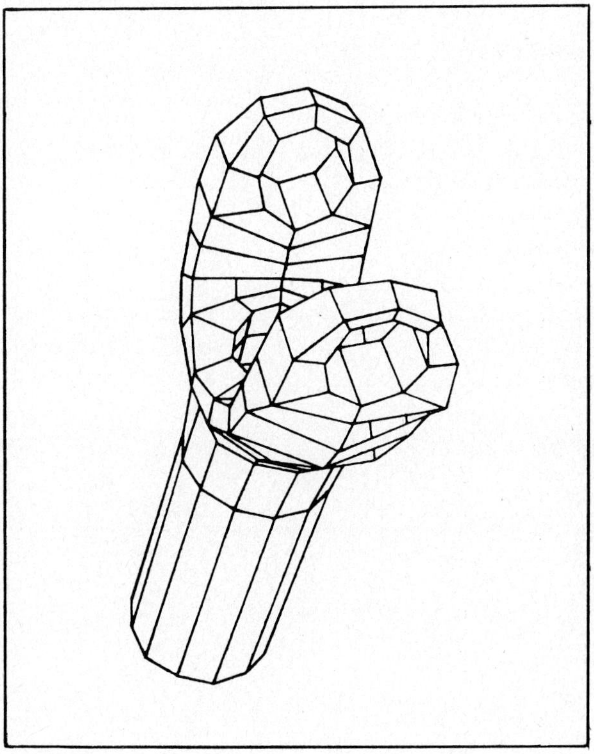

CONTENTS

ABSTRACT

The finite element method is a numerical method originally invented by engineers out of necessity to solve complex practical problems. It later attracted the interest of mathematicians who recognised its place amongst known mathematical procedures. This allowed to generalise the method and to use it with confidence in other fields. Fast and large computers made possible applications to more and more complicated practical problems at reasonable cost. General purpose computer codes written by engineers appeared allowing the solution of a wide range of problems by one code and with minimum theoretical knowledge required by the user. The finite element method became a tool for an engineer and scientist comparable with other experimental procedures used in the practice.
The success unfortunately (but naturally) often led to an overconfidence of the users who forgot that the computer cannot replace clear understanding of the fundamentals of the method, experience and skill in modelling real structures and to coupling techniques and theory correctly and efficiently together.
The correct usage for the future progress of the method is in close cooperation of mathematicians, scientists, computer programmers and users. This paper tries to demonstrate how an engineer understands the finite element method and its application.

INTRODUCTION

It is essential to distinguish four areas of interests in the finite element methods, each on its own very important for the future development and usage of the technique.

Firstly, mathematical theory of the method, which has to work with idealised models of reality in order to achieve generalisation, is essential for mathematical validity of specific approaches and for future direction of the technique as a whole.

Secondly, development of specific element formulations and efficient numerical methods for specific tasks by scientists cannot always be based on precise mathematical foundations. It often has to rely on intuition and validation by computer or material experiments in order to satisfy practical needs and in such a way creates problems of interest to mathematicians.

Thirdly, implementation of the method into an efficient computer code requires cooperation of mathematicians, scientists, computer programmers and users in order to arrive at a correct, efficient and useful code.

Fourthly, the application of the method to solving practical problems assumes a correct understanding of the method with respect to approximations used. This is vital because a model of real complex structures in real media is based on many simplifications and results obtained have to be interpreted in view of these simplifications.

It is surprising how often the importance and inter-relationship between these four areas is disregarded. There are many computer codes which were designed and programmed by engineers without a sufficient knowledge and experience in efficient programming (usability, maintenance, data-base design, sorting procedures, etc.). It becomes virtually impossible for such codes to keep up to date with new element formulations and efficient solution techniques.

Some research codes are prematurely released as commercial codes without proper consideration of quality assurance. For example, they do not include features essential for practical usage (beam off-set, multiple constraints in user friendly form etc.) and they produce wrong results (incorrect procedures for solving non-linear problems, incorrect contour plots due to using wrong function points for interpolation, etc.).

Scientists often test new elements on an assembly of a few elements only, assuming that some results will be obtained with larger assemblies. They do not take into consideration the difference between "laboratory" scale and "production" scale. Some of their elements are not working for practical aspect ratios (element size to element thickness), some are not useful for vibration analysis because of the hour-glass modes, etc.

It is also interesting to observe how many engineering departments at universities do not consider practical application of the finite element method "scientific" enough and produce papers based on more or less mental exercises, as if the object of engineering investigations were the abstract model and not the reality itself.

The author of this paper uses the finite element method as en engineering tool. The computer program has a place in his laboratory similar to that of any other test rig. He learns by using it. He recognises its reliability by comparing results with natural expectations or with physical tests. If the program does not work properly or is missing important features, he gets rid of it and looks for a more suitable program. He certainly tries to understand basic ideas behind the "black-box" he is using but he has no time or interest in developing better or new features. He initiates new developments in the finite element methods by his needs leaving the details of formulation and programming to experienced specialists in relevant scientific areas.

This paper, presented to applied mathematicians, aims at demonstrating how practical engineers use the finite element method. The author also had in his mind his students, to whom he would like to show that there are many things they should know in detail, other things they should have heard about and be careful when using them, and many other things which are too complex to be understood so that they must believe methematical proofs which are absolutely true - but only in one of many mathematical spaces.

ENGINEERING CONCEPT - DISCRETE SYSTEMS

The finite element method evolved from various origins and along different paths. The engineering concept of the method will be presented first as it corresponds to the historical development of the method. The mathematical concept will be discussed later.

Engineers have been solving truss and beam structures for more than a hundred years. The repetition of identical elements and the "pointwise" interconnection of elements make the application of basic principles of mechanics very simple. Although matrix algebra is not essential for the formulation and solution, it is a very convenient method of book-keeping and computer programming. It will therefore be used throughout this paper.

Let us consider a two-dimensional structure from Fig. 2.1. It consists of three members which can resist tension or compression but not bending. The individual members (elements) are numbered 1, 2, 3. The elements are connected with each other at node 2 and with the frame at nodes 1, 2, 4. The structure is loaded by a force F in the node 2.

The displacement of node 2 and internal forces in truss elements are to be calculated.

The usual procedure in engineering analysis is to consider the deformation of each element separately and then to satisfy the equilibrium of internal and external forces at each node.

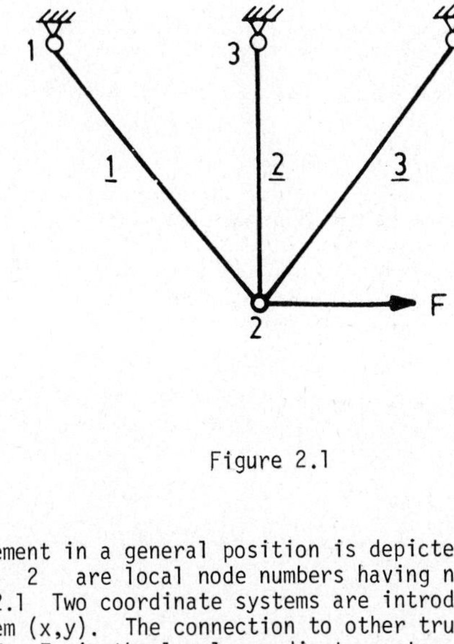

Figure 2.1

An individual element in a general position is depicted in Fig. 2.2. The node numbers 1 and 2 are local node numbers having no relation to the node numbers in Fig. 2.1 Two coordinate systems are introduced, a global system (X,Y) and a local system (x,y). The connection to other truss elements is replaced by internal forces F_1, F_2 in the local coordinate system. A linear elastic analysis with small displacements is assumed.

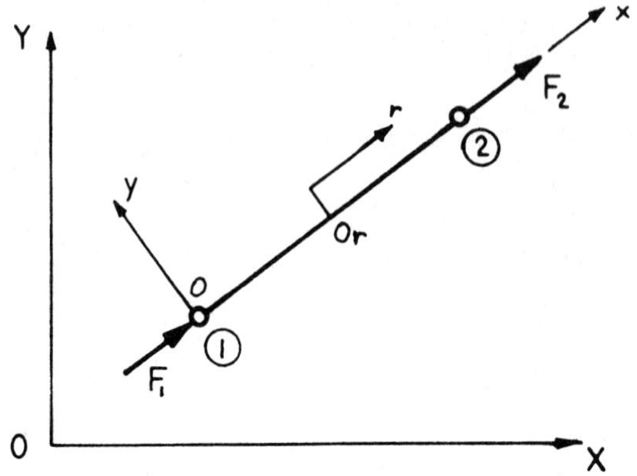

Figure 2.2

The action of internal forces F_1 and F_2 (Fig. 2.2) will result in displacements u_1 and u_2 in the direction of x.

The strain ε of the element is $(u_2 - u_1)/l$.

The stress $\sigma = \varepsilon E$, and the force in the truss is given by

$$F_2 = -F_1 = \sigma A \text{ or } F_2 = (AE/l) \cdot (u_2 - u_1).$$

This relationship can be written in a matrix form as

$$\begin{bmatrix} F_1 \\ F_2 \end{bmatrix} = \frac{A.E}{l} \begin{bmatrix} 1 & -1 \\ -1 & 1 \end{bmatrix} \cdot \begin{bmatrix} u_1 \\ u_2 \end{bmatrix} \quad (2.1)$$

or

$$\underset{\sim}{F}_\ell = \underset{\approx}{K}_\ell \, \underset{\sim}{u}_\ell \quad (2.2)$$

The matrix $\underset{\approx}{K}_\ell$ is called the local element stiffness matrix.

The subscript ℓ is introduced to indicate quantities in the local system in order to distinguish them from those in the global system (X, Y), which will be indicated by a subscript g.

Local stiffness matrices can be calculated for each element. To be able to calculate displacements of the global nodes (Fig. 2.1) the local displacements $\underset{\sim}{u}_\ell$ must be transformed into the global coordinate system (X,Y).

The transformation matrix is easily obtained using Fig. 2.3. The displacement u_1

in the x direction is given in terms of global components u_{1x} and u_{1y} by

$$u_1 = u_{1x} \cos \alpha + u_{1y} \cos \beta \, .$$

A similar expression exists for u_2 - the displacement of the second node of the element. In a matrix form this can be written as

$$\begin{bmatrix} u_1 \\ u_2 \end{bmatrix} = \begin{bmatrix} \cos \alpha & \cos \beta & 0 & 0 \\ 0 & 0 & \cos \alpha & \cos \beta \end{bmatrix} \cdot \begin{bmatrix} u_{1x} \\ u_{1y} \\ u_{2x} \\ u_{2y} \end{bmatrix} \qquad (2.3)$$

or $$\underset{\sim}{u}_\ell = \underset{\approx}{T} \, \underset{\sim}{u}_g$$

where T is the transformation matrix.

Inserting equation (2.3) into equation (2.2) results in

$$\underset{\sim}{F}_\ell = \underset{\approx}{K}_\ell \, \underset{\approx}{T} \, \underset{\sim}{u}_g \qquad (2.4)$$

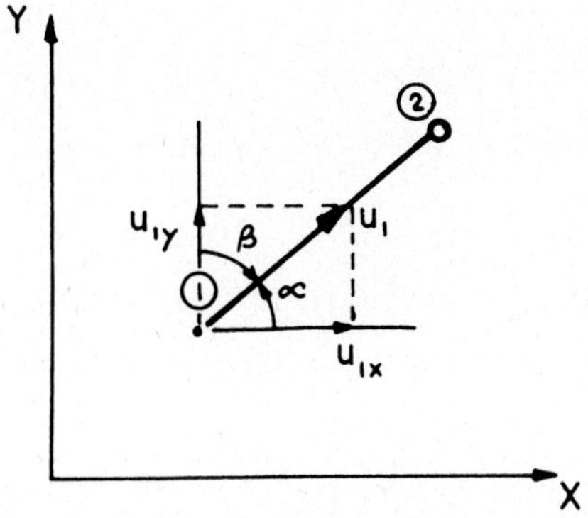

Figure 2.3

This relation is used for calculation of internal forces (specified in local co-ordinate systems of each truss) after the global displacements u_g have been calculated. Most of the computer programs precalculate the matrix $\underset{\approx}{K}.\underset{\approx}{T}$ for each element and store them on a tape for later use.

The equilibrium of internal and external forces in each node requires the transformation of internal forces $\underset{\sim}{F}_\ell$ into the global coordinate system.

It follows from Fig. 2.2, $F_{1x} = F_1 \cos \alpha$, $F_{2x} = F_2 \cos \beta$, or in a matrix form (for both nodes)

$$\begin{bmatrix} F_{1x} \\ F_{1y} \\ F_{2x} \\ F_{2y} \end{bmatrix} = \begin{bmatrix} \cos \alpha & 0 \\ \cos \beta & 0 \\ 0 & \cos \alpha \\ 0 & \cos \beta \end{bmatrix} \begin{bmatrix} F_1 \\ F_2 \end{bmatrix}$$

i.e.

$$\underset{\sim}{F}_g = \underset{\approx}{T}^T \underset{\sim}{F}_\ell \tag{2.5}$$

After multiplying equation (2.4) from left by the transposed transformation matrix $\underset{\approx}{T}^T$ the equation in the global coordinate system is obtained for each element as

$$\underset{\sim}{F}_g = \underset{\approx}{T}^T \underset{\approx}{K}_\ell \underset{\approx}{T} \underset{\sim}{u}_g$$

or

$$\underset{\sim}{F}_g = \underset{\approx}{K}_g \underset{\sim}{u}_g \tag{2.6}$$

where the matrix $\underset{\approx}{K}_g = \underset{\approx}{T}^T \underset{\approx}{K}_\ell \underset{\approx}{T}$ is the element stiffness matrix in the global coordinate system.

Until now, the individual elements have been considered. The forces $\underset{\sim}{F}_g$ in equation (2.6) are internal forces at the nodes 1 and 2 of each element. The elements are connected together and global points are considered (numbered 1,2,3,4 in Fig. 2.1). Each element contributes by its internal force to the global node to which it is connected. There are also external forces acting on the global nodes (some of them will normally be zero). Equilibrium requires that the sum of internal forces on each global node balance the external force on this node.

Considering equation (2.6) for each element, the vector of internal forces on the left side will be replaced by the force vector of external forces in the global nodes and the elements of $\underset{\approx}{K}_g$ matrices corresponding to the nodal point entries and displacement entries will be added together.

The result will be the equation for the complete structure.

$$\underset{\sim}{F} = \underset{\approx}{K} \underset{\sim}{u} \tag{2.7}$$

where $\underset{\sim}{F}$ is the vector of external forces, $\underset{\approx}{K}$ is the system stiffness matrix, $\underset{\sim}{u}$ is the vector of global displacements. Symbolically, this assembly procedure of the stiffness matrix $\underset{\approx}{K}$ is depicted in Fig. 2.4. The displacement notation u_1x, u_1y, ...u_4x, u_4y has been replaced by u_1, ...u_8, where the subscripts 1,...,8 indicate equation numbers.

The force vector is a vector of external forces corresponding to each displacement (equation number). Zero entry corresponds to zero external force component for the relevant equation.

The stiffness matrix $\underset{\approx}{K}$ in equation (2.7) is singular. This fact can be easily interpreted physically. Our assembly procedure in Fig. 2.4 assumes free nodes 1,3,4; the structure is free, so that large motions are possible without any deformations (the so called rigid body modes).

There are two ways to handle geometric (fixed) boundary conditions. The first one is to eliminate from equation (2.7) the rows and columns corresponding to fixed boundaries. In Fig. 2.4, this would be the rows and columns numbered 1,2,5,6,7,8. An identification vector relating the equation numbers of the reduced system to the equation numbers of the original system has to be created and stored in order to relate displacements (now only available for free node 2) to the components of the original vector $\underset{\sim}{u}$ after solution.

The second method uses so called boundary elements. It is of interest to mention that this method is equivalent to the treatment of geometric boundary conditions suggested by Courant (2), which is in fact the penalty method. The connections between nodes 1,3,4 and the frame are replaced by very stiff springs (Fig. 2.5).

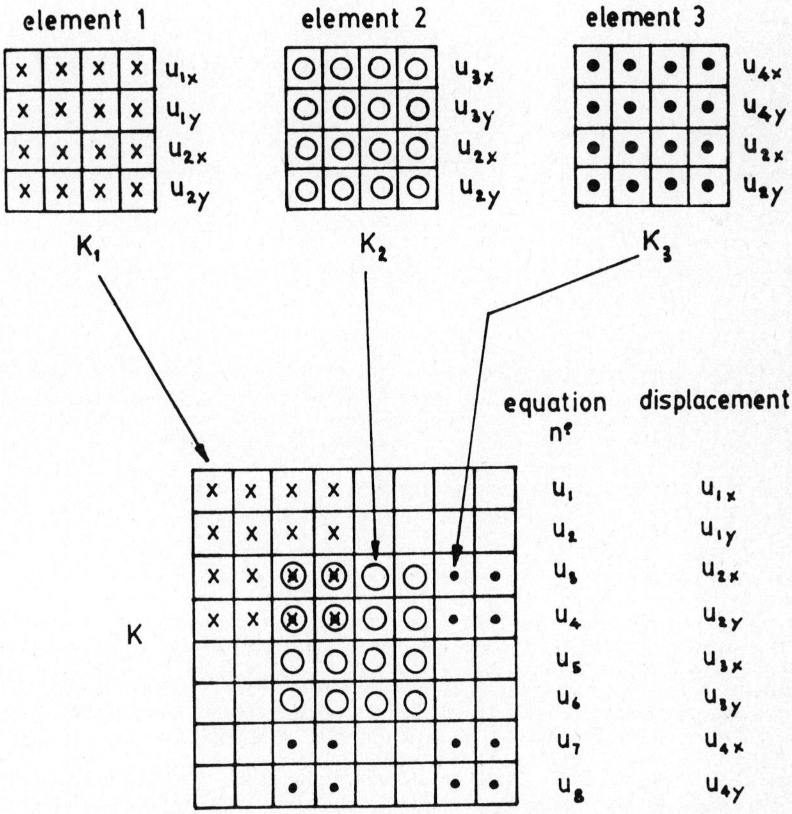
element 1
element 2
element 3
u_{1x}
u_{1y}
u_{2x}
u_{2y}
u_{3x}
u_{3y}
u_{2x}
u_{2y}
u_{4x}
u_{4y}
u_{2x}
u_{2y}
K_1
K_2
K_3
K
equation n°
displacement
u_1 u_{1x}
u_2 u_{1y}
u_3 u_{2x}
u_4 u_{2y}
u_5 u_{3x}
u_6 u_{3y}
u_7 u_{4x}
u_8 u_{4y}

Figure 2.4

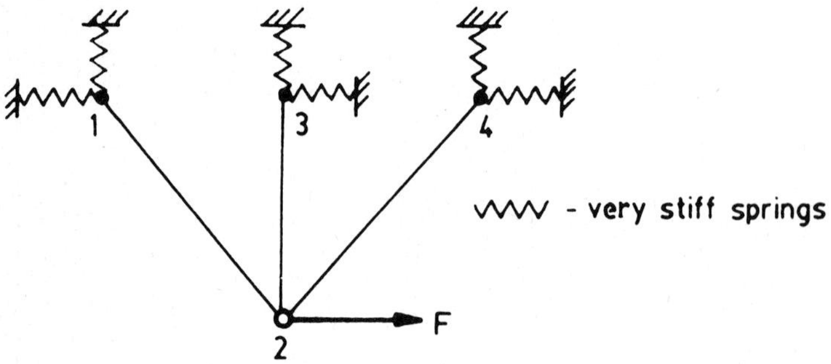

Figure 2.5

The stiffness matrix of the spring is only a single element which is added to the corresponding diagonal term of the system stiffness matrix. When equation (2.7) is solved by the Gaussian elimination procedure, all row matrix terms and the force component are divided by the corresponding diagonal term.

This will result in the equation for the row i

$$F_i/k_{ii}=(\ldots)/k_{ii} + u_{ii} + (\ldots)/k_{ii} \qquad (2.8)$$

which gives the result $u_{ii} = 0$ when k_{ii} is very large.

This method can also be used to introduce prescribed displacements at some nodes. A stiff spring is added to the node with prescribed deflection d. The deflection is applied to the system through this spring. The force corresponding to the deflection d is equal to $k_s.d$. Equation (2.8) will have on the left side $F_1=k_s.d$. Because the difference between k_s and k_{ii} is negligible (k_{ii} is the original k_{ii} plus $k_s >> k_{ii}$) the result will be $u_{ii} \simeq d$ as required.

Programmes as CAL 78, SAP4, ADINA use the first methods; programs MARC and ABAQUS use the second method. ABAQUS uses a front-wave solution method in which the system matrix is not completely assembled so that the boundary element method does not represent any drawback and is very comfortable from the programming point of view.

After the geometric boundary conditions have been accommodated in equation (2.7), equation (2.7) is solved.

All of the programs with which the author is familiar use the Gaussian elimination method. Since the matrix $\underset{\sim}{K}$ is symmetrical and banded, an economical storage scheme is possible. SAP and MARC store the complete upper half-band, ADINA stores the upper half-band in the so-called skyline fashion (column-wise up to the last non-zero element above the diagonal), ABAQUS uses a front-wave solution method in which an element matrix is condensed, keeping only equations which are common to elements not yet included, the next element matrix is added and condensed again, and so on.

It is clear that the order of numbering nodes is important for minimizing band width, while the order of numbering elements is important for keeping the number of equations required for condensation at any time to the minimum in the front-wave solution method.

After equation (2.7) has been solved, the displacements of relevant nodes forming an element are extracted from the vector $\underset{\sim}{u}$ and the internal forces (stresses) are calculated from equation (2.4).

Although it is arguable whether the above procedure of solving the truss system should be called finite element method, nevertheless it contains all of the basic features of the true finite element method for the solution of continua. The difference is that the stiffness matrix is calculated exactly for trusses while for continua it can only be approximated. However, in both cases the forces can be applied only at nodes and the displacements obtained are the displacements of the nodes.

In a similar way, beam elements can be introduced. If elementary beam theory is assumed (normals to the neutral beam axis remain normal) the stiffness matrix can again be calculated exactly. The unknown variables will now be not only displacements of the nodes, but also their derivatives (slopes) to allow bending.

ENGINEERING CONCEPT - CONTINUOUS SYSTEMS

Structural problems of solid mechanics can be formulated for the finite element method directly without using variational principles or weighted residual methods. This fact is probably the main reason for the success of the method among engineers. Every third year mechanical or civil engineering student understands that the work of external forces during deformation is converted into internal energy of the body. This balance is valid for any part of the body, and so for a finite element. The finite element has a physical meaning, it can be drawn on the body, or cut-out and seen from all sides. On the other hand, formulation of equilibrium of a point by means of a partial differential equation and subsequent numerical solution by the finite difference method represents for an engineer a double-artificial process having no connection with the reality.

The principle of virtual work and the principle of minimum potential energy will be used in this section to derive the stiffness matrix of a continuum. Unlike the stiffness matrix of truss or beam elements, the stiffness matrix of a continuum solid element can be evaluated only approximately.

Any real component or structure is a three-dimensional continuum. In many cases some of the dimensions can be neglected and either one-dimensional models, such as trusses or beams, or two-dimensional plates, axi-symmetric solids or shells can be constructed. The main practical gain is a substantial reduction in computer requirements.

Historically, the first attempt to solve a continuum by a method related to the finite element approach followed the procedure outlined in the previous section. The continuum was replaced by a framework of trusses or beams arranged in a definite pattern and having elastic properties selected in a way that equivalent displacements and strains of the framework converged to the exact values for the continuum if the size of the individual bars was made infinitesimally small (3). This approach was abandoned when Argyris (4) and Turner et al (5) developed a more direct method of replacing a continuum by equivalent small subdivisions called later by Clough (6) finite elements.

Let us demonstrate the method with a simple problem - a two-dimensional plate loaded in its middle plane (the so-called plane stress problem (Fig. 3.1).

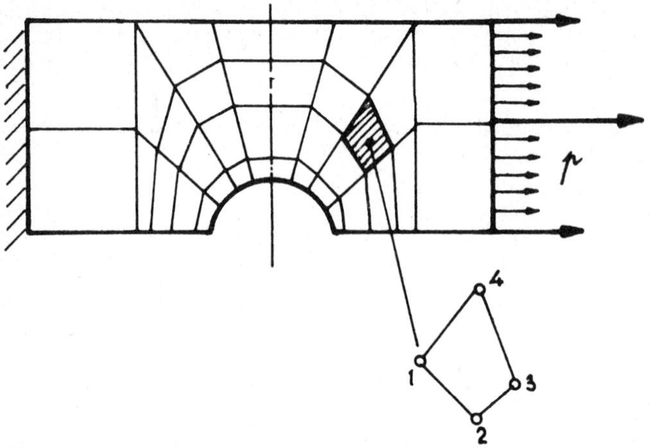

Figure 3.1

The continuum is first discretised into finite elements. There is an unlimited number of ways of constructing the element mesh. One of the possibilities is shown in Fig. 3.1. The coarseness of the mesh and the precision of the solution are directly related as will be shown later. Intuitively, the mesh should be finer in areas of large displacement gradients.

There are also many possible element shapes. Four-node quadrilateral elements have been chosen in Fig. 3.1, but other types, such as triangular elements with three or six nodes or quadrilateral elements with eight or twelve nodes could equally have been selected. The nodes have a function similar to that described earlier for trusses: their displacements are the basic unknowns of the problem, and all loads must be replaced by equivalent forces applied at the nodes.

The displacement within each element is approximated by a shape function (also called a displacement, trial, approximation, basis or interpolation function) which has the unknown displacements of the nodes as parameters.

This displacement function allows strains within the element to be expressed in terms of the nodal displacements so that stresses throughout the element and also on its boundaries may be calculated.

The principle of virtual work is then applied to an isolated element expressing the balance between the virtual work of internal forces at the nodes and the virtual strain energy within the element for any combination of virtual displacements. The result will be a set of linear algebraic equations similar to the system of equations (2.7).

Clearly, the displacement function has to satisfy certain requirements in order to ensure continuity of displacements along boundaries between adjacent elements and to allow convergence to the exact solution with mesh refinement.

Let us cut out one element from the structure in Fig. 3.1 and number its nodes 1,2,3,4 (local nodal numbers) as shown in Fig. 3.1).

The displacement component u (in the x direction) at any point (x,y) within the element is approximated as

$$u = \sum_{i=1}^{4} N_i u_i , \tag{3.1}$$

where u_i, i = 1, ..., 4 are as yet unknown nodal displacements in the x direction.

Plane strain ε is defined in terms of displacements u and v as

$$\underset{\sim}{\varepsilon}^T = \left[\varepsilon_x \; \varepsilon_y \; \varepsilon_{xy}\right] = \left[\frac{\partial u}{\partial x} \; \frac{\partial v}{\partial y} \; \frac{\partial u}{\partial x} + \frac{\partial u}{\partial y}\right] \tag{3.2}$$

The approximate value of the strain can be easily expressed in terms of nodal point displacements u_i , v_i by inserting (3.1) into (3.2). We then have

$$\underset{\sim}{\varepsilon} = \underset{\approx}{B} \underset{\sim}{u_i} , \quad \underset{\sim}{u_i} = \begin{bmatrix} u_i \\ v_i \end{bmatrix} , \quad i = 1,\ldots,4. \tag{3.3}$$

The strain displacement matrix $\underset{\approx}{B}$ is obtained by appropriate differentiation of the functions N_i.

The stress components can be expressed in terms of the strains. For an isotropic material, the linear elastic behaviour is described by relations (if initial strains and stresses are not considered).

$$\varepsilon_x = (\sigma_x - \nu\sigma_y)/E$$

$$\varepsilon_y = \sigma_y - \nu\sigma_x)/E$$

$$\varepsilon_{xy} = 2(1 + \nu)\sigma_{xy}/E$$

which after solving for stresses gives

$$\underset{\sim}{\sigma} = \underset{\approx}{D} \underset{\sim}{\varepsilon} . \tag{3.4}$$

The matrix $\underset{\approx}{D}$ is called the material matrix. The action of adjacent elements on the element under consideration is replaced by equivalent nodal forces. These forces deform the element by shifting the nodal points by displacements $\underset{\sim}{u_i}$ and thus strain the element throughout.

The natural procedure for deriving the conditions of equilibrium is the principle of virtual work.

A virtual displacement at the nodes δ_{u_i} results in virtual strains within the element equal to

$$\delta_{\underset{\sim}{\varepsilon}} = \underset{\approx}{B} \cdot \delta_{\underset{\sim}{u_i}} . \tag{3.5}$$

The internal work per unit volume done by the stresses is

$$\delta_{W_i} = \delta_{\underset{\sim}{\varepsilon}}{}^T . \underset{\sim}{\sigma} = \delta_{\underset{\sim}{u_i}}{}^T \underset{\approx}{B}{}^T \underset{\sim}{\sigma} . \tag{3.6}$$

The external work of nodal forces (external with respect to the element, internal with respect to the structure) is

$$\delta_{W_e} = \delta_{\underset{\sim}{u_i}}^T \underset{\sim}{f_i} . \tag{3.7}$$

Integrating the internal work (3.6) over the volume of the element gives the total internal work, and this must be equal to the virtual work of external forces:

$$\delta \underset{\sim}{u}_i^T \cdot \underset{\sim}{f}_i = \delta \underset{\sim}{u}^T \int_V \underset{\approx}{B}^T \underset{\sim}{\sigma}\, dV \quad . \qquad (3.8)$$

Equation (3.8) must be satisfied for any values of virtual displacements $\delta \underset{\sim}{u}_i$ so that

$$\underset{\sim}{f}_i = \int_V \underset{\approx}{B}^T \underset{\sim}{\sigma}\, dV \quad . \qquad (3.9)$$

This is valid for any stress-strain relationship.

The linear elastic behaviour of the material, as expressed by the relationship (3.4), gives

$$\underset{\sim}{f}_i = \int_V \underset{\approx}{B}^T \underset{\approx}{D}\, \underset{\sim}{\varepsilon}\, dV = \int_V \underset{\approx}{B}^T \underset{\approx}{D}\, \underset{\approx}{B}\, \underset{\sim}{u}_i\, dV$$

or

$$\underset{\sim}{f}_i = \underset{\approx}{K}_e\, \underset{\sim}{u}_i \qquad (3.10)$$

where the element stiffness matrix is given by

$$\underset{\approx}{K}_e = \int_V \underset{\approx}{B}^T \underset{\approx}{D}\, \underset{\approx}{B}\, dV \qquad (3.11)$$

Equation (3.10) is formally identical to equation (2.6) derived for the truss structure in the previous Section.

The assembly procedure of the element stiffness matrices into the global system matrix is identical to that described in the previous Section for the truss structure. Formally, it can be derived by applying the principle of virtual work to the whole system of elements. Internal forces will be replaced by external forces acting at the nodes. The virtual work of a given external force is balanced by part of the strain energy only in those elements connected to the node at which the force is applied. This means that only these displacements enter the equation having the force on the left side. The system of equations (2.7) will be banded, and a substantial saving of computer storage can be achieved by proper nodal point numbering.

The principle of virtual work is not the only method available for the finite element formulation.

The same result can be achieved by the principle of minimization of total potential energy.

For a linear elastic material the strain energy U of a solid of volume V is given by

$$U = \tfrac{1}{2} \int \underset{\sim}{\varepsilon}^T \underset{\approx}{D}\, \underset{\sim}{\varepsilon}\, dV \quad . \qquad (3.12)$$

The work W of external forces is

$$W = \underset{\sim}{u}_i^T \cdot \underset{\sim}{f}_i \qquad (3.13)$$

Replacing $\underset{\sim}{\varepsilon}$ by the expression (3.3), the total potential energy P = U - W can be written as

$$P = \tfrac{1}{2} \int \underset{\sim}{u}^T \underset{\approx}{B}^T \underset{\approx}{D}\, \underset{\approx}{B}\, \underset{\sim}{u}_i dV - \underset{\sim}{u}_i^T \cdot \underset{\sim}{f}_i \quad . \qquad (3.14)$$

The minimum of P is achieved by setting

$$\frac{\partial P}{\partial \underset{\sim}{u}_i} = 0 \ . \tag{3.15}$$

Applying this procedure to equation (3.14) yields

$$(\int \underset{\approx}{B}^T \underset{\approx}{D} \underset{\approx}{B} \, dV) \underset{\sim}{u}_i = \underset{\sim}{f}_i \ ,$$

which is identical to equation (3.10).

Recognition of the fact that the finite element formulation requires the total potential energy to be a minimum, was of critical importance for the further development of the finite element method as will be demonstrated in the next Section.

MATHEMATICAL CONCEPT

We have seen in the previous section that the system of equations (2.7) can be derived either by the principle of virtual work or by minimization of the total potential energy. The integral of the total potential energy is called a functional. There are many other mathematical and physical problems in which functionals whose stationary value represents the solution of the problem can be defined this means that the finite element method, originally developed for the solution of structural problems of solid mechanics, can be used for the solution of many other fields.

Further, engineers and physicists have been using the Rayleigh-Ritz method since its discovery at the beginning of this century for a direct numerical technique for optimizing functionals. The principle of the Rayleigh-Ritz method is summarised as follows: a) a finite number of trial functions $\phi_1, \phi_2, \ldots \phi_n$ are chosen; b) these functions are combined in the linear form $u = \Sigma a_i \phi_i$; c) the unknown coefficients a_i are determined in such a way that the functional $I(u)$ is minimized. If this idea is compared with the minimization of the potential energy from the last Section, it is clear that nodal displacements correspond to the unknown parameters a_i and the shape functions N_i to the trial functions ϕ_i. However, there is a fundamental difference between the classical Rayleigh-Ritz method and the finite element method. In the Rayleigh-Ritz method the trial functions are defined over the whole region and have to satisfy fixed boundary conditions along the whole boundary. The more complex the boundary or the greater the accuracy required, the more trial functions of greater accuracy have to be included. This limits the Rayleigh-Ritz method substantially for engineering application.

On the other hand, the finite element method subdivides the region into smaller subdivisions (elements) and within each element very simple trial (shape) functions are defined, usually polynomials of a low degree. Accurate approximation of complex boundaries is achieved by mesh refinement.

This equivalence of the finite element method and the Rayleigh-Ritz method gives the method a respectable mathematical basis. Error estimates and convergence criteria already existing for the classical Rayleigh-Ritz method can be applied to the finite element method.

It is worth mentioning a mathematical paper of Courant (2) from 1941. When solving a plane torsion problem, he realized that the application of the Rayleigh-Ritz method to the whole domain would be very cumbersome and therefore he divided the domain in a net of small triangular subdomains. He then defined linear trial functions over each subdomain separately and replaced the functional by a finite element sum of subdomain contributions. The results were function values in nodal points. We can see that this procedure contains all basic features of the finite element technique. Unfortunately, its potential for numerical solution of practical problems was not recognised for more than two decades. This is reflected in mathematical textbooks on numerical methods for the solution of partial differential equations which only now include the finite element method after thirty years later (see e.g. (7) from 1969 with no reference at all, new edition (8) from 1977 has a brief chapter only).

The Rayleigh-Ritz method can be applied only to functionals. Unfortunately, there are many non-structural problems where the functionals do not exist or are difficult to find. The partial differential equations governing the problem are usually known. In such cases, the classical methods of weighted residuals can be introduced in a similar way to the classical Rayleigh-Ritz method and a finite element approximation can be then evaluated.

Let us demonstrate both procedures mentioned above on an example of a steady-state heat transfer problem in two dimensions.

The problem is governed by the differential equation.

$$L(u) = \frac{\partial}{\partial x}(k\frac{\partial u}{\partial x}) + \frac{\partial}{\partial y}(k\frac{\partial u}{\partial y}) + Q(u) = 0 \quad (4.1)$$

The variable u corresponds to the temperature inside a body Ω.

Let us assume for simplicity that $u = 0$ on the boundary Γ .

If the variable k (conductivity) in equation (4.1) is constant, the problem of solving the differential equation (4.1) is equivalent to the minimization of the functional

$$I(u) = \int_{\Omega} (\tfrac{1}{2}k(\frac{\partial u}{\partial x})^2 + \tfrac{1}{2}k(\frac{\partial u}{\partial y})^2 - Qu)\, d\Omega \quad (4.2)$$

The proof is simple. Let us calculate a small change (variation) in I due to a change in u. Following the rules of calculus of variation, the variation in I is

$$\delta I = \int_{\Omega}(k\frac{\partial u}{\partial x}\,\delta(\frac{\partial u}{\partial x}) + k\frac{\partial u}{\partial y}\,\delta(\frac{\partial u}{\partial y}) - Q\,\delta u)\,d\Omega \quad (4.3)$$

Since $\delta(\partial u/\partial x) = \partial(\delta u)/\partial x$, equation (4.3) can be integrated by parts. We then have

$$\delta I = -\int_{\Omega} \delta u\,(\frac{\partial}{\partial x}(k\frac{\partial u}{\partial x}) + \frac{\partial}{\partial y}(k\frac{\partial u}{\partial y}) + Q)\,d\Omega + \int_{\Gamma}\delta u(k\frac{\partial u}{\partial n})d\Gamma = 0 \quad (4.4)$$

The boundary integral over Γ is identically equal zero for $u = 0$. This follows from the given boundary condition $u = 0$. Inside the domain Ω δu is arbitrary. To satisfy equation (4.4), where $\delta I = 0$ for any δu, the expression in brackets must be zero. The result is identical to the original equation (4.1).

Instead of solving the differential equation (4.1) together with necessary boundary conditions, the function (4.2) can be solved directly by approximating the unknown temperature u by the function v.

$$u \simeq v = \Sigma N_i v_i \quad , \quad (4.5)$$

where N_i are shape functions and v_i are approximate temperatures at the nodes.

Inserting (4.5) into (4.2) gives

$$I = \int_\Omega \tfrac{1}{2} k \left(\Sigma \frac{\partial N_i}{\partial x} v\right)^2 d\Omega + \int_\Omega \tfrac{1}{2} k \left(\Sigma \frac{\partial N_i}{\partial y} v\right)^2 d\Omega$$

$$- \int_\Omega Q \, \Sigma N_i v_i \, d\Omega = 0 . \tag{4.6}$$

The stationary value of the approximated function (4.6) can be obtained by differentiation of (4.6) with respect to the parameters v_j, $j = 1, \ldots, n$. The system of equations can be written in matrix form as

$$\underset{\approx}{K} \, \underset{\sim}{v} = \underset{\sim}{f} \tag{4.7}$$

where $K_{ij} = K_{ji} = \int_\Omega k \left(\frac{\partial N_i}{\partial x} \frac{\partial N_j}{\partial x} + \frac{\partial N_i}{\partial y} \frac{\partial N_j}{\partial y} \right) d\Omega$,

$$f_j = \int_\Omega N_j \, Q \, d\Omega.$$

The system (4.7) is again formally equivalent to the system of equations (2.7).

The functional (4.2) assumes constant k. If k is not constant, the Rayleigh-Ritz method cannot be used because there is no functional related to equation (4.1). A Galerkin method of weighted residuals must then be applied.

The unknown u is again approximated by (4.5). Assuming that the boundary conditions are satisfied, the approximate solution v can be made to satisfy (4.1) in the mean by making the residual L(v) orthogonal to a set of weighting functions w, that is

$$\int_\Omega w L(v) \, d\Omega = 0 . \tag{4.8}$$

If the weighting functions are chosen as trial functions N, the well-known Galerkin procedure is obtained. Applied to the heat transfer problem we obtain

$$\int_\Omega N_i \left(\frac{\partial}{\partial x}\left(k \frac{\partial u}{\partial x}\right) + \frac{\partial}{\partial y}\left(k \frac{\partial u}{\partial y}\right) - Q \right) d\Omega = 0 . \tag{4.9}$$

Integrating by parts and realizing that $N_i = 0$ at Γ we obtain

$$\int_\Omega k \left(\frac{\partial N_i}{\partial x} \frac{\partial u}{\partial x} + \frac{\partial N_i}{\partial y} \frac{\partial u}{\partial y} \right) d\Omega - \int_\Omega N_i \, Q \, d\Omega = 0 . \tag{4.10}$$

Expressing the derivatives of u by (4.5) will result again in the equation (4.7). This is not surprising because the Rayleigh-Ritz method and Galerkin method are identical for the problems where a functional exists.

If the original differential equation (4.1) is compared with the variational forumulation (4.2) or the Galerkin formulation (4.9) a very important difference regarding the continuity of v can be immediately recognised. The differential equation (4.1) requires the second derivatives of u to exist throughout the domain. The variational or Galerkin formulation contain only the first order derivatives of u. A lower order approximation v may be used for the approximation in (4.5). Also the function u has to satisfy all boundary conditions (geometric and natural) while the approximation v has to satisfy only geometric conditions because the natural boundary conditions are satisfied implicitly by the formulation.

CONVERGENCE

The continuum has an infinite number of degrees of freedom. It is approximated by a finite number of shape functions so that the minimum of the potential energy (functional) is only approximated. As in the classical Rayleigh-Ritz method, the approximation of the true displacement distribution can be considered as an additional constraint of the system. Thus the true minimum of energy may be reached only in the limit when the size of the elements goes to zero. The consequence is that the finite element model of a structural problem is always stiffer than the true mathematical model.

It certainly is desirable that the true displacement distribution can be approximated as closely as necessary. This requirement can be achieved only if the chosen shape functions satisfy special criteria of completeness and conformity.

Mathematical definition of conformity can be found elsewhere (e.g. (9) or (10)). We will use a physical definition applied to the displacement functions in structural analysis.

Completeness requires that the displacement function is able to represent the condition of constant strain including zero strain.

Conformity (also called compatibility) requires continuity of displacements between elements so that the strains at the interface between elements remain finite.

Both criteria need to be satisfied in the limit as the size of the elements tends to zero. However, an improved accuracy is achieved if these criteria are satisfied already on elements of finite size.

If we consider that in the plane-stress problems (linear elastic) the strains are defined by first derivatives of displacements, then conformity requires C_0 continuity of the displacement functions. In beam, plate or shell problems, the strains depend on the second derivative of displacements, so that C_1 continuity is required for the displacement functions.

There are many mathematical studies analysing error bounds and rate of convergence of the finite element approximations. Although they have not yet proved to be particularly useful for quantitative estimates because they contain terms depending on the unknown solution, they nevertheless allow us to assess the order of convergence.

Again, the reader is referred for mathematical discussion elsewhere. A heuristic proof of convergence is now given in the manner presented in (11).

A polynomial expansion is assumed to approximate displacements in an element. If this expansion fits the correct solution exactly, the approximation will yield the exact answer. This is the case of beam elements, for instance, where a cubic polynomial corresponds to the exact solution of the beam model based on the elementary beam theory.

Let us now assume that the exact solution is expanded in the vicinity of a point (x_i, y_i) within the element by use of the Taylor series:

$$u = u_i + \left(\frac{\partial u}{\partial x}\right)_i x + \left(\frac{\partial u}{\partial y}\right)_i y + \ldots \tag{5.1}$$

where x and y are sides of the element, both of "size" h. If within an element of "size" h a polynomial expansion of order p is employed, this will fit locally the Taylor expansion up to that order and the error in displacements will be of the order of neglected terms, i.e. $O(h^{p+1})$. The strains (and stresses) are given by m-th derivative of displacements, so that they would converge with an error of $O(h^{p+1-m})$.

In the case of the plane stress, a linear polynomial (p = 1) satisfies the compatibility criteria. The rate of convergence of displacements is then $O(h^2)$. The strain is defined as the first derivative of displacement, m = 1. Then the rate of convergence in strains and stresses is O(h). Suppose that in a given mesh the elements are halved in size. Then the error in displacements is reduced to 25% of the original error, but the error in strains (and stresses) is reduced only to 50%. This is certainly a very practical result: by solving the same problem with two different meshes, an almost exact solution in any point can be extrapolated, in the case of a monotonic convergence.

Other errors are introduced by computers. The round-off errors depend on the number of significant digits carried in the computer. In some machines double or higher precision is necessary in order to keep the errors below reasonable limits.

Evaluation of the integrals in equation (4.7) is performed numerically. This introduces another error. Surprisingly, this error has an influence on the minimum value of the energy functional opposite to that introduced by the finite element discretization, that is, it makes the structure softer. This fact is exploited very efficiently in many element formulations by using reduced integration to make element behaviour closer to that of the real structure. Convergence is preserved; however it will not be monotonic. This method of improvement is discussed in the next section.

SHAPE FUNCTIONS AND ELEMENT MATRICES OF CONFORMING ELEMENTS

As stated in the Section on Engineering Concept, the displacement within a finite element is approximated in terms of the displacements of the element nodes, i.e.

$$\underset{\sim}{u} = \underset{\sim}{N}^T \underset{\sim}{u}_i \ . \tag{6.1}$$

The functions N_i must be chosen in an appropriate way in order to achieve convergence with mesh refinement. As shown in the last Section, the conditions for the convergence are those of completeness and compatibility of the shape functions (6.1).

Compatibility is automatically achieved in beam and truss elements where the adjacent elements are connected in one point. The truss element requires C_0 continuity of u, while the beam requires C_1 continuity (continuity in slopes and displacements).

Compatibility is also relatively easily achieved in two-and three-dimensional solids because only C_0 continuity of $\underset{\sim}{u}$ is required.

Plate bending and thin shell analyses present a problem because of the C_1 compatibility requirement, and strain being defined as the second order derivative of displacement. In fact, very few fully compatible shell elements have been developed because of their impracticability (computationally expensive, higher order derivatives as parameters, internal nodes required etc.). Non-compatible elements (see next Section) are preferred in modern computer codes.

Number of elements have been developed for 2-D and 3-D solids. However, the user convenience has become an increasingly important consideration to program developers and almost all of the modern programs use one type of the solid element. This is the so-called isoparametric solid element. Its formulation allows any distorted configuration or degeneration into lower order elements (for example, quadrilateral into triangle by simply merging two nodes together). Its edges can be curved so that more complicated boundaries can be easily approximated.

A two-dimensional plane-stress isoparametric quadrilateral will be briefly discussed. An extension of the formulation into three-dimensions or axi-symmetric types is straightforward and can be found in the literature (e.g. (11)).

The basic idea is quite simple. The region is subdivided into elements which may be quadrilaterals (or triangles) with straight or curved edges.

Cartesian (global) coordinates are transformed into a new cartesian (r-s) system in which element edges become straight lines with r or s equal -1 or +1 (Fig. 6.1).

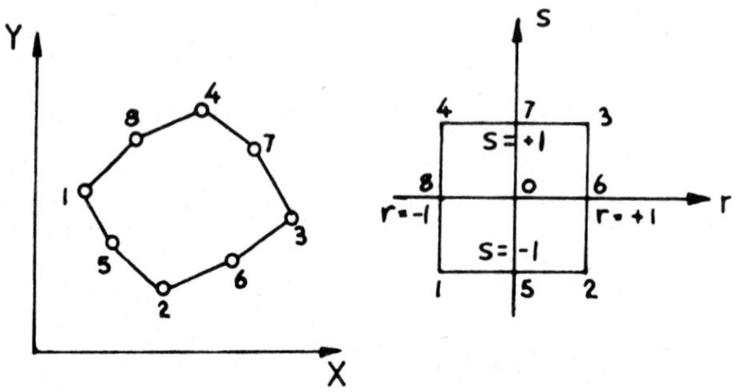

Figure 6.1

The nodes are transformed into corresponding nodes in the r-s plane. Fig. 6.1 shows this transformation for an eight node quadrilateral. The four node quadrilateral is obtained as a special case by deleting nodes 5 to 8. Higher approximations are also possible but above the cubic (4 nodes to each edge) internal nodes are required in order to satisfy compatibility, which is very impractical and so rarely used in computer programs.

The bilinear approximation of x and y will transform the r-s square into x-y quadrilateral with straight edges. The biquadratic approximation is required for the eight node element:

$$x(r,s) = a_1 + a_2 r + a_3 s + a_4 rs + a_5 r^2 + a_6 s^2 + a_7 r^2 s + a_8 rs^2 \tag{6.2}$$

We express y(r,s) in a similar fashion.

The unknown parameters a_i are calculated from the conditions that vertices in (x,y) plane must become vertices in (r,s) plane.

The resulting transformation for the eight node element is

$$x = \sum_{i=1}^{8} N_i x_i$$

$$y = \sum_{i=1}^{8} N_i y_i \tag{6.3}$$

where x_i, y_i are global coordinates of the nodes (known) and N_i are the shape functions. By inserting x_i (y_i) and corresponding (r_i, s_i) into equation (6.2) for all eight nodes, the eight unknown parameters a_i can be evaluated and shape functions N_i obtained for both x and y.

The result is

$$N_1 = \tfrac{1}{4}(1-r)(1-s) - \tfrac{1}{2}N_5 - \tfrac{1}{2}N_8$$

$$N_2 = \tfrac{1}{4}(1+r)(1-s) - \tfrac{1}{2}N_5 - \tfrac{1}{2}N_6$$

$$N_3 = \tfrac{1}{4}(1+r)(1+s) - \tfrac{1}{2}N_6 - \tfrac{1}{2}N_7$$

$$N_4 = \tfrac{1}{4}(1-r)(1+s) - \tfrac{1}{2}N_7 - \tfrac{1}{2}N_8$$

$$N_5 = \tfrac{1}{2}(1-r^2)(1-s)$$

$$N_6 = \tfrac{1}{2}(1-s^2)\ (1+r)$$

$$N_7 = \tfrac{1}{2}(1-r^2)(1+s)$$

$$N_8 = \tfrac{1}{2}(1-s^2)(1-r) \tag{6.4}$$

It is easy to prove that after inserting the appropriate (r,s) coordinates of a nodal point, the expansion (6.3) gives (x,y) coordinates of this node.

Also the expansion of equation (6.3) for an edge in (r,s) will contain only the coordinates of the nodes located on the edge (e.g. for s = -1, equation (6.3) will contain the nodes 1,5,2 only). This means that adjacent elements will have common edges after the transformation given by equation (6.3).

The isoparametric approximation is obtained by using the same shape functions (6.4) for the approximation of displacements, i.e.

$$\underset{\sim}{u} = \sum_{i=1}^{8} N_i \underset{\sim}{u}_i \ , \tag{6.5}$$

where $\underset{\sim}{u}_i = \begin{bmatrix} u_i \\ v_i \end{bmatrix}$ are the nodal point displacements,

$\underset{\sim}{u} = \begin{bmatrix} u \\ v \end{bmatrix}$ is the displacement of a point (r,s)

$N_i = N_i$ (r,s,) is the corresponding shape function from equation (6.4). Following our discussion about coordinates of points along the edges, the displacement function (6.5) will have C_0 continuity across element boundaries, this is the same displacements will be obtained along the edge of both adjacent elements from nodal point displacements because only displacements of nodal points common to both elements enter the function (6.5) along the edge. Hence the isoparametric element is compatible.

To prove completeness let us consider a linear displacement field applied to the element

$$u = a_1 + a_2x + a_3y \ ,$$

$$v = b_1 + b_2x + b_3y \ . \tag{6.6}$$

At the nodes the displacements would be

$$u_i = a_1 + a_2 x_i + a_3 y_i \quad ,$$

$$v_i = b_1 + b_2 x_i + b_3 y_i \quad . \tag{6.7}$$

In the isoparametric formulation the displacements are approximated as

$$u = \sum_{i=1}^{8} N_i \, u_i \quad ,$$

$$v = \sum_{i=1}^{8} N_i \, v_i \quad . \tag{6.8}$$

Inserting (6.7) into (6.8) gives

$$u = a_1 \Sigma N_i + a_2 \Sigma N_i x_i + a_3 \Sigma N_i y_i \quad , \quad v = \,.....$$

However, the coordinate transformation (6.3) states that $x = \Sigma N_i x_i$, $y = \Sigma N_i y_i$ so that (6.9) yields

$$u = \Sigma N_i + a_2 x + a_3 y \quad ,$$

$$v = \Sigma N_i + b_2 x + b_3 y \quad . \tag{6.10}$$

This is equivalent to (6.6) providing $\Sigma N_i = 1$.

It can easily be shown by inspection of expressions (6.4) that the condition $\Sigma N_i = 1$ is satisfied for four and eight node isoparametric element. Therefore isoparametric elements are complete. This is valid for any isoparametric element.

The element stiffness matrix is given by equation (3.11), that is

$$\underset{\approx}{K_e} = \int_V \underset{\approx}{B}^T \underset{\approx}{B} \, \underset{\approx}{D} \, dV \quad . \tag{6.11}$$

The matrix $\underset{\approx}{B}$ can be written as

$$\underset{\approx}{B} = [\underset{\approx}{B_1} \; \underset{\approx}{B_2} \; \underset{\approx}{B_3} \; ...... \; \underset{\approx}{B_n}] \quad . \tag{6.12}$$

n is the number of element nodes and submatrices $\underset{\approx}{B_i}$ are given as

$$\underset{\approx}{B_i} = \begin{bmatrix} \frac{\partial N_i}{\partial x} & 0 \\ 0 & \frac{\partial N_i}{\partial y} \\ \frac{\partial N_i}{\partial y} & \frac{\partial N_i}{\partial x} \end{bmatrix} \tag{6.13}$$

We can see that the components of the $\underset{\approx}{B}$ matrix are given by partial derivatives of the shape function N_i with respect to the global coordinates (x,y). However the same shape functions N_i are specified in local coordinates r,s as defined in equation (6.4).

The evaluation of the integral (6.11) in local coordinates r,s requires two transformations: one for the global derivatives of the N_i's in the $\underset{\approx}{B}$ matrices, and a second for the elementary volume dV = dx.dy.

Partial differentiation of N_i gives

$$\begin{bmatrix} \frac{\partial N_i}{\partial r} \\ \\ \frac{\partial N_i}{\partial s} \end{bmatrix} = \underset{\approx}{J} \begin{bmatrix} \frac{\partial N_i}{\partial x} \\ \\ \frac{\partial N_i}{\partial y} \end{bmatrix} , \qquad (6.14)$$

where the matrix $\underset{\approx}{J}$ is the Jacobian matrix

$$\text{given as } \underset{\approx}{J} = \begin{bmatrix} \frac{\partial x}{\partial r} & \frac{\partial y}{\partial r} \\ \\ \frac{\partial x}{\partial s} & \frac{\partial y}{\partial s} \end{bmatrix} = \begin{bmatrix} \Sigma \frac{\partial N_i}{\partial r}.x_i & \Sigma \frac{\partial N_i}{\partial r}.y_i \\ \\ \Sigma \frac{\partial N_i}{\partial s}.x_i & \Sigma \frac{\partial N_i}{\partial s}.y_i \end{bmatrix} \qquad (6.15)$$

Assuming that the inverse of the Jacobian matrix exists, the global derivatives of N_i's can be evaluated from (6.14).

The transformation of the elementary volume is

$$dV = dx.dy = \det(\underset{\approx}{J}).dr.ds \quad . \qquad (6.16)$$

The sign of the Jacobian determinant must remain unchanged throughout the element if one-to-one mapping as defined by equation (6.2) should exist. It can be shown that the necessary condition for isoparametric elements is that element angles are less than 180° and for eight-node elements the middle nodes are in the middle third of the side on which they are located (11).

As a result of the transformation, the integral (6.11) is written as

$$\int_{-1}^{1} \int_{-1}^{1} F(r,s)\, dr\, ds \quad . \qquad (6.17)$$

The function F(r,s) cannot be integrated analytically. Numerical integration has to be used for the evaluation.

NUMERICAL INTEGRATION OF ELEMENT MATRICES

The integral (6.17) can be solved analytically for only a very few simple elements. Almost invariably, element matrices are evaluated using numerical integration technique.

Principles and details of various methods of numerical integration can be found elsewhere (for example, (12)).

Two integration methods mostly used are: a) Newton-Cotes quadrature, b) Gauss quadrature.

Newton-Cotes quadrature such as the trapezoidal or Simpson rule is mainly used for the evaluation of integrals over the cross-section of beams in large displacement analysis (13).

Gauss quadrature is used for the evaluation of matrices of isoparametric elements because such a scheme requires the least number of evaluations of functions in equation (6.17).

Numerical integration introduces an additional error into the calculation by replacing the integral (6.17) by a weighted sum of a few function values. It also requires a substantial amount of computer time. Therefore, it is important to determine the minimum integration order necessary for convergence. There is no reason to expect that exact integration is required in order to achieve convergence. As already shown in the Section, Convergence, a condition on the displacement (shape) function is to achieve a state of constant strain in the limit of zero element size. The energy functional (3.12) is, in case of elements with C_0 continuity, a function of first derivatives of the displacement functions N. The rate of convergence of strains is $O(h^{p-m+1})$, if the strains are given by m-th derivatives of displacements. The energy is given by the square of strains, so that the rate of convergence in energy is $O(h^{2(p-m)+1})$.

The minimum rate of convergence is O(h). This will be achieved if p = m, or p = 1 for C_0 elements. Such approximation is sufficient for convergence and will result in a constant value of F in the integral (6.17). This means that convergence will occur if the volume of the element

$\int_V dV = \int_{-1}^{1} \int_{-1}^{1} \det(\underset{\approx}{J})\, dr\, ds$ is correctly evaluated by numerical integration.

It can easily be shown that for bilinear approximation, the det $(\underset{\approx}{J})$ is a linear function of r and s because the terms with the rs-product cancel. This means that one point integration is sufficient for this type of element.

Although bilinear approximation with one integration point is sufficient to assure convergence, higher order approximation will result in a higher rate of convergence. Quadratic elements have p = 2 so that the error is $O(h^3)$ if elements are not too much distorted, i.e. det $(\underset{\approx}{J})$ const. If quadratic elements are distorted, det $(\underset{\approx}{J})$ is a polynomial of the 4th order so that p = 3 is required.

The above values are related to approximations at element level. The element stiffness matrices are then assembled and the system of linear algebraic equations

$$\underset{\approx}{K}\ \underset{\sim}{u}_i = \underset{\sim}{f}_i \tag{7.1}$$

has to be solved after the boundary conditions have been inserted. With exact integration the matrix $\underset{\approx}{K}$ will not be singular and the system (7.1) can be solved.

The question is how the replacement of the exact integration by the numerical integration can influence the singularity of $\underset{\approx}{K}$.

The independent relations are expressed by the strain-displacement relationship which expresses strain components in terms of a linear form of the nodal point displacements $\underset{\sim}{u}_i$. In C_0 plane problems, there are three such equations (see equation 3.3)).

When numerical integration is used, the strains are evaluated at each of the integration points. If the number of degrees of freedom (components of $\underset{\sim}{u}_i$) is larger than the number of independent strain components, the element matrix $\underset{\approx}{K}_e$ will be singular.

Let us consider four and eight node quadrilateral elements. Fig. 7.1 shows a structure modelled by one element only. The rigid body modes are removed by appropriate supports.

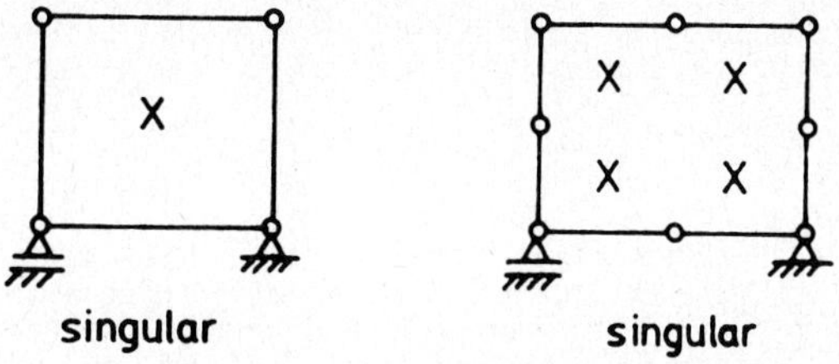

Figure 7.1

There are two degrees of freedom per node (translations in x and y direction) and three independent relations per integration point. In both cases, the matrix $\underset{\sim}{K}$ will be singular because the number of unknown parameters exceeds the number of independent relations.

Fig. 7.2 shows a structure modelled by two elements. The four node quadrilateral is still singular, the eight node quadrilateral is non-singular.

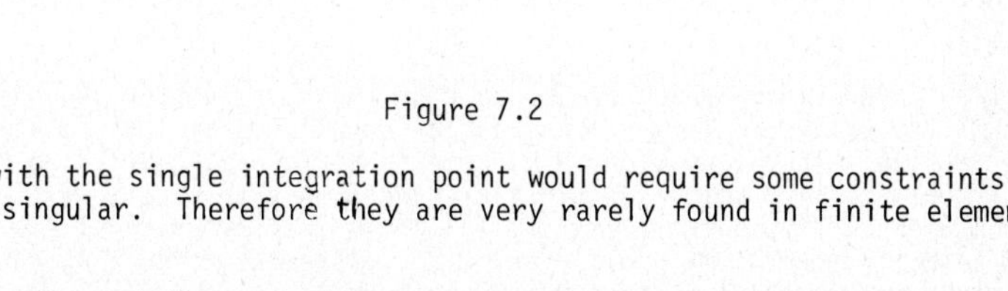

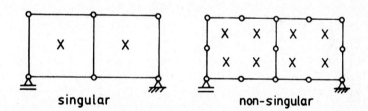

Figure 7.2

Elements with the single integration point would require some constraints in order to be non-singular. Therefore they are very rarely found in finite element programs.

The eight node quadrilateral elements with four integration points are today the most commonly used plane elements and twenty node solid elements with 2 x 2 x 2 integration points are the most commonly used three-dimensional elements.

However, as mentioned above, one integration point is sufficient for convergence with an error O(h) for any C_0 element. In fact, it was found by computer experiments that reduced integration makes elements correspond more closely to the behaviour of real solids. As already mentioned, any methods based on the

Rayleigh-Ritz or Galerkin processes result in a model (in structural analysis) which is stiffer than the real object. Reduced integration has an opposite effect and the convergence is still guaranteed (at a lower rate). This is the explanation for the success of such elements in practice. However, the user of finite element programs should be aware of some problems which can result from reduced integration, such as singularity for insufficiently constrained structures.

NON-CONFORMING ELEMENTS

Plates and shells require C_1 continuity (displacements and their derivatives must be continuous). Determination of shape functions satisfying convergence requirements is now much more difficult. Although conforming elements for plates and shells have been derived, they are not only impractical (second order derivatives are required at some nodes; they are very difficult to combine with other elements) but their performance is very poor (a very fine mesh is required for reasonable convergence).

During the "prehistoric" period of finite element development, when the conditions of convergence and compatibility were not clearly recognised, many non-conforming elements were designed and used. Surprisingly enough, some of them gave much better results than conforming elements. It was important to explain this fact.

If strains are defined by second order derivatives of displacements (as in beams or shells), a discontinuity in first derivatives of displacements will cause infinite strains at the element interfaces. However, the convergence criteria need only be satisfied in the limit as element size tends to zero. Therefore, if in the limit the continuity is restored, the nonconforming element will work correctly.

A test for the acceptability of nonconforming elements has been invented as a computer test performed on an assembly of elements - a patch test (14). A patch of elements of arbitrary geometry is subject to an arbitrary constant strain condition by an appropriate deformation of the patch boundary. The patch is solved on the computer and stresses in all elements are printed. They should be identical. Strictly speaking, the patch should have infinitesimal element sizes but a fine mesh will do just as well.

The patch test was later defined mathematically, and it proved that the intuitive experimental approach as described above was correct (see, eg. (10) or (15)).

Recognition that nonconformity improves element performance suggested the idea of improving the performance of conforming elements by either adding nonconforming shape functions or by using substitute shape functions with a smoothing effect on the derivatives.

An example of improving element performance by adding nonconforming shape functions is Wilson's element (16). It is a four node isoparametric rectangular element (Section - Shape Functions and Element Matrices of Conforming Elements) to which two nonconforming shape functions are added in order to improve the element's performance in bending. Fig. 8.1(a) shows how the original element deforms if bending moments are applied, Fig. 8.1(b) demonstrates that the behaviour after two nonconforming shape functions have been added, closely corresponds to the reality.

The two new shape functions are defined as

$$N_5 = (1 - r^2) ,$$

$$N_6 = (1 - s^2) . \qquad (8.1)$$

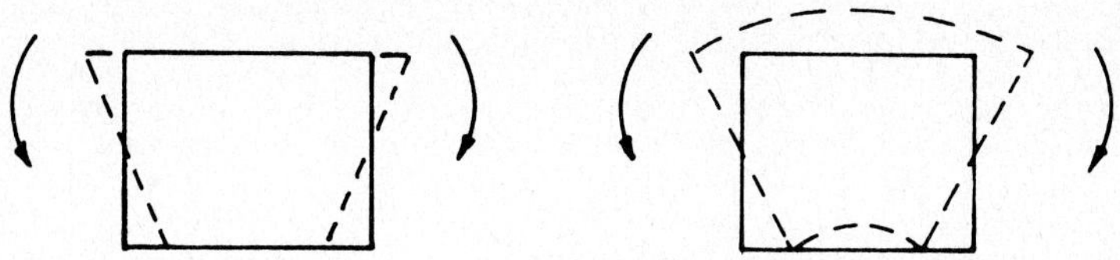

Figure 8.1

They assume the values according to equation (8.1) inside the element. Outside of the element they are zero, therefore they are not continuous at the boundary. The two parameters u_5 and u_6 have no physical meaning. They are evaluated as functions of the remaining parameters (nodal displacements) from the condition of minimum total energy and they are then eliminated from the element stiffness matrix by static condensation.

This element passes the patch test for rectangular configuration (11). A general isoparametric element does not pass the patch test. Taylor (17) satisfied the conditions of the test by altering the nonconforming shape functions and by modifying the numerical integration.

Substitute shape function can be demonstrated using again the example of four node isoparametric quadrilateral element. The shape function for node 1 ($r = -1$, $s = 1$) has the form $N = (1 - r)(1 - s)/4$. It has one term of the second order (rs) which provides twist but does not contribute to convergence. Fig. 8.2(a) shows the shape function which is clearly not a plane. We can expect that the same results (or better if the overstiffening effect of the original method in general is considered) will be obtained by averaging the N surface by a linear function.

$$\bar{N}_1 = a_1 + a_2 r + a_3 s \tag{8.2}$$

which would replace the derivatives $\partial N_1/\partial r, \partial N_1/\partial s$ by their constant averages Fig. 8.2(b)). The surface $\bar{N}_1$ (r,s) will then remain planar. This method has been successfully applied to plate bending elements (11). The performance of the nonconforming element is much better than that achieved by the original confirming elements.

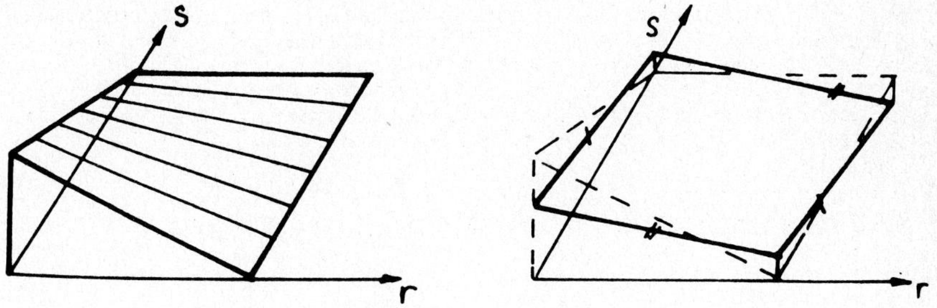

Figure 8.2

One can intuitively expect tthat minimizing the energy functional (3.12), that is

$$U = \tfrac{1}{2} \int_V \underset{\sim}{\varepsilon} \underset{\approx}{D} \underset{\sim}{\varepsilon} \, dV + W \tag{8.3}$$

using the exact strains $\underset{\sim}{\varepsilon}$ will be equivalent to inserting a least square fit to the exact strains. This is successful when the energy integral is evaluated exactly. Mathematical proof of this can be found in (11).

Using numerical integration, the strains are calculated in a few points only. It is extremely important in terms of the order of the error that correct points are used. Fig. 8.3 shows exact strain (full-line) and piece-wise linear least square approximations (dotted lines). It is evident that there are points within each element where the exact and approximated strains coincide. It was shown by (18) that these points are very close to Gauss integration points (the result is exact

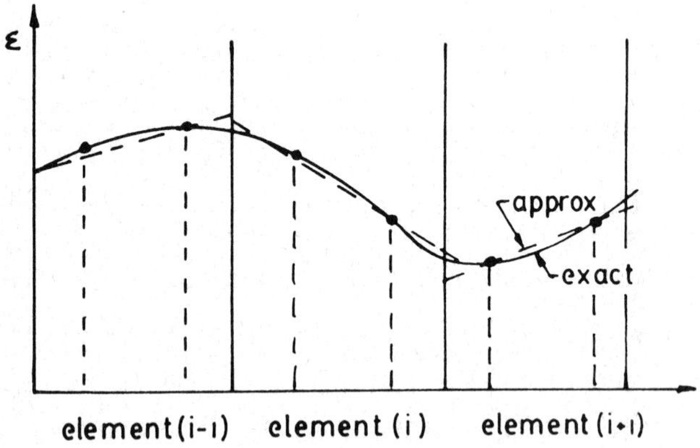

Figure 8.3

for one-dimensional integration), which are sufficient to integrate exactly the polynomial function under the energy functional. This means that by using these points for the evaluation of strains in equation (8.3) the approximation will be almost one order better than sampling them elsewhere.

The practical result is that the best and most economical plane element is the eight node quadrilateral (Fig. 8.4) with four Gaussian integration points.

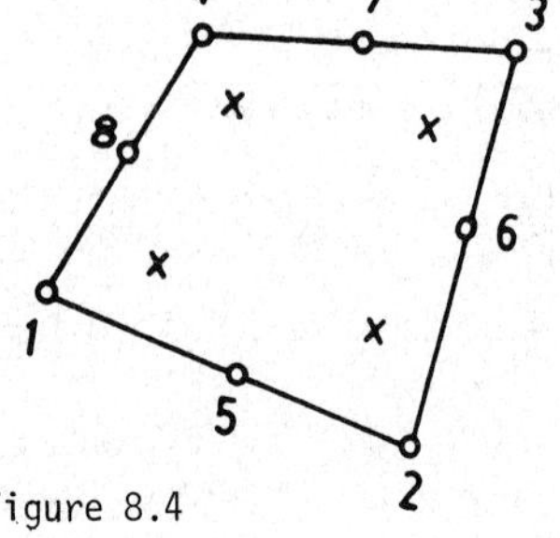

Figure 8.4

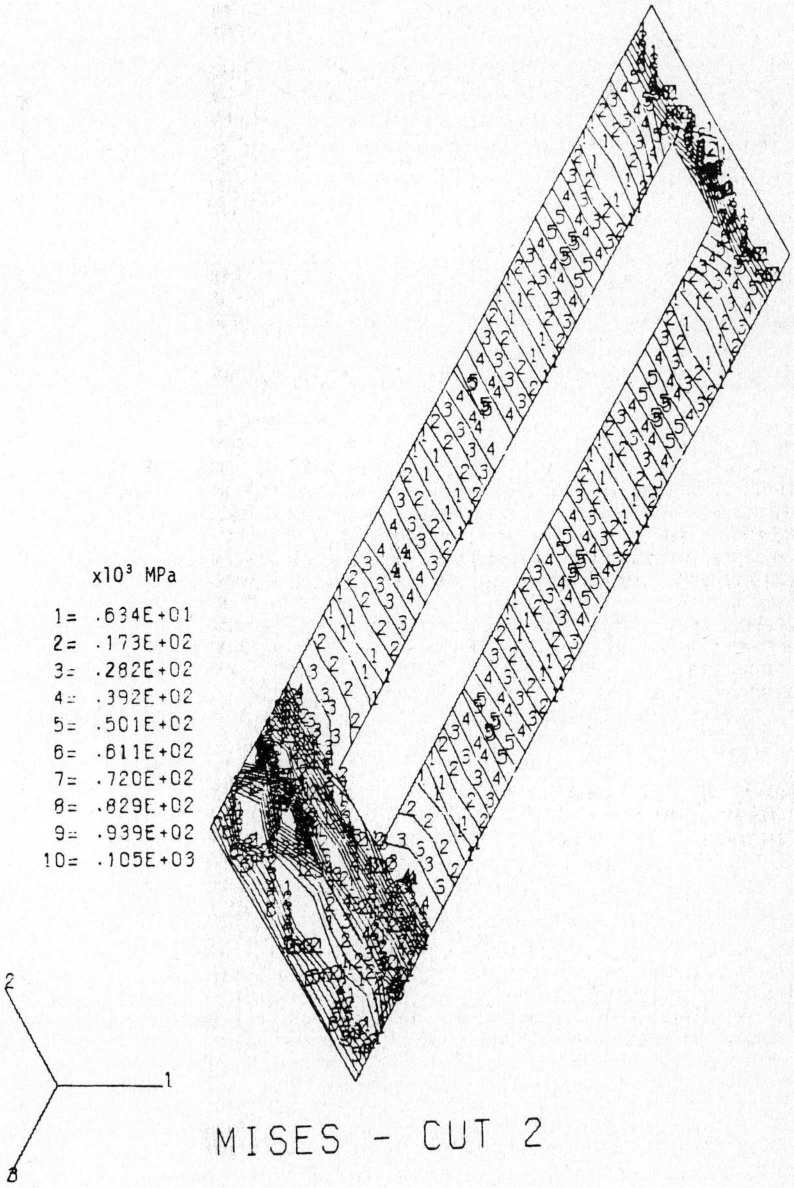
x10³ MPa
1= .634E+01
2= .173E+02
3= .282E+02
4= .392E+02
5= .501E+02
6= .611E+02
7= .720E+02
8= .829E+02
9= .939E+02
10= .105E+03
2
1
3
MISES - CUT 2

Figure 8.5

The shape functions N_i are complete polynomials of the second order (p = 2), strains are proportional to the first derivative of N_i (m = 1), that is a polynomial of order p - m = 1. The integrand in the energy integral will be a polynomial of order 2(p - m) requiring two Gaussian points (in each direction). At these points the strains are evaluated nearly one order better, that is the convergence rate of order $O(h^{p-m+1}) = O(h^2)$ is improved to $O(h^{p-m+2}) = O(h^3)$.

The stresses are calculated from the strains after the solution of the system (2.7). The result of the previous discussion shows that the stresses should never be calculated at nodes but at the Gauss points (1 Gauss point for a four-node quadrilateral or 8-node 3-D solid element; 4 Gauss points for a 8 node quadrilateral; 8 Gauss points for a 20-node 3-D solid element).

Evidence that not observing this rule can lead to useless results is demonstrated in Fig. 8.5. The shaft of a seat-belt retractor was modelled by 20-node brick elements and 3 x 3 x 3 integration points were used for calculation and evaluation of stresses. The Mises stress levels plotted in Fig. 8.5 make no sense for the given loading. If the stresses are calculated at 2 x 2 x 2 = 8 integration points the results are correct.

IMPOSITION OF CONSTRAINTS

Difficulties with continuity requirements in the case of beam and shell elements or difficulties in overcoming other problems such as incompressibility, have led to the formulation of various types of nonconforming elements where the coupled variables are approximated independently by separate shape functions and the relation between them is imposed as a constraint.

There are basically three methods used: a) penalty function, b) exact constraints in internal points, c) Lagrange multipliers.

The <u>penalty function</u> method is based on the fact that the minimization of a functional with constraints is equivalent to the minimization of the original functional with a penalty function added to it (2).

Let us consider a function V(u) such as that in Equation (4.2) and a constraint u = g. The solution of this problem can be achieved by minimizing the functional

$$V(u) + \tfrac{1}{2}\alpha \int_{\Omega} (u - g)^2 d\Omega \qquad (9.1)$$

for $\alpha \to \infty$

The Ritz method can be applied to equation (9.1) . It is clear that only an approximate solution can be obtained because there is a limit on the coefficient α. For sufficiently large values of α ill-conditioned matrices will appear. However, it is reasonable to assume that there is a relationship between the error introduced by discretization (element size) and the error due to a finite value of α. Babuska (19) determined the error for geometric boundary conditions (boundary element in the second Section) where he found that

$$\alpha = c \cdot h^{-p} , \qquad (9.2)$$

where h is the element "size", and p is the degree of the complete approximation polynomial.

Introducing shape functions into equation (9.1) and minimizing the functional with respect to the unknown parameters will result in a system of equations.

$$(\underset{\approx}{K_1} + \alpha \underset{\approx}{K_2})\, \underset{\sim}{u_i} = \underset{\sim}{f_i} + \alpha \underset{\sim}{g_i} \qquad (9.3)$$

Comparing this equation with boundary element method in second Section earlier on, we see that they are identical, $\underset{\approx}{K}_2$ being the stiffness matrix of stiff springs simulating the rigid boundary. With α very large, the $\underset{\sim}{u}_i$'s with the constraint will be practically equal zero or $\underset{\sim}{g}_i$. This was required in the case of fixed boundaries.

If the matrix $\underset{\approx}{K}_2$ is made to be singular, then

$$\underset{\approx}{K}_2 . \underset{\sim}{u}_i \to 0, \text{ while } \underset{\sim}{u}_i \neq 0 \text{ with } \alpha \to \infty$$

The singularity is introduced into $\underset{\approx}{K}_2$ by the use of a low order integration scheme as discussed in the Section - Numerical Integration of Element Matrices. The higher the degree of singularity, the greater the influence on α.

As an example let us consider the beam element (11). C_1 continuity is required (displacements and slopes). Normally Hermitian polynomials for shape functions are used to satisfy continuity requirements:

$$\underset{\sim}{u} = \begin{bmatrix} u \\ \\ \frac{du}{dx} \end{bmatrix} = \Sigma N_i \underset{\sim}{u}_i + \Sigma L_i \frac{d\underset{\sim}{u}_i}{dx} \tag{9.4}$$

Using a penalty function allows us to approximate displacements and slopes independently, that is

$$u = \Sigma N_i u_i \;, \qquad \theta = \Sigma N_i \theta_i \tag{9.5}$$

and relate them by a constraint

$$C = \frac{du}{dx} - \theta = 0 \;. \tag{9.6}$$

The original functional $V(u,\theta)$ is now extended by the penalty function

$$f = \tfrac{1}{2}\alpha \int \left(\frac{du}{dx} - \theta\right)^2 dx \tag{9.7}$$

The influence of the singularity of the matrix $\underset{\approx}{K}_2$ is demonstrated in (11 - p.291). It shows the ratio of approximate to exact solution for a cantilever beam with point load at its free end. The results show that the exact integration tends to zero with large α and that remarkable results are achieved by making matrix $\underset{\approx}{K}_2$ singular by means of reduced integration.

A similar approach has been successful in applications in the field of incompressible materials (20). A constraint has been imposed on the volumetric strain which is zero for such materials.

The constraints can also be introduced in an approximate manner by imposing them at a limited number of points within the element. This introduces additional equations which allow us to eliminate some of the element variables at element level. It is clear that this procedure can violate even the C_0 continuity. A patch test is necessary to assure that the element is converging.

Let us demonstrate the method again on the beam problem. The displacements and slopes are approximated independently as in equation (9.5). The constraint (9.6) is enforced exactly at two internal points with coordinates r_1, r_2. The beam has an additional node in the middle (Fig. 9.1.)

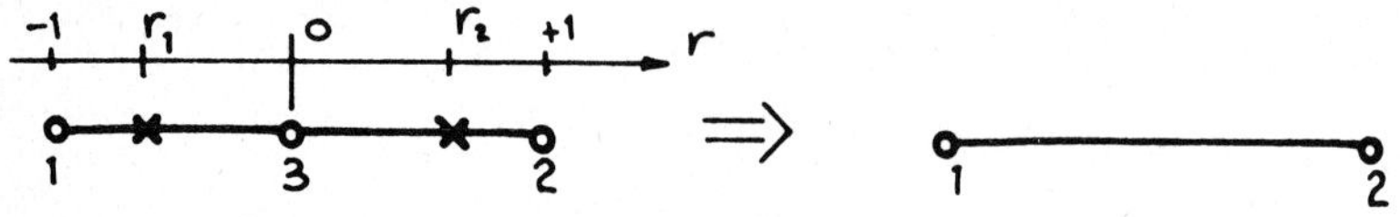

Figure 9.1

Introducing the shape functions (9.5) into the constraint (9.6) gives

$$\sum_{i=1}^{3} \frac{dN_i}{dx}(r_1)\, u_i - \sum_{i=1}^{3} N_i(r_1)\, \theta_i = 0 \,,$$

$$\sum_{i=1}^{3} \frac{dN_i}{dx}(r_2)\, u_i - \sum_{i=1}^{3} N_i(r_2)\, \theta_i = 0 \,. \tag{9.8}$$

The two parameters u_3 and θ_3 of the internal node can be eliminated from the original shape functions using the relationship (9.8).

It is of interest to realise that if the points r_1 and r_2 are Gauss points, then the stiffness matrix obtained after elimination of the internal node is precisely that obtained by Hermitian polynomials (slope and displacement continuity).

This method has been used for design of a series of plate and shell elements. Since a relatively large number of internal constraint points are now used, the elimination process is very complicated and often requires matrix inversion at element level. Clearly, the question of economy of such elements has to be considered.

An interesting example is the famous semi-loof element of Irons (14) (Fig. 9.2).

The original element has 27 degrees of freedom. This number is reduced by imposing 10 internal constrains and one integral constraint to an element with 16 degrees of freedom. The integral constraint requires that a condition in the mean on the total perimeter be satisfied.

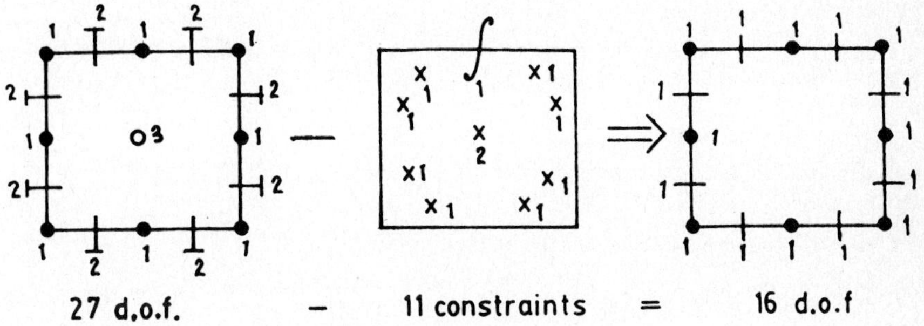

Figure 9.2

The introduction of constrains into a variational formulation using Lagrange multipliers is a well-known method. It can also be used in the finite element formulation. The disadvantage is that additional unknowns are introduced which in the case of a large number of constraints may be prohibitive.

The Lagrangian multipliers often have physical meaning, and lead to so-called mixed variational principles and hybrid formulations. This approach allows to define compatibility inside the element and to enforce inter-element compatibility by Lagrange multipliers defined on the inter-element boundary. The interested reader can find a detailed explanation elsewhere (11).

NONLINEAR PROBLEMS - STATIC ANALYSIS

So far, we have applied the finite element method to linear problems. The stiffness matrix and the force vector in equation (3.10) were constant. In the mathematical concept, the differential equation (4.1) was linear and the functional (4.2) had the standard quadratic form.

The application of the finite element method is not limited to linear problems. It can equally well be used for the solution of problems involving very large deformations of elastic bodies, and can include nonlinear material properties (elastic-plastic analysis) nonlinear field problems (diffusion, potential flow) and time dependant problems (creep, dynamic problems, impact).

Let us demonstrate the general procedure for the solution of nonlinear problems in the case of nonlinear elasticity.

The equation (3.9)

$$\underset{\sim}{f}_i = \int_V \underset{\approx}{B}^T \underset{\sim}{\sigma} \, dV \qquad (10.1)$$

is still valid.

The strain-displacement relationship (3.3) is now nonlinear

$$\underset{\sim}{\varepsilon} = \underset{\sim}{\varepsilon}(\underset{\sim}{u}) \ , \tag{10.2}$$

as is also the stress-strain relationship (3.4)

$$\underset{\sim}{\sigma} = \underset{\sim}{\sigma}(\underset{\sim}{\varepsilon}) \ . \tag{10.3}$$

Let us sumplify the problem and assume that the strain $\underset{\sim}{\varepsilon}$ is small and the material is linear. Then equation (3.4) remains in the original form

$$\underset{\sim}{\sigma} = \underset{\approx}{D} \, \underset{\sim}{\varepsilon} \ . \tag{10.4}$$

This assumption is true for most structural problems. Large displacements occur due to rotations without causing large strains.

Let us rewrite equation (10.1) in the following form

$$\underset{\sim}{F}_i(\underset{\sim}{u}_j) = \int_V \underset{\approx}{B}^T(\underset{\sim}{u}_j) \cdot \underset{\sim}{\sigma} \cdot dV - \underset{\sim}{f}_i(\underset{\sim}{u}_j) = 0 \tag{10.5}$$

There is no general method which would solve equation (10.5) exactly. Numerical methods must be used, and so the solution can be obtained only to a prescribed degree of accuracy. Such a solution is not unique, and the one obtained depends on the initial estimate of the solution and on the increment size used in the numerical method.

The most commonly used solution scheme in finite element programs is the combination of the incremental loading procedure and the Newton-Raphson iterative procedure (9).

<u>The Newton-Raphson method</u> is used as follows: The equation (10.5) is written for <u>the $(n + 1)^{th}$ iteration</u> step assuming that the solution for the step n is known. Then a curtailed Taylor series gives an improved solution

$$\underset{\sim}{F}_i(\underset{\sim}{u}_j^{(n+1)}) = \underset{\sim}{F}_i(\underset{\sim}{u}_j^{(n)}) + \left(\frac{\partial \underset{\sim}{F}_i}{\partial \underset{\sim}{u}_j}\right)_n \Delta\underset{\sim}{u}_j^{(n)} = 0$$

The increment $\Delta\underset{\sim}{u}_j^{(n)}$ is calculated from the linear system

$$\underset{\approx}{K}_T \cdot \Delta\underset{\sim}{u}_j^{(n)} = \underset{\sim}{F}_i(\underset{\sim}{u}_j^{(n)}) \ , \tag{10.6}$$

where $\underset{\approx}{K}_T = (\partial \underset{\sim}{F}_i / \partial \underset{\sim}{u}_j)_n$ is called a tangential matrix (or Jacobian matrix).

The incremental loading procedure is a direct iteration method. The equation (10.1) is written in the form

$$\underset{\approx}{K}(\underset{\sim}{u}_j)\,\underset{\sim}{u}_i = \underset{\sim}{f}_i \ . \tag{10.7}$$

If the solution at the iteration step n is $\underset{\sim}{u}_i^{(n)}$ the improved solution $\underset{\sim}{u}_i^{(n+1)}$ can be found from equation (10.7), since after inserting $\underset{\sim}{u}_i^{(n)}$ into $\underset{\approx}{K}$, it becomes a linear system of equations:

$$\underset{\approx}{K}(\underset{\sim}{u}_i^{(n)})\ \underset{\sim}{u}_i^{(n+1)} = \underset{\sim}{f}_i^{(n)} . \tag{10.8}$$

Both the Newton-Raphson method and the incremental loading method can be used independently for the complete solution. They both have serious drawbacks. The incremental loading method often fails to converge and leads to accumulation of errors (see Fig. 10.1).

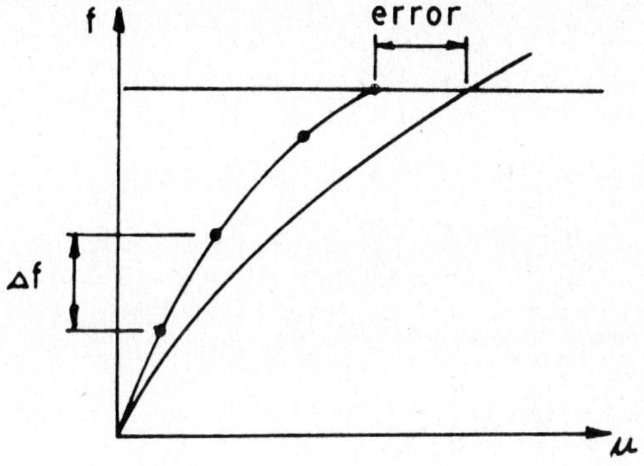

Figure 10.1

If the Newton-Raphson method is applied directly to the system (10.1) it may well diverge whenever the difference between the initial solution $u_i^{(o)}$ (found from the linearized equation) and the final solution $u_i^{(n)}$ is large. Also a quite different solution than that required may be obtained because of nonlinearity.

A combination of both procedures seems to be the best and most economical way of solving equation (10.1).

The loads are applied in increments which are sufficiently small so that we may assume a linear solution. The result is then corrected by the Newton-Raphson method. The corrected solution is subsequently used as the starting point for the next incremental step (Fig. 10.2) (9).

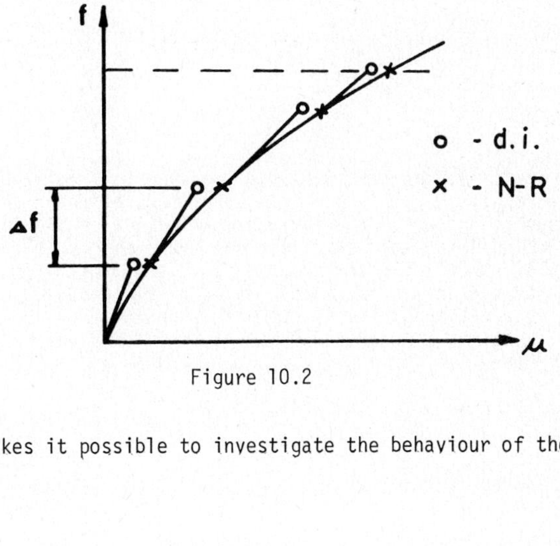

Figure 10.2

This procedure makes it possible to investigate the behaviour of the system during loading. The intermediate states can be tested for stability, bifurcation points can be found and, if necessary, multi-valued solutions may be obtained.

Calculation of the Jacobian matrix in equation (10.7) is computationally expensive. Most of the programs use various modifications such as the modified Newton-Raphson scheme (9) or BFGS-method (22). The trade-off is that more iteration steps are required for convergence than if the full Newton-Raphson method is used. The more non-linear the problem, the smaller the saving (13).

Let us now derive the tangential matrix $\underset{\approx}{K}_T$, which is required for the Newton-Raphson solution process (10.7) in the case of the large displacement analysis.

Equation (10.1) is first written in an incremental form

$$\Delta \underset{\sim}{F} = \int_V \Delta \underset{\approx}{B}^T \underset{\sim}{\sigma}\, dV + \int_V \underset{\approx}{B}^T \Delta \underset{\sim}{\sigma}\, dV - \Delta \underset{\sim}{f}_i = 0 \tag{10.9}$$

For small strain problems we have from (10.4)

$$\Delta \underset{\sim}{\sigma} = \underset{\approx}{D} \,.\, \Delta \underset{\sim}{\varepsilon} = \underset{\approx}{D} . \underset{\approx}{B} . \Delta \underset{\sim}{u}_i \ , \tag{10.10}$$

so that

$$\Delta \underset{\sim}{F} = \int_V \Delta \underset{\approx}{B}^T \underset{\sim}{\sigma}\, dV + \int_V \underset{\approx}{B}^T \underset{\approx}{D}\, \underset{\approx}{B}\, dV\, \Delta \underset{\sim}{u}_i - \Delta \underset{\sim}{f}_i = 0 \,. \tag{10.11}$$

The first matrix in equation (10.11) can be written as

$$\frac{\partial \underset{\approx}{B}^T}{\partial \underset{\sim}{u}_i} \Delta \underset{\sim}{u}_i \ ,$$

where the matrix $\frac{\partial \underset{\approx}{B}^T}{\partial \underset{\sim}{u}_i} = \underset{\approx}{K}_\sigma$

is called the initial stress matrix. The second matrix in equation (10.11) can be resolved into a linear and non-linear part $\underset{\approx}{K}_L + \underset{\approx}{K}_N$. The matrix $\underset{\approx}{K}_L$ is called the small displacement matrix, and $\underset{\approx}{K}_N$ is called the initial displacement matrix. The increment in the displacement-dependant load vector $\Delta \underset{\sim}{f}_i$ can also be expressed as $\Delta \underset{\sim}{f}_i = \underset{\approx}{R} . \Delta \underset{\sim}{u}_i$, where the matrix $\underset{\approx}{R}$ is called the initial-load matrix. If loads do not depend on displacements, the matrix $\underset{\approx}{R}$ is zero.

The tangential matrix $\underset{\approx}{K}_T$ is then

$$\underset{\approx}{K}_T = \underset{\approx}{K}_L + \underset{\approx}{K}_\sigma + \underset{\approx}{K}_N + \underset{\approx}{R} \ . \tag{10.12}$$

The tangential stiffness matrix (10.12) can be used to predict buckling loads and modes in a modal analysis. This can be done at the beginning of the solution process or at any point during the analysis. Hibbit uses the following procedure (13): At some point in the analysis, the stiffness matrix is stored. After one or more loading increments the stiffness of the structure is stored again. At this point, the change in stiffness associated with the added loads that would cause collapse is evaluated. The assumption is that the change in stiffness will be proportional to the change in added loads. This is a valid assumption for stiff shells or beams.

The magnification factor λ for the collapse is calculated from the eigen-problem

$$(\underset{\approx}{K} + \lambda \Delta \underset{\approx}{K})\, \underset{\sim}{v} = 0 \ , \tag{10.13}$$

where $\underset{\approx}{K}$ is the stiffness under the "dead" load

$\Delta\underset{\approx}{K}$ is the change in stiffness caused by the "live" loads

λ is the "live" load magnification factor (the eigen-value)

$\underset{\sim}{v}$ is the buckling mode shape (the eigen-vector).

There are two possible ways to proceed after each load increment.

The Total Lagrangian formulation relates the displacements throughout all load increments to the original local (or global system). The transformation matrices remain the same, the nodal points and element geometry are not changed.

The updated Lagrangian formulation recalculates the nodal points and the element geometry after each load increment. The transformation matrix $\underset{\approx}{T}$ has to be updated because of the changed angle α . The advantage of this approach lies in the decrease in the nonlinear terms. If load increments are sufficiently small, the initial displacement matrix can be neglected (23).

NONLINEAR PROBLEMS - FLUID FLOW

In general, the finite element formulation of fluid flow proceeds along the lines shown in the Section - Mathematical Concept. The governing differential equation is inserted into the residual integral and the Galerkin method is used to derive the matrix equation. The solution of this equation is more difficult because the system matrix is non-symmetrical, non-linear and often ill-conditioned so that instability in the iteration procedure is much more likely to occur.

Let us briefly demonstrate the formulation process on an example of two-dimensional plane flow of an incompressible viscid fluid as it was used in (24). The governing differential equations are the Navier-Stoke equations.

$$\rho u\frac{\partial u}{\partial x} + \rho v\frac{\partial u}{\partial y} - \rho f_1 - \frac{\partial \tau_{11}}{\partial x} - \frac{\partial \tau_{12}}{\partial y} = 0 ,$$

$$\rho u\frac{\partial v}{\partial x} + \rho v\frac{\partial v}{\partial y} - \rho f_2 - \frac{\partial \tau_{21}}{\partial x} - \frac{\partial \tau_{22}}{\partial y} = 0 , \qquad (11.1)$$

where $\tau_{ij} = -p\delta_{ij} + \mu(\frac{\partial u}{\partial x_j} + \frac{\partial v}{\partial x_i})$, $x_1 = x$, $x_2 = y$,

where u, v are the velocity components in the x and y direction, p is the pressure, ρ is the density, f is the body force, μ is the viscosity, τ_{ij} is the stress tensor and δ_{ij} the Kronecker delta.

The condition of incompressibility is given in the form

$$\frac{\partial u}{\partial x} + \frac{\partial v}{\partial y} = 0 . \qquad (11.2)$$

To complete the problem a sufficient set of boundary conditions has to be specified. For example, on part of the boundary the velocity components can be prescribed, on the other part the pressure may be specified.

The region of interest is then divided into finite elements. In contrast with the formulation used in solid mechanics, the fluid elements are fixed and cannot change shape. This implies that the Eulerian description of the fluid motion must be used.

The unknown parameters are the two velocity components u_i, v_i and the pressure p_i at each of element nodal points.

Within each element, the velocity and pressure fields are approximated by shape functions with u_i, v_i, p_i ; $i = 1,2, \ldots, n$ at the nodal points as parameters:

$$\begin{bmatrix} u \\ v \end{bmatrix} = \sum_{i=1}^{n} N_i \begin{bmatrix} u_i \\ v_i \end{bmatrix} ,$$

$$p = \sum_{i=1}^{m} M_i \, p_i \tag{11.3}$$

The different indices n and m indicate that in general the velocity and pressure functions do not have to use the same number of nodal points for definition.

The approximations (11.3) are inserted into equations (11.1) and (11.2) and subsequently made orthogonal to the interpolation functions N_i, M_i from equation (11.3).

After carrying out these operations and removing the nodal point values from integrals, a system of nonlinear algebraic equations is obtained

$$(\underset{\approx}{K}_L + \underset{\approx}{K}_N(\underset{\sim}{u}_i)\,)\,\underset{\sim}{u}_i = \underset{\sim}{f}_i \ , \tag{11.4}$$

where

$$\underset{\sim}{u}_i = \begin{bmatrix} u_i \\ v_i \\ p_i \end{bmatrix} .$$

The method of introduction of boundary conditions into equation (11.4) in terms of velocities and pressure is the same as that used in solid mechanics,

The matrices in equation (11.4) will be banded for a proper numbering of nodal points.

The system (11.4) is solved in (24) by the iterative method from

$$(\underset{\approx}{K}_L + \underset{\approx}{K}_N(\underset{\sim}{u}_i^{(n)})\,)\,\underset{\sim}{u}_i^{(n+1)} = \underset{\sim}{f}_i \ . \tag{11.5}$$

The matrix $\underset{\approx}{K}$ is not symmetric. In (24), the transformation to a symmetric system was not successful, so that the asymmetric form has been solved. The convergence behaviour depends on the Reynolds number

$$Re = \frac{v_{ref}.h}{\mu} < C \ . \tag{11.6}$$

This indicates that for a particular mesh size there is a limit on viscosity for convergence. The mesh size must decrease with decreasing viscosity. The value of the constant C in expression (11.6) is unfortunately not known, so that the user must rely on computer experiments.

Other formulations of the above problem are possible, some of which use the stream function (C_1 continuity is required) or incorporate the incompressibility condition by a penalty function.

EDUCATION OF ENGINEERS

One important aspect of engineering education is to demonstrate to the students that mathematics cannot be applied to real objects. A mathematical model has to be constructed for a given task, and only this idealized object can be mathematically analysed. To construct such a model sometimes requires extraordinary skill, intuition and experience.

The results from the computer analysis are valid only for the mathematical model used. Extrapolation to reality is a practical problem again requiring experience, imagination and a refined criticism.

In between is the finite element program used as a tool. As in any profession, the deeper the understanding of the tool, the higher the quality of the products.

These three steps - modelling, solving, extrapolation of results - are closely interrelated as can be clearly demonstrated in two case studies where a twenty-node three-dimensional solid element was used.

A machine part, a universal joint, has been modelled using 20-node 3-D elements. One quarter of the hardware was measured manually and nodal point coordinates and element connectivities were input into the computer. By double mirroring the geometry of the complete component was generated. The plotting program MOVIE-BYU was used to check the geometry. Fig. 12.1 shows the model.

The seemingly correct model was submitted to the finite element program. This program also has plotting facilities. A plot from the finite element program was required as a further check before the analysis. Fig. 12.2 shows the model.

We can now see that the model which seemed to be correct is worthless.

There are many edges which are completely wrong. If such model were submitted for solution, completely wrong results would be obtained.

The reason for the errors is simple. Some of the middle nodes were not in the middle of the edges. An isoparametric element approximates 3-node edges by parabolas. To obtain a close approximation to a given curve, the middle node must be at the correct position (in the middle for a circular arc). The errors could not be discovered by the graphics program because straight lines were plotted between the nodes in order to increase speed of display.

The second example is related again to the 20-node solid element. A shaft of a seat-belt retractor mechanism was modelled. This time the model was correct. The structure was analysed for given loads and Mises stresses were plotted in prescribed planes throughout the structure. The result is on Fig. 8.5

It is clear that the results are wrong. A simple analysis by using beam elements reveals this.

What is the cause of this error ? The stiffness matrices are integrated numerically. In this case, 3 x 3 x 3 integration points have been used. The stresses have been calculated in these 27 integration points and extrapolated for contour plotting. As discussed in the Section - Non-Conforming Elements, the correct stresses are obtained only in 2 x 2 x 2 integration points. This means that the contours in Fig. 8.5 have been extrapolated from quite wrong values. The user again did not know enough about the tool. He had a choice of both integration schemes and selected the wrong one.

However, there are codes which calculate stresses in 3 x 3 x 3 integration points for 20-node solids. As a consequence, the results they plot are then incorrect.

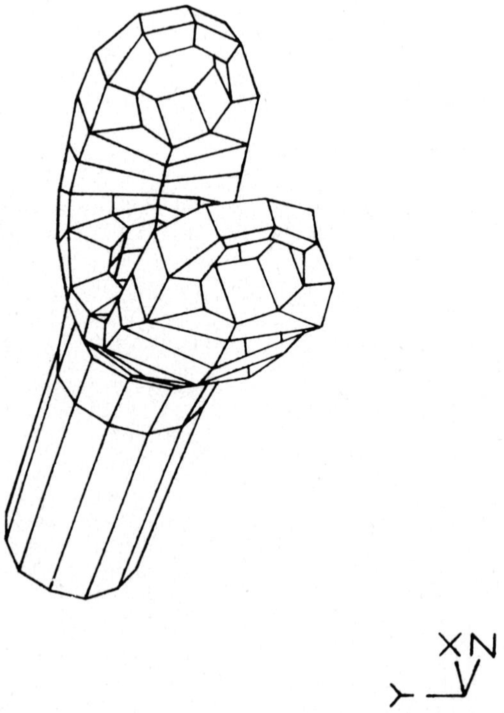

Figure 12.1

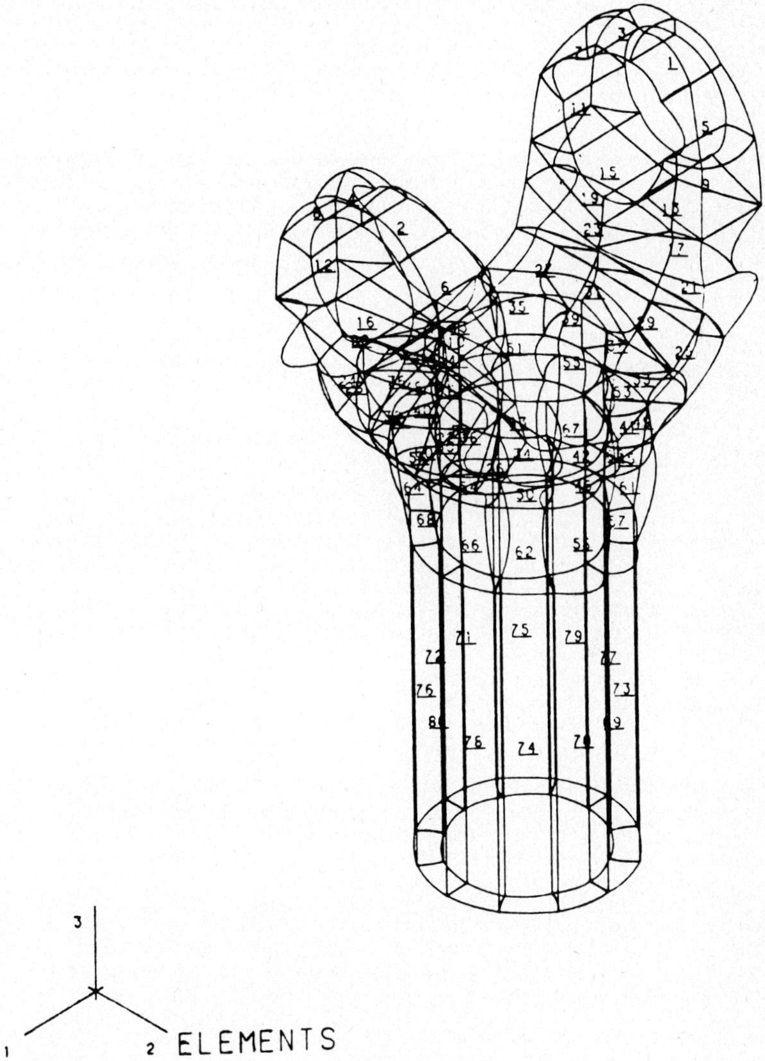

Figure 12.2

The user should be interested in the quality of computer graphics software he is using.

These examples demonstrate quite clearly that the finite element method can be a dangerous weapon in the hands of under-qualified users.

At RMIT, the potential of the finite element method as a practical instrument was recognised several years ago. A group consisting of lecturers from various departments and fatulties was formally established. The group objectives for the next two years have been set (25). They include hardware, software and man-power resources and their appropriate usage in graduate and postgraduate teaching and applied research.

A finite element course was designed and introduced in the second semester of 1980. The course is offered as an optional subject in mathematics to third year degree students in civil, mechanical and aeronautical engineering. It is taught by four members of the group (two applied mathematicians and two engineers - civil and mechanical).

Two finite element programs are used in the course: CAL78 (26) and TWODEPEP (27). CAL78 has three levels of user involvement. At the first level, the user uses only simple matrix operations, such as loading, adding, multiplication and solving. He has to specify all matrices including transformation matrices and solve the problem step by step. The second level allows the user to specify the order of assembly of element matrices into the global stiffness matrix by using an integer matrix. It also calculates automatically stiffness matrices of beams in global coordinates.

The third level is close to the finite element programs as used in engineering practice. Nodal point coordinates, element connectivities, material properties, loads and boundary conditions are specified in a normal way. Plane stress elements from 3 to 8 nodes are available with a different number and position of integration points. The effect of reduced integration and precision on stresses can be studied. Computer time is printed out for each operation so that the user can clearly see the cost difference in using more nodes or more integration points.

The computer graphics program M3D (28) is available for model display. The importance of careful checking of correctness of the model is emphasised again and again.

The program TWODEPEP is used in the second half of the course. It requires a quite different approach. The problem to be solved has to be described by a partial differential equation. The students are briefly informed of the method of weighted residuals on which TWODEPEP is based. Simple heat-transfer and potential fluid flow problems are formulated and solved.

The student assignments must include a discussion about the development of the model, the graphical plot of the model and validation or results by means of any suitable and available method (e.g. handbook formula, experiment, analytical solution of a simplified system).

Fourth year degree students who are working on industrial projects requiring finite element method use the nonlinear program ABAQUS (29). This program, together with mesh generation program FEMGEN and plotting program MOVIE.BYU, represent the most advanced level of finite element application. The mesh is generated on the video-screen from a basic geometric description of the component. A digitizing tablet is available for input of point coordinates from drawings.

When using FEMGEN to generate finite element mesh of a model, the basic geometric parts, such as points, lines, surfaces and bodies, are named. A data base containing the detailed information about the geometry of the component is created and saved. It contains not only coordinates of nodal points and element connectivities, but also coordinates and names of major points and tables with the definitions of the geometric parts.

In the next stage, an ABAQUS input file is created interactively in conversation mode. The communication is by means of FEMGEN point, line, surface and body names, so the user has no knowledge about nodal point or element numbers. This permits the user to generate a correct input file. This is important when such large programs as ABAQUS are used. The computer usage is limited to overnight runs for ABAQUS because of the large number of time-sharing computer users of the system. This means that one turn-around per day is the maximum.

More detailed information about student projects and postgraduate projects can be found in (30) or (31).

SIGFEM is also organising introductory and advanced courses in the finite element method for practising engineers from industry. Practical application of the method again is the main objective.

CONCLUSIONS

The finite element method is today used in so many areas of engineering that it is not physically possible to present all applications in a single paper.

Such diverse problems as elastoplastiticy, creep, phase change, crack propagation, impact, soil mechanics, friction, and many more, can be relatively easily accommodated in the finite element formulation. Most of these features already have been incorporated in major general-purpose finite element computer programs and can be used in a routine way.

The engineer, who uses finite element codes developed by specialists at universities or professional software companies, faces two major difficulties when solving more complex problems (32).

The first difficulty is the cost of computing, including model development. This can become prohibitive, being 10 to 100 times more than for solving simple linear problems. The second is the reliability of the formulation used in the code and the proper interpretation and application of it.

Many novice users of the finite element method become over-confident in the power and capabilities of modern computer, as if the computer could think for them, and solve problems before they clearly understood and formulated the problems.

The approach by some managers who are introducing the method into their companies reflects this dangerous way of thinking. They do not hesitate to spend vast amounts of money for modern computer hardware and sophisticated software, but they neglect basic training of their engineers.

The area of engineering education in modern computer methods, including the finite element method, is an exicting opportunity for co-operation between mathematicians and engineers. The use of modern methods and technology would seem to require an integrated presentation of theory and practice.

ACKNOWLEDGEMENT

The author would like to acknowledge the value of co-operation of SIGFEM members in helping to develop new ideas in education and applied research at RMIT. It would have been difficult to write this paper without this background.

GLOSSARY OF SYMBOLS

Symbol	Meaning
A	Matrix *or* Area
a	Unknown coefficients
B	Strain displacement matrix
C	Continuity *or* Constant
D	Material matrix
d	Deflection value
E	Constant
e	Subscript (refers to external)
F	Force
f	Force component
g	Subscript (refers to global)
h	"Size" of element
I	Functional (Integral of total Potential Energy)
i	Subscript (refers to internal)
J	Jacobian matrix
j	Subscript
$\underset{\approx}{K}$	Local element stiffness matrix
k	Element of matrix $\underset{\approx}{K}$
L	An operator
l	Subscript (refers to local)
M	Interpolation function
m	Number of derivatives
N	Shape or interpolation functions
n	Counter
P	Potential energy
p	Order of polynomial *or* Pressure
Q	A function
R	Initial load matrix
Re	Reynolds number
r	Radial coordinate

Symbol	Meaning
s	Coordinate
T	Transformation matrix
U	Strain energy
u	Displacement component *or* Temperature value *or* Velocity component
V	Volume *or* Functional
v	Displacement component *or* Temperature value *or* Velocity component
W	Work
w	Weighting functions
X	Coordinate axis (global)
x	Coordinate axis (local)
Y	Coordinate axis (global)
y	Coordinate axis (local)

Greek Alphabet

α	Angle
β	Angle
Γ	Boundary
Δ	Small change in value
δ	Virtual displacement *or* Kronecker delta
ε	Strain
λ	Magnification factor
μ	Viscosity
ρ	Density
Σ	Sum
σ	Stress
τ	Stress tensor
ϕ	Trial functions
Ω	Domain

Other Symbols

$\sim$	Indicates a vector
$\approx$	Indicates a matrix

REFERENCES

(1) Poincare, E., "Science and Method", Dover 1952.

(2) Courant, R., "Variational Method for the Solution of Problems of Equilibrium and Vibration", Bull. Am. Math. Soc., Vol. 49, 1-23, 1945.

(3) Hrennikoff, A., "Solution of Problems in Elasticity by the Framework Method", J.Appl.Mech., Vol. 8, A169-A175, 1941.

(4) Argyris, J.R., "Energy Theorems and Structural Analysis", Aircraft Eng., Vol.26 (1954), also Butterworth 1960.

(5) Turner, M.J., Clough, R.W., Martin, H.C., Topp, L.J., "Stiffness and Deflection Analysis of Complex Structures", J.Aero.Sci., Vol. 23, 805-23, 1956.

(6) Clough, R.W., "The Finite Element in Plane Stress Analysis", Proc. 2nd A.S.C.E. Conf. on Electronic Computation, Pittsburgh, Pa., Sept. 1960.

(7) Ames, W.F., "Numerical Methods for Partial Differential Equations", First Edition, Nelson, 1969.

(8) Ames, W.F., "Numerical Methods for Partial Differential Equations", Second Edition, Academic Press, 1977.

(9) Oden, J.T., "Finite Elements of Nonlinear Continua", McGraw-Hill, 1972.

(10) Strang, G., Fix, G.J., "An Analysis of the Finite Element Method", Prentice-Hall, 1973.

(11) Zienkiewicz, O.C., "The Finite Element Method", McGraw-Hill, 1977.

(12) Bathe, K.J., Wilson, E.L., "Numerical Methods in Finite Element Analysis", Prentice Hall, 1976.

(13) Hibbit, H.D., Karlsson, B.I., Sorensen, E.P., "ABAQUS - A Finite Element Code for Nonlinear Dynamic Analysis", Theory Manual, 1980.

(14) Irons, B., Ahmad, S., "Techniques of Finite Elements", John Wiley & Sons, 1980.

(15) Mitchell, A.R., Wait, R., "The Finite Element Method in Partial Differential Equations", John Wiley & Sons, 1977.

(16) Wilson, E.L., Taylor, R.L., Doherty, W.P., Ghaboussi, J., "Incompatible Displacement Models", Num. & Comp. Methods in Structural Mechanics, pp. 43-51, Academic Press, 1973.

(17) Taylor, R.L., Beresford, P.J., Wilson, E.L., "A Non-Conforming Element for Stress Analysis", Int. J. Num. Meth. Eng., Vol. 10, pp. 1211-1220, 1976.

(18) Barlow, J., "Optional Stress Locations in Finite Element Models", Int. J. Num. Meth. Eng., Vol. 10, pp. 243-251, 1976.

(19) Babuska, I., "Finite Element Method with Penalty", Rept. BN-910, University of Maryland.

(20) Oden, J.T., Kikuchi, N., "Finite Element Methods for Constrained Problems in Elasticity", TICOM Report, The University of Texas, Austin, 1981.

(21) Irons, B.M., "The Semiloof Shell Element", pp. 197-222 of "Finite Elements for Thin Shells and Curved Members", Wiley & Sons, 1976.

(22) Bathe, K.J., "ADINA - A Finite Element Program for Automatic Dynamic Incremental Nonlinear Analysis", Report 82448-1, MIT, Cambridge, 1978.

(23) Murray, D.W., Wilson, E.L., "Finite Element Post Buckling Analysis of Thin Elastic Plates", Proc. 2nd Conf. Matrix Methods in Struct. Mechanics, AFFDL, Oct. 1968.

(24) Garthing, D.K., "Texas Fluid Analysis Program - User Manual", TICOM Report 75-2, Austin, Texas, 1974.

(25) "Finite Element Methods at RMIT" - SIGFEM Report 001, RMIT, Melbourne, 1979.

(26) Wilson, E.L., "CAL78 - Computer Analysis Language", University of Berkeley, 1978.

(27) "TWODEPEP - User Manual", IMSL, Houston, 1980.

(28) Trinder, S., "M3D - Mesh Display and Editing Program", SIGFEM Report 005, RMIT, Melbourne, 1980.

(29) Hibbitt, H.D., Karlsson, B.I., Sorensen, E.P., "ABAQUS - User Manual, 1981.

(30) Tomas, J.A., "SIGFEM & Application Case Studies in Industry", ACADS International Symposium, ACADS Publication No. U214, pp. 150-154, Melbourne, 1980.

(31) Robinson, K.J, Tomas, J.A., Trinder, S.M., "Applied Research for Industry Using ADINA", ADINA Conference, MIT, 1981.

(32) Oden, J.T., Bathe, K.J., "A Commentary on Computational Mechanics", Applied Mechanics Review, Vol. 31, No. 8, 1978, pp. 1053-1058.

Computational Techniques for Differential Equations
J. Noye (Editor)

AN INTRODUCTION TO THE BOUNDARY ELEMENT METHOD

LEIGH WARDLE
CSIRO Division of Applied Geomechanics, Melbourne, Australia

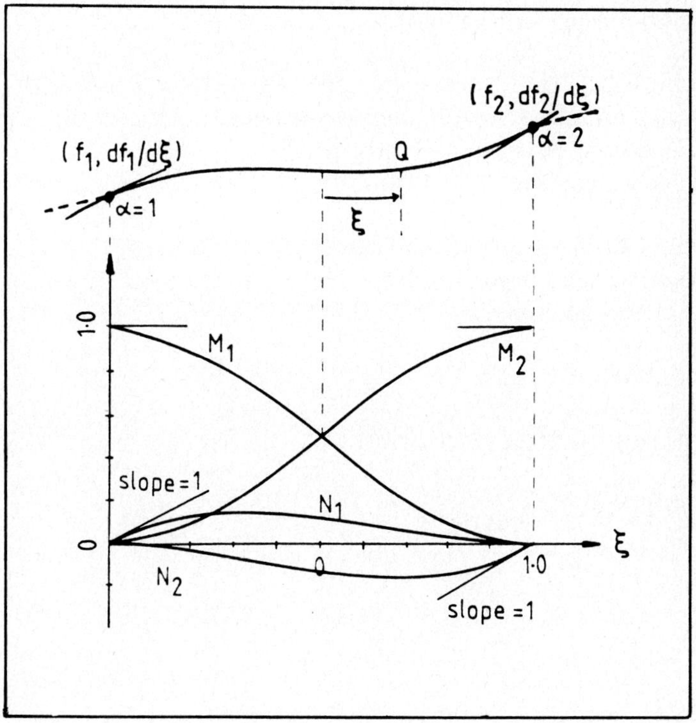

CONTENTS

Reprinted from
Numerical Solutions of Partial Differential Equations
J. Noye (Editor)

AN INTRODUCTION TO THE BOUNDARY ELEMENT METHOD

L.J. Wardle

Commonwealth Scientific and Industrial Research Organization
Division of Applied Geomechanics
Melbourne Australia

The problem of solving a partial differential equation within a given domain can be transformed into one solving an equivalent integral equation on the boundary of the domain. This reformulation is particularly advantageous for numerical applications as the interior of the domain does not have to be discretized. Alternative methods of formulating and solving the integral equations are described. Techniques for considering non-homogeneous and non-linear problems are discussed.

1. INTRODUCTION

This paper introduces the mathematical background to boundary integral equation methods i.e., methods by which the problem of the solution of a partial differential equation valid in a given domain is recast into the solution of an integral equation which applies only to the boundary of the domain. The principal advantage of such a reformulation is that the dimensionality of the problem is reduced by one. For example, a two-dimensional partial differential equation is replaced by a one-dimensional integral equation. In practical terms this means that two-dimensional problems involve discretization into line segments on the boundary, in contrast to finite element and finite difference procedures which require meshes over the plane domain area within the boundary (Fig. 1). For problems involving an infinite domain the boundary integral formulation is particularly advantageous because the behaviour at infinity is usually automatically included without having to discretize an artificial 'remote' boundary as with other methods.

In this paper details are given of various alternative boundary integral equation formulations. Having arrived at an integral equation there are many methods of numerical solution possible. For example, for a rectangular region it may be possible to express the unknown boundary quantities in terms of Fourier series or power series. For arbitrary geometries, however, resort must be made to numerical methods that allow the boundary geometry of the problem to discretized into 'elements' over which the boundary variables have a given polynomial order of variation, i.e. we use a finite element method on the boundary surface. To contrast this procedure with the conventional domain-type finite element method, the boundary version is now commonly called *the boundary element method*.

For many problems, the boundary element method possesses a similar level of generality to the finite element method, e.g. no limitations on the shape of the boundary or on the connectivity of the domain it encloses.

The use of integral equation formulations for partial differential equations dates back to the mid nineteenth century investigations involving the representation of harmonic functions (i.e. solutions to Laplace's equation) by single-layer and double-layer potentials. The existence of solutions to such equations was first demonstrated by Fredholm [1] using a discretization procedure. However, the use

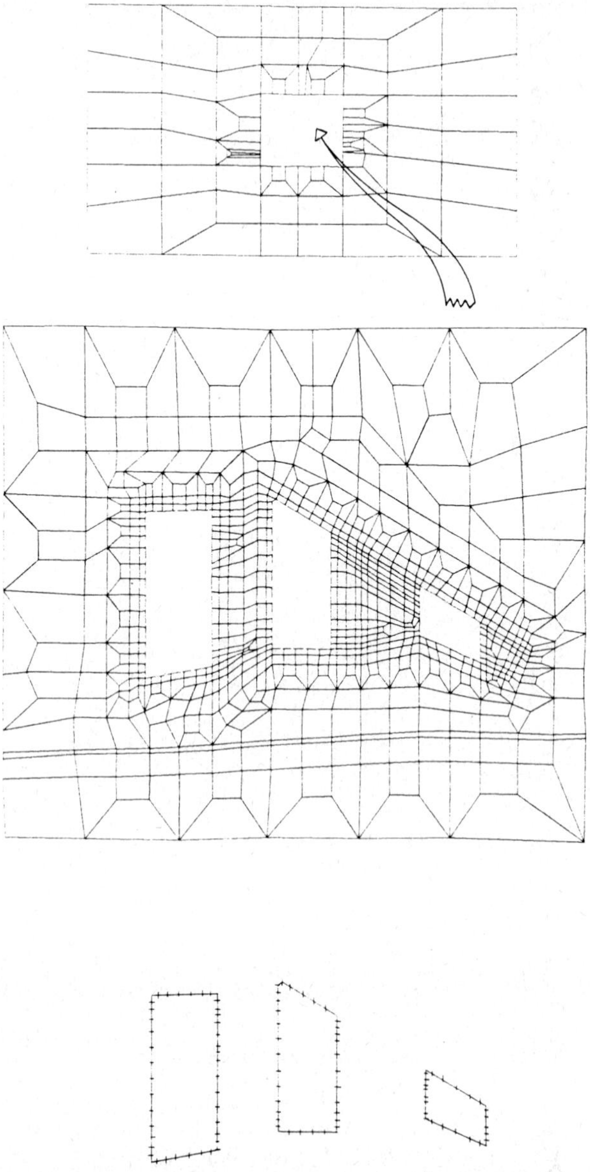

Figure 1
Boundary Element and Finite Element Meshes for Same Domain.

of a discretization procedure to actually construct solutions did not become feasible until the advent of electronic digital computers (circa 1960).

In this paper, Laplace's equation has been chosen as an example to demonstrate the basic features of boundary integral formulations and associated numerical procedures for their solution. Laplace's equation arises in a wide range of physical phenomena such as gravitational fields, electrostatic fields, steady-state heat conduction, incompressible flow, torsion and so on. For completeness, all of the essential classical results relating to surface potential representation of harmonic functions are introduced in Section 2. Using these results a variety of alternative boundary integral equations can be formulated.

In Section 3 methods used to obtain numerical solutions to boundary integral equations are illustrated by some specific examples. Section 4 gives an overview of how the basic methods can be extended to treat Poisson's equation, non-homogeneous material properties and non-linear behaviour. Section 5 discusses how the techniques have been used for the solution of other partial differential equations.

2. INTEGRAL EQUATION FORMULATIONS FOR LAPLACE'S EQUATION

To illustrate the basic features of boundary integral formulations consider Laplace's equation

$$\nabla^2\phi = 0 \qquad (2.1)$$

in some domain V. (Fig. 2).

The boundary conditions on the surface S are of two types

$$\phi(P) = f(P) \qquad \textit{Dirichlet conditions} \qquad (2.2)$$

or

$$\frac{\partial\phi}{\partial n}(P) = g(P) \qquad \textit{Neumann conditions} \qquad (2.3)$$

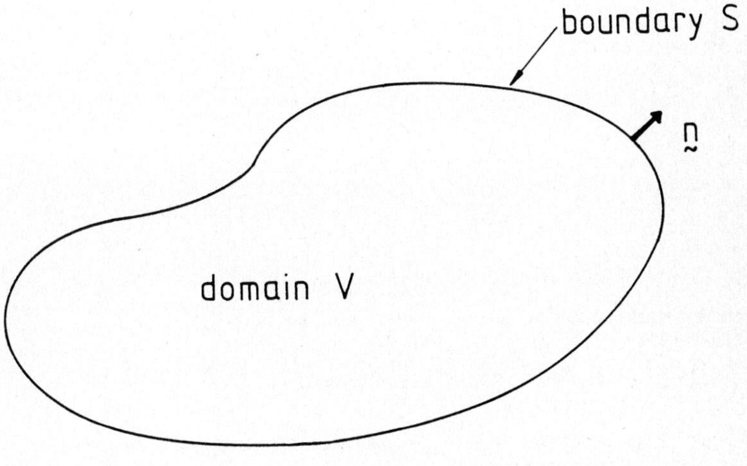

Figure 2
Definitions

2.1 Fundamental Solutions

The starting point of an equivalent integral equation formulation is to consider the appropriate fundamental singular solution for Laplace's equation, i.e. a solution of Laplace's equation that depends merely on the distance r of the point p from a fixed point q.

For such functions the Laplacian operator in three dimensions is $\partial^2/\partial r^2 + (2/r)\,\partial/\partial r$ and in two dimensions is $\partial^2/\partial r^2 + (1/r)\,\partial/\partial r$. It is easily verified by direct substitution that the functions

$$1/r(p,q) \qquad \text{in three dimensions}$$

and

$$\log[1/r(p,q)] \qquad \text{in two dimensions}$$

satisfy the respective Laplace's equation for $r(p,q)\neq 0$, where $r(p,q)$ is the scalar distance between points p and q.

Furthermore, if the point $r(p,q) = 0$ is included it can be shown that for three dimensions

$$\nabla^2 \frac{1}{r(p,q)} = -4\pi\delta(p,q) \qquad (2.4)$$

and for two dimensions

$$\nabla^2 \log[1/r(p,q)] = -2\pi\delta(p,q) \qquad (2.5)$$

where the Dirac delta function is defined to have the following properties

$$\delta(p,q) = 0, \qquad \underset{\sim}{x}_p \neq \underset{\sim}{x}_q$$
$$\int_V \rho(p)\delta(p,q)dV(q) = \rho(p) \qquad (2.6)$$

we infer the function

$$\psi(p,q) = \frac{1}{4\pi r(p,q)} \qquad (2.7)$$

to be the fundamental solution (the factor 4π is included for convenience) for three dimensions. The fundamental solution in two dimensions can be taken to be

$$\psi(p,q) = \frac{1}{2\pi} \log \left(\frac{1}{r(p,q)}\right) \qquad (2.8)$$

In the remainder of this work $\psi(p,q)$ denotes the fundamental solution defined by either equations (2.7) or (2.8) appropriate to the dimensionality of a given problem. The use of this notation in most cases permits two and three dimensions to be treated simultaneously.

2.2 Single and Double Layer Potentials

A continuous distribution of simple sources extending over the surface S and of surface density $\sigma(Q^{\dagger})$ at $Q \in S$ generates the potential

$$\phi(p) = \int_S \sigma(Q)\psi(p,Q)dS(Q) \qquad (2.9)$$

and is called a single-layer (or, sometimes, simple-layer) potential.

† In all that follows the symbols p, q will represent the field points in the domain V; the symbols P, Q will represent the field points on the surface S of the domain.

From the single-layer distribution we can generate another distribution by considering two surfaces separated by a small distance ε. Assume each surface has a distribution of attraction of equal magnitude but of opposite sign for neighbouring points and take the magnitude of the attraction to be inversely proportional to ε. The potential associated with the point $Q(\underset{\sim}{x})$ and the point $Q(\underset{\sim}{x} + d\underset{\sim}{x})$ is given by the limit

$$\lim_{\varepsilon\to 0} \left\{ \frac{1}{\varepsilon} \left[(p,Q+\varepsilon) - (p,Q) \right] \right\} \tag{2.10}$$

which can be seen to be the derivative of the potential $\psi(p,Q)$ in the direction normal to the surface S. Integrating over all the surface points gives the double-layer potential

$$\phi(p) = \int_S \mu(Q) \frac{d\psi(p,Q)}{dn(Q)} dS(Q) \tag{2.11}$$

The function $\mu(Q)$ is the surface density of the double layer potential.

2.3 Continuity of the Surface Potentials

The single-layer and double-layer potentials obey Laplace's equation everywhere except on the surface S, in which case the integrands involve singularities when $r(p,Q) = 0$. Briefly, to investigate the behaviour of the surface potentials, the boundary is broken into two surfaces : a small disc tangent to the surface at a point $P(\underset{\sim}{x})$ and the remaining surface which will contain no singularities because $P(\underset{\sim}{x}) \neq Q(\underset{\sim}{x})$.

Provided that σ is Holder continuous at P, the single layer potential is continuous when crossing S.[2]. The double-layer potential is discontinuous when crossing S. The limiting form of equation (2.11) as $p(\underset{\sim}{x}) \to P(\underset{\sim}{x})$ from the inside is

$$\phi^+(P) = -\frac{1}{2}\mu(P) + \int_S \mu(Q) \frac{d\psi(P,Q)}{dn(Q)} dS(Q) \tag{2.12}$$

and from the outside is

$$\phi^-(P) = \frac{1}{2}\mu(P) + \int_S \mu(Q) \frac{d\psi(P,Q)}{dn(Q)} dS(Q) \tag{2.13}$$

Subtracting equation (2.12) from (2.13) shows that the double-layer potential has a discontinuity or jump of $-\mu(P)$ as the point $p(\underset{\sim}{x})$ passes from outside the region to inside the region.

2.4 Indirect Boundary Integral Equations

A variety of boundary integral equations can be obtained by considering single-layer and double-layer distributions separately.

For example, for the Drichlet boundary conditions (ϕ prescribed on S), the unknown function $\phi(p)$ may be expressed solely as a double-layer potential of unknown density $\mu(Q)$.

$$\phi(p) = \int_S \mu(Q) \frac{d\psi(p,Q)}{dn(Q)} dS(Q) \tag{2.14}$$

Using the limiting behaviour of the double-layer potential as $p(\underset{\sim}{x})$ approaches $P(\underset{\sim}{x})$ from inside S (equation 2.12) gives

$$\phi(P) = -\frac{1}{2}\mu(P) + \int_S \mu(Q) \frac{d\psi(P,Q)}{dn(Q)} dS(Q) \tag{2.15}$$

Equation (2.15) is a Fredholm equation of the second kind with $\mu(Q)$ being the only unknown.

To obtain an integral equation for Neumann boundary conditions ($d\phi/dn$ prescribed on S), the unknown function $\phi(p)$ can be expressed solely as a single-layer potential with unknown density $\sigma(Q)$

$$\phi(p) = \int_S \sigma(Q)\psi(p,Q)dS(Q) \qquad (2.16)$$

For points $p(\underset{\sim}{x})\varepsilon V$ we may take the derivative of $\phi(p)$ in the ξ-direction

$$\frac{d\phi}{d\xi} = \int_S \sigma(Q)\frac{d\psi(p,Q)}{d\xi(p)}dS(Q) \qquad (2.17)$$

If ξ is taken to be the direction of the outer normal to S as $p(\underset{\sim}{x})$ is taken to $P(\underset{\sim}{x})$ then as $d\phi/d\xi$ is continuous up to and including S,

$$\frac{d\phi}{dn}(P) = -\frac{1}{2}\sigma(P) + \int_S \sigma(Q)\frac{d\psi(P,Q)}{dn(P)}\,dS(Q) \qquad (2.18)$$

which is a Fredholm equation of the second kind with $d\phi/dn(P)$ known and $\sigma(P)$ unknown.

The two integral equations (2.15) and (2.18) are not the only possible ones that can be formulated using the single and double layer potentials. For example, instead of using the double-layer potential for the Dirichlet boundary conditions we could have used a single-layer potential. This leads to an integral equation of the first kind. For mixed boundary value problems, the use of either single-layer or double-layer potentials alone leads to coupled first and second kind equations. An alternative is to use a double-layer distribution on the part of the boundary where ϕ is defined and single-layer on the remainder. This formulation gives an integral equation of the second kind.

2.5 Direct Boundary Integral Equation

In principle, the solution to Laplace's equation can be represented by a *combination* of single-layer and double-layer singularities,

$$\phi(p) = \int_S \{\sigma(Q)\psi(p,Q) + \mu(Q)\frac{d\psi(p,Q)}{dn(Q)}\}\,dS(Q) \qquad (2.19)$$

Since there is only one boundary condition and there are two functions, there is a "degree of freedom" in the problem. Thus the singularity distributions that give rise to a particular potential are not unique. The remainder of this sub-section shows how by combining our sought after solution ϕ and the fundamental solution ψ through Green's theorem, we can arrive at a representation of the form (2.19) in which the single and double layer densities $\sigma(Q)$ and $\mu(Q)$ are expressed in terms of the actual boundary values of ϕ and its normal derivative.

If ϕ and ψ are two continuous functions with continuous first derivatives, the application of the divergence theorem to the region V with surface S gives the following Green's theorem:

$$\int_V(\phi\nabla^2\psi - \psi\nabla^2\phi)dV = \int_S(\phi\frac{d\psi}{dn} - \psi\frac{d\phi}{dn})dS \qquad (2.20)$$

The normal direction to S is taken outwards from V.

We take $\phi(q)$ to be our *unknown* harmonic function (satisfies $\nabla^2\phi = 0$) and $\psi(p,q)$ as the *fundamental* solution to Laplace's equation (from Section 2.1).

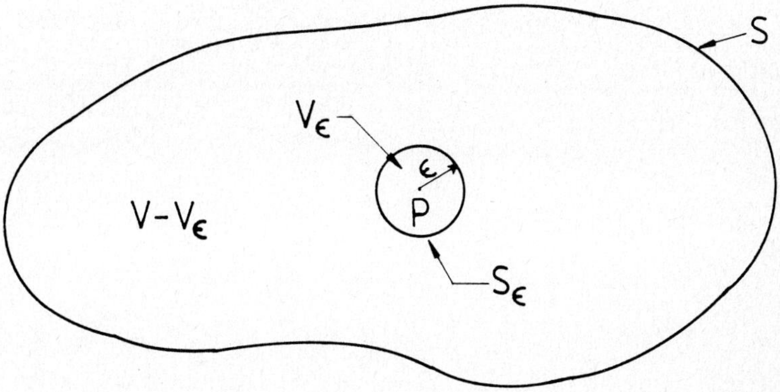

Figure 3
Domain for which Green's Theorem is written, if p is inside V.

As ψ does not obey Laplace's equation at p(x), we surround this point by a small sphere (or circle) of radius ε, denoted V_ε with surface S_ε (Fig. 3). In the region $V-V_\varepsilon$, $\nabla^2\psi = 0$. Taking Green's theorem for this region

$$\int_{V-V_\varepsilon} (\phi\nabla^2\psi - \psi\nabla^2\phi)dV = \int_{S+S_\varepsilon} (\phi \frac{d\psi}{dn} - \psi \frac{d\phi}{dn})dS \tag{2.21}$$

Because of the assumed continuity of $\phi(p)$

$$\lim_{\varepsilon\to 0} \int_{S_\varepsilon} \phi \frac{d\psi}{dn} dS = \phi(p) \, . \lim_{\varepsilon\to 0} \int_{S_\varepsilon} \frac{d\psi}{dn} dS \tag{2.22}$$

On the surface S_ε, $d/dn = -d/dr$. Considering the three-dimensional case,

$$\frac{d\psi}{dn} = -\frac{d}{dr}\left(\frac{1}{4\pi r}\right) = \frac{1}{4\pi r^2} \tag{2.23}$$

giving

$$\int_{S_\varepsilon} \frac{d\psi}{dn} dS = \frac{1}{4\pi\varepsilon^2} \int_{S_\varepsilon} dS = 1 \tag{2.24}$$

For the two-dimensional case

$$\frac{d\psi}{dn} = -\frac{1}{2\pi}\frac{d}{dr}\ell n\left(\frac{1}{r}\right) = \frac{1}{2\pi r} \tag{2.25}$$

giving

$$\int_{S_\varepsilon} \frac{d\psi}{dn} dS = \frac{1}{2\pi\varepsilon} \int_{S_\varepsilon} dS = 1 \tag{2.26}$$

In the limit as $\varepsilon \to 0$, equation (2.21) becomes

$$\phi(p) = \int_S \{-\phi(Q)\frac{d\psi(p,Q)}{dn(Q)} + \frac{d\phi(Q)}{dn(Q)} \psi(p,Q)\} dS(Q) \tag{2.27}$$

This is the important integral identity of potential theory, that a harmonic function ϕ may be expressed as the sum of a single-layer potential with density $d\phi/dn$ and a double-layer potential with density ϕ.

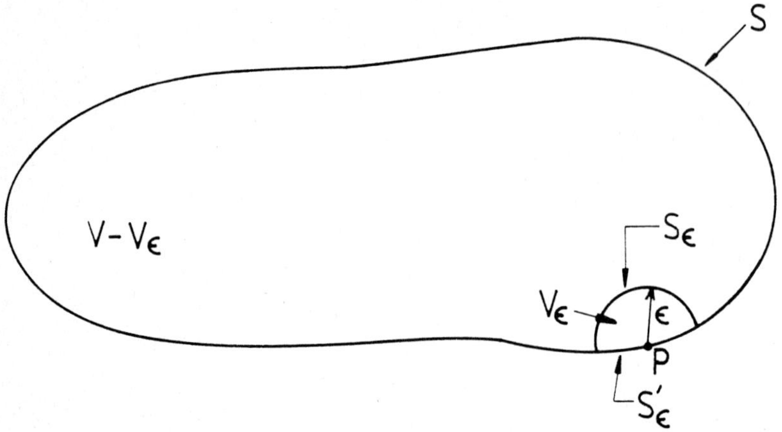

Figure 4
Domain for which Green's Theorem is written, if P is on S.

We now consider the case when the interior point p is taken to the boundary point P. The analysis is the same as for the interior case, except that instead of using the entire sphere (or circle), we only consider the part of the sphere (or circle) that is contained in V (Fig. 4). Instead of equations (2.24) or (2.26) we get

$$\lim_{\varepsilon \to 0} \int_{S_\varepsilon} \frac{d\psi}{dn}\, dS = c(P) \tag{2.28}$$

where c(P) is the proportion of surface area of S_ε contained within the domain. For a smooth boundary $c = \frac{1}{2}$.

Thus, in the limit, equation (2.27) becomes

$$c(P)\phi(P) + \int_S \phi(Q) \frac{d\psi(P,Q)}{dn(Q)}\, dS(Q) = \int_S \psi(P,Q)\frac{d\phi(Q)}{dn(Q)}\, dS(Q) \tag{2.29}$$

Equation (2.29) represents a constraint equation between the Dirichlet boundary conditions (ϕ defined) and the Neumann ($d\phi/dn$ defined) boundary conditions. For Neumann boundary conditions the right hand side of equation (2.29) is known, giving a Fredholm equation of the second kind for the unknown boundary values of the function $\phi(Q)$. For the Dirichlet problem equation (2.29) becomes a Fredholm equation of the first kind for the unknown boundary values of $d\phi/dn$. The mixed boundary value problem leads to a mixed integral equation for the unknown boundary data.

2.6 Direct vs Indirect Formulations

As the unknowns in the integral equation are physical boundary quantities (either ϕ or $d\phi/dn$), equation (2.29) is called the *direct* boundary integral equation to distinguish it from integral equations that involve a 'fictitious' source density, for example equations (2.15) and (2.18). Formulations based on the source density approach are called indirect because the 'complementary' boundary data can only be obtained after the source density distribution has been solved for.

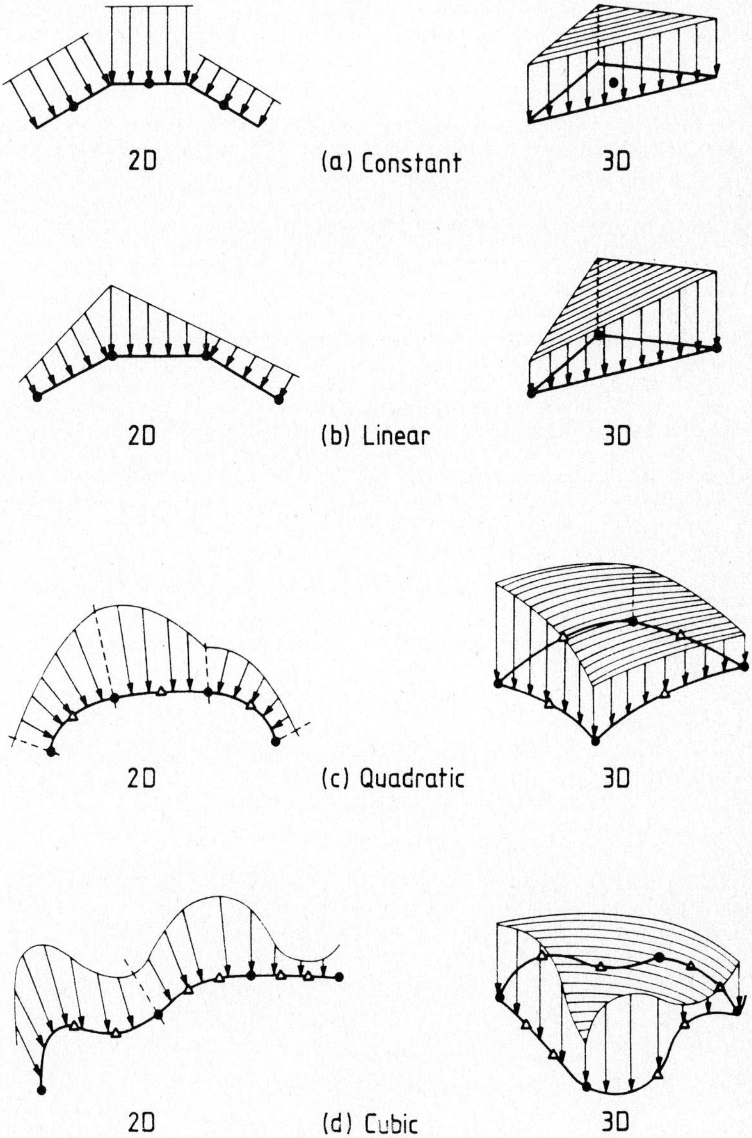

Figure 5
Common Boundary Elements

Almost without exception, authors of research papers on boundary element methods use only one of the two alternative methods of formulation.

It is fair to say that proponents of the direct method have developed the method further in terms of numerical techniques and mathematical sophistication than the supporters of the indirect methods have. However, this leaves aside the question of the numerical effectiveness of the various formulations. A proper numerical comparison of the methods still needs to be undertaken. It is possible that no one formulation is best for all problems but that each formulation has a class of problems, depending on geometry, boundary conditions etc. for which it is suited.

In the remainder of this paper, whenever a specific formulation needs to be considered, the direct formulation will be used. In general, however, the remarks apply to either method of formulation.

3. NUMERICAL SOLUTION OF BOUNDARY INTEGRAL EQUATIONS

Except for very simple geometries, analytical solutions for the boundary integral equations are unavailable and numerical methods must be used. By assuming that the boundary is divided into elements over which the data has a prescribed polynomial order of variation we can reduce the problem to that of solving a matrix system for the unknown coefficients. A variety of elements have been used (see Fig. 5). To illustrate how we can arrive at a system of linear equations we look at the so-called 'constant' element. We also discuss a number of higher-order elements of the isoparametric type (called isoparametric because the same order of polynomial variation is used for both the geometry and boundary data). These elements are particularly useful for curved geometries. Some of the factors to be considered when solving large systems of equations are pointed out. Brief mention is also made of the use of weighted residual formulations such as the Galerkin scheme.

3.1 Constant Elements

For purposes of illustration, consider the two-dimensional direct boundary integral equation (equation (2.29) with $\psi = (1/2\pi)\log(1/r)$). Divide the boundary of the body into N straight line segments (see Fig. 6). The function and its

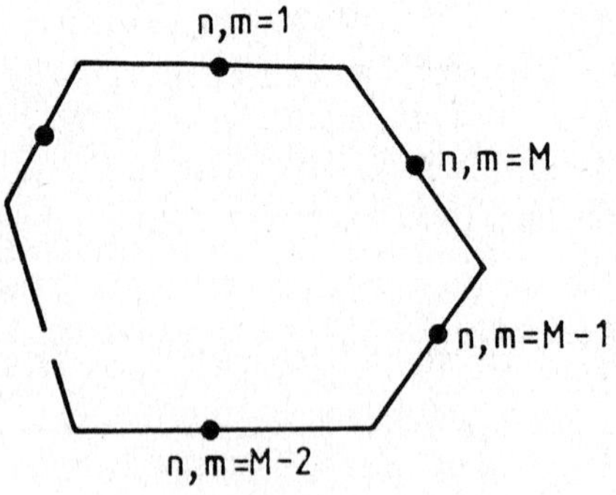

Figure 6
Constant Boundary Elements

normal derivative are assumed constant on each segment. The boundary integral equation is written at the mid-point of each segment and the values of ϕ and $d\phi/dn$ can be extracted from the integration over each segment.

$$\pi\phi(P_n) + \sum_{m=1}^{N} \phi(Q_m) \int_{S_m} \frac{d}{dn(Q)} \log \frac{1}{r(P_n,Q)} dS(Q) =$$
$$\sum_{m=1}^{N} \frac{d\phi}{dn}\Big|_{Q_m} \int_{S_m} \log \frac{1}{r(P_n,Q)} dS(Q) \quad (n = 1,N). \tag{3.1}$$

The integrals can be expressed in closed form [3]. Care must be taken to handle the segment where the mid-point $P_n = Q_m$. This integral must be evaluated in the Cauchy Principal Value sense, i.e., by deleting a small segment of length 2ε at the mid-point and taking the limit

$$\lim_{\varepsilon\to 0} \{ \int_{Q_1}^{-\varepsilon} (\ )dS(Q) + \int_{\varepsilon}^{Q_2} (\ )dS(Q)\}. \tag{3.2}$$

Denoting the results of the integrals by $A^*(P_n,Q_m)$ and $B(P_n,Q_m)$, equation (3.1) becomes

$$\pi\phi(P_n) + \sum_{m=1}^{N} \phi(Q_m)A^*(P_n,Q_m) = \sum_{m=1}^{N} \frac{d\phi}{dn}\Big|_{Q_m} B(P_n,Q_m) \qquad (n = 1,N) \tag{3.3}$$

Equation (3.3) may be rewritten in matrix form (A^* becomes A by adding π to the diagonal terms of A)

$$[A]\ \{\phi\} = [B]\ \{\frac{d\phi}{dn}\}\ . \tag{3.4}$$

For any well-posed boundary value problem, equation (3.4) involves just N knowns and N unknowns. This equation is normally very well conditioned and can be solved by standard matrix reduction schemes.

Values of ϕ at any interior point can be obtained using a discretized version of equation (2.27).

3.2 Ergatoudis Isoparametric Elements

One of the most efficient methods for numerical solution of boundary integral equations is based on the parametric representation of geometry and functions in terms of the Ergatoudis shape functions which are widely used in the finite element method [4]. This section introduces the quadratic element for two-dimensional problems [5]. The analogous element for three-dimensional problems is described in Ref. [6].

Assume that the boundary data at point Q within a typical element can be represented by mapping from the line $(-1<\xi<1)$ according to

$$f(Q) = \sum_{\alpha=1}^{3} M^{\alpha}(\xi)\ f_{\alpha} \tag{3.5}$$

where f_α is the value of the function at one of the three nodes associated with the element, (Fig. 7a). $M_\alpha(\xi)$ is a set of three 'shape functions' (Fig. 7b)

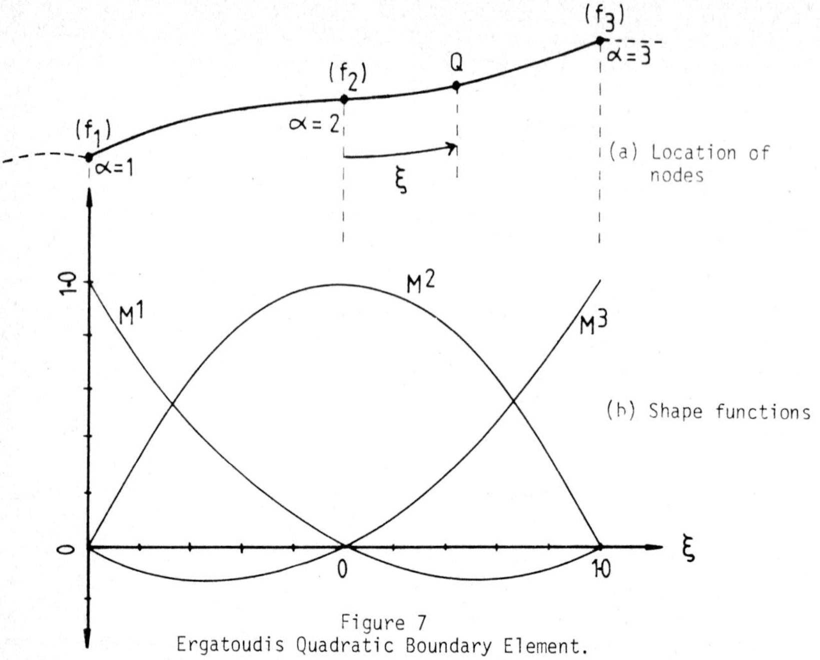

Figure 7
Ergatoudis Quadratic Boundary Element.

$$
\begin{aligned}
M^1(\xi) &= -\frac{1}{2}\xi + \frac{1}{2}\xi^2 \\
M^2(\xi) &= 1 - \xi^2 \\
M^3(\xi) &= \frac{1}{2}\xi + \frac{1}{2}\xi^2
\end{aligned}
\tag{3.6}
$$

The actual boundary curve is assumed to be described by the same form as the boundary data (hence the name 'isoparametric')

$$
\begin{aligned}
x(Q) &= \sum_{\alpha=1}^{3} M^{\alpha}(\xi)x_{\alpha} \\
y(Q) &= \sum_{\alpha=1}^{3} M^{\alpha}(\xi)y_{\alpha}
\end{aligned}
\tag{3.7}
$$

where x_{α}, y_{α} are the coordinates of the nodes.

Based on the above discretizing assumptions, equation (2.29) may be written:

$$
\begin{aligned}
&\pi c(P_n)\phi(P_n) + \sum_{m=1}^{(N/2)} \sum_{\alpha=1}^{3} \phi_m^{\alpha} \int_{S_m} M^{\alpha}(\xi) \frac{d}{dn(Q)} \log \frac{1}{r(P_n,Q(\xi))} J(\xi)d\xi \\
&= \sum_{m=1}^{N/2} \sum_{\alpha=1}^{3} \left(\frac{d\phi}{dn}\right)_m^{\alpha} \int_{S_m} M^{\alpha}(\xi) \log \frac{1}{r(P_n,Q(\xi))} J(\xi)d\xi \qquad (n = 1,N)
\end{aligned}
\tag{3.8}
$$

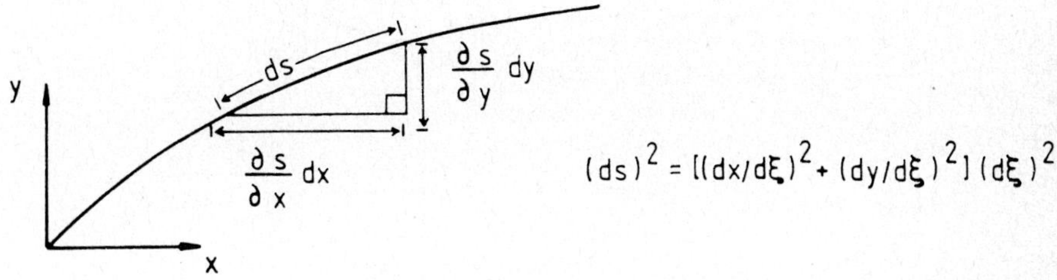

Figure 8
Notation for Curved Boundary.

In equation (3.8), $J(\xi)$ is the Jacobian given by (see Fig. 8)

$$J(\xi) = \frac{dS(Q)}{d\xi} = \left| \left(\frac{dx}{d\xi}\right)^2 + \left(\frac{dy}{d\xi}\right)^2 \right|^{\frac{1}{2}} \tag{3.9}$$

Substituting equation (3.7) gives

$$J(\xi) = \left[\left(\sum_{\alpha=1}^{3} \frac{dM^\alpha}{d\xi} x_\alpha \right)^2 + \left(\sum_{\alpha=1}^{3} \frac{dM^\alpha}{d\xi} y_\alpha \right)^2 \right]^{\frac{1}{2}} \tag{3.10}$$

The quantities ϕ_m^α, $(d\phi/dn)_m^\alpha$ represent values at local node α of element m. In general because the end nodes are associated with the elements on either side, $\phi_1^3 = \phi_2^1$, $(d\phi/dn)_1^3$ $(d\phi/dn)_2^1$ etc. The exception is at a point where discontinuous values are to be considered (say at a corner). In this case a pair of nodes can be used, each one associated with a different element.

The system of equations (3.8) becomes

$$[A] \; \{\phi\} = [B] \left\{\frac{d\phi}{dn}\right\} \tag{3.11}$$

The square matrices [A] and [B] contain integrals of the form

$$I_{nm}^\alpha = \int_{-1}^{1} M^\alpha(\xi)\psi(P_n, Q(\xi))J(\xi)d\xi \tag{3.12}$$

where ψ is one of the kernel functions $\log(\frac{1}{r})$ or $\frac{d}{dn}(\log(\frac{1}{r}))$.

The presence of the Jacobian (a consequence of the curved boundary) necessitates the use of numerical quadrature. For further details see Reference [5].

3.3 Hermitian Cubic Elements

Watson [7,8] has used Hermitian cubic elements for two-dimensional elasticity problems. These elements feature interelement continuity of the tangent plane and also of tangential derivatives of the boundary data. There are no midside nodes, thus simplifying data preparation.

The Hermitian cubic element has two nodes (Fig. 9a). The boundary data at point Q is representing by mapping from intrinsic coordinates $(-1<\xi<1)$ according to

$$f(Q) = \sum_{\alpha=1}^{2} \{M^\alpha(\xi)f_\alpha + N^\alpha(\xi)(df/d\xi)_\alpha\} \tag{3.13}$$

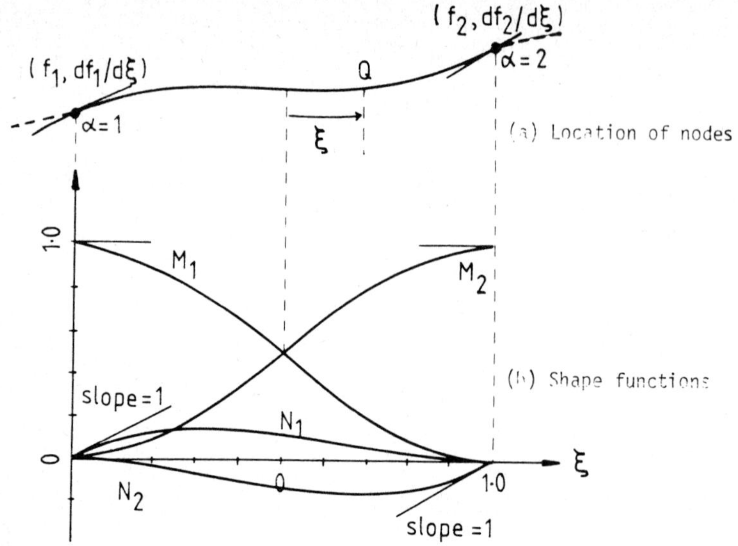

Figure 9
Hermitian Cubic Boundary Element.

where f_α (α =1,2) are the values of the function at the nodes and $(df/d\xi)_\alpha$ (α=1,2) represents the tangential derivatives of the function at the nodes.

The shape functions are given by (see Fig. 9b):

$$\begin{aligned} M^1(\xi) &= \frac{1}{4}(1+\xi)^2(2-\xi) \\ M^2(\xi) &= \frac{1}{4}(1-\xi)^2(2+\xi) \\ N^1(\xi) &= -\frac{1}{4}(1+\xi)^2(1-\xi) \\ N^2(\xi) &= \frac{1}{4}(1-\xi)^2(1+\xi) \end{aligned} \qquad (3.14)$$

Again, as for the Ergatoudis elements, the Hermitian element is assumed to be isoparametric, i.e. the boundary geometry is modelled by the same shape functions.

The implementation of Hermitian cubic elements requires the derivation of an extra integral equation because of the introduction of the tangential derivatives. This is obtained by taking the tangential derivative of the basic integral equation [7,8].

3.4 Considerations of Computational Efficiency

In terms of computer effort, the two principal tasks are firstly the calculation of the coefficients and secondly the solution of the resulting linear equations. If for arguments sake we assume that a given problem is modelled by N constant elements, N^2 coefficients need to be computed. Contrasting with the symmetric banded coefficient matrices arising from the finite element method, the coefficient matrix is nonsymmetric and fully populated. The systems of equations are commonly solved by Gaussian elimination or iterative techniques. As is well known the computing effort involved in direct elimination is proportional to N^3 and is essentially independent of the number of right hand sides involved (these would arise if multiple sets of boundary conditions are considered in a single analysis). For an iterative solution the computing effort is proportional to IRN^2, where I is the number of iterations required for convergence and R is the number of right hand sides being considered.

For exterior problems (i.e. infinite domain) Hess [9] reports that iterative solutions converge in a number of iterations that is independent of N but dependent on the shape of the domain. In many applications I is of the order of 10 - 20. As a consequence iteration becomes more efficient than elimination for problems involving 100 - 300 elements. On the other hand, for interior problems (i.e. finite domain) I increases linearly with N, so in general, iteration is not competitive with direct elimination.

The computer effort involved in computing the terms in the coefficient matrix is of the order N^2. These terms involve integrating the field at the centre of one element (the 'field point') due to a prescribed source distribution within another element (the 'source element'). For flat elements these integrals are expressible analytically. The resulting expressions, which involve logarithms and inverse trignometric functions, depend on the orientation and dimensions of the source element and the relative coordinates of the field point. When a field point is far from the source element, such precision is inappropriate, and approximate expressions should be used to reduce computing costs. Hess [9] suggests that if the field point is further away than four times the maximum source element dimension, the effect of the distributed source can be replaced by that of a point source with no significant loss of accuracy. For large problems over 90% of the N^2 coefficients can be evaluated using the point source formulae, reducing the computational effort by an order of magnitude. Going further than this, for three-dimensional problems, computer time can be dramatically reduced by replacing the effects of a remote block of elements by an equivalent point source [10].

3.5 Weighted Residual Formulations

The numerical methods discussed so far are based on what is commonly known as 'nodal collocation', or point-matching. That is we have chosen a number of discrete points at which our integral equation is assumed to be satisfied so as to supply the same number of algebraic equations as we have unknowns. Alternative numerical methods can be based on assuming that the integral equation is satisfied in some sort of 'best fit' sense over the entire boundary rather than attempting to have zero error at discrete points and unknown error in between. This class of methods is commonly known as 'weighted residual methods'. Better known methods in this class are the Galerkin method, Method of moments and the Rayleigh-Ritz method. A discussion of the application of these methods applied to the integral equation for Laplace's equation is given by Reference [11]. The Galerkin scheme has been used more widely than the others. Computationally the Galerkin scheme involves double surface integrals over the boundary, as opposed to single surface integrals for nodal collocation. For a given surface discretization the Galerkin scheme will be correspondingly more costly than nodal collocation. The question is whether superior accuracy for a given discretization will allow sufficiently fewer elements to be used to offset this disadvantage.

4. EXTENSIONS TO THE BASIC BOUNDARY ELEMENT METHOD

The basic boundary integral formulations are based on the use of fundamental solutions of the governing partial differential equation over an infinite domain with homogeneous material properties. The boundary integral equations were constructed by superposition of fundamental solutions with their centres on the boundary. By distributing centres of the fundamental solution throughout the domain we show how to develop integral equation formulations for non-homogeneous partial differential equations such as Poisson's equation.

For non-homogeneous material properties in general an analytical fundamental solution does not exist. However we show how the boundary element method can be extended to solve piecewise homogeneous geometries.

Although the integral equations are based on the superposition property of linear partial differential equations, we demonstrate how various types of non-linear behaviour can be treated by iterative methods.

4.1 Non-homogeneous Partial Differential Equations

The non-homogeneous counterpart of Laplace's equation is Poisson's equation:

$$\nabla^2\phi\ (\underset{\sim}{x}) = \rho(\underset{\sim}{x}) \tag{4.1}$$

In steady flow problems $\rho(\underset{\sim}{x})$ represents the distribution of sources within the domain.

If the term $\rho(\underset{\sim}{x})$ is included in Green's theorem (Equation 2.20), the boundary integral equation becomes

$$\begin{aligned} c(P)\phi(P) + \int_S \phi(Q)\, \frac{d\psi(P,Q)}{dn(Q)}\, dS(Q) &= \\ \int_S \psi(P,Q)\, \frac{d\phi(Q)}{dn(Q)}\, dS(Q) &+ \int_V \rho(q)\psi(P,q)dV(q) \end{aligned} \tag{4.2}$$

In addition to the usual surface integrals, we now have a known volume integral term involving $\rho(q)$, and for arbitrary $\rho(q)$ we must resort to a volume discretization scheme to evaluate the volume integral.

Assume that the domain is subdivided into M suitably sized 'volume elements' for which ρ can be taken as constant (Fig. 10). The volume integral then becomes

$$\int_V \rho(q)\psi(P,q)dV(q) = \sum_{\ell=1}^{M} \rho(q_\ell) \int_{V_\ell} \psi(P,q)dV(q) \tag{4.3}$$

The volume integrations can be performed using Gaussian quadrature.

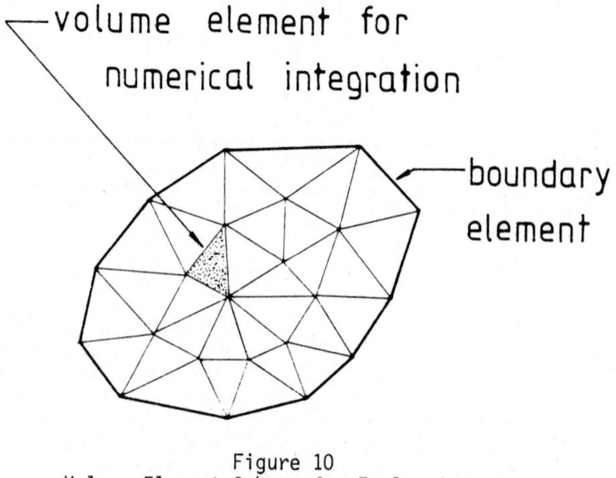

Figure 10
Volume Element Scheme for Evaluation of
Volume Integrals for Domain with Interior Source Density.

The matrix system obtained by the use of the simple constant boundary elements for Laplace's equation (Equation (3.4)) now generalizes to

$$[A]\,\{\phi\} \;=\; [B]\,\{\frac{d\phi}{dn}\} \;+\; [C]\,\{\rho\} \tag{4.4}$$

where, as before, matrices A and B are NXN (N being the number of boundary elements) and matrix C is an NXM matrix used to multiply the M known values of ρ within the volume elements.

Although the volume discretization has the appearance of that used in the finite element method, no unknowns have been introduced into the interior of the domain. In section 4.3 it will be shown how the volume discretization scheme can be used to solve non-linear problems using integral formulations.

The volume discretization scheme is unnecessary if $\rho(x)$ has certain properties. For example, if $\rho(x)$ is harmonic, i.e. $\nabla^2\rho(x) = 0$, reference [5] shows that for the two-dimensional case the volume integral can be expressed as a simple surface integral.

4.2 Non-homogeneous Material Properties

An implicit assumption used in the formulation of boundary integral equations is that the material properties are uniform throughout the domain. This stems from the fact that the fundamental solutions involved are for uniform material properties. Boundary element methods can be used to analyse a piecewise homogeneous system, for example a steady-state fluid flow problem involving distinct zones, each with constant permeability.

Figure 11 shows a problem with two different permeabilities. On the interface between the two zones the potential and the normal velocity component must be continuous, i.e.

$$\begin{aligned} \phi_1 &= \phi_2 \\ C_1\,\frac{\partial\phi_1}{\partial n_1} &= -\,C_2\,\frac{\partial\phi_2}{\partial n_2} \end{aligned} \tag{4.5}$$

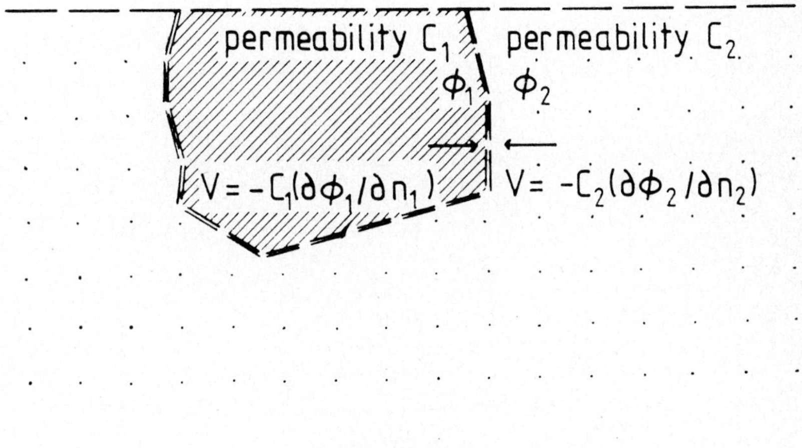

Figure 11
Boundary Element Method for Piecewise Homogeneous System

Assume that constant boundary elements are used and zone 1 has N_1 nodes, zone 2 has N_2 nodes and there are M nodes on the interface. Then there are N_1+N_2+2M unknowns of either ϕ or $d\phi/dn$ on the exterior boundaries and both ϕ and $d\phi/dn$ on the interface. Equation (3.4) applied in each zone yields N_1+N_2 equations and equations (4.5) the remaining 2M equations needed. By counting points on the interface only once, the system can be assembled into N_1+N_2 equations and unknowns.

4.3 Non-linear Problems

Non-linear effects often need to be taken into account in order to realistically model many physical phenomena. For the sake of illustration we consider steady-state flow of the type that arises in problems of heat conduction, seepage through porous media, fluid flow and so on. To solve these problems it is often necessary to obtain a numerical solution to the differential equation[+]

$$\frac{\partial}{\partial x_i} \left(C \frac{\partial \phi(\underset{\sim}{x})}{\partial x_i}\right) = q(\underset{\sim}{x}) \tag{4.6}$$

where ϕ is the potential at a point
C is the conductivity or permeability
and q is the specified source distribution in the interior of the domain

The boundary conditions are usually specified in terms of the potential ϕ or the velocity normal to the surface, V_n, given by

$$V_n = V_i n_i = -C \frac{\partial \phi}{\partial x_i} n_i \tag{4.7}$$

If C is constant or piecewise constant within the domain, equation (4.6) reduces to Poisson's equation or Laplace's equation (when $q = 0$) and the integral formulations discussed earlier apply.

Non-linear behaviour that may have to be taken into account in practical applications can come from one or more of the following sources:

i) part of the boundary S may not be known initially,

ii) the source q may depend on ϕ or V_i,

iii) the material property C may be dependent on ϕ or V_i.

Non-linear boundary conditions

In many flow problems such as groundwater flow, seepage through dams etc. (Fig. 12), the free surface of flow is not known *a priori*. Although on the known parts of the boundary either ϕ or V_n is known, on the unknown free surface both conditions must be satisfied simultaneously. On the free surface we require

$$\phi = Z \quad \text{and} \quad V_n = 0 \tag{4.8}$$

where Z is the height above some reference datum.

[+] In this section the summation convention is used, i.e. repeated subscripts are summed over the number of spatial dimensions.

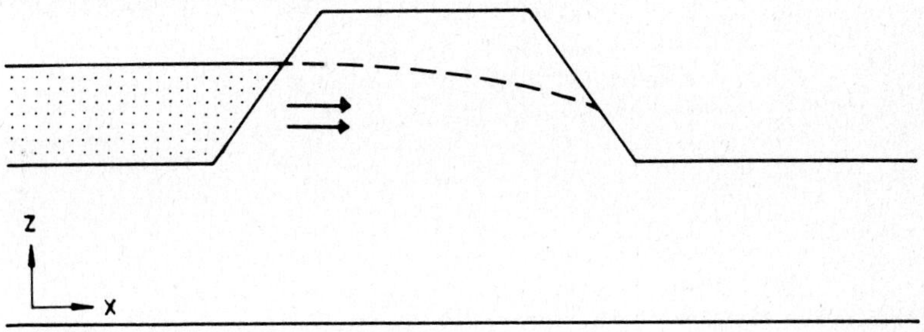

Figure 12
Example of Flow with Free Surface.

The boundary element procedure used to solve this type of problem is as follows [12]:

i) An initial guess is used for the location of the free surface.

ii) For each of the nodes on the free surface, the boundary condition $V_n = 0$ is prescribed. Of course the boundary conditions on the remainder of the boundary are prescribed unambiguously.

iii) The unspecified boundary data is solved for using the conventional boundary element method. This solution provides a set of values for ϕ on the free surface.

iv) From the relationship $\phi = Z$, a fresh estimate of the location of the free surface is obtained.

Steps (ii) and (iv) are repeated until successive locations of the free surface agree to within a prescribed tolerance. For problems solved by Liggett [12] one percent accuracy was obtained in 5 iterations.

Non-linear source

If, for example, the source $q(\underset{\sim}{x})$ is dependent on the potential ϕ, we can use the volume discretization introduced for Poisson's equation. This will lead to the final system of equations (see Equation (4.4.)).

$$[K] \ \{x\} \ = \ \{c\} \ + \ [C] \ \{q\} \tag{4.9}$$

where the vector $\{x\}$ contains the unknown boundary data and the vector $\{c\}$ represents the effects of the prescribed boundary conditions.

The computational scheme is as follows:

i) take an initial guess for $\{q\}$ (zero, say)

ii) solve equation (4.9) and determine the interior values of ϕ

iii) determine new values of $\{q\}$ from the interior values of ϕ (such a relationship has been assumed)

iv) repeat the process until the changes in $\{q\}$ between successive iterations are less than a prescribed tolerance.

Material non-linearities

In many applications the material parameter C is dependent on the values of ϕ or V_i. Assume that C is expressed in the form:

$$C = C^0 + C^d \tag{4.10}$$

where C^0 is the constant part
and C^d is the part dependent on ϕ or V_i.

The differential equation (4.6), with q = 0, becomes

$$C^0\nabla^2\phi + \frac{\partial}{\partial x_i}\left(C^d \frac{\partial\phi}{\partial x_i}\right) = 0 \tag{4.11}$$

or

$$\nabla^2\phi = \rho \tag{4.12}$$

where

$$\rho = -\frac{1}{C^0}\frac{\partial}{\partial x_i}\left(C^d \frac{\partial\phi}{\partial x_i}\right) \tag{4.13}$$

On the boundary the value of ϕ remains unaltered, but the normal velocity must be modified to

$$V_n' = V_n + C^d \frac{\partial\phi}{\partial x_i} n_i \tag{4.14}$$

The problem is reduced to the solution of equation (4.12) subject to the modified normal velocity V_n'.

We can now use the boundary integral formulation (Equation 4.2) introduced for Poisson's equation to obtain

$$\begin{aligned} c(P)\phi(P) + \int_S \phi(Q)\frac{d\psi(P,Q)}{dn(Q)}\,dS(Q) &= \\ \frac{1}{C^0}\int_S \psi(P,Q)\,V_n'(Q)dS(Q) - \frac{1}{C^0}\int_V \frac{\partial}{\partial x_i}&\left[C^d\frac{\partial\phi(q)}{\partial x_i}\right]\psi(P,q)dV(q) \end{aligned} \tag{4.15}$$

By applying the divergence theorem to the volume integral we obtain [13]:

$$\begin{aligned} c(P)\phi(P) + \int_S \phi(Q)\frac{d\psi(P,Q)}{dn(Q)}\,dS(Q) &= \\ \frac{1}{C^0}\int_S \psi(P,Q)\,V_n(Q)dS(Q) + \frac{1}{C^0}\int_V C^d&\frac{\partial\phi(q)}{\partial x_i}\frac{\partial\psi(P,q)}{\partial x_i}\,dV(q) \end{aligned} \tag{4.16}$$

Use of this equation does not require any modification to the normal velocity at the boundary.

The numerical solution of equation (4.16) can be obtained using the iterative procedure outlined in the previous section.

5. OTHER PARTIAL DIFFERENTIAL EQUATIONS

5.1 Steady-state Phenomena

The basic concepts of boundary element methods have been illustrated in the previous sections within the context of Laplace's equation.

Other partial differential equations arising from steady state phenomena (i.e. elliptical equations), can be transformed to boundary integrals once the appropriate fundamental solution is found. Going further than this, a direct formulation can be derived using an appropriate reciprocal relationship analogous to Green's theorem. Using these concepts all of the well known 'classical' partial differential equations, e.g. Helmholz equation, isotropic elasticity etc. have been solved by numerical techniques similar to those described in Sections 3 and 4. That is to say that to use the techniques one usually doesn't have to derive the integral equations from scratch; they are already in the literature. Some work has been done on establishing methods of formulation for quite general forms of differential equations, for example Clements and Rizzo [14] derive a direct method for two-dimensional problems governed by quite general second order elliptic systems. Such systems arise in anisotropic elasticity, for example.

It is worth pointing out that it is not guaranteed that an *analytical* fundamental solution can be found for a given system. Such examples arise in three dimensional anisotropic elasticity, which with the lowest order of symmetry requires 21 independent elastic constants. Even for cubic symmetry (3 independent constants), the resulting characteristic equation cannot be solved analytically [15]. The fundamental solution can still be derived *numerically* however [16].

5.2 Transient Phenomena

When transient phenomena are considered the complexity of the system is increased by the independent variable of time.

The most commonly used technique is to apply the Laplace transform to the partial differential equations and boundary conditions, thus temporarily eliminating time as an independent variable. Using this process, parabolic and hyperbolic systems generally reduce to elliptic systems. In the transform space the elliptic system can be reformulated in terms of a boundary integral equation which can then be solved for a sequence of values of the transform parameter. The time solution can then be obtained by numerical Laplace transform inversion. A survey of the application of this technique to transient heat conduction, viscoelasticity and wave propagation is given in Reference [17].

An alternative method is to work directly in terms of the fundamental solution for the original partial differential equation. This solution will, of course, have time dependence and have a singularity in the time domain as well as in the spatial domain. This approach has been used to solve transient heat conduction problems [18]. Integration with respect to time is performed step-wise. In contrast with the Laplace transform method, the resultant integral equations involve a volume integral.

Shaw [19] gives a survey of boundary integral techniques applied to the wave equation, and suggests that time-stepping methods are appropriate for early time and the Laplace transform method for longer time.

6. CONCLUSIONS

The boundary element method has become an important alternative to the conventional finite element method, i.e. the finite element method applied to the partial differential equations in the domain. The simplicity of the input data requirements, particularly for three-dimensional and infinite domain problems, makes boundary element computer programs easier to use than programs based on finite element or finite difference methods.

Although the basic boundary element method was developed for linear homogeneous material properties, non-homogeneous problems involving distinct homogeneous regions can also be solved by the method. Problems involving a continuous spatial variation in material properties are probably best left to the finite element method. As demonstrated in Section 4 the boundary element method can be adapted to non-linear problems, but the efficiency of the method compared to finite element or finite difference methods remains an open question. Indeed, it may be more efficient to use a combination of methods. The development of methods of coupling boundary element methods with the finite element method is an active research topic [20, 21].

The basic features of the method have been illustrated in the context of Laplace's equation. Over the last decade the method has been applied successfully to partial differential equations arising in such diverse physical applications as acoustics, electromagnetism, aerodynamics, elasticity and so on. Apart from journal papers, the interested reader is referred to the proceedings of conferences on boundary elements [22-25] and proceedings of two other conferences than contain a large proportion of papers on boundary element methods [26,27]. Also worth mentioning is a monograph of surveys of various application areas [28]. A number of books that give an introductory treatment of boundary element methods applied to potential theory and elasticity are available [29-31].

APPENDIX : GLOSSARY OF SYMBOLS

A	An NxN matrix
B	An NxN matrix
C	An NxM matrix (referring to conductivity or permeability)
c	Function describing the proportion of surface area contained within the domain
I	Number of iterations required for convergence
i	Subscript
J	Jacobean matrix
K	An NxM matrix
M	Shape function or Volume elements
m	Coordinate of midpoint of element
N	Number of divisions of boundary line segments
P	Field point on surface S
p	Field point in domain V
Q	Field point on surface S
q	Field point in domain V
R	Number of right hand sides being considered
r	Distance
S	Surface or boundary

V	A domain
x	Unknown boundary data (a vector)
Z	Height above some reference datum

α	Numbered coordinates
Δ	Laplacian operator
δ	Dirac delta function
ε	Some small distance
μ	Function referring to surface density of double layer potential
ξ	Direction of outward normal to S
π	3.1416.......
ρ	Referred to Poisson's equation
ϕ	Function referring to a single layer potential
ψ	Function determining fundamental solution

REFERENCES

[1] Fredholm, I. Solution d'un probleme fundamental de la theorie de l'elasticte Arkiv fur Mathematik Astonomie und Fysik, (1905).

[2] Smirnov, V.I., Integral equations and partial differential equations, a course of higher mathematics, Vol. IV, Pergamon, London, (1964).

[3] Cruse, T.A., Mathematical foundations of the boundary-integral equation method in solid mechanics. U.S. Air Force Office of Scientific Research, Report TR-77-1002, (1977).

[4] Zienkiewicz, O.C., Finite element methods in engineering science, 3rd ed. McGraw Hill, London, (1977).

[5] Wu, Y.S., Rizzo, R.J., Shippy, D.J. and Wagner, J.A., An advanced boundary integral equation method for two-dimensional electromagnetic field problems, Electrical Machines and Electromechanics; Vol, 1, No. 4, pp. 301-313, (1977).

[6] Lachat, J.C. and Watson, J.O., Effective numerical treatment of boundary integral equations: A formulation for three-dimensional elastostatics, Int. J. Numerical Methods in Engineering, Vol. 10, pp. 991-1005 (1976).

[7] Watson, J.O., Hermitian cubic boundary elements for plane elastostatics, Proc. 2nd Int. Symp. on Innovative Numerical Analysis in Applied Engineering Sciences, Montreal, pp. 403-412 (see Ref. [27]), (1980).

[8] Watson, J.O., Hermitian cubic boundary elements for plane problems of fracture mechanics, Res. Mechanica, (to appear)(1981).

[9] Hess, J.L., Review of integral-equation techniques for solving potential-flow problems with complicated boundaries. Proc. 2nd Int. Symp. on Innovative Numerical Analysis in Applied Engineering Sciences, Montreal, pp. 131-143 (see Ref. [27]), (1980).

[10] Diest, F.H. and Georgiadis, E., A computer system for three-dimensional elastic analysis using a boundary element approach, Chamber of Mines of South Africa, Res. Rept. No. 43/76, (December, 1976), (1976).

[11] Lean, M.H., Friedman, M. and Wexler, A., Application of the boundary element method in electrical engineering problems, Chapter 7 in Developments in Boundary Element Methods - I, (P.K. Banerjee & R. Butterfield, eds.) Applied Science Publishers, Barking, Essex, (1979).

[12] Liggett, J.A., Location of free surface in porous media, Proc. A.S.C.E. Hydraulics Division, Vol. 103, pp. 353-365, (1977).

[13] Banerjee, P.K., Non-linear problems of potential flow, Chapter 2 in Developments in Boundary Element Methods - I, (P.K. Banerjee & R. Butterfield, eds.) Applied Science Publishers, Barking, Essex, (1979).

[14] Clements, D.L. and Rizzo, F.J., Method for numerical-solution of boundary-volume problems governed by 2nd-order elliptic systems, J. Inst. Math. Appl. Vol. 22, pp. 197-202, (1978).

[15] Head, A.K., The Galois unsolvability of the sextic equation of anisotropic elasticity, J. Elasticity, Vol. 9, pp. 9-20, (1979).

[16] Wilson, R.B. and Cruse, T.A., Efficient implementation of anisotropic three-dimensional boundary-integral equation stress analysis, Int. J. for Numerical Methods in Engineering, Vol. 12, pp. 1383-1397, (1978).

[17] Shippy, D.J., Application of the boundary-integral equation method to transient phenomena in solids, Boundary Integral Equation Method : Computational applications in Applied Mechanics, (T.A. Cruse & F.J. Rizzo eds), ASME, New York, 1975, pp. 15-30, (1975).

[18] Chang, Y.P., Kang, C.S. and Chen, D.J., The use of fundamental Green's functions for the solution of problems of heat conduction in anisotropic media, Int. J. Heat & Mass Transfer, Vol. 16, pp. 1905-1918 (1973).

[19] Shaw, R.P., Boundary integral equation methods applied to wave problems, Chapter 6 in Developments in Boundary Element Methods - I. (P.K. Banerjee & R. Butterfield, eds), Applied Science Publishers, Barking, Essex, (1979).

[20] Zienkiewicz, O.C., Kelly, D.W. and Bettess, P., The coupling of the finite element method and boundary solution methods, Int. J. Numerical Methods in Engng, Vol. 11, No. 2, pp. 355-375, (1977).

[21] Kelly, D.W., Mustoe, G.G.W. and Zienkiewicz, O.C., Coupling boundary element methods with other numerical methods, Chapter 10 in Developments in Boundary Element Methods - I, (P.K. Banerjee & R. Butterfield, eds) Applied Science Publishers, Barking, Essex, (1979).

[22] Cruse, T.A. and Rizzo, F.J. (eds), Boundary-integral equation methods: computational applications in applied mechanics, AMD-11, Applied Mechanics Division, ASME, New York, 141 pp, (1975).

[23] Brebbia, C.A. (editor), Recent Advances in Boundary Element Methods, (Proc. 1st Int. Seminar, Southampton, 1978), Pentech Press, Plymouth, (1978).

[24] Brebbia, C.A. (editor), New Developments in Boundary Element Methods, (Proc. 2nd Int. Seminar, Southampton, 1980), CML Publications, Southampton, (1980).

[25] Brebbia, C.A. (editor), Further Developments in Boundary Element Methods, (Proc. 3rd Int. Seminar, California, 1981), Springer-Verlag, Berlin, (1981).

[26] Cruse, T.A., Lachat, J.C., Rizzo, F. and Shaw, R.P. (eds), Proceedings Int. Symposium on Innovative Numerical Analysis in Applied Engineering Science, (Versailles, France), C.E.T.I.M., Senlis, France (1977).

[27] Shaw, R., Pilkey, W., Pilkey, B., Wilson, R., Lakis, A., Chaudouet, A. and Marino, C. (editors), Innovative numerical analysis for the engineering sciences, (Proc. 2nd Int. Symposium on Numerical Analysis in Applied Engineering Sciences), Montreal, 1980, University Press of Virginia, Charlottesville, (1980).

[28] Banerjee, P.K. and Butterfield, R. (eds), Developments in Boundary Element Methods - I, Applied Science Publishers, Essex, (1979).

[29] Brebbia, C.A., The Boundary Element Method for Engineers, Pentech Press, Plymouth, (1978).

[30] Brebbia, C.A. and Walker, S., Boundary Element Techniques in Engineering, Newnes-Butterworth, London, (1980).

[31] Banerjee, P.K. and Butterfield, R., Boundary Element Methods in Engineering Science, McGraw-Hill, London, (1981).

Computational Techniques for Differential Equations
J. Noye (Editor)

DIRECT SOLUTION AND STORAGE OF LARGE SPARSE LINEAR SYSTEMS

KEN MANN
Chisholm Institute of Technology, Victoria, Australia

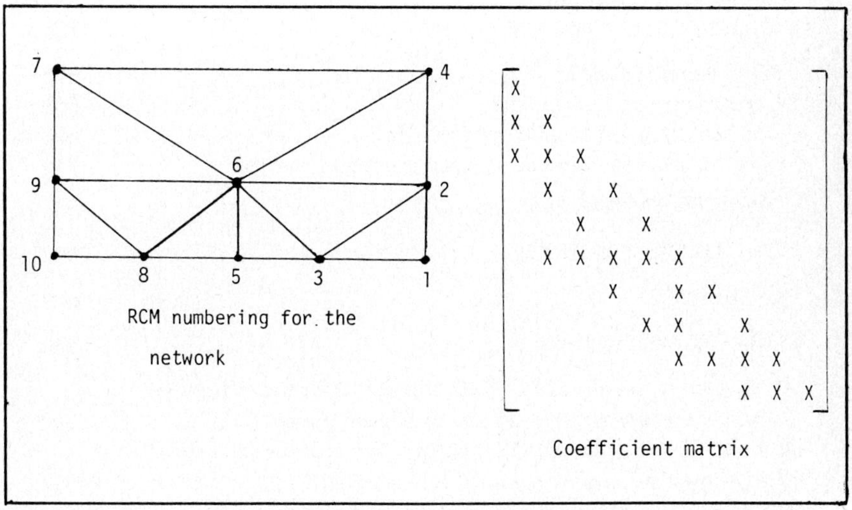

RCM numbering for the network

Coefficient matrix

CONTENTS

1. INTRODUCTION

1.1 General Introduction

Linear systems of equations occur in many diverse areas of mathematical modelling. The analysis of experimental data by linear regression and statistical estimation techniques, and linear economic models, are some areas; others include mechanical or electrical networks, structural analysis and surveying. Many of these linear systems are large and sparse. The sparsity in management science problems is due to the fact that, although the operations of a large company might entail several thousand variables, individual constraints such as those on the production of a single factory, involve only a very few variables. In physical problems, the equation associated with an individual node will involve the variable or variables belonging to that node and the variables belonging to nodes to which it is connected in the network. Because each node is likely to be connected to only a few neighbouring nodes, the matrix is sparse.

The material in this paper has been presented as a first reading for recent graduates in science and engineering to assist them in the initial stages of efficiently solving large sparse linear systems on the digital computer.

There are numerous texts which supply various aspects of this paper in great detail with regards to derivation, error analysis, etc..., through the functional analysis approach or other procedures. However, these are not always easy to read. This paper sets out as simply as possible, using numerical examples where appropriate, some of the fundamental techniques. The reader may then approach a more rigorous analysis with a better understanding. It should be noted that even when an advanced knowledge of the area has been attained, there are many variations from the standard approaches which have been found most suitable for particular modelling problems. These would be found in the appropriate applied journals.

A survey of classical matrix theory may be found in Faddeev and Faddeeva (1963) and Householder (1964). Fox (1964) and Forsythe and Moler (1967) give an elementary approach to computational linear algebra with Forsythe (1967) presenting a basic survey. Wilkinson (1961, 1963,

1965) and Wilkinson and Reinsch (1971) present an in-depth study of numerical methods in linear algebra. The basics of direct methods for the solution of linear systems are given in Young and Gregory (1973), including an outline of several misconceptions often encountered. Reviews of direct methods for large sparse linear systems are presented by George (1977), Reid (1977 a,b) and Duff (1981) while Tewarson (1973) and Jennings (1977) are texts covering this area. Some special sparse linear systems are reviewed by Fox (1977). Many researchers are currently involved in devising efficient techniques for large sparse matrix systems.

Jennings (1977) considers that matrix programs that are unnecessarily complicated, highly efficient or incapable of producing accurate solutions may be the result of their author's not having a sufficient background in large matrices. He states that a matrix program should attain a balance between simplicity, economy and versatility. It should also include the characteristic of being well-documented: to enable others, as well as the author, to follow the program readily and, perhaps, add to, or update, it.

It is always desirable to obtain a better understanding of the numerical algorithm by first practising some manual examples. With the understanding obtained through this, it is then easier to proceed to the coding of a FORTRAN program. To this end, some solutions are presented, and practice exercises, with answers, are then set. The appendices of this paper contain FORTRAN subroutines, well documented to enable easy usage and effective understanding of the procedures involved. Adherence to this computer language has been guided by the fact that most applied journals, texts, etc. produce their listings in this form and most practising engineers/scientists in this field know this language. In writing one's own subroutines for some of these methods it is useful to have available a selection of matrices to use for testing the programs. Westlake (1968) and Gregory and Karney (1969) have examples appropriate for this purpose, as well as the manual exercises.

A treatment of computational linear algebra with particular reference to large sparse linear systems must include a study of the economization of computer storage, input/output procedures and central processor unit (c.p.u.) time. It also often involves a compromise between accuracy and cost.

1.2 Mathematical introduction

A general method of representing a physical system by a mathematical model is to choose a number of points, called nodes, within the system at which equations are set up describing that aspect of the system which is of interest. We can say then, that the equations for large systems are usually associated with individual nodes and involve variables belonging to that node and to a few connected nodes, so that although the total number of variables employed may be large, the number used in any particular equation will be relatively small; i.e. each equation may contain all the variables but for all but a few the multiplying coefficients will be zero.

The system of equations considered here is a linear system which may be represented in matrix form as

$$AX = B \tag{1.2.1}$$

where A is an nxn coefficient matrix, X is the solution vector and B is the "constant" vector. If A is invertible, a solution may be found by setting

$$X = A^{-1}B \tag{1.2.2}$$

The contents of a matrix are referred to as entries rather than elements in order to avoid confusion in finite element analysis, where much discussion centres on a study of the matrices as well as on the finite elements.

Direct Methods lead to an exact solution in a finite number of steps if roundoff error is not present. Most direct methods are based on Gaussian elimination which produces a system of equations with a triangular coefficient matrix which can then be easily solved. Variants arise from the different methods of intermediate storage and utilisation of special properties of certain matrices. For a system of m linear equations, Birkhoff and Fix (1974) indicate that for m less than about 2000, direct solutions are preferred, but for m greater than 2000, iterative solutions are preferred.

A matrix is considered to be spares when it has sufficiently many zero entries to make it advantageous to employ special techniques that avoid storing or operating with the zeros. A 100 x 100 matrix with ten zero entries would simply be treated as a dense matrix. A 1000 x 1000 matrix with 4000 non-zero entries would be considered sparse.

In the analysis of a special technique for sparse matrices, it is necessary to consider "fill-ins", those zero entries that become non-zero. Direct methods lead to fill-ins being created during solution. An efficient algorithm will limit the number of fill-ins.

In considering the computer implementation of algorithms it is important to note those arithmetic processes which are more significant in the use of central processor unit time than others, whereas in accuracy considerations all the arithmetic processes are important. Multiplication and division are more time consuming than addition.

It is advantageous to the reader to be already familiar with
(i) Gaussian elimination
(ii) LU decomposition of the Crout, Doolittle, Cholesky type
(iii) positive-definite, symmetric matrices
(iv) partitioning of matrices
(v) laws of matrix algebra.

Symmetric, positive-definite matrix systems arise in situations such as the application of the finite element least squares and classic variational criteria. The large systems that occur are usually sparse. A useful introduction to the fundamentals of symmetric, positive-definite matrices is given in Dahlquist and Bjorck (1974).

George (1977) states than in comparing two different approaches to solving large sparse systems of equations there are factors such as how many problems having the same structure must be solved, computer charging algorithms etc. which need be considered. He also says that because these algorithms will be applied on a computer, any comparison must involve the computer implementation of them: some data structures exact a higher price in execution and storage than others.

Quite often, the constraint of cost is real. With unlimited finance, a very good result may be obtained. With less finance, a less accurate solution results. In the computer solution of these large systems, the cost factors are essentially

(i) central processor unit time,

(ii) storage - core and backing store,

(iii) input/output time.

The many variants of Gaussian elimination that are designed for particular structures of the sparse matrix are implemented to minimize these cost factors and minimize the accumulation of round-off error. To achieve the latter, a technique is sought to

(i) keep the number of arithmetic processes to a minimum,

(ii) avoid involving the zero entries,

(iii) avoid fill-in.

It is desirable that the zero entries do not occupy storage space.

If the mathematical model being solved is that of a continuum problem and is not data-based and there is access to a high speed digital computer, with large core, which stores its numbers to a large number of significant digits and there is no restraint on cost, then the subject of this paper probably is of little interest. However, most engineers usually do not have the benefit of this combination.

2. FORMS OF COEFFICIENT MATRICES

The coefficient matrix A may be classed as dense or sparse depending on the number of non-zero elements and in each case it can be symmetric or unsymmetric. A matrix is symmetric if $a_{ij} = a_{ji}$ for all i, j, where i is the row number and j is the colum number, i.e. it is symmetric about the main diagonal; $A = A^T$. It is also said to be <u>positive-definite</u> if $X^TAX > 0$ for all $X \neq 0$, where X^T is the transpose of X. This definition is not particularly helpful; however, it can be shown that if C is a non-singular matrix then the product $C^TC = A$ is symmetric positive-definite for any C. From this, it follows that if A is symmetric positive-definite there is a unique lower-triangular matrix L, with positive diagonal elements, such that

$$A = LL^T \tag{2.1}$$

This factorization is used in the Choleski method of solution, to be described later.

Example 1

An example of $C^TC = A$ is

$$\begin{bmatrix} 3 & 4 \\ -1 & 2 \end{bmatrix} \begin{bmatrix} 3 & -1 \\ 4 & 2 \end{bmatrix} = \begin{bmatrix} 25 & 5 \\ 5 & 5 \end{bmatrix}$$

and this is then symmetric, positive-definite. It is factorised into LL^T as

$$\begin{bmatrix} 25 & 5 \\ 5 & 5 \end{bmatrix} = \begin{bmatrix} 5 & . \\ 1 & 2 \end{bmatrix} \begin{bmatrix} 5 & 1 \\ . & 2 \end{bmatrix}$$

Although there is no clear boundary at which a matrix becomes dense or sparse we will only be dealing with cases where the non-zero entries represent, at most, a few per cent of the total so that all that follows will clearly pertain to sparse matrices and in particular to large sparse matrices although, for convenience, small matrices may be shown in the examples.

A banded matrix is one in which all the non-zeros lie in a relatively narrow region about the main diagonal. The bandwidth is derived from the maximum number of non-zero entries to any one side of the diagonal. If this number is M then the bandwidth is 2M + 1. Tridiagonal and pentadiagonal matrices are two types of banded matrices for which special algorithms exist enabling solutions to be found efficiently. Figure 2.1 provides a typical example of a tridiagonal matrix.

$$\begin{bmatrix} 1 & 2 & & & \\ 2 & 1 & 3 & & \\ & 1 & 2 & 3 & \\ & & 3 & 1 & 2 \\ & & & 1 & 2 \end{bmatrix}$$

Figure 2.1 Tridiagonal matrix

The form of the matrix may depend upon the numbering of the nodes. For the ring system shown in Figure 2.2, a matrix map of the nodal interconnections can be made in many ways. Of the two maps shown Figure 2.2(a) has a bandwidth of 6, Figure 2.2(b) has a bandwidth of 5.

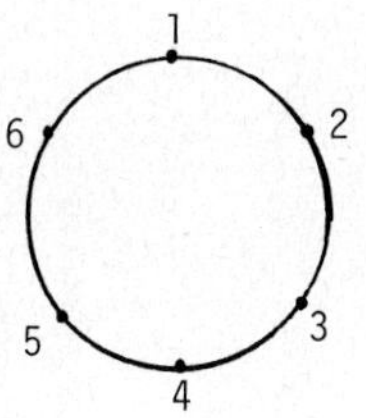

$$\begin{bmatrix} X & X & & & & X \\ X & X & X & & & \\ & X & X & X & & \\ & & X & X & X & \\ & & & X & X & X \\ X & & & & X & X \end{bmatrix} \quad \text{(a)}$$

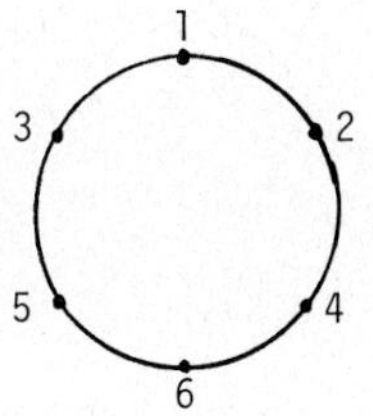

$$\begin{bmatrix} X & X & X & & & \\ X & X & & X & & \\ X & & X & & X & \\ & X & & X & & X \\ & & X & & X & X \\ & & & X & X & X \end{bmatrix} \quad \text{(b)}$$

Digraph Matrix

Figure 2.2 Bandwidths

The X in row i, column j, denotes a coefficient a_{ij} resulting from a connection from node i to node j. The bandwidth allotted to the matrix in Figure 2.2(a) does not fit the formula for a banded matrix but is the maximum bandwidth of the interconnections.

In the symmetric case shown in Figure 2.2, the directions are not shown for the connections but in the unsymmetric case directional arrows not included in the digraph (directed graph).

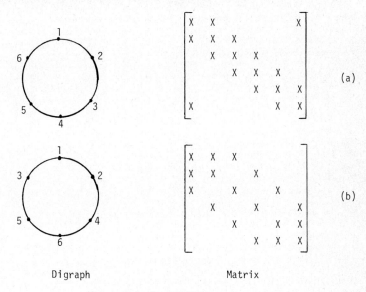

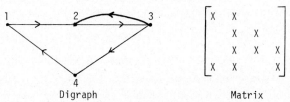

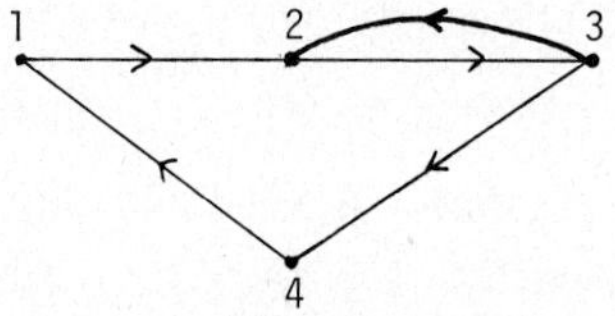

$$\begin{bmatrix} X & X & & \\ & X & X & \\ & X & X & X \\ X & X & & X \end{bmatrix}$$

Digraph Matrix

Figure 2.3 Unsymmetric matrix system

A banded matrix may have a constant bandwidth or variable bandwidth. This is, in some cases, a matter of definition, since, by including zero entries a variable bandwidth can be treated as constant. In Figure 2.4, the matrix is symmetrical, so that only the main diagonal and sub-diagonal entries need be shown to fully describe it. This corresponds with the fact that these are usually the only entries held in storage.

$$\begin{bmatrix}
X & & & & & & & & \\
X & X & & & & & & & \\
X & X & X & & & & & & \\
X & X & X & X & & & & & \\
0 & X & X & X & X & & & & \\
 & 0 & 0 & X & X & X & & & \\
 & & 0 & 0 & 0 & X & X & & \\
 & & & 0 & 0 & X & X & X & \\
 & & & & X & X & X & X & X
\end{bmatrix}$$

Figure 2.4 Banded matrix store

The non-zero entries form a matrix in which the bandwidth is not regular from row to row. Including the zeros shown forms a matrix with a regular bandwidth. Variable bandwidth matrices are also classifiable into those with re-entrant rows and those without. A re-entrant row is defined as one in which the column number of the first entry is less than the column number of the first entry of the previous row. In the variable bandwidth matrix shown in Figure 2.4, the last row is re-entrant.

When dealing with symmetric matrices, requiring only a knowledge of the diagonal and sub-diagonal entries, we often speak of the bandwidth in terms of these entries. Jennings (1977) would say that the third row of the matrix in Figure 2.4 had a semibandwidth of 3. George (1977) uses a term $B_i(A)$ which defines the bandwidth of the ith row and a term $B(A)$ as the maximum $B_i(A)$, $1 \leq i \leq n$. For the above matrix $B_3(A) = 3$ and $B(A) = 4$. The profile of A is then defined as $\sum_{i=1}^{n} B_i(A)$.

Given a matrix A with i rows and j columns, permutation matrices can operate on it to interchange rows or columns. The permutation PAP^T of the rows and columns of A can be used to arrange it into a form better suited to the solution process. It should be noted that, with this type of permutation, all the original entries are retained on the main diagonal.

<u>Example 2</u>

A permutation of a 3x3 matrix is presented as follows

$$\begin{bmatrix} 1 & 0 & 0 \\ 0 & 0 & 1 \\ 0 & 1 & 0 \end{bmatrix} \begin{bmatrix} a & b & c \\ d & e & f \\ g & h & i \end{bmatrix} \begin{bmatrix} 1 & 0 & 0 \\ 0 & 0 & 1 \\ 0 & 1 & 0 \end{bmatrix}$$

$$= \begin{bmatrix} 1 & 0 & 0 \\ 0 & 0 & 1 \\ 0 & 1 & 0 \end{bmatrix} \begin{bmatrix} a & c & d \\ d & f & e \\ g & i & h \end{bmatrix}$$

$$= \begin{bmatrix} a & c & d \\ g & i & h \\ d & f & e \end{bmatrix}$$

In this type of operation, if the ith and r^{th} columns are interchanged so also are the ith and r^{th} rows. This need not always be the case; a permutation PAQ could be used where Q is not the transpose of P.

3. STORAGE

3.1 Introduction

In the 2-dimensional array commonly used for A, each variable is allotted one column and each equation, corresponding to one node, is allotted one line. Hence any line of a large sparse matrix will consist mostly of zeros. This distribution may not be a static one, however. In the process of solution, additional non-zeros may be inserted. The extent of these fill-ins depends on the type of system and the manner of solution.

To carry out a solution, the elements of A are stored in computer core and accessed when required. If there is insufficient space in core, a backing (disc) store is used. Variables are transferred between core and backing store during the solution process. For reasons of economy, and because of limitations in core space, it is useful if it can be arranged that only non-zeros be stored or that, at least, as few zeros as possible are retained. If the system can be broken up into smaller self-consistent systems it may only be necessary to store a small part of it at any one time. As variables are eliminated during the progressive solution of the smaller sub-systems they can be transferred to disc and the storage space re-allocated to new variables.

The method of storage depends on many factors; there is one method best for all cases. A method of storage for very sparse matrices may not be as satisfactory for matrices still classifiable as sparse but with greater numbers of non-zeros.

In the following two subsections on storage, a vector is real or integer according to the Fortran convention for the first letter of its name.

3.2 Storage for a random sparse symmetric matrix

In some applications such as finite element calculations one encounters symmetric, positive-definite matrices, where the location of the nonzero elements differs greatly from row to row. In the use of the finite element least squares and finite element classic variational criteria the resultant linear system is symmetric, positive-definite. In other large scale applications, the matrices are simply symmetric. In such cases it can be more efficient to use the "profile-storage" scheme devised by Jennings (1966).

The nxn symmetric matrix A is replaced by two vectors, E and R. In E, the entries of the lower triangle of A are stored row by row starting with the first non-zero in each row and including all subsequent entries, whether zero or non-zero, as far as the diagonal. The pointer vector, R, indicates the entries of E which are on the diagonal. R(i) points to the location in E of the diagonal entry of the ith row of A.

<u>Example 3</u>

$$A = \begin{bmatrix} 7 & 3 & 0 & 0 & 0 \\ 3 & 1 & -2 & 4 & 0 \\ 0 & -2 & 2 & 0 & 0 \\ 0 & 4 & 0 & 8 & -1 \\ 0 & 0 & 0 & -1 & 6 \end{bmatrix}$$

Using "profile storage", the vectors E and R are formed as follows:

$$E = (7,3,1,-2,2,4,0,8,-1,6)$$
$$R = (1,3,5,8,10)$$

Advantage or disadvantage?

In this example the lower triangle submatrix of A has 15 entries whereas the combined E and R vectors have 15 entries, so there has been no gain in storage. This is typical of small matrices in which storage is never of concern. However, for a large sparse matrix of this type the gain would be considerable. A 100x100 matrix with 400 nonzero entries has a lower triangular submatrix of 5050 entries, but, using "profile storage", vector E would have, at most, 300 entries and vector R would have exactly 100 entries, making a total of 400 in comparison to the 5050.

Exercise 1

1. Use the "profile-storage" scheme for symmetric matrices to formulate the E and R arrays from each of the following:

(a) The 100x100 matrix

$$\begin{bmatrix} 2 & -1 & & & & & \\ -1 & 2 & -1 & & & & \\ & -1 & 2 & -1 & & & \\ & & & \cdots\cdots & & & \\ & & & & -1 & 2 & -1 \\ & & & & & -1 & 1 \end{bmatrix}$$

(b) $$\begin{bmatrix} 38 & 19 & -21 \\ 19 & 21 & -12 \\ -21 & -12 & 14 \end{bmatrix}$$

(c) $$\begin{bmatrix} 5 & 3 & 0 & 1 & -6 \\ 3 & 7 & 0 & 2 & 0 \\ 0 & 0 & 2 & 0 & 4 \\ 1 & 2 & 0 & 6 & -1 \\ -6 & 0 & 4 & -1 & 11 \end{bmatrix}$$

(d)
$$\begin{bmatrix} 2 & 0 & -3 & 0 & 0 \\ 0 & -4 & -1 & 2 & 0 \\ -3 & -1 & 1 & 0 & 4 \\ 0 & 2 & 0 & -6 & 0 \\ 0 & 0 & 4 & 0 & 2 \end{bmatrix}$$

Answers

(a) E = (2,-1,2,-1,2,-1,2,-1,2..........2,-1,1) : 199 entries.
R = (1,3,5,7,9,...................,197,199) : 100 entries.

(b) E = (38,19,21,-21,-12,14)
R = (1,3,6)

(c) E = (5,3,7,2,1,2,0,6,-6,0,4,-1,11)
R = (1,3,4,8,13)

(d) E = (2,-4,-3,-1,1,2,0,-6,4,0,2)
R = (1,2,5,8,11)

3.3 Storage for a random sparse unsymmetric matrix

In applications such as the computational design of electrical networks, the matrices are unsymmetric with highly irregular sparseness. In the use of the finite element Galerkin criterion for nonlinear problems, the resultant linear system is unsymmetric. Hence the "profile-storage" scheme is not appropriate.

The Gustavson (1972) method, describing the data structure of an unsymmetric matrix, A, with irregular sparseness by means of three vectors, is outlined as follows. Three vectors AH, IH, IA replace matrix A. Vector AH contains the non-zero entries of A row by row; the zero entries are not stored. Vector IH contains the column number of the corresponding entries of AH such that IH(j) contains the column number of the entry AH(j). Vector IA is constructed so that IA(i) gives the position in the vectors AH and IH of the first entry of the ith row of A and the last additional entry in IA is equal to the total number of elements in AH plus one.

Example 4

$$A = \begin{bmatrix} 3 & 0 & -4 & 1 & 0 \\ 0 & -2 & 3 & 0 & 0 \\ 0 & -1 & 4 & 0 & -1 \\ -2 & 0 & 0 & -3 & 0 \\ 0 & -5 & 6 & 0 & 2 \end{bmatrix}$$

Using the Gustavson scheme, vectors AH, IH and IA are formed as follows:

AH = (3,-4,1,-2,3,-1,4,-1,-2,-3,-5,6,2)

IH = (1,3,4,2,3,2,3,5,1,4,2,3,5)

IA = (1,4,6,9,11,14)

Advantage or disadvantage?

In this example, matrix A has 25 entries whereas the combined AH, IH and IA vectors have 32 entries. Therefore for this example the system has proven to be a disadvantage. It is reasonable to expect that there will be little gain in the application of this scheme to small matrices or to matrices with only a small percentage of the entries as zeros. This particular example falls into both categories. However, it can readily be seen that for a large sparse unsymmetric matrix of order 1000x1000 with a random sparseness giving, at most, 30 non-zeros in each row that, instead of storing 10^6 entries, the maximum for the total of the three vectors AH, IH and IA would be $3.10^4 + 3.10 + 1001 \doteqdot 6.10^4$ entries: a significant reduction. This reduction would be more pronounced for even larger systems.

Exercise 2

1. Use the Gustavson storage scheme for unsymmetric matrices to formulate arrays AH, IH and IA from each of the following:

(a)
$$\begin{bmatrix} 5 & 0 & -2 & 0 & 0 \\ 0 & 7 & 0 & 1 & 0 \\ 3 & 0 & 2 & 0 & 1 \\ 0 & -5 & 6 & 0 & 0 \\ 0 & 0 & 1 & 0 & 4 \end{bmatrix}$$

(b)
$$\begin{bmatrix} 7 & 0 & -3 & 0 & -1 & 0 \\ 2 & 8 & 0 & 0 & 0 & 0 \\ 0 & 0 & 1 & 0 & 0 & 0 \\ -3 & 0 & 0 & 5 & 0 & 0 \\ 0 & -1 & 0 & 0 & 4 & 0 \\ 0 & 0 & 0 & -2 & 0 & 6 \end{bmatrix}$$

(c)
$$\begin{bmatrix} -2.4 & 7.6 & 0 & 0 & 0 & 8.9 \\ 3.8 & -0.4 & 0 & 2.9 & 0 & 0 \\ 0 & 0 & 3.1 & -5.6 & 11.3 & 0 \\ -1.4 & 0 & 0 & 6.7 & 0 & 4.9 \\ 3.8 & 1.4 & 0 & 0 & -2.9 & 0 \\ 7.2 & 0 & 0 & 0 & -1.6 & 5.3 \end{bmatrix}$$

Answers

(a) AH = (5,-2,7,1,3,2,1,-5,6,1,4)
IH = (1,3,2,4,1,3,5,2,3,3,5)
IA = (1,3,5,8,10,12)

(b) AH = (7,-3,-1,2,8,1,-3,5,-1,4,-2,6)
IH = (1,3,5,1,2,3,1,4,2,5,4,6)
IA = (1,4,6,7,9,11,13)

(c) AH = (-2.4,7.6,8.9,3.8,-0.4,2.9,3.1,-5.6,11.3,-1.4,6.7,4.9,3.8, 1.4,-2.9,7.2,-1.6,5.3)
IH = (1,2,6,1,2,4,3,4,5,1,4,6,1,2,5,1,5,6)
IA = (1,4,7,10,13,16,19)

A similar scheme has been incorporated by Fletcher (1976) in a subroutine called SPAREL which solves a system of linear algebraic equations using Gaussian elimination. A documented and ordered listing of SPAREL has been given in appendix B of this paper. Fletcher has arrays functioning identically to vectors AH and IH but, instead of the vector IA giving the position in AH and IH of the first entry of each row, he has a vector JEND such that JEND(i) indicates the number of non-zero entries in the ith row, and a separate variable LCMAX contains the total number of non-zeros in matrix A. The overall effect of the Fletcher scheme is the same as the Gustavson scheme. There may be other similar schemes.

Example 5

For the model example 4, the Fletcher scheme forms AH, IH, and JEND and LCMAX as follows:

AH = (3,-4,1,-2,3,-1,4,-1,-2,-3,-5,6,2)
IH = (1,3,4,2,3,2,3,5,1,4,2,3,5)
JEND = (3,2,3,2,3)
LCMAX = 13

Thus the procedure is to indicate the column number of each non-zero entry in one vector, which has a one-to-one ratio with a vector containing the non-zero entries; then a further array is needed to indicate when each new row commences. This is necessary because of problems occurring in situations such as those in which the only non-zero entry in row 3 is in column 2 and the only non-zero entry in row 4 is in column 3. It would be an easy mistake to consider that they belong to the same row unless there is this further constraint.

In using SPAREL, it is necessary to set a further vector B to store the right-hand side of the system of equations and the identifier N indicating the number of equations to be solved. The solution returns in vector X.

Exercise 3

1. Arrange the system AX = B into the appropriate identifiers and arrays for solution by subroutine SPAREL (in Appendix C), where matrices A and B are given as

$$A = \begin{bmatrix} 5 & -3 & 0 & 0 & 2 \\ 0 & 2 & 4 & -1 & 0 \\ -1 & 0 & -3 & 0 & 0 \\ 2 & 0 & 0 & -5 & 4 \\ 0 & 0 & 0 & 3 & 6 \end{bmatrix} \qquad B = \begin{bmatrix} 11 \\ 21 \\ 37 \\ 14 \\ 5 \end{bmatrix}$$

Answers

AH = (5,-3,2,2,4,-1,-1,-3,2,-5,4,3,6)
IH = (1,2,5,2,3,4,1,3,1,4,5,4,5)
JEND = (3,3,2,3,2)
B = (11,21,37,14,5)
LCMAX = 13
N = 5

3.4 Storage for banded matrices

A method used for storing a symmetric banded matrix is called diagonal storage. All the diagonals of the lower triangle which contain non-zero entries are stored in a rectangular array with n rows and a number of columns equal to the maximum $B_i(A) + 1$. The A_{ij} are submatrices of the symmetric matrix and $A_{ij} = A_{ji}$.

$$\begin{bmatrix} A_{11} & A_{12} & & A_{14} & & & \\ A_{21} & A_{22} & & & & & \\ & & A_{33} & & A_{35} & A_{36} & \\ A_{41} & & & A_{44} & A_{45} & & \\ & & A_{53} & & A_{55} & & A_{57} \\ & & A_{63} & & & A_{66} & A_{67} \\ & & & & A_{75} & A_{76} & A_{77} \end{bmatrix}$$

A (symmetric)

$$\begin{bmatrix} & & & A_{11} \\ & & A_{21} & A_{22} \\ & & & A_{33} \\ A_{41} & & & A_{44} \\ & A_{53} & A_{54} & A_{55} \\ A_{63} & & & A_{66} \\ & A_{75} & A_{76} & A_{77} \end{bmatrix}$$

Figure 3.1 Diagonal storage of A

If there is wide variation in the value of $B_i(A)$ this method may be quite wasteful of storage space; in which case the method of profile storage, previously described, could be used.

4. METHODS OF SOLUTION

4.1 Gaussian elimination

Direct methods of solution are based on the method of Gaussian elimination.

Let A be a given square matrix of order n and B a given n-vector.

The linear system of n equations

$$\begin{aligned} a_{11}x_1 + a_{12}x_2 + a_{13}x_3 + \dots + a_{1n}x_n &= b_1 \\ a_{21}x_1 + a_{22}x_2 + a_{23}x_3 + \dots + a_{2n}x_n &= b_2 \\ \dots\dots\dots\dots\dots\dots\dots\dots\dots\dots \\ \dots\dots\dots\dots\dots\dots\dots\dots\dots\dots \\ a_{n1}x_1 + a_{n2}x_2 + a_{n3}x_3 + \dots + a_{nn}x_n &= b_n \end{aligned} \qquad (4.1.1)$$

can be put in matrix form

$$\begin{bmatrix} a_{11} & a_{12} & a_{13} & \dots & a_{1n} \\ a_{21} & a_{22} & a_{23} & \dots & a_{2n} \\ \dots & \dots & \dots & \dots & \dots \\ \dots & \dots & \dots & \dots & \dots \\ a_{n1} & a_{n2} & a_{n3} & \dots & a_{nn} \end{bmatrix} \begin{bmatrix} x_1 \\ x_2 \\ . \\ . \\ x_n \end{bmatrix} = \begin{bmatrix} b_1 \\ b_2 \\ . \\ . \\ b_n \end{bmatrix} \qquad (4.1.2)$$

or, more briefly, as

$$AX = B \qquad (4.1.3)$$

In order to solve the system, the matrix A is reduced to upper triangular form and the solution found by back-substitution. The method depends on the fact that the solution is unaffected by

(i) interchanging equations.

(ii) adding a scalar multiple of one equation to another.

The whole process can be outlined as

$$AX = LUX = LY = B \qquad (4.1.4)$$

$$UX = Y = L^{-1}B \qquad (4.1.5)$$

where L is lower triangular and U is upper triangular. X can be found by back-substitution commencing with $x_n = y_n/u_{nn}$.

In more detail, the elimination process leading to the desired upper-triangular form can be summarised as follows

$$a_{ij}^{(k+1)} = a_{ij}^{(k)} - \frac{a_{ik}^{(k)}}{a_{kk}^{(k)}} . a_{kj}^{(k)} \qquad (i,j > k) \qquad (4.1.6)$$

$$b_i^{(k+1)} = b_i^{(k)} - \frac{a_{ik}^{(k)}}{a_{kk}^{(k)}} . b_k^{(k)} \qquad (i > k) \tag{4.1.7}$$

expressing the operations at the kth step, $k = 1,2,\ldots,n-1$ commencing with $A^{(1)} = A$, $B^{(1)} = B$. This leads to the desired upper triangular system, which, on back substitution, gives

$$x_k = \frac{1}{a_{kk}^{(k)}}\left(b_k^{(k)} - \sum_{j=k+1}^{n} a_{kj}^{(k)} x_j \right) \qquad k = n,n-1,\ldots,1 \tag{4.1.8}$$

During the process each $a_{ij}^{(k+1)}$ can replace $a_{ij}^{(k)}$ in storage and the multipliers $\ell_{ik} = a_{ik}^{(k)}/a_{kk}^{(k)}$ may overwrite $a_{ik}^{(k)}$. We thus have obtained the triangular factorisation

$$A = LU \tag{4.1.9}$$

where L is unit lower triangular with off-diagonal elements ℓ_{1k} and U is the upper triangular matrix $[a_{ij}^{(i)}]$, $i \leq j$. The first few terms of L are

$$\begin{bmatrix} 1 & & & \\ (a_{21}/a_{11})^{(1)} & 1 & & \\ (a_{31}/a_{11})^{(1)} & (a_{32}/a_{22})^{(2)} & 1 & \\ \cdots\cdots & \cdots\cdots & \cdots\cdots & \end{bmatrix} \tag{4.1.10}$$

When a term such as $a_{ij}^{(k)}$ is a zero, $a_{ij}^{(k+1)}$ may be a new non-zero for which storage space needs to be found if, say, a Gustavson type storage method is used. This problem is dealt with by Reid (1977). Gustavson storage requires that a fresh copy be made of the row (or column) thereby temporarily wasting the space occupied by the old row (or column). A means of avoiding this is to use a linked list. The non-zeros are stored as triples (a_{ij},i,j) held contiguously, in any order, in a real array and two integer arrays. The integers indicate where the non-zero belongs in the matrix. Two more integers are held with each non-zero to give the address of the next non-zero in the row and the address of the next non-zero in the column. It is these latter two integers which expand the structure into a linked list. Pointers are also required to the first elements in the rows and columns. Thus rows and/or columns can be accessed without scanning the whole array and non-zeros can be deleted or inserted without performing a major operation involving the whole row or column.

4.2 Block-triangular form

The number of fill-ins can be limited by interchanging rows and columns. A simple case is illustrated below

$$\begin{bmatrix} X & X & X & X & X \\ X & X & & & \\ X & & X & & \\ X & & & X & \\ X & & & & X \end{bmatrix} \qquad \begin{bmatrix} X & & & & X \\ & X & & & X \\ & & X & & X \\ & & & X & X \\ X & X & X & X & X \end{bmatrix}$$

(a) (b)

Figure 4.1 Sparse matrices

The matrix in Figure 4.1(b) is the matrix in Figure 4.1(a) permuted so that the first and last rows and the first and last columns are interchanged. Elimination applied to the former results in the creation of a totally dense matrix, whereas no new non-zeros would be created in the latter.

It is obvious that fill-in is avoided with a triangular matrix, since forward or back substitution can be commenced immediately, but a more general case is for a matrix to be block-triangular as illustrated in (4.2.1), where each diagonal block A_{kk} is a square matrix

$$A = \begin{bmatrix} A_{11} & & & \\ A_{21} & A_{22} & & \\ \cdot & & A_{33} & \\ \cdot & & \cdot & \\ \cdot & & & \ddots \\ A_{n1} & A_{n2} & \cdots & A_{nn} \end{bmatrix} \qquad (4.2.1)$$

The equation $AX = B$ is now broken up into a sequence of smaller problems

$$A_{kk}X_k = B_k - \sum_{j=1}^{k-1} A_{kj}X_j \qquad k = 1,2,\ldots,n \qquad (4.2.2)$$

Having solved for A_{11}, for instance, all the variables in A_{21} are known so we need only solve for the unknowns in A_{22} at the next step and so on. Any fill-in occurring will always be confined to the diagonal blocks A_{kk}.

There are several algorithms for forming a block-triangular matrix, where it is possible to do so. The first step is to find a transversal i.e. set non-zeros on the diagonal. Reid (1977a) describes a method due to Hall (1956) and also outlines methods of finding the block-triangular form due to Sargent and Westerberg (1964) and Tarjan (1972). Duff and Reid (1976) have provided a Fortran program to implement the method of Tarjan.

If a matrix has t non-zeros, no more than O(t) operations are needed to place one of them on the main diagonal and it follows that no more than O(nt) operations are needed to find a transversal.

The above-mentioned algorithms will not be described here, but the following example may give some idea of what is involved in the process of block-triangularisation. Starting with a matrix A which has zeros on the main diagonal we show it after a transversal has been found, as illustrated in Figure 4.2. A digraph is then constructed which clearly shows that there are three composite nodes. Renumbering the ordinary nodes of the composite nodes is equivalent to a permutation of A which can result in the block-triangular form, as shown in Figure 4.3.

<u>Exercise 4</u>

1. Write down a matrix for an 8x8 system which replaces columns 3,5,6 and 8 with columns 6,8,5 and 3 respectively.

2. Write down a matrix which permutes a square matrix of order 7 such that it replaces rows 1,3,4 and 7 with rows 4,7,1 and 3 respectively.

Answers

(1)
$$\begin{bmatrix} 1 & & & & & & & \\ & 1 & & & & & & \\ & & \cdot & & & 1 & & \\ & & & 1 & & & & \\ & & & & \cdot & & & 1 \\ & & & & 1 & \cdot & & \\ & & & & & & 1 & \\ & & 1 & & & & & \cdot \end{bmatrix}$$

(2)
$$\begin{bmatrix} \cdot & & & 1 & & & \\ & 1 & & & & & \\ & & \cdot & & & & 1 \\ 1 & & & \cdot & & & \\ & & & & 1 & & \\ & & & & & 1 & \\ & & 1 & & & & \cdot \end{bmatrix}$$

$$\begin{bmatrix} X & X & X & & & & & \\ X & X & X & & & & & \\ & X & & X & & & & X \\ & & & X & & & & X \\ & & & X & X & X & X & \\ & & & & X & & X & \\ & & & & X & X & & \\ X & X & X & & & & & \end{bmatrix} \qquad \begin{bmatrix} X & X & & & & & & X \\ X & X & & & & & & X \\ & X & X & X & & & & \\ & & X & X & & & & \\ & & & X & X & X & X & \\ & & & & X & X & & \\ & & & & X & & X & \\ X & X & & & & & & X \end{bmatrix}$$

(a) Before transversal (b) After transversal

Figure 4.2

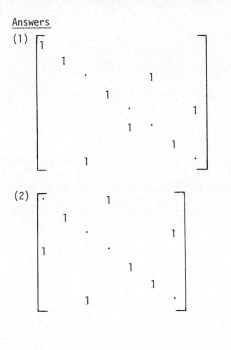

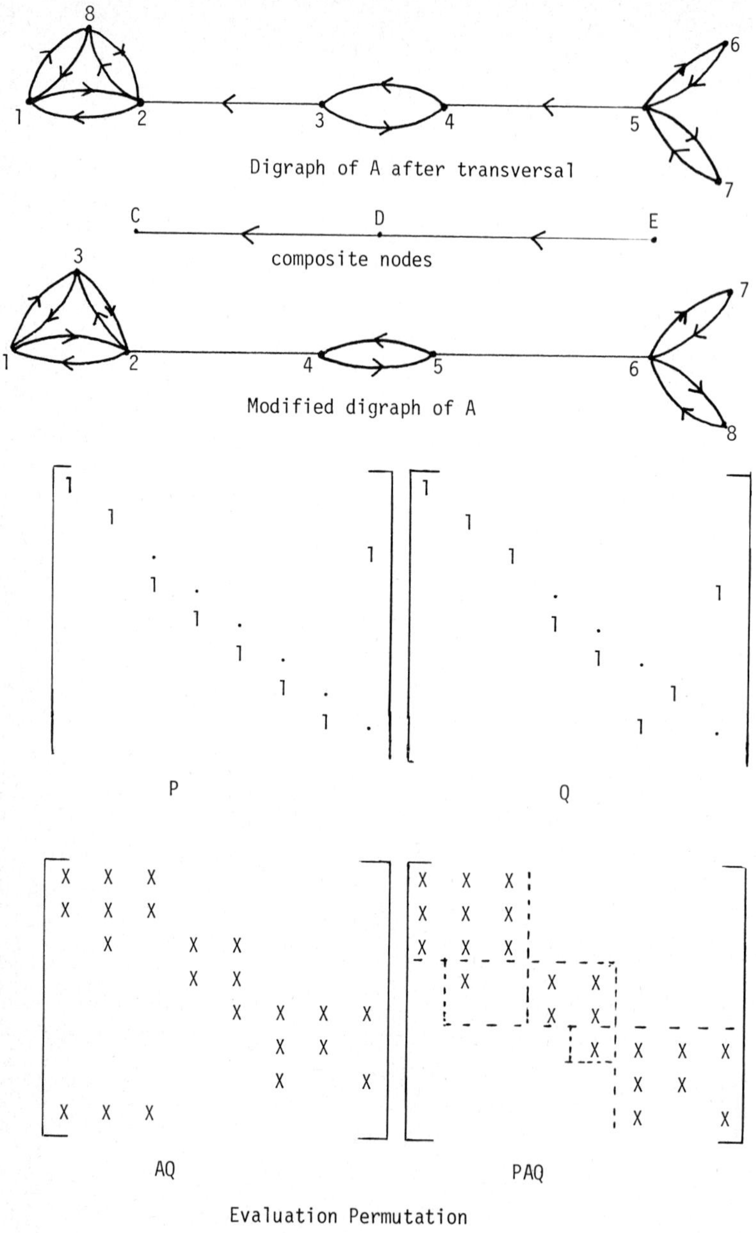

Figure 4.3 Digraph and Permutation

4.2.1 Pivoting

Numerical instability can arise when large numbers are added to small numbers in the elimination process as information present in the smaller number may be lost. For example in 4 decimal arithmetic

$$1.248 + 201.1 = 202.3 \tag{4.2.3}$$

and the last two digits of 1.248 have been lost. The large number could arise when one number is divided by a much smaller number. Pivoting is a strategy used to avoid this and to obtain stability in the process of solution by Gaussian elimination. Rows or columns are interchanged to ensure that the pivoting entry $a_{kk}^{(k)}$ of equations (4.1.6) and (4.1.7) satisfies one of the conditions

$$|a_{kk}^{(k)}| \geq u.\max_{k\leq i\leq n} |a_{ik}^{(k)}| \quad \text{(column search)} \tag{4.2.4}$$

$$|a_{kk}^{(k)}| \geq u.\max_{k\leq j\leq n} |a_{kj}^{(k)}| \quad \text{(row search)} \tag{4.2.5}$$

where u is a parameter chosen in the range $0 < u \leq 1$. Then, from equation (4.1.6), the inequality

$$|a_{ij}^{(k+1)}| \leq (1 + \frac{1}{u})\max_{i,j}|a_{ij}^{(k)}| \tag{4.2.6}$$

may be deduced, and we see that the growth in size of matrix entries is limited at each stage to the factor $1 + \frac{1}{u}$. Ordinary partial pivoting, as used with a dense matrix, corresponds to a choice of $u = 1$. If this were the case with a sparse matrix, the severe restriction in choice of pivots would greatly reduce the possibility of block-triangulation. Fortunately, this is not the case; u may be chosen less than 1, thus allowing exploitation of the sparsity. Curtis and Reid (1976) recommend $u = \frac{1}{4}$ on the grounds of numerical experiments. For linear programming matrices, Tomlin (1972) recommends $u = 1/100$.

From equation (4.1.6) it will be seen that the entries in a column only change if there is a non-zero in that column in the pivot row. Thus, overall growth of an entry in column j is limited by a factor

$$(1 + \frac{1}{u})^{p_j - 1} \tag{4.2.7}$$

where p_j is the total number of non-zeros, including fill-ins, in column j. For a sparse matrix, p_j may be small, so the factor may not be enormous, in contrast to the case of a dense matrix if values of $u \ll 1$ were used. However, even for a sparse matrix the growth factor as calculated above may be too large. In actuality, growth is likely to be much smaller than the value calculated in this manner and it may be advisable to insert a routine in the program to monitor the size of the entries. If the growth is too great, a fresh start with a new value of u is necessary.

Finally, although pivoting involves row or column interchanges, they need not be physically interchanged. Permutation vectors of length n can be used in which the interchanges can be recorded. The row permutation vector is initially $(1,2,3,\ldots n)^T$ corresponding to the n rows. To exchange rows i and j, the ith and j^{th} entries of the permutation vector are exchanged.

In practice the criterion of Markowitz (1957) has been found generally satisfactory in choosing a pivot entry for a sparse matrix. A non-zero in $A^{(k)}$ is found which satisfies (4.2.4) and (4.2.5) and has the smallest product of number of other non-zeros in its row and number of other non-zeros in its column and this is then brought into the pivotal position (k,k) by interchanging rows and columns. For example, if two suitable entries are found, one of which has 5 other non-zeros in the row and 3 in the column and the other has 4 and 4 respectively, the first one will be chosen as a pivot since the product 5 x 3 = 15 is less than 4 x 4 = 16.

A fresh search is made at each stage of the elimination, requiring that the sparsity pattern be known at each stage, so it must be held in a data structure and updated at each stage. This is a disadvantage which can be overcome by ordering the columns and processing them one by one, i.e. all operations for the jth column are performed before operations on the $(j + 1)^{st}$ column. This can be done if the multipliers $a_{ik}^{(k)}/a_{kk}^{(k)}$ are stored. Duff and Reid (1974) have found experimentally that it is satisfactory for the columns to be chosen in order of increasing numbers of non-zeros. The pivot $a_{kk}^{(k)}$ is chosen by searching the kth column of the rearranged matrix for an entry $a_{ik}(i \geq k)$ satisfying the stability criterion and whose row has least non-zeros in the original matrix. The sparsity pattern is not updated. This produces more fill-ins than the original Markowitz criterion but against this must be weighed the simplicity of the method. The following figures are given by Reid (1977b)

Matrix order		54	57	199
Non-zeros in A		291	281	701
Non-zeros in L/U	Markowitz	381	315	1387
	a priori using A	475	408	2767

Table 4.1

Wilkinson (1974a) provides some basic ideas on the introduction of pivoting.

4.2.2 Scaling

If the ith row of the coefficient matrix and right-hand side vector is multiplied by a number k_i, the solution of the system is not affected, so we could choose k_i such that a_{ii} becomes the largest entry in the ith column and if a pivot entry is to be selected purely on the basis of it being the largest number then, for partial pivoting by rows, this would seem a suitable choice. However, multiplying the ith row by k_i does not change the error propagation as can be shown by multiplying each row by some number. From (4.1.6)

$$a_{ij}^{(k+1)} = \left[a_{ij}^{(k)} - \frac{a_{ik}^{(k)}}{k_k a_{kk}^{(k)}} \cdot k_k a_{kj}^{(k)}\right] (i,j > k) \qquad (4.2.8)$$

$$= \left[a_{ij}^{(k)} - \frac{a_{ik}^{(k)}}{a_{kk}^{(k)}} \cdot a_{kj}^{(k)}\right] \qquad (4.2.9)$$

and the relative error due to the subtraction is unchanged.

In fact, leaving the equations unchanged and selecting the largest pivot is no different from the above process since we can consider each row as having already been multiplied by some constant. In order to avoid what amounts to an arbitrary choice of pivot, the matrix is scaled so that the rows are comparable in some way. Various methods of scaling have been suggested. The rows could be normalised by dividing each row by its largest entry or dividing by the average magnitude of its coefficients or by the square root of the sum of the squares of its coefficients. The scaled values should then by used in choosing the pivot entry. The scheme of Hamming (1971) is quite effective.

The process described above implies that an explicit scaling of the rows is carried out but this need not be so. Having calculated the factor by which each row should be multiplied we use this when choosing a pivot; i.e. we compare the magnitudes of the products of the entries under consideration and their associated scale factors but we leave the original matrix entries unchanged for the solution process, as scaling them could introduce an error due to the operation of division, and as we have seen scaling does not of itself reduce the error.

Scaling by columns is also a possible process but this requires re-ordering the solution entries and row pivoting is generally preferred. Full pivoting would involve interchanges of rows and columns but it is generally considered that the gain achieved is usually not sufficient to warrant the extra work involved. Again, cost is an important factor.

Whilst scaling is essential when working with a full matrix, its necessity in the case of a sparse matrix depends on the degree of sparsity. For very sparse matrices it could be omitted.

It would, however, always seem advisable to inspect the linear system to see if there is any obvious disparity in the magnitudes of the entries in the matrices.

Scaling across a row is a reasonably easy procedure to incorporate into a program. Scaling down a column requires more accounting as this then alters the value of the unknown relevant to that column in the resulting system. The following simple example illustrates this.

Example 6

$$356\ x_1 + 5.4\ x_2 = -\ 7.3$$
$$-\ 724\ x_1 + 6.7\ x_2 = 8.9$$

For this system, divide column 1 by 100 to give a system where entries are of the same order as follows

$$3.56\ x_1^* + 5.4\ x_2 = -\ 7.3$$
$$-\ 7.24\ x_1^* + 6.7\ x_2 = 8.9$$

It is necessary to note that, on solution, the x_1^* must be divided by the division factor of 100 to obtain the required x_1.

Scaling matrices automatically still remains an area of research with quite a deal of support for the scheme of Hamming (1971). This technique will not be described here. Curtis and Reid (1972) researched the area of automatic scaling of matrices for Gaussian elimination and were in favour of the method of Hamming.

4.3 Cholesky method

Symmetric positive-definite matrices usually arise as a result of the finite element method particularly with the least squares and classic-variational criteria. If

$$A^{(k)} = a_{ij}^{(k)} \qquad i \geq j,\ j \geq k \tag{4.3.1}$$

is symmetrical, then so also is

$$A^{(k+1)} = a_{ij}^{(k+1)} \qquad i \geq k+1,\ \ j \geq k+1 \tag{4.3.2}$$

and only the lower triangular part need be computed and stored. For this case it has been shown by Wilkinson (1967) that

$$\max_{i,j}|a_{ij}^{(k+1)}| \leq \max_{i,j}|a_{ij}^{(k)}| \qquad (4.3.3)$$

i.e. there is no growth in the size of the matrix entries and the system is stable. Consequently, symmetric interchanges can be made without regard to the required sparsity pattern. The resulting matrix PAP^T has a triangular factorisation LL^T and L can be computed by the Cholesky algorithm. The Markowitz pivoting strategy of choosing the diagonal non-zero with least other non-zeros in its row can be used, so minimising fill-in.

If triangular decomposition is possible, a matrix, A, can be expressed as the product of two matrices

$$A = LU \qquad (4.3.4)$$

where L is lower triangular and U is upper triangular. If A is symmetric positive-definite then U can be the transpose of L and

$$A = LL^T \qquad (4.3.5)$$

Only the coefficients of L need be found and stored and these can be calculated by formula. A further condition on the L is that it has positive entries on the diagonal.

The method can be illustrated by means of the following 4x4 system.

$$\begin{bmatrix} a_{11} & a_{12} & a_{13} & a_{14} \\ a_{21} & a_{22} & a_{23} & a_{24} \\ a_{31} & a_{32} & a_{33} & a_{34} \\ a_{41} & a_{42} & a_{43} & a_{44} \end{bmatrix} = \begin{bmatrix} \ell_{11} & & & \\ \ell_{21} & \ell_{22} & & \\ \ell_{31} & \ell_{32} & \ell_{33} & \\ \ell_{41} & \ell_{42} & \ell_{34} & \ell_{44} \end{bmatrix} \cdot \begin{bmatrix} \ell_{11} & \ell_{12} & \ell_{13} & \ell_{14} \\ & \ell_{22} & \ell_{23} & \ell_{24} \\ & & \ell_{33} & \ell_{34} \\ & & & \ell_{44} \end{bmatrix} \qquad (4.3.6)$$

where the u_{ij} have been replaced with the ℓ_{ij}, and $\ell_{ij} = \ell_{ji}$. Multiplying the two matrices on the RHS and equating the resulting coefficients to the coefficients of the LHS give

$$\left.\begin{aligned}
\ell_{11} &= \sqrt{a_{11}} \\
\ell_{i1} &= a_{i1}/\ell_{11} \qquad\qquad i = 2,3,\ldots,n \\
\ell_{jj} &= \sqrt{a_{jj} - \sum_{k=1}^{j-1} \ell_{jk}^2} \\
\ell_{ij} &= \frac{1}{\ell_{jj}}\left(a_{ij} - \sum_{k=1}^{j-1} \ell_{ik}\ell_{jk}\right) \qquad i = j+1,j+2,\ldots,n \\
\ell_{ij} &= 0 \quad \text{for} \quad i < j
\end{aligned}\right\} j = 2,3,\ldots,n \qquad (4.3.7)$$

Then

$$LL^TX = B \qquad (4.3.8)$$

and putting $L^TX = Y$ gives

$$LY = B \qquad (4.3.9)$$

from which Y can be found by forward substitution and then from

$$L^TX = Y \qquad (4.3.10)$$

X can be found by back substitution.

Example 7

Form A as the product C^TC and then derive L, given

$$C = \begin{bmatrix} 1 & 3 & 2 \\ 2 & 2 & 1 \\ 1 & 0 & 1 \end{bmatrix}$$

Solution

$$\begin{bmatrix} 1 & 2 & 1 \\ 3 & 2 & 0 \\ 2 & 1 & 1 \end{bmatrix} \begin{bmatrix} 1 & 3 & 2 \\ 2 & 2 & 1 \\ 1 & 0 & 1 \end{bmatrix} = \begin{bmatrix} 6 & 7 & 5 \\ 7 & 13 & 8 \\ 5 & 8 & 6 \end{bmatrix}$$

(Note that the largest number in A is always on the main diagonal)

$$\ell_{11} = \sqrt{a_{11}} = \sqrt{6} = 2.4495$$

$$\ell_{21} = \frac{a_{21}}{\ell_{11}} = \frac{7}{\sqrt{6}} = 2.8577$$

$$\ell_{31} = \frac{a_{31}}{\ell_{11}} = \frac{5}{\sqrt{6}} = 2.0412$$

$$\ell_{22} = \sqrt{a_{22} - \ell_{21}{}^2} = \sqrt{13 - (2.8577)^2} = 2.1985$$

$$\ell_{32} = \frac{1}{\ell_{22}}(a_{32} - \ell_{31}\ell_{21}) = \frac{1}{2.1985}(8 - 2.0412 \times 2.8577) = 0.9856$$

$$\ell_{33} = \sqrt{a_{33} - \ell_{31}{}^2 - \ell_{32}{}^2} = \sqrt{6 - (2.0412)^2 - (0.09856)^2} = 0.9285$$

$$L = \begin{bmatrix} 2.4495 & & \\ 2.8577 & 2.1985 & \\ 2.0412 & 0.9856 & 0.9285 \end{bmatrix}$$

<u>Exercise 5</u>

1. Find the lower-triangular matrix L with positive diagonal entries, such that

$$A = LL^T$$

for the following symmetric, positive-definite matrices, A.

(a) $\begin{bmatrix} 1 & 2 & 6 \\ 2 & 5 & 15 \\ 6 & 15 & 46 \end{bmatrix}$ (b) $\begin{bmatrix} 14 & 16 & 16 \\ 16 & 21 & 19 \\ 16 & 19 & 29 \end{bmatrix}$

<u>Answers</u>

(a) $\begin{bmatrix} 1 & & \\ 2 & 1 & \\ 6 & 3 & 1 \end{bmatrix}$ (b) $\begin{bmatrix} 3.74 & & \\ 4.28 & 1.65 & \\ 4.28 & 0.41 & 3.24 \end{bmatrix}$

2. Solve, using Cholesky factorisation, AX = B where

$$A = \begin{bmatrix} 14 & 16 & 16 \\ 16 & 21 & 19 \\ 16 & 19 & 29 \end{bmatrix} \qquad B = \begin{bmatrix} -1 \\ 2 \\ -3 \end{bmatrix}$$

given that A is symmetric, positive-definite.

Answer

$$L = \begin{bmatrix} 3.74 & & \\ 4.28 & 1.65 & \\ 4.28 & 0.41 & 3.24 \end{bmatrix}$$

$Y = (-0.27,\ 1.91,\ -0.82)$
$X = (1.38,\ 1.22,\ -0.25)$

The Cholesky method is sometimes written as

$$A = S^T S$$

where S is an upper-triangular matrix with positive entries on the diagonal. The algorithm is then written slightly differently; the effect, of course, is the same.

5. BAND MATRICES

5.1 Introduction

The simplest type of band matrix is one with constant bandwidth. Although they seem to be good candidates for iterative methods of solution, special algorithms exist which make direct methods preferable. The algorithms are much more efficient than Gaussian elimination on a full matrix.

Algorithms for band symmetric and unsymmetric decompositions are available in Wilkinson and Reinsch (1971) and Meis and Marcowitz (1981): the former contains Algol programs, the latter contains Fortran programs.

5.2 The tridiagonal matrix system

The tridiagonal algorithm, which is usually described as the Thomas Tridiagonal algorithm, may be derived in several ways. The procedure employed here is that of LU-decomposition of the Doolittle form

$$\begin{bmatrix} b_1 & c_1 & & & & \\ a_2 & b_2 & c_2 & & & \\ & a_3 & b_3 & c_3 & & \\ & & \cdots & \cdots & \cdots & \\ & & & a_{n-1} & b_{n-1} & c_{n-1} \\ & & & & a_n & b_n \end{bmatrix} \begin{bmatrix} x_1 \\ x_2 \\ x_3 \\ \cdots \\ x_{n-1} \\ x_n \end{bmatrix} = \begin{bmatrix} d_1 \\ d_2 \\ d_3 \\ \cdots \\ d_{n-1} \\ d_n \end{bmatrix} \tag{5.2.1}$$

$$\text{i.e.} \qquad AX = D \tag{5.2.2}$$

$$\text{i.e.} \quad a_i x_{i-1} + b_i x_i + c_i x_{i+1} = d_i$$

$$\text{for} \quad 1 \leqslant i \leqslant n \tag{5.2.3}$$

$$\text{with} \quad a_1 = c_n = 0$$

If this coefficient matrix has an LU-decomposition, let it be

$$LU \equiv \begin{bmatrix} 1 & & & & \\ \alpha_2 & 1 & & & \\ & \alpha_3 & 1 & & \\ & & \cdots & \cdots & \\ & & & \alpha_n & 1 \end{bmatrix} \begin{bmatrix} \beta_1 & c_1 & & & \\ & \beta_2 & c_2 & & \\ & & \beta_3 & c_3 & \\ & & & \cdots & \cdots \\ & & & \beta_{n-1} & c_{n-1} \\ & & & & \beta_n \end{bmatrix} \tag{5.2.4}$$

This is an example of DOOLITTLE decomposition.

Equating entries of $A \equiv LU$ will produce the unknowns β_i, $i = 1,\ldots,n$; α_i, $i = 2,\ldots,n$.

1st row: $\quad b_1 = \beta_1 \qquad \rightarrow \beta_1 = b_1$

2nd row: $\quad a_2 = \alpha_2\beta_1 \qquad \rightarrow \alpha_2 = \dfrac{a_2}{\beta_1}$

$\qquad b_2 = \alpha_2 c_1 + \beta_2 \quad \rightarrow \beta_2 = b_2 - \alpha_2 c_1$

3rd row: $\quad a_3 = \alpha_3\beta_2 \qquad \rightarrow \alpha_3 = \dfrac{a_3}{\beta_2}$

$\qquad b_3 = \alpha_3 c_2 + \beta_3 \quad \rightarrow \beta_3 = b_3 - \alpha_3 c_2$

to lead to the generalization

$$\beta_1 = b_1 \, , \quad \boxed{\alpha_i = \frac{a_i}{\beta_{i-1}} \, , \quad \beta_i = b_i - \alpha_i c_{i-1}} \quad i = 2,3,\ldots,n \tag{5.2.5}$$

Then $$AX = D \tag{5.2.6}$$

may be written $$LUX = D \tag{5.2.7}$$

i.e. $$\begin{aligned} LY &= D \\ UX &= Y \, . \end{aligned} \tag{5.2.8}$$

Now $LY = D$ is written as

$$\begin{bmatrix} 1 & & & & \\ \alpha_2 & 1 & & & \\ & \alpha_3 & 1 & & \\ & & \cdots\cdots & & \\ & & & \alpha_n & 1 \end{bmatrix} \begin{bmatrix} y_1 \\ y_2 \\ y_3 \\ \cdot\cdot \\ y_n \end{bmatrix} = \begin{bmatrix} d_1 \\ d_2 \\ d_3 \\ \cdot\cdot \\ d_n \end{bmatrix} \tag{5.2.9}$$

to give, on forward substitution, $y_1 = d_1$
$y_2 = d_2 - \alpha_2 y_1$
$y_3 = d_3 - \alpha_3 y_2$

to lead to the generalization

$$y_1 = d_1 \, , \quad \boxed{y_i = d_i - \alpha_i y_{i-1}} \quad \text{for } i = 2,3,\ldots,n \tag{5.2.10}$$

Hence we solve $UX = Y$ which is written as

$$\begin{bmatrix} \beta_1 & c_1 & & & & \\ & \beta_2 & c_2 & & & \\ & & \beta_3 & c_3 & & \\ & & & \cdots\cdot & & \\ & & & & \beta_{n-1} & c_{n-1} \\ & & & & & \beta_n \end{bmatrix} \begin{bmatrix} x_1 \\ x_2 \\ x_3 \\ \cdot\cdot \\ x_{n-1} \\ x_n \end{bmatrix} = \begin{bmatrix} y_1 \\ y_2 \\ y_3 \\ \cdot\cdot \\ y_{n-1} \\ y_n \end{bmatrix} \tag{5.2.11}$$

to give, on back substitution,

$$x_n = \frac{y_n}{\beta_n}$$

$$x_{n-1} = \frac{y_{n-1} - c_{n-1}x_n}{\beta_{n-1}}$$

to lead to the generalization

$$x_n = \frac{y_n}{\beta_n}, \quad \boxed{x_i = \frac{y_i - c_i x_{i+1}}{\beta_i}} \quad \text{for } i = (n-1),\ldots,2,1. \tag{5.2.12}$$

So the algorithm may be written

$$\beta_1 = b_1$$

$$y_1 = d_1$$

$$\left.\begin{aligned} \alpha_i &= \frac{a_i}{\beta_{i-1}} \\ \beta_i &= b_i - \alpha_i c_{i-1} \\ y_i &= d_i - \alpha_i y_{i-1} \end{aligned}\right\} \quad i = 2,3,\ldots,n \tag{5.2.13}$$

$$x_n = \frac{y_n}{\beta_n}$$

$$x_i = \frac{y_i - c_i x_{i+1}}{\beta_i} \qquad i = (n-1),\ldots,2,1.$$

<u>Example 7</u>

Use the tridiagonal algorithm to solve the system

$$\begin{bmatrix} 2 & -1 & & \\ -1 & 2 & -1 & \\ & -1 & 2 & -1 \\ & & -1 & 1 \end{bmatrix} \begin{bmatrix} x_1 \\ x_2 \\ x_3 \\ x_4 \end{bmatrix} = \begin{bmatrix} 10 \\ 6 \\ -8 \\ 4 \end{bmatrix}$$

Solution: Given: $b_1 = b_2 = b_3 = 2$, $b_4 = 1$

$a_2 = a_3 = a_4 = -1$

$c_1 = c_2 = c_3 = -1$

$d_1 = 10,\ d_2 = 6,\ d_3 = -8,\ d_4 = 4$

$\beta_1 = b_1 = 2$

$y_1 = d_1 = 10$

$\alpha_2 = a_2/\beta_1 = -1/2$

$\beta_2 = b_2 - \alpha_2 c_1 = 2 - (-\tfrac{1}{2})(-1) = 3/2$

$y_2 = d_2 - \alpha_2 y_1 = 6 - (-\tfrac{1}{2}).10 = 11$

$\alpha_3 = a_3/\beta_2 = -1/(3/2) = -2/3$

$\beta_3 = b_3 - \alpha_3 c_2 = 2 - (-2/3).(-1) = 4/3$

$y_3 = d_3 - \alpha_3 y_2 = -8 - (-2/3).11 = -2/3$

$\alpha_4 = a_4/\beta_3 = -1/(4/3) = -3/4$

$\beta_4 = b_4 - \alpha_4 c_3 = 1 - (-3/4)(-1) = \tfrac{1}{4}$

$y_4 = d_4 - \alpha_4 y_3 = 4 - (-3/4)(-2/3) = 7/2$

$x_4 = y_4/\beta_4 = (7/2)/(\tfrac{1}{4}) = 14$

$x_3 = (y_3 - c_3 x_4)/\beta_3 = (-2/3 - (-1)(14))/(4/3) = 10$

$x_2 = (y_2 - c_2 x_3)/\beta_2 = (11 - (-1)(10))/(3/2) = 14$

$x_1 = (y_1 - c_1 x_2)/\beta_1 = (10 - (-1)(14))/2 = 12$

Answer: $(x_1,x_2,x_3,x_4) = (12,14,10,14)$.

Exercise 6

1. Use the tridiagonal algorithm to solve

$$\begin{bmatrix} 3 & -5 & & & \\ -3 & 2 & -1 & & \\ & 6 & -4 & 3 & \\ & & -2 & 5 & 4 \\ & & & 7 & -2 \end{bmatrix} \begin{bmatrix} x_1 \\ x_2 \\ x_3 \\ x_4 \\ x_5 \end{bmatrix} = \begin{bmatrix} 2 \\ -6 \\ 3 \\ -3 \\ 4 \end{bmatrix}$$

Answer: $(x_1,x_2,x_3,x_4,x_5) = (1129/486.161/162,55/54,10/27,-19/27)$.

2. Use the tridiagonal algorithm to solve, correct to 3D, the system

$$\begin{bmatrix} 4 & -1 & & & \\ -1 & 4 & -1 & & \\ & -1 & 4 & -1 & \\ & & 1- & 4 & -1 \\ & & & -2 & 1 \end{bmatrix} \begin{bmatrix} x_1 \\ x_2 \\ x_3 \\ x_4 \\ x_5 \end{bmatrix} = \begin{bmatrix} 0.4 \\ 0.8 \\ 1.2 \\ 1.6 \\ 1.6 \end{bmatrix}$$

Answer: $(x_1,x_2,x_3,x_4,x_5) = (0.223,0.491,0.940,2.070,5.740)$.

3. Use the tridiagonal algorithm to solve, correct to 3D, the system

$$\begin{bmatrix} 2 & 1 & & & \\ 3 & -2 & -3 & & \\ & 4 & -4 & 2 & \\ & & -1 & 3 & -2 \\ & & & 5 & 4 \end{bmatrix} \begin{bmatrix} x_1 \\ x_2 \\ x_3 \\ x_4 \\ x_5 \end{bmatrix} = \begin{bmatrix} 3 \\ 2 \\ 1 \\ 0 \\ -1 \end{bmatrix}$$

Answer: $(x_1,x_2,x_3,x_4,x_5) = (1.244,0.511,0.237,-0.048,-0.190)$.

Arithmetic operations

For the whole process of LU decomposition followed by forward and backward substitution - all of which constitutes the tridiagonal algorithm - the number of arithmetic operations is represented in Table 5.1.

Operations	Tridiagonal	Gaussian Elimination	Gauss-Seidel for 1 iteration
Multiplication	3(n - 1)	$\frac{n^3}{3} + \frac{n^2}{2} + \frac{5n}{6}$	2n
Division	2n - 1	n(n + 1)/2	n
Addition	3(n - 1)	$\frac{n^3}{3} + \frac{n^2}{2} - \frac{5n}{6}$	2n

Table 5.1

From Table 5.1,

(i) the total number of multiplications and divisions $= 5n - 4$. Compare this with Gaussian elimination $= \frac{n^3}{3} + n^2 - \frac{n}{3}$; Gauss-Seidel for 1 iteration $= 3n$.

(ii) the total number of arithmetic operations $= 8n - 7$. Compare with Gaussian elimination $= \frac{2n^3}{3} + \frac{3n^2}{2} - \frac{7n}{6}$; Gauss-Seidel for 1 iteration $= 5n$.

Recall that for c.p.u. cost, (i) is important, while for error analysis (ii) is important.

For a tridiagonal system, Table 5.2 presents a comparison illustrating the effectivenss of the tridiagonal algorithm. For this table a system of 1000 equations has been chosen.

Operations	Tridiagonal	Gaussian Elimination	Gauss-Seidel for 1 iteration
Multiplication and Division	5.10^3	3.10^8	3.10^3
Arithmetic Operations	8.10^3	6.10^8	5.10^3

Table 5.2 Comparison for a tridiagonal system of 1000 equations

As usual, for systems such as that for full Gaussian elimination, only the term with the highest power of n is significant.

A FORTRAN listing of the algorithm for the tridiagonal matrix system is given in Appendix A. Because the α_i are only used to obtain the corresponding β_i and y_i, they need not be stored as an array but simply occupy a variable location.

In iteration usage of this scheme, through such as time-dependent problems or degradation of nonlinear systems, there are certain terms which need be computed once only as they are independent of the particular step of the iterative scheme. This comment is mainly applicable to constant coefficient matrix problems.

It may be readily shown that the algorithm for the tridiagonal matrix may also be written as follows:

$$\left.\begin{aligned}
\beta_1 &= b_1 \\
\gamma_1 &= \frac{d_1}{b_1} \\
\beta_i &= b_i - \frac{a_i c_{i-1}}{\beta_{i-1}} \\
\gamma_i &= \frac{d_i - a_i \gamma_{i-1}}{\beta_i}
\end{aligned}\right\} \quad i = 2,3,\ldots,n \tag{5.2.14}$$

$$x_n = \gamma_n$$

$$x_i = \gamma_i - \frac{c_i x_{i+1}}{\beta_i} \qquad i = n-1, n-2, \ldots, 2, 1.$$

For matrices that are symmetric, positive-definite, a unique L is formed where L is lower-triangular with positive diagonal entries

$$\text{i.e.} \quad \begin{aligned} A &\equiv LU \\ &\equiv LU^T \end{aligned} \tag{5.2.15}$$

So for tridiagonal systems of this type, only the L need be found and so storage is conserved, realising that the $U = L^T$.

<u>Exercise 7</u>

In Appendix A, there is a listing of a FORTRAN program for the tridiagonal algorithm. Implement this program on your computer using the following example:

$$\begin{bmatrix} 2 & -1 & & & & & \\ -1 & 2 & -1 & & & & \\ & -1 & 2 & -1 & & & \\ & & \cdots & \cdots & \cdots & & \\ & & & & -1 & 2 & -1 \\ & & & & & -1 & 1 \end{bmatrix} \begin{bmatrix} x_1 \\ x_2 \\ x_3 \\ \cdots \\ x_{n-1} \\ x_n \end{bmatrix} = \begin{bmatrix} 0 \\ 0 \\ 0 \\ \cdots \\ 0 \\ 1 \end{bmatrix}$$

Answers: $(x_1, x_2, x_3, \ldots, x_n) = (1, 2, 3, \ldots, n)$.

5.3 The pentadiagonal matrix system

The pentadiagonal algorithm may be derived through LU-decomposition similarly to that of the derivation of the tridiagonal algorithm.

$$\begin{bmatrix} c_1 & d_1 & e_1 & & & & & \\ b_2 & c_2 & d_2 & e_2 & & & & \\ a_3 & b_3 & c_3 & d_3 & e_3 & & & \\ & a_4 & b_4 & c_4 & d_4 & e_4 & & \\ & & \cdots & \cdots & \cdots & \cdots & & \\ & & & a_{n-2} & b_{n-2} & c_{n-2} & d_{n-2} & e_{n-2} \\ & & & & a_{n-1} & b_{n-1} & c_{n-1} & d_{n-1} \\ & & & & & a_n & b_n & c_n \end{bmatrix} \begin{bmatrix} x_1 \\ x_2 \\ x_3 \\ x_4 \\ \cdot\cdot \\ x_{n-2} \\ x_{n-1} \\ x_n \end{bmatrix} = \begin{bmatrix} f_1 \\ f_2 \\ f_3 \\ f_4 \\ \cdot\cdot \\ f_{n-2} \\ f_{n-1} \\ f_n \end{bmatrix} \tag{5.3.1}$$

The equations for the pentadiagaonl matrix system may be written

$$a_i x_{i-2} + b_i x_{i-1} + c_i x_i + d_i x_{i+1} + e_i x_{i+2} = f_i$$

$$\text{for} \quad 1 \leqslant i \leqslant n \tag{5.3.2}$$

$$\text{with} \quad a_1 = b_1 = a_2 = e_{n-1} = d_n = e_n = 0$$

The algorithm for this system is as follows:

$$\delta_1 = \frac{d_1}{c_1}$$

$$\lambda_1 = \frac{e_1}{c_1}$$

$$\gamma_1 = \frac{f_1}{c_1}$$

$$\mu_2 = c_2 - b_2\delta_1$$

$$\delta_2 = (d_2 - b_2\lambda_1)/\mu_2$$

$$\lambda_2 = e_2/\mu_2$$

$$\gamma_2 = (f_2 - b_2\gamma_1)/\mu_2$$

$$\left.\begin{aligned}
\beta_i &= b_i - a_i\delta_{i-2} \\
\mu_i &= c_i - \beta_i\delta_{i-1} - a_i\lambda_{i-2} \\
\delta_i &= (d_i - \beta_i\lambda_{i-1})/\mu_i \\
\lambda_i &= e_i/\mu_i \\
\gamma_i &= (f_i - \beta_i\gamma_{i-1} - a_i\gamma_{i-2})/\mu_i
\end{aligned}\right\} \quad i = 3,4,\ldots,n-2$$

$$\beta_{n-1} = b_{n-1} - a_{n-1}\delta_{n-3}$$

$$\mu_{n-1} = c_{n-1} - \beta_{n-1}\delta_{n-2} - a_{n-1}\lambda_{n-3}$$

$$\delta_{n-1} = (d_{n-1} - \beta_{n-1}\lambda_{n-2})/\mu_{n-1}$$

$$\gamma_{n-1} = (f_{n-1} - \beta_{n-1}\gamma_{n-2} - a_{n-1}\gamma_{n-3})/\mu_{n-1}$$

$$\beta_n = b_n - a_n\delta_{n-2}$$

$$\mu_n = c_n - \beta_n\delta_{n-1} - a_n\lambda_{n-2}$$

$$\gamma_n = (f_n - \beta_n\gamma_{n-1} - a_n\gamma_{n-2})/\mu_n$$

$$x_n = \gamma_n$$

$$x_{n-1} = \gamma_{n-1} - \delta_{n-1}x_n$$

$$x_i = \gamma_i - \delta_i x_{i+1} - \lambda_i x_{i+2} \qquad i = n-2,n-3,\ldots,2,1 \qquad (5.3.3)$$

Storage space may be conserved during the execution of this algorithm by realising that β_i and μ_i are used only to compute δ_i, λ_i and γ_i and need not be stored once this latter group have been evaluated.

Example 8

Use the pentadiagonal algorithm to solve, correct to 2D, the system

$$\begin{bmatrix} 2 & -3 & 4 & & & \\ -5 & -2 & 1 & 2 & & \\ 1 & -3 & 3 & -4 & -1 & \\ & 2 & 4 & -1 & 3 & -2 \\ & & -1 & 3 & -2 & 5 \\ & & & -5 & 3 & -4 \end{bmatrix} \begin{bmatrix} x_1 \\ x_2 \\ x_3 \\ x_4 \\ x_5 \\ x_6 \end{bmatrix} = \begin{bmatrix} 2 \\ -3 \\ -2 \\ 4 \\ -1 \\ 5 \end{bmatrix}$$

Solution: Given: $c_1 = 2,\ c_2 = -2,\ c_3 = 3,\ c_4 - -1,\ c_5 = -2,\ c_6 = -4.$
$b_2 = -5,\ b_3 = -3,\ b_4 = 4,\ b_5 = 3,\ b_6 = 3.$
$a_3 = 1,\ a_4 = 2,\ a_5 = -1,\ a_6 = -5.$
$d_1 = -3,\ d_2 = 1,\ d_3 = -4,\ d_4 = 3,\ d_5 = 5.$
$e_1 = 4,\ e_2 = 2,\ e_3 = -1,\ e_4 = -2.$
$f_1 = 2,\ f_2 = -3,\ f_3 = -2,\ f_4 = 4,\ f_5 = -1,\ f_6 = 5.$

$\delta_1 = d_1/c_1 = (-3)/2 = -3/2$
$\lambda_1 = e_1/c_1 = 4/2 = 2$
$\gamma_1 = f_1/c_1 = 2/2 = 1$

$\mu_2 = c_2 - b_2\delta_1 = -2 - (-5)(-3/2) = -9.5$
$\delta_2 = (d_2 - b_2\lambda_1)/\mu_2 = 1 - (-5)(2)\ /(-9.5) = -1.1579$
$\lambda_2 = e_2/\mu_2 = 2/(-9.5) = -0.2105$
$\gamma_2 = (f_2 - b_2\gamma_1)/\mu_2 = (-3 - (-5)(1))/(-9.5) = -0.2105$

$\beta_3 = b_3 - a_3\delta_1 = -3 - (1)(-3/2) = -3/2.$
$\mu_3 = c_3 - \beta_3\delta_2 - a_3\lambda_1 = 3 - (-3/2)(-1.1579) - (1)(2) = -0.7368$
$\delta_3 = (d_3 - \beta_3\lambda_2)/\mu_3 = (-4 - (-3/2)(-0.2105))/(-0.7368) = 5.8574$
$\lambda_3 = e_3/\mu_3 = (-1)/(-0.7368) = 1.3572$
$\gamma_3 = (f_3 - \beta_3\gamma_2 - a_3\gamma_1)/\mu_3 = (-2 - (-3/2)(-0.2105)-(1)(1))/(-0.7368)$
$= 4.5094$

$\beta_4 = b_4 - a_4\delta_2 = 4 - (2)(-1.1579) = 6.3158$
$\mu_4 = c_4 - \beta_4\delta_3 - a_4\lambda_2 = -1 - (6.3158)(5.8574)-(2)(-0.2105) = -37.5732$
$\delta_4 = (d_4 - \beta_4\lambda_3)/\mu_4 = (3 - (6.3158)(1.3572))/(-37.5732) = 0.1483$
$\lambda_4 = e_4/\mu_4 = -2/(-37.5732) = 0.0532$
$\gamma_4 = (f_4 - \beta_4\gamma_3 - a_4\gamma_2)/\mu_4 = (4-(6.3158)(4.5094)-(2)(-0.2105))/$
$(-37.5732) = 0.6403$

$\beta_5 = b_5 - a_5\delta_3 = 3 - (-1)(5.8574) = 8.8574$

$\mu_5 = c_5 - \beta_5\delta_4 - a_5\lambda_3 = -2 - (8.8574)(0.1483)-(-1)(1.3572) \doteq -1.9564$

$\delta_5 = (d_5 - \beta_5\lambda_4)/\mu_5 = (5 - (8.8574)(0.0532))/(-1.9564) = -2.3149$

$\gamma_5 = (f_5 - \beta_5\lambda_4 - a_5\gamma_3)/\mu_5 = (-1 - (8.8574)(0.6403) -(-1)(4.5094)/$
$(-1.9564) = 1.1051$

$\beta_6 = b_6 - a_6\delta_4 = 3 - (-5)(0.1483) = 3.7415$

$\mu_6 = c_6 - \beta_6\delta_5 - a_6\lambda_4 = -4 - (3.7415)(-2.3149) - (-5)(0.0532)$
$= 4.9271$

$\gamma_6 = (f_6 - \beta_6\gamma_5 - a_6\gamma_4)/\mu_6 = (5 - (3.7415)(1.0953) - (-5)(0.6403))/$
$(4.9271) = 0.8254$

$x_6 = 0.8254$

$x_5 = \gamma_5 - \delta_5 x_6 = 1.0953 - (-2.3149)(0.8254) = 3.0138$

$x_4 = \gamma_4 - \delta_4 x_5 - \lambda_4 x_6 = 0.6403 - (0.1483)(3.0138) - (0.0532)$
$(0.8254) = 0.1479$

$x_3 = \gamma_3 - \delta_3 x_4 - \lambda_3 x_5 = 4.5094 - (5.8574)(0.1479) - (1.3572)(3.0138)$
$= -0.4566$

$x_2 = \gamma_2 - \delta_2 x_3 - \lambda_2 x_4 = -0.2105 - (-1.1579)(-0.4566) - (-0.2105)$
$(0.1479) = -0.7081$

$x_1 = \gamma_1 - \delta_1 x_2 - \lambda_1 x_3 = 1 - (-3/2)(-0.7081) - (2)(-0.4566) = 0.8511$

Answer: $(x_1,x_2,x_3,x_4,x_5,x_6) = (0.85,-0.71,-0.46,0.15,3.01,0.82)$

Exercise 8

1. Use the pentadiagonal algorithm to solve, correct to 2D, the system

(a)

$$\begin{bmatrix} 3 & -4 & 2 & & & \\ -2 & 5 & -3 & 1 & & \\ 4 & 2 & -3 & 5 & 1 & \\ & -3 & 4 & 1 & -2 & 2 \\ & & -5 & 2 & -4 & 3 \\ & & & -2 & 5 & 4 \end{bmatrix} \begin{bmatrix} x_1 \\ x_2 \\ x_3 \\ x_4 \\ x_5 \\ x_6 \end{bmatrix} = \begin{bmatrix} -0.78 \\ 2.29 \\ -2.71 \\ 2.88 \\ -1.57 \\ 0.44 \end{bmatrix}$$

(b)

$$\begin{bmatrix} 2 & -1 & & & & & & & \\ -1 & 2 & -1 & & & & & & \\ -1 & -1 & 4 & -1 & -1 & & & & \\ & -1 & -1 & 4 & -1 & -1 & & & \\ & & \cdot & \cdot & \cdot & \cdot & \cdot & \cdot & \\ & & & -1 & -1 & 4 & -1 & -1 \\ & & & & & -1 & 2 & -1 \\ & & & & & & -1 & 1 \end{bmatrix} \begin{bmatrix} x_1 \\ x_2 \\ x_3 \\ x_4 \\ \cdot \\ x_{n-2} \\ x_{n-1} \\ x_n \end{bmatrix} = \begin{bmatrix} 0 \\ 0 \\ 0 \\ 0 \\ \cdot \\ 0 \\ 0 \\ 1 \end{bmatrix}$$

Answers: (a) $(x_1,x_2,x_3,x_4,x_5,x_6) = (2.64,3.12,1.89,-2.36,-2.04,$ $1.48)$.

(b) $(x_1,x_2,\ldots,x_n) = (1,2,3,\ldots,n)$.

A FORTRAN listing of the algorithm for the pentadiagonal matrix system is given in Appendix B.

There are other standard systems such as the bi-tridiagonal matrix and tri-tridiagonal matrix. The algorithms for these are listed in the appendix of Von Rosenberg (1969).

The higher order of algorithms for the wider bands become progressively more complicated.

5.4 General variable-bandwidth matrix system

If the nodes of a particular triangular element in a finite element mesh are numbered 5, 19 and 37 and if there is one variable at each node then the semi-bandwidth, n, of the symmetric, positive-definite matrix is $n \geqslant 37 - 5 = 32$. With the bandwidth given, only a triangle of coefficients is retained on core and, as elimination proceeds, this core moves diagonally downwards. When the first variable is eliminated using the first equation, all the other coefficients in the triangle are modified, and as they are modified they are shifted diagonally upwards. The second equation now takes first place in the coefficients on core. And so the system proceeds. Usually the band is sparsely filled and frequently the bandwidth varies from one position to another. Often then only the coefficients within the $n_t \times n_t$ submatrix are modified, where n_t is the temporary semi-bandwidth but this is not always satisfactory.

The above description is equally appropriate for unsymmetric systems.

5.5 Block-tridiagonal matrix system

Isaacson and Keller (1966), Varah (1972) and others give details on the block-tridiagonal matrix system.

In the finite-difference solution of partial differential equations a block tridiagonal matrix is commonly obtained. The layout is for the one above but with sub-matrices substituted for the entries. The same formula is used but there are now matrix additions and multiplications, so that where there was a division by an entry b_i, there is now a multiplication by a matrix $[b_i]^{-1}$. This requires calculation of an inverse, which can be done by Gaussian elimination. The method is still more efficient than simple iteration or full Gaussian elimination.

For symmetric positive-definite matrices, Jennings (1977) gives a method of solution which is a type of Cholesky triangular decomposition expressed in terms of submatrices and involving $7(n - 1)b^3/6$ multiplications, where the submatrices are bxb. In this case the tridiagonal matrix is better shown as

$$\begin{bmatrix} A_{11} & A_{21}^T & & & \\ A_{21} & A_{22} & A_{32}^T & & \\ & A_{32} & A_{33} & A_{43}^T & \\ & & \cdots\cdots & & \\ & & & \cdot & A_{nn} \end{bmatrix} \begin{bmatrix} x_1 \\ x_2 \\ x_3 \\ \vdots \\ x_n \end{bmatrix} \begin{bmatrix} d_1 \\ d_2 \\ d_3 \\ \vdots \\ d_n \end{bmatrix} \qquad (5.5.1)$$

where the A_{ii} are submatrices.

The lower triangular matrix L_{11} is obtained by Cholesky decomposition of A_{11}, then premultiplying the first submatrix equation by L_{11}^{-1} gives

$$L_{11}^T x_1 + L_{21}^T x_2 = h_1 \qquad (5.5.2)$$

The inverse L_{11}^{-1} need not be calculated since L_{21}^T and h_1 are obtained by forward substitution in the equations

$$L_{11}L_{21}^T = A_{21}^T \qquad (5.5.3)$$

and

$$L_{11}h_1 = d_1 \qquad (5.5.4)$$

From the first of these we obtain

$$A_{21} = L_{21}L_{11}^T \qquad (5.5.5)$$

so that premultiplying the new first rows (5.5.2), by L_{21} and subtracting from row 2 gives a second line as

$$A_{22}^*x_2 + A_{32}^Tx_3 = b_2^* \qquad (5.5.6)$$

where

$$A_{22}^* = A_{22} - L_{21}L_{21}^T \qquad (5.5.7)$$

and

$$b_2^* = d_2 - L_{21}h_1 \qquad (5.5.8)$$

Putting $A_{22}^* = L_{22}L_{22}^T$, the second line, (5.5.6), is pre-multiplied by L_{22}^{-1} and the process repeated as for the first line.

The final result is

$$\begin{bmatrix} L_{11}^T & L_{21}^T & & & \\ & L_{22}^T & L_{32}^T & & \\ & & L_{33}^T & L_{43}^T & \\ & & & \cdots & \\ & & & & L_{nn}^T \end{bmatrix} \begin{bmatrix} x_1 \\ x_2 \\ x_3 \\ \cdot \\ \cdot \\ x_n \end{bmatrix} = \begin{bmatrix} h_1 \\ h_2 \\ h_3 \\ \cdot \\ \cdot \\ h_n \end{bmatrix} \qquad (5.5.9)$$

which can be solved by back substitution. Only a part of the matrix is held in core at any one time, the submatrices being transferred in and out as required.

6. SPECIAL TECHNIQUES

6.1 Backing store

A simple example of the use of backing store for a symmetric band matrix is given by Jennings (1977) and a modified version is shown below for Cholesky decomposition of a matrix with a semibandwidth of 3.

Referring to the formulae for the Cholesky method, if only the first three rows of A are in main (core) store, the a_{ij} can be replaced with the ℓ_{ij}.

$$\left[\begin{array}{lllll} a_{11} & & & & \\ a_{21} & a_{22} & & & \\ a_{31} & a_{32} & a_{33} & & \\ \hline & a_{42} & a_{43} & a_{44} & \\ & & a_{53} & a_{54} & a_{55} \\ & & & \cdots & \end{array}\right. \longrightarrow \left[\begin{array}{lllll} \ell_{11} & & & & \\ \ell_{21} & \ell_{22} & & & \\ \ell_{31} & \ell_{32} & \ell_{33} & & \\ \hline & a_{42} & a_{43} & a_{44} & \\ & & a_{53} & a_{54} & a_{55} \\ & & & \cdots & \end{array}\right. \qquad (6.1.1)$$

The first column can then be transferred to backing (disc) store and the fourth line of A brought into main store with all entries moved up one row and across one column. The fourth row of A (now the third in store) can then be operated on.

$$\left[\begin{array}{lll} \ell_{22} & & \\ \ell_{32} & \ell_{33} & \\ a_{42} & a_{43} & a_{44} \\ \hline \end{array}\right. \longrightarrow \left[\begin{array}{lll} \ell_{22} & & \\ \ell_{32} & \ell_{33} & \\ \ell_{42} & \ell_{43} & \ell_{44} \\ \hline \end{array}\right. \qquad (6.1.2)$$

The process can be repeated until the whole matrix L is stored column by column in backing store. It can then be recalled to main store, column by column, for the operation LY = B and returned column by column to backing store. When y_i is found ($3 \leq i \leq n$), column i - 2 is returned to backing store. Finally L can be recalled in reverse order for the operation $L^T X = Y$, where X is found by back substitution (Y having been found by forward substitution).

The total number of multiplications and divisions in the formation of the L matrix is $\simeq nb^2/2$.

With such a narrow bandwidth, there is no necessity for such frequent transfers between main and backing store. It would be preferable to hold more lines in the main store and transfer several columns at a time.

The elimination scheme just described can be used with a variable bandwidth store. If there are no re-entrant rows, the extension is quite straight-forward. If re-entrant rows are present the number of rows, held in main store before decomposition can commence, is increased as the diagram in Figure 6.1 shows.

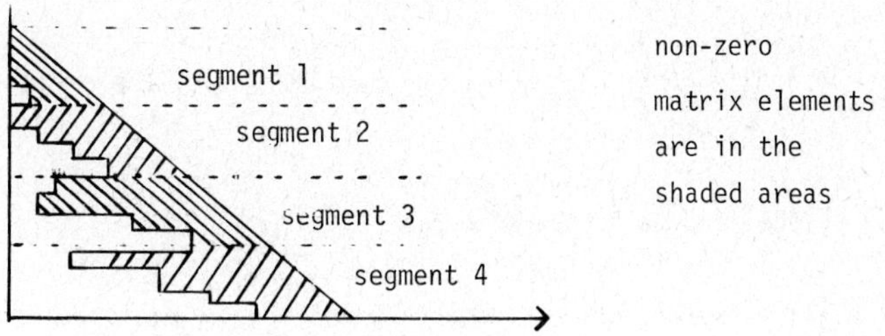

Figure 6.1

For elimination of the first column segment 1 and 2 are required to be held in core. For elimination of the first columns of segment 4, all four segments are required. Jennings and Tuff (1971) have discussed this method.

6.2 Cuthill-McKee algorithm

If a matrix has a variable bandwidth, with or without re-entrant rows, a variable band algorithm may be more efficient than a diagonal band elimination. It allows more effective use of the property that, for elimination without pivoting, zero elements before the first non-zero in any row always remain zero.

The main disadvantage is that additional storage is required for an address sequence. For very large matrices, this sequence may have to be generated by a special program segment which scans the input data. For an efficient, economical program it may be necessary to rearrange the input data by renumbering the variables. The Cuthill-McKee algorithm (1969) provides an efficient numbering scheme. Jennings (1977) gives a simple description as follows -

"(a) Choose a node to be relabelled 1. This should be located at an extremity of the graph and should have, if possible, few connections to other nodes.

(b) The nodes connected to the new node are relabelled 2, 3, etc., in the order of their increasing degree (the degree of a node is the number of nodes to which it is connected).

(c) The sequence is extended by relabelling the nodes which are directly connected to the new node 2 and which have not previously been relabelled. The nodes are again listed in the order of their increasing degree.

(d) The last operation is repeated for the new nodes 3, 4, etc., until the renumbering is complete. "

The proper choice of starting node is important for the effectiveness of the algorithm, and this is often selected using the algorithm of Gibbs, Poole and Stockmeyer (1976).

The example, illustrated in Figure 6.2, shows how a network may be renumbered using the Cuthill-McKee algorithm.

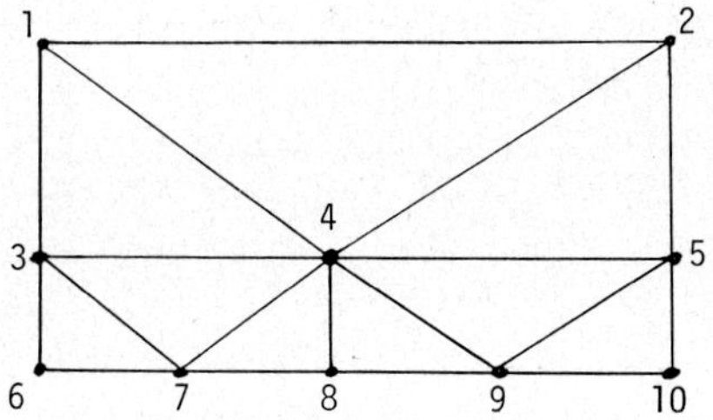

Initial numbering scheme for the network

X									
X	X								
X		X							
X	X	X	X						
	X		X	X					
		X			X				
		X	X		X	X			
			X			X	X		
			X	X			X	X	
				X				X	X

Coefficient matrix

Table 6.2

The node connection list for the network of Figure 6.2 is given in Table 6.1.

Node	No. of Connections	Connection List
1	3	2,3,4
2	3	1,4,5
3	4	1,4,6,7
4	7	1,2,3,5,7,8,9
5	4	2,4,9,10
6	2	3,7
7	4	3,4,6,8
8	3	4,7,9
9	4	4,5,8,10
10	2	5,9

Table 6.1
Node Connection List for Network of Figure 6.2.

Going through the list, node 6 seems to be a suitable starting node, and this is labelled 1. 3 and 7 are next renumbered as 2 and 3. The connection list for 3 now shows that 1 and 4 are next to be renumbered as 4 and 5, and since 1 has fewer connections than 4, it must be renumbered first. The process continues until all the nodes have been renumbered.

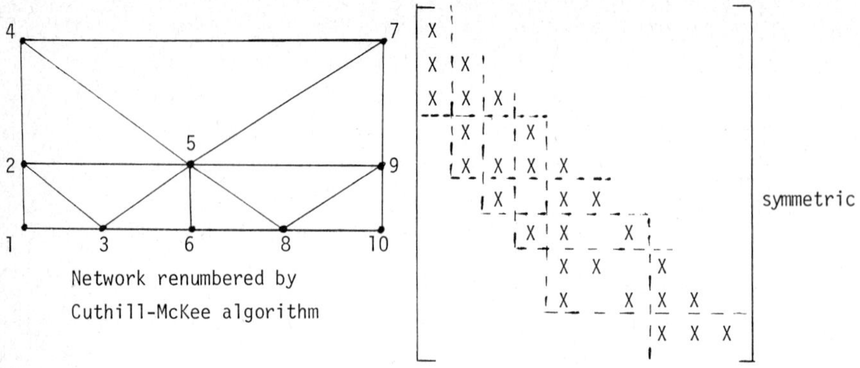

Network renumbered by Cuthill-McKee algorithm

Coefficient matrix

Figure 6.3 Network and matrix using Cuthill-McKee algorithm

The Cuthill-McKee algorithm results in a variable bandwidth store which cannot have re-entrant rows. It may be considered to be made up of overlapping triangles as shown by the dashed lines in Figure 6.3. If the sides of the triangles span $p_1, p_2, \ldots, p_r$ rows and the sides of the overlaps span $q_1, q_2, \ldots, q_r$ rows, the number of storage locations is

$$s = \sum_{i=1}^{r} \frac{p_i(p_i + 1)}{2} - \sum_{i=1}^{r-1} \frac{q_i(q_i + 1)}{2} \qquad (6.2.1)$$

Inserting figures in formula (6.2.1), we obtain, for the above matrix

$$s = \frac{3.4}{2} + \frac{4.5}{2} + \frac{4.5}{2} + \frac{4.5}{2} + \frac{5.6}{2} + \frac{3.4}{2} - \left(\frac{2.3}{2} + \frac{3.4}{2} + \frac{3.4}{2} + \frac{3.4}{2} + \frac{2.3}{2}\right)$$

$$= 57 - 24 = \underline{\underline{33}}$$

The original matrix store required 40 locations so we have reduced this by 7.

The total number of multiplications and divisions for triangular decomposition is

$$m = \sum_{i=1}^{r} \frac{p_i(p_i - 1)(p_i + 4)}{6} - \sum_{i=1}^{r-1} \frac{q_i(q_i - 1)(q_i + 4)}{6} \qquad (6.2.2)$$

6.3 The reverse Cuthill-McKee algorithm (RCM)

It was recognised by George (1971) that if a Cuthill-McKee numbering scheme is reversed a more efficient scheme often results. To reverse the numbers, (i,j) is replaced by (n-j+1,n-i+1). This merely reverses the triangles and the storage and computational requirements remain the same. However, the storage requirement can often be lessened by taking advantage of the re-entrant rows which may occur. It has been shown by George and also by Lin and Sherman (1976), that the Reverse Cuthill-McKee algorithm cannot give a less efficient bandwidth store than the direct method and will often be more efficient. In the example in Figure 6.4, use of a re-entrant row reduces the storage requirement to 32 locations, a gain of 1 over the direct method. With a larger system, the gain could be much greater.

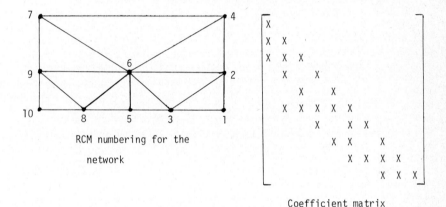

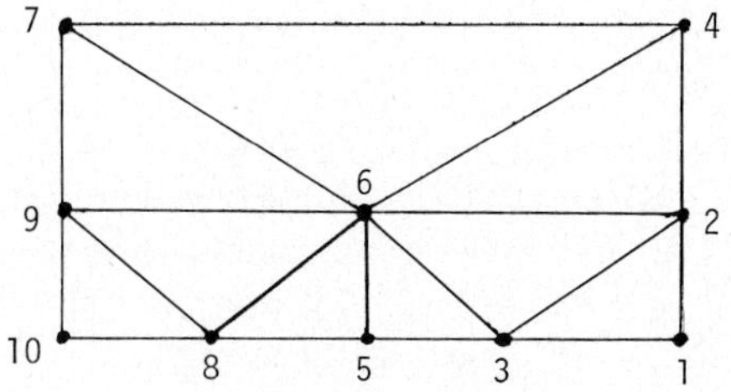

RCM numbering for the network

$$
\begin{bmatrix}
X & & & & & & & & & \\
X & X & & & & & & & & \\
X & X & X & & & & & & & \\
 & X & & X & & & & & & \\
 & & X & & X & & & & & \\
 & X & X & X & X & X & & & & \\
 & & & X & & X & X & & & \\
 & & & & X & X & & X & & \\
 & & & & & X & X & X & X & \\
 & & & & & & & X & X & X
\end{bmatrix}
$$

Coefficient matrix

Figure 6.4 Network and matrix using Reverse Cuthill-McKee algorithm

7. FRONTAL METHODS

7.1 Introduction

Irons (1970) developed a Fortran program for the assembly and solution of symmetric positive-definite equations arising in the use of the finite element method, when using criteria such as the "least squares". He aimed to improve on the standard band-matrix algorithms. There is some doubt as to whether or not he developed the concept of the "frontal approach" but he was the first to develop a program based on this approach.

The frontal approach uses computer peripherals such as disc or magnetic tape to store numbers that are not immediately required for computation. As a result of this the core of the computer need not be particularly large for such programs provided this continuous Input/Output access is available.

Hood (1977a,b) developed a frontal solution program for the assembly and solution of unsymmetric matrix equations arising in certain applications of the finite element method. This program was also coded in Fortran.

The Hood program is now discussed before that of Irons because of its generality. Both programs use Gaussian elimination.

7.2 Frontal solution for unsymmetric matrices

As an illustration of the Hood subroutine, a 3-element mesh with 5 nodes, as in Figure 7.1, will be used although, typically, a finite element system might have 5000 variables with a front width of 200 (i.e. the order of 10^6 non-zeros). All coefficients are kept in backing store, from which they are brought into core as required during assembly. With an unsymmetric matrix, the choice of pivot entry is important but a portion of the solution without pivoting will first be considered.

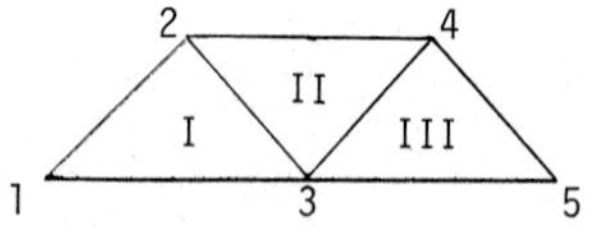

Figure 7.1 Three element mesh

Element I is first assembled into core. At this stage, the row and column associated with node 1 are fully summed and a_{11} may be used to eliminate the variable x_1, even though nodes 2 and 3 are not fully summed.

The total coefficient matrix for this three-element mesh is given in (7.2.1) with the submatrix, enclosed within dashed lines, being that position in core after assembly of the first element.

$$\begin{bmatrix} a_{11} & a_{12} & a_{13} & & \\ a_{21} & a_{22} & a_{23} & a_{24} & \\ a_{31} & a_{32} & a_{33} & a_{34} & a_{35} \\ & a_{42} & a_{43} & a_{44} & a_{45} \\ & & a_{53} & a_{54} & a_{55} \end{bmatrix} \tag{7.2.1}$$

After assembly of the first element, the matrix equation, where superscript I denotes the element number from which the matrix entry was derived, is as follows:

$$\begin{bmatrix} a_{11}^{I} & a_{12}^{I} & a_{13}^{I} \\ a_{21}^{I} & a_{22}^{I} & a_{23}^{I} \\ a_{31}^{I} & a_{32}^{I} & a_{33}^{I} \end{bmatrix} \begin{bmatrix} x_1 \\ x_2 \\ x_3 \end{bmatrix} = \begin{bmatrix} b_1^{I} \\ b_2^{I} \\ b_3^{I} \end{bmatrix} \tag{7.2.2}$$

After eliminating x_1 and transferring row 1 to backing store the system becomes, in core,

$$\begin{bmatrix} a_{22}^{I} - \dfrac{a_{21}^{I}}{a_{11}^{I}} a_{12}^{I} & a_{23}^{I} - \dfrac{a_{21}^{I}}{a_{11}^{I}} a_{13}^{I} \\ a_{32}^{I} - \dfrac{a_{31}^{I}}{a_{11}^{I}} a_{12}^{I} & a_{33}^{I} - \dfrac{a_{31}^{I}}{a_{11}^{I}} a_{13}^{I} \end{bmatrix} \begin{bmatrix} x_2 \\ x_3 \end{bmatrix}$$

$$= \begin{bmatrix} b_2^{I} - \dfrac{a_{21}^{I}}{a_{11}^{I}} b_1^{I} \\ b_3^{I} - \dfrac{a_{31}^{I}}{a_{11}^{I}} b_1^{I} \end{bmatrix} \tag{7.2.3}$$

Assembly of the next elements makes node 2 fully summed. Once again there are three equations in core and the matrix system is as follows:

$$\begin{bmatrix} a_{22}^{I} - \dfrac{a_{21}^{I}}{a_{11}^{I}} a_{12}^{I} + a_{22}^{II} & a_{23}^{I} - \dfrac{a_{21}^{I}}{a_{11}^{I}} a_{13}^{I} + a_{23}^{II} & a_{24}^{II} \\ a_{32}^{I} - \dfrac{a_{31}^{I}}{a_{11}^{I}} a_{12}^{I} + a_{32}^{II} & a_{33}^{I} - \dfrac{a_{31}^{I}}{a_{11}^{I}} a_{13}^{I} + a_{33}^{II} & a_{34}^{II} \\ a_{42}^{II} & a_{43}^{II} & a_{44}^{II} \end{bmatrix} \begin{bmatrix} x_2 \\ x_3 \\ x_4 \end{bmatrix}$$

$$= \begin{bmatrix} b_2^{I} - \dfrac{a_{21}^{I}}{a_{11}^{I}} b_1^{I} + b_2^{II} \\ b_3 I - \dfrac{a_{31}^{I}}{a_{11}^{I}} b_1^{I} + b_3^{II} \\ b_4^{II} \end{bmatrix} \qquad (7.2.4)$$

Again, elimination occurs using the first entry of the first row as pivot and then the first row is transferred to backing store.

Note that the nodes of element I would have been dealt with in the same way regardless of what numbers were allotted to them.

In practice more fully summed equations would be entered to a predetermined level and a full pivotal search instituted. If a_{11}, for instance, were zero, the above solution would fail. Total pivoting, as compared with partial pivoting, involves a higher pivotal-search time and often causes a larger growth of non-zero entries with the associated increase in computation time and round-off error.

Hood indicated that in unsymmetric linear systems an entry should be used as a pivot only if its row and column are fully assembled. Therefore, assembly is continued until the front exceeds a critical size, at which stage the pivot is chosen to be the maximal entry in the fully assembled rows and columns. His reason for choosing total pivoting rather than partial pivoting for the assembled submatrix was that he had already restricted the pivotal choice by taking a small submatrix and that he felt it would be undesirable to further reduce the choice by a partial pivoting scheme.

After all the variables have been eliminated in the described manner, solution can proceed by back substitution, recalling the equations from backing store in reverse order.

The order in which the nodal solutions are evaluated is independent of their numbering but dependent on the order in which the elements are assembled to give the fully assembled equations. It is possible for a fully assembled equation to remain in core for a large part of the elimination process, simply due to its not having been chosen as a pivotal equation.

In band-solution routines the numbering of the nodes is very important. In frontal solutions, the order of the nodal numbering is completely irrelevant to the procedure; the technique is dependent on the order of the element numbering. In the assembly and solution of the equations, the only equations which must remain in core are those which are not yet fully summed. The variables represented by these equations constitute the "front". Equations behind the front have either been chosen as a pivotal equation and placed in backing store or are awaiting their selection for this. The "front width" is kept as small as possible to minimize core requirements: efficient element numbering achieves this.

Table 7.1 contains a selection from a comparison table constructed by Hood in support of his assertion that the frontal approach was an improvement on the band solution routines. A rectangular mesh was used for the finite elements and the nodes were numbered in the optimum manner for the band-solution routines. The equations arose in the finite element formulation of the Navier-Stokes equations for fluid flow. The results of the comparison indicate that for large scale problems there is a considerable advantage in the use of the frontal approach.

	FRONTAL SOLUTION				BAND SOLUTION		
No. of Equations	Type of Pivoting	Front Width	Computer Time(sec)	Core (words)	Band Width	Computer Time(sec)	Core (words)
876	Diagonal	58	156	6900	163	344	23,000
	Total		181	6900			

Table 7.1 Comparison between solution routines

Apart from its using less storage and less arithmetic than band routines, the frontal approach permits an easier refining of the mesh if it is considered too coarse in some region.

Reid (1977b), commenting on Hood's assembly and solution criteria, suggests that Hood is unreasonably cautious and that it would be preferable to follow each assembly with as many eliminations as possible subject to each pivot satisfying the relative tolerance as given in (4.2.4) and (4.2.5). Lee (1980) found that, in his use of the Hood subroutine in the study of convection, using the Navier-Stokes equations for this field flow, total pivoting was not necessary nor in fact was partial pivoting but purely diagonal pivoting was sufficient. This comment is really moving into variations for special cases. Mann (1981a) contains a Fortran program which includes the frontal approach for the solution of the equations resulting from a model of atmospheric convection using the finite element Galerkin criterion.

7.3 Frontal solution for symmetric, positive-definite matrices

The Irons solution technique is similar to that described in the Hood approach. A knowledge of the basic properties of symmetric, positive-definite matrices would assist in a clear understanding of the mathematics involved.

Because of the symmetry, both core and backing store are conserved by simply using either the upper or lower triangular submatrix.

Irons (1980) states that with positive-definite matrices it makes little difference in taking the diagonal pivots in a different order whereas the use of non-diagonal pivots is highly dangerous. In his program each fully assembled equation is taken as a pivotal equation and eliminated from core to backing store as soon as it is fully assembled. There is no pivotal search. The diagonal entries are the pivots.

Because of the lack of any pivotal-search in the Irons program, it is decidely unwise to use this program for symmetric matrices that are other than positive-definite.

Mann (1981b) contains a Fortran program which includes the frontal solution of the symmetric, positive-definite equations resulting from a model of atmospheric convection using the finite element least squares criterion.

8. COMPUTER PACKAGES

There are many routines for the basic techniques for dense systems available as packages. Some of these occur, in varying degree, in the International Mathematical and Statistical Libraries (IMSL), Numerical Algorithms Group (NAG) and LINPACK, with a lesser emphasis on sparse routines. From time to time computer listings of relevant packages for sparse systems are available in journals such as

International Journal for Numerical Methods in Engineering
Computers and Fluids
Computers and Structures
ACM Transactions on Mathematical Software

Usually these are written in Fortran, occasionally in Algol.

The commercial packages are usually available in magnetic tape, etc., in a form immediately applicable to specific computers. This eliminates the disadvantage of having to adapt the package to the particular computer.

The frontal approach for solving large sparse matrix systems is used in the recently available finite element package TWODEPEP.

9. CONCLUDING REMARKS

Some of the standard solution techniques, and other important considerations, involved in the solution of large sparse linear systems by direct methods have been considered in this paper. However, there are many refinements and alternatives to these approaches available for specific systems. Fox (1977) reviews some of these. The well-known applied numerical journals regularly provide these specific approaches.

Duff (1981) provides a recent addition to this area of applications. This contains many invited papers from a recent conference on sparse matrices and its uses and would seem to be ideal for further reading from this paper.

10. REFERENCES

Birkhoff, G. and Fix, G.J. (1974) "Higher Order Linear Finite Element Methods", Report AD-779-341, Harvard University.

Bjorck, A., Plemmons, R.J. and Schneider, H. (1980) "Large Scale Matrix Problems", North-Holland.

Curtis, A.R. and Reid, J.K. (1972) "On the Automatic Scaling of Matrices for Gaussian Elimination", J. Inst. Math. Appl., 10, 118-124.

Curtis, A.R. and Reid, J.K. (1971) "The Solution of Large Sparse Unsymmetric Systems of Linear Equations", J. Inst. Math. Appl., 8, 344-353.

Cuthill, E. and McKee, J. (1969) "Reducing the Bandwidth of Sparse Symmetric Matrices", ACM Proc. 24th National Conf., New York.

Dahlquist, G. and Bjorck, A. (1974) "Numerical Methods", Prentice-Hall.

Duff, I.S. (Ed.) (1981) "Sparse Matrices and Their Uses", Academic.

Duff, I.S. and Reid, J.K. (1976) "An Implementation of Tarjan's Algorithm for the Block Triangularisation of a Matrix", Harwell Report CSS 29.

Dufort, E.C. and Frankel, S.P. (1953) "Stability Conditions in the Numerical Treatment of Parabolic Differential Equations", Math. Comp., 7, 135-152.

Evans, D.J. (Ed.) (1974) "Software for Numerical Mathematics", Academic.

Faddeev, D.K. and Faddeeva, V.N. (1963) "Computational Methods of Linear Algebra", Freeman, San Francisco.

Fletcher, C.A.J. (1976) "The Application of the Finite Element Method to Two-Dimensional Inviscid Flow", WRE. Tech. Note 1606.

Forsythe, G.E. (1967) "Today's Computational Methods of Linear Algebra", SIAM Review, 9, 489-515.

Forsythe, G. and Moler, C.B. (1967) "Computer Solution of Linear Algebraic Systems", Prentice-Hall.

Fox, L. (1964) "An Introduction to Numerical Linear Algebra", Clarendon Press, Oxford.

Fox, L. (1977) "Finite Difference Methods for Elliptic Boundary-Value Problems" in The State of the Art in Numerical Analysis (Ed. Jacobs, D.A.H.) Academic.

Fox, L. and Mayers, D.F. (1968) "Computing Methods for Scientists and Engineers", Clarendon Press, Oxford.

George, J.A. (1972) "Block Elimination of Finite Element Systems of Equations" in Sparse Matrices and Their Applications (Eds. Rose, D.J. and Willoughby, R.A.) 41-52, Plenum, New York.

George, J.A. (1971) "Computer Implementation of the Finite Element Method", Stanford Comp. Sc. Dept., Tech. Report STAN-CS-71-208, California.

George, J.A. (1977) "Solution of Linear Systems of Equations: Direct Methods for Finite Element Problems" in Sparse Matrix Systems (Ed. Barker, V.A.) Springer-Verlag.

Gibbs, N.E., Poole, W.G. and Stockmeyer, P.K. (1976) "An Algorithm for Reducing the Bandwdith and Profile of a Sparse Matrix", SIAM J. Num. Anal., 13, 236-250.

Gregory, R.T. and Karney, L.K. (1969) "A Collection of Matrices for Testing Computational Algorithms", Wiley.

Gustavson, F.G. (1972) "Some Basic Techniques for Solving Sparse Systems of Linear Equations" in Sparse Matrices and Their Applications (Eds. Rose, D.J. and Willoughby, R.A.), 41-52, Plenum, New York.

Hall, M. (1956) "An Algorithm for Distinct Representatives", Amer. Math. Monthly, 63, 716-717.

Hamming, R.W. (1971) "Introduction to Applied Numerical Analysis", McGraw-Hill.

Hamming, R.W. (1973) "Numerical Methods for Scientists and Engineers", 2nd. Ed., McGraw-Hill.

Hood, P. (1977a) "Frontal Solution Program for Unsymmetric Matrices", Int. J. Num. Meth. Eng., 10, 379 - 399.

Hood, P. (1977b) "Note on Frontal Solution Program for Unsymmetric Matrices", Int. J. Num. Meth. Eng., 11, 1202.

Householder, A.S. (1964) "The Theory of Matrices in Numerical Analysis", Blaisdell, New York.

Irons, B.M. (1970) "A Frontal Solution Program for Finite Element Analysis", Int. J. Num. Meth. Eng., 2, 5-32.

Irons, B.M. (1980) Private Communication.

Isaacson, E.A. and Keller, H.B. (1966) "Analysis of Numerical Methods", Wiley.

Jacobs, D.A.H. (Ed.) (1977) "The State of the Art in Numerical Analysis", Academic.

Jennings, A. (1966) "A Compact Storage Scheme for the Solution of Simultaneous Equations", Comput. J., 9, 281-285.

Jennings, A. (1977) "Matrix Computation for Engineers and Scientists", Wiley.

Jennings, A. and Tuff, A.D. (1971) "A Direct Method for the Solution of Large Sparse Symmetric Simultaneous Equations" in Large Sparse Sets of Linear Equations (Ed. Reid, J.K.), Academic.

Lee, R.L. (1980) Private Communication.

Lowenthal, F. (1975) "Linear Algebra with Differential Equations", Wiley.

Mann, K.J. (1981a) "Finite Element Galerkin Program for Atmospheric Convection", CIT Publication.

Mann, K.J. (1981b) "Finite Element Least Squares Program for Atmospheric Convection", CIT Publication.

Martin, H.C. and Carey, G.F. (1973) "Introduction to Finite Element Analysis", McGraw-Hill.

Meis, T. and Marcowitz, U. (1981) "Numerical Solution of Partial Differential Equations", Springer-Verlag.

Noble, B. (1969) "Applied Linear Algebra", Prentice-Hall.

Pizer, S.M. (1975) "Numerical Computing and Mathematical Analysis", Science Research Associates (SRA), Chicago.

Ralston, A. (1965) "A First Course in Numerical Analysis", McGraw-Hill.

Reid, J.K. (Ed.) (1971) "Large Sparse Sets of Linear Equations", Academic.

Reid, J.K. (1977a) "Solution of Linear Systems of Equations: Direct Methods (General)" in Sparse Matrix Systems (Ed. Barker, V.A.), Springer-Verlag.

Reid, J.K. (1977b) "Sparse Matrices" in The State of the Art in Numerical Analysis (Ed. Jacobs, D.A.H.), Academic.

Rose, D.J. and Willoughby, R.A. (Eds.) (1972) "Sparse Matrices and Their Applications", Plenum.

Sargent, R.W.H. and Westerberg, A.W. (1964) "Speed up in Chemical Engineering Design", Trans. Inst. Chem. Eng., 42, 190-197.

Tarjan, R.E. (1972) "Depth First Search and Linear Graph Algorithms", SIAM J. Comput., 1, 146-160.

Tewarson, R.P. (1973) "Sparse Matrices", Academic.

Varah, J.M. (1972) "On the Solution of Block-Tridiagonal Systems Arising from Certain Finite-Difference Equations", Math. Comp., 26, 859-868.

Von Rosenberg, D.U. (1969) "Methods for the Numerical Solution of Partial Differential Equations", Elsevier.

Westlake, J. (1968) "A Handbook of Numerical Matrix Inversion and Solution of Linear Equations", Wiley.

Wilkinson, J.H. (1974a) "Error Analysis and Related Topics. The Classical Error Analyses for the Solution of Linear Systems", Bull I.M.A., Vol. 10, no. 9/10.

Wilkinson, J.H. (1961) "Error Analysis of Direct Methods of Matrix Inversion", J. Assoc. Comput. Mach., 8, 281-330.

Wilkinson, J.H. (1974b) "Numerical Linear Algebra on Digital Computers", Bull I.M.A., Vol. 10, no. 9/10.

Wilkinson, J.H. (1963) "Rounding Errors in Algebraic Processes", Prentice-Hall.

Wilkinson, J.H. (1965) "The Algebraic Eigenvalue Problem", Oxford University Press.

Wilkinson, J.H. and Reinsch, C. (1971) "Handbook for Automatic Computation", Vol. II, Linear Algebra, Springer-Verlag.

Young, D.M. and Gregory, R.T. (1973) "A Survey of Numerical Mathematics", Vol. II., Chpt. 12, 779-834, Addison-Wesley.

11. COMMERCIAL PACKAGES

IMSL Available through International Mathematical and Statistical Libraries, Inc., 6th Floor - NBC Building, 7500 Bellaire Boulevard, Houston, Texas 77036, USA.

LINPACK Available through IMSL Inc. (as above).

NAG Available through NAG Central Office, 7 Banbury Road, Oxford OX2 6NN, England.

TWODEPEP Available through IMSL Inc. (as above).

Appendix A   Listing of Fortran program for the solution of large linear systems with a tridiagonal coefficient matrix

```
C *****************************************************************
C
      SUBROUTINE TRIDIA(N,A,B,C,D,X,BETA,Y)
C
C     This subroutine solves the matrix equation:-
C                A * X = D          (1)
C     where  A is a tridiagonal matrix.
C
      DIMENSION A(N),B(N),C(N),D(N),X(N)
      DIMENSION BETA(N), Y(N)
C
C
C -----------------------------------------------------------------
C     GLOSSARY OF TERMS
C -----------------------------------------------------------------
C
C     A(N)    array of entries of 'diagonal' on the left of
C         the leading diagonal. (Note: A(1)=0.0 must be
C         read in with the data).
C     AMULT   is set equal to constant for each row.
C     B(N)   array of diagonal entries.
C     B(I)    component of B(N), it contains the entry
C         from the 'Ith' row.
C     BETA(N)    an intermediate array arising from the
C            LU decomposition of A.
C     C(N)    array of entries of 'diagonal' on the right of
C         the leading diagonal.(Note: C(N)=0.0 must be
C         read in with data).
C     D(N)    array of entries from the R.H.S of equation 1
C     X(N) array of the solution to tridiagonal matrix.
C     X(I)  contains the solution to the 'Ith' equation
C     Y(N)    an intermediate array arising from the
C           LU decomposition af A.
C
C -----------------------------------------------------------------
C
C
      BETA(1) = B(1)
      Y(1) = D(1)
C
C     Forward substitution begins.
      DO 1 I=2,N
           AMULT = A(I)/BETA(I-1)
           BETA(I) = B(I) - AMULT*C(I-1)
           Y(I) = D(I) - AMULT*Y(I-1)
    1 CONTINUE
C
C     Backward substitution begins.
      X(N) = Y(N)/BETA(N)
      DO 2 J=2,N
           I=N-J+1
           X(I) = (Y(I) - C(I)*X(I+1))/BETA(I)
    2 CONTINUE
C
C     Outputing the data.
      WRITE(1,3)
    3 FORMAT(//,5X,'For the TRIDIAGONAL matrix',///,

     *5X,'A(I)',6X,'B(I)',6X,'C(I)',9X,'D(I)',8X,'THE SOLUTION',
     *2X,'X(I)',/,4X,25('-'),8X,5('-'),8X,18('-'))
      DO 5 K=1,N
           WRITE(1,4) A(K),B(K),C(K),D(K),X(K)
    4      FORMAT(3(3X,F7.3),6X,F7.3,17X,F8.4)
    5 CONTINUE
      RETURN
      END
```

Appendix B   Listing of Fortran program for the solution of large linear systems with a pentadiagonal coefficient matrix.

```
      SUBROUTINE PENTA(N,A,B,C,D,E,F,X)
C * * * * * IMPLEMENTS THE PENTADIAGONAL MATRIX ALGORITHM
C
C  GLOSSARY:
C           N = ORDER OF COEFFICIENT MATRIX
C           G(N)=COEFFICIENT MATRIX
C           F(N)=CONSTANTS VECTOR
C           X(N)=SOLUTION VECTOR TO SYSTEM OF EQUATIONS GX=F.
C           A,B,C,D,E(I)= NON-ZERO ENTRIES,IN ORDER,OF ITH ROW OF MATRIX
C                G. C(I) IS THE DIAGONAL ENTRY OF G.
C                (SEE NOTE IN MAIN SECTION)
C           DELTA(N),LAMDA (N),GAMMA(N)       =INTERMEDIATE ARRAYS
C                GENERATED IN THE PENTADIAGONAL ALGORITHM
C           BETA,MU =INTERMEDIATE PARAMETERS USED ONLY IN EACH ROW TO
C                COMPUTE INTERMEDIATE ARRAYS,AND NOT REQUIRED FOR
C                SUBSEQUENT ROWS.
C
C
      DIMENSION A(1),B(1),C(1),D(1),E(1),F(1),X(1)
      REAL MU
      DIMENSION DELTA(40),GAMMA(40)
      REAL LAMDA(40)
C
      DELTA(1)=D(1)/C(1)
      LAMDA(1)=E(1)/C(1)
      GAMMA(1)=F(1)/C(1)
      MU=C(2)-B(2)*DELTA(1)
      DELTA(2)=(D(2)-B(2)*LAMDA(1))/MU
      LAMDA(2)=E(2)/MU
      GAMMA(2)=(F(2)-B(2)*GAMMA(1))/MU
C
      NM2=N-2
      DO 10 I=3,NM2
            BETA    =B(I)-A(I)*DELTA(I-2)
            MU= C(I)-BETA*DELTA(I-1)-A(I)*LAMDA(I-2)
            DELTA(I)=(D(I)-BETA*LAMDA(I-1))/MU
            LAMDA(I)=E(I)/MU
            GAMMA(I)=(F(I)-BETA*GAMMA(I-1)-A(I)*GAMMA(I-2))/MU
   10 CONTINUE
C
      BETA=B(N-1)-A(N-1)*DELTA(N-3)
      MU=C(N-1)-BETA*DELTA(N-2)-A(N-1)*LAMDA(N-3)
      DELTA(N-1)=(D(N-1)-BETA*LAMDA(N-2))/MU
      GAMMA(N-1)=(F(N-1)-BETA*GAMMA(N-2)-A(N-1)*GAMMA(N-3))/MU
      BETA=B(N)-A(N)*DELTA(N-2)
      MU=C(N)-BETA*DELTA(N-1)-A(N)*LAMDA(N-2)
      GAMMA(N)=(F(N)-BETA*GAMMA(N-1)-A(N)*GAMMA(N-2))/MU
      X(N)=GAMMA(N)
      X(N-1)=GAMMA(N-1)-DELTA(N-1)*X(N)
C
      DO 20 I=1,NM2
            K=N-1-I
            X(K)=GAMMA(K)-DELTA(K)*X(K+1)-LAMDA(K)*X(K+2)
   20 CONTINUE
C
      RETURN
      END
```

Appendix C Listing of Fortran program for the solution of large linear systems with random sparseness

```
C      SUBROUTINE FOR SPARSE LINEAR SYSTEMS
C      ------------------------------------
C
C                         DESIGNED BY:   C. FLETCHER
C                                 Weapons Research 1976
C
C                    IMPLIMENTED BY:   C. HALL  1981
C
C                          SUPERVISOR   : K. J. MANN
C                                Mathematics Dept.   C. I. T.
C
C                            COMPUTER:    PRIME (C. I. T. )
C
C      SOLVES SYSTEMS OF N LINEAR EQUATIONS STORING ONLY NON ZERO
C      ELEMENTS. USES SPARSE GUASS ELIMINATION
C
       SUBROUTINE SPAREL(AH, IH, N, LCMAX, JEND, IDIM, B, X,
     *IELIM, JPT, AHP, IHP, IHPDIM)
C
C -------------------------------------------------------------
C      GLOSSARY OF TERMS
C -------------------------------------------------------------
C      AH        :   Non-zero elements of Sparse matrix.
C      AHP       :   Temporary location of 'fill-in' terms.
C      B         :   Array containing right-hand side of equation.
C      I         :   Current work row.
C      IA        :   Is the position of the largest coefficient in
C                    the Ith row.
C      IDIM      :   Is the increased dimensions of AH and IH.
C      IELIM(I):     Contains the position of largest coefficient
C                :   in row I.
C      IH        :   Column position of corresponding terms in AH.
C      IHOLD     :   Temparary storage location of coefficient
C                    in row I for finding  IELIM.
C      IHP       :   Column position of corresponding terms in AHP
C      IHPDIM    :   The dimensions of arrays IHP and AHP.
C      J         :   No. of equation in second sweep.  (DO 13).
C      JCBIG     :   Is the maximum value of JCP.
C      JCM       :   Column counter for deleting the location of PIVOT
C                    in row J.
C      JCMH      :   Temporary stores JCM during row  J.
C      JCP       :   Column position of term deleted from row J.
C      JEND(i)   :   Contains the No. of non-zeros in row  i.
C      JPT       :   No. of non-zeros in each column.
C      JSA       :   Counter in the reverse sweep.
C      KEND      :   No. of non-zero's in row  I.
C      L         :   Column position in the Ith equation.  (DO 8).
C      LA        :   Column counter in row  J.
C      LCBIG     :   Is the maximum No. of terms in AH.
C      LCMAX     :   Is the location of the last entry in  AH.
C      LCT       :   Column counter for finding IHOLD.
C      LHOLD     :   Temporary location of LCT. (for row I).
C      M         :   Column position in the Jth equation.  (DO 9).
C      MCT       :   LCT - KEND.
C      MEND      :   No. of non-zeros in row J.
C      MHOLD     :   Temporary storage of  MCT + 1.
C      N         :   No. of equations.
C      NCT       :   Column counter for eliminating X(IHOLD) from Jth row
C      PIVOT     :   Is the multiplying factor applied to all rows
C                    about the pivotal row.
C      X         :   Array containing the solution.
C -------------------------------------------------------------
C
       DIMENSION B(N), X(N), JEND(N), AH(IDIM), IH(IDIM)
       DIMENSION IELIM(N), JPT(N), AHP(IHPDIM), IHP(IHPDIM)
C
C
       NM1=N-1
       LCBIG=0
       JCBIG=0
       LCT=0
```

```
C
C     I is the equation from which  X(IHOLD) will be eliminated.
      DO 26 I=1,NM1
         LCT=LCT+1
         KEND =JEND(I)
         PIVOT =AH(LCT)
         IHOLD=IH(LCT)
         DO 1 K=2,KEND
            LCT=LCT +1
            IF(ABS(AH(LCT)).LT.ABS(PIVOT)) GO TO 1
            PIVOT  =AH(LCT)
            IHOLD=IH(LCT)
    1    CONTINUE
C
C        Normalize the Ith equation with the largest coefficient.
         MCT =LCT-KEND
         IELIM(I)=IHOLD
         DO 2 K=1,KEND
            MCT=MCT+1
            AH(MCT)=AH(MCT)/PIVOT
    2    CONTINUE
         B(I)=B(I)/PIVOT
C
C        Eliminate  X(IHOLD) from all subsequent equations.
         MCT=LCT
         IP1=I+1
         JCP=0
         JCM=0
         DO 13 J=IP1,N
            JCPH=JCP
            JCMH=JCM
            MEND=JEND(J)
            MHOLD=MCT+1
            NHOLD=MCT+MEND
            NCT=MCT
            IF(IH(MHOLD).GT.IHOLD)GO TO 10
            LHOLD=LCT-KEND
            DO 9 M =1,MEND
               MCT=MCT+1
               IF(IH(MCT).GT.IHOLD)GO TO 10
               IF(IH(MCT).LT.IHOLD)GO TO 9
               PIVOT=AH(MCT)
               NCT=MHOLD-1
               B(J)=B(J)-B(I)*PIVOT
               DO 8 L=1,KEND
                  LA=LHOLD+L
    3             NCT=NCT+1
                  IF(NCT.GT.NHOLD)GO TO 6
                  IF(IH(LA).LT.IH(NCT))GO TO 6
                  IF(IH(LA).EQ.IH(NCT))GO TO 4
                  JD=NCT-JCM
                  AH (JD)=AH(NCT)
                  IH(JD)=IH(NCT)
                  GO TO 3
    4             IF(IH(LA).NE.IHOLD)GO TO 5
                  JCM=JCM+1
                  GO TO 8
C
C                 Drop XIHOLD) from the Jth equation.
    5             AH(NCT)=AH(NCT)-AH(LA)*PIVOT
                  GO TO 7
C
C                 Include   X(IH(LA)) in the Jth equation
C                 and store temporarily in  AHP/IHP.
    6             JCP =JCP +1
                  IHP(JCP) = IH(LA)
                  AHP(JCP) = - AH(LA) * PIVOT
                  NCT =NCT - 1
                  GO TO 8
    7             JD = NCT -JCM
                  AH(JD) = AH(NCT)
                  IH(JD) = IH(NCT)
    8          CONTINUE
    9       CONTINUE
```

```
10          MCT = MHOLD - 1 + JEND(J)
11          NCT = NCT + 1
            IF(NCT .GT. NHOLD) GO TO 12
            JD = NCT - JCM
            AH(JD) = AH(NCT)
            IH(JD) = IH(NCT)
            GO TO 11
12          JEND(J) = JEND(J) - JCM + JCMH
            JPT(J) = JCP - JCPH
13      CONTINUE
        LCMAX =LCMAX -JCM
        MCT = LCMAX + 1
        MHOLD = MCT
        LCMAX = LCMAX + JCP
        IF(LCMAX .LE. LCBIG) GO TO 14
        LCBIG = LCMAX
14      IF(JCP .LE. JCBIG) GO TO 15
        JCBIG =JCP
15      CONTINUE
C
C       Test that the Dimensions of AHP/IHP and AH/IH
C       have not been exceeded.
        IF (JCP .LE. IHPDIM) GO TO 17
        WRITE(1,16) JCP,I
16      FORMAT('  JCP TOO LARGE, = ',I8,' I = ',I5,)
        RETURN
17      IF(LCMAX .LE. IDIM) GO TO 19
        WRITE(1,18) LCMAX,I,J
18      FORMAT('  LCMAX TOO LARGE  = ',I8,' I = ',I5,' J = ',I5)
        RETURN
C
C       Do a reverse pass thru  AH/IH to fill in new terms which
C       have been stored temporarily in  AHP/IHP
19      DO 25 JA =IP1,N
            J = IP1 + N - JA
            MEND = JEND(J)
            JCPH = JCP
            JSA = 0
            MCT = MHOLD
            IF(JCP .LE. 0) GO TO 24
            DO 22 M = 1,MEND
                MCT = MCT -1
                IF(JSA .GE. JPT(J)) GO TO 21
20              IF(IH(MCT) .GE. IHP(JCP)) GO TO 21
                JSA = JSA + 1
                JD = MCT +JCP
                AH(JD) = AHP(JCP)
                IH(JD) = IHP(JCP)
                JCP = JCP - 1
                IF(JCP .LE. 0) GO TO 24
                IF(JSA .LT. JPT(J)) GO TO 20
21              JD = MCT + JCP
                AH(JD) = AH(MCT)
                IH(JD) = IH(MCT)
22          CONTINUE
23          IF(JSA .GE. JPT(J)) GO TO 24
            JSA = JSA + 1
            JD = MCT + JCP - 1
            AH(JD) = AHP(JCP)
            IH(JD) = IHP(JCP)
            JCP = JCP - 1
            GO TO 23
24          MHOLD = MHOLD - JEND(J)
            JEND(J) = JEND(J) -JCP + JCPH
25      CONTINUE
26 CONTINUE
```

```
C
C      Backsubstitution to obtain the  X's.
       IELIM(N) = IH(LCMAX)
       B(N) = B(N)/AH(LCMAX)
       LCT = LCMAX + 1
       DO 28 I = 1,N
             NA = N + 1 -I
             KEND = JEND(NA)
             IA = IELIM(NA)
             X(IA) =B(NA)
             DO 27 K = 1,KEND
                   LCT = LCT -1
                   IF(IH(LCT) .EQ. IA) GO TO 27
                   IB = IH(LCT)
                   X(IA) = X(IA) - X(IB) * AH(LCT)
   27        CONTINUE
   28  CONTINUE
       WRITE(1,29) LCT,LCBIG,JCBIG
   29  FORMAT('  LCT = ',I5,'  LCBIG = ',I5,' JCBIG = ',I5)
       RETURN
       END
```

Computational Techniques for Differential Equations
J. Noye (Editor)

ITERATIVE METHODS FOR SOLVING LARGE SPARSE SYSTEMS OF LINEAR ALGEBRAIC EQUATIONS

LEONARD COLGAN
South Australian Institute of Technology, Adelaide, South Australia

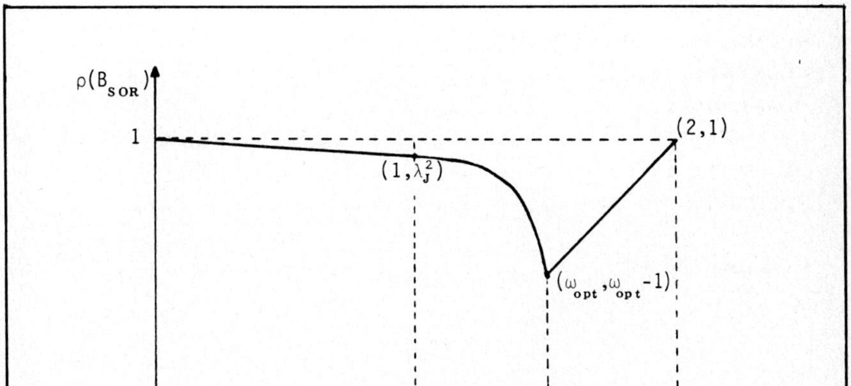

CONTENTS

1. INTRODUCTION

Because of the substantial increase in speed and efficiency of modern digital computers, numerical methods involving iterative processes have regained much popularity. For example, the discretization of partial differential equations, using either finite differences or finite elements, invariably leads to the problem of solving a large system of linear equations. In particular, elliptic partial differential equations in two dimensions, or more so in three dimensions, yield linear systems which, because of the very large number of equations with very few non-zero coefficients, suggest an iterative method rather than a direct method. The available literature on iterative techniques is vast, while the number of substantially different methods is equally vast. Hence a suitable selection has been made, taking into account the simplicity of description, ease of implementation on a computer, efficiency in the use of storage and time, availability of software, and most importantly, personal preference. Consequently, rather than detailing a succession of methods and their many variants, a small representative selection has been included with a list of reference material for those interested enough to seek further information.

A brief summary of the contents of this article is as follows: The first section introduces two model problems for subsequent discussion, and this is followed by the necessary matrix preliminaries. Linear stationary iterative methods of the first degree are then described, along with their corresponding convergence criteria. The particular methods detailed are the basic Jacobi, Gauss-Seidel, Successive Overrelaxation (SOR), and Symmetric Successive Overrelaxation (SSOR) methods. Of these, it is fair to say that only Successive Overrelaxation has any practical claim for direct implementation. However, the Jacobi and SSOR methods are included because they can be accelerated substantially using Chebyshev or conjugate gradient acceleration procedures, which comprise the next two sections. Then follows a description of how the size of the system can possibly be reduced using a so-called "red-black" ordering on rectangular domains. Alternating-Direction Implicit (ADI) methods are represented by the Peaceman-Rachford method. Throughout these discussions, it should be understood that variants of these basic methods can yield further improvements. For example it is possible to use block-iterative methods (or line-iterative methods) in certain circumstances, and these can produce worthwhile time savings when compared to the fundamental point-iterative methods. These are described extensively in many references in the bibliography. Finally, a brief comparison of selected iterative methods is included. Appendix A contains a list of the symbols used in this chapter, and subsequent Appendices list possible FORTRAN subroutines (and outputs) to implement a number of the methods discussed.

A special acknowledgement must be made to Professor David M. Young of the Center for Numerical Analysis, at the University of Texas at Austin. In preparing a review article such as this, it was necessary to select notation, definitions, particular methods and examples from the list of reference material. Professor Young's influence in this area is especially significant and much of the following material can be attributed directly to him.

2. THE MODEL PROBLEMS

Consider the two-dimensional Poisson equation

$$\frac{\partial^2 u}{\partial x^2} + \frac{\partial^2 u}{\partial y^2} = -1, \text{ with } u(x,y) = 0$$

on the boundary of the unit square $0 \leq x \leq 1$, $0 \leq y \leq 1$.

Using a uniform grid of mesh-size h, we can approximate the given elliptic partial differential equation at a point (x,y) by the usual finite difference formula

$$\frac{u(x,y-h)+u(x-h,y)-4u(x,y)+u(x+h,y)+u(x,y+h)}{h^2} = -1.$$

<u>"Model Problem 1"</u> corresponds to $h = \frac{1}{5} = 0.2$. Then there would be 16 unknowns corresponding to the function values at the internal nodes. Let these be denoted by u_i, $i=1,\ldots,16$, where the <u>natural ordering</u> is used.

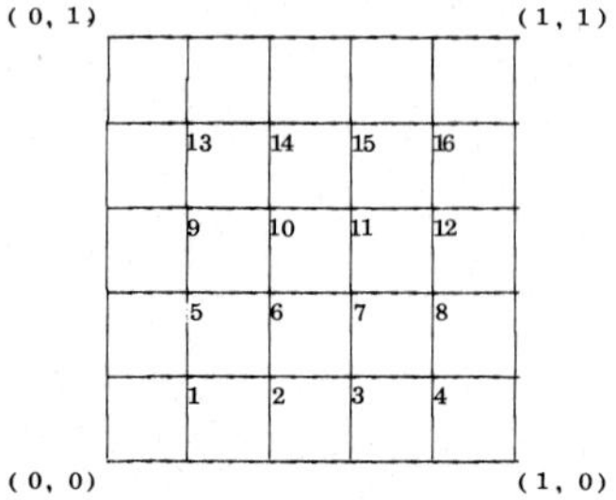

This then leads to the matrix equation

$$\begin{pmatrix}
4 & -1 & 0 & 0 & -1 & 0 & 0 & 0 & 0 & 0 & 0 & 0 & 0 & 0 & 0 & 0\\
-1 & 4 & -1 & 0 & 0 & -1 & 0 & 0 & 0 & 0 & 0 & 0 & 0 & 0 & 0 & 0\\
0 & -1 & 4 & -1 & 0 & 0 & -1 & 0 & 0 & 0 & 0 & 0 & 0 & 0 & 0 & 0\\
0 & 0 & -1 & 4 & 0 & 0 & 0 & -1 & 0 & 0 & 0 & 0 & 0 & 0 & 0 & 0\\
-1 & 0 & 0 & 0 & 4 & -1 & 0 & 0 & -1 & 0 & 0 & 0 & 0 & 0 & 0 & 0\\
0 & -1 & 0 & 0 & -1 & 4 & -1 & 0 & 0 & -1 & 0 & 0 & 0 & 0 & 0 & 0\\
0 & 0 & -1 & 0 & 0 & -1 & 4 & -1 & 0 & 0 & -1 & 0 & 0 & 0 & 0 & 0\\
0 & 0 & 0 & -1 & 0 & 0 & -1 & 4 & 0 & 0 & 0 & -1 & 0 & 0 & 0 & 0\\
0 & 0 & 0 & 0 & -1 & 0 & 0 & 0 & 4 & -1 & 0 & 0 & -1 & 0 & 0 & 0\\
0 & 0 & 0 & 0 & 0 & -1 & 0 & 0 & -1 & 4 & -1 & 0 & 0 & -1 & 0 & 0\\
0 & 0 & 0 & 0 & 0 & 0 & -1 & 0 & 0 & -1 & 4 & -1 & 0 & 0 & -1 & 0\\
0 & 0 & 0 & 0 & 0 & 0 & 0 & -1 & 0 & 0 & -1 & 4 & 0 & 0 & 0 & -1\\
0 & 0 & 0 & 0 & 0 & 0 & 0 & 0 & -1 & 0 & 0 & 0 & 4 & -1 & 0 & 0\\
0 & 0 & 0 & 0 & 0 & 0 & 0 & 0 & 0 & -1 & 0 & 0 & -1 & 4 & -1 & 0\\
0 & 0 & 0 & 0 & 0 & 0 & 0 & 0 & 0 & 0 & -1 & 0 & 0 & -1 & 4 & -1\\
0 & 0 & 0 & 0 & 0 & 0 & 0 & 0 & 0 & 0 & 0 & -1 & 0 & 0 & -1 & 4
\end{pmatrix}
\begin{pmatrix} u_1\\ u_2\\ u_3\\ u_4\\ u_5\\ u_6\\ u_7\\ u_8\\ u_9\\ u_{10}\\ u_{11}\\ u_{12}\\ u_{13}\\ u_{14}\\ u_{15}\\ u_{16} \end{pmatrix}
= h^2 \begin{pmatrix} 1\\1\\1\\1\\1\\1\\1\\1\\1\\1\\1\\1\\1\\1\\1\\1 \end{pmatrix}
= \begin{pmatrix} 0.04\\0.04\\0.04\\0.04\\0.04\\0.04\\0.04\\0.04\\0.04\\0.04\\0.04\\0.04\\0.04\\0.04\\0.04\\0.04 \end{pmatrix}$$

i.e. $A\underset{\sim}{u} = \underset{\sim}{b}$.

The coefficient matrix A is symmetric and positive definite, sparse, and has a band-width of 9.

<u>"Model Problem 2"</u> corresponds to $h = \frac{1}{40} = 0.025$.

In this second model problem, we have a large sparse linear system $A\underset{\sim}{u} = \underset{\sim}{b}$, where the matrix A is of order 1521, and the vector $\underset{\sim}{u}$ contains the unknown values u_i, $i=1,\ldots,1521$, numbered in the natural ordering. A has a band-width of 79, and there are at most five non-zero entries in each row of A.

Apart from equations derived from points adjoining a boundary, we have to solve a system (h = 1/40) of the form

$$-u_{i-39} - u_{i-1} + 4u_i - u_{i+1} - u_{i+39} = h^2 = 0.000625. \qquad (2.1)$$

In general, numerical solutions of (elliptic) partial differential equations lead to linear systems with the coefficient matrix having a structure somewhat similar to the ones generated by the model problems. However, finite element schemes often produce a matrix which is less sparse and with less pattern.

In the subroutines listed in the Appendix, the matrix A is to be input in an unsymmetric sparse form, although for the second problem it would clearly be more economical to use a symmetric sparse storage scheme.

For the first model problem, this would require the data to be input as

A |4, -1, -1, -1, 4, -1, -1, -1, 4, -1, -1, -1, 4, -1, -1, 4,
-1, -1, -1, -1, 4, -1, -1, -1, -1, 4, -1, -1, -1, -1, 4, -1,
-1, 4, -1, -1, -1, -1, 4, -1, -1, -1, -1, 4, -1, -1, -1, -1,
4, -1, -1, 4, -1, -1, -1, 4, -1, -1, -1, 4, -1, -1, -1, 4|

JA|1, 2, 5, 1, 2, 3, 6, 2, 3, 4, 7, 3, 4, 8, 1, 5, 6, 9, 2, 5, 6,
7, 10, 3, 6, 7, 8, 11, 4, 7, 8, 12, 5, 9, 10, 13, 6, 9, 10, 11,
14, 7, 10, 11, 12, 15, 8, 11, 12, 16, 9, 13, 14, 10, 13, 14, 15,
11, 14, 15, 16, 12, 15, 16|

ISTART|1, 4, 8, 12, 15, 19, 24, 29, 33, 37, 42, 47, 51, 54, 58, 62, 65|

Although the solution of the second model problem, $h = 1/40$, is more typical of "large sparse systems of linear algebraic equations", the first problem, with only 16 variables, has been introduced to enable some outputs to be shown in the Appendix.

3. MATRIX PRELIMINARIES

Let B be a real $N \times N$ matrix, and $\underset{\sim}{x}$ an N-dimensional real vector.

(a) Then $\rho(B)$, the <u>spectral radius</u> of B, is the maximum of the moduli of the eigenvalues of B. Moreover, let $m(B)$ and $M(B)$ (or, if unambiguous, merely m and M) represent the smallest and largest eigenvalues of B, so that $m(B) \le \lambda \le M(B)$ whenever all the eigenvalues λ are real.

(b) For <u>any</u> matrix norm $\| \;\|$, $\rho(B) \le \|B\|$.

(c) We define the <u>2-norm</u> of $\underset{\sim}{x}$ by

$$\|\underset{\sim}{x}\|_2 = (\underset{\sim}{x},\underset{\sim}{x})^{\frac{1}{2}} = \left(\sum_{i=1}^{N} x_i^2\right)^{\frac{1}{2}}.$$

The 2-norm of B is given by

$$\|B\|_2 = \{\rho(B^T B)\}^{\frac{1}{2}},$$

and this matrix norm is <u>subordinate</u> to the vector norm in that

$$\|B\|_2 = \max_{\underset{\sim}{x} \ne \underset{\sim}{0}} \frac{\|B\underset{\sim}{x}\|_2}{\|\underset{\sim}{x}\|_2}.$$

Clearly $\|B\underset{\sim}{x}\|_2 \le \|B\|_2 \|\underset{\sim}{x}\|_2$. Also, if B is symmetric, then $\|B\|_2 = \rho(B)$.

(d) For any nonsingular matrix W, we define the <u>W-norm</u> of the vector $\underset{\sim}{x}$ and of the matrix B by

$$\|\underset{\sim}{x}\|_W = \|W\underset{\sim}{x}\|_2$$

and

$$\|B\|_w = \|WBW^{-1}\|_2.$$

Hence, if WBW^{-1} is symmetric, then $\|B\|_w = \rho(B)$.

(e) B is <u>symmetric positive definite</u> if and only if B is symmetric and $\underset{\sim}{x}^T B\underset{\sim}{x}>0$ for all $\underset{\sim}{x}\neq\underset{\sim}{0}$.

Then all the eigenvalues of B are positive, and there exists a unique symmetric positive definite matrix, denoted by $B^{1/2}$, whose square is equal to B. Also, B is symmetric positive definite if and only if there exists a non-singular matrix X such that $B = X^T X$.

(f) B is <u>irreducible</u> (indecomposable) if it cannot be put into the form $\begin{pmatrix} F & 0 \\ G & H \end{pmatrix}$, where F and H are square, by simultaneous row and column permutations.

(g) A matrix B is said to have "<u>Property A</u>" if, by appropriate simultaneous row and column permutations, B can be written in the form

$$\begin{pmatrix} D_1 & G \\ K & D_2 \end{pmatrix},$$

where D_1 and D_2 are square diagonal matrices.

(h) $B = \{b_{ij}\}$ has <u>weak diagonal dominance</u> if

$$|b_{ii}| \geq \sum_{\substack{j=1 \\ j\neq i}}^{N} |b_{ij}| \qquad (1\leq i\leq N),$$

and for at least one i, the inequality is strict.

Then if B is a real symmetric matrix which is irreducible, $b_{ii}>0$ for $i=1,\ldots,N$, and B has weak diagonal dominance then B is symmetric positive definite.

The matrices A in the linear systems $A\underset{\sim}{u} = \underset{\sim}{b}$ derived from the model problems are irreducible, have "Property A", have weak diagonal dominance and are symmetric positive definite.

4. BASIC ITERATIVE METHODS

Consider a large sparse linear system

$$A\underset{\sim}{u} = \underset{\sim}{b}, \tag{4.1}$$

where A is a given real, nonsingular N×N matrix, and $\underset{\sim}{b}$ is a given real N-component vector. Such systems that arise from discretizations of partial differential equations yield a matrix A that is usually symmetric positive definite, and so this assumption shall be made. In any case, such a condition is often necessary to guarantee convergence of the iterative methods considered.

<u>A linear stationary iterative method of the first degree</u> can be constructed (theoretically) in the following manner. Suppose there exists a nonsingular matrix Q which in some sense can be imagined as being an approximation to A, but, unlike A, is readily invertible. More particularly, it is possible to solve conveniently any linear system of the form $Q\underset{\sim}{x} = \underset{\sim}{c}$. For example, Q may be

a diagonal, tri-diagonal, or lower triangular matrix, or the product of two or more such matrices. Q is called a <u>splitting matrix</u>.

The system (4.1) is clearly equivalent to

$$Q\underset{\sim}{u} = (Q-A)\underset{\sim}{u} + \underset{\sim}{b}$$

which leads to the iterative process

$$Q\underset{\sim}{u}^{(n+1)} = (Q-A)\underset{\sim}{u}^{(n)} + \underset{\sim}{b}.$$

Although Q^{-1} is not usually found explicitly, this can be written as

$$\underset{\sim}{u}^{(n+1)} = B\underset{\sim}{u}^{(n)} + \underset{\sim}{k}, \tag{4.2}$$

where $B = I - Q^{-1}A$ and $\underset{\sim}{k} = Q^{-1}\underset{\sim}{b}$.

Because the <u>iteration matrix</u> B is constant, this process is called stationary. (Information concerning nonstationary methods, such as Richardson's method, can be found in Forsythe and Wasow (1960); or Young (1954), (1971).)

The basic method (4.2) is <u>convergent</u> if the sequence $\underset{\sim}{u}^{(0)}, \underset{\sim}{u}^{(1)}, \underset{\sim}{u}^{(2)}, \ldots$ converges to the true solution $\underset{\sim}{u}$ for all initial vectors $\underset{\sim}{u}^{(0)}$, and this limit is independent of the choice of $\underset{\sim}{u}^{(0)}$.

A necessary and sufficient condition for convergence is that $\rho(B)<1$.

The <u>error vector</u> after n iterations is

$$\underset{\sim}{\varepsilon}^{(n)} = \underset{\sim}{u}^{(n)} - \underset{\sim}{u}.$$

The <u>residual vector</u> is

$$\underset{\sim}{r}^{(n)} = \underset{\sim}{b} - A\underset{\sim}{u}^{(n)}, \quad \text{(from (4.1))}$$

and the <u>pseudo-residual vector</u> is

$$\underset{\sim}{\delta}^{(n)} = B\underset{\sim}{u}^{(n)} + \underset{\sim}{k} - \underset{\sim}{u}^{(n)}$$

$$= \underset{\sim}{u}^{(n+1)} - \underset{\sim}{u}^{(n)}, \quad \text{(from (4.2)), } \underline{\text{in this case.}}$$

Since $\underset{\sim}{u} = B\underset{\sim}{u} + \underset{\sim}{k}$, then from (4.2) we have

$$\underset{\sim}{\varepsilon}^{(n+1)} = B\,\underset{\sim}{\varepsilon}^{(n)} \text{ and hence}$$

$$\underset{\sim}{\varepsilon}^{(n)} = B^n\,\underset{\sim}{\varepsilon}^{(0)}.$$

Consequently, $\dfrac{\|\underset{\sim}{\varepsilon}^{(n)}\|_2}{\|\underset{\sim}{\varepsilon}^{(0)}\|_2} \le \|B^n\|_2$, for any $\underset{\sim}{\varepsilon}^{(0)} \ne \underset{\sim}{0}$.

The <u>average rate of convergence</u> for n iterations is defined by

$$R_n(B) = -\frac{1}{n}\ln \|B^n\|_2.$$

If $R_n(B_1) < R_n(B_2)$, then B_2 is iteratively faster for n iterations than B_1.

Then, to reduce an initial error by a prescribed factor, say 10^{-6}, it would require approximately

$$n \simeq \frac{\ln 10^6}{R_n(B)} \text{ iterations.}$$

However, the average rate of convergence, $R_n(B)$, is rarely available.

The <u>asymptotic rate of convergence</u>, denoted by $R(B)$, is defined by

$$R(B) = \lim_{n\to\infty} R_n(B),$$

and it can be shown that

$$\underline{R(B) = -\ln \rho(B).} \qquad (4.3)$$

(Since $\|B_n\|_2 \geq [\rho(B)]^n$ for all $n\geq 1$, then $R(B) \geq R_n(B)$ for any positive n for which $\|B^n\|_2 < 1$.)

This asymptotic rate of convergence $R(B)$, for a convergent matrix B, is certainly the simplest practical way in modern usage of measuring the rapidity of convergence. Hence, a crude estimate of the number of iterations required to reduce the norm of the error by a factor 10^{-6} is given by

$$n \simeq \frac{\ln 10^6}{R(B)}. \qquad (4.4)$$

However, this estimate can often yield misleadingly low values for n. [For example, see Varga (1962).]

4.1 Jacobi Method

For the first model problem, the Jacobi iterative method is represented by

$$u_1^{(n+1)} = \tfrac{1}{4}(u_2^{(n)} + u_5^{(n)} + 0.04)$$

$$u_2^{(n+1)} = \tfrac{1}{4}(u_1^{(n)} + u_3^{(n)} + u_6^{(n)} + 0.04)$$

$$u_3^{(n+1)} = \tfrac{1}{4}(u_2^{(n)} + u_4^{(n)} + u_7^{(n)} + 0.04)$$

$$u_4^{(n+1)} = \tfrac{1}{4}(u_3^{(n)} + u_8^{(n)} + 0.04)$$

$$u_5^{(n+1)} = \tfrac{1}{4}(u_1^{(n)} + u_6^{(n)} + u_9^{(n)} + 0.04)$$

$$u_6^{(n+1)} = \tfrac{1}{4}(u_2^{(n)} + u_5^{(n)} + u_7^{(n)} + u_{10}^{(n)} + 0.04)$$

$$u_7^{(n+1)} = \tfrac{1}{4}(u_3^{(n)} + u_6^{(n)} + u_8^{(n)} + u_{11}^{(n)} + 0.04)$$

$$u_8^{(n+1)} = \tfrac{1}{4}(u_4^{(n)} + u_7^{(n)} + u_{12}^{(n)} + 0.04)$$

$$u_9^{(n+1)} = \tfrac{1}{4}(u_5^{(n)} + u_{10}^{(n)} + u_{13}^{(n)} + 0.04)$$

$$u_{10}^{(n+1)} = \tfrac{1}{4}(u_6^{(n)} + u_9^{(n)} + u_{11}^{(n)} + u_{14}^{(n)} + 0.04)$$

$$u_{11}^{(n+1)} = \tfrac{1}{4}(u_7^{(n)} + u_{10}^{(n)} + u_{12}^{(n)} + u_{15}^{(n)} + 0.04)$$

$$u_{12}^{(n+1)} = \tfrac{1}{4}(u_8^{(n)} + u_{11}^{(n)} + u_{16}^{(n)} + 0.04)$$

$$u_{13}^{(n+1)} = \tfrac{1}{4}(u_9^{(n)} + u_{14}^{(n)} + 0.04)$$

$$u_{14}^{(n+1)} = \tfrac{1}{4}(u_{10}^{(n)} + u_{13}^{(n)} + u_{15}^{(n)} + 0.04)$$

$$u_{15}^{(n+1)} = \tfrac{1}{4}(u_{11}^{(n)} + u_{14}^{(n)} + u_{16}^{(n)} + 0.04)$$

$$u_{16}^{(n+1)} = \tfrac{1}{4}(u_{12}^{(n)} + u_{15}^{(n)} + 0.04)$$

Using an initial estimate $\underset{\sim}{u}^{(0)} = \underset{\sim}{0}$, subroutine JAC (see Appendix B), a corresponding calling program, and a suitable stopping criterion (see later), the output in Appendix B is obtained. The spectral radius (SR in the subroutine) is estimated, and stopping is invoked when

$$\frac{SR}{1-SR}\,\frac{\|\underset{\sim}{\delta}^{(n)}\|_2}{\|\underset{\sim}{u}^{(n)}\|_2} < \zeta \qquad (\zeta = 10^{-6} \text{ in this case}).$$

For the second model problem, the Jacobi method is equivalent to rewriting (2.1) in the iterative form

$$u_i^{(n+1)} = \tfrac{1}{4}(u_{i-39}^{(n)} + u_{i-1}^{(n)} + u_{i+1}^{(n)} + u_{i+39}^{(n)} + h^2).$$

We can express the matrix A as the matrix sum

$$A = D + L + U,$$

where D is the diagonal matrix such that $d_{ii} = a_{ii}$, and L and U are respectively strictly lower and upper triangular N×N matrices, whose entries are those respectively below and above the main diagonal of A. We can then rewrite (4.1) as

$$D\underset{\sim}{u} = -(L+U)\underset{\sim}{u} + \underset{\sim}{b},$$

which in turn yields the Jacobi iteration method

$$\underset{\sim}{u}^{(n+1)} = -D^{-1}(L+U)\underset{\sim}{u}^{(n)} + D^{-1}\underset{\sim}{b}.$$

Hence our general notation corresponds to a splitting matrix $Q_J \equiv D$, an iteration matrix

$$B_J \equiv -D^{-1}(L+U) = I - D^{-1}A, \text{ and } \underset{\sim}{k}_J \equiv D^{-1}\underset{\sim}{b}.$$

It should be pointed out that it is possible to normalize the initial matrix problem so that the diagonal elements are unity (see Varga (1962)), and this avoids numerous division operations during the iterative process.

The Jacobi method will converge providing $\rho(B_J) < 1$. However, this is not guaranteed, even though the matrix A might be symmetric positive definite, because, although it can be shown that the eigenvalues of B are real and < 1, it is possible that $m(B_J) < -1$. (The Jacobi method will converge if both A and 2D-A are symmetric positive definite.)

A useful sufficient condition for convergence is that A is irreducible and has weak diagonal dominance with $d_{ii} > 0$ (a stronger condition than positive definiteness).

Another sufficient extra condition besides positive definiteness is for the matrix A to have "Property A". In this case the eigenvalues of B_J are either zero or exist as real pairs $\pm\lambda_i$ for $|\lambda_i| < 1$. This would imply that $\rho(B_J) = M(B_J) < 1$.

Because of its importance in the iterative methods still to be discussed, we denote $\lambda_J = M(B_J)$ to be the largest positive eigenvalue of B_J.

The Jacobi iteration matrices for the model problems possess all these properties. In fact, it can be shown that $\lambda_J = \cos\pi h$ in each case.

For $h = 0.2$, $\lambda_J \simeq 0.809017$. For $h = 0.025$, $\lambda_J \simeq 0.996917$.

Hence the approximate number of iterations required to reduce the error by a factor 10^{-6} is

$$n \simeq \frac{\ln 10^6}{-\ln \cos \pi h}.$$

For $h = 0.2$, this implies 66 iterations, and is verified in Appendix B. With $h = 1/40$, this yields $n \simeq 4475$. [With $\underset{\sim}{u}^{(n)} = \underset{\sim}{0}$, it actually takes 4479.]

We note that the convergence rate of the Jacobi method appears to be very slow. It can be shown for any generalized Dirichlet problem that the number of iterations required to achieve a specified level of convergence is proportional to $1/h^2$ for a small mesh-size h. The fact that the Jacobi method can be accelerated, using Chebyshev acceleration, to yield a proportionality of $1/h$, is adequate reason for its inclusion here. The next two basic methods, Gauss-Seidel and Successive Overrelaxation, although individually faster than Jacobi, cannot be accelerated in the same way.

NOTE: Suppose our basic Jacobi iteration was of the form

$$u_i^{(n+1)} = a_1 u_{i-k}^{(n)} + a_2 u_{i-1}^{(n)} + a_3 u_{i+1}^{(n)} + a_4 u_{i+k}^{(n)} + c, \text{ for } a_i > 0,\ i=1,..,4.$$

This might correspond to a general elliptic partial differential equation where, say,

$$k = \frac{1}{h} - 1,$$

and $u(x,y)$ is given on the boundary of the unit square. Then it can be shown that

$$\lambda_J = 2(\sqrt{a_1 a_4} + \sqrt{a_2 a_3}) \cos\pi h.$$

Our model problems correspond to $a_i = \frac{1}{4}$, $i=1,..,4$.

4.2 Gauss-Seidel Method

For the first model problem, the Gauss-Seidel iterative method is represented by

$$u_1^{(n+1)} = \tfrac{1}{4}(u_2^{(n)} + u_5^{(n)} + 0.04)$$

$$u_2^{(n+1)} = \tfrac{1}{4}(u_1^{(n+1)} + u_3^{(n)} + u_6^{(n)} + 0.04)$$

$$u_3^{(n+1)} = \tfrac{1}{4}(u_2^{(n+1)} + u_4^{(n)} + u_7^{(n)} + 0.04)$$

$$u_4^{(n+1)} = \tfrac{1}{4}(u_3^{(n+1)} + u_8^{(n)} + 0.04)$$

$$u_5^{(n+1)} = \tfrac{1}{4}(u_1^{(n+1)} + u_6^{(n)} + u_9^{(n)} + 0.04)$$

$$u_6^{(n+1)} = \tfrac{1}{4}(u_2^{(n+1)} + u_5^{(n+1)} + u_7^{(n)} + u_{10}^{(n)} + 0.04)$$

$$u_7^{(n+1)} = \tfrac{1}{4}(u_3^{(n+1)} + u_6^{(n+1)} + u_8^{(n)} + u_{11}^{(n)} + 0.04)$$

$$u_8^{(n+1)} = \tfrac{1}{4}(u_4^{(n+1)} + u_7^{(n+1)} + u_{12}^{(n)} + 0.04)$$

$$u_9^{(n+1)} = \tfrac{1}{4}(u_5^{(n+1)} + u_{10}^{(n)} + u_{13}^{(n)} + 0.04)$$

$$u_{10}^{(n+1)} = \tfrac{1}{4}(u_6^{(n+1)} + u_9^{(n+1)} + u_{11}^{(n)} + u_{14}^{(n)} + 0.04)$$

$$u_{11}^{(n+1)} = \tfrac{1}{4}(u_7^{(n+1)} + u_{10}^{(n+1)} + u_{12}^{(n)} + u_{15}^{(n)} + 0.04)$$

$$u_{12}^{(n+1)} = \tfrac{1}{4}(u_8^{(n+1)} + u_{11}^{(n+1)} + u_{16}^{(n)} + 0.04)$$

$$u_{13}^{(n+1)} = \tfrac{1}{4}(u_9^{(n+1)} + u_{14}^{(n)} + 0.04)$$

$$u_{14}^{(n+1)} = \tfrac{1}{4}(u_{10}^{(n+1)} + u_{13}^{(n+1)} + u_{15}^{(n)} + 0.04)$$

$$u_{15}^{(n+1)} = \tfrac{1}{4}(u_{11}^{(n+1)} + u_{14}^{(n+1)} + u_{16}^{(n)} + 0.04)$$

$$u_{16}^{(n+1)} = \tfrac{1}{4}(u_{12}^{(n+1)} + u_{15}^{(n+1)} + 0.04)$$

See Appendix C for a possible subroutine GS, with an output corresponding to $\underset{\sim}{u}^{(0)} = \underset{\sim}{0}$, and a stopping criterion similar to that for the Jacobi method. It required 34 iterations.

For the second model problem, this is equivalent to rewriting (2.1) in the iterative form

$$u_i^{(n+1)} = \tfrac{1}{4}(u_{i-39}^{(n+1)} + u_{i-1}^{(n+1)} + u_{i+1}^{(n)} + u_{i+39}^{(n)} + h^2);$$

i.e. the latest estimates for all components of $\underset{\sim}{u}$ are always used in subsequent computations. In matrix form,

$$\underset{\sim}{u}^{(n+1)} = D^{-1}(-L\underset{\sim}{u}^{(n+1)} - U\underset{\sim}{u}^{(n)} + \underset{\sim}{b})$$

$$(D+L)\underset{\sim}{u}^{(n+1)} = -U\underset{\sim}{u}^{(n)} + \underset{\sim}{b}$$

$$\underset{\sim}{u}^{(n+1)} = -(D+L)^{-1}U\underset{\sim}{u}^{(n)} + (D+L)^{-1}\underset{\sim}{b}.$$

Hence the splitting matrix is $Q_{GS} \equiv D+L$, the iteration matrix is $B_{GS} \equiv -(D+L)^{-1}U = I - (D+L)^{-1}A$, and $\underset{\sim}{k}_{GS} \equiv (D+L)^{-1}\underset{\sim}{b}$.

It can be shown that if A is symmetric positive definite, then the Gauss-Seidel iteration <u>will</u> converge. (The eigenvalues need not be real, and complex eigenvalues tend to produce an irregular convergence.)

However, if A is "consistently ordered" (see Kincaid and Young (1978); or Varga (1962), for a precise definition) and has "Property A", then the eigenvalues of B_{GS} are related to those of B_J in the following way: the eigenvalues of B_{GS} will be real, and for each nonzero pair $\pm\lambda_i$ of B_J, there corresponds eigenvalues $0, \lambda_i^2$ for B_{GS}.

Hence, in this case,

$$\rho(B_{GS}) = [\rho(B_J)]^2 = \lambda_J^2,$$

and the Gauss-Seidel iteration converges about twice as fast as the Jacobi iteration. It would require about half as many iterations to achieve the accuracy specified in (4.4); i.e. $R(B_{GS}) = 2\,R(B_J)$.

This implies about 33 iterations for $h = 0.2$, and about 2240 iterations for $h = 1/40$. (Using subroutine GS, the 2240 iterations have been verified.)

The number of iterations is again, of course, proportional to $1/h^2$.

The Gauss-Seidel method is included here mainly to introduce the next procedure, Successive Overrelaxation. Before doing so, a brief comment on a stopping criterion is warranted. Ideally, we might wish to guarantee that

$$\frac{\|\underset{\sim}{\varepsilon}^{(n)}\|_2}{\|\underset{\sim}{u}\|_2} = \frac{\|\underset{\sim}{u}^{(n)}-\underset{\sim}{u}\|_2}{\|\underset{\sim}{u}\|_2} < 10^{-6}$$

say, with an added safeguard condition taken $n \le n_{max}$, for some prescribed n_{max}.

However, we cannot measure either $\|\underset{\sim}{\varepsilon}^{(n)}\|_2$ or $\|\underset{\sim}{u}\|_2$ precisely. We can find the pseudo-residual $\underset{\sim}{\delta}^{(n)}$ at each step, but this can only be used effectively if we have a good estimate for the spectral radius of the iteration matrix. It should be pointed out that a stopping test of the form

$$\frac{\|\underset{\sim}{u}^{(n)} - \underset{\sim}{u}^{(n-1)}\|_2}{\|\underset{\sim}{u}^{(n)}\|_2} < 10^{-6}$$

does <u>not</u> imply the accuracy specified above.

4.3 Successive Overrelaxation (SOR) Method

For the second model problem, this is equivalent to rewriting (2.1) in the iterative form

$$u_i^{(n+1)} = \omega\{\tfrac{1}{4}(u_{i-39}^{(n+1)} + u_{i-1}^{(n+1)} + u_{i+1}^{(n)} + u_{i+39}^{(n)} + h^2)\} + (1-\omega)u_i^{(n)}.$$

The parameter ω is called a relaxation parameter. The new estimate $u_i^{(n+1)}$ can be regarded as a weighted average of a term obtained in a manner similar to the Gauss-Seidel concept and the current estimate $u_i^{(n)}$. This iteration is linear, stationary and completely consistent with (4.1), in the sense that if it converges, it gives the true solution $\underset{\sim}{u}$.

In matrix form,

$$\underset{\sim}{u}^{(n+1)} = \omega\{D^{-1}(-L\underset{\sim}{u}^{(n+1)} - U\underset{\sim}{u}^{(n)} + \underset{\sim}{b})\} + (1-\omega)\underset{\sim}{u}^{(n)}$$

$$(D+\omega L)\underset{\sim}{u}^{(n+1)} = [-\omega U + (1-\omega)D]\underset{\sim}{u}^{(n)} + \omega\underset{\sim}{b}$$

$$\underset{\sim}{u}^{(n+1)} = (D+\omega L)^{-1}[-\omega U + (1-\omega)D]\underset{\sim}{u}^{(n)} + (D+\omega L)^{-1}\omega\underset{\sim}{b}.$$

The splitting matrix is $Q_{SOR} \equiv \frac{1}{\omega}D + L$, the iteration matrix is

$$B_{SOR} \equiv (D+\omega L)^{-1}[-\omega U+(1-\omega)D] = I - (\tfrac{1}{\omega}D + L)^{-1}A,$$

and

$$\underset{\sim}{k}_{SOR} \equiv (D+\omega L)^{-1}\omega\underset{\sim}{b} = (\frac{1}{\omega}D + L)^{-1}\underset{\sim}{b}.$$

For $\omega = 1$, the method is merely Gauss-Seidel. It can be shown that if A is symmetric positive definite and irreducible, then $\rho(B_{SOR})<1$ if and only if $0 < \omega < 2$. (See Young (1971).) i.e. if A is symmetric positive definite, this iterative method converges if and only if the relaxation parameter $\omega \in (0,2)$. Moreover, $\rho(B_{SOR}) \geq |\omega-1|$, with equality only if all eigenvalues of B_{SOR} are of modulus $|\omega-1|$. (See Kahan (1958), or Varga (1962).)

For $0 < \omega < 1$, the method is termed underrelaxation.

For $1 < \omega < 2$, the method is termed overrelaxation, and this is usually the case.

Suppose the matrix A is symmetric positive definite and consistently ordered. Then we know the eigenvalues of B_J are real with modulus less than one. Then the eigenvalues λ of B_J are related to the eigenvalues μ of B_{SOR} as follows:

The eigenvalues of B_J are $\pm\lambda_1, \pm\lambda_2, \ldots, \pm\lambda_{N1}$ $(\lambda_i > 0)$ and N2 zero eigenvalues $(2N1+N2 = N)$.

Corresponding to $\pm\lambda$, $\lambda > 0$, there are two eigenvalues μ of B_{SOR} satisfying the relationship

$$\mu + \omega - 1 = \omega\lambda\sqrt{\mu}, \tag{4.3.1}$$

or equivalently,

$$\mu = \left(\frac{\omega\lambda \pm \sqrt{\omega^2\lambda^2-4(\omega-1)}}{2}\right)^2.$$

For $\omega > 1$, this usually produces complex eigenvalues μ. The remaining N2 eigenvalues of B_{SOR} are all equal to $(1-\omega)$.

In particular, we know that $\omega = 1$ corresponds to the Gauss-Seidel iteration. Then equation (4.3.1) confirms the earlier statement concerning the rates of convergence of the Jacobi and Gauss-Seidel methods under the conditions being assumed; i.e. $\rho(B_{GS}) = [\rho(B_J)]^2 = \lambda_J^2$.

There is an optimal value of ω, denoted by ω_{opt}, that minimises the spectral radius of B_{SOR}, depending on λ_J;

i.e. ω_{opt} minimises $\max\limits_{-\lambda_J \leq \lambda_i \leq \lambda_J} \left|\frac{\omega\lambda_i \pm \sqrt{\omega^2\lambda_i^2-4(\omega-1)}}{2}\right|^2.$

It can be shown that

$$\omega_{opt} = \frac{2}{1+\sqrt{1-\lambda_J^2}}, \tag{4.3.2}$$

and for a general relaxation parameter ω,

$$\rho(B_{SOR}) = \begin{cases} \left(\frac{\omega\lambda_J + \sqrt{\omega^2\lambda_J^2-4(\omega-1)}}{2}\right)^2, & \text{if } 0 < \omega < \omega_{opt} \\ \omega - 1, & \text{if } \omega \geq \omega_{opt}. \end{cases}$$

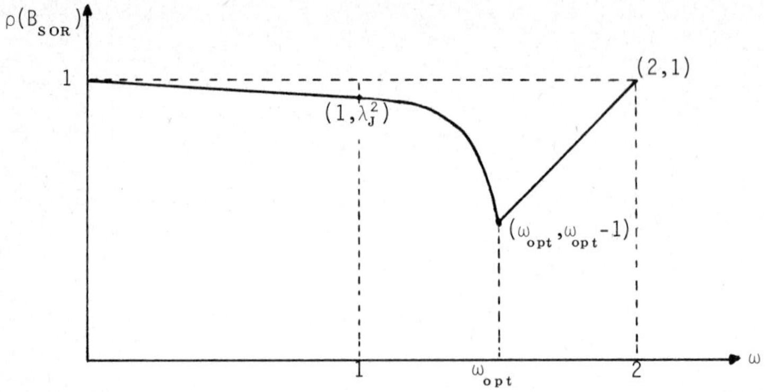

This graph shows how $\rho(B_{SOR})$ varies with choices of ω between 0 and 2, and how the rate of convergence would improve dramatically as ω approaches ω_{opt} from below.

For convenience of notation, when $\omega = \omega_{opt}$, let $\rho(B_{SOR}) = \bar{\mu}$. If we were to know ω_{opt}, which we can certainly find via (4.3.2) if we know λ_J, then the rate of convergence of the SOR method is about $2\sqrt{2}$ times the square root of the rate of convergence of the Jacobi method (and hence about twice the square root of the rate of convergence of the Gauss-Seidel method). In particular, the number of iterations required to obtain the error reduction previously prescribed is now proportional to $1/h$.

For the model problems, $\lambda_J = \cos \pi h$.

Hence, from (4.3.2), $\omega_{opt} = \dfrac{2}{1+\sin \pi h}$, which is $\simeq 1.2596$ for $h = 0.2$, and $\simeq 1.8545$ for $h = 1/40$. The spectral radius of B_{SOR} with $\omega = \omega_{opt}$ is

$$\bar{\mu} = \omega_{opt} - 1 = \frac{1 - \sin\pi h}{1 + \sin\pi h}$$

which is $\simeq 0.2596$ for $h = 0.2$, and $\simeq 0.8545$ for $h = 1/40$.

Using

$$n \simeq \frac{\ln 10^6}{-\ln(\omega_{opt} - 1)},$$

we expect to require 11 iterations for $h = 0.2$ and 88 iterations for $h = 1/40$.

However, using the true value of ω_{opt} as inputs into the subroutine SOR of Appendix D, it is found that 15 and 114 iterations respectively are actually required. This is because the iteration matrix B_{SOR}, corresponding to ω_{opt}, is not able to be diagonalised, but has a Jordan Canonical Form which leads to a smaller reduction in the errors than predicted by the spectral radius alone.

For effective use of the SOR method, it is essential that we have an accurate estimate for λ_J, and thereby ω_{opt}. If we overestimate ω_{opt}, the decrease in the asymptotic rate of convergence is not too serious. However, an underestimate of ω_{opt} produces a considerable reduction in the convergence rate. If the matrix A has "Property A", it is interesting that the value of ω_{opt} is independent of the order in which the unknowns are actually determined for most practical orderings. Estimates for λ_J can sometimes be obtained a priori, or else an adaptive procedure can be employed for the automatic determination of λ_J as the actual iterative procedure is being carried out. (See Young (1971) or Hageman (1972).)

A possible adaptive procedure then can easily be incorporated into an SOR code using the fact that given either λ_J (the spectral radius of B_J) or $\bar{\mu}$ (the spectral radius of B_{SOR} with $\omega = \omega_{opt}$), all three quantities λ_J, $\bar{\mu}$ and ω_{opt} can be determined via relation (4.3.2) as well as

$$\lambda_J = \frac{\bar{\mu} + \omega_{opt} - 1}{\omega_{opt}\sqrt{\bar{\mu}}} \tag{4.3.3}$$

which is derivable from (4.3.1) and the formula for $\bar{\mu} = \rho(B_{SOR})$.

The following adaptive procedure is not rigorous in that it requires $\omega < \omega_{opt}$ for all values of ω used while we seek the optimum value. However, in practice, it has been quite successful.

(i) Choose a value ω so that $\omega < \omega_{opt}$ is certain; for example, we could start with $\omega = 1$.

(ii) Apply a few steps of SOR, and then use the ratio of consecutive pseudo-residuals

$$R^{(n)} = \frac{\|\underset{\sim}{\delta}^{(n)}\|_2}{\|\underset{\sim}{\delta}^{(n-1)}\|_2},$$

where $\underset{\sim}{\delta}^{(n)} = \underset{\sim}{u}^{(n+1)} - \underset{\sim}{u}^{(n)}$, as an estimate for $\bar{\mu}$.

(iii) Use this estimate for $\bar{\mu}$, and the current ω, to calculate a corresponding estimate for λ_J via (4.3.3)

(iv) Use this λ_J estimate to calculate an improved value for ω_{opt} via (4.3.2). Then return to (ii).

Provided we are sensible in deciding how often to apply this adaptive process, the value of ω_{opt} can be obtained fairly easily without adding much cost to the overall Successive Overrelaxation method. This process theoretically relies on the estimate ω approaching ω_{opt} from below. As it does so, the convergence rate of the actual SOR method improves noticeably.

A <u>stopping test</u> which is very nearly equivalent to

$$\frac{\|\underset{\sim}{\varepsilon}^{(n)}\|_2}{\|\underset{\sim}{u}\|_2} < 10^{-6}, \text{ say, is as follows:}$$

After the nth iteration, define

$$H = \begin{cases} R^{(n)}, & \text{if } \omega - 1 < R^{(n)} < 1, \\ \omega - 1, & \text{if } R^{(n)} < \omega - 1. \end{cases}$$

Then the SOR process terminates when

$$\frac{\|\delta^{(n)}\|_2}{(1-H)\|u^{(n+1)}\|_2} < 10^{-6}.$$

This test is found to be satisfied within a couple of iterations after the desired accuracy has actually been achieved.

Appendix D shows the output corresponding to using an initial input of $\omega = 1$, and then adapting to find ω_{opt}. It takes 17 iterations, which is only 2 more than needed if the true ω_{opt} was known in advance. (Obviously a better initial estimate for ω would be supplied, if available.) The subroutine SOR does not allow the value of ω to change too often, and in this case there is only one change from the initial $\omega = 1$ to $\omega = 1.28213$ after 5 iterations. Suppose we make it even more difficult to change the value of ω by altering the 115th line of the subroutine (with 0.01) to read

```
IF(ABS(SR-PSNORM/PSPREV).GT.0.0001)THEN
```

i.e. change 0.01 to 0.0001. Then the only change in ω is after 14 iterations to yield a new $\omega = 1.25967$. Although this is almost exactly ω_{opt}, it takes a total of 24 iterations before convergence is obtained. Basically, the more iterations performed with a given ω, the better will be the next estimate for ω_{opt}. However, we cannot afford to use a poor value of ω too long. Obviously, a compromise is required.

Similarly, for the second model problem with $h = 1/40$ and 1521 variables, the adaptive SOR method (with 0.0001) requires 154 iterations, compared to 114 if ω_{opt} was known. Using an initial $\omega = 1$, the adaptive method changes ω to $\omega = 1.8000$ after 26 iterations, to $\omega = 1.8500$ after 34 iterations, and to $\omega = 1.8549$ after 130 iterations, thereby converging in 154 iterations. (Compare with true $\omega_{opt} = 1.8545$.)

4.4 Symmetric Successive Overrelaxation (SSOR) method

Each iteration of the SSOR method can be regarded as two distinct operations. Firstly, it uses a forward SOR iteration in which the unknowns are computed successively in their natural ordering, and then a backward SOR iteration in which the unknowns are computed in the opposite order. The same value of the relaxation factor ω, $0 < \omega < 2$, is used in both parts of each SSOR iteration.

Let B_{BSOR} be the iteration matrix associated with the backward sweep. Then each SSOR iteration, depending upon a relaxation factor ω, can be written in a matrix form such as

$$u^{(n+\frac{1}{2})} = B_{SOR}u^{(n)} + k_1$$

$$u^{(n+1)} = B_{BSOR}u^{(n+\frac{1}{2})} + k_2$$

where

$$B_{SOR} = (D+\omega L)^{-1}[-\omega U+(1-\omega)D]$$

and

$$B_{BSOR} = (D+\omega U)^{-1}[-\omega L+(1-\omega)D].$$

Hence, the iteration matrix for the Symmetric Successive Overrelaxation method is given by

$$B_{SSOR} \equiv (D+\omega U)^{-1}[-\omega L+(1-\omega)D](D+\omega L)^{-1}[-\omega U+(1-\omega)D].$$

It can be shown that the corresponding splitting matrix is

$$Q_{SSOR} \equiv \frac{1}{\omega(2-\omega)}(D+\omega L)D^{-1}(D+\omega U).$$

Interestingly, if A is a symmetric positive definite matrix (i.e. $L^T=U$), and $0 < \omega < 2$, then the splitting matrix Q is itself symmetric positive definite. This fact can be used to prove that, if $0 < \omega < 2$, all the eigenvalues λ of B_{SSOR} are real and lie in the interval $0 \le \lambda < 1$. (See Young (1971).) The rate of convergence of this SSOR method is not particularly sensitive to the precise choice of the relaxation factor ω, and so the exact optimum value of ω is not necessary. A number of good values for ω have been expounded, depending upon the information available. In any case, it is known that the SSOR method converges satisfactorily provided

$$\rho(D^{-1}LD^{-1}U) \le \tfrac{1}{4}.$$

This particular condition is satisfied for the model problem, and for most examples of generalised Dirichlet problems provided the natural ordering of the mesh points is employed.

If we know that $\rho(D^{-1}LD^{-1}U) \le \frac{1}{4}$, and we also know λ_J, the largest eigenvalue of the corresponding Jacobi iteration matrix, then a good value for ω would be

$$\omega^* = \frac{2}{1+\sqrt{2(1-\lambda_J)}}. \quad \text{(See Young (1971).)}$$

Moreover, for this value ω^*, it can be shown that

$$\rho(B_{SSOR}) \le \frac{1-\sqrt{\dfrac{1-\lambda_J}{2}}}{1+\sqrt{\dfrac{1-\lambda_J}{2}}}.$$

In a more general situation, we might be able to find an upper bound for $\rho(D^{-1}LD^{-1}U)$, say $\bar{\beta}(\ge\frac{1}{4})$, and an <u>upper bound</u> for λ_J, say $\bar{\lambda}_J$. In this case, a good value for ω would be

$$\omega^* = \frac{2}{1+\sqrt{1-2\bar{\lambda}_J+4\bar{\beta}}}$$

with a corresponding bound for $\rho(B_{SSOR})$.

As a typical comparison between the effectiveness of the SOR method using ω_{opt}, and the SSOR method using ω^*, consider the model problems. Here $\lambda_J = \cos\pi h$, and we know $\bar{\beta} \le \frac{1}{4}$.

Hence

$$\omega^* = \frac{2}{1+\sqrt{2(1-\cos\pi h)}} = \frac{2}{1+2\sin\frac{\pi h}{2}}.$$

which is very close to

$$\omega_{opt} = \frac{2}{1 + \sin\pi h} .$$

Also, $\rho(B_{SOR}) = \frac{1 - \sin\pi h}{1 + \sin\pi h}$, whereas

$$\rho(B_{SSOR}) \le \frac{1 - \sin\frac{\pi h}{2}}{1 + \sin\frac{\pi h}{2}} .$$

Comparing these, we can see that the SOR Method converges about twice as quickly as the SSOR method. This condition is true for most linear systems derived from generalized Dirichlet problems for which $\bar{\beta} \le \frac{1}{4}$, and the number of iterations necessary is approximately proportional to $1/h$, for small h. Besides needing twice as many iterations, the SSOR method requires twice as much work per iteration. However, these disadvantages can be overcome in two ways. Firstly, it is possible to extrapolate the SSOR method, which can square the spectral radius of the iteration matrix. But most importantly, since the eigenvalues λ of B_{SSOR} are real with $0 \le \lambda < 1$, we can accelerate the convergence of the SSOR method by employing either Chebyshev acceleration or conjugate gradient acceleration (see ahead).

A substantial improvement in the convergence rate can be achieved in this manner. This is not possible with the SOR method. Similarly, we can accelerate the Jacobi method, but not Gauss-Seidel.

5. CHEBYSHEV ACCELERATION (See Hageman and Young (1981); Kincaid and Young (1978); or Varga (1962).)

Consider a basic stationary iterative method of the form

$$\underset{\sim}{u}^{(n+1)} = B\,\underset{\sim}{u}^{(n)} + \underset{\sim}{k},$$

where the eigenvalues of the iteration matrix B are real.

For example, we might use the Jacobi iteration with $B_J = I - D^{-1}A$, or the Symmetric Successive Overrelaxation iteration with

$$B_{SSOR} = (D+\omega U)^{-1}[-\omega L+(1-\omega)D](D+\omega L)^{-1}[-\omega U+(1-\omega)D].$$

Let $m = m(B)$ and $M = M(B)$ represent respectively the smallest and largest eigenvalues λ of B, so that

$$m \le \lambda \le M, \text{ for all } \lambda.$$

The Chebyshev acceleration technique does not assume that the matrix A of the original large sparse linear system is necessarily positive definite. Instead, it merely requires that the iteration matrix B is similar to a symmetric matrix with real eigenvalues less than one.

i.e. $M < 1$.

Even if the basic method diverges because $m < -1$, this acceleration process converges and is still very effective. In all cases, it produces a considerable improvement in the rate of convergence.

Fundamentally, we require the existence of a nonsingular symmetrization matrix W such that WBW^{-1} is symmetric with eigenvalues $\lambda < 1$.

Equivalently, we require W such that

$$W(I-B)W^{-1} = I - WBW^{-1}$$

is symmetric positive definite.

For example, suppose A is symmetric positive definite, and let $W = D^{\frac{1}{2}}$. (The basic Jacobi method will diverge if $m(B_J) < -1$.) Then

$$W(I-B_J)W^{-1} = D^{\frac{1}{2}}(D^{-1}A)D^{-\frac{1}{2}}$$

$$= C^T C$$

where $C = A^{\frac{1}{2}}D^{-\frac{1}{2}}$.

Similarly, for the SSOR method, let

$$W = \frac{1}{\sqrt{\omega(2-\omega)}} D^{-\frac{1}{2}}(D+\omega U), \quad 0 < \omega < 2.$$

Then $W(I-B_{SSOR})W^{-1}$ is symmetric positive definite. In fact, we could also choose $W = Q^{\frac{1}{2}}_{SSOR}$ or $W = A^{\frac{1}{2}}$ in this case.

Even though it is not essential, it is convenient to assume that A is symmetric positive definite in order to analyze the effectiveness of Chebyshev acceleration. It is a polynomial acceleration, and is termed a semi-iterative method with respect to the corresponding basic iterative method. An outline of the theory behind Chebyshev acceleration is as follows:

Suppose we have a basic iterative method that generates a sequence of vectors $\underset{\sim}{v}^{(i)}$; i.e. we have an initial vector $\underset{\sim}{v}^{(0)}$, and then

$$\underset{\sim}{v}^{(1)} = B\,\underset{\sim}{v}^{(0)} + \underset{\sim}{k} \text{ etc, such that}$$

$$\underset{\sim}{v}^{(n+1)} = B\,\underset{\sim}{v}^{(n)} + \underset{\sim}{k}.$$

At the nth iteration, we wish to choose constants α_{ni}, i=0,1,..,n and then define

$$\underset{\sim}{u}^{(0)} = \underset{\sim}{v}^{(0)}; \quad \underset{\sim}{u}^{(n)} = \sum_{i=0}^{n} \alpha_{ni}\,\underset{\sim}{v}^{(i)},$$

in such a way that the sequence of vectors $\underset{\sim}{u}^{(n)}$ tends towards the true solution $\underset{\sim}{u}$ of $A\underset{\sim}{u} = \underset{\sim}{b}$ in some optimal fashion. If the initial vector $\underset{\sim}{v}^{(0)}$ was in fact $\underset{\sim}{u}$, then clearly $\underset{\sim}{v}^{(n)} = \underset{\sim}{u}$ for all $n \geq 0$, and so we require in this case that $\underset{\sim}{u}^{(n)} = \underset{\sim}{u}$ for all $n \geq 0$.

Hence we have the added condition that $\sum_{i=0}^{n} \alpha_{ni} = 1$, for each $n \geq 0$.

Since $\underset{\sim}{u} = \sum_{i=0}^{n} \alpha_{ni}\underset{\sim}{u}$, we have

$$\underset{\sim}{\varepsilon}^{(n)} = \underset{\sim}{u}^{(n)} - \underset{\sim}{u} = \sum_{i=0}^{n} \alpha_{ni}(\underset{\sim}{v}^{(i)} - \underset{\sim}{u})$$

$$= \sum_{i=0}^{n} \alpha_{ni} B^i\,\underset{\sim}{\varepsilon}^{(0)} \quad \text{since } \underset{\sim}{u}^{(0)} = \underset{\sim}{v}^{(0)}$$

$$= \left(\sum_{i=0}^{n} \alpha_{ni} B^i\right) \underset{\sim}{\varepsilon}^{(0)}$$

i.e. $\underset{\sim}{\varepsilon}^{(n)} = P_n(B)\,\underset{\sim}{\varepsilon}^{(0)}$, where the polynomial $P_n(x)$ is given by

$$P_n(x) = \sum_{i=0}^{n} \alpha_{ni} x^i, \text{ for } n \geq 0.$$

The only restriction we have imposed so far on the polynomials $P_n(x)$ is that $P_n(1) = 1$.

Let W be the symmetrization matrix corresponding to the matrix B. Since WBW^{-1} is symmetric, then $WP_n(B)W^{-1}$ is also symmetric, and hence $\|P_n(B)\|_W = \rho(P_n(B))$ (see Matrix Preliminary (d)).

Thus,

$$\|\underset{\sim}{\varepsilon}^{(n)}\|_W \leq \|P_n(B)\|_W \|\underset{\sim}{\varepsilon}^{(0)}\|_W$$

implies

$$\frac{\|\underset{\sim}{\varepsilon}^{(n)}\|_W}{\|\underset{\sim}{\varepsilon}^{(0)}\|_W} \leq \rho(P_n(B)).$$

At the nth iteration, our objective is therefore to choose the constants α_{ni} so as to minimise $\rho(P_n(B))$.

Let $\lambda_1 = m, \lambda_2, \ldots, \lambda_{N-1}, \lambda_N = M < 1$ be the eigenvalues of B. The eigenvalues of $P_n(B)$ are $P_n(\lambda_1), \ldots, P_n(\lambda_N)$. Hence, for each n, we wish to minimise

$$\max_{i=1,\ldots,N} |P_n(\lambda_i)|.$$

Instead of this problem, consider a similar minimisation problem which determines the virtual spectral radius $\bar{\rho}(P_n(B))$; we minimise

$$\max_{m \leq \lambda \leq M} |P_n(\lambda)|,$$

given that $P_n(1) = 1$.

The solution of this problem is classical, and can be expressed in terms of Chebyshev polynomials, $T_n(x)$. Because of a linear transformation, and the condition $P_n(1) = 1$, we find that

$$P_n(x) = \frac{T_n\left(\frac{2x-M-m}{M-m}\right)}{T_n\left(\frac{2-M-m}{M-m}\right)}, \quad n \geq 0.$$

Clearly this expression could be used to generate the coefficients α_{ni}, but this would be cumbersome, and, besides, the necessary storage of all the vectors $\underset{\sim}{v}^{(0)}, \underset{\sim}{v}^{(1)}, \ldots, \underset{\sim}{v}^{(n)}$ would be unmanageable. This Chebyshev acceleration process can be reduced to a second degree iteration by making use of the second order recurrence relation for Chebyshev polynomials; viz.

$$T_0(x) = 1$$

$$T_1(x) = x$$

$$T_{n+1}(x) = 2x\,T_n(x) - T_{n-1}(x), \quad n \geq 1.$$

Using this recurrence, and a certain amount of algebra, the Chebyshev acceleration process can be expressed simply as the second-degree semi-iterative method:

Let m, $M < 1$ be the smallest and largest eigenvalues respectively of the general iteration matrix B. Define constants

$$\gamma = \frac{2}{2-M-m}; \quad \sigma = \frac{M-m}{2-M-m}.$$

Then, given an arbitrary $\underset{\sim}{u}^{(0)}$, for $n \geq 0$ let

$$\underset{\sim}{u}^{(n+1)} = \omega_{n+1}\{\gamma(B\underset{\sim}{u}^{(n)} + \underset{\sim}{k}) + (1-\gamma)\underset{\sim}{u}^{(n)}\} + (1-\omega_{n+1})\underset{\sim}{u}^{(n-1)},$$

where the sequence of acceleration parameters ω_n is defined by

$$\omega_1 = 1; \quad \omega_2 = \frac{1}{1-\frac{1}{2}\sigma^2}; \quad \omega_{n+1} = \frac{1}{1-\frac{1}{4}\sigma^2\omega_n} \text{ for } n \geq 2.$$

If m and M are correctly chosen, then Chebyshev acceleration converges considerably faster than the basic method. One possible disadvantage is that it requires storage of two iteration vectors at each stage. A first-degree form does exist, but the acceleration parameters then tend to become large, possibly making the iteration unstable due to rounding error.

It can be shown that the sequence of ω_n parameters tends towards a limit ω_∞, which is given by

$$\omega_\infty = \frac{2}{1+\sqrt{1-\sigma^2}}.$$

Let $r = \omega_\infty - 1 = \dfrac{1-\sqrt{1-\sigma^2}}{1+\sqrt{1-\sigma^2}}$.

From the derivation of the acceleration procedure, and knowing the maximum value of the Chebyshev polynomials over the relevant interval, it can then be shown that the virtual spectral radius is

$$\bar{\rho}(P_n(B)) = \frac{2r^{n/2}}{1+r^n}.$$

Of course, $\rho(P_n(B)) \leq \dfrac{2r^{n/2}}{1+r^n} \simeq 2r^{n/2}$ for large n. Consider the second model problem, where $M = -m = \cos\pi h$, $h = 1/40$, when we apply Chebyshev acceleration to the basic Jacobi method. Then $\gamma = 1$, $\sigma = M$, and so

$$r = \frac{1-\sin\pi h}{1+\sin\pi h}.$$

To achieve the accuracy prescribed earlier, we require

$$n \simeq \frac{2\ln(2\times10^6)}{-\ln r} \quad 185 \text{ iterations},$$

which can be confirmed by experiment. (Using subroutine JSI of Appendix E, with the true values of M, m input, and a suitable stopping test, 187 iterations were required.)

The number of iterations is proportional to 1/h for small h, and there are approximately 1½ to 2 times as many as for the Successive Overrelaxation method using ω_{opt}.

If we were to use Symmetric Successive Overrelaxation, and apply Chebyshev acceleration, a similar analysis shows that the number of iterations is now proportional to $1/\sqrt{h}$ for small h, and in the case of the second model problem, only 26 iterations are required. This seems to be very few, considering we are, in effect, looking for 6 significant figures for the 1521 variables.

However, this needs to be balanced by taking into account the amount of work involved in each iteration, along with the total time and storage requirements.

It should be noted that the formulae for the constants γ and σ, as well as for the acceleration parameters ω_n, involve the smallest and largest eigenvalues m, M of the basic iteration matrix B.

Also, a stopping test that is frequently used in order to terminate the Chebyshev acceleration procedure at about the correct iteration is of the form

$$\frac{1}{1-M}\,\frac{\|\underset{\sim}{\delta}^{(n)}\|_2}{\|\underset{\sim}{u}^{(n)}\|_2} < 10^{-6}, \text{ or similar, where } \underset{\sim}{\delta}^{(n)} = B\underset{\sim}{u}^{(n)} + \underset{\sim}{k} - \underset{\sim}{u}^{(n)}.$$

Hence, if the values of m, M are not known, estimates will have to be found. Suppose these are $\bar{m}$, $\bar{M}$ respectively. It can be shown that Chebyshev acceleration is not really sensitive to errors involved in $\bar{m}$, especially if $\bar{m} \le m$, but it is very sensitive to errors in $\bar{M}$. In particular, as $\bar{M}$ approaches M from below, the rate of converges improves dramatically. Adaptive procedures have been developed which generate accurate estimates for $\bar{M}$ while the actual acceleration process is in operation. In the case of Jacobi with Chebyshev acceleration, it is often sufficient to let $\bar{m} = -\bar{M}$, whilst if SSOR is the basic method, $\bar{m}$ is usually left as zero throughout. Of course, in this latter case, a good value ω^* of the overrelaxation parameter also needs to be found adaptively. Details of these various adaptive procedures can be found in Hageman and Young (1981); or Grimes, Kincaid, MacGregor and Young (1978).

A possible subroutine JSI to implement the Jacobi method with Chebyshev acceleration, and which adapts to find M, is listed in Appendix E. It requires an initial estimate for M (called SRJ in the subroutine), which is often zero, and then adapts. There are inbuilt restrictions to prevent the parameters being changed too regularly.

[Briefly, if n-s is the number of iterations performed since the estimate of M was last changed at iteration number s, then define, using some of the notation in the previous two mentioned references,

$$QT = \frac{2r^{(n-s)/2}}{1+r^{(n-s)}}$$

$$QA = \frac{\|\underset{\sim}{\delta}^{(n)}\|_2}{\|\underset{\sim}{\delta}^{(s)}\|_2}$$

$$Z = \frac{(1+r)^{(n-s)}}{2}\left(QA + (QA^2 - QT^2)^{\frac{1}{2}}\right)$$

$$X = Z^{1/(n-s)}$$

$$Y = (X + r/X)/(1 + r)$$

and then a new value for $\bar{M}$ is given by

$\frac{1}{2}(\bar{M} + \bar{m} + Y(2-\bar{M}-\bar{m}))$,

where $\bar{m}$, $\bar{M}$ are the previous estimates, with the option of letting a new $\bar{m} = -\bar{M}$.

(All the above variables are, in effect, the most straightforward way of solving the "Chebyshev equation".)]

For the first model problem, it requires 22 iterations if m, M are known, and 26 iterations if they are to be found adaptively (see Appendix E). Starting with $\bar{M} = 0$, $\bar{m} = -1$, and thereafter letting $\bar{m} = -\bar{M}$, the subroutine changes $\bar{M}$ to 0.7624 after 2 iterations, and then to 0.8087 after 6 iterations. (Compare with true M = 0.8090.)

For the second model problem, it requires 187 and 218 iterations respectively.

For more information on Chebyshev acceleration, and its comparison with the Successive Overrelaxation iterative method, one can refer to Golub (1959); or Golub and Varga (1961).

6. CONJUGATE GRADIENT ACCELERATION

Hestenes and Stiefel (1952), in "Methods of conjugate gradients for solving linear systems", J. Res.Nat.Bur.Standards, introduced the classic conjugate gradient concept. It is a particular case of earlier conjugate-direction methods, and can be derived by modifying the "steepest descent" optimisation technique. Moreover, for an Nth order system $A\underset{\sim}{x} = \underset{\sim}{b}$, it would converge theoretically in at most N iterations. However, after a certain amount of research flurry that immediately followed, little practical implementation of this conjugate gradient method was used until recently. Renewed interest has been sparked by broadening the scope of the original method of Hestenes and Stiefel. Rather than regarding the conjugate gradient method as an isolated one, it can be considered as possibly the simplest case of accelerating a basic iterative method using "conjugate gradient acceleration". More specifically, consider the basic stationary Richardson iterative technique, where the splitting matrix Q and the symmetrization matrix W (if A is symmetric positive definite) are both the identity matrix I, and the iteration matrix is I-A. Then the original conjugate gradient method can be regarded as conjugate gradient acceleration applied to this Richardson iteration. Moreover, a conjugate gradient acceleration can be applied under the same conditions as required for Chebyshev acceleration, and invariably it converges in fewer iterations. Also, the acceleration parameters do not involve eigenvalues, and hence can be generated automatically, although the largest eigenvalue M of the iteration matrix may eventually be required for a suitable stopping test similar to the ones outlined in earlier methods. The fact that the method theoretically terminates in N iterations is unimportant for large sparse problems, because any practical iterative technique necessarily must converge in considerably fewer steps than this.

For simplicity, we shall assume that A is symmetric positive definite. Actually, as is the situation with Chebyshev acceleration, it is sufficient merely to require that a symmetrization matrix W exists for the iteration matrix B. e.g. Richardson (W=I), Jacobi ($W=D^{1/2}$), SSOR ($W = 1/\sqrt{\omega(2-\omega)}\, D^{-1/2}(D+\omega U)$), under suitable assumptions.

Let $\underset{\sim}{r}^{(n)} = \underset{\sim}{b} - A\,\underset{\sim}{u}^{(n)}$ be the residual vector after n iterations. Suppose $\underset{\sim}{u}^{(0)}$ is arbitrary. Then the original <u>conjugate gradient method</u> is of the form

$$\underset{\sim}{u}^{(n+1)} = \underset{\sim}{u}^{(n)} + \alpha_n \underset{\sim}{p}^{(n)},$$

where the $p^{(n)}$ are suitably chosen direction vectors, and are generated iteratively by the relation

$$\underset{\sim}{p}^{(n)} = \underset{\sim}{r}^{(n)} + \beta_{n-1}\underset{\sim}{p}^{(n-1)}, \text{ given } \underset{\sim}{p}^{(0)} = \underset{\sim}{r}^{(0)}.$$

The parameters α_n, β_n are determined by

$$\alpha_n = \frac{(\underset{\sim}{r}^{(n)},\underset{\sim}{r}^{(n)})}{(\underset{\sim}{p}^{(n)},A\underset{\sim}{p}^{(n)})}, \text{ or equivalently } \frac{(\underset{\sim}{p}^{(n)},\underset{\sim}{r}^{(n)})}{(\underset{\sim}{p}^{(n)},A\underset{\sim}{p}^{(n)})},$$

and

$$\beta_n = \frac{(\underset{\sim}{r}^{(n+1)},\underset{\sim}{r}^{(n+1)})}{(\underset{\sim}{r}^{(n)},\underset{\sim}{r}^{(n)})}, \text{ or equivalently } \frac{-(\underset{\sim}{r}^{(n+1)},A\underset{\sim}{p}^{(n)})}{(\underset{\sim}{p}^{(n)},A\underset{\sim}{p}^{(n)})}.$$

Hestenes and Stiefel suggest that the first expressions for α_n and β_n are simpler to compute, but the second values for α_n and β_n tend to yield better results. In practice, the iterations would proceed thus:

given $\underset{\sim}{u}^{(n)}$, $\underset{\sim}{p}^{(n-1)}$, compute $\underset{\sim}{r}^{(n)}$, followed by, in order, β_{n-1}, $\underset{\sim}{p}^{(n)}$, α_n and hence $\underset{\sim}{u}^{(n+1)}$.

The validity of the classic conjugate gradient method lies in the fact that the residuals are orthogonal,

i.e. $(\underset{\sim}{r}^{(i)},\underset{\sim}{r}^{(j)}) = 0$ for $i \neq j$; and the direction vectors are "$A^{1/2}$ - orthogonal", or "conjugate",

i.e. $(A^{1/2}\underset{\sim}{p}^{(i)},A^{1/2}\underset{\sim}{p}^{(j)}) = 0$ for $i \neq j$, or equivalently, $(\underset{\sim}{p}^{(i)},A\underset{\sim}{p}^{(j)}) = 0$ for $i \neq j$.

The property of orthogonal residuals has as an immediate consequence the theoretical termination property mentioned earlier.

Now, suppose we decide to eliminate the direction vectors $\underset{\sim}{p}^{(n)}$ from the formulae above. This yields a second-degree iterative process for $\underset{\sim}{u}^{(n+1)}$ involving the residual vectors;

$$\underset{\sim}{u}^{(n+1)} = \omega_{n+1}\{\gamma_{n+1}\underset{\sim}{r}^{(n)} + \underset{\sim}{u}^{(n)}\} + (1-\omega_{n+1})\,\underset{\sim}{u}^{(n-1)},$$

where

$$\gamma_{n+1} = \frac{(\underset{\sim}{r}^{(n)},\underset{\sim}{r}^{(n)})}{(\underset{\sim}{r}^{(n)},A\underset{\sim}{r}^{(n)})},$$

and the parameters ω_n are defined by the sequence

$$\omega_1 = 1$$

$$\omega_{n+1} = \frac{1}{1 - \dfrac{(\underset{\sim}{r}^{(n)},\underset{\sim}{r}^{(n)})\gamma_{n+1}}{(\underset{\sim}{r}^{(n-1)},\underset{\sim}{r}^{(n-1)})\gamma_n\omega_n}}, \text{ for } n \geq 1.$$

As can be seen, this has a form somewhat similar to the Chebyshev acceleration process. In fact, this can be regarded as a special case of the following more general conjugate gradient acceleration method.

Consider a general linear stationary iterative method of the first degree to solve $A\underset{\sim}{u} = \underset{\sim}{b}$, of the form

$$\underset{\sim}{u}^{(n+1)} = B\,\underset{\sim}{u}^{(n)} + \underset{\sim}{k},$$

corresponding to a splitting matrix Q, where $B = I - Q^{-1}A$. Let W be a non-singular symmetrization matrix such that $W(I-B)W^{-1}$ is symmetric positive definite. Let $\underset{\sim}{\delta}^{(n)} = B\,\underset{\sim}{u}^{(n)} + \underset{\sim}{k} - \underset{\sim}{u}^{(n)}$ be the pseudo-residual vector.

Then the conjugate gradient acceleration procedure as applied to this basic method is given by:

Given an arbitrary $\underset{\sim}{u}^{(0)}$, for $n \geq 0$ let

$$\begin{aligned}\underset{\sim}{u}^{(n+1)} &= \omega_{n+1}\{\gamma_{n+1}(B\underset{\sim}{u}^{(n)}+\underset{\sim}{k}) + (1-\gamma_{n+1})\underset{\sim}{u}^{(n)}\} + (1-\omega_{n+1})\underset{\sim}{u}^{(n-1)}\\ &= \omega_{n+1}\{\gamma_{n+1}\underset{\sim}{\delta}^{(n)} + \underset{\sim}{u}^{(n)}\} + (1-\omega_{n+1})\underset{\sim}{u}^{(n-1)},\end{aligned} \tag{6.1}$$

where the sequences of acceleration parameters γ_n and ω_n are defined by

$$\gamma_{n+1} = \cfrac{1}{1 - \cfrac{\underset{\sim}{\delta}^{(n)T}W^TWB\underset{\sim}{\delta}^{(n)}}{\underset{\sim}{\delta}^{(n)T}W^TW\underset{\sim}{\delta}^{(n)}}} \qquad \text{and}$$

$$\omega_1 = 1$$

$$\omega_{n+1} = \cfrac{1}{1 - \cfrac{\underset{\sim}{\delta}^{(n)T}W^TW\underset{\sim}{\delta}^{(n)}\;\gamma_{n+1}}{\underset{\sim}{\delta}^{(n-1)T}W^TW\underset{\sim}{\delta}^{(n-1)}\;\gamma_n\omega_n}}.$$

It can be shown that this procedure minimises the $[W^TW(I-B)]^{1/2}$ norm of the error vector $\underset{\sim}{\varepsilon}^{(n)}$ when compared with any general polynomial acceleration procedure applied to the same basic iterative method. In particular, it would often require fewer iterations than Chebyshev acceleration. It is worthwhile to point out a few relevant properties of this conjugate gradient acceleration process:

(a) If we consider the basic Richardson iteration with $Q = I$, $B = I-A$, $\underset{\sim}{k} = \underset{\sim}{b}$, $\underset{\sim}{\delta}^{(n)} = \underset{\sim}{r}^{(n)}$, $W = I$ etc., it can be seen immediately that the conjugate gradient acceleration reduces to the second-degree form of the classic conjugate gradient method.

(b) Unlike Chebyshev acceleration, the symmetrization matrix W is needed in this process. More precisely, we require W^TW, and upon inspecting W for the Jacobi and SSOR methods, this is usually simple or can be eliminated by a suitable preconditioning (scaling) of the original linear system. It can be shown that the pseudo-residuals are W orthogonal

i.e. $(W\underset{\sim}{\delta}^{(i)}, W\underset{\sim}{\delta}^{(j)}) = 0$, for $i \neq j$.

(c) The acceleration parameters do not require knowledge of the eigenvalues m(B), M(B). However, since a stopping test is of the form

$$\frac{1}{1-M}\,\frac{\|\underset{\sim}{\delta}^{(n)}\|_2}{\|\underset{\sim}{u}^{(n)}\|_2} \quad 10^{-6} \text{ say,}$$

an estimate for M(B) must be found. Adaptive procedures to find M(B) have been developed. Again, details can be found in Hageman and Young (1981); or Grimes, Kincaid, MacGregory and Young (1978).

(d) Since $B\underset{\sim}{\delta}^{(n)}$ is needed in the evaluation of the parameter γ_{n+1}, it is beneficial if we can avoid having to determine $B\underset{\sim}{u}^{(n)}$ also.

This can be achieved by using the second form (6.1) for the definition of $\underset{\sim}{u}^{(n+1)}$, and generating the pseudo-residual vectors $\underset{\sim}{\delta}^{(n)}$ by means of the second degree recursion

$$\underset{\sim}{\delta}^{(n+1)} = \omega_{n+1}\{\gamma_{n+1}B\underset{\sim}{\delta}^{(n)} + (1-\gamma_{n+1})\underset{\sim}{\delta}^{(n)}\} + (1-\omega_{n+1})\underset{\sim}{\delta}^{(n-1)},$$

which can be derived quite simply.

(e) There appears to be more work necessary in the calculation of the acceleration parameters on each iteration of the conjugate gradient acceleration than for the corresponding Chebyshev acceleration. Also, more storage is required. Nevertheless, for a wide range of problems, conjugate gradient acceleration uses less computer time, even when the eigenvalues for Chebyshev acceleration are known in advance. However, counter-examples to this can be constructed.

(f) Suppose the matrix A is not necessarily symmetric, but $A^T + A$ is symmetric positive definite. We can then use a splitting matrix

$$Q = \tfrac{1}{2}(A^T+A)$$

and so obtain a basic iterative method

$$Q\underset{\sim}{u}^{(n+1)} = \tfrac{1}{2}(A^T-A)\underset{\sim}{u}^{(n)} + \underset{\sim}{b},$$

where Q is symmetric positive definite. For general problems, Q may not be easily invertible, and consequently it may require an internal iteration procedure within each basic iteration before $\underset{\sim}{u}^{(n+1)}$ can be computed. Furthermore, the eigenvalues of the iteration matrix $B = \tfrac{1}{2}Q^{-1}(A^T-A)$ are purely imaginary. However, a generalized conjugate gradient acceleration procedure has been developed, corresponding to this basic iterative process. [See Concus and Golub (1976).]

Let $\underset{\sim}{\delta}^{(n)} = Q^{-1}[\tfrac{1}{2}(A^T-A)\underset{\sim}{u}^{(n)} + \underset{\sim}{b}] - \underset{\sim}{u}^{(n)}$ be the pseudo-residual. Then this method can be expressed in the form

$$\underset{\sim}{u}^{(n+1)} = \omega_{n+1}(\underset{\sim}{\delta}^{(n)}+\underset{\sim}{u}^{(n)}) + (1-\omega_{n+1})\underset{\sim}{u}^{(n-1)}, \text{ for } n \geq 0,$$

where the acceleration parameters are given by

$$\omega_1 = 1$$

$$\omega_{n+1} = \cfrac{1}{1 + \cfrac{(\underset{\sim}{\delta}^{(n)},Q\underset{\sim}{\delta}^{(n)})}{(\underset{\sim}{\delta}^{(n-1)},Q\underset{\sim}{\delta}^{(n-1)})\omega_n}}, \text{ for } n \geq 1.$$

Appendix F contains a simplified subroutine JCG to implement Conjugate Gradient Acceleration applied to the basic Jacobi iterative method. As with JSI in the previous section, it can adapt to find M, the largest eigenvalue of B , which is required in the stopping test. To do this, Hageman and Young (1981) have shown that it is convenient to form (theoretically) a symmetric tridiagonal matrix, of order n, after n iterations, represented by

$$[(\frac{\omega_i - 1}{\gamma_{i-1}\gamma_i\omega_{i-1}\omega_i})^{\frac{1}{2}}, \frac{\gamma_i - 1}{\gamma_i}, (\frac{\omega_{i+1} - 1}{\gamma_i\gamma_{i+1}\omega_i\omega_{i+1}})^{\frac{1}{2}}], \quad 1 \le i \le n.$$

Then the largest eigenvalue of this tridiagonal is the new estimate for M. The subroutine NEWTON is sufficient to accomplish this. Once M is found to a required accuracy, no further estimates are sought. Because M is only required in the stopping test and not in the actual iterations, the number of iterations for convergence of the JCG method is independent of this. The output for the first model problem indicates that only 3 iterations were necessary, and at the time of stopping, the adaptive technique had achieved M = 0.809017 (precisely!). For the second model problem, it required only 63 iterations, during which M was found correctly on the 27th iteration, and no further estimates for M were sought after that.

7. REDUCED SYSTEMS (RED-BLACK SYSTEMS)

Consider a simpler version of the model problems where we shall use $h = \frac{1}{4}$. Then A is of order 9. Suppose we number the internal nodes in the "red-black" ordering, as follows:

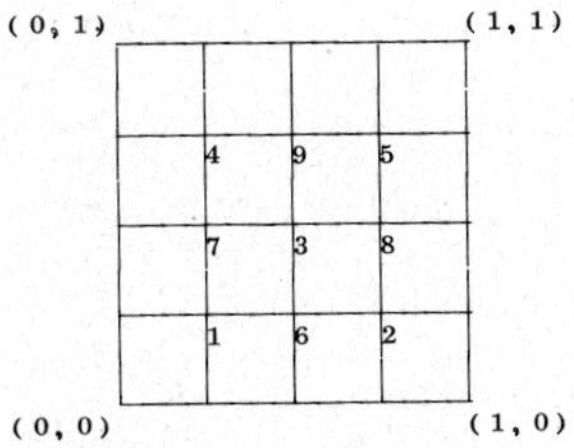

This will lead to the matrix equation

$$\left(\begin{array}{ccccc|cccc}
4 & 0 & 0 & 0 & 0 & -1 & -1 & 0 & 0 \\
0 & 4 & 0 & 0 & 0 & -1 & 0 & -1 & 0 \\
0 & 0 & 4 & 0 & 0 & -1 & -1 & -1 & -1 \\
0 & 0 & 0 & 4 & 0 & 0 & -1 & 0 & -1 \\
0 & 0 & 0 & 0 & 4 & 0 & 0 & -1 & -1 \\
\hline
-1 & -1 & -1 & 0 & 0 & 4 & 0 & 0 & 0 \\
-1 & 0 & -1 & -1 & 0 & 0 & 4 & 0 & 0 \\
0 & -1 & -1 & 0 & -1 & 0 & 0 & 4 & 0 \\
0 & 0 & -1 & -1 & -1 & 0 & 0 & 0 & 4
\end{array}\right)
\left(\begin{array}{c} u_1 \\ u_2 \\ u_3 \\ u_4 \\ u_5 \\ \hline u_6 \\ u_7 \\ u_8 \\ u_9 \end{array}\right)
= h^2 \left(\begin{array}{c} 1 \\ 1 \\ 1 \\ 1 \\ 1 \\ \hline 1 \\ 1 \\ 1 \\ 1 \end{array}\right)$$

In general, a system $A\underset{\sim}{u} = \underset{\sim}{b}$ is said to be a "red-black system" if it can be partitioned so that

$$A = \begin{pmatrix} D_1 & G \\ K & D_2 \end{pmatrix}, \quad \underset{\sim}{u} = \begin{pmatrix} \underset{\sim}{u}_1 \\ \underset{\sim}{u}_2 \end{pmatrix}, \quad \underset{\sim}{b} = \begin{pmatrix} \underset{\sim}{b}_1 \\ \underset{\sim}{b}_2 \end{pmatrix},$$

where D_1, D_2 are diagonal matrices of order N_1, N_2 respectively such that N_1, N_2 are the numbers of "red" and "black" points, and $N = N_1 + N_2$. If we assume A is symmetric positive definite, then $K^T = G$. Clearly, whenever finite difference discretizations are used over rectangular domains, we can arrange for the large sparse linear systems to be "red-black" systems.

Then $A\underset{\sim}{u} = \underset{\sim}{b}$ can be rewritten as

$$\begin{cases} D_1 \underset{\sim}{u}_1 + G\, \underset{\sim}{u}_2 = \underset{\sim}{b}_1 \\ K\, \underset{\sim}{u}_1 + D_2 \underset{\sim}{u}_2 = \underset{\sim}{b}_2, \end{cases}$$

which suggests an iteration of the form:

Given $\underset{\sim}{u}_2^{(0)}$, perform

$$\begin{cases} D_1\, \underset{\sim}{u}_1^{(n)} = -G\, \underset{\sim}{u}_2^{(n)} + \underset{\sim}{b}_1 \\ D_2\, \underset{\sim}{u}_2^{(n+1)} = -K\, \underset{\sim}{u}_1^{(n)} + \underset{\sim}{b}_2. \end{cases} \tag{7.1}$$

Since D_1^{-1}, D_2^{-1} are trivially found, the sequence of $\underset{\sim}{u}_2^{(n)}$ vectors can be computed along with $\underset{\sim}{u}_1^{(n)}$. Eliminating $\underset{\sim}{u}_1^{(n)}$ from the two equations, the above iteration process can theoretically be regarded as

$$\underset{\sim}{u}_2^{(n+1)} = D_2^{-1} K D_1^{-1} G\, \underset{\sim}{u}_2^{(n)} + (D_2^{-1} \underset{\sim}{b}_2 - D_2^{-1} K D_1^{-1} \underset{\sim}{b}_1).$$

The size of this system is N_2, which is about half the size of the original system. Using the general notation for basic iterative methods, this <u>Reduced System</u> can be expressed in the form

$$\underset{\sim}{u}_2^{(n+1)} = B_{RS}\, \underset{\sim}{u}_2^{(n)} + \underset{\sim}{k}_{RS},$$

where $B_{RS} \equiv D_2^{-1} K D_1^{-1} G$, $\underset{\sim}{k}_{RS} \equiv (D_2^{-1} \underset{\sim}{b}_2 - D_2^{-1} K D_1^{-1} \underset{\sim}{b}_1)$. In practice however, the form (7.1) would be used for the computations during each iteration, and it is clearly possible to store the $\underset{\sim}{u}_1^{(n)}$ vectors in the same computer locations as the $\underset{\sim}{u}_2^{(n)}$. Moreover, the amount of work involved in each step of the Reduced System iterative method is about the same as for the basic Jacobi method.

As $\underset{\sim}{u}_2^{(n)} \to \underset{\sim}{u}_2$, then $\underset{\sim}{u}_1^{(n)} \to \underset{\sim}{u}_1$.

Suppose $W = D_2^{1/2}$. Then

$$\begin{aligned} W B_{RS} W^{-1} &= D_2^{1/2} (D_2^{-1} K D_1^{-1} G) D_2^{-1/2} \\ &= D_2^{-1/2} K D_1^{-1} G D_2^{-1/2} \\ &= C^T C \text{ where } C = D_2^{-1/2} G D_2^{-1/2}, \text{ since } K^T = G. \end{aligned}$$

Since $W B_{RS} W^{-1} = C^T C$ is symmetric, then $\|B_{RS}\|_W = \rho(B_{RS})$, and the eigenvalues of B_{RS} are real and non-negative. Also, using this "red-black" ordering, the basic Jacobi iterative method can be written in the form

$$\begin{pmatrix} \underset{\sim}{u}_1^{(n+1)} \\ \underset{\sim}{u}_2^{(n+1)} \end{pmatrix} = \begin{pmatrix} 0 & -D_1^{-1} G \\ -D_2^{-1} K & 0 \end{pmatrix} \begin{pmatrix} \underset{\sim}{u}_1^{(n)} \\ \underset{\sim}{u}_2^{(n)} \end{pmatrix} + \begin{pmatrix} D_1^{-1} & \underset{\sim}{b}_1 \\ D_2^{-1} & \underset{\sim}{b}_2 \end{pmatrix},$$

and so

$$B_J \equiv \begin{pmatrix} 0 & -D_1^{-1} G \\ -D_2^{-1} K & 0 \end{pmatrix}.$$

Now,

$$B_J^2 = \begin{pmatrix} D_1^{-1}GD_2^{-1}K & 0 \\ 0 & D_2^{-1}KD_1^{-1}G \end{pmatrix} = \begin{pmatrix} B^*_{RS} & 0 \\ 0 & B_{RS} \end{pmatrix}, \text{ say.}$$

i.e. the $\underset{\sim}{u}_2$ values after two Jacobi iterations are the same as after one Reduced System iteration. It can be shown that $\rho(B_J^2) = \rho(B^*_{RS}) = \rho(B_{RS})$. When the matrix A is symmetric positive definite and has "Property A", we know that $\rho(B_J) < 1$. Hence, under these conditions, $\rho(B_{RS}) = [\rho(B_J)]^2 < 1$.

Because of these properties, we can clearly accelerate this Reduced System method using either Chebyshev acceleration or conjugate gradient acceleration. In all cases, the number of iterations is about one half as many as needed in the corresponding Jacobi process to achieve the same accuracy. The gains in time and storage requirements are of a similar order.

For example, the Chebyshev acceleration procedure applied to the reduced system could be carried out in the following way:

Given $\underset{\sim}{u}_2^{(0)}, \ldots, \underset{\sim}{u}_2^{(n)}$,

find $\underset{\sim}{u}_1^{(n)}$ using $D_1\underset{\sim}{u}_1^{(n)} = -G\,\underset{\sim}{u}_2^{(n)} + \underset{\sim}{b}_1$, and the pseudo-residual $\underset{\sim}{\delta}_2^{(n)}$ by means of the expression

$$D_2\underset{\sim}{\delta}_2^{(n)} = -K\,\underset{\sim}{u}_1^{(n)} + \underset{\sim}{b}_2 - \underset{\sim}{u}_2^{(n)}.$$

Then the Chebyshev acceleration is of the form

$$\underset{\sim}{u}_2^{(n+1)} = \omega_{n+1}\{\gamma\underset{\sim}{\delta}_2^{(n)} + \underset{\sim}{u}_2^{(n)}\} + (1-\omega_{n+1})\underset{\sim}{u}_2^{(n-1)},$$

where constants $\gamma = \frac{2}{2-\lambda_J^2}$, $\sigma = \frac{\lambda_J^2}{2-\lambda_J^2}$, and the acceleration parameters ω_n are given recursively by

$$\omega_1 = 1;\ \omega_2 = \frac{2}{2-\lambda_J^2};\ \omega_{n+1} = \frac{1}{1-\frac{1}{4}\sigma^2\omega_n}, \text{ for } n \geq 2.$$

This is, of course, equivalent to the general form, because $m = 0$ and $M = \lambda_J^2$. To obtain the prescribed accuracy for $\underset{\sim}{u}$, a suitable stopping test might be

$$\frac{\sqrt{2}}{1-\lambda_J^2}\,\frac{\|\underset{\sim}{\delta}_2^{(n)}\|_2}{\|\underset{\sim}{u}_2^{(n)}\|_2} < 10^{-6}, \text{ or similar.}$$

Again, if λ_J is not known, adaptive processes have been developed.

Other methods similar to a Reduced System with Chebyshev acceleration are also available, and are obtainable as modifications of the fundamental "red-black" ordering concept.

8. ALTERNATING-DIRECTION IMPLICIT (ADI) METHODS

Peaceman-Rachford Method

There are a number of different ADI methods. Only the Peaceman-Rachford method will be considered here (see Birkhoff, Varga and Young (1962); or Varga (1962)).

Consider the simpler model problem corresponding to $h = \frac{1}{4}$, and the resulting linear system $A\underset{\sim}{u} = \underset{\sim}{b}$. Assume we are using the natural ordering.

Then the matrix A can be written as $A = A_1 + A_2$, where

$$A_1 = \begin{pmatrix} 2 & -1 & 0 & 0 & 0 & 0 & 0 & 0 & 0 \\ -1 & 2 & -1 & 0 & 0 & 0 & 0 & 0 & 0 \\ 0 & -1 & 2 & 0 & 0 & 0 & 0 & 0 & 0 \\ 0 & 0 & 0 & 2 & -1 & 0 & 0 & 0 & 0 \\ 0 & 0 & 0 & -1 & 2 & -1 & 0 & 0 & 0 \\ 0 & 0 & 0 & 0 & -1 & 2 & 0 & 0 & 0 \\ 0 & 0 & 0 & 0 & 0 & 0 & 2 & -1 & 0 \\ 0 & 0 & 0 & 0 & 0 & 0 & -1 & 2 & -1 \\ 0 & 0 & 0 & 0 & 0 & 0 & 0 & -1 & 2 \end{pmatrix}$$

and

$$A_2 = \begin{pmatrix} 2 & 0 & 0 & -1 & 0 & 0 & 0 & 0 & 0 \\ 0 & 2 & 0 & 0 & -1 & 0 & 0 & 0 & 0 \\ 0 & 0 & 2 & 0 & 0 & -1 & 0 & 0 & 0 \\ -1 & 0 & 0 & 2 & 0 & 0 & -1 & 0 & 0 \\ 0 & -1 & 0 & 0 & 2 & 0 & 0 & -1 & 0 \\ 0 & 0 & -1 & 0 & 0 & 2 & 0 & 0 & -1 \\ 0 & 0 & 0 & -1 & 0 & 0 & 2 & 0 & 0 \\ 0 & 0 & 0 & 0 & -1 & 0 & 0 & 2 & 0 \\ 0 & 0 & 0 & 0 & 0 & -1 & 0 & 0 & 2 \end{pmatrix}.$$

The elliptic partial differential equation from which this arose was

$$\frac{\partial^2 u}{\partial x^2} + \frac{\partial^2 u}{\partial y^2} = -1,$$

when $u(x,y) = 0$ on the boundary of the unit square $0 \le x \le 1$, $0 \le y \le 1$.

After using the finite difference discretization and multiplying by $(-h^2)$, we can see that A_1 is derived from the $\partial^2 u/\partial x^2$ term, and A_2 is derived from $\partial^2 u/\partial y^2$.

In a more general situation, suppose we discretize

$$a \frac{\partial^2 u}{\partial x^2} + b \frac{\partial^2 u}{\partial y^2} + c \frac{\partial u}{\partial x} + d \frac{\partial u}{\partial y} + e\, u = f,$$

on, say, a rectangular region.

Let the matrices H, V, Σ respectively contain the terms corresponding to

$$a \frac{\partial^2 u}{\partial x^2} + c \frac{\partial u}{\partial x}, \quad b \frac{\partial^2 u}{\partial y^2} + d \frac{\partial u}{\partial y} \quad \text{and } e\, u,$$

after multiplying by $(-h^2)$. Then $A = H + V + \Sigma$. Assume Σ is a non-negative diagonal matrix, and H, V are symmetric positive definite matrices with non-positive off-diagonal elements.

Let $A_1 = H + \tfrac{1}{2}\Sigma$, $A_2 = V + \tfrac{1}{2}\Sigma$, so that $A = A_1 + A_2$.

It is worth noting that most of the <u>rigorous</u> justification of the following method relies on H, V being symmetric positive definite, Σ being a multiple of the identity matrix, and $HV = VH$. This then implies that A_1 and A_2 are symmetric positive definite, and $A_1A_2 = A_2A_1$.

The equation $A\underset{\sim}{u} = \underset{\sim}{b}$ is clearly equivalent to each of the equations

$$(A_1 + \omega_{n+1} I)\underset{\sim}{u} = \underset{\sim}{b} - (A_2 - \omega_{n+1} I)\underset{\sim}{u}$$

and

$$(A_2 + \omega_{n+1} I)\underset{\sim}{u} = \underset{\sim}{b} - (A_1 - \omega_{n+1} I)\underset{\sim}{u},$$

for any ω_{n+1}.

This suggests the Peaceman-Rachford ADI method:

$$\begin{cases} (A_1 + \omega_{n+1} I)\underset{\sim}{u}^{(n+\frac{1}{2})} = \underset{\sim}{b} - (A_2 - \omega_{n+1} I)\underset{\sim}{u}^{(n)} \\ (A_2 + \omega_{n+1} I)\underset{\sim}{u}^{(n+1)} = \underset{\sim}{b} - (A_1 - \omega_{n+1} I)\underset{\sim}{u}^{(n+\frac{1}{2})}. \end{cases}$$

The ω_n are acceleration parameters. Since $A_1 + \omega_{n+1} I$ is usually a tridiagonal matrix, and $A_2 + \omega_{n+1} I$ can be made tridiagonal after a suitable permutation, each of the implicit steps in the iterations can be carried out using the well-known Gaussian elimination algorithm for tridiagonal systems. In effect, the first equation is equivalent to horizontal sweeps, whereas the second equation should be regarded as vertical sweeps; hence the category "Alternating-Direction Implicit Methods". The auxiliary vector $\underset{\sim}{u}^{(n+\frac{1}{2})}$ is not retained from one complete iteration to the next.

In practice, it is efficient to use acceleration parameters in the manner $\omega_1, \omega_2, \ldots, \omega_k, \omega_1, \omega_2, \ldots, \omega_k, \omega_1, \omega_2, \ldots, \omega_k, \ldots$ for a suitable k. If k=1, the process is stationary. The Peaceman-Rachford ADI method can theoretically be expressed in the form

$$\underset{\sim}{u}^{(n+1)} = B_{n+1}\underset{\sim}{u}^{(n)} + \underset{\sim}{k}_{n+1}, \quad n \geq 0,$$

where the iteration matrix B_{n+1} is given by

$$B_{n+1} = (A_2 + \omega_{n+1} I)^{-1}(A_1 - \omega_{n+1} I)(A_1 + \omega_{n+1} I)^{-1}(A_2 - \omega_{n+1} I).$$

Then $\underset{\sim}{\varepsilon}^{(n)} = \prod_{i=1}^{n} B_i \underset{\sim}{\varepsilon}^{(0)}$.

If $\rho(\prod_{i=1}^{n} B_i)$ is the spectral radius of this matrix product, then the average rate of convergence after n iterations is $R_n = -\frac{1}{n} \ln(\rho(\prod_{i=1}^{n} B_i))$.

Although optimal acceleration parameters ω_n to minimise this average rate of convergence are not available, good ones, that are simple to calculate, have been found to work quite satisfactorily.

Let α, β be found so that $0 < \alpha \leq \mu \leq \beta$ and $0 < \alpha \leq \lambda \leq \beta$ for all eigenvalues μ of A_1 and eigenvalues λ of A_2. Such bounds on the eigenvalues can easily be determined using the relevant theories.

Then two commonly used parameters are the following:

(1) Peaceman-Rachford Parameters
Find the smallest integer k such that $(\sqrt{2}-1)^{2k} \leq \alpha/\beta$. Then, for i=1,2,..,k, let $\omega_i = \alpha(\alpha/\beta)^{(2i-1)/2k}$. These parameters are clearly not evenly spaced, but are geometrically spaced. (In the special case of the stationary Peaceman-Rachford method, k = 1, and so $\omega = \sqrt{\alpha\beta}$.)

It can then be shown that

$$\rho(\prod_{i=1}^{k} B_i) \leq \left(\frac{1-(\alpha/\beta)^{1/2k}}{1+(\alpha/\beta)^{1/2k}}\right)^2 = \delta^2,$$

say, and so the average rate of convergence

$$R_k \simeq -\frac{2}{k}\ln\delta.$$

The condition $(\sqrt{2}-1)^{2k} \leq \alpha/\beta$ is used to determine k because it is a consequence of the result that R_k is maximised when $\delta = \sqrt{2}-1$.

For example, consider the second model problem. It can be shown that $\alpha = 4\sin^2\pi h/2$ and $\beta = 4\cos\pi^2 h/2$. Hence,

$$\frac{\alpha}{\beta} = \tan^2\frac{\pi h}{2},$$

and so we require $(\sqrt{2}-1)^k \leq \tan\frac{\pi h}{2}$. For h = 1/40, this yields k = 4. Then we obtain

$$\omega_1 \simeq 0.01385,\ \omega_2 \simeq 0.06986,\ \omega_3 \simeq 0.35245,\ \omega_4 \simeq 1.7781.$$

(2) Wachspress Parameters
Find the smallest integer $k \geq 2$ such that $(\sqrt{2}-1)^{2(k-1)} \leq \frac{\alpha}{\beta}$. Then

$$\omega_i = \alpha\left(\frac{\beta}{\alpha}\right)^{(i-1)/(k-1)},\ \text{for } i=1,2,\ldots,k.$$

For the second model problem with h = 1/40, this yields k = 5. Then we obtain

$$\omega_1 = \alpha \simeq 0.006165,\ \omega_2 \simeq 0.03110,\ \omega_3 \simeq 0.1569,\ \omega_4 \simeq 0.7916,$$

$$\omega_5 = \beta \simeq 3.9938.$$

Numerical experiments suggest that the Wachspress parameters are superior to the Peaceman-Rachford parameters by a factor of about 2, provided that k is determined as described.

For both sets of parameters, the number of iterations required to obtain the specified accuracy varies according to ln(1/h) as the mesh-size h tends to zero. This is to be compared with 1/h for both

(i) the Successive Overrelaxation method, and
(ii) the Jacobi method with Chebyshev acceleration

and with $1/\sqrt{h}$ for the accelerated SSOR method. However, these other methods are perhaps applicable over a wider range of problems, and also the extra computations involved in the implicit solutions needed in each iteration of the Peaceman-Rachford method are appreciable. Hence, this ADI method is really only advantageous provided h is very small. In fact, experimentation suggests that, providing the ADI method is applicable, there exists a value of h, say h*, such that for h > h* the SOR-type methods are preferable, but for h < h*, the Peaceman-Rachford method uses less computer time.

9. COMPARISON OF ITERATIVE METHODS

Comments made about the convergence rates of the various iterative methods described in the preceding sections give a reasonable indication as to the suitability of each. Of course, the amount of time and machine storage space necessary to perform the iterations themselves must be taken into account. A number of references in the bibliography quote numerical results for a range of problems, and there is a reasonable degree of uniformity in their conclusions. For example, see Birkhoff, Varga and Young (1962); or Westlake (1968).

Most of the iterative methods considered rely on the matrix A being symmetric positive definite, or else require (theoretically) the existence of a symmetrization matrix W. If A is not symmetric, newer methods that are basically extensions or variations of the classic methods have been formulated. Details can be found in Hageman and Young (1981), or Manteuffel (1975), or Axelsson and Gustafsson (1977), or Widlund (1978).

Of the methods outlined in the preceding sections, those that can seriously be considered for implementation to solve large sparse linear systems are

(1) Successive Overrelaxation method, which may need to be adaptive if ω_{opt} is unknown.

(2) Chebyshev acceleration of the basic Jacobi or SSOR methods. Again, adaptive techniques will be needed if the eigenvalues or relaxation parameters are not known beforehand.

(3) Conjugate gradient acceleration of the Jacobi or SSOR methods.

(4) The reduced system concept, using a "red-black" ordering, in conjunction with either Chebyshev or conjugate gradient acceleration.

(5) Peaceman-Rachford Alternating Direction method.

If the reduced system methods are applicable, numerical results indicate that they frequently use fewer iterations, less time and less storage than the other techniques. In particular, the reduced system combined with conjugate gradient acceleration has often proven to be optimal in each of these classifications. The SSOR method, with either Chebyshev or conjugate gradient acceleration also produces favourable results, and covers a wider class of problems. However, the actual programming implementation is more complicated. The Successive Overrelaxation method is preferred by many because of its simplicity in this regard, but usually considerably more iterations, and somewhat more execution time, are required. The Peaceman-Rachford method is really only superior when the region is rectangular (or similar), the mesh-size is small, and $k > 1$. It is advisable to select a value of k, the number of iteration parameters, which is too large than to use one which is too small. Savings have been found by actually choosing a value of k larger than the ones produced by the formulae described in that section.

APPENDIX A

LIST OF SYMBOLS

A	: coefficient matrix of linear system
A_1, A_2	: $A = A_1 + A_2$ in Peaceman-Rachford Method
B	: general iteration matrix
B_J	: Jacobi iteration matrix
B_{GS}	: Gauss-Seidel iteration matrix
B_{SOR}	: Successive Overrelaxation iteration matrix
B_{BSOR}	: Backward Successive Overrelaxation iteration matrix
B_{SSOR}	: Symmetric Successive Overrelaxation iteration matrix
B_{RS}	: Reduced system iteration matrix
B_n	: Peaceman-Rachford iteration matrices
D	: diagonal part of A
G	: submatrix of A in "red-black" ordering
H	: estimate for spectral radius of B_{SOR}
K	: submatrix of A in "red-black" ordering
L	: lower triangular part of A
M	: largest eigenvalue of B
$\bar{M}$	: estimate of M
N	: order of matrix A
$P_n(x)$	: transformed Chebyshev polynomials
Q	: general splitting matrix
QA	: variable used in solving Chebyshev equation
QT	: variable used in solving Chebyshev equation
$R(B)$	: asymptotic rate of convergence
$R_n(B)$	: average rate of convergence after n iterations
$R^{(n)}$	: ratio of consecutive pseudo-residuals in SOR method
$T_n(x)$	: Chebyshev polynomials
U	: upper triangular part of A
W	: symmetrization matrix
X	: variable used in solving Chebyshev equation
Y	: variable used in solving Chebyshev equation
Z	: variable used in solving Chebyshev equation
$\underset{\sim}{b}$	: right-hand side of linear system
h	: mesh size
$\underset{\sim}{k}$	: vector in general iteration
m	: smallest eigenvalue of B
$\bar{m}$	: estimate of m
n	: the number of the iteration

$\underset{\sim}{p}^{(n)}$	: direction vectors in conjugate gradient method
r	: parameter in Chebyshev acceleration
$\underset{\sim}{r}^{(n)}$	: residual vector after n iterations
$\underset{\sim}{u}$	: solution of linear system
$\underset{\sim}{u}^{(n)}$	: estimate of u after n iterations
α	: lower eigenvalue bound in Peaceman-Rachford method
α_n	: conjugate gradient parameters
β	: upper eigenvalue bound in Peaceman-Rachford method
β_n	: conjugate gradient parameters
$\bar{\beta}$	: upper bound for $\rho(D^{-1}LD^{-1}U)$ in SSOR method
γ	: Chebyshev acceleration constant
γ_n	: conjugate gradient acceleration parameters
$\underset{\sim}{\delta}^{(n)}$	: pseudo-residual vector
$\underset{\sim}{\varepsilon}^{(n)}$	: error vector
λ	: general eigenvalue
λ_J	: largest positive eigenvalue of B_J
$\bar{\lambda}_J$	: upper bound for λ_J, in SSOR method
μ	: eigenvalue of B_{SOR}
$\bar{\mu}$	: spectral radius of B_{SOR}
$\rho(B)$	: spectral radius of B
σ	: Chebyshev acceleration constant
ω	: relaxation parameter
ω_{opt}	: optimal value for ω in SOR method
ω^*	: good value for ω in SSOR method
ω_n	: Chebyshev acceleration parameters
	: conjugate gradient acceleration parameters
	: Peaceman-Rachford acceleration parameters
ω_∞	: limiting value of ω_n in Chebyshev acceleration

APPENDIX B

SUBROUTINE JAC

```
C      SUBROUTINE TO SOLVE LARGE SPARSE LINEAR
C      SYSTEMS USING JACOBI ITERATION
C              WRITTEN BY L.COLGAN       1983
C      ------------------------------------------------------------
C      SOLVES SYSTEMS OF N SPARSE LINEAR EQUATIONS WHERE
C      THE COEFFICIENT MATRIX IS STORED IN UNSYMMETRIC
C      SPARSE FORM
       SUBROUTINE JAC(A,JA,ISTART,N,NP1,IADIM,B,U,ZETA,
      +SR, IADAPT,ITMAX,NUMITS,V,D)
C      ------------------------------------------------------------
C
C      GLOSSARY OF TERMS
C
C      ------------------------------------------------------------
C      A( )        :NON-ZERO ELEMENTS OF SPARSE COEFF. MATRIX (INPUT)
C      JA( )       :COLUMN POSITIONS OF ELEMENTS IN A (INPUT)
C      ISTART(I)   :POSITION IN A & JA OF THE FIRST ENTRY FROM
C                  :ROW I. LAST ENTRY IS IADIM+1 (INPUT)
C      N           :NO. OF EQUATIONS (INPUT)
C      NP1         :N+1 (INPUT)
C      IADIM       :DIMENSION OF A & JA (INPUT)
C      B( )        :RIGHT HAND SIDE OF EQUATIONS (INPUT)
C      U( )        :SOLUTION. ON INPUT CONTAINS INITIAL GUESS
C                  :WHICH MAY BE ZERO VECTOR (INPUT,OUTPUT)
C      ZETA        :STOPPING TEST. RELATIVE ERROR<ZETA (INPUT)
C      SR          :SPECTRAL RADIUS OF ITERATION MATRIX
C                  :(MUST BE INPUT IF ADAPT=0)
C      IADAPT      :=0 DOES NOT ADAPT TO ESTIMATE SR
C                  :=1 WILL ADAPT (INPUT)
C      ITMAX       :MAX NO. OF ITERATIONS (INPUT)
C      NUMITS      :NO. OF ITERATIONS PERFORMED (OUTPUT)
C      V( )        :TEMPORARY STORAGE WORK AREA OF SIZE N
C      D( )        :DIAGONAL STORAGE WORK AREA OF SIZE N
C      PSPREV      :PREVIOUS PSEUDO-RESIDUAL NORM
C      PSNORM      :CURRENT PSEUDO-RESIDUAL NORM
C      VNORM       :CURRENT SOLUTION NORM
C      TEST        :STOPPING TEST
C      ------------------------------------------------------------
       REAL A(IADIM),B(N),U(N),ZETA,SR
       INTEGER JA(IADIM),ISTART(NP1),N,NP1,IADIM,IADAPT,
      +NUMITS,ITMAX
       REAL V(N),PSPREV,PSNORM,D(N),VNORM,SUM,TEST
C      ****
C      SET UP DIAGONAL ELEMENTS
C      FIND ORIGINAL PSEUDO-RESIDUAL NORM
C      CALCULATE FIRST ITERATES
C      ****
       PSPREV=0.
       DO 10 I=1,N
       SUM=B(I)
       DO 20 J=ISTART(I),ISTART(I+1)-1
       IF(JA(J).EQ.I)THEN
       D(I)=A(J)
       ELSE
       SUM=SUM-A(J)*U(JA(J))
       ENDIF
20     CONTINUE
```

```
      V(I)=SUM/D(I)
      PSPREV=PSPREV+(V(I)-U(I))**2
10    CONTINUE
      PSPREV=SQRT(PSPREV)
      NUMITS=1
      DO 30 I=1,N
      U(I)=V(I)
30    CONTINUE
C     ****
C     PERFORM THE NEXT ITERATION
C     CALCULATE THE NORMS OF THE ESTIMATE
C     AND THE PSEUDO-RESIDUAL
C     ****
40    NUMITS=NUMITS+1
      IF(NUMITS.GT.ITMAX)RETURN
      PSNORM=0.
      VNORM=0.
      DO 50 I=1,N
      SUM=B(I)
      DO 60 J=ISTART(I),ISTART(I+1)-1
      IF(JA(J).NE.I)SUM=SUM=A(J)*U(JA(J))
60    CONTINUE
      V(I)=SUM/D(I)
      PSNORM=PSNORM+(V(I)-U(I))**2
      VNORM=VNORM+V(I)**2
50    CONTINUE
      PSNORM=SQRT(PSNORM)
      VNORM=SQRT(VNORM)
      DO 70 I=1,N
      U(I)=V(I)
70    CONTINUE
C     ****
C     STOPPING TEST
C     IF IADAPT=1,USE RATIO OF CONSECUTIVE PSEUDO-RESIDUALS
C     AS AN ESTIMATE FOR THE SPECTRAL RADIUS OF THE
C     ITERATION MATRIX
C     ****
      IF(IADAPT.EQ.1)SR=PSNORM/PSPREV
      TEST=SR*PSNORM/(1.-SR)/VNORM
      IF(ABS(TEST).LT.ZETA)RETURN
C     ****
C     PERFORM ANOTHER ITERATION
C     ****
      PSPREV=PSNORM
      GOTO40
      END
```

<u>OUTPUT</u>

The subroutine JAC, with appropriate PRINT statements, $u^{(0)} = \underset{\sim}{0}$, and a calling PROGRAM to implement the solution of the first model problem (N=16), yields the following output: (showing the number of the iteration, NUMITS; progressive estimate of the spectral radius, SR; and the 16 variable values, $U_1, \ldots, U_{16}$. Only some of the iterations have been exhibited.)

NUMITS = 1
.0100000 .0100000 .0100000 .0100000 .0100000 .0100000 .0100000 .0100000
.0100000 .0100000 .0100000 .0100000 .0100000 .0100000 .0100000 .0100000
SR = 0 (has not changed)

```
NUMITS = 2
.0150000  .0175000  .0175000  .0150000  .0175000  .0200000  .0200000  .0175000
.0175000  .0200000  .0200000  .0175000  .0150000  .0175000  .0175000  .0150000
SR = 770552

NUMITS = 3
.0187500  .0231250  .0231250  .0187500  .0231250  .0287500  .0287500  .0231250
.0231250  .0287500  .0287500  .0231250  .0187500  .0231250  .0231250  .0187500
SR = .805001

NUMITS = 4
.0215625  .0276563  .0276563  .0215625  .0276563  .0359375  .0359375  .0276563
.0276563  .0359375  .0359375  .0276563  .0215625  .0276563  .0276563  .0215625
SR = .808623

NUMITS = 5
.0238281  .0312891  .0312891  .0238281  .0312891  .0417969  .0417969  .0312891
.0312891  .0417969  .0417969  .0312891  .0238281  .0312891  .0312891  .0238281
SR = .808978

   .
   .
   .

NUMITS = 10
.0300404  .0413385  .0413385  .0300404  .0413385  .0580456  .0580456  .0413385
.0413385  .0580456  .0580456  .0413385  .0300404  .0413385  .0413385  .0300404
SR = .809017

   .
   .
   .

NUMITS = 15
.0321921  .0448201  .0448201  .0321921  .0448201  .0636789  .0636789  .0448201
.0448201  .0636789  .0636789  .0448201  .0321921  .0448201  .0448201  .0321921
SR = .809017

   .
   .
   .

NUMITS = 20
.0329378  .0460267  .0460267  .0329378  .0460267  .0656312  .0656312  .0460267
.0460267  .0656312  .0656312  .0460267  .0329378  .0460267  .0460267  .0329378
SR = .809017

   .
   .
   .

NUMITS = 60
.0333333  .0466665  .0466665  .0333333  .0466665  .0666665  .0666665  .0466665
.0466665  .0666665  .0666665  .0466665  .0333333  .0466665  .0466665  .0333333
SR = .809017

NUMITS = 66
.0333333  .0466666  .0466666  .0333333  .0466666  .0666666  .0666666  .0466666
.0466666  .0666666  .0666666  .0466666  .0333333  .0466666  .0466666  .0333333
SR = .809017
```

APPENDIX C

SUBROUTINE GS

```
C       SUBROUTINE TO SOLVE LARGE SPARSE LINEAR
C       SYSTEMS USING GAUSS-SEIDEL ITERATION
C                 WRITTEN BY L.COLGAN       1983
C       ---------------------------------------------------------------
C       SOLVES SYSTEMS OF N SPARSE LINEAR EQUATIONS WHERE
C       THE COEFFICIENT MATRIX IS STORED IN UNSYMMETRIC
C       SPARSE FORM
         SUBROUTINE GS(A,JA,ISTART,N,NP1,IADIM,B,U,ZETA,
      +SR,IADAPT,ITMAX,NUMITS,D)
C       ---------------------------------------------------------------
C
C       GLOSSARY OF TERMS
C
C       ---------------------------------------------------------------
C       A( )       :NON-ZERO ELEMENTS OF SPARSE COEFF. MATRIX (INPUT)
C       JA( )      :COLUMN POSITIONS OF ELEMENTS IN A (INPUT)
C       ISTART( )  :POSITION IN A & JA OF THE FIRST ENTRY FROM
C                  :ROW I. LAST ENTRY IS IADIM+1 (INPUT)
C       N          :NO. OF EQUATIONS (INPUT)
C       NP1        :N+1 (INPUT)
C       IADIM      :DIMENSION OF A & JA (INPUT)
C       B( )       :RIGHT HAND SIDE OF EQUATIONS (INPUT)
C       U( )       :SOLUTION. ON INPUT CONTAINS INITIAL GUESS
C                  :WHICH MAY BE ZERO VECTOR (INPUT,OUTPUT)
C       ZETA       :STOPPING TEST. RELATIVE ERROR<ZETA (INPUT)
C       SR         :SPECTRAL RADIUS OF ITERATION MATRIX
C                  :(MUST BE INPUT IF IADAPT=0)
C       IADAPT     :=0 DOES NOT ADAPT TO ESTIMATE SR
C                  :=1 WILL ADAPT (INPUT)
C       ITMAX      :MAX NO. OF ITERATIONS (INPUT)
C       NUMITS     :NO. OF ITERATIONS PERFORMED (OUTPUT)
C       D( )       :DIAGONAL STORAGE WORK AREA OF SIZE N
C       PSPREV     :PREVIOUS PSEUDO-RESIDUAL NORM
C       PSNORM     :CURRENT PSEUDO-RESIDUAL NORM
C       UNORM      :CURRENT SOLUTION NORM
C       TEST       :STOPPING TEST
C       TEMP       :TEMPORARY STORAGE OF NEW SOLUTION ELEMENT
C       ---------------------------------------------------------------
        REAL A(IADIM),B(N),U(N),ZETA,SR
        INTEGER JA(IADIM),ISTART(NP1),N,NP1,IADIM,IADAPT,
      +NUMITS,ITMAX
        REAL PSPREV,PSNORM,D(N),UNORM,SUM,TEST,TEMP
C       ****
C       SET UP DIAGONAL ELEMENTS
C       FIND ORIGINAL PSEUDO-RESIDUAL NORM
C       CALCULATE FIRST ITERATES
C       ****
        PSPREV=0.
        DO 10 I=1,N
        SUM=B(I)
        DO 20 J=ISTART(I),ISTART(I+1)-1
        IF(JA(J).EQ.I)THEN
        D(I)=A(J)
        ELSE
        SUM=SUM-A(J)*U(JA(J))
        ENDIF
 20     CONTINUE
```

```
      TEMP=SUM/D(I)
      PSPREV=PSPREV+(TEMP-U(I))**2
      U(I)=TEMP
10    CONTINUE
      PSPREV=SQRT(PSPREV)
      NUMITS=1
C     ****
C     PERFORM THE NEXT ITERATION
C     CALCULATE THE NORMS OF THE ESTIMATE
C     AND THE PSEUDO-RESIDUAL
C     ****
40    NUMITS=NUMITS+1
      IF(NUMITS.GT.ITMAX)RETURN
      PSNORM=0.
      UNORM=0.
      DO 50 I=1,N
      SUM=B(I)
      DO 60 J=ISTART(I),ISTART(I+1)-1
      IF(JA(J).NE.I)SUM=SUM-A(J)*U(JA(J))
60    CONTINUE
      TEMP=SUM/D(I)
      PSNORM=PSNORM+(TEMP-U(I))**2
      UNORM=UNORM+TEMP**2
      U(I)=TEMP
50    CONTINUE
      PSNORM=SQRT(PSNORM)
      UNORM=SQRT(UNORM)
C     ****
C     STOPPING TEST
C     IF IADAPT=1,USE RATIO OF CONSECUTIVE PSEUDO-RESIDUALS
C     AS AN ESTIMATE FOR THE SPECTRAL RADIUS OF THE
C     ITERATION MATRIX
C     ****
      IF(IADAPT.EQ.1)SR=PSNORM/PSPREV
      TEST=SR*PSNORM/(1.-SR)/UNORM
      IF(ABS(TEST).LT.ZETA)RETURN
C     ****
C     PERFORM ANOTHER ITERATION
C     ****
      PSPREV=PSNORM
      GOTO40
      END
```

OUTPUT

The subroutine GS yields the following output, under the same conditions as in Appendix B. It is found that about half as many iterations are required for the first model problem being considered.

```
NUMITS=1
.0100000  .0125000  .0131250  .0132813  .0125000  .0162500  .0173438  .0176563
.0131250  .0173438  .0186719  .0190820  .0132813  .0176563  .0190820  .0195410
SR=0.000000

NUMITS=2
.0162500  .0214063  .0230078  .0201660  .0214063  .0293750  .0321777  .0278564
.0230078  .0321777  .0356299  .0307568  .0201660  .0278564  .0307568  .0253784
SR=.684742
```

```
NUMITS=3
.0207031  .0282715  .0301538  .0245026  .0282715  .0402246  .0434662  .0346814
.0301538  .0434662  .0471115  .0367928  .0245026  .0346814  .0367928  .0283964
SR=.700422

NUMITS=4
.0241357  .0336285  .0353993  .0275202  .0336285  .0485474  .0514349  .0389370
.0353993  .0514349  .0541139  .0403618  .0275202  .0389370  .0403618  .0301809
SR=.696357

NUMITS=5
.0268143  .0376902  .0391613  .0295246  .0376902  .0545626  .0566937  .0416450
.0391613  .0566937  .0585277  .0425884  .0295246  .0416450  .0425884  .0312942
SR=.686504

    :

NUMITS=10
.0324713  .0455369  .0457519  .0328758  .0455369  .0651866  .0654688  .0460676
.0457519  .0654688  .0656973  .0461820  .0328758  .0460676  .0461820  .0330910
SR=.655921

    :

NUMITS=15
.0332296  .0465308  .0465568  .0332784  .0465308  .0664888  .0665228  .0465947
.0465568  .0665228  .0665503  .0466085  .0332784  .0465947  .0466085  .0333042
SR=.654544

    :

NUMITS=20
.0333209  .0466503  .0466535  .0333267  .0466503  .0666453  .0666494  .0466580
.0466535  .0666494  .0666527  .0466597  .0333267  .0466580  .0466597  .0333298
SR=.654509

    :

NUMITS=25
.0333318  .0466647  .0466651  .0333325  .0466647  .0666641  .0666646  .0466656
.0466651  .0666646  .0666650  .0466658  .0333325  .0466656  .0466658  .0333329
SR=.654509

    :

NUMITS=30
.0333332  .0466664  .0466665  .0333332  .0466664  .0666664  .0666664  .0466665
.0466665  .0666664  .0666665  .0466666  .0333332  .0466665  .0466666  .0333333
SR=.654508

    :

NUMITS=34
.0333333  .0466666  .0466666  .0333333  .0466666  .0666666  .0666666  .0466666
.0466666  .0666666  .0666666  .0466666  .0333333  .0466666  .0466666  .0333333
SR=.654508
```

APPENDIX D

SUBROUTINE SOR

```
C      SUBROUTINE TO SOLVE LARGE SPARSE LINEAR
C      SYSTEMS USING SOR ITERATION
C               WRITTEN BY L.COLGAN       1983
C      ---------------------------------------------------------------
C      SOLVES SYSTEMS OF N SPARSE LINEAR EQUATIONS WHERE
C      THE COEFFICIENT MATRIX IS STORED IN UNSYMMETRIC
C      SPARSE FORM
        SUBROUTINE SOR(A,JA,ISTART,N,NP1,IADIM,B,U,ZETA,
      +SR,OMEGA,IADAPT,ITMAX,NUMITS,D)
C      ---------------------------------------------------------------
C
C      GLOSSARY OF TERMS
C
C      ---------------------------------------------------------------
C      A( )      :NON-ZERO ELEMENTS OF SPARSE COEFF. MATRIX (INPUT)
C      JA( )     :COLUMN POSITIONS OF ELEMENTS IN A (INPUT)
C      ISTART(I) :POSITION IN A & JA OF THE FIRST ENTRY FROM
C                :ROW I. LAST ENTRY IS IADIM+1 (INPUT)
C      N         :NO. OF EQUATIONS (INPUT)
C      NP1       :N+1 (INPUT)
C      IADIM     :DIMENSIONS OF A & JA (INPUT)
C      B( )      :RIGHT HAND SIDE OF EQUATIONS (INPUT)
C      U( )      :SOLUTION. ON INPUT CONTAINS INITIAL GUESS
C                :WHICH MAY BE ZERO VECTOR (INPUT,OUTPUT)
C      ZETA      :STOPPING TEST. RELATIVE ERROR<ZETA (INPUT)
C      SR        :SPECTRAL RADIUS OF ITERATION MATRIX
C                :(MUST BE INPUT IF IADAPT=0)
C      OMEGA     :OVER-RELAXATION PARAMETER
C                :(MUST BE INPUT IF IADAPT=0)
C      IADAPT    :=0 DOES NOT ADAPT TO ESTIMATE SR,OMEGA
C                :=1 WILL ADAPT (INPUT)
C      ITMAX     :MAX NO. OF ITERATIONS (INPUT)
C      NUMITS    :NO. OF ITERATIONS PERFORMED (OUTPUT)
C      D( )      :DIAGONAL STORAGE WORK AREA OF SIZE N
C      PSPREV    :PREVIOUS PSEUDO-RESIDUAL NORM
C      PSNORM    :CURRENT PSEUDO-RESIDUAL NORM
C      UNORM     :CURRENT SOLUTION NORM
C      TEST      :STOPPING TEST
C      H         :USED IN STOPPING TEST
C      TEMP      :TEMPORARY STORAGE OF NEW SOLUTION ELEMENT
C      SRJ       :SPECTRAL RADIUS OF JACOBI MATRIX
C      T(I)      :UPPER BOUND FOR THE I'TH ESTIMATE OF OMEGA
C      IEST      :NUMBER OF DIFFERENT ESTIMATES OF OMEGA
C      NEWITS    :NUMBER OF ITERATIONS USING THE NEW ESTIMATE
C      CHANGE    :TEST IF SR IS REASONABLE ESTIMATE
C      ---------------------------------------------------------------
       REAL A(IADIM),B(N),U(N),ZETA,SR,OMEGA
       INTEGER JA(IADIM),ISTART(NP1),N,NP1,IADIM,IADAPT,
      +NUMITS,ITMAX
       REAL PSPREV,PSNORM,D(N),UNORM,SUM,TEST,TEMP,H,SRJ,T(9),CHANGE
       INTEGER IEST,NEWITS
       DATA T/1.5,1.8,1.85,1.9,1.94,1.96,1.975,1.985,1.992/
C      ****
C      SET UP DIAGONAL ELEMENTS
C      FIND ORIGINAL PSEUDO-RESIDUAL NORM
C      CALCULATE FIRST ITERATES
C      ****
```

```
      IF(IADAPT.EQ.1)THEN
      IEST=1
      OMEGA=MIN(OMEGA,T(1))
      ENDIF
      PSPREV=0.
      DO 10 I=1,N
      SUM=B(I)
      DO 20 J=ISTART(I),ISTART(I+1)-1
      IF(JA(J).EQ.I)THEN
      D(I)=A(J)
      ELSE
      SUM=SUM-A(J)*U(JA(J))
      ENDIF
20    CONTINUE
      TEMP=OMEGA*SUM/D(I)+(1.0-OMEGA)*U(I)
      PSPREV=PSPREV+(TEMP-U(I))**2
      U(I)=TEMP
10    CONTINUE
      PSPREV=SQRT(PSPREV)
      NUMITS=1
      NEWITS=1
C     ****
C     PERFORM THE NEXT ITERATION
C     CALCULATE THE NORMS OF THE ESTIMATE
C     AND THE PSEUDO-RESIDUAL
C     ****
40    NUMITS=NUMITS+1
      IF(NUMITS.GT.ITMAX)RETURN
      NEWITS=NEWITS+1
      PSNORM=0.
      UNORM=0.
      DO 50 I=1,N
      SUM=B(I)
      DO 60 J=ISTART(I),ISTART(I+1)-1
      IF(JA(J).NE.I)SUM=SUM-A(J)*U(JA(J))
60    CONTINUE
      TEMP=OMEGA*SUM/D(I)+(1.0-OMEGA)*U(I)
      PSNORM=PSNORM+(TEMP-U(I))**2
      UNORM=UNORM+TEMP**2
      U(I)=TEMP
50    CONTINUE
      PSNORM=SQRT(PSNORM)
      UNORM=SQRT(UNORM)
C     ****
C     STOPPING TEST
C     IF IADAPT=1,USE RATIO OF CONSECUTIVE PSEUDO-RESIDUALS
C     AS AN ESTIMATE FOR THE SPECTRAL RADIUS OF THE
C     ITERATION MATRIX
C     MUST PERFORM AT LEAST TWO ITERATIONS WITH
C     NEW VALUE OF OMEGA BEFORE STOPPING
C     ****
      IF(IADAPT.EQ.0)THEN
      TEST=PSNORM/(1.-SR)/UNORM
      IF(ABS(TEST).LT.ZETA)RETURN
      GOTO 80
      ENDIF
      IF(NEWITS.LT.2)GOTO 80
      IF(ABS(SR-PSNORM/PSPREV).GT.0.01)THEN
      CHANGE=0.
      ELSE
```

```
      CHANGE=1.
      ENDIF
      SR=PSNORM/PSPREV
      IF(SR.GE.1.0)GOTO80
      IF(SR.LT.OMEGA-1.0)THEN
      H=OMEGA-1.0
      ELSE
      H=SR
      ENDIF
      TEST=PSNORM/(1.-H)/UNORM
      IF(ABS(TEST).LT.ZETA)RETURN
C     ****
C     CHANGE OMEGA IF NECESSARY
C     MUST BE AT LEAST FIVE ITERATIONS USING PREVIOUS OMEGA
C     ****
      IF(NEWITS.LT.5.OR.SR.LT.(OMEGA-1.)**0.75.OR.CHANGE.EQ.0.)GOTO 80
      IEST=IEST+1
      NEWITS=0
      SRJ=(SR+OMEGA-1.0)/OMEGA/SQRT(SR)
      OMEGA=2.0/(1.0+SQRT(1.0-SRJ**2))
      IF(IEST.LE.9)THEN
      OMEGA=MIN(OMEGA,T(IEST))
      ELSE
      OMEGA=MIN(OMEGA,T(9))
      ENDIF
C     ****
C     PERFORM ANOTHER ITERATION
C     ****
80    PSPREV=PSNORM
      GOTO40
      END
```

OUTPUT

The subroutine SOR yields the following output for the first model problem. Initially the relaxation parameter ω is given the value 1, and the subroutine changes it at appropriate times only. This occurs after the fifth iteration only. The value of H, used in the stopping test, is also shown.

```
NUMITS=1
ω=1.000000
.0100000  .0125000  .0131250  .0132813  .0125000  .0162500  .0173438  .0176563
.0131250  .0173438  .0186719  .0190820  .0132813  .0176563  .0190820  .0195410
H not yet applicable

NUMITS=2
ω=1.000000
.0162500  .0214063  .0230078  .0201660  .0214063  .0293750  .0321777  .0278564
.0230078  .0321777  .0356299  .0307568  .0201660  .0278564  .0307568  .0253784
H=.684742

NUMITS=3
ω=1.000000
.0207031  .0282715  .0301538  .0245026  .0282715  .0402246  .0434662  .0346814
.0301538  .0434662  .0471115  .0367928  .0245026  .0346814  .0367928  .0283964
H=.700422
```

```
NUMITS=4
ω=1.000000
.0241357  .0336285  .0353993  .0275202  .0336285  .0485474  .0514349  .0389370
.0353993  .0514349  .0541139  .0403618  .0275202  .0389370  .0403618  .0301809
H=.696357

NUMITS=5
ω=1.000000
.0268143  .0376902  .0391613  .0295246  .0376902  .0545626  .0566937  .0416450
.0391613  .0566937  .0585277  .0425884  .0295246  .0416450  .0425884  .0312942
H=.686504

   :
   :

NUMITS=7
ω=1.282128
.0312275  .0441691  .0449078  .0326485  .0441691  .0638370  .0648243  .0460275
.0449078  .0648243  .0656558  .0464129  .0326485  .0460275  .0464129  .0332224
H=.534703

   :

NUMITS=9
ω=1.282128
.0329194  .0462823  .0464766  .0332860  .0462823  .0663611  .0665409  .0466257
.0464766  .0665409  .0665959  .0466478  .0332860  .0466257  .0466478  .0333289
H=.434433

   :

NUMITS=11
ω=1.282128
.0333109  .0466583  .0466609  .0333322  .0466583  .0666587  .0666648  .0466671
.0466609  .0666648  .0666674  .0466675  .0333322  .0466671  .0466675  .0333338
H=.282128

   :

NUMITS=13
ω=1.282128
0.333320  .0466667  .0466671  .0333335  .0466667  .0666673  .0666673  .0466669
.0466671  .0666673  .0666670  .0466668  .0333335  .0466669  .0466668  .0333334
H=.282128

   :

NUMITS=15
ω=1.282128
.0333335  .0466668  .0466668  .0333334  .0466668  .0666668  .0666667  .0466667
.0466668  .0666667  .0666667  .0466667  .0333334  .0466667  .0466667  .0333333
H=.367594

   :

NUMITS=17
ω=1.282128
.0333333  .0466667  .0466667  .0333333  .0466667  .0666667  .0666667  .0466667
.0466667  .0666667  .0666667  .0466667  .0333333  .0466667  .0466667  .0333333
H=.300329
```

APPENDIX E

SUBROUTINE JSI

```
C       SUBROUTINE TO SOLVE LARGE SPARSE LINEAR
C       SYSTEMS USING JACOBI ITERATION WITH CHEBYSHEV ACCELERATION
C                WRITTEN BY L.COLGAN       1983
C       -----------------------------------------------------------
C       SOLVES SYSTEMS OF N SPARSE LINEAR EQUATIONS WHERE
C       THE COEFFICIENT MATRIX IS STORED IN UNSYMMETRIC
C       SPARSE FORM
         SUBROUTINE JSI(A,JA,ISTART,N,NP1,IADIM,B,U,ZETA,
      +SRJ,SRNJ,ICASE,IADAPT,ITMAX,NUMITS,V,W,D)
C       -----------------------------------------------------------
C
C       GLOSSARY OF TERMS
C
C       -----------------------------------------------------------
C       A( )       :NON-ZERO ELEMENTS OF SPARSE COEFF. MATRIX (INPUT)
C       JA( )      :COLUMN POSITIONS OF ELEMENTS IN A (INPUT)
C       ISTART(I)  :POSITION IN A & JA OF THE FIRST ENTRY FROM
C                  :ROW I. LAST ENTRY IS IADIM+1 (INPUT)
C       N          :NO. OF EQUATIONS (INPUT)
C       NP1        :N+1 (INPUT)
C       IADIM      :DIMENSION OF A & JA (INPUT)
C       B( )       :RIGHT HAND SIDE OF EQUATIONS (INPUT)
C       U( )       :SOLUTION. ON INPUT CONTAINS INITIAL GUESS
C                  :WHICH MAY BE ZERO VECTOR (INPUT,OUTPUT)
C       ZETA       :STOPPING TEST. RELATIVE ERROR<ZETA (INPUT)
C       SRJ        :SPECTRAL RADIUS OF BASIC JACOBI ITERATION MATRIX
C                  :(MUST BE INPUT)
C       SRNJ       :LEAST EIGENVALUE;MOST NEGATIVE (INPUT)
C       ICASE      :=0 DO NOT CHANGE SRNJ
C                  :=1 LET SRNJ=-SRJ (INPUT)
C       IADAPT     :=0 DOES NOT ADAPT TO ESTIMATE SRJ,SRNJ
C                  :=1 WILL ADAPT (INPUT)
C       ITMAX      :MAX NO. OF ITERATIONS (INPUT)
C       NUMITS     :NO. OF ITERATIONS PERFORMED (OUTPUT)
C       V( )       :TEMPORARY STORAGE WORK AREA OF SIZE N
C       W( )       :STORAGE FOR PREVIOUS ITERATION OF SIZE N
C       D( )       :DIAGONAL STORAGE WORK AREA OF SIZE N
C       PSORIG     :ORIGINAL PSEUDO-RESIDUAL NORM WITH CURRENT SRJ
C       PSNORM     :CURRENT PSEUDO-RESIDUAL NORM
C       UNORM      :CURRENT SOLUTION NORM
C       TEST       :STOPPING TEST
C       GAMMA      :PARAMETER
C       SIGMA      :PARAMETER
C       R          :PARAMETER
C       OMEGA      :ACCELERATION PARAMETER
C       TEMP       :VARIOUS TEMPORARY STORAGES
C       QA         :ACTUAL RESIDUAL QUOTIENT
C       QT         :THEORETICAL RESIDUAL QUOTIENT
C       NEWITS     :NUMBER OF ITERATIONS USING THE NEW ESTIMATE
C       -----------------------------------------------------------
        REAL A(IADIM),B(N),U(N),ZETA,SRJ,SRNJ
        INTEGER JA(IADIM),ISTART(NP1),N,NP1,IADIM,ICASE,IADAPT,
      +NUMITS,ITMAX
        REAL PSORIG,PSNORM,V(N),W(N),D(N),UNORM,SUM,TEST,TEMP
        REAL GAMMA,SIGMA,R,OMEGA,QA,QT
        INTEGER NEWITS
```

```
C       ****
C       SET UP DIAGONAL ELEMENTS
C       ****
        DO 10 I=1,N
        DO 20 J=ISTART(I),ISTART(I+1)-1
        IF(JA(J).EQ.I)THEN
        D(I)=A(J)
        GOTO 10
        ENDIF
20      CONTINUE
10      CONTINUE
        NUMITS=0
C       ****
C       CALCULATE OTHER PARAMETERS
C       ****
25      GAMMA=2.0/(2.0-SRJ-SRNJ)
        SIGMA=(SRJ-SRNJ)/(2.0-SRJ-SRNJ)
        R=SQRT(1.0-SIGMA**2)
        R=(1.0-R)/(1.0+R)
        NEWITS=0
C       ****
C       CALCULATE THE ACCELERATION PARAMETER
C       PERFORM THE NEXT ITERATION
C       IF NEWITS=1,FIND ORIGINAL PSEUDO-RESIDUAL
C       NORM USING THE NEW VALUES OF SRJ,SRNJ
C       ****
30      NUMITS=NUMITS+1
        IF(NUMITS.GT.ITMAX)RETURN
        NEWITS=NEWITS+1
        PSNORM=0.
        UNORM=0.
        IF(NEWITS.EQ.1)GOTO40
        IF(NEWITS.EQ.2)THEN
        OMEGA=1.0/(1.0-0.5*SIGMA**2)
        ELSE
        OMEGA=1.0/(1.0-0.25*OMEGA*SIGMA**2)
        ENDIF
40      DO 50 I=1,N
        SUM=B(I)
        DO 60 J=ISTART(I),ISTART(I+1)-1
        IF(JA(J).NE.I)SUM=SUM-A(J)*U(JA(J))
60      CONTINUE
        TEMP=SUM/D(I)-U(I)
        PSNORM=PSNORM+TEMP**2
        IF(NEWITS.EQ.1)THEN
        V(I)=GAMMA*TEMP+U(I)
        ELSE
        V(I)=OMEGA*(GAMMA*TEMP+U(I))+(1.0-OMEGA)*W(I)
        ENDIF
        UNORM=UNORM+V(I)**2
50      CONTINUE
        UNORM=SQRT(UNORM)
        PSNORM=SQRT(PSNORM)
C       ****
C       STOPPING TEST
C       ****
        TEST=PSNORM/(1.0-SRJ)/UNORM
        IF(ABS(TEST).LT.ZETA)RETURN
        IF(NEWITS.EQ.1)PSORIG=PSNORM
```

```
C       ****
C       INTERCHANGE ARRAYS
C       ****
        DO 70 I=1,N
        W(I)=U(I)
        U(I)=V(I)
70      CONTINUE
C       ****
C       CHANGE PARAMETERS IF NECESSARY
C       ****
        IF(IADAPT.EQ.0.OR.NEWITS.EQ.1)GOTO30
        QA=PSNORM/PSORIG
        TEMP=R**(NEWITS-1)
        QT=2.0*SQRT(TEMP)/(1.0+TEMP)
        IF(QA.LT.QT**0.75)GOTO30
C       ****
C       SOLVE CHEBYSHEV EQUATION TO FIND A
C       NEW ESTIMATE FOR SRJ
C       ****
        TEMP=0.5*(1.0+TEMP)*(QA+SQRT(QA*QA-QT*QT))
        TEMP=TEMP**(1.0/(NEWITS-1.0))
        TEMP=(TEMP+R/TEMP)/(1.0+R)
        SRJ=0.5*(SRJ+SRNJ+TEMP*(2.0-SRJ-SRNJ))
        IF(ICASE.EQ.1)SRNJ=-SRJ
        GOTO25
        END
```

OUTPUT

The subroutine JSI yields the following output for the first model problem. It uses the notation SRJ, SRNJ for M, m respectively, starting with SRJ=0, SRNJ=-1, adapting when convenient to find a better SRJ, and thereafter letting SRNJ=-SRJ. This occurred after the second and sixth iterations only.

```
NUMITS=1
.0066667  .0066667  .0066667  .0066667  .0066667  .0066667  .0066667  .0066667
.0066667  .0066667  .0066667  .0066667  .0066667  .0066667  .0066667  .0066667
SRJ=0.000000

NUMITS=2
.0117467  .0129412  .0129412  .0117647  .0129412  .0141176  .0141176  .0129412
.0129412  .0141176  .0141176  .0129412  .0117647  .0129412  .0129412  .0117647
SRJ=0.000000

NUMITS=3
.0164706  .0197059  .0197059  .0164706  .0197059  .0235294  .0235294  .0197059
.0197059  .0235294  .0235294  .0187059  .0164706  .0197059  .0197059  .0164706
SRJ=.762438

NUMITS=4
.0231671  .0298375  .0298375  .0231671  .0298375  .0387883  .0387883  .0298375
.0298375  .0387883  .0387883  .0298375  .0231671  .0298375  .0298375  .0231671
SRJ=.762438

NUMITS=5
.0270955  .0363603  .0363603  .0270955  .0363603  .0496681  .0496681  .0363603
.0363603  .0496681  .0496681  .0363603  .0270955  .0363603  .0363603  .0370955
SRJ=.762438
```

```
NUMITS=6
.0293013  .0401694  .0401694  .0293013  .0401694  .0561959  .0561959  .0401694
.0401694  .0561959  .0561959  .0401694  .0293013  .0401694  .0401694  .0293013
SRJ=.762438

NUMITS=7
.0300847  .0414167  .0414167  .0300847  .0414167  .0581826  .0581826  .0414167
.0414167  .0581826  .0581826  .0414167  .0300847  .0414167  .0414167  .0300847
SRJ=.808666

NUMITS=8
.0313919  .0435149  .0435149  .0313919  .0435149  .0615504  .0615504  .0435149
.0435149  .0615504  .0615504  .0435149  .0313919  .0435149  .0435149  .0313919
SRJ=.808666

NUMITS=9
.0322941  .0449798  .0449798  .0322941  .0449798  .0639283  .0639283  .0449798
.0449798  .0639283  .0639283  .0449798  .0322941  .0449798  .0449798  .0322941
SRJ=.808666

NUMITS=10
.0327923  .0457924  .0457924  .0327923  .0457924  .0652538  .0652538  .0457924
.0457924  .0652538  .0652538  .0457924  .0327923  .0457924  .0457924  .0327923
SRJ=.808666

NUMITS=11
0.330548  .0462177  .0462177  .0330548  .0462177  .0659432  .0659432  .0462177
.0462177  .0659432  .0659432  .0462177  .0330548  .0462177  .0462177  .0330548
SRJ=.808666

NUMITS=12
.0331913  .0464371  .0464371  .0331913  .0464371  .0662957  .0662957  .0464371
.0464371  .0662957  .0662957  .0464371  .0331913  .0464371  .0464371  .0331913
SRJ=.808666

NUMITS=13
.0332610  .0465493  .0465493  .0332610  .0465493  .0664762  .0664762  .0465493
.0465493  .0664762  .0664762  .0465493  .0332610  .0465493  .0465493  .0332610
SRJ=.808666

NUMITS=14
.0332963  .0466065  .0466065  .0332963  .0466065  .0665691  .0665691  .0466065
.0466065  .0665691  .0665691  .0466065  .0332963  .0466065  .0466065  .0332963
SRJ=.808666

NUMITS=15
.0333142  .0466358  .0466358  .0333142  .0466358  .0666167  .0666167  .0466358
.0466358  .0666167  .0666167  .0466358  .0333142  .0466358  .0466358  .033142
SRJ=.808666

NUMITS=16
.0333235  .0466508  .0466508  .0333235  .0466508  .0666411  .0666411  .0466508
.0466508  .0666411  .0666411  .0466508  .0333235  .0466508  .0466508  .0333235
SRJ=.808666

NUMITS=17
.0333283  .0466585  .0466585  .0333283  .0466585  .0666535  .0666535  .0466585
.0466585  .0666535  .0666535  .0466585  .0333283  .0466585  .0466585  .0333283
SRJ=.808666
```

```
NUMITS=18
.0333308  .0466625  .0466625  .0333308  .0466625  .0666599  .0666599  .0466625
.0466625  .0666599  .0666599  .0466625  .0333308  .0466625  .0466255  .0333308
SRJ=.808666

NUMITS=19
.0333320  .0466645  .0466645  .0333320  .0466645  .0666632  .0666632  .0466645
.0466645  .0666632  .0666632  .0466645  .0333320  .0466645  .0466645  .0333320
SRJ=.808666

NUMITS=20
.0333326  .0466656  .0466656  .0333326  .0466656  .0666649  .0666649  .0466656
.0466656  .0666649  .0666649  .0466656  .0333326  .0466656  .0466656  .0333326
SRJ=.808666

NUMITS=21
.0333330  .0466661  .0466661  .0333330  .0466661  .0666657  .0666657  .0466661
.0466661  .0666657  .0666657  .0466661  .0333330  .0466661  .0466661  .0333330
SRJ=.808666

NUMITS=22
.0333332  .0466664  .0466664  .0333332  .0466664  .0666662  .0666662  .0466664
.0466664  .0666662  .0666662  .0466664  .0333332  .0466664  .0466664  .0333332
SRJ=.808666

NUMITS=23
.0333332  .0466665  .0466665  .0333332  .0466665  .0666664  .0666664  .0466665
.0466665  .0666664  .0666664  .0466665  .0333332  .0466665  .0466665  .0333332
SRJ=.808666

NUMITS=24
.0333333  .0466666  .0466666  .0333333  .0466666  .0666665  .0666665  .0466666
.0466666  .0666665  .0666665  .0466666  .0333333  .0466666  .0466666  .0333333
SRJ=.808666

NUMITS=25
.0333333  .0466666  .0466666  .0333333  .0466666  .0666666  .0666666  .0466666
.0466666  .0666666  .0466666  .0466666  .0333333  .0466666  .0466666  .0333333
SRJ=.808666
```

APPENDIX F

SUBROUTINE JCG

```
C       SUBROUTINE TO SOLVE LARGE SPARSE LINEAR SYSTEMS
C       USING JACOBI ITERATION WITH CONJUGATE GRADIENT ACCELERATION
C                 WRITTEN BY L.COLGAN        1983
C       ----------------------------------------------------------------
C       SOLVES SYSTEMS OF N SPARSE LINEAR EQUATIONS WHERE
C       THE COEFFICIENT MATRIX IS STORED IN UNSYMMETRIC
C       SPARSE FORM
        SUBROUTINE JCG(A,JA,ISTART,N,NP1,IADIM,B,U,ZETA,
       +SRJ,IADAPT,ITMAX,NUMITS,W,D,PSU,PSV,PSW)
C       ----------------------------------------------------------------
C
C       GLOSSARY OF TERMS
C
C       ----------------------------------------------------------------
C       A( )       :NON-ZERO ELEMENTS OF SPARSE COEFF. MATRIX (INPUT)
C       JA( )      :COLUMN POSITIONS OF ELEMENTS IN A (INPUT)
C       ISTART(I)  :POSITION IN A & JA OF THE FIRST ENTRY FROM
C                  :ROW I. LAST ENTRY  IS IADIM+1 (INPUT)
C       N          :NO. OF EQUATIONS (INPUT)
C       NP1        :N+1 (INPUT)
C       IADIM      :DIMENSION OF A & JA (INPUT)
C       B( )       :RIGHT HAND SIDE OF EQUATIONS (INPUT)
C       U( )       :SOLUTION. ON INPUT CONTAINS INITIAL GUESS
C                  :WHICH MAY BE ZERO VECTOR (INPUT,OUTPUT)
C       ZETA       :STOPPING TEST. RELATIVE ERROR<ZETA (INPUT)
C       SRJ        :SPECTRAL RADIUS OF BASIC JACOBI ITERATION MATRIX
C                  :(MUST BE INPUT)
C       IADAPT     :=0 DOES NOT ADAPT TO ESTIMATE SRJ
C                  :=1 WILL ADAPT (INPUT)
C       ITMAX      :MAX NO. OF ITERATIONS (INPUT)
C       NUMITS     :NO. OF ITERATIONS PERFORMED (OUTPUT)
C       W( )       :STORAGE FOR PREVIOUS ITERATION OF SIZE N
C       D( )       :DIAGONAL STORAGE WORK AREA OF SIZE N
C       PSNORM     :CURRENT PSEUDO-RESIDUAL NORM
C       UNORM      :CURRENT SOLUTION NORM
C       TEST       :STOPPING TEST
C       PSU( )     :STORAGE FOR CURRENT PSEUDO-RESIDUAL VECTOR
C       PSV( )     :STORAGE FOR NEW PSEUDO-RESIDUAL VECTOR
C       PSW( )     :STORAGE FOR PREVIOUS PSEUDO-RESIDUAL VECTOR
C       PSUPSV     :INNER PRODUCT
C       PSUPSU     :INNER PRODUCT
C       PSWPSW     :INNER PRODUCT
C       GAMMAU     :PARAMETER
C       GAMMAW     :PARAMETER
C       OMEGAU     :CURRENT ACCELERATION PARAMETER
C       OMEGAW     :PREVIOUS ACCELERATION PARAMETER
C       ALPHA( )   :DIAGONAL ELEMENTS OF TRIDIAGONAL MATRIX
C       BETA( )    :SQUARE OF OFF-DIAGONAL ELEMENTS
C       CHANGE     :=0 IF SRJ IS ACCURATE ENOUGH
C       TEMP       :VARIOUS TEMPORARY STORAGES
C       ----------------------------------------------------------------
        REAL A(IADIM),B(N),U(N),ZETA,SRJ
        INTEGER JA(IADIM),ISTART(NP1),N,NP1,IADIM,IADAPT,
       +NUMITS,ITMAX
        REAL PSNORM,W(N),D(N),PSU(N),PSV(N),PSW(N),UNORM,SUM,
       +TEST,TEMP,PSUPSV,PSUPSU,PSWPSW,GAMMAU,GAMMAW,
       +OMEGAU,OMEGAW,ALPHA(200),BETA(200)
```

```
      INTEGER CHANGE
C     ****
C     SET UP DIAGONAL ELEMENTS
C     FIND ORIGINAL PSEUDO-RESIDUAL
C     AND ITS INNER PRODUCT
C     ****
      PSUPSU=0.0
      DO 10 I=1,N
      SUM=B(I)
      DO 20 J=ISTART(I),ISTART(I+1)-1
      IF(JA(J).EQ.I)THEN
      D(I)=A(J)
      ELSE
      SUM=SUM-A(J)*U(JA(J))
      ENDIF
20    CONTINUE
      PSU(I)=SUM/D(I)-U(I)
      PSUPSU=PSUPSU+D(I)*PSU(I)**2
10    CONTINUE
      CHANGE=1
      NUMITS=0
      GAMMAU=0.0
      GOTO50
C     ****
C     ESTIMATE SRJ
C     ****
30    IF(CHANGE.EQ.0.OR.IADAPT.EQ.0)GOTO40
      IF(NUMITS.GT.2)THEN
      CALL NEWTON(SRJ,CHANGE,ZETA,ALPHA,BETA,NUMITS)
      ELSEIF(NUMITS.EQ.1)THEN
      SRJ=ALPHA(1)
      ELSE
      SRJ=ALPHA(1)+ALPHA(2)
      TEMP=ALPHA(1)-ALPHA(2)
      SRJ=0.5*(SRJ+SQRT(TEMP**2+4.0*BETA(2)))
      ENDIF
      PRINT 35,NUMITS,SRJ
35    FORMAT(1X,'NUMITS =',I4/1X,'SRJ =',E12.6/)
C     ****
C     STOPPING TEST
C     ****
40    TEST=PSNORM/(1.0-SRJ)/UNORM
      IF(ABS(TEST).LT.ZETA)RETURN
C     ****
C     CALCULATE NEW PARAMETERS
C     NEEDS TO FIND (ITERATION MATRIX)*(PSU)
C     ****
50    NUMITS=NUMITS+1
      IF(NUMITS.GT.ITMAX)RETURN
      PSUPSV=0.0
      DO 60 I=1,N
      SUM=0.0
      DO 70 J=ISTART(I),ISTART(I+1)-1
      IF(JA(J).NE.I)SUM=SUM-A(J)*PSU(JA(J))
70    CONTINUE
      PSV(I)=SUM/D(I)
      PSUPSV=PSUPSV+PSU(I)*D(I)*PSV(I)
60    CONTINUE
```

```
      GAMMAW=GAMMAU
      GAMMAU=1.0/(1.0-PSUPSV/PSUPSU)
C     ****
C     CALCULATE NEW ACCELERATION PARAMETERS
C     ****
      IF(NUMITS.EQ.1)THEN
      OMEGAU=1.0
      ELSE
      TEMP=1.0/(1.0-PSUPSU*GAMMAU/(OMEGAU*GAMMAW*PSWPSW))
      OMEGAW=OMEGAU
      OMEGAU=TEMP
      ENDIF
C     ****
C     FIND NEW SOLUTION ESTIMATE AND NORM
C     FIND NEW PSEUDO-RESIDUAL AND NORM
C     INTERCHANGE ARRAYS FOR NEXT ITERATION
C     ****
      UNORM=0.0
      PSNORM=0.0
      PSWPSW=PSUPSU
      PSUPSU=0.0
      IF(NUMITS.EQ.1)THEN
      DO 80 I=1,N
      TEMP=GAMMAU*PSU(I)+U(I)
      W(I)=U(I)
      U(I)=TEMP
      UNORM=UNORM+TEMP**2
      TEMP=GAMMAU*PSV(I)+(1.0-GAMMAU)*PSU(I)
      PSW(I)=PSU(I)
      PSU(I)=TEMP
      TEMP=TEMP**2
      PSNORM=PSNORM+TEMP
      PSUPSU=PSUPSU+D(I)*TEMP
80    CONTINUE
      ALPHA(1)=1.0-1.0/GAMMAU
      ELSE
      DO 90 I=1,N
      TEMP=OMEGAU*(GAMMAU*PSU(I)+U(I))+(1.0-OMEGAU)*W(I)
      W(I)=U(I)
      U(I)=TEMP
      UNORM=UNORM+TEMP**2
      TEMP=OMEGAU*(GAMMAU*PSV(I)+(1.0-GAMMAU)*PSU(I))
     ++(1.0-OMEGAU)*PSW(I)
      PSW(I)=PSU(I)
      PSU(I)=TEMP
      TEMP=TEMP**2
      PSNORM=PSNORM+TEMP
      PSUPSU=PSUPSU+D(I)*TEMP
90    CONTINUE
      ALPHA(NUMITS)=1.0-1.0/GAMMAU
      BETA(NUMITS)=(OMEGAU-1.0)/(GAMMAW*GAMMAU*OMEGAW*OMEGAU)
      ENDIF
      UNORM=SQRT(UNORM)
      PSNORM=SQRT(PSNORM)
      GOTO30
      END
C     ****
      SUBROUTINE TO FIND NEW SRJ ESTIMATE
C     LARGEST EIGENVALUE OF A TRIDIAGONAL MATRIX
C     USES NEWTON'S METHOD
C     ****
```

```
      SUBROUTINE NEWTON(SRJ,CHANGE,ZETA,ALPHA,BETA,NUMITS)
      REAL SRJ,ZETA,ALPHA(200),BETA(200)
     +,X,FX(200),DFX(200),DELTAX
      INTEGER CHANGE,NUMITS
      X=SRJ
100   FX(1)=ALPHA(1)-X
      DFX(1)=-1.0
      FX(2)=FX(1)*(ALPHA(2)-X)-BETA(2)
      DFX(2)=DFX(1)*(ALPHA(2)-X)-FX(1)
      DO 110 I=3,NUMITS
      FX(I)=FX(I-1)*(ALPHA(I)-X)-FX(I-2)*BETA(I)
      DFX(I)=DFX(I-1)*(ALPHA(I)-X)-FX(I-1)-DFX(I-2)*BETA(I)
110   CONTINUE
      DELTAX=FX(NUMITS)/DFX(NUMITS)
      X=X-DELTAX
      IF(ABS(DELTAX)GT.ZETA)GOTO100
      IF(ABS(X-SRJ).LT.ZETA)CHANGE=0
      SRJ=X
      RETURN
      END
```

OUTPUT

The subroutine JCG yields the following output for the first model problem. It does not require any parameters during the actual iterations, but must estimate M (called SRJ in the subroutine) for the stopping test. This is done inside the additional subroutine NEWTON.

```
NUMITS=1
.0400000  .0400000  .0400000  .0400000  .0400000  .0400000  .0400000  .0400000
.0400000  .0400000  .0400000  .0400000  .0400000  .0400000  .0400000  .0400000
SRJ=.750000

NUMITS=2
.0320000  .0480000  .0480000  .0320000  .0480000  .0640000  .0640000  .0480000
.0480000  .0640000  .0640000  .0480000  .0320000  .0480000  .0480000  .0320000
SRJ=.806186

NUMITS=3
.0333333  .0466667  .0466667  .0333333  .0466667  .0666667  .0666667  .0466667
.0466667  .0666667  .0666667  .0466667  .0333333  .0466667  .0466667  .0333333
SRJ=.809017
```

BIBLIOGRAPHY

Axelsson, O., (1977) "Solution of linear systems of equations: iterative methods", in Lecture Notes in Math.: Sparse matrix techniques 572 (V.A. Barker, ed.) Springer-Verlag, N.Y.

Axelsson, O. and Gustafsson, I., (1977) "A modified upwind scheme for convective transport equation and the use of the conjugate gradient method for the solution of nonsymmetric systems of equations", Computer Sciences Dept. Report 77-12R, Chalmers Univ.Tech., Goteborg, Sweden.

Bartels, R. and Daniel, J.W., (1973) "A conjugate gradient approach to nonlinear elliptic boundary value problems in irregular regions", Lecture Notes in Math.363, Springer-Verlag, N.Y.

Birkhoff, G. and Varga, R.S., (1959) "Implicit alternating direction methods", Trans.Amer.Math.Soc. 92.

Birkhoff, G., Varga, R.S. and Young, D., (1962) "Alternating direction implicit methods", Advances in Computers 3.

Buzbee, B.L., Golub, G.H. and Nielson, C.W., (1970) "The method of odd/even reduction and factorization with application to Poisson's equation", Siam J.Numer. Anal. 7.

Carré, B.A., (1961) "The determination of the optimal accelerating factor for successive overrelaxation", the Computer Journal 4.

Chandra, R., (1977) "Conjugate gradient methods for partial differential equations", doctoral thesis, Yale University, New Haven, Connecticut.

Concus, P. and Golub, G.H., (1976) "A generalized conjugate gradient method for nonsymmetric systems of linear equations", Stan-CS-76-536, Computer Science Dept., Stanford University, Palo Alto, California.

Concus, P., Golub, G.H. and O'Leary, D.R., (1976) "A generalized conjugate gradient method for the numerical solution of elliptic partial differential equations", in Sparse Matrix Computation (J.R. Bunch and D.J. Rose, eds.), Academic Press.

Douglas, J. Jr., (1955) "On the numerical integration of $\frac{\partial^2 u}{\partial x^2} + \frac{\partial^2 u}{\partial y^2} = \frac{\partial u}{\partial t}$ by implicit methods", J.Soc.Ind.Appl.Math.3.

Douglas, J. Jr., (1961) "A survey of numerical methods for parabolic differential equations" Advances in Computers, Vol.2, (F.L. Alt, ed.), Academic Press, N.Y.

Douglas, J. Jr. and Rachford, H.H. Jr., (1956) "On the numerical solution of heat conduction problems in two or three space variables", Trans.Amer.Math.Soc.82.

Ehrlich, L.W., (1963) "The Block symmetric successive overrelaxation method", doctoral thesis, Univ.Texas at Austin, Austin, Texas.

Engeli, M., Ginsburg, T., Rutishauser, H. and Stiefel, E., (1959) "Refined iterative methods for the computation of the solution and the eigenvalues of self-adjoint boundary value problems", Mitt.Inst.f.angew. Math.ETH Zurich, No. 8.

Evans, D.J., (1967) "The use of preconditioning in iterative methods for solving linear equations with symmetric positive definite matrices", J.Inst.Math.Appl.4.

Forsythe, G.E. and Wasow, W.R., (1960) "Finite Difference Methods for Partial Differential Equations", Wiley, N.Y.

Ginsburg, T., (1963) "The conjugate gradient method", Numer.Math.5.

Golub, G.H., (1959) "The use of Chebyshev matrix polynomials in the iterative solution of linear systems compared with the methods of successive overrelaxation", doctoral thesis, Univ.Illinois, Urbana, Illinois.

Golub, G.H. and Varga, R.S., (1961) "Chebyshev semi-iterative methods, successive overrelaxation methods, and second-order Richardson iterative methods, Part I and Part II", Numer.Math.3.

Grimes, R.G., Kincaid, D.R., MacGregor, W.I. and Young, D.M., (1978) "ITPACK Report: Adaptive Iterative Algorithms using Symmetric Sparse Storage", CNA-139, Center for Numerical Analysis, Univ. of Texas at Austin, Austin, Texas.

Habetler, G.J. and Wachspress, E.L., (1961) "Symmetric successive overrelaxation in solving diffusion difference equations", Math.Comp.15.

Hageman, L.A., (1972) "The estimation of acceleration parameters for the Chebyshev polynomials and the successive overrelaxation methods", WAPD-TM-1038, Bettis Atomic Power Laboratory, Pittsburgh, Pennsylvania.

Hageman, L.A. and Young, D.M., (1981) "Applied iterative methods", Academic Press, N.Y.

Hageman, L.A., Luk, F. and Young, D.M., (1977) "The acceleration of iterative methods: preliminary report", CNA-129, Center for Numerical Analysis, Univ. of Texas at Austin, Austin, Texas.

Heller, J., (1960) "Simultaneous successive and alternating direction schemes", J. Soc. Indust.Appl.Math.8.

Henrici, P., (1962) "Discrete variable methods in Ordinary Differential Equations", John Wiley & Sons Inc., N.Y.

Hestenes, M.R. and Stiefel, E.L., (1952) "Methods of conjugate gradients for solving linear systems", NBS J.Res. 49.

Kahan, W., (1958) "Gauss-Seidel methods of solving large systems of linear equations", doctoral thesis, Univ. of Toronto, Toronto.

Kincaid, D.R., (1971) "An analysis of a class of norms of iterative methods for systems of linear equations", doctoral thesis, Univ. of Texas at Austin, Austin, Texas.

Kincaid, D.R. and Young, D.M. (1978) "Survey of iterative methods", CNA-135, Center for Numerical Analysis, Univ. of Texas at Austin, Austin, Texas.

Manteuffel, T.A., (1975) "An iterative method for solving nonsymmetric linear systems with dynamic estimation of parameters", doctoral thesis, Univ. Illinois, Urbana, Illinois.

Peaceman, D.W. and Rachford, H.H. Jr., (1955) "The numerical solution of parabolic and elliptic differential equations" J.Soc.Indus.Appl. Math. 3.

Price, H. and Varga, R.S., (1962) "Recent numerical experiments comparing successive overrelaxation iterative methods with implicit alternating direction methods", Report No. 91, Gulf Research and Development Co., Pittsburgh, Pennsylvania.

Schwarz, H.R., Rutishauser, H. and Stiefel, E., (1973) "Numerical analysis of Symmetric Matrices", Prentice-Hall Inc., N.J.

Stone, H.L., (1968) "Iterative Solutions of implicit approximations of multi-dimensional partial differential equations", Siam J.Num.Anal. 5.

Tee, G.J., (1963) "Eigenvectors of the successive overrelaxation process and its combination with Chebyshev semi-iteration", Comput.J. 6.

Varga, R.S., (1962) "Matrix Iterative Analysis", Prentice-Hall, N.J.

Wachspress, E.L., (1966) "Iterative solution of elliptic systems and applications to the Neutron diffusion equations of reactor physics", Prentice-Hall. N.J.

Westlake, J.R., (1968) "A handbook of numerical matrix inversions and solution of linear equations", John Wiley & Sons Inc., N.Y.

Widlund, O., (1978) "A Lanczos method for a class of nonsymmetric systems of linear equations", Siam J.Numer.Anal. No. 4.

Wrigley, H.E., (1963) "On accelerating the Jacobi method for solving simultaneous equations by Chebyshev extrapolation when the eigenvalues of the iteration matrix are complex", Comput.J. 6.

Young, D.M. (1950) "Iterative methods for solving partial difference equations of elliptic type", doctoral thesis, Harvard Univ., Cambridge, Massachusetts.

Young, D.M. (1954) "On Richardson's method for solving linear systems with positive definite matrices", J.Math.Phys. XXXII.

Young, D.M. (1970) "Convergence properties of the symmetric and unsymmetric successive overrelaxation methods and related methods", Math.Comp. 24.

Young, D.M. (1971) "Iterative solution of large linear systems" Academic Press, N.Y.

Young, D.M. (1977) "On the accelerated SSOR method for solving large linear systems", in Advances in Math. 23.